INSTITUTION OF CIVIL ENGINEERS

Advances in site investigation practice

Proceedings of the international conference held in
London on 30–31 March 1995

¶¶ Thomas Telford

Conference organized by the Institution of Civil Engineers and co-sponsored by the Association of Geotechnical Specialists (AGS), the British Geotechnical Society (BGS) and the International Society for Soil Mechanics and Foundation Engineering (ISSMFE).

Organizing committee: C. Craig, Chairman; C. H. Adam, Fugro, Scotland; B. G. Clarke, University of Newcastle upon Tyne; EurIng Professor C. R. I. Clayton, University of Surrey; S. M. Herbert, TBVScience; J. P. Love, Geotechnical Consulting Group; J. J. M. Powell, Building Research Establishment; N. A. Trenter, Sir William Halcrow & Partners Ltd.

Published by Thomas Telford Publishing, Thomas Telford Services Ltd, 1 Heron Quay, London, E14 4JD.

First published 1996

Distributors for Thomas Telford books are
USA: American Society of Civil Engineers, Publications Sales Department, 345 East 47th Street, New York, NY 10017-2398
Japan: Maruzen Co. Ltd, Book Department, 3–10 Nihonbashi 2-chome, Chuo-ku, Tokyo 103
Australia: DA Books and Journals, 648 Whitehorse Road, Mitcham 3132, Victoria

A CIP catalogue record for this book is available from the British Library.

ISBN: 0 7277 2513 0

Classification
Availability: Unrestricted
Content: Collected papers
Status: Refereed
User: Academic and practising engineers

Printed and bound in Great Britain by Redwood Books, Trowbridge, Wiltshire.

Contents

SESSION 1: PLANNING, CONTROL, MANAGEMENT AND QA

Aspects of the site investigation for the Waterloo International Terminal.
A. PICKLES and S. EVERTON   1

Recent developments in planning and execution of nearshore site investigation.
A. J. HODGSON, C. H. ADAM and M. SNEDDON   13

Terrain systems mapping and geomorphological studies for the Channel Tunnel
rail link. A. M. WALLER and P. PHIPPS   25

Development of a knowledge-based system for ground investigation in soil and rock.
M. G. WINTER, G. D. MATHESON and P. McMILLAN   39

Expert system for geotechnical testing. X. XUE and P. SMART   51

Managing site investigation fieldwork using ISO 9000. T. G. BALL and
A. P. WHITTLESTONE   63

The practical application of quality management to site investigation.
J. C. WOODWARD   75

Site investigation for lime stabilisation of highway works. J. PERRY,
R. A. SNOWDEN and P. E. WILSON   85

Investigation and assessment of cohesive soils for lime stabilisation. L. J. BARBER   97

Spatial dependence in site investigation design. I. W. FARMER, G. C. XIAO
and P. G. CHALLINOR   109

Overview of initial ground investigations for the Channel Tunnel high speed
rail link. R. C. BECKWITH, W. J. RANKIN, I. C. BLIGHT and I. HARRISON   119

Historic maps in the investigation of land use and the identification of possible
contamination. B. RIDEOUT and D. SMITH   145

Moderator's report on Session 1. N. A. TRENTER   155

Discussion on Session 1. J. COUTTS   165

SESSION 2: DATA MANAGEMENT

From paper to silicon chip — the growing art of computerised data management.
R. A. NICHOLLS, A. S. PYCROFT, M. J. PALMER and J. A. FRAME   186

A computer system for site investigation data management and interpretation.
A. J. OLIVER and D. G. TOLL   198

Management of spatial data for the heavy foundations of a blast furnace.
P. NATHANAIL and M. ROSENBAUM   210

Management of geologic and instrumentation data for landslide remediation.
L. W. ABRAMSON and T. S. LEE   218

Investigation of London Underground earth structures. B. T. McGINNITY
and D. RUSSELL   230

Moderator's report on Session 2. L. THREADGOLD   243

Discussion on Session 2. D. G. TOLL   250

SESSION 3: DRILLING, BORING, SAMPLING AND DESCRIPTION

Dynamic (window) sampling — a review. C. S. ECCLES and R. P. REDFORD   257

Deep rotary cored boreholes in soils using wireline drilling. P. HEPTON   269

Sample disturbance in rotary core tube sampling of softrock. F. TATSUOKA,
Y. KOHATA, T. TSUBOUCHI, K. MURATA, K. OCHI and L. WANG   281

Rock and soil description and classification — a view from industry.
D. R. NORBURY and R. C. GOSLING   293

A new core orientation device. I. P. WEBBER and D. GOWANS   306

The use of down-hole CCTV for collection of quantitative discontinuity data.
P. McMILLAN, A. R. BLAIR and I. M. NETTLETON   312

Advances in survey techniques for contaminated land. P. C. ROBERY,
W. L. BARRETT and S. M. HERBERT   324

Moderator's report on Session 3. D. W. HIGHT   337

Discussion on Session 3.   361

SESSION 4(a): PENETROMETERS

Development of automated dilatometer and comparison with cone penetration
tests at the University of Adelaide, Australia. W. S. KAGGWA, M. B. JAKSA
and R. K. JHA   372

Dynamic probing and its use in clay soils. A. P. BUTCHER, K. McELMEEL
and J. J. M. POWELL   383

The development of the seismic cone penetration test and its use in
geotechnical engineering. P. A. JACOBS and A. P. BUTCHER   396

Applications of penetration tests for geo-environmental purposes.
P. K. ROBERTSON, T. LUNNE and J. POWELL   407

The multifunctional Envirocone® test system. D. A. O'NEILL, G. BALDI
and A. DELLA TORRE   421

Use of piezocone tests in non-textbook materials. T. LUNNE, J. POWELL
and P. ROBERTSON   438

Cone penetration testing in volcanic soil deposits. K. TAKESUE, H. SASAO
and Y. MAKIHARA   452

Moderator's report on Session 4(a). P. K. ROBERTSON   464

Discussion on Session 4(a). C. H. ADAM   470

SESSION 4(b): PRESSUREMETER, PERMEABILITY AND PLATE TESTS

A new field test for the 'degree of compaction'. D. W. COX   487

A new technique for in situ stress measurement by overcoring.
A. P. WHITTLESTONE and C. LJUNGGREN   499

A pore water pressure probe for the in situ measurement of a wide range of
soil suctions. A. M. RIDLEY and J. B. BURLAND   510

Groundwater monitoring — the Sellafield approach. C. ELDRED,
J. SCARROW, M. AMBLER and A. SMITH   521

Recent developments in the cone pressuremeter. R. W. WHITTLE   533

Ménard pressuremeter test to foundation design — an integrated concept.
J. NUYENS, F. BARNOUD and M. GAMBIN   547

A practical guide to the derivation of undrained shear strength from
pressuremeter tests. B. G. CLARKE and J. A. SADEEQ   559

Reliable parameters from imperfect SBP tests in clay. D. A. SHUTTLE
and M. G. JEFFERIES   571

Pressuremeter tests in unsaturated soils. F. SCHNAID, G. C. SILLS
and N. C. CONSOLI   586

The determination of deformation and shear strength characteristics of Trias
and Carboniferous strata from in situ and laboratory testing for the Second
Severn Crossing. J. D. MADDISON, S. CHAMBERS, A. THOMAS
and D. B. JONES   598

M5 Avonmouth bridge strengthening — in situ testing of Mercia Mudstone.
N. V. PARRY, A. J. JONES and M. R. DYER   610

Moderator's report on Session 4(b). B. G. CLARKE   623

Discussion on Session 4(b). J. J. M. POWELL   642

SESSION 5: GEOPHYSICAL TESTING

Better correlation between geophysical and geotechnical data from improved
offshore site investigations. J. F. NAUROY, J. L. COLLIAT, A. PUECH,
D. POULET, J. MEUNIER and F. LAPIERRE   659

Confidence in seismic characterisation of the ground. G. A. RICKETTS,
J. SMITH and B. O. SKIPP   673

Geophysical surveying methods in a site investigation programme. D. M. McCANN
and C. A. GREEN   687

Practical considerations for field geophysical techniques used to assess
ground stiffness. A. P. BUTCHER and J. J. M. POWELL   701

Site investigation for seismically designed structures. P. D. DAVIS,
P. J. L. ELDRED, J. D. BENNELL, D. W. HIGHT and M. S. KING   715

The selection and interpretation of seismic geophysical methods for site investigation. M. A. GORDON, C. R. I. CLAYTON, T. C. THOMAS and M. C. MATTHEWS 727

The use of acoustic emission to monitor the stability of soil slopes. N. DIXON, R. HILL and J. KAVANAGH 739

The use of slimline logging techniques in engineering ground investigation. D. PASCALL, A. D. J. LAW and D. C. CURTIS 750

Use of the resistivity dipmeter for discontinuity assessment and stability design of a major cutting. D. G. GUY and D. A. O'CALLAGHAN 762

Moderator's report on Session 5. C. R. I. CLAYTON and V. S. HOPE 774

Discussion on Session 5. A. P. BUTCHER 789

SESSION 6: LABORATORY TESTING

Investigation of the fabric of engineering soil using high resolution X-ray densimetry. M. A. PAUL, G. C. SILLS, L. A. TALBOT and B. F. BARRAS 805

Considerations in the geotechnical testing of contaminated samples. R. G. CLARK and G. P. KEETON 816

A hydraulic fracturing test based on radial seepage in the Rowe consolidation cell. C. H. de A. C. MEDEIROS and A. I. B. MOFFAT 828

Quick, accurate, consistent measurements of permeability of clays. J. T. ARARUNA Jnr, B. G. CLARKE and A. H. HARWOOD 840

The shear strength and deformation behaviour of a glacial till. A. CHEGINI and N. A. TRENTER 851

The use of local strain measurements in triaxial testing to investigate brittleness of residual soil. L. A. BRESSANI, A. V. D. BICA and F. B. MARTINS 867

Recent Japanese practice for investigating elastic stiffness of ground. S. SHIBUYA, T. MITACHI, S. YAMASHITA and H. TANAKA 875

The measurement of strength, stiffness and in situ stress in the Thanet Beds using advanced techniques. J. G. A. JOHNSON, R. L. NEWMAN, T. S. PAUL and D. S. PENNINGTON 887

Moderator's report on Session 6. R. J. JARDINE 900

Discussion on Session 6. J. P. LOVE 916

CLOSING ADDRESS

The role of in situ testing in geotechnical engineering — thoughts about the future. M. JAMIOLKOWSKI 929

Aspects of the site investigation for the Waterloo International Terminal

ANDREW PICKLES - Chief Geotechnical Engineer
Sir Alexander Gibb & Partners Ltd, Reading, England

STEVEN EVERTON - Geotechnical Engineer
Sir Alexander Gibb & Partners Ltd, Reading, England

ABSTRACT
A major site investigation has been performed on the site of the new Waterloo International Terminal in London. The investigation provided parameters for general design, and for finite element modelling of ground movements and underground tunnel distortions due to construction. This paper presents details of the investigation and interpretation of design parameters.

INTRODUCTION
The new Waterloo International rail terminal (WIT) in London was officially opened in May 1993. The WIT was constructed over part of the existing British Rail (BR) domestic station at Waterloo. The site of the WIT is underlain by the station and running tunnels of London Underground Limited's (LUL) Bakerloo and Northern lines, and is directly adjacent to the existing rail terminal. A plan on the WIT is shown in Figure 1 and a cross section through the terminal is shown in Figure 2.

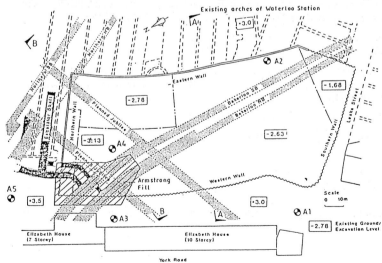

Figure 1 : Plan of site

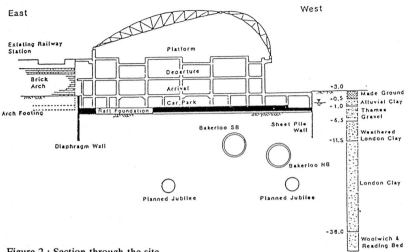

Figure 2 : Section through the site

The construction of the WIT required demolition of part of the existing station and excavation to an average depth of approximately 6 m across the site. The LUL underground lines and BR station were to remain in operation throughout the construction of the WIT. Accurate predictions of ground movements and tunnel and structure movements were therefore required to assure both LUL and BR that the construction would not affect the operation of either the underground or the remaining Waterloo station. Extensive soil/structure finite element modelling incorporating non linear soil stiffness properties was carried out for the design of the terminal and to develop appropriate design and construction methodologies.

This paper describes the ground investigation that was carried out at the site to determine the geotechnical parameters required for design. The paper compares the results obtained using various investigation techniques and sets out the geotechnical parameters adopted for design. The in situ stress state and non linear stiffness properties of the London Clay and Thames Gravel are discussed in detail by Hight et al (1993) and these aspects are only dealt with in outline in this paper. General comments are made with respect to the applicability of various investigation techniques on this site.

THE WIT DEVELOPMENT
At the time of the ground investigation the proposed development comprised the new international terminal with two levels of below ground parking, and a commercial development comprising five high rise office blocks, two of which would be constructed over the terminal and 3 adjacent to the terminal. The total area of the WIT and associated commercial development was approximately 3 Ha. Subsequent to the ground investigation the commercial development was cancelled and only one level of basement was required for the WIT. The basement comprises a reinforced concrete raft with approximate dimensions of 170 m by 60 m. Demolition and excavation followed by

construction resulted in a net long term unloading of approximately 100 kN/m². The basement was constructed within a retaining/cut-off wall, a diaphragm wall being used for the sides directly adjacent to the existing station and a sheet pile wall on the remaining sides. The terminal itself comprises a reinforced concrete substructure with a steel and glass arched roof (Hunt et al, 1994).

SITE HISTORY

The site originally formed part of the low level marshes within the floodplain alongside the River Thames. The area had been filled to a level of approximately 3 mOD prior to construction of the station. The soil profile at the site is typical of this part of London with the alluvial clay being underlain successively by Thames Gravel, London Clay, Woolwich and Reading Bed clay, Thanet Sands and Chalk. Scour hollows, where the Thames Gravel locally penetrates the London Clay to considerable depth, are often found in this area of London (Berry, 1979).

The history of construction at the site has an implication on the in situ stresses in the ground, and hence on the response of the ground to the new construction. Waterloo station was constructed in three stages between 1848 and 1880, with the construction of the Southern Railway office block on part of the site in 1886. The Waterloo and City Line was completed in 1898, the Bakerloo Lines were completed below the site in 1902 and the Northern Lines in 1926. The 7 to 15 storey Elizabeth House was constructed on the site of the Southern Railway office block in 1965.

A number of investigations have been carried out at the site in connection with the various developments at the station. The majority of these investigations comprised only descriptions of the soil encountered in boreholes, with a limited amount of index testing. No information was available on the in situ stresses and the non linear stress strain properties of the ground.

SUMMARY OF INVESTIGATION TECHNIQUES

The presence of the LUL tunnels beneath the site and of the remaining BR station directly adjacent to the site meant that accurate predictions of ground movements would be an important element of the design. In order to make accurate predictions it was essential that the investigation established both the in situ stress state and the non linear stiffness characteristics of the Thames Gravel and London Clay.

The fieldwork was carried out by Soil Mechanics Limited between May and October 1989 under the supervision of Sir Alexander Gibb and Partners Ltd. The majority of the drilling and testing was performed beneath the existing brick arches of Waterloo Station which had not yet been demolished.

Detailed investigation was concentrated at five locations, designated A1 to A5 on Figure 1. At each location a combination of methods was used, comprising; a triple tube rotary cored borehole to obtain high quality samples for descriptive logging and laboratory testing; a percussion borehole with pushed thin wall sampling to obtain high quality samples for laboratory testing; a piezocone profile; a Marchetti plate dilatometer profile; and a series of self boring pressuremeter tests at various levels. This methodology allowed direct comparison of a suite of tests to be made, and aided correlation and interpretation of the results. A series of four piezocone profiles and four dilatometer profiles were carried out at a radial distances of 2 m, 6 m, 11 m and 18 m from the side of the Bakerloo Line (southbound station tunnel) in an attempt to measure the extent of the influence of the tunnel construction on the ground conditions.

In addition to the 5 "A" sites, 18 boreholes were drilled to a depth of approximately 40 m at an average spacing of 50 m across the site to prove the consistency and depth of the London Clay. U100 and disturbed samples were recovered from these boreholes and SPTs taken at approximately 4 m centres. Nine boreholes were taken to approximately 10 m to prove the thickness and consistency of the Made Ground, Alluvium and Thames Gravel overlying the London Clay. Standpipe piezometers were installed in boreholes in the Thames Gravel and vibrating wire piezometers were installed in the boreholes in the London Clay.

The presence of scour hollows would have serious implications for the design and construction of the retaining/cut-off wall around the site, and therefore required careful investigation. The boreholes were supplemented by dynamic probing along the proposed line of the cut-off wall using both the Flow Through Sampler and the Dynamic Cone Penetrometer.

Trial pits were excavated to inspect the shallow foundations of the existing station substructure, and rotary core samples were taken of the brickwork piers and mass concrete strip footings for laboratory testing. The thickness of the cast iron lining to the LUL tunnels and the tensile strength capacity of the tunnel lining bolts was measured.

In addition to standard index tests the following were carried out in the laboratory:

* Filter paper suction tests on both thin wall and rotary core samples of the London Clay, to allow an assessment of the in situ horizontal stress to be made.

* Consolidated and unconsolidated undrained triaxial tests on the London Clay and Woolwich and Reading Bed clay. All tests included the measurement of porewater pressure and the measurement of initial sample suction prior to testing. These tests allowed both the undrained and drained strength characteristics to be measured and an assessment of the in situ horizontal stress to be made.

* Triaxial stress path testing with small strain measurement on thin wall and rotary core samples of the London Clay and reconstituted samples of the Thames Gravel, to allow the non linear stress strain characteristics to be measured.

* Conventional load/unload oedometer tests on the London Clay, and stress controlled K_0 (zero lateral strain) swelling tests in the triaxial cell to low vertical effective stress (approximately 10 kN/m^2). These tests allowed the unloading stiffness of the London Clay to be measured.

The typical soil profile encountered on the site is summarised on Figure 2. The strength parameters adopted for design are shown in Table 1. The non linear stress strain characteristics of the Thames Gravel and London Clay and the estimated in situ stress profile in the London Clay are presented by Hight et al (1993).

Table 1 : Design Parameters

STRATUM	γ (kN/m³)	c' (kN/m²)	ϕ' (°)	v (°)
MADE GROUND	18	0	30	0
ALLUVIUM	17	0	24	0
THAMES GRAVELS	20	0	35	17.5
LONDON CLAY :				
i) Weathered	20	5	23	11.5
ii) Unweathered	20	15	25	12.5
WOOLWICH AND READING	22	200	27	0

v : Dilation Angle

PARTICULAR ASPECTS OF THE SITE INVESTIGATION
Rotary Coring of the London Clay
Sampling of the London Clay was carried out to obtain standard U100 open-drive samples, 100 mm diameter pushed thin wall samples and 100 mm diameter rotary core samples. The majority of laboratory testing was carried out on thin wall and rotary core samples. U100 samples were only included for comparison of strength test results with data from previous investigation.

The rotary core samples allowed a detailed visual inspection of the soil fabric. This enabled the presence of sand and silt seams in the London Clay and Woolwich and Reading Beds to be logged, and the presence of claystones to be established. The distribution of sand/silt seams and claystones in the five rotary cored boreholes is shown in Figure 3. The sand/silt seams have an important influence on the following aspects:

* The pattern of ground water flow to the tunnels and therefore the pore pressure profile at the site.

* Base stability of deep excavations.

* The amount of immediate heave and the rate of longer term swelling of the clay following excavation.

* The suction measured in a sample. Where sand/silt seams are present in a sample the sample can desaturate as the sand/silt gives up water to the surrounding clay. This will lead to the measurement of both lower suction pressures and an under-estimation of undrained shear strength.

It can be seen in Figure 3 that sand/silt seams were generally not apparent in the upper weathered London Clay. The use of rotary coring proved extremely useful in gaining a better understanding of the potential behaviour of the London Clay and was thus valuable when interpreting the results of the in situ and laboratory testing.

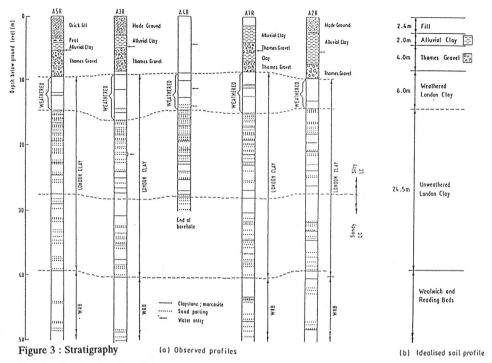

Figure 3 : Stratigraphy   (a) Observed profiles   (b) Idealised soil profile

Suction in the London Clay

The suction existing in a sample of clay taken from the ground is related to the mean effective stress of the clay in situ. Suction may be modified by the sampling process and by the procedure followed in making the measurement (Hight, 1985). In stiff overconsolidated clays it might be expected that prevention of dilation of the clay whilst pushing a sample tube into the ground would lead to an increase in the suction measured in a sample. Conversely, during rotary coring with a water, polymer mud or foam flush the soil core has some access to water, even when using a triple tube drilling system. This access to water is likely to lead to a loss of suction in the sample.

In the WIT investigation filter paper suction tests were carried out on both pushed thin wall samples and rotary core samples. Care was taken in selecting only thin wall samples with no evidence of damage to the tubes. Rotary core samples were selected for testing on site immediately upon retrieval of the core from the borehole. The sample liner was split and the surface of the sample was lightly dried using tissue paper and any evidence of softened material on the surface of the sample scraped off. Samples were then wrapped in clingfilm and waxed.

The results of the filter paper suction tests on both the rotary core and thin wall samples at three of the "A" sites are shown in Figure 4. The suction measured on the rotary core samples is, as expected, significantly lower than measured on the pushed thin wall samples. Also shown is the estimated mean effective stress profile calculated using the procedure described by Burland et al (1979) making an allowance for one dimensional loading of the clay by the Made Ground, Alluvium and Thames Gravel.

6

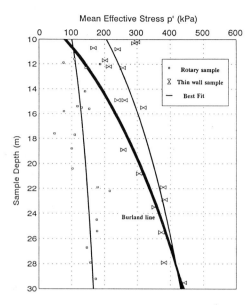

Figure 4 : Mean effective stress

A comparison between the K_0 profiles measured using the suction measurements, the pressuremeter, the dilatometer and also based on the procedure described by Burland et al is given in Hight et al (1993) for two of the "A" sites. Based on all the results it is considered that the use of the filter paper method on samples recovered by pushed thin walled samples is an appropriate method of estimating the in situ mean effective stress in stiff clays, although at shallow depths it may overestimate K_0. It should be noted however that great care must be taken in the selection of samples for testing. Samples from damaged tubes or those containing sand/silt seams should not be used.

The loss of suction measured on the rotary core samples would lead to an under-prediction of the in situ horizontal stress. Rotary coring was generally carried out using polymer mud flush (GS 550). Foam flush was used for one of the boreholes in an attempt to reduce the loss of suction. Comparison of the filter paper test results between the boreholes drilled using polymer mud and foam did not show any significant difference in the measured suction. In addition there was no noticeable difference in the sample quality for visual logging.

Probing the Thames Gravel-London Clay Interface
The possibility of a scour hollow beneath the foundation raft would have had serious implications on its design. Dynamic probing was used as a cheaper, quicker alternative to conventional boring to locate the gravel-clay interface. Probing was carried out using both the Flow Through sampler (FTS) and the Dynamic Cone penetrometer (DCP). The Flow Through Sampler consists of a hydraulic percussion hammer which drives a sampling tube continuously into the ground during which the penetration rate is measured. The Dynamic Cone penetrometer consists of a 50kg weight which is repeatedly dropped through 500mm to drive an oversize cone of base area 15cm² through the ground. The number of drops of the hammer to achieve penetration is recorded.

Typical results from the DCP and FTS profiles are presented in Figure 5, showing that both techniques were able to differentiate the boundary with acceptable precision.

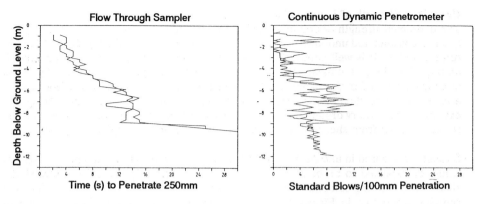

Figure 5 : Probing of the Gravel-Clay Interface

Pore Water Pressure Profile in the London Clay

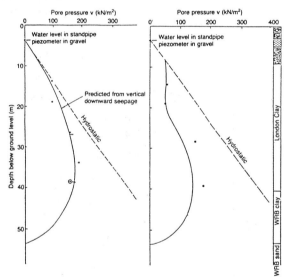

Figure 6 : Piezometric profile

Vibrating wire piezometers were installed throughout the London Clay to measure the pore water pressure profile. The results are summarised on Figure 6. It is clear that the presence of the tunnels has a significant effect on the pore water pressure profile with the tunnels acting as drains. The sand and silt layers result in a relatively high horizontal permeability compared to the vertical permeability. The effect of the tunnels on the pore pressure distribution is therefore apparent at relatively large horizontal distances from the tunnel. Further details on the pore pressure distribution are given in Hight et al (1993).

Piezocone dissipation tests were performed in order to obtain a measure of the permeability of the ground. It was considered that the in situ equilibrium pore pressure could potentially be established from the dissipation test results. Dissipation monitoring continued overnight in order to establish the equilibrium pore pressure. Despite continuing some of the dissipation tests for up to 18 hours it was not possible to measure the in situ equilibrium pore pressure using the piezocone.

Occasionally both the piezocone probe and the Marchetti dilatometer were obstructed by claystones and the probes needed to be advanced by pre-boring through the claystone.

Undrained Strength of London Clay

The undrained strength of London Clay has traditionally been measured by carrying out quick unconsolidated undrained triaxial tests on samples recovered using the U100 open-drive sampler. It is well known, however, that the undrained strength is a function of, among other things, the mean effective stress (p') in the sample prior to shearing. It has already been noted in this paper that the initial mean effective stress in a sample can be affected by the method of sampling. Changes in p' can occur as a result of swelling (for example during rotary drilling or as a result of sand/silt seams which desaturate following stress relief) or from shear strains which occur as a result of sample tube penetration.

Strengths measured in unconsolidated undrained (UU) triaxial tests on thin wall samples would be expected to be more reliable than those measured on U100 samples as the shear strains during sampling are smaller. The results of the UU tests on thin wall samples are shown in Figure 7. The lower values of strength in Figure 7 tend to correspond with sand/silt seams which would have lead to a reduction in p' prior to testing.

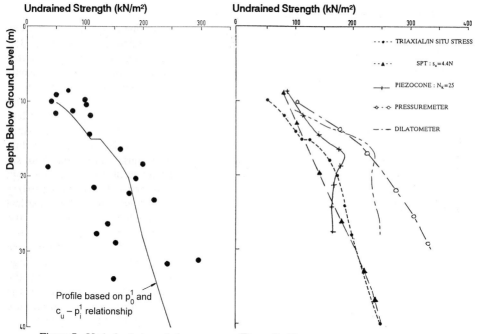

Figure 7 : Undrained strength from thin wall triaxial samples

Figure 8 : Undrained strength - various methods

Also shown in Figure 7 is the s_u profile determined from consolidated undrained (CU) triaxial tests. The stress paths showed that the samples exhibited shear at approximately constant effective stress (i.e. vertical in t-s' space), and the effective stresses were converted to depths using the profiles determined from the suction measurements.

The undrained strength was also estimated from the results of the piezocone, SPT, pressuremeter and dilatometer tests. A comparison of these profiles with that calculated from the triaxial test results is shown in Figure 8. The following methods have been used to calculate the various profiles:

*   Piezocone - using the method in Meigh (1987) and adopting a cone factor $N_k=25$

*   SPT - using the method proposed by Stroud (1988) and a correlation factor of 4.4

*   Pressuremeter - using the method proposed by Gibson and Anderson (1961)

*   Dilatometer using the method proposed by Powell and Uglow (1988)

Both the piezocone and SPT provide a reasonable agreement with the triaxial data, whereas both the pressuremeter and dilatometer tend to predict a higher undrained strength. On the basis of all the test data measured at Waterloo it is considered that measurement of the undrained strength using the results of triaxial tests where the samples are consolidated to their estimated in situ mean effective stress is the most appropriate method of establishing the undrained strength profile. The estimate should also be compared with the results of the SPT. The use of triaxial tests without the measurement of pore pressure makes assessment of the quality and reliability of the test results extremely difficult.

Effective Stress Strength Parameters

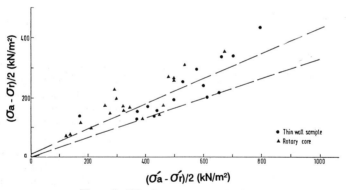

Figure 9 : Effective stress strength - London Clay

All of the triaxial tests were performed with pore pressure measurement to enable drained strength parameters to be measured. The peak strengths measured in tests on unweathered London Clay are shown in Figure 9. It can be noted that a failure envelope corresponding approximately to the critical state strength for London Clay provides a lower bound to the data. It is also worth noting that the results of the tests carried out on the rotary core samples are generally greater than those for the thin wall samples, suggesting perhaps that less disturbance is caused by the coring process.

The peak strengths measured on the tests carried out on the Woolwich and Reading Bed clay are shown in Figure 10. All tests were performed on rotary core samples of the clay. As can be seen the Woolwich and Reading Bed clay exhibits a significant drained cohesive intercept with results showing a strength of c' = 200 kN/m² and ϕ' = 27°

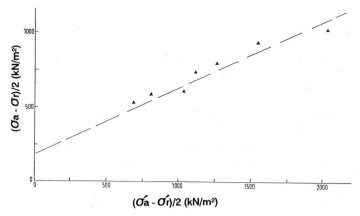

Figure 10 : Effective stress strength - Woolwich and Reading beds

Swelling of London Clay

Incremental swelling tests were performed on samples of London Clay in the oedometer. The results of these tests are summarised in Figure 11a. All the tests were loaded in increments to the initial starting stress for the swelling stages. The results of the tests have been replotted in Figure 11b by normalising the vertical stress for each swelling increment by the maximum stress prior to swelling.

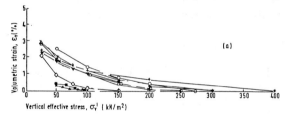

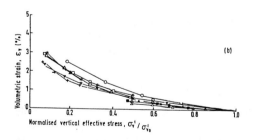

In addition to the oedometer tests the swelling characteristics of two samples of London Clay were tested in a triaxial cell. The radial stress applied to the samples was controlled so as to maintain zero radial strain during the swelling. The initial vertical stress in both samples was 400 kN/m². The results of the tests are shown in Figure 11c. The test results are also compared to the average line obtained from the oedometer tests. It can be seen that the oedometer tests on small samples are comparable to the one dimensional results on the larger triaxial samples.

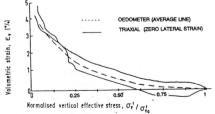

Figure 11 : Swelling from triaxial and oedometer

SUMMARY

A detailed site investigation has been performed at Waterloo which has added to the data available for these soils. High quality sampling methods and modern laboratory tests have been performed to determine non-linear design parameters and to estimate the in situ stress state with confidence. This work has allowed realistic assessments of ground and structural movements to be made with complex numerical analyses. At the project feasibility stage it was envisaged that costly protection works would be required to the underground tunnels and escalators. The combination of an extensive site investigation and detailed numerical modelling demonstrated that these works were not necessary, resulting in a significant cost saving to the client. Successful correlation of simple techniques such as dynamic probing, conventional oedometer and standard penetration tests with more detailed techniques has confirmed that these tests are still useful for routine design purposes.

REFERENCES

BERRY F G (1979) "Late Quarternary scour hollows and related features in central London" Quarterly Journal of Engineering Geology 12

BURLAND J B, SIMPSON B, ST JOHN H D (1979) "Movements around excavations in London Clay" Proceedings 7th European Conference on Soil Mechanics, Brighton

GIBSON R E, ANDERSON W F (1961) "In situ measurement of soil properties with the pressuremeter" Civil Engineering and Public Works Review, Vol 56, No 658

HIGHT D W (1985) "Assessing BS 5930" Geological Society of London

HIGHT D W, HIGGINS KG, JARDINE R J, POTTS D M, PICKLES A R, DE MOOR E K, NYIRENDA Z M (1983) "Predicted and measured tunnel distortions associated with the construction of Waterloo International Terminal" Proceedings Wroth Memorial Symposium, Oxford

HUNT A J, JONES A C, OTLET M, DEXTER D I (1994) "Waterloo International Rail Terminal trainshed roof structure" The Structural Engineer Vol 72 No 8

MEIGH A C (1987) "Cone penetration testing" CIRIA

POWELL J J M, UGLOW I M (1988) "The interpretation of the Marchetti Dilatometer test in UK clays" Proceedings Penetration Testing in the UK

STROUD M A (1988) State of the Art Report, Proceedings Penetration testing in the UK

Recent developments in planning and execution of nearshore site investigation

A.J. HODGSON, C.H. ADAM AND M. SNEDDON
Fugro Ltd, Hemel Hempstead, England, Fugro Scotland, Glasgow, Scotland and Seacore Ltd, Gweek, England.

1.   INTRODUCTION

A wide variety of civil engineering projects in coastal waters have a demanding site investigation requirement. This paper discusses the range of techniques currently available for such works and also recent developments in sampling and insitu testing methods.

Furthermore, the marine environment dictates careful consideration with respect to craft selection and safe methods of working. An outline of factors concerning the selection and approval process is also discussed.

While this paper is primarily concerned with marine plant and investigation techniques applicable in coastally exposed waters, extending from the high water line to 25m water depth, general observations regarding shallow and sheltered water work are included where relevant.

2.   POLARISED DEVELOPMENT

Traditionally, nearshore site investigations were generally carried out as extensions to onshore investigations. This led to the use of static, land based drilling plant usually mounted on locally chartered barges or pontoons, the combination of which, with no alternative available, became the accepted standard 'marine spread'.

The principal method of site investigation operated on land and therefore extended overwater is cable-tool boring, a low cost, low technology method which uses percussive techniques to advance casings, sampling and insitu testing equipment into the soils. It is assumed that these traditional techniques are familiar to the reader and need not be detailed further.

Rotary methods of drilling were rarely considered for soils investigation, and overwater, rotary applications were generally limited to the proof coring of obstructions or bedrock. Rotary coring tools and plant do not permit dynamic action and therefore required the drilling plant to be isolated from the movement of the barge. As floating barges were the common drilling craft, at the simplest level, rotary plant was generally of the 'pendant' type, i.e. a lightweight, easy to handle power swivel that could be mounted directly onto the casing or be suspended from the cable tool mast with some suitable form of lateral rigging providing torque reaction to the power swivel. Such typical arrangements, which avoided any sophistication, were slow to set up and operate, but required little skill and were attractively inexpensive.

In the move from land to overwater drilling, the land based traditions accompanied the plant. Traditionally the soils 'boring' foreman and the rotary 'drilling' foreman were usually quite separate. Today, it is still quite normal for soils work to be undertaken by a soils foreman using cable tool plant while any coring - other than simple pendant work - is undertaken by a separate rotary foreman using more sophisticated rotary plant, each foreman competent in their respective skills. Indeed if one examines the recent list of BDA accredited drillers, few of them will be seen to be accredited in both cable percussive and rotary techniques.

As can be seen, the combined use of land based plant and accompanying traditions, generally discriminate against overwater rotary methods while greatly favouring percussive methods.

Many other factors have ranged themselves against overwater rotary soils drilling methods:
a)   lack of market confidence due to poor, ill-considered attempts at using rotary soils drilling on land
b)   the high cost of development of plant and techniques
c)   the greater skill levels required of the personnel.

Such factors have left the percussive method dominating coastal investigation works.

Unfortunately, the cable percussive rig is little more than a simple vertical, static lifting frame with a free fall winch. As such it requires a stable platform base from which to operate. The heavy drop tools, suspended as they are by a single wire from the crown of the high mast, rapidly become dangerous and uncontrollable behemoths when deployed from any vessel which begins to pitch and roll. This makes free-fall cable tool methods generally incompatible with floating plant - other than in the most sheltered of applications.

By the late 1970s, the offshore oil industry, faced with having to install large and very expensive structures in water depths of up to several hundred metres, realised that alternative drilling systems were required for deep water site investigations. Existing methods, cable tool boring in particular, were entirely inappropriate offshore due to the depth of water, the poor drilling production, inhospitable environment, and the associated large costs of offshore vessels, support craft etc. Subsequently, research and development focused on three main aspects:
a)   rotary soils drilling with heave compensation
b)   high quality sampling
c)   insitu testing.

With these objectives in mind, offshore/nearshore geotechnical equipment development and methodology became highly polarised. As a result and for a considerable period of time, geotechnical investigations offshore were, and to a great extent remain, generally more sophisticated and advanced than those nearshore or on land (Ref. 1 Power et al 1994). However, these advanced methods for overwater investigations are now available for the nearshore environment. They have been further improved by the separate development of jack-ups specifically for coastal site investigations.

Unfortunately, the revised BS 5930, Code of Practise for Site Investigation (Ref. 2) pays little attention to the problems associated with overwater investigations and gives the practising engineer little direction as to the selection of the most suitable marine plant and investigatory techniques. It is hoped that the following sections of this paper will in many ways rectify this omission.

3. CRAFT SELECTION AND APPROVAL

The selection of a suitable principal craft, appropriate tender vessels and an efficient method of working with which to safely execute overwater investigations depend on many compounding factors that may apply to each particular site. Only a detailed desk study by a competent marine drilling engineer will permit an accurate evaluation of the various relevant parameters in order to select craft and methods of working that are entirely appropriate to each project and which will balance quality assurance against cost-effectiveness without compromising safety of the crews, plant, works, other mariners and any existing installations.

Site Location/Environmental Factors

The environmental conditions, to a large extent, dominate most other factors in craft selection. The type of location (harbour, river, estuary, coastline, etc.,) where the investigation is to be conducted gives a first indication of the likely general environmental considerations (currents, tide, winds, sea-states, visibility, etc.,) that may be relevant. Wave height, wave period and wave shape can be of particular significance depending on the prevailing wind direction, distance of fetch and water depth.

Water Depths/Bed Conditions

The range of water depths, the tidal regime, the extent of any intertidal drying out, and the bed conditions (smooth/hard, regular/irregular, clear/foul, etc.,) also play a major role in the plant selection process.

Navigational/Marine Restrictions

Other ship movements, existing sea bed services, fishing interests, moorings, and moored craft, the confines of navigable waterways or channels, contribute to craft selection. The proximity/size of the drilling vessel in relation to operational quays, jetties, locks, docks, etc., or the approaches thereto, and in particular, areas of special hazard such as petroleum or gas handling facilities, are also relevant when choosing the most suitable craft.

Timing of the Work

Weather/environmental conditions are of particular significance on coastally exposed sites, and hence the time of the year in which the works are to be executed can bear significantly on plant selection. Similarly, rivers liable to suffer sudden floods, or the effects of tidal bores need also to be considered in the work programme and plant selection.

Nature of the Site Investigation Works

Often neglected considerations in plant selection are the technical requirements of the investigation and the depths of the exploratory works, for the plant must not only be suitable for the assurance of the data acquisition, but must be capable of remaining on station throughout the exploration works without interruption.

Resource Availability/Competitiveness

In making the final decision as to an appropriate craft, at this point resource availability, cost sensitivity and competitiveness may jointly become the over-riding factor. It is here that experience and professionalism play one of their greatest roles; in that it is essential in this final adjudication that the forces of competitiveness do not compromise or be allowed to displace the vital principal factors.

4.   ADVANTAGES/DISADVANTAGES OF VARIOUS CRAFT

Floating Craft

Whereas a small anchored or spud leg pontoon barge with a simple static boring rig may be entirely adequate for a sheltered shallow soils boring project in, say, a non-tidal harbour, its suitability rapidly decreases in relationship to the exposure and remoteness of the site, the depth and technical requirements of the boreholes, and the water depths. As these considerations become increasingly demanding, they can rarely be satisfied by simply providing a larger barge. For conventional cable-tool boring, a heave of 0.5m is likely to be the 'maximum' vertical movement at the drilling position for the safety of operating personnel.

Anchors are generally inappropriate in irregular/boulder strewn waters since the wires are likely to become snared in the obstructions preventing sensible deployment and requiring the highest level of skilled seamanship if long delays are to be avoided.

Spud Leg Barges

Where the bed is irregular or where anchor wires are likely to be a potential hazard to other mariners, spudded barges may be more appropriate. Spud leg barges have the added advantage in that whereas it can take many hours to deploy an adequate anchor pattern, modern spudded craft have a very fast move capability. Spudded craft are generally only suitable for sheltered water working, since they are simply floating vessels which use legs as seabed gravity 'pins' in lieu of anchors to hold them on location.

Jack-ups

Jack-ups, (Fig. 1) as the name implies, having lowered their legs to the sea bed, can elevate the hull above the water. The stable, elevated deck of the jack-up is no longer subject to the environmental forces as are associated with all floating vessels, the effects of which are incompatible with conventional static drilling techniques.

Once elevated clear of the water, jack-ups are the ideal solution to shallow water drilling works. Typically they can operate in water depths of up to 25m. However, in gaining access to the site and when moving between exploratory hole locations, jack-ups must navigate as floating craft, and as such are subject to the same forces as with all floating craft. When afloat and moving in shallow water, the legs are raised high above deck level.

The effects of wind on the raised legs can be a major limiting condition. Conversely, in deep water, with the legs suspended at depth beneath the platform, the effects of tidal streams and currents may well be a major consideration, possibly restricting the craft to moving only at slack water.

Whereas much emphasis is placed on the leg length and elevating height capability of jack-ups, little consideration may be given to hull size, gross tonnage and leg bearing values, or the leg deployment and elevating speed capability. These factors can be an important consideration in plant selection as the following sections illustrate.

Jack-ups are possibly at their most sensitive when partially spudded but afloat. This condition arises when the craft's legs first make contact with the seabed, for example, upon arrival at a borehole location. The hardness of the seabed is a significant consideration, and the leg deployment, jacking speed and elevating method are also significant. The craft which can deploy its legs quickly and rise out of the water, is clearly preferable to slower craft. It should be noted that sand can be 'as hard as rock' to a spudding jack-up or to any craft that is required to 'take the bottom'.

Jack-ups come in a range of sizes and shapes and have varying numbers of legs. In selecting the appropriate craft, the risk of toppling failure is a dominant factor. While it costs significantly less to produce, maintain and therefore operate 3 leg jack-ups, it has been expensively learned that a jack-up having a minimum of 4 legs is desirable, not only to safeguard the works and the personnel but also with due regard to other mariners and marine installations. Jack-ups that carry out inshore geotechnical works generally only occupy each site for a brief period of time. However, weather can prolong the occupancy time at each site, and in either case, jack-up craft integrity is essential.

Excluding collision damage, toppling can be induced in many ways:
a) waves striking the elevated platform
b) scour of the foundation soils
c) sudden load bearing failure of the foundation soils beneath one or more of the jack-up legs (punch through failure).

Whereas the dangers of toppling failures mentioned should be avoidable with due diligence, the dangers of punch through failures - which are frequently sudden and are likely to be catastrophic - can be reduced by preloading of the foundation soils. To permit this, a minimum of 4 legs is required, which is why this configuration is the minimum now employed by all leading operators and manufacturers of inshore jack-ups. Preloading is achieved by alternately raising each diagonally opposite leg pair. This allows the foundation soils to be quickly loaded close to twice the normal bearing pressure.

Where soft, deep alluvial deposits are likely to be present, the jack-up which offers the lowest leg bearing pressure may prove to be the most effective. The heavier the leg bearing pressure of the jack-up, the greater the leg penetration will be, therefore reducing the effective water depth capability of the craft. Furthermore, it can be very time consuming to push long legs into deep alluvium, and more so when endeavouring to withdraw them.

Given that all jack-ups are most sensitive to environmental forces when afloat, active consideration must be given to the overall size of the jack-up. While operating costs increase generally in direct proportion to the size of the principal craft, the weather risk costs associated with different size craft require careful balancing in craft selection and tender approval. A small floating jack-up platform with towering, heavy legs can remain weather-bound for unacceptably long periods even in marginal weather, thereby discrediting the many general benefits of jack-ups as inshore drilling platforms. The more exposed and remote the site and the more demanding the environmental considerations, the greater the likelihood that a larger platform will prove to be more cost effective, and also that data acquisition and programme completion dates can be achieved without undue delay.

Extreme weather and gale force winds, will prevent every jack-up from moving. But extreme weather generally tends to be very limited in duration, the frequency dependent upon the time of year and the vagaries of the climate. By far the greatest demurrage costs are generally incurred as a result of 'marginal' sea conditions. It is here, while waiting to move, that wave height and wave period gain recognition as the principal critical factors in craft selection. Even where wind speeds are negligible or absent, and bright, clear weather prevails, running seas can continue without let up. Once more, the larger craft capable of moving in marginal conditions should have the clear advantage.

Regrettably, the general method of measurement in current tendering procedures, i.e. TBA (Ref. 3 and 4) specifies a set number of days for the Contractor to price for weather delays. This procedure actively discriminates against larger jack-ups, even where ideally suited, and denies the Engineer, with the responsibility for tender evaluation, appropriate flexibility for adjudication. To allow the selection and approval of craft to be based on the true merits of the many compound principal factors, it is the author's experience that a pre-set, lump sum for demurrage linked to a daily "rate only" item in the Bill is one method whereby the Engineer, in making his adjudication, can sensibly allocate suitable delay periods as appropriate to differing crafts.

With the advances that have been made in nearshore investigation techniques, just as it is inappropriate to place a traditional static, land based drilling rig on a weather dependent, low cost barge, it is equally inappropriate to accept such a drilling arrangement on a state of the art jack-up. The performance of the low cost drilling unit is incompatible with the high standing cost of the advanced jack-up.

A summary of potential marine craft and their general suitability given various criteria established from a desk study is presented in Table 1 below:

CRAFT	SPUD LEG BARGE	ANCHORED BARGE	JACK-UP
Intertidal			✓
Irregular Seabed	✓		✓
Smooth Seabed	✓	✓	✓
Deep Borings			✓
Insitu Testing			✓
High Tidal Range			✓
Shallow Water - Sheltered	✓	✓	✓
Deep Water - Sheltered	✓	✓	✓
Sheltered Water - Exposed		✓	✓
Deep Water - Exposed			✓

TABLE 1 - SUITABILITY OF MARINE PLANT

5.   SELECTION OF EXPLORATORY TECHNIQUES
General
It is vitally important that the most appropriate exploratory techniques are selected to investigate the soil/rock conditions pertaining to a particular site. The costs of having to repeat work or adopt conservative design parameters due to inappropriate or inadequate investigation techniques being used can be substantial particularly in the marine environment. As part of the selection process a desk study is essential to obtain all available geotechnical data from whatever source to assist in this selection process. The assessment of the geological conditions which exist at a site can be obtained from Geological maps and BGS databases as well as previous surveys from the same area.

The exploratory techniques which can be adopted fall into three main categories namely:
  a)   Geophysical/Remote Sensing
  b)   Borings with sampling/coring and down hole testing i.e. SPT, Pressuremeter
  c)   Insitu Testing using independent techniques i.e. cone penetrometer, or vane.

Geophysical Surveys
For certain projects it may be prudent to perform a geophysical survey prior to commencing the geotechnical investigation. Geophysical surveys can provide information on the following:

  a)   Bathymetry
  b)   Seabed topography and seabed obstruction if any
  c)   Subsurface soil/rock conditions

Using the data from a geophysical survey the selection of borehole locations and, if appropriate, insitu tests can be optimised.

Drilling Techniques

Although for many nearshore site investigations, conventional light cable percussive boring methods are still used, in recent years there has been an increasing use of rotary systems for coring soils. These systems i.e. Geobore S or SK6L have been used to establish the stratification of sites particularly in areas where there are complicated geological structures. The core is generally taken in a thin transparent plastic liner which enables it to be transported to the testing laboratory in a reasonably undisturbed state for detailed description and analysis.

Typically these systems are used in soils/rocks which are unlikely to contain significant proportions of coarse grained material. Cobbles and boulders in a weaker matrix can lead to core loss and poor recovery. In these soils which are typically glacial deposits, the light cable tool boring method is often still the most appropriate method of obtaining representative samples of coarse grained soils. The size of boulders sampled is often dictated by the diameter of casing used to drill the borehole.

Sampling Techniques

A similar range of sampling tools are available for use in nearshore site investigation to those associated with onshore work. Wireline operated systems are available for sampling in deep water but generally percussive techniques are still widely used. Piston samplers operated either by hydraulic or water pressure are also available for sampling the softer soils.

Insitu Testing

Insitu testing in the nearshore environment using various techniques is becoming increasingly accepted as a reliable means of obtaining accurate soils data efficiently and at reasonable cost. The most commonly used insitu techniques at the present time are:-

a)   Cone Penetration Testing (including the piezocone)
b)   Vane testing
c)   Pressuremeter testing
d)   Nuclear Density and other in hole logging tools.

Cone Penetration Testing (CPT)

The most widely used insitu testing technique in nearshore investigations is the electric static cone penetration test. This is due to the following factors:-

a)   The recognition of the effectiveness of the cone penetration test for soil profiling where the soil type is identified from the combined measurement of cone end resistance, local sleeve friction, and more recently pore water pressure.

b)   The ability to incorporate additional sensors in the cone to measure other soil parameters such as shear wave velocity, soil temperature and electrical conductivity, etc.

c)   The use of cone end resistance data in the design of foundations, particularly in granular soils.

In the shallower water depth range, up to about 25 metres, CPT's can be performed directly from the deck level of a jack-up using static land based CPT equipment. A jack-up platform is generally preferred, since the movement of a floating vessel can lead to problems with rod buckling and can adversely interrupt data acquisition and the thrust that can be applied.

A cone penetrometer rig can be readily mounted and fixed to the deck of the jack-up to provide up to 20 tonnes reaction. Prior to testing, a stabilising casing pipe has to be installed between the jack-up deck and seabed level together with inner sets of casing to provide adequate lateral resistance to the sounding rods during penetration.

There are also light seabed units available for testing the top few metres of soil for pipeline and dredging surveys. These units have a single fixed stroke length and are incorporated in a frame similar to that of a vibrocorer. Typically penetrations of up to 6 metres are obtained.

The CPT is the only available insitu testing technique which provides an accurate and continuous profile of soil stratification with depth. The repeatability of the electric cone penetration test is excellent due to the automated performance of the tests and the use of data loggers to record and process the results. The continuous recording, results in a more accurate profile allowing thin soil layers to be identified which may be overlooked in a conventional borehole. A typical CPT performed from a jack-up platform is presented in Fig. 2.

When pore water pressure is being measured using a piezocone the insitu permeability can be estimated by temporarily halting penetration and allowing the excess pore pressure to dissipate with time. Clearly, no vertical movements, however small, can be accommodated, and this criteria dictates the use of a jack-up platform.

Vane Testing
Vane shear tests can be carried out down the borehole using conventional land-type equipment mounted on the deck of the jack-up. Whilst the vane equipment can be fixed to the casing pipe when working on a floating barge this is rarely acceptable and a jack-up platform is the preferred solution. Whilst borehole vane equipment is the most commonly used, the penetration "Geonor" vane can be used in conjunction with a cone penetrometer rig.

The limitations of the vane test in the nearshore environment is that generally the vane equipment is normally calibrated for a penetration of up to 10m. If the depth of water is significant then the vane data can be less accurate due to the extended string of rods used.

Cone Pressuremeter
Although the self boring pressuremeter and Menard type can be used in the marine environment, the cone pressuremeter is now available for nearshore projects. This device is quicker, more simple in its operation and hence more economical than other types of pressuremeter. The use of the cone pressuremeter enables both the standard cone

parameters to be measured as well as deriving parameters from the pressuremeter. The cone data can be used to ensure that suitable depths are chosen for the pressuremeter tests.

Borehole Logging

Borehole logging techniques are being increasingly used in geotechnical investigations to provide a complete stratification profile. Generally the tool is run up and down the boring when the borehole has reached final penetration. A variety of wireline logging tools can be used which record selected soil and rock parameters continuously as the sonde's travel the depth of the borehole.

Most of the tools provide more detailed data if used in an open-hole i.e. if the drill pipe is withdrawn to the mudline. However if hole collapse is likely without casing, certain tools can be used inside the drill pipe, although some attenuation in the signals will occur. If operated from a floating craft, the wireline tool must be compensated to eliminate the vertical movement during the logging operation.

Site Supervision

Direct full-time supervision of all marine works by a geotechnical engineer is essential.

Due to the high cost of equipment spreads for nearshore geotechnical investigations compared to onshore, the scope of work for the near shore investigation is sometimes limited by cost. It is therefore essential that a geotechnical engineer is on site at all times to supervise the work. This enables the drilling and testing programme to be varied on site, and can considerably cut the costs of additional borings which may be necessary if supervision is inadequate.

The engineer is also able to make on site descriptions of the soil samples and carry out routine soil strength tests. This enables field records of the boring to be issued on site, with a degree of confidence. Characteristics of the drilling can also be monitored and recorded by the engineer. In particular if rotary soil drilling is performed then a record of mud pressure, bit load and drilling speed can assist with an interpretation of the soil conditions between the sampling intervals.

6.   CONCLUSIONS

- A nearshore site investigation should not be looked upon as merely an extension to a land-based investigation and appropriate equipment and techniques for working in the nearshore environment should be selected.

- Equipment and techniques developed for offshore oil related geotechnical investigations are now available for the shallow water nearshore environment.

- A detailed desk study is necessary to ensure that the most suitable plant is selected to suit the site environmental conditions.

- Significant advances have been made in insitu testing techniques, particularly cone penetration testing for nearshore investigations.

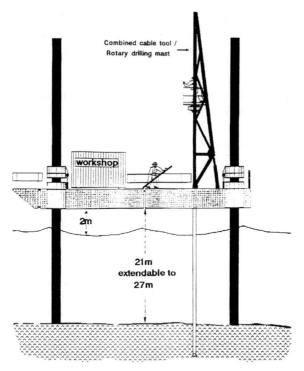

Figure 1. - Typical Jack-up Platform

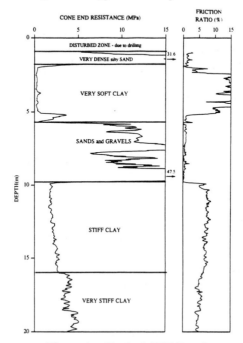

Figure 2. - Typical CPT Result

REFERENCES

1.   Tjelta T.I. (1992) Historical Overview of Geotechnical Design in the North Sea International Conference on Behaviour of Offshore Structures. Boss '92 London July 1992.

2.   BS 5930 (1981):   Code of Practice for Site Investigations, British Standards Institution, London.

3.   Specification and Method of Measurement for Ground Investigation.   HMSO, 1987.

4.   1993 Site Investigation Steering Group Specification for Ground Investigation, Thomas Telford, 1993.

Terrain systems mapping and geomorphological studies for the Channel Tunnel rail link

A M WALLER and P PHIPPS
Union Railways Limited, Croydon, UK

1.0 BACKGROUND

The Union Railways' Channel Tunnel Rail Link (CTRL) is to be constructed between St Pancras and the Channel Tunnel Railway Terminal at Cheriton near Folkestone. This high speed rail link is some 108 km in length and incorporates three major tunnels, numerous viaducts, bridges and cut and cover tunnels and large sections of cuttings and embankments. The route also traverses a wide range of ground features and solid and drift geology. A geomorphological approach has been used to assist in identifying and predicting the ground conditions relevant to the 'structures, earthworks and trackbed of the high speed rail link.

A Geotechnical Management Unit (GMU) was set up in 1993 by Union Railways. The unit has designed, planned and implemented:-

- a £2.4m Phase I and a £3.6m Phase II package of 'deep', 'shallow' and contaminated land investigations;
- a £0.30m package of geological and geomorphological field and desk studies; and,
- collated and computerised a geodata reference room with over 4000 relevant data items.

This paper describes the role in particular of the geomorphological studies and the use of terrain systems mapping in formulating and assisting the development of ground models for the route.

2.0 INTRODUCTION

A range of Geomorphological and Geological studies were undertaken to provide relevant information, in a structured system, for support of a Parliamentary Bill and for use by future designers and constructors of a high speed railway. The strategy was to develop ground models which would assist future design work and the planning of a cost effective and appropriate Phase II Ground Investigation. The work also focused on identifying types and forms of possible adverse ground conditions, thus enabling an assessment of potential geo-hazards and a quantification of the risks associated with the uncertainty of variable ground.

The CTRL studies have covered a wide corridor and have involved the collection of a large amount of information from ground investigations. However, the point data inevitably covers only a small fraction of the route. Two key concerns were identified from this observation and previous experience; firstly, ground models are essential to enable the interpolation and extrapolation of values of selected point data, and secondly a structured approach is required to managing the large volumes of data. The first has been addressed through terrain system mapping and geological studies, the latter by the storage and referencing of information by geosegments (a combination of up to 3 ground models which have definable boundaries, see Table 1). These two concerns are outlined in the following sections.

Table 1 Geosegment Summary Table

GEOSEGMENT CODE	GEOSEGMENT NAME	LENGTH (m)	CONSTRUCTION FEATURE	TERRAIN SYSTEM	TERRAIN UNIT
STP	ST PANCRAS TERMINUS	400	INTERNATIONAL TERMINUS	LONDON BASIN TERTIARIES	LONDON BASIN (I A) (CENTRAL LONDON)
NLL	NORTH LONDON LINE	7850	SURFACE AND TUNNEL		
STD	STRATFORD	2000	INTERNATIONAL STATION	OTHER MAJOR FLOODPLAIN	RIVER LEA FLOODPLAIN (IIA)
TSB	STRATFORD TO BARKING	7170	TUNNEL	LONDON BASIN TERTIARIES	LONDON BASIN (IB) (EAST END)
BDA	BARKING, DAGENHAM	5650	EMBANKMENTS	THAMES FLOODPLAIN	BARKING CREEK - LONDON ROAD VIADUCT (II B)
HRW	HORNCHURCH, RAINHAM, WENNINGTON	5780	EMBANKMENTS/ VIADUCTS		
PWT	PURFLEET, WEST THURROCK	4370	CUTTINGS/ VIADUCTS/(TUNNEL)	CHALK LOWLANDS -------------------	LONDON ROAD VIADUCT - LONDON ROAD PURFLEET (IVA) ------ --------------
TRC	RIVER THAMES CROSSING	3000	TUNNEL	THAMES FLOODPLAIN	LONDON ROAD - SWANSCOMBE MARSHES (IIC)
EBV	EBBSFLEET VALLEY	2050	EMBANKMENTS & CUTTINGS	CHALK LOWLANDS	GREEN MANOR WAY - EBBSFLEET RIVER (IV B)
GPH	GRAVESEND, PEPPER HILL	2050	SURFACE VIADUCT	TERTIARY OUTLIERS -------------------	PEPPER HILL (III A) -------------------
ISR	ISTEAD RISE, COBHAM	3640	AT GRADE PLUS EMBANKMENTS	CHALK DIP SLOPES (a)	PEPPER HILL - WROTHAM ROAD (VI A)
				CHALK VALLEY SIDE SLOPES	WROTHAM ROAD DRY VALLEY (V A)
				CHALK DIP SLOPES (a)	WROTHAM ROAD - BASE OF SCALERS HILL (VI B)
SCH	SCALERS HILL	3310	CUTTING	TERTIARY OUTLIERS	SCALERS HILL (III B)
CXC	COBHAM, CUXTON	3400	AT GRADE PLUS EMBANKMENTS	CHALK DIP SLOPES (b)	COBHAM - CUXTON (VI C)

Table 2 Terrain Unit Divisions of Identified Terrain Systems

TERRAIN SYSTEM	TERRAIN UNITS					
	A	B	C	D	E	F
LONDON BASIN TERTIARIES I	LONDON BASIN (CENTRAL LONDON)	LONDON BASIN (EAST END)				
THAMES FLOOD PLAIN II	RIVER LEA FLOOD PLAIN	BARKING CREEK LONDON ROAD VIADUCT	LONDON ROAD SWANSCOMBE MARSHES			
TERTIARY OUTLIERS III	PEPPER HILL	SCALERS HILL	COLEWOOD RESERVOIRS			
CHALK LOWLANDS IV	LONDON ROAD VIADUCT-LONDON ROAD, PURFLEET	GREEN MANOR WAY-EBBSFLEET RIVER				
CHALK VALLEY SIDE SLOPES V	WROTHAM ROAD DRY VALLEY	WEST R.MEDWAY VALLEY SIDE	EAST R.MEDWAY VALLEY SIDE			
CHALK DIP SLOPES VI	PEPPER HILL - WROTHAM ROAD	WROTHAM ROAD - BASE OF SCALERS HILL	COBHAM - CUXTON	NASHENDEN VALLEY	BLUE BELL HILL	

TERRAIN SYSTEM

CLASSIFICATION OF LAND DEPENDENT ON CHARACTERISTIC TOPOGRAPHY, DRAINAGE, VEGETATION ETC., CORRELATED TO GEOLOGY AND GEOMORPHOLOGY

TERRAIN UNIT

SPECIFIC AREA WHERE THE DEFINED GENERAL TERRAIN SYSTEM IS APPROPRIATE.

Table 3 Geomorphological And Geological Prioritisation For Each Terrain System To Assist Developing Ground Models And Identifying Adverse Ground Conditions

TERRAIN SYSTEMS	Approx Length (km)	Occurances	Bibliographic Dbase Compilation	BH DBase Compilation	Computer Ground Modelling and Geostatistics	Geomorphological Field Mapping	Detailed Aerial Photograph Interpretation	Specialist Database Searches	Input from Specialist Advisors relating to proposed scope of works			
									SR	B	FW	DI
London Basin Tertiaries	15	1	4	5	5	1	1	5	2	1	1	5
Thames Floodplain	15.5	2	5	5	5	1	2	5	2	1	1	5
Tertiary Outliers	3.9	2	4	5	4	4	4	3	5	1	4	3
Chalk Lowlands	4	2	5	4	4	1	4	4	4	4	3	3
Chalk Valley Side Slopes	1.7	4	5	3	2	5	3	5	4	4	3	3
Chalk Dip Slopes	12.1	5	4	5	4	5	4	5	2	4	3	4
Chalk Scarp Slopes	2	1	3	2	2	5	5	4	2	4	3	4
Chalk Forelands	1.3	5	3	3	3	1	1	3	4	5	1	2
Other Major Floodplains	2.7	2	5	2	2	5	3	3	2	5	3	5
Gault Clay Vale	9	6	4	2	1	4	4	3	3	3	3	3
Dissected Folkestone Beds Surface	19.8	2	3	2	2	5	4	3	4	3	3	4
Hythe Beds Surface	4	1	5	2	1	5	5	3	4	3	3	4
Hythe Beds / Atherfield Clay Dissected Surface	3.2	4	4	2	1	3	4	3	4	3	3	3
Sandgate Beds Lowlands	8	4		2		3		3	4	3	3	3

KEY

5 Most assistance in developing ground models and identifying adverse ground conditions
1 Least assistance in developing ground models and identifying adverse ground conditions

SR Specialist Report
B Bibliography Supply
F Fieldwork
DI Data Interpretation

28

3.0   GROUND MODEL DEVELOPMENT BY TERRAIN SYSTEMS

3.1   <u>Background to Terrain Systems</u>

At a macro-scale, terrain can be inferred from information on geology and structure, such as 1:50000 geological mapping. The CTRL traverses three distinct areas, the Thames basin, the North Downs and the northern limb of the Weald. At a scale of 1:10000, the combination of the geological materials, structural development and also past and contemporary environmental processes can be seen to have developed different features. Classification and assessment of these features have identified preliminary ground models for the route. The features manifest themselves in distinct land form assemblages (terrain systems), such as the Hythe Beds Escarpment, and it's these assemblages that repeat as separate terrain units along the route corridor, e.g. Hythe Bed Escarpments between Mersham and Smeeth, and between Ashford Station and Godinton Park.

At this scale of applied studies, engineering properties are also a function of geological materials, structural development and former and current environmental processes. The classification of the terrain can therefore provide boundaries to an appropriate ground model and assist in a systematic identification of relevant engineering properties of soils and rocks of significance in future design and soil-parameter models.

Table 2 presents an example of terrain unit divisions for some terrain system that were preliminary identified, from available map information, to be relevant to the CTRL - in all 14 generic terrain systems were adopted. Subsequently, these were confirmed to repeat as a sequence of 40 individual terrain units along the route. The identified terrain units are presented on maps such as indicated in Figure 1.

The use of physical characteristics to distinguish the ground model of an area has been used at a variety of scales for a variety of applications. However, the main difference between many previous studies and the CTRL is its linear form and the fact that the route traverses important terrain units in very short distances. Consequently, to be able to make a sensible assessment of these units requires that a variable corridor width be defined for different aspects of the studies. Although the boundaries to these areas represent a suitable division of the route ground model for field and desk studies, testing of the accuracy of these boundaries also forms an important part of the desk studies in checking the validity of the ground models.

3.2   <u>Task Prioritisation</u>

From the outset of the Geomorphological and Geological Studies, it was realised that there was a significant time restraint on the amount of work that could be undertaken as part of a Phase I study. Therefore, all available resources had to be managed in a controlled manner within the budgetary limits and within a justifiable geomorphological and geological framework. There were a number of specific tasks identified. Those selected were undertaken in the context of the generic ground model of the terrain system. The tasks comprised:

- specialist advisor input;
- bibliographic database compilation;
- borehole database compilation;
- computer ground modelling and geostatistics;
- geomorphological field mapping;
- detailed aerial photograph interpretation; and
- specialist database searches.

Figure 1 Geosegments And Terrain Units: Ground Model Uncertainties

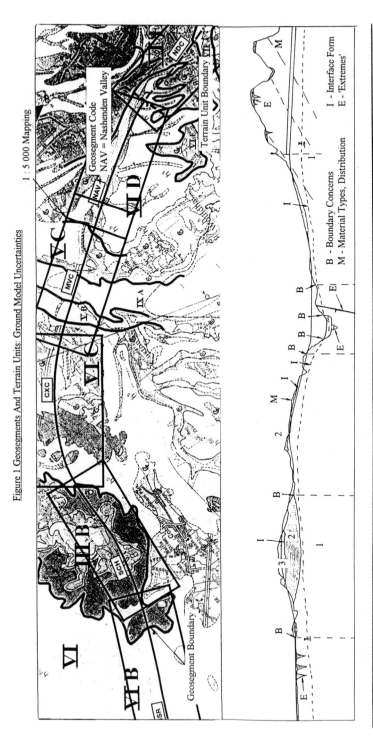

In order to prioritise resources, a simple five-point rating was developed on the basis of multidisciplinary experience to target which tasks would be most useful in which terrain system, see Table 3. From early on in development of the methodology, the GMU recognised the importance of framework reports on the geology, hydrogeology, stratigraphy and geomorphology and the use of the experience and judgement of Specialist Advisors in guiding fieldwork and reporting, particularly for products to be used by future engineering and construction users. The authors used this knowledge to identify the key aspects of uncertainty in the ground models (boundaries, interfaces, lithologies and geological 'extremes') and so focus the aims of the desk studies, fieldwork and reporting. Other than the geomorphological studies, the tasks undertaken include:-

- **terrain boundaries**; initially geological maps were used to assess terrain units, however geological mapping is an interpretation process and as such the base maps are not consistent for the whole route. Remote sensing imagery was examined to assist the assessment of boundaries (1:50000 scale), field mapping to refine the detail of the terrain in the Medway to Thames area, (1:10000 scale).
- **lithological materials**; the full range of lithologies and the local and regional controls on the distribution are not yet been fully identified. Bibliographic data searches, reference core logging and the collection of historical data and case study information are all being used to address this uncertainty by providing context for the GI information.
- **ground surface instability**; data on the type, location on a terrain basis and frequency of certain features are being collated to assess the potential hazards and risk. Data collated includes information on:-
    - Mass movement features - landslips; and
    - cambering and valley bulging.
- **weathering features**; data on the type, location on a terrain basis and frequency of dissolution features, particularly in the chalk is also being gathered and assessed together with geophysical field work to establish the potential hazards and risk. The features include:-
    - natural cavities, e.g. solution pipes; and
    - man-made cavities, e.g. deneholes.
- **quaternary features**; borehole data and known features are being examined along with computer generated geo-statistical plots for the potential for other features, such as:-
    - buried channels;
    - scour features; and
    - drift-filled hollows.
- **man-made ground**; made ground and fill sites are being examined in a structured approach that includes the recording of the quality and certainty of the data to enable hazard and risk assessments to be made. Map inspections and aerial photographic interpretations of morphological and tonal features are being undertaken on a route wide basis, together with historical map inspection on a site by site basis.

4.0 GEOMORPHOLOGICAL FIELD MAPPING

4.1 Field Studies

Both geomorphological and geological field studies have been carried out; these include:-

- geomorphological field mapping of 'type sites' of terrain systems by Union Railways;
- geological field mapping between the Thames and the Medway by the British Geological Survey (BGS); and
- stratigraphical field mapping of key chalk marker horizons by Dr R Mortimore from the Medway and across the North Downs.

To develop the desk based generic ground models for each terrain system the technique of field morphological mapping was adopted. The aim was to carry out mapping of each terrain system and then a geomorphological interpretation at a scale appropriate to the engineering of the CTRL. Morphological mapping enabled the ground surface form to be described on an objective and consistent basis, and provided a basis for numerous field observations. The geomorphological maps provided a synthesis of, the ground surface form, the properties of the near surface materials, (in particular superficial deposits but also geological strata) and the type, magnitude and products of contemporary or relict environmental processes that contribute to the surface forms.

4.2   **Geomorphological Field Mapping of "Type-Sites"**

Type sites were chosen for each terrain system and a detailed geomorphological mapping exercise undertaken at scales of 1:1250 or 1:2500. Two to three days were spent to cover areas up to 0.75 km². For the Phase I geomorphological fieldwork, a total of nine areas (covering over 20 km of the CTRL route) were visited as type-sites in a 50 man-day programme. The sites were:-

-       Scalers Hill NW (Tertiary Outliers)
-       Nashenden Valley Tunnel Portal (Chalk Dip-Slopes)
-       River Medway Valley Sides (Chalk Valley Side-Slopes)
-       North Downs Crossing Tunnel Portal and Approaches (Chalk Forelands)
-       Boxley Abbey and Longham Wood Areas(Gault Clay Vale)
-       Warren Wood Area (Dissected Folkestone Beds Surface)
-       South of Smeeth (Hythe Beds Cuesta and Dissected Atherfield Clay Surface)

Two or more terrain systems were mapped in some cases for convenience. Others were selected to provide data at boundaries, so as to address railway trackbed support concerns at transition zones between different terrain units. Following the morphological mapping, each terrain system type-site was subdivided into terrain facets from the geomorphological observations and interpretation. The facets included interfluve surfaces, valley bottoms, steep side slopes, etc.

For each of the terrain facets the engineering geomorphologist assessed the characteristic slopes, drainage, materials, associated relict and contemporary processes, and engineering concerns. Borehole data, field observations of drift and solid exposures and hand augering in the type site areas all contributed to the process of defining the terrain facet types. The specialists then reviewed, in the field, the proposed facets for their accuracy of description and appropriateness to engineering and construction.

4.3   **Extension Work For Each "Type-Site" and Development of Ground Models**

Various techniques were employed to map accurately, but quickly the remaining part of the CTRL route with the identified terrain facets. As many of the boundaries were morphological it was possible to extend the identification of the various facets away from the type-sites using aerial photographic interpretation of colour stereoscopic pairs at 1:2,500 scale. Most of the route is in open fields so that this process proved very fast and accurate to <5m. Even in wooded areas the morphology was generally identifiable, although the accuracy is accordingly reduced. Ground truthing was and continues to be undertaken. Other slope classification methods have been examined, together with the use of MOSS programmes to classify the available digital ground model. In some instances reconnaissance mapping has been carried out to identify the distribution of the facets. The final maps produced identify for each terrain system a number of repeatable terrain facets and consequently provide the following:-

-       a statement of the contemporary landscape;

- an insight to the development of the landscape through recognition of relatively older and younger surfaces;
- a basis for structuring the location and setting of natural features such as solution pipes, cambering observations, drift filled hollows, sarsen stones etc.;
- a basis for identifying obvious limits to any ground model and engineering descriptions of the model;
- a means to interpolate and extrapolate near surface point data with a qualified degree of certainty;
- a robust and objective statement on the characteristics of the terrain in each facet;
- a preliminary interpretation and set of assumptions for the ground model on the drift and solid strata and geological structure which underlie the surface terrain; and,
- a synthesis of the ground model concerns likely to be relevant to the engineering of a high speed rail link.

Figure 2 presents an example of the completed facet maps, comprehensive keys and summary tables for the type-sites as identified.

4.4 **Adverse Ground Conditions**
All tonal and textural features identified during an objective route long aerial photograph inspection, were overlain on the facet maps. In many cases these features were directly explained by the natural morphology. For example, dry valleys in chalk lands tended to be darker on the aerial photos than the surrounding slopes reflecting the thicker superficials that hold more moisture, etc. Some features have a morphological expression but are not part of the natural land form assemblages. All such features that were still lacking an obvious geological or geomorphological cause were subsequently checked against other existing data sources, consisting of:-

- Historical Land Use Maps;
- Archaeological Maps; and
- Services Plans.

Any features still unidentified were then reported on with possible causes highlighted to ensure that all such anomalies were recorded for consideration in future design and construction.

4.5 **Reporting of Terrain Systems and Ground Models**
For each terrain system, a schematic 3-D ground model has been prepared. These show how the surface features relate to processes, underlying geological materials and structure. These were used to raise the awareness of the type of ground conditions expected in each terrain system at an early stage of the studies, see Figure 3. The main Phase I report comprises the type site morphological mapping, the facet maps and summary tables and a series of specialist stratigraphic, geological and geomorphological framework reports.

In addition for a Phase II report, a series of composite drawings are being produced along the centre line of the route for each geosegment. These present the ground models related to the specific terrain units that have been identified from field work and aerial photograph extension work. They comprise 1:5000 horizontal plans of the terrain facets, 1:500 vertical sections with borehole logs and geohazard information on the plans, sections and in table form, see Figure 4.

Figure 2 Example Geomorphological Map, Key And Table

KEY: TERRAIN SYSTEM X : GAULT CLAY VALE

FACET	NAME
I	Interfluve Plateau Surface
II	Gault Clay Vale Side Slope
III	Steep Gault Clay Valley Side Slope
IV	Valley Bottoms
V	Dry Valley Bottoms

Facet Code	Facet Name	Slopes	Materials	Drainage	Contemporary Processes	Relict Processes	Engineering Concerns
II = □	Gault Clay Valley Side Slopes	7°–4°	See I.

Soliflucted zone is likely to be nearer to maximum proposed thickness of 7m owing to accumulation processes on these lower angled slopes. | No perennial drainage features are present except where influences of man is present. Ditches, ponds and mainly piped ground are associated. | Wash and creep process are ongoing with accumulation occurring lower on the lowest angled slopes. This is mainly applicable to fines. Slow landsliding may be ongoing in specific locations, mainly triggered by fluctuations in ground water levels or disturbance by mans activities. The types of movement will tend to be translational and shallow (<3m) affecting the solifluced Gault (where present) and/or deeper rotational movements that could involve material from the ground surface down into the weathered zone. | See I.

The actual Gault Clay sub-surface profile gives evidence for past processes that have effected the landscape. The presence of solifluced materials is more likely on these valley side slopes where successive phases of material moved towards the valley bottoms would have been more likely to occur than on the plateau surface. Slope surfaces that are oriented sub-parallel to the ground surface are probable in these areas providing a distinct boundary between the transported material above and the cryoturbated weathered Gault Clay below. | See I.

The presence of potentially increased shear surfaces sub-parallel to the ground surface with reduced parameters or residual soil has serious implications for any excavations which exposes these and may lead to reactivation of the movements. The above problem also exists for embankments that are constructed on the sloping ground. Any type of foundation may have to deal with differing responses to any applied loads owing to the variability of the ground profile both in depth and spatial. |
| III = □ | Steep Gault Clay Valley Side Slopes | >4° | See II.

The thickness of the solifluced zone would tend to be nearer its lower required limit of 1m than on the side slopes that are of less than 4° | See II. | See II.

Potential for contemporary instability is greater than in facet II as slopes are generally inclined at angles greater than the reported angle of repose for Gault Clay slopes of 4°. There is more morphological evidence for deeper rotational movements in the Gault Clay, which may extend down into the weathered zone. If surface morphology is not diagnostic of instability then the area is probably OK under contemporary conditions. | See II.

Ploughing, for example, may have destroyed the morphological evidence of past instability even though the associated shear surfaces at depth may exist. | See II.

Potential for re-activating relict instability is greater than the lower angled side slopes. |

Figure 3 3-D Diagram

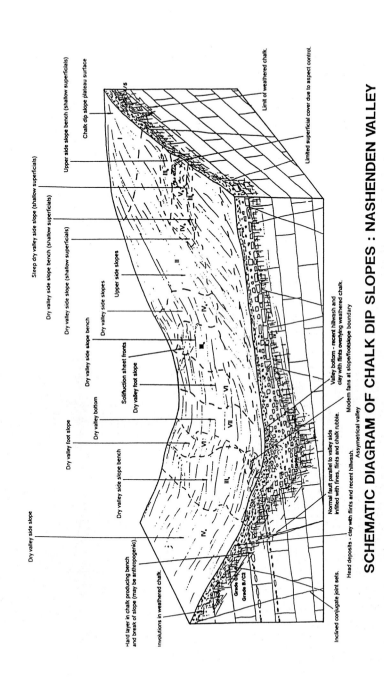

SCHEMATIC DIAGRAM OF CHALK DIP SLOPES : NASHENDEN VALLEY

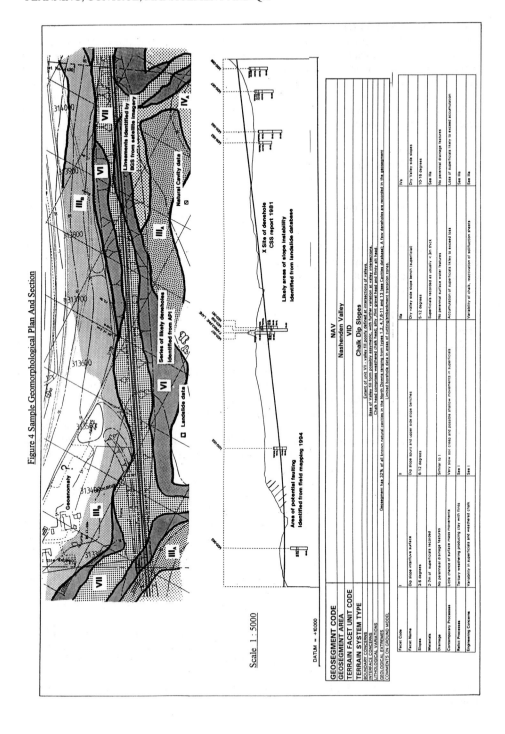

Figure 4 Sample Geomorphological Plan And Section

5.0   REFERENCING GEOLOGICAL DATA

As noted in the introduction, the second objective of the geological and geomorphological studies was to provide an information system, structured to support the geotechnical design process and the Parliamentary Bill. The approach adopted was to divide the route into geo-segments, where each geosegment is a package of up to 3 terrain units. These provide the means of sorting, referencing and presenting the relevant geo-information for each ground model on a single composite plan and section drawing. The geosegments were determined by considering the following aspects in a descending order of priority:

-    terrain units;
-    major infrastructure items; and
-    length of route (3 km maximum for presentation at 1:5000 scale A1 drawing).

All geological and geotechnical data acquired by URL has been classified by geosegment with key information being entered onto computerised databases. A reference room has been developed which holds over 3500 relevant borehole records; 350 ground investigation reports; 200 geotechnical case studies, theses and publications; 150 specialist geotechnical papers and all the reports and drawings prepared by Union Railways covering the geotechnical works since 1989. All the key information is coded by geosegment but can also be searched by a variety of location criteria through the computerised databases or visual inspection of a simple GIS system.

6.0   CONCLUSIONS

There are a number of ways that the geomorphological uses of terrain systems mapping for ground model development have benefited the CTRL project.

-    The geomorphological approach of generic terrain systems has provided a rapid and sensible hierarchical framework for the whole route ground conditions to be set into context.
-    The terrain unit mapping has provided appropriate ground model boundaries in terms of the geological and geomorphological concerns to engineering and construction.
-    As geology, structure and weathering are the key controls on terrain and engineering properties, so the geomorphological models in areas of drift deposits and past periglacial processes are likely to be a main part of any engineering soil parameter's model developed for near surface structures.
-    The ground models are based on identification of terrain units. These translate directly or in combinations to a simpler geosegment basis. These provided the means to present a large range and volume of information in packages suited to design and construction requirements.
-    The ground models provide the basis for prediction of dispersed point data, e.g. boreholes and trial excavations, through a structured framework and a set of assumptions on boundaries between ground models, interfaces between strata in the models, lithological variations within the strata and the presence or absence of geological extremes.
-    From an understanding of the environmental processes the geomorphologist has identified features that may be indicative of significant active contemporary geomorphological processes or past relict ones and provide structure to the assessment of specific hazards such as natural cavities.
-    The identification of a robust set of terrain unit boundaries and geosegments, have enabled a structured approach to the assessment of geological uncertainty and guided specialist studies and ground investigation to areas of greatest risk in the design and construction of the CTRL.

The geomorphological studies have also directly benefited the ground investigations, through the following:-

- They provided a robust terrain classification to assist, and ensure a consistent and appropriate approach, to the ground investigation designs in areas of similar terrain;
- The findings from the Phase I Geomorphological and Geological studies were used to prioritise areas for Phase II Ground Investigations as the uncertainties are intrinsically linked to the soil-parameter concerns and objectives of the geotechnical design of the CTRL.
- The summary report for each of the type-sites provided recommendations for the type of ground investigation, that are required to obtain suitable information in refining the generic ground models; these recommendations provided the basis of ground investigation. They were added to, so as to address design concerns other than geological, e.g. structures, foundations, earthwork materials, contamination conditions, etc.
- The actual numbers, types of investigation methods and specific locations were developed from the findings of Phase I ground investigations and studies. It was possible, with close liaison with the engineers involved with the design of the Phase II ground investigations, to target locations in a way that decisions on methodology were backed by sensible considerations derived from the existing ground models, facet maps and geomorphological mapping.

The results of the recent ground investigations have been integrated to update the individual ground models. This approach of hierarchical ground modelling with the review of historical and recent GI data to revise the ground model, has enhanced the certainty in the assumption on boundaries, interfaces, lithological variation and geological extremes. It has also clarified the ground model features that are relevant for predicting the extent of the units in soil-parameter models in design. Other relevant studies to be completed are aiming to enhance the potential use of the terrain based ground models. These are:-

- incorporating the geological materials and their depths to provide a full three dimensional ground model for future geotechnical design; and,
- preparation of semi-quantitative assessments of the uncertainty within the different route ground models.

The geomorphological approach based on geo-segments terrain systems, units and facets will continue to at least mid 1995. The studies will then have provided a major contribution to, identification, assessment and quantification of the significant ground conditions associated with the design and construction of the CTRL.

Acknowledgements
We are grateful to the staff of the Union Railways and the Geotechnical Management Unit for the assistance in the course of these studies and preparation of the paper, in particular the assistance of David Harris in the field mapping and production of the figures and Professor R. Mortimore (Brighton University), Dr A. Thompson (Travers Morgan Consulting Group), Dr J. Griffith (Plymouth University), Dr David Bridgland (Durham University) and R. Ellison and B. Lake (BGS) as specialist advisors. The information is presented with the kind permission of Union Railways, however the opinions expressed in this paper are those of the Authors and do not necessarily represent the views of these organisations.

Development of a knowledge-based system for ground investigation in soil and rock

M. G. WINTER, G. D. MATHESON & P. McMILLAN
Transport Research Laboratory Scotland, Craigshill West, Livingston, West Lothian EH54 5DU, UK.

SYNOPSIS: A knowledge-based system for ground investigation on Scottish trunk road projects is being developed. The objective is to increase the cost-effectiveness of such investigations. The initial structure is common to both soil and rock and reflects the common division of a ground investigation into preliminary, main and additional phases. The paper describes the initial structure and gives further information on more detailed aspects of ground investigation procedures in both soil and rock. Applications include the design of ground investigations, the checking of procedures followed in existing or planned investigations, and training.

INTRODUCTION

A knowledge-based system is being researched by the Transport Research Laboratory, Scotland for the Scottish Office Roads Directorate. The project is divided into soil and rock components and early work was carried out in collaboration with the Departments of Civil Engineering at the Universities of Glasgow and Strathclyde respectively (Winter & Matheson 1992; Thomas, Kor & Matheson 1992). A common initial structure, incorporating the concept of phasing in ground investigations, and an interactive menu screen have been adopted. Thereafter the structure and logic increasingly diverge as the differing principles of soil and rock mechanics become dominant in the ground investigation.

The present paper outlines the structure and logic of the initial structure and compares and contrasts the approaches to ground investigation in soil and rock.

OBJECTIVES

The main objectives of a ground investigation for a highway project are to provide information on the conditions likely to be encountered during construction and to acquire data which can be used in design. These allow an optimum route to be chosen, appropriate construction techniques to be selected, and structures to be designed in accordance with the host ground conditions. A ground investigation is therefore an essential requirement in all highway projects.

Successful projects are often those which have accurately forecast the out-turn costs of the construction contract. In practice, out-turn costs generally exceed those forecast by significant amounts (Matheson & Keir 1978; Anon 1991). All cost increases during the construction contract contribute to higher out-turn costs, including inflation, variations and claims. Many claims relating to earthworks have been specifically caused by deficiencies in the ground investigation, often in highly variable and moisture sensitive soils and where there has been incorrect interpretation of in-situ rock conditions. There is therefore considerable scope for cost savings by increasing the effectiveness of ground investigations.

Deficiencies in ground investigation have been related not to the shortcomings in state-of-the-art but to poor planning, to the selection of inappropriate investigatory techniques, and to the acquisition of unrepresentative data on ground conditions (Matheson & Keir 1978). Although problems have on occasion arisen from inaccurate test data, many are caused by shortcomings in the expertise and experience of the engineer designing the investigation.

Clearly a system which would assist the efficient planning of a ground investigation and help select the most appropriate techniques would be of benefit. This could result in more effective investigations, cost-effective design, better estimation of construction costs and a decrease in the level of claims.

The objectives in developing a knowledge-based system are therefore to summarise expert knowledge and to present it in an easily accessible format. The user can thus be advised, guided and assisted in the planning and implementation of effective ground investigations.

SYSTEM STRUCTURE AND LOGIC

The system structure is based on the need to divide a ground investigation into preliminary, main and additional phases (Matheson 1979a). This allows information to be collected in a planned and systematic manner. During the investigation the nature of the data collected changes from being qualitative to semi-quantitative in the early phase to being quantitative in later phases. The preliminary phase is thus intended to indicate rather than define the conditions likely to be encountered and the main phase to define the conditions important to engineering construction. The additional phase is used to augment information obtained during the earlier phases and generally takes the form of specific studies. To encourage systematic and progressive collection of data and avoid duplication, the user should be given the opportunity of reviewing previous work before embarking on any of the later investigatory phases.

Although a large and complex ground investigation would be expected to include all three phases, it is recognised that an additional ground investigation will not always be required. Further, for small and simple projects constructed in geologically simple regimes, a small scale investigation may be adequate. Typically, aspects of both the preliminary and main ground investigation are incorporated into such investigations.

The system is therefore structured (Figure 1) in terms of an introduction, in which the basic principles are explained, and three possible phases of ground investigation each of which logically and progressively collects information on ground conditions. Reports listing the techniques selected can be generated for each investigatory phase or for the full investigation.

Introduction

Within the introduction the knowledge-based system is described in terms of Objectives, Scope and Method. Objectives sets out the basic intentions of a ground investigation; Scope introduces the system hierarchy and extent, and the type of advice and assistance provided; Method explains the interactive screen layout (Figure 2) and its use. A single main screen has been developed which allows access to the system through user selection of prompted options, and which displays text information and guidance. The interactive screen is described in detail by Thomas et al. 1992.

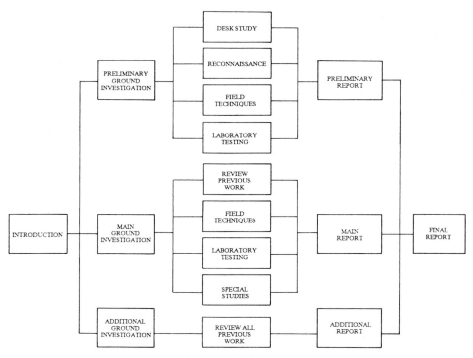

Figure 1 - Structure of the knowledge-based system (soil and rock).

GIMAIN   MAIN GROUND INVESTIGATION   LEVEL 2
                    Gimain/Field techniques/Exploratory holes
Surface excavations Percussion boring Rotary coring Open hole drilling

This will provide surface excavations procedures

Various methods may be employed to explore the subsurface conditions in rock.

There are no hard and fast rules about the depth and spacing of exploratory holes. However, general guidelines may be applied.

Boreholes should be placed and of sufficient depth to identify the distribution, nature and variability of rock materials and structure relevant to the proposed construction.

A view must be taken on the likelihood of variations to the horizontal and vertical alignment and whether these should be catered for in the ground investigation.

PgDn

F1Help   F2Info   F3Quit   F4Exit   F5Prog str   F6Menu str   F9Log

Figure 2 - Example interactive screen for ground investigation in rock.

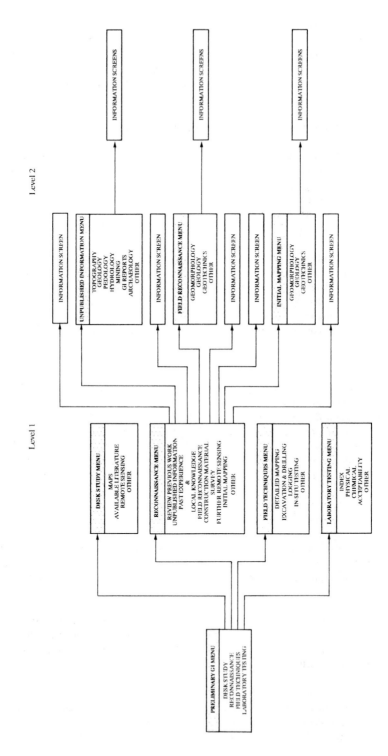

Figure 3 - Example menu structure for preliminary ground investigation in soil.

Preliminary Ground Investigation

The main objectives of a preliminary ground investigation (Matheson 1979b) are to provide an overview of ground conditions, to highlight potential problems, to facilitate route selection, and to evaluate the need for, and the content of, a main ground investigation. The data should be qualitative or semi-quantitative. At this stage a high level of detail is neither necessary nor desirable and it is sufficient to indicate rather than define the ground conditions. The information acquired allows the main investigation to be planned and effective methods and techniques of acquiring quantitative data chosen.

The first stage in a preliminary ground investigation (Matheson 1979b) is a desk study to collect and collate all published and unpublished information pertaining to the site. The menu structure is therefore organised accordingly (Figure 3) and includes the study of Maps, Available Literature and the results of Remote Sensing, including air photographs. The association of menu items reflects the fact that all information in this stage can be readily obtained from within an office environment, without site visits, and without external contact.

The second stage involves a reconnaissance to acquire data on the site and the menu items include Unpublished Information, Past Experience & Local Knowledge, Field Reconnaissance, Construction Material Survey, Further Remote Sensing and Initial Mapping. Included in this stage are subjects which require direct contact with either the proposed site itself or with other organisations. The idea is to quickly survey a corridor of interest, make observations on field conditions, highlight potential problems and collect local knowledge and experience.

In the third and fourth stages those aspects which had been identified for further study by the reconnaissance survey are investigated in more detail. These can be categorised in terms of Field Techniques and Laboratory Testing respectively. Field Techniques include Detailed Mapping, Excavation & Drilling, Logging, and In-Situ Testing. An "Other" menu category is included for subsidiary subjects such as Geophysics and Instrumentation. Laboratory Testing includes Index, Physical, Chemical and Acceptability tests. The test techniques actually used will depend on the data required and the ground conditions. The results act as a guide for specifying a more detailed testing programme in the main investigation phase.

The menu structures for desk study and reconnaissance are very similar for both soil and rock. Figure 3 shows the menu structure for preliminary ground investigation and expands fully the reconnaissance stage for soil. In this instance the menu structure for rock is very similar, with only two changes from the soil structure: in the unpublished information menu the collection of data on pedology and archaeology are excluded. In other categories, such as topography, the emphasis will be different for soil and rock but the basic logical structure remains the same.

It is at a detailed level in field techniques and laboratory tests that the differences between ground investigation in soil and rock become apparent. The methods and techniques available and their logical organisation reflect these differences. This is discussed below, in the context of the main ground investigation.

Main Ground Investigation

The purpose of the main ground investigation is to provide the detailed geological and geotechnical data necessary for the design of the construction works, including the engineering structures. This phase may be carried out a considerable time after the

43

preliminary investigation, be designed by a different consultant, and be conducted by a different contractor. It is therefore essential that, prior to designing the main investigation, all previous work is reviewed and evaluated, and possible shortcomings identified.

The main investigation draws on most of the methods and techniques available at the preliminary investigation phase, the difference being the increased scope and level of detail. The same menu structure is therefore used. However a category has been introduced to cater for situations where special studies, such as field or other trials, may be necessary to evaluate, for example, the full-scale behaviour of the ground to structural loading. It should be noted however that special studies are considered to be atypical of routine investigations.

The laboratory test menu expansions for soil and rock (Figures 4 & 5) show subtle differences. These reflect the fact that a far greater emphasis is placed on laboratory testing in the investigation of soils than of rock. In particular, the focus in the soil structure on index tests is not apparent in the rock structure. There is also a much broader spectrum of chemical tests for soil than there is for rock. In rock investigation the emphasis is placed far more on in-situ tests and logging to determine the rock mass structure.

Additional ground investigations

Additional ground investigations are intended to supplement or augment existing information on ground conditions. These may be necessary because of a route alteration, after changes in proposed structures or their design, to remove uncertainty regarding an evaluation based on limited or suspect data, or to confirm a proposed course of action.

Reporting

A record of the route taken through the system can be kept and reported in terms of the individual phases or of the investigation as a whole. The intention is to allow sequential progression though the system without automatically recording the route through the system structure or the investigatory techniques considered. Positive action will be required to record a selection; this will permit browsing whilst still retaining a list of the techniques chosen.

IMPLEMENTATION

Two levels of detail have been defined within the present knowledge-based system. The first, Level 1, includes the basic organisation (Figures 1 & 3 to 5) and extends to giving general advice on the purpose of each phase and stage. Level 1 is common to both soil and rock system components. The second, Level 2, gives more detailed advice on the planning of investigations (Figure 3 to 5) and assists the selection of appropriate methods and techniques. It is in Level 2 that the differences between ground investigation in soil and rock become apparent, in terms of the methods and techniques available, and their logical organisation.

A user interface flowchart showing the facilities available and the technique of interrogation is given in Figure 6. The interactive screen (Figure 2) is central to the interrogation technique and its design is common to all menus at both levels of detail. This simplifies data exchange and standardises the user interface. Full help facilities are available and basic guidance and assistance appropriate to the menu option can be accessed. References are provided when appropriate. Additional information is available to the user in the form of Further Guidance and Contract Experience. The intention is to regularly update this knowledge-base with experiences from other highway projects. The user will thus be kept in touch with problems and developments.

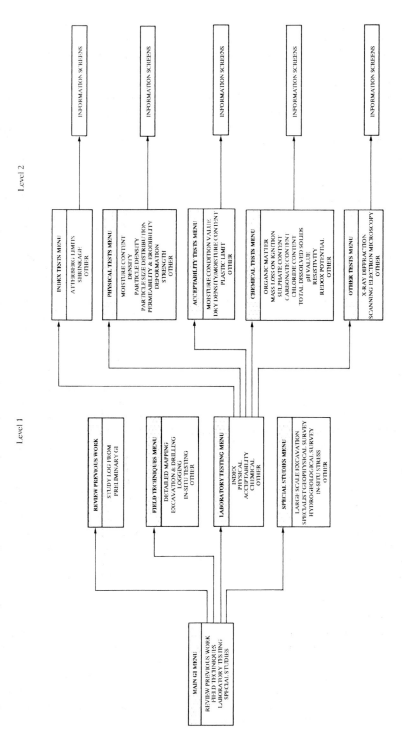

Figure 4 - Example menu structure for main ground investigation in soil.

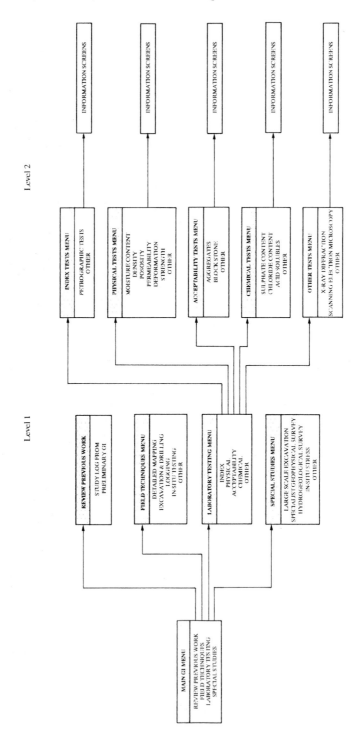

Figure 5 - Example menu structure for main ground investigation in rock.

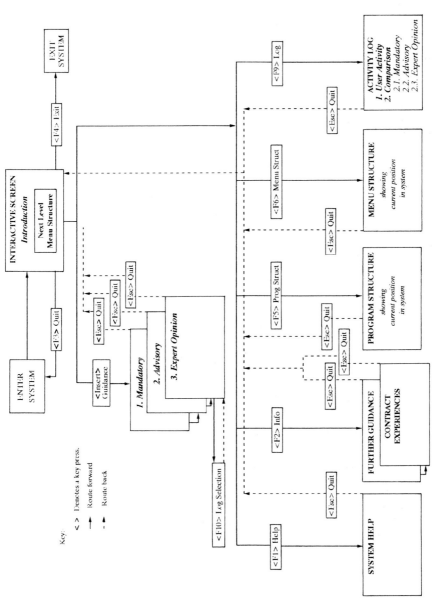

Figure 6 - Knowledge-based system for ground investigation - user interface.

A facility to view graphically the structure of either the program and the menu, and highlight the current position under interrogation, is available. This allows the user to monitor progress though the system.

The system uses a logical approach to tailor the advice proffered to the problem in hand. Questions are asked of the user at stages relevant to the ground investigation process. For example, in the laboratory testing section, soil type will determine the type of permeability test which will be most appropriate; falling head tests are relevant to cohesive soils while constant head tests are relevant to granular soils. This input of relatively simple data by the user thus allows the system to guide the selection of appropriate ground investigation techniques according to the prevailing geotechnical conditions and the structural unit under consideration.

A facility within Level 2 permits the user to log the investigatory techniques selected and report on the methods and techniques chosen. The resultant activity log can also be interrogated on completing each phase and compared to a list of mandatory and advisory procedures, and to expert opinion. The mandatory and advisory procedures are drawn from Specifications (Anon 1987), the Design Manual for Roads and Bridges (Anon 1993a; 1993b; 1993c) and British Standards (Anon 1981). Expert opinion is derived from journal and conference papers, textbooks and personal expertise and experience. The content of existing or proposed ground investigations can therefore be checked against specifications, established practice, and expert opinion. The idea is to detect omissions which might impair the effectiveness of an investigation.

HARDWARE REQUIREMENTS

The system has been programmed using the Leonardo© expert system shell running under MSDOS© using conventional PC-type microcomputers. The final system requires a minimum 20286 microcomputer with hard disc, a VGA screen and a standard printer. However, speed of interrogation is greatly improved by the use of faster processors.

POTENTIAL FOR EXPANSION

The present system is intended for use on trunk road projects in Scotland. However it could be modified and expanded to encompass other geographical areas and other types of construction project. The addition of a further level, Level 3, would allow the system to evaluate the data obtained in a ground investigation and guide its use in design. Further expansion could thus take the system into the design process. Such design-based systems are at present being developed in the UK to deal with, for instance, the design of retaining walls. It is thus possible to envisage a system that will guide the engineer through the processes of ground investigation, the evaluation of the data obtained, the selection of design data, and the eventual design of the construction works.

SUMMARY

This paper concerns the development of a knowledge-based system for ground investigations in highway projects in Scotland. The objectives are to increase the effectiveness of such investigations, improve cost forecasting and reduce claims arising from unforseen ground conditions.

Guidance has been taken from the mandatory or advisory requirements of British Standards and the Design Manual for Roads and Bridges, good practice contained in professional publications, and expert opinion. The system conforms with the principles of phasing of a ground investigation. This involves a systematic and progressive collection of data on ground conditions during preliminary, main and additional phases.

A common initial structure and interactive menu screen has been developed. Menu screens and menu trees have been flowcharted to allow the systematic and progressive collection of data. The help, guidance and advisory facilities available are comprehensive and permit a check against mandatory and advisory procedures, and expert opinion.

The advice offered by the system can be tailored to suit the prevailing geotechnical and structural conditions and so guide the selection of appropriate ground investigation techniques.

Applications of the system include the design of ground investigations, checking the content of existing and proposed investigations, and training.

ACKNOWLEDGEMENTS: The Authors are grateful to the following for useful discussions and suggestions: Mr A P Gunning (formerly TRL Scotland); Professor D Muir-Wood, Mrs X Xue, Dr A H Chan and Dr P Smart (Glasgow University); and Drs K C Yeo, P R Thomas and F H Kor, Professors A McGown and K Z Andrawes (University of Strathclyde).

The work described in this paper forms part of a Scottish Office funded research programme conducted by the Transport Research Laboratory, Scotland (Director, Dr G D Matheson), and the paper is published by permission of the Scottish Office and the Chief Executive of TRL.

REFERENCES
ANON 1981. Code of practice for site investigations (Formerly CP2001), BS5930, *British Standards Institution*, London.

ANON 1987. Specification and method of measurement for ground investigation, Department of Transport, Department of the Environment, HMSO, London.

ANON 1991. Inadequate site investigation, *The Institution of Civil Engineers*, London.

ANON 1993a. Documentation requirements for ground investigation contracts, Highways, Safety and Traffic Departmental Standard HD 13/87, Design Manual for Roads and Bridges, Volume 4, Section 1, Department of Transport, Scottish Office Industry Department, The Welsh Office, The Department of the Environment for Northern Ireland, HMSO, London. (Including Scottish Addendum.)

ANON 1993b. Ground Investigation Procedure, Highways and Traffic Advice Note HA 34/87, *Ibid.* (Including Scottish Addendum.)

ANON 1993c. Earthworks; design and preparation of contract documents, Highways, Safety and Traffic Departmental Advice Note HA 44/91, *Ibid.* (Including Scottish Addendum.)

MATHESON G. D. 1979a. Phasing of site investigations for highways, *Transport & Road Research Laboratory*, Department of the Environment, Department of Transport, TRRL Leaflet LF890, Crowthorne.

MATHESON G. D. 1979b. Preliminary site investigations for highways: Suggested contents, *Transport & Road Research Laboratory*, Department of the Environment, Department of Transport, TRRL Leaflet LF891, Crowthorne.

MATHESON, G. D. & KEIR, W. G. 1978. Site investigation in Scotland, *Transport & Road Research Laboratory*, Department of the Environment, Department of Transport, TRRL Laboratory Report LR828, Crowthorne.

THOMAS, P. R., KOR, F. H. & MATHESON, G. D. 1992. Development of a knowledge-based system for ground investigation in rock, *Rock Characterization*. pp. 159-162, Thomas Telford, London.

WINTER, M. G. & MATHESON, G. D. 1992. Development of a knowledge-based system for ground investigation, *Géotechnique et Informatique*, pp. 623-630, Presses de L'École Nationale des Ponts et Chaussées.

Expert system for geotechnical testing

XIAONIAN XUE[1] and PETER SMART[2]
[1]L & M Geotechnic Pte. Ltd., Singapore; formerly ICE Visiting Research Fellow, University of Glasgow. [2]University of Glasgow, Glasgow G12 8QQ, Scotland.

Abstract
A pair of computer programs have been established to help young engineers when ordering and checking Geotechnical testing. One produces a schedule of test; the other provides verbal help and a draft specification of method. The system is flexible and can be tailored to specific jobs. The programs run jointly under MS DOS in a PC.

INTRODUCTION

Correct conceptual design requires a correct conceptual understanding of the geotechnical situation, which in turn depends mainly on the desk studies, geological mapping, and preliminary investigations. However, correct detailed design requires the correct details of the geotechnical situation. In particular, if the correct tests have been made and the correct parameters have been measured, it will be difficult for the designer not to make the correct calculations; but if the requisite parameters have not been measured, these calculations will be impossible. Yet is it precisely at the point of choosing the correct tests that junior geotechnical engineers experience the greatest difficulty. The system described here gives guidance on how to choose field and laboratory tests for a particular design and construction. It can either help junior engineers to make decisions or help a senior engineer to give instructions to his assistant engineers.

In practice, which methods of tests are chosen should depend on the type of works and the type of soil. The present version of the main program, which is called F-7 TESTS, can handle up to 120 tests, 12 types of works, and 20 types of soil simultaneously; currently, 87 tests are installed arranged in groups as indicated in Table 1; different sets of tests, works and soils, can be substituted easily. A geotechnical score-base (numerically-based rule base) is used as the primary source of advice. The user is permitted to take his own decisions in the light of this primary advice.

To support the score-base, a conventional help system, which is based on BS 1377:1990 has been compiled to advise when taking these decisions; and this help system has been extended to enable the user to extract a draft specification for ground investigation contracts.

Objectives
Fig. 1 illustrates the objectives of the system. In order to construct the works the contractor will require to know what is to be built and how it is to be built; this implies a flow of queries from the contractor about the design and about the ground treatment. The ground treatment specialist will also need to know about the design; and all three bodies will need to know about the soil. The successful geotechnical specialist will have anticipated all these

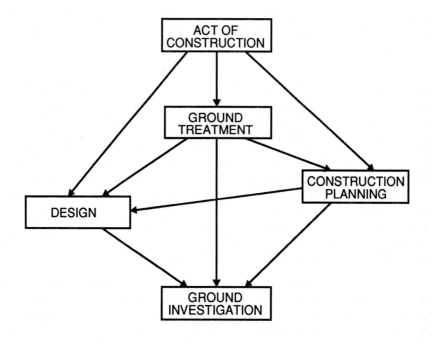

Fig. 1   Requests for information

Table 1   Groups of tests in the prototype score-base

Group	Sub-group	Number of tests
Classification	–	11
Compaction	–	4
Strength and deformation	Triaxial tests	10
	Shear box tests	5
	Static and dynamic probings	12
	Miscellaneous tests	11
Consolidation	–	4
Ground water levels	–	4
Permeability	Indirect estimation	2
	Field tests	6
	Laboratory tests	4
Chemical tests	–	8
Miscellaneous tests	–	6
Total		87

queries and provided all answers in advance. The immediate objective is to decide what testing will be needed to provide the information to enable design, ground treatment, and construction to proceed efficiently.

Pre-requisites

The basic concept of F7-TESTS is that the number of samples to be tested by each method will depend on: (a) the type of works; (b) the type of soil; and (c) the scale of the works. The procedure adopted here is to divide the site into geotechnical units in each of which the type of soil, the type of works, and consequently the geotechnical treatment, are all essentially uniform (Smart and Xue, 1992). For example, if a light foundation on carse clay spans two sandy levees, a peat-filled creek between them, and a clayey back-swamp beyond them, the site will be divided into four geotechnical units, each of which may require a different foundation design. The concept of a piece of ground which may be considered to be essentially uniform for the purpose in hand was adopted from Webster and Beckett (1965). As in the original, it may be necessary in practice to use units of uniform variability, i.e. all that may usefully be said of them is the expected scale of the variability within them. Unlike the original, three-dimensional units may be involved here.

Fig. 2 illustrates the general scheme of site investigation based on BS 5930:1981. Geotechnical testing is concentrated in the Preliminary Ground Investigation, in the Field and Laboratory parts of the Main Ground Investigation, and in the Additional (syn. Supplementary) Ground Investigation. F7-TESTS is designed for use when planning and checking these parts of the work. For each geotechnical unit in turn, the program reads the intended types of works and the expected types of soils from a file which has been prepared in advance. Without this information the program will not run; and, in this respect, use of the program puts some pressure on the engineer to complete the conceptual study correctly.

The scale of works is represented in the program by 'total number of samples'; this is a short-hand method of saying 'total number of replicates which must be made by whichever method of test is used most often'. A rule-of-thumb is that this number should be proportional to the product of the logarithm of the area of the geotechnical unit, the coefficient of variation of the soil properties, the cost of construction, and the consequences of failure. For the present purpose, it is regarded as data; but, as explained below, a provisional scheme of testing may be drawn up to assist in deciding what it should be.

OPERATION

Overview

Figs. 3 and 4 illustrate the operation of the main program, F7-TESTS. The heart of this is the operation, Adjust by Tests, in the centre of Fig. 3. Here, the program is looping over the geotechnical units one-by-one (outer loop in Fig. 3) and requiring the user to accept or adjust its recommendations for testing. The program obtains details of the type of works and type of soils from the file Input Details in Fig. 4. Using the numerical rules which it has obtained from the Score Base, it calculates a recommended schedule of tests, presents these to the user, and permits him to amend the schedule in the light of any information he has which is unknown to the system. If the user is satisfied, the schedule of tests is written to the decision file. Finally, all the data from Input Details is written to Updated Details after altering a status flag to 'complete' if appropriate to do so. The number of samples may also have been updated. The program normally then searches for the next incomplete

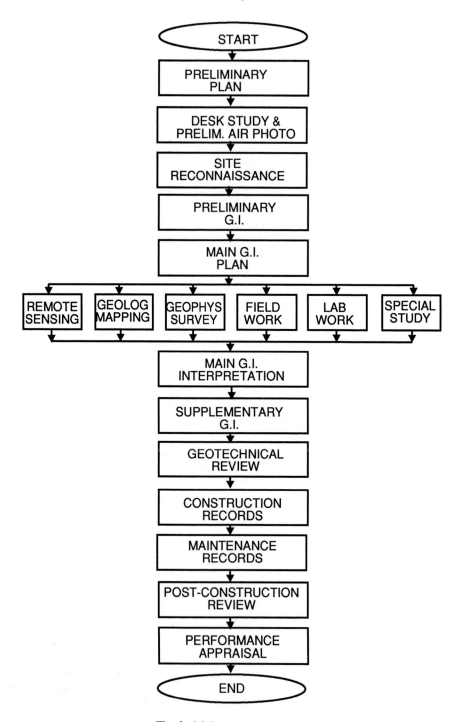

Fig. 2   Major stages in ground investigation

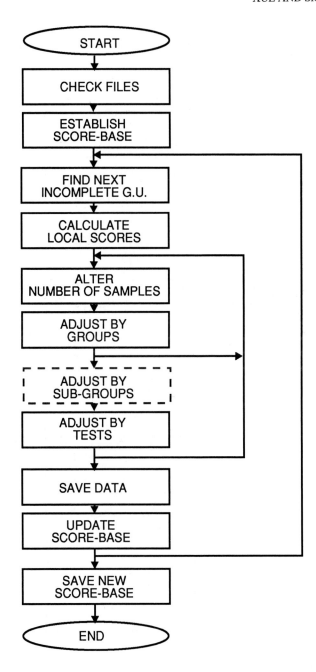

Fig. 3 Operation of F7-TESTS

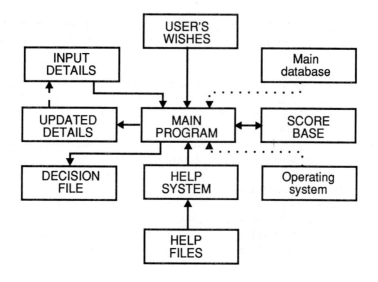

Fig. 4   Structure of F7-TESTS

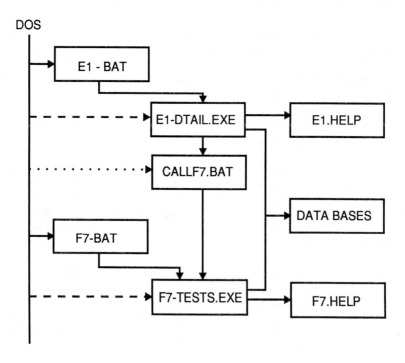

Fig. 5   Control pathways for running F7-TESTS

geotechnical unit, copying the details of complete units from file to file as it does so. At the end of a successful run, the user should copy the Updated Details over the Input Details ready for the next run.

Because the Score Base may vary from job to job, and the input Details certainly will do so, it is necessary for the proper versions of these files to be made available to F7-TESTS. Therefore, although F7-TESTS may be called directly from the operating system, it is preferable to call it from a batch file, F7.BAT, see Fig. 5; this file then has the task of arranging all the files properly, and, by its nature, it is entirely under the control of the user. Fig. 5 also shows an alternative way of using F7-TESTS. There is another program, E1-DTAIL, which takes a broader view of detailed ground investigation; in essence, this program monitors the arrangements for borehole location, geotechnical testing, ground water investigation, and geotechnical hazards. If desired, E1-DTAIL can call F7-TESTS as if it were a sub-program, passing the data for one geotechnical unit to it, and setting a flag which informs F7 that it has been called by E1. In this case, F7-TESTS assumes that the geotechnical unit is incomplete, and the outer loop in Fig. 3 is disabled.

Details
On starting the program, F7-TESTS, the user checks that the proper files have been found, and the program reads in the score base. Next, the incomplete geotechnical units are scanned until the user selects one on which to work; (the flow chart is simplified here).

For the selected unit, the types of works and the types of soils must have been specified in advance, and the total number of samples may have been. Regardless of whether this number has been decided or not, the task is that of deciding what proportion is to be tested by each particular method. The program's recommendations are calculated in the step called Calculate Local Scores in Fig. 3; the method is explained below.

If only one type of works and one type of soil is present; then the scores in the score base are taken as the recommended proportions.

If several types of works are superimposed in one geotechnical unit, then for each test, F7-TESTS checks through the types of works concerned, extracts the maximum of the individual recommendations, and suggests this maximum to the user. For example, if the recommended procedure for ground sulphates is to test 25% of the samples for a cut and 100% of the samples for a foundation for a major structure, then 100% of the samples should be tested when a major structure is superimposed on a cut.

If several types of soil are included in the same geotechnical unit, then for each test, F7-TESTS checks through the types of soil concerned, extracts the individual recommendations, and suggests using the average weighted by the proportion of each soil present. For example, if field permeability is required for 70% of the samples from lodgement till and for 50% of samples from carse, and if these two soil types are present in equal proportions, then 60% of the total number of 'samples' should be tested for field permeability.

Before considering the program's recommendations, the user is given a chance to alter (or enter) the total number of samples.

```
Program F7-TESTS ======================================= 93.11.25
```

	Geotechnical Unit Number 012. Fill	Glacial sands and gravels		
		Total Number of Samples =		75
ref	Classification tests	orig. %	user %	number
1	Field Description	50	50	38
2	Laboratory Description	100	100	75
3	Moisture Content	100	100	75
4	Wet Sieve Test	100	100	75
5	Clay Content	50	50	38
6	Liquid Limit	5	0	0
7	Plastic Limit	5	5	4
8	Organic Carbon	10	10	8
9	Dry Density	100	100	75
10	Porosity	80	80	60
11	Specific Gravity	100	100	75

Enter P = change %, N = change No, Q = quit, else OK:

Fig. 6   Screen-dump from Adjust by Tests.

It will be clearer to the reader to defer discussing Adjust by Groups and Adjust by Sub-groups and to explain Adjust by Tests next, see Fig. 3.   Fig. 6 is a copy of a screen-dump taken whilst this part of the program was processing the first group of tests shown in Table 1.   The data is shown at the top.   For the eleven tests being considered, the program's recommendations are shown in the column headed 'orig'; this column cannot be altered by the user.   Initially, the program copies the percentages in 'orig' to 'user' and calculates the values in 'number'.   The user may then alter either 'user' or 'number',   the program automatically keeping these two columns in agreement.   In the example shown, the user has changed the values for Test No. 6; either the user reduced the percentage and the program reduced the number, or vice versa.

After working through all of the sub-groups of tests, the program checks whether an attempt has been made to increase the total number of samples.   If so, the user is alerted and obliged to reconsider the whole series of tests.

When the check above has been passed, the user is given a chance to work through the whole series of tests again.   In addition to catering for doubt and error, this gives the user who is unfamiliar with the version of the score base which is in use, an opportunity to scroll through all the recommendations before starting to take decisions.

After working through a geotechnical unit, the user is given a chance in Save Data in Fig. 3 to write the final choice for this geotechnical unit to the Decision File, see Fig. 4.   The essential data written out consists of a list of all tests required, giving for each test, the reference number of the test, the name of the test, and either the number or proportion of samples to be tested, see Table 2.

Table 2   Extract from a Decision File, un = unchecked.

un	Geotech Unit	012
ok	Map Ref	AB 120450
ok	Chainage	30
ok	Type of Works	Fill
ok	Strata	Glacial sand and gravel
ok	No. of samples	75
..	File Number	...
un	Progress	Specifying main GI
un	Compiled by	X.X
un	Date	18.10.93

Index	Test	Number
110100	Field Description	38
110300	Moisture Content	75
110400	Wet Sieve Test	30
110500	Clay Content	4
110600	Liquid Limit	4
110700	Plastic Limit	4
110800	Organic Carbon	8
110900	Dry Density	38
111100	Specific Gravity	4
330300	Dynamic probe static record vertical	38
510100	Ground Water Level in borehole	4
710100	Soil sulphate	4
710500	pH Value	4
710600	Chloride Content	4

In Update Score-Base in Fig. 3, the user is given a chance to save the new set of percentages into the version of the score base held within the program.   In simple cases, the program copies back from the 'user' column into the appropriate part of the score base;   but there are also options to create a new type of works and a new type of soils.

As the run proceeds, the program copies the data read from the progress file into the Updated Details file in Fig. 4, updating the states of geotechnical units to 'complete but unchecked' if appropriate to do so.   A small but important point is that F7-TESTS cannot raise a status to 'ok';   an independent check at the E1-DTAIL Level is required.

To return to Adjust by Groups in Fig. 3, before starting the test-by-test adjustment, the program extracts the maximum score for each group and permits the user to adjust the scores for whole groups of tests, i.e. all the scores in one group may be scaled up, scaled down, or set to zero in a single operation.   If necessary, the user is taken back to adjust the total number of samples.   A similar operation on sub-groups may be added to the program if it seems useful.

Table 3   Extract from a draft specification prepared by the Help System

---

@$810130   Linear shrinkage

BS 1377: Part 2: 1990: Section 6.5   Determination of shrinkage characteristics: Linear shrinkage.

Particles > 425 microns shall be removed by wet sieving.

Soil in the natural state shall be used without air drying.

@$810300   Frost Heave

Frost heave shall be determined in accordance with BS 1377: Part 5 : 1900;   Section 7 Determination of frost heave and with BS 812 Part 124.

Undisturbed cylindrical specimens shall be prepared in accordance with BS 1377: Part 1: 1900: Section 8.5   Preparation of cylindrical specimen from undisturbed block sample.

Comment
The Particle size distribution, the plasticity index of the fine fraction (if appropriate) and the optimum moisture content and maximum dry density must all be obtained, see Section 7.2.1.

At the end of the run, there is an option to update the version of the score-base held in file by writing the program's version over it.   This feature gives this rule-using system a rule-making capacity, especially in large organisations where modified score-bases can be collected from several engineers working on several different projects.   This feature also permits a senior engineer to tailor the system exactly to fit one particular job and then to leave a junior engineer to finish drawing up all the schedules of tests.

HELP SYSTEM
In addition to providing the user of the system with recommendations in numerical form, F7-TESTS is provided with a help system which gives access to the non-numerical rules which have to be taken into account.   These are available to the user when running the system.

The rules provided in the first instance run to 55 double-spaced A4 pages.   There were compiled almost entirely from BS 1377:1990 ignoring other sources for the time being. Extracting these rules took far longer than had been expected, mainly because of the large number of variations which had to be considered.   They are intended as a preliminary draft with which to test the system and to serve as a basis for future expansion and development. They are held as a set of 9 files shown as a single box in Fig. 4.

An existing program, MARI, described by Smart and Xue (1992), was adapted and linked to the main program of F7-TESTS in such a way that the appropriate help file can be read;   this program is called Help System in Fig. 4. It enables the user to navigate through a text file by following a menu, provided that relatively simple cross-referencing has been added to the file. (This cross-referencing doubles the length of the file, and makes print-outs difficult to read, so an ancillary program has been written to strip all the cross-referencing out of these files.)

In addition to specifying how many tests are to be made, it is necessary to specify exactly what those tests are and which variations are to be used. In general, this is done by quoting the 'British Standard test number' and appending other relevant details. Most of the relevant wording has been put into the help-system rule-base. Therefore, MARI was extended to enable selected wording to be extracted and recorded in an ASCII file when necessary, see Table 3; the user can add non-standard wording as well. In this way, the F7-TESTS module provides the user with a draft specification which will require only minor editing by a word processor before being issued.

When writing these help files, consideration was given to listing those properties which are to be measured. It appeared that, variations excepted, this information is included explicitly in BS 1377:1900, and it was decided not to complicate the issue by re-specifying it. After the help files had been drafted, the Authors were shown one specification for geotechnical testing which did state explicitly and clearly what was to be reported for every test. However, for any test, the properties which are to be measured and the procedure which is to be used can both vary. Setting up a system to handle both sets of variation simultaneously and to maintain consistency between these sets of variations, would require a special program to be written; and this has not been attempted at this stage.

MISCELLANEOUS POINTS
From time-to-time, it will be necessary to reconsider geotechnical units which have been flagged as complete. This is potentially dangerous, especially if the geotechnical unit has been checked and upgraded to 'ok'. Therefore, this reconsideration has been made difficult; the status flag in the Input Details file must be edited in a separate step before the file is used. An ordinary editor can be used, but a special editor is available which minimises the risk of error.

The Input Details files have been designed to hold only the data which is needed to enable F7-TESTS to run. This data is an abstract of what is known about the site. In case the user wishes to access the full details or to make calculations or to refer to other files, two facilities are provided, see Fig. 4. First the program will call a .BAT file which is intended to call the main data base; it is left to the user to set this up, and it could be used to call any program which could be run in the available memory. Secondly, the user may exit temporarily to the operating system.

As indicated earlier, the program is designed to be used (a) when planning an investigation, and (b) when checking what has been done. In the interim, more has been learned about the site. In particular, where a geomorphological classification may have been appropriate initially, a material classification may be appropriate finally. This situation can be catered for by having two versions of the score base.

In some cases, it may be appropriate to start from separate score bases for design, ground treatment, and construction. In this case, a preliminary step would be required to combine whichever scorebases were to be used for any particular case.

E.O.A. Awoleye (Per. comm. 1993) suggested multiplying the schedule of tests by unit prices to obtain a preliminary costing. The Authors' present opinion is that this idea should be extended to include operator time, machine time, and time to completion, all of which are needed when comparing alternative schedules of tests.

The system runs on a PC under MS DOS; use of a Ram Disk is useful but not essential.

SUMMARY
Having been told what types of works are intended, what types of soil to expect, and if available the total number of 'samples' to be used, the program F7-TESTS calculates a draft schedule of geotechnical tests which are to be or should have been performed and permits the user to make any necessary modifications to the standard recommendations. A subsidiary program gives, for each test, the British Standard reference, a list of variations, and some comments on these; this program can also print out a draft specification of methods to indicate precisely which test and which variations are intended. The whole system is very flexible and can be adapted to individual jobs and extended relatively simply. This flexibility and capability for extension has been achieved by keeping all the rules in ordinary files within the computer. The textual rule base is almost a conventional branching-tree; but some extra cross-referencing has been found to be useful. The use of a numerical rule base (score base) in a less usual feature of the system and adds considerably to its capabilities.

Acknowledgements
The work was made possible by the ICE R & D Fund. Mrs. I. Duncan and Mrs. T. Bryden prepared the figures and text, respectively.

References
BS 1377: 1990. *Methods of test for soil for civil engineering purposes.* British Standards Institution, London.
BS 5930: 1981. *Code of practice for site investigation.* British Standards Institution, London.
Smart, P., and Xue, X. 1992. *Expert systems for soil description.* Actes du Colloque Geotechnique et Informatique, Paris. pp. 599-606.
Webster, R., and Becket, P.H.T. 1965. *A physiographic map of the Oxford Region.* D. Survey, Ministry of Defence, London.

Managing site investigation fieldwork using ISO 9000

T G BALL and A P WHITTLESTONE
Soil Mechanics Limited, Wokingham, United Kingdom

INTRODUCTION

Over the last few years British industry appears to have eagerly adopted ISO 9000 (BSI, 1994) an international standard for quality assurance management systems. Companies from a wide variety of industries have set up management systems and sought accreditation for their achievement. The widespread usage of such systems and market forces have, in recent years encouraged the site investigation industry to follow suit and European Community directives are adding to the impetus (Grover & Laver, 1993). The application of ISO 9000 management systems to site investigation practise is widely seen by both the contractors and their clients as beneficial, but the practical operation of these management systems within the civil engineering industry is often problematic.

The ISO 9000 standard was written for the manufacturing industry and is directed towards companies providing a range of products on a repetitive basis. Many service industries which have successfully applied the standard often provide a similarly repetitive service. Whilst site investigation techniques are generally routine, ground conditions are unique to each site and poorly defined before the investigation. The investigative approach has to be site specific and is often modified as the investigation progresses hence some contractual requirements are only loosely defined. ISO 9000 guidance documents provide little constructive assistance for interpretation of the standard in these respects by the site investigation industry or independent assessors; differences in interpretation may lead to conflict.

The successful adoption of an ISO 9000 management system in site investigation depends on the interpretation of several key clauses. This paper discusses the clauses relating to contract review, process control and the control of non-conforming product. The interpretations presented below have been tested over a number of years and have been shown to meet the requirements of the industry and the standard.

AN INTERPRETATION OF QUALITY MANAGEMENT

At the outset it is important to clarify what is meant by quality management , before trying to relate the concept to a specific activity such as site investigation. There are many misconceptions about the almost meaningless phrase "Quality Assurance" ("QA") frequently exacerbated by cavalier use. The management systems advocated by the ISO 9000 standard are nothing more than formalised methods of describing good company and project management, the implication being that good practices lead to good products. There is nothing new about the principles, they have been used by managers for many years and certainly predate the 1960's and 70's when written standards first started to appear.

Quality in the terms of the ISO 9000 standard should not necessarily be taken to imply a product of the highest specification and workmanship. In contractual situations quality can only mean compliance with the customer's specified requirements. If the customer specifies the grade of product with standards for components, manufacturing and reliability he is defining the quality required, and thus almost anything could comply with ISO 9000.

The clearest definition of the phrase "quality management" is probably that based on Feddersen (1991):

> "systematic management to achieve compliance of the product or service with the requirements of the customer as specified in the contract documents."

The Management Loop (Fig. 1) illustrates the important concepts of compliance management for any company which are:

1   A management system which includes:

- Definition of overall control and responsibilities.

- Planning and review procedures at a company/department level.

- Standard operating procedures.

- Monitoring systems, eg auditing.

2   Contract specific activities are iterative and necessarily repetitive. The basic principles required for one contract should be applied to all contracts eg:

- Specify with clarity.
  The specification and order documents should define what is required clearly without ambiguity. Aspects of this responsibility are shared by both client and supplier.

- Plan to regulations and standards.
  The work should be planned to comply with the relevant regulations and standards including legal, industry, client and internal requirements.

- Design for objectives.
  Client and contractual objectives must be clearly understood at all levels so that the work is performed in the way that meets the objectives most efficiently.

- Purchase for quality (compliance).
  All materials equipment and services purchased to fulfil a contract should comply with the specified requirements.

- Supply the right product and service.
  Having carried out these actions a company should be in a position to supply the right product or service and be capable of learning from the experience.

Fig. 1: The Management Loop

QUALITY MANAGEMENT AND SITE INVESTIGATION

ISO 9000 principles have been used on many of the largest site investigation projects in the UK since the mid 1980's particularly those relating to the nuclear and transport industries. Despite this experience with the standard very little seems to have been written on the subject except for generalised encouragements to use "QA". The available publications tend to cover the construction phases of projects rather than the approach used for site investigation (Feddersen 1991, Vu Hong 1986). The Site Investigation in Construction publications (SISG 1993 a & b) provide some guidance and conclude that the application of ISO 9000 style management principles is desirable without addressing the questions of interpretation.

As mentioned above, ISO 9000 is geared towards managing compliance and, as a result, it concentrates almost exclusively on making sure that the product or service meets the specification. There are few problems identifying the product of a company making widgets or even a car hire company but what are the products of site investigation?

BS 5930 (BSI, 1981) defines site investigation as:

> "the investigation of sites for the purposes of assessing their suitability for the construction of civil engineering and building works and of acquiring knowledge of the characteristics of a site that affect the design and construction of such work and the securing of neighbouring land and property."

The full range of site investigation activities is very broad and includes design and interpretation and numerous drilling, excavation, sampling, monitoring and testing techniques which encompass many of the natural sciences. Whilst a number of the practical techniques are well established and the subject of international standards, novel and innovative procedures are frequently required. A formalised ISO 9000 management system therefore needs to be designed to provide adequate controls for the range of activities and standards without stifling creative thinking. Similarly the investigation contracts should be flexible and allow for the changes in the specification which will almost certainly result from the investigation process itself.

A UK site investigation project generally involves the following parties:

Client   The company or organisation commissioning the investigation.

Consultant   A company providing technical expertise to the Client, and whose main duties include designing and managing the investigation and interpreting its results. Generally only one consultant is used, although occasionally more than one may be employed to provide the necessary range of skills. The consultant may also fulfil the roles of the contractor.

Contractor   A company carrying out practical investigative work to obtain information on the site. Results are generally presented in a factual manner, ie without interpretation. Any number of contractors may be used on a single project.

The consultant's design and interpretation products are relatively easy to control, in quality management terms, using some form of peer review. In contrast some of the factors which affect the excellence of the contractor's products (data and samples) are not under his control. The quality of both data and samples depends on:

-    The skill and care of the operator and his supervisors.

-    The test or sampling method selected.

-    The equipment used to do the test or take the sample.

-    The material being tested.

No one has a controlling influence on the material being tested or sampled; after all lack of knowledge about the material is the sole reason for carrying out the work in the first place. Natural materials are not homogenous, they are in engineering terms imperfect and most techniques have a "success" rate which is at least partially independent of the equipment, method and operator. In these circumstances the contractor has to do what he can within the limits of his contract to minimise failures.

This lack of control over the materials being tested is one reason why it is inappropriate to simply transfer quality assurance principles directly from manufacturing industry to site investigation. The constraints imposed by geotechnical and environmental uncertainties must be taken into account; off-the-shelf management systems are rarely adequate for clients, consultants or contractors. ISO 9000 will never assure excellent site investigation data every time, it is after all a standard for **management of contract compliance** not **management of perfection**.

Successful management of site investigation fieldwork depends on a clear understanding of the investigation objectives, feasible methods and how any problems which arise can be dealt with constructively. These factors need to be understood by all parties, so clearly communication is vitally important.

CONTRACT REVIEW

The client is ultimately responsible for project management, he communicates his requirements to his supplier, ie consultant or contractor, using some form of specification. Clause 4.3 states that the supplier shall review the contract documents and decide how he is going to meet the requirements and how they relate to his normal way of working. Contract review already forms part of most companies' administration procedures as suppliers need to reassure themselves that the client has supplied an adequate specification and they have the capacity and ability to meet all its requirements.

The ISO 9000 definition of contract includes accepted tenders, orders and any statement of requirements between a supplier and a customer. The contract review process is consequently not just a single stage operation which happens before mobilisation; it also needs to deal with written and verbal instructions given during the course of the work. It may well be appropriate to use two levels of review, an initial detailed review of the main contract documents supported by an abbreviated system for dealing with variation orders including site instructions.

A company's management system belongs to that company and should not be affected by contractual arrangements, but it is in all the parties' interests to address potential problems in a reasonably consistent manner. The supplier's management system should be robust enough to deal with the unusual and the contract review process must be able to identify areas where the client requires alternative methods to be used, particular care must be taken to ensure that these requirements are highlighted so that the staff carrying out the work know exactly what should be done. If the supplier considers that the specification is technically inadequate he must raise the problem with the client, agree a revised specification, or accept the original requirements or refuse to take on the work.

Suppliers have to remember that, in the terms of a contract, the ultimate document is the contract not their management system. A site investigation contractor's procedural document will probably say that they log samples in accordance with BS 5930 (BSI, 1981). If a client asks them to use a different system then the contractor should use that method, the contractor can not insist on supplying BS 5930 logs as by doing so he would fail to meet the specification even if the required method is technically inadequate. Whilst it would be onerous, there is nothing to stop the contractor compiling additional BS 5930 logs for his own records and technical satisfaction, they might even help him if things do go wrong. On the other hand clients have to remember that suppliers are likely to have good reasons for suggesting alternative methods so consultation is often advisable.

A Quality Plan can be produced as a result of this initial stage of contract review; this should then act as the interface between the client's project management system and the supplier's internal company management system.

Site investigation contracts frequently run on the basis of a plethora of site instructions and any review procedure will be unworkable if much paperwork is generated when the day's instructions are reviewed. The review of these variation orders has to be adequate to deal appropriately with both routine instructions like moving a borehole two metres from A to B and more significant instructions which effectively result in a change to the specification and trigger a more extensive review process.

Oral instructions have always been a problem, and there is little prospect of this changing. The standard puts the onus on the supplier to record spoken instructions. In practice common sense has to prevail, do we really want a JCB driver (supplier) to record the geologist's (customer) instruction to start digging the trial pit? A successful solution depends on sensible cooperation and communication between all the parties concerned.

PROCESS CONTROL

Clause 4.9.1 of the standard states:

> "The supplier shall identify and plan the production and, where applicable, installation processes which directly affect quality and shall ensure that these processes are carried out under controlled conditions."

The standard subsequently sets out the ways in which a company should control the processes it performs and requirements for inspecting and testing the product to verify compliance. The principles embodied in the standard do not work particularly well for site investigation practices because the product is rarely something which can be checked by physical inspection. The following are examples of site investigation "products" which can not be verified at a later date:

-       sample depths: these can not be checked after the borehole has been advanced.

-       depths of water strikes.

-       laboratory test results: these can not be checked fully, even the results of tests performed on two halves of the same sample will not be identical.

-       in situ measurement results such as standard penetration test (SPT) results.

All these things can be checked for gross errors, but even minor mistakes which are impossible to detect could have a marked influence on the engineering interpretation. We consider that the majority of site investigation activities are actually Special Processes in the terms of ISO 9000. The standard stipulates that they shall be carried out using approved methods, appropriate well maintained equipment and qualified personnel. The management system should therefore concentrate on ensuring that the people and resources are adequately controlled.

<u>Methods</u>   Documented method statements are an essential requirement for a formal management system, but they should not be viewed as irritating bureaucratic paperwork which a company has because the standard says it should. Perhaps the most frequently used QA slogan is "Do it right first time", this is a laudable aim but its implications are overlooked all too often. " Doing it right" implies:

-    There is a right way.

-    The worker knows what the right way is.

-    The worker does the work in the right way.

Effective method statements are written in a way which recognises these points and help the worker to select the most effective *feasible* procedure. A company can learn a great deal by going through the process of compiling method statements which consider the possibilities of alternative methods and contingency planning.   If a company approaches this task positively it will learn a lot about itself and gain a valuable set of training documents which can be used to address the last two of these points and enhance all aspects of its work.

Personnel frequently have to check their own work and supervisors must rely on the professionalism and honesty of their staff. This self checking does not conflict with the standard, however, it should be organised in a way which helps the worker rather than hindering him.

Records are required which demonstrate that the work has been carried out in the most effective feasible way. The records need to identify working methods selected, standard drillers logs can fulfil this purpose for most drilling and boring operations. Some form of check list is often used to check work completion, these should not insult the worker's intelligence and should strike a balance between *aide-mémoire* and a check that contract requirements have been met.

<u>Equipment</u> All equipment, particularly those items used to obtain samples or take technical measurements should be fully controlled, regularly inspected and calibrated to national standards where appropriate. Detailed equipment control procedures are fairly standard for all industries and are not covered in detail here.

<u>Personnel</u> The quality of site investigation fieldwork depends primarily on the efficiency, care and experience of the staff doing the work. ISO 9000 requires staff to be adequately trained and the people doing the work must be able to demonstrate that they know what they are doing.   The industry now has established standards for accredited drillers, chartered geologists and engineers, and geotechnical specialists and advisers. Clients, engineers and contractors should not lack suitably qualified staff and there can be no excuse for reputable companies relying on substandard personnel, consultants or contractors.

There can be no doubt that a company's most significant asset is its staff. Poor workers produce bad work even if they are using the right equipment and procedures.

NON-CONFORMANCE

The 1987 version of ISO 9000 only included explicit requirements for dealing with non-conformances when these affected the product, a non-conforming product being one that does not meet its specification. The scope of these requirements was widened for the 1994 revision which now includes the words:

> "initiate action to prevent occurrence of any non-conformances relating to product, process and quality system"

The 1987 standard resulted in problems with the definition of what the product is and hence what constitutes a non-conformance which merits formal identification, investigation, review and preventative action. The revised text has reduced, but not totally removed, the problems of identification; the practical difficulties with how to deal with non-conformances remain.

The primary products of site investigation are raw and interpreted data and samples, does this imply that inadequate data or poor samples are non-conforming? Unfortunately the answer to this question is both yes and no. Geotechnical and environmental uncertainties and constraints cause problems with both test and sampling procedures, it is perfectly feasible for experienced operators to carry out the work using the right equipment and excellent methods and still end up with inadequate data or poor samples. The contractor may have complied with the full requirements of the specification but the right "product" has not been supplied. In many instances the validity of results can only checked by interpretation but the contractor is only responsible for using the "right" method and reporting the results obtained. This scenario raises a number of questions:

- Did the specification address the full range of potential geotechnical and environmental possibilities adequately?

- Did the client, consultant and contractor all understand the implications of the potential conditions for the equipment, methods and operators specified?

- Did the specification allow for the use of alternative methods or procedures?

- Should the problems have been anticipated?

It is not appropriate to discuss all the ramifications here however this simplistic scenario should be enough to demonstrate that there is understandable uncertainty about what is or is not a non-conforming product for the site investigation industry. Arguments can only be avoided by applying ISO 9000 non-conformance and corrective action management principles to all problems, whether or not they could affect the quality of the product.

All personnel are responsible for monitoring their own work and the activities of others they are responsible for. Problems need to be reported so that they can be dealt with constructively. The way in which problems are handled may differ for different categories of problem but the management principles will be the same.

Problem categories may include:

- Equipment failures.

- Procedural problems.

- Customer Complaints.

- Safety and environmental problems.

Safety and environmental problems fall outside the remit of ISO 9000 however there is no reason why they should not be included in a company wide management system, and controlling them within the general umbrella of quality management is often easier than having a number of totally divorced systems.

Consistent management principles should apply to the reporting and investigative processes for these different types of problem. A problem should be dealt with in a way that ensures that all possible lessons are learnt from it. Some companies have taken this positive attitude to the point where the jargon starts to hide the intention, some no longer call errors "mistakes" they call them "opportunities for improvement" instead. Our experience suggests that this style of jargon is self defeating however good its intentions.

ISO 9000 requirements for dealing with problems are clear and include the following stages:

1   Identify the problem.

2   a   Take immediate action to prevent it causing further problems.

     b   Report it.

     The order in which these two actions are carried out depends on the severity of the problem and the responsibility/experience of the person discovering it.

3   Investigate the problem, what happened, why it happened and what the actual or potential effects are.

4   a   Specify actions to deal with the immediate problem, if actions to beyond 2a are necessary.

     b   Specify actions to stop it happening again in the future.

5   Check on the success of all actions taken to deal with the problem.

6   Report details of all aspects of the problem and how it was handled.

7   Review; this should be part of an overall review of all problems encountered so that trends can be identified particularly where problems might have common or related causes.

Preventative and corrective actions are dictated by what is possible, a company may have to live with an ongoing problem which it can not currently solve fully. This type of decision should be made by choice not by default, and making it does not remove the obligation to continue reporting its repeat occurrences. A full set of problem reports will enable you to decide whether the status quo is being maintained or the situation is deteriorating to the point where action has to be taken despite the cost or practical difficulties.

Individual companies do not work in isolation, relationships with clients and contractors have significant psychological effects on how they deal with problems. Confrontational contracts have a tendency to result in problems being swept under the carpet; clients do not admit mistakes because they may lead to claims against them; contractors do not admit mistakes because this would reduce the chances of full payment or repeat orders. This attitude is all too common in the site investigation industry and clearly conflicts with ISO 9000 management standards. There is no easy answer to this psychological problem, even a perfect documented system can only be a small part of the solution, success depends on changing the attitudes of individuals who would be greatly encouraged by less adversarial contracts.

A WAY FORWARD

If the essence of ISO 9000 is a common sense approach to compliance management, why does its implementation within the site investigation industry appear to be so problematic? The answer seems to lie in a combination of current contract management practices and confusion about how to apply ISO 9000 control principles to site investigation products. The entire industry needs to take on board three basic lessons if it is going to progress and use quality management principles successfully on site investigation projects:

1   Quality begins with the client.

The Site Investigation Steering Group makes the valid comment that "Quality begins with the client" (SISG 1993b). The client is responsible for:

-      managing the contract.

-      the contract specification.

-      maintaining open lines of communication with all relevant parties.

The consultant may have delegated responsibilities and hence influence in these areas, but there is little a contractor can do about them if the client is unsympathetic. Confrontational contracts and inadequate communication undermine all the good intentions of ISO 9000 management systems and are particularly detrimental to planning and the way in which problems are dealt with.

Successful contract management depends on the establishment of effective interfaces between the management systems of the individual parties. The most successful projects are clearly those where the client recognises his responsibilities and establishes good practices at every level of the project. A number of reports like those of the Site Investigation Steering Group (SISG 1993 a & b) have stressed that successful site investigation depends on consideration of all the possibilities and careful planning. ISO 9000 is an ideal tool for formalising these systems and reducing errors; careful implementation of its principles should help the industry move forward.

2   The technical expertise of the contractor has to be recognised.

Contracts which neglect the contractors' experience ignore the possibility that they might know something which could assist the way in which the work is carried out and the quality of the information which can be obtained. If the contractor is expected to do what he is told without question then he is unlikely to accept responsibility for any resulting problems, and whilst this attitude may be understandable it can not be justified because it only increases the likelihood of further confrontation. Contractors have a responsibility to raise problems with the client at as early a stage as possible, and for dealing with them in a responsible manner.

Natural materials are inherently variable and many site investigation tests are unrepeatable, furthermore the results can not be checked fully. Stringent controls are necessary to ensure that appropriate staff are employed and that they use the best methods and appropriate equipment. Staff carrying out the work have a very high level of responsibility, if they do their work poorly the errors can be impossible to identify and may have a significant effect on the engineering design.

Engineering and site investigation design procedures together with all aspects of data acquisition and interpretation have to be carefully supervised and controlled yet remain flexible so that changes can be made in response to the conditions actually identified during the investigation. The influences exercised by the client, consultant and contractor have to be fully appreciated when the management system is established, documented and audited.

3   Teamwork is the new order of the day.

Successful scientific investigations depend on all the parties involved working together as a team, if information does not flow freely between them then efficiency is impaired and mistakes are made. Confrontational contracts do everything but lead to constructive teamwork. The New Engineering Contract (ICE 1993) may help to reduce confrontation, and the early indications suggest that it is helping to foster constructive contract management attitudes on all sides of the industry.

The oil industry has a long history of adversarial contractual arrangements which has many parallels in site investigation and civil engineering. A number of the oil and associated service companies are currently learning the advantages of cooperative working within various quality management frameworks. ISO 9000, Total Quality Management, and Continuous Improvement systems are all being implemented by various companies and there is a growing body of evidence that this is reducing costly confrontations and improving industry wide efficiency (Johnstone et al 1993, Kadaster et al 1992, King 1993, Robins and Carlisle 1993, Sandison and Kern 1992). Evidence for similar trends in the civil engineering industry is much harder to find; maybe we should be learning from their experience.

It is unfortunate that the requirements of ISO 9000 are frequently misconstrued and the standard poorly implemented because its value to site investigation practice is undeniable but the benefits depend on potential problems being recognised early and tackled positively by all the parties concerned.

ACKNOWLEDGEMENTS

Acknowledgement is given to the Directors of Soil Mechanics Limited for their support in publishing this paper.

REFERENCES

BRITISH STANDARDS INSTITUTION. 1981. BS 5930. Code of practice for site investigation. British Standards Institution.

BRITISH STANDARDS INSTITUTION. 1994. BS EN ISO 9000 Quality systems. British Standards Institution.

FEDDERSEN, M. A. 1991. Project management and quality management hand-in-hand. Proceedings of innovation and economics in building conference, Institute of Engineers, Australia.

GROVER, R. & LAVERS, A. 1993. The impact of European Communities' policy on quality management in construction. CIRIA Special Publication 89.

INSTITUTION OF CIVIL ENGINEERS (ICE). 1993. New Engineering Contract. London.

JOHNSTONE, J. A., MORRISON, C., & FERNANDEZ, R. L. 1993. Directional well planning in the oil and gas industry to quality standard ISO 9001. Quality Forum vol. 19.

KADASTER, A. G., TOWNSEND, C. W. & ALBAUGH, E. K. 1992. Drilling time analysis: A Total Quality Management tool for drilling in the 1990's. Paper SPE 24559, 67th Annual Technical Conference and Exhibition of the Society of Petroleum Engineers.

KING, G. W. 1993. The development of a competency-based training programme for drilling engineers using Quality Assurance principles. Paper SPE/IADC 25763, 1993 Society of Petroleum Engineers / International Association of Drilling Contractors Drilling Conference.

ROBINS, K.E. & CARLISLE, J. A. 1993. Customer/supplier cooperation: Working together for quality. Paper SPE/IADC 25679, 1993 Society of Petroleum Engineers / International Association of Drilling Contractors Drilling Conference.

SANDISON, G. F., & KERN, A. M. 1992. Drilling for quality. Paper SPE 24559, 67th Annual Technical Conference and Exhibition of the Society of Petroleum Engineers.

SITE INVESTIGATION STEERING GROUP (SISG). 1993a. Without site investigation ground is a hazard. Site investigation in construction, Part 1. Thomas Telford, London.

SITE INVESTIGATION STEERING GROUP (SISG). 1993b. Planning, procurement and quality management. Site investigation in construction, Part 2. Thomas Telford, London.

VU HONG, L. 1986. Quality Assurance as a management tool for large projects. In Safety and quality assurance of civil engineering structures - Preliminary report. 1986 Symposium of the International Association for Bridge and Structural Engineers.

The practical application of quality management to site investigation

J C WOODWARD

Geotechnical Consultant, Princes Risborough, England

SUMMARY

This paper is written from the site investigation contractor's viewpoint to deal with current concerns on how to "assure" the quality of an investigation to an agreed standard. The following key points are reviewed:
- impact of geological variations
- contract terms which are incompatible with the concept of self-checking quality plans
- appropriate assessment and certification.

Quality is not optional - it is essential, but does not happen by chance. To assist the contractor, detailed guidance is given on preparing a relevant Quality Management System (QMS) to BS 5750 which will avoid the pitfalls of over-documentation and the complications of third party involvement. It is concluded that the industry should adopt a uniform QMS embodying 'self-certification' as the cost-effective means of applying quality management.

INTRODUCTION

A Quality Management System is the means of giving the client confidence that the agreed requirements are met. Following the changes to BS 5750, 'Quality Systems'[1] in recent years, most major construction companies have successfully developed the principles of quality management to improve business and benefit their clients. Site investigation (SI) contractors should follow this lead, making use of the 'special process' controls provided in BS 5750.

There are, however, two key factors which require special consideration when applying a QMS as described in BS 5750 to SI:
- the "uncertainty" of the geology of a site
- the current need for repeated referral of all aspects of the investigation to a third party "for approval" before the contractor provides the "end product", ie the Report.

On this first point, while innate variability exists in any process, the significant local variations and uncertainties in the natural materials which have to be dealt with in site investigations, do not exist to the same degree in manufacturing (for which BS 5750 was designed) nor in general construction. It is not feasible to apply the process manufacturing strictures of BS 5750 to SI without some modification. Neither would it be reasonable to expect a QMS to "assure" the geology of a site, bearing in mind the practicalities which always limit the number of boreholes and tests carried out on any site. Furthermore the QMS, on its own, will not prevent the technical problems high-lighted in part 1 of the ICE document[2], Site Investigation in Construction, 'Without site investigation ground is a hazard'. What a well-prepared QMS will provide is documented evidence that the specification and other agreed procedures have been complied with to produce the end-product.

Regarding the second point, the concept of quality assurance puts the onus of providing an acceptable end product totally on the supplier. Enterprising construction companies have shown that their QMS can only be effectively applied if they are in full control of the entire construction process through a comprehensive, single responsibility, 'design and construct' form of contract. This requires the designer, internal or commissioned, to produce specifications which provide all the necessary acceptance criteria without the need for additional approval by others as the work progresses.

The SI contractor is in fact already well-placed to adopt this single responsibility approach. He has the track record, the knowledge, the equipment and the qualified people under one management to enable him to take full control of an investigation and develop innovative and cost-effective solutions to geological variations. If a relevant QMS and compatible contract terms are then properly applied to these skills, the image and credibility of the industry will be restored and the client can again have full confidence in allowing the contractor proper control of his work.

Despite these benefits there are still several hurdles which have to be overcome to convince the typical SI contractor that it will pay him to implement a QMS; frequently quoted are:
- cost of third party assessment and up-keep of registration
- potential legal implications
- mountains of new paperwork
- unfamiliar terminology in BS 5750
- confusion of duties.

Much has been written on these points in applying quality management to construction, but most of the guidance available on QMSs for the SI industry, (eg part 2 of the ICE document[2], and the AGS paper[3] produced following their QA symposium in 1990), tends to concentrate on the role of the overseeing Engineer or the 'Principal Geotechnical Adviser', rather than address the problems faced by the specialist contractor. The following review and proposals set out to answer these concerns.

PROCUREMENT

Two systems of procurement are proposed in part 2 of the ICE document[2]:
"System 1" - provides for an Engineer for design, supervision and approval and a contractor to undertake fieldwork, testing and reporting
"System 2" - requires a single contractor to undertake a design and execute ground investigation contract, including a factual/ interpretive report.

System 2 is stated as being "attractive.....because it provides single-point responsibility", and "reduces administration". It is submitted that for the full benefits of quality management to be gained by the parties, System 2 is the only method of procurement for most SI contracts

The "Design and Investigate" form of SI contract proposed by Uff and Clayton[4,5] to operate System 2 would appear to provide an excellent basis for integrating the contractor's QMS with his obligations to produce a limited warranty Report. Under these terms "the Specialist [contractor] is in full control of production of an adequate and appropriate investigation as far as is needed to provide the information, data or other objects required". It follows that the contractor has to develop his QMS to address the design control requirements of BS 5750

Part 1, and provide the essential acceptance criteria not currently included in the prescribed "recognised specification". The contractor would therefore employ the geotechnical specialists to design, manage and interpret the work directly from start to finish; he would provide the professional indemnity. With the operation of rigorous internal non-conformance and concession procedures a true quality managed project is then in place. Other benefits which flow from the System 2 contract are quicker responses to unexpected conditions and less double-handling of documents leading to shorter reporting times.

One could speculate on why this non-adversarial System 2 contract has not been adopted for SI, but the increasing success of the ICE New Engineering Contract[6] could provide the alternative means of making quality management for SI compatible with contract terms.

Although System 2 provides for the effective integration of specification and quality management, most SI contractors are still employed under System 1, using the ICE Conditions of Contract for Ground Investigations[7] and the Specification for Ground Investigation, (part 3 of ICE document[2]). Both these and similar documents require the intervention of the Engineer to issue instructions/acceptance criteria during the work, usually through full-time on site representation. With the introduction of the new "Principal Technical/Geotechnical Adviser"[2] and an auditor for the QMS, the client's procurement of external management of the contractor has become critical to the contractor's performance.

While it is likely that on small SI contracts the duties of the three parties (Engineer, Adviser and auditor) will be carried out by the same person, it has to be made clear in all cases (and agreed with the contractor) what powers are delegated and to whom, bearing in mind only the Engineer has contractual authority to communicate with the contractor. It is also essential that all the parties involved have QMSs compatible with each other and the contract documents. If this is not done, it is just one of the potential areas for generating costly delays, confusion and layers of unnecessary paperwork which will render any QMS unwieldy - if not inoperable.

CONTRACTUAL CONSIDERATIONS
Contract Clause
A clear clause in the contract which commits both sides to effective quality management appropriate to the work will be of considerable benefit in improving understanding of the duties imposed. The following clause is developed from Barber[8] and is proposed for routine insertion into the ICE Contract[7]:

> "The contractor will be required to develop and operate a Quality Management System complying with BS 5750, Part 2. He will also produce a Quality Manual and Project Quality Plan in compliance with his Quality Management System and the project requirements stated in the Annex. The Project Quality Plan shall state that the laboratory tests specified for conformance with BS 1377[9] shall be carried out in a NAMAS accredited laboratory".

No obligation is imposed on the contractor by this clause to provide third party accreditation (except for standard lab testing). The Engineer would be required in the Annex to audit the contractor's QMS and Project Plan in accordance with his own quality management plan as the second party assessor (see below). This would ensure compatibility for the surveillance of the contractor's QMS with the requirement in the Specification for the Engineer to intervene and give technical approvals.

Surveillance

This is currently the most confused area of quality management in construction with yet further potential for creating considerable bureaucracy to hamper progress, particularly when the acceptance criteria under surveillance are not absolutely clear from the designers.

While most forms of contract do not provide for non-compliance with specification, it is recognised both in the SI Contract and through a BS 5750 QMS that the work can be varied, either by the Engineer giving his judgment/approval, or by the QMS auditor granting/advising concessions in respect of non-conformance with the management system. In a System 1 contract operated with a QMS as set out below, both these variation processes have to apply, but should be kept separate by the parties. As the Engineer usually acts also as auditor on short-term SI contracts, two points should be clear:
- the Engineer does not need extra powers conferred on him through either his own or the contractor's QMS to operate the Specification, reject defective work or withhold money for defective work
- a QMS does not negate the need for the Engineer to have in place a qualified representative to issue timely instructions to the contractor on his decisions in respect of the geological uncertainties exposed (or suspected) during the work.

Some simple examples from actual cases illustrate these points:
Core drilling - if the contractor does not keep the Engineer informed of the ground conditions and actions he is taking to optimise core recovery as required in his QMS, this is a *non-conformance* both with the Specification and the QMS, and re-work could be required;
- if, on the other hand, the information is duly given and the prescribed actions are taken but the core recovery continues to be low, this is *not* a non-conformance. The Engineer has to intervene promptly, give his decision in "the prevailing conditions" as specified and vary the work as appropriate.
In-situ tests - if the contractor sets up and carries out the test as specified or in accordance with an approved method statement, but the results are not as anticipated, this is not a non-conformance. The Engineer has to investigate and consider the significance of the results, and if necessary vary the test;
- if the specified/approved method was not followed, this is a clear non-conformance.
Purchasing - if the auditor finds that the contractor had not procured equipment or subcontracts in accordance with the QMS, the auditor's non-conformance report would be passed to the Engineer who would decide if the results of the investigation were jeopardised by the non-compliance and issue a re-work order or a concession.

The confusion of when the Engineer has to step in and give a decision and when to issue a non-conformance report, is avoided in a System 2 contract. Here the Engineer as "Supervisor" would be the auditor of the contractor's comprehensive design and investigate QMS and could only intervene when the QMS is not being operated.

Surveillance of a System 1 contract can only be rationalised if the Specification is re-written to eliminate the "for approval" referrals and fully define the acceptance criteria. While this would be costly initially, it would improve definition of responsibilities and save contractors considerable time and expense by avoiding repeated submissions.

Duty of Skill and Care

Although, as Barber[8] states, Courts are reluctant to imply a duty of skill and care in a construction contract, SI contractors are specialists employed for their skill, and it is wise for them to assume that such a duty will be implied if not expressly stated. The proposed contract clause above does not of itself impose additional liabilities on the contractor (other than a requirement to produce a QMS). But failure by the contractor to comply with his own QMS could amount to failure to show he has exercised the due skill and care levels he intended to apply to the work, thereby possibly compromising the final Report - leading to potential fault-based liability. A contractor who writes a sensible QMS and then fails to employ the right people to operate his system, inevitably leaves himself exposed.

Fit for Purpose

An SI Final Report which is "fit for purpose" is not a requirement of the current forms of contract. It is difficult to visualise how invoking a carefully written QMS to BS 5750 would extend the existing implied duty of skill and care to an obligation to produce a Report that was fit for purpose. (*Note* the contractor has already accepted in the Contract that the testing operations and processes he uses will be fit for purpose to meet the Specification).

ASSESSMENT OF QUALITY MANAGEMENT

First Party Assessment

Self-assessment, (that is the company's internal checks and audits), is the most important part of preparing, maintaining and developing a QMS, and should cover the four main aspects of the business: marketing, production, administration and strategy. The first vital step is to unearth all the documents and systems currently used by the contractor. Inevitably gaps or missing links in the communication process will be discovered and will need addressing, or existing sound, but un-documented, procedures will need writing up. This process will almost certainly produce an excess of documentation and too much detail in the procedures; a thorough review and streamlining is therefore essential, making sure that the business performance is not affected.

The objectives of this prime assessment and documentation are to demonstrate that investigations will be run with a management system showing how the specification will be met and that the appropriate skill and care will be used to produce the required final Report; it is not to produce documents which simply conform with BS 5750 manufacturing terminology.

Standard check lists have been produced for manufacturers to prepare procedures and attempts have been made to adapt them for service industries. As they usually retain the process jargon, they give little help to the SI contractor. He must start by setting down how his own existing business is run. Only when this is fully appreciated can compliance with BS 5750 headings as below be considered.

Second Party Assessment

This is defined as audits of the supplier's management systems carried out by the client/purchaser. Any management which leaves the review of its performance to an outside agency on an occasional basis is not providing quality management within its own organisation and is unlikely to be contributing to quality on behalf of the client. However, external auditing is an essential part of demonstrating capabilities to the client, and in the construction industry, which is largely project based, the benefits of external auditing will be best achieved by second party assessment, either directly by the client, his Engineer or other client appointed auditor.

The reasons are twofold; firstly the client can ensure that the contractor's QMS and specifically his Project Quality Plan are compatible with the project quality management procedures and specifications, and secondly the responsibilities on both sides of the contract are properly delegated to provide the collaborative approach so essential to good SI.

Another important factor in recommending second party assessment in construction, (as highlighted by Baden Hellard[10]) is that quality management begins at the top - and this is the client. He and his Engineer have to steer the project from initiation to conclusion and only they have the overview on how to achieve successfully the desired, unique, end-product. They must have in place an overall QMS which effectively pulls together all the interlocking parts and decide if a "sub QMS" meets their project objectives.

Unfortunately, the contractor could be faced with repeated trials of strength with auditors wanting to impose new corrective action procedures on each job. A standard QMS for SI contractors agreed by the industry on a 'self-certification'[11] basis (see below), would avoid the costly one-off changes to the basic management procedures and provide a uniform approach to Project Quality Plans.

Third Party Assessment
This is defined as audits of the supplier carried out by an independent body established for the purpose. The objective in manufacturing of having third party (eg BSI) certification of a supplier's systems and processes is to enable the purchaser to know, without inspecting for himself, that the system and procedures exist and can be used to assure satisfaction of the multiple/repetitive end-product against the specification. Site investigation is not a stand-off process like this. The Engineer and designer will want to see for themselves the results of the investigation as it proceeds to minimise uncertainties and ensure that the optimum design can be prepared. Seeing a certificate that a QMS is in place is no substitute for this hands-on participation.

The fact that the SI contractor is one of the earliest specialists on site is frequently used as the reason for needing a third party certificate - because the client or Engineer does not yet have the Project Plan in place. This must be wrong - the client committed to requiring quality management on his project must operate a quality assessment process to select designers and management teams at the start. Once the client has his quality team in place, assessment and on-going surveillance will be project related and rightly be second party based.

It is a general principle of assessment is that the auditor should not be one of the persons involved in the process, but, in the companies being considered in this paper where 'matrix' management has to be relied upon, this is not always possible. This would appear to lead only to third party assessment. However, as Baden Hellard explains, the current thinking on this difficulty in the construction industry is that the more formal BS 5750 principles are of less importance than the regular review of procedures - the main objectives being the response to the client and the improvement of the company's performance through the use of sound procedures. It is therefore considered acceptable for one of the contractor's staff to be trained to undertake internal audits and the project-specific variations to the QMS to satisfy BS 5750 self-certification and also function in routine unrelated line management.

Technical Systems. There are two additional activities which the SI contractor has to document - the laboratory testing and the technical procedures for writing the Report.

In view of the formalised nature of the soils testing methods and the unambiguous acceptance criteria detailed in BS 1377, together with the controlled laboratory environment necessary, it is recommended that all soils labs should be formally accredited. As BSI will not certify laboratories against technical standards, calibrations, techniques, etc., it is necessary to bring in the National Measurement and Accreditation Service (NAMAS) to assess and certify the lab for routine tests required in the Specification. Any sub-contracted tests should be similarly accredited. For special tests, project-specific method statements will be required in the Project Quality Plan and be assessed by the second party auditor for the Engineer's approval.

The technical manual of an SI contractor is a confidential document and should not be offered for formal assessment. The relevant extracts required for the Project Quality Plan will be made available to the second party auditor, eg the reporting proformas, method statements, etc., but the manual itself should only be available as appropriate to the auditor for examination to satisfy him that the contractor is meeting his technical obligations.

Safety Manual. The relevant instructions (eg relating to contaminated ground) will be given in the Project Quality Plan, and the manual made available for review by an auditor as requested.

Costs. In addition to the initial assessment costs, a BSI certified company has to pay for two follow-up surveillance visits each year during the three year registration period. NAMAS also require fees for two checks on the soils lab each year to maintain its certificate. Such interruptions four times a year are an unwarranted costly burden on a small contractor's business - hence the proposal to omit BSI involvement in a self-certification scheme.

Self-Certification. At the AGS symposium in 1990, concerns were raised at the confusion and fragmentation of surveillance and assessment as well the cost, and requests were made for guidance. Following the line of the above arguments, the additional cost of BSI assessment/registration and the current excessive bureaucracy are unnecessary if the industry is prepared to operate collectively the recommended project-based first and second party assessments of the QMS as the cost-effective way forward to sound quality management within the BS 5750 framework. This amounts to formal self-certification either by a group of SI companies or on an individual basis. The Engineer/auditor would then only need to check the specific content of the Project Plan to ensure the fully specified fieldwork, testing and reporting requirements were being met.

OUTLINE QUALITY MANAGEMENT SYSTEM
An effective and practical QMS has two requirements:
     - it must satisfy the needs and expectations of the client and
     - satisfy the needs and interests of the company.

The most practical and concise published guide for the SI contractor's QMS is the Building Employers Confederation document[12], setting out the sequence of events and giving clear examples of procedures needed to fit in with BS 5750 Part 2 requirements. The following detailed guidelines are drawn from experience to satisfy a System 1 Contract with second party assessment, once the basic company documentation has been rationalised.

The QMS contains three main documents: The Quality Manual (company policies); the Management Procedures (details of operations); the Project-Specific Plan (instructions for the works). Commercial data is not included.

The Quality Manual - Clauses 4.1, 4.2

This contains the company policies covering:

- how the QMS is to be controlled, ie numbered copies to named holders, restrictions on use and copying, up-dating and issues
- company objectives
- company organisation, including structure and reporting routes and summary of responsibilities
- summary of the management, technical and laboratory procedures cross referenced to the BS 5750 clauses
- method of management review

The Management Procedures

These are the core of the QMS and describe the basic way the company is run. Each section will describe personal responsibilities, have a brief narrative under headings covering the points made below, and have the relevant company proformas and recording documents attached:-

Section 1 Document Control - Clause 4.4 describes arrangements for the receipt, distribution and custody, up-dating and dispatch of documents. It should state how enquiries, tenders, awards, drawings, instructions, field and lab test results, reports etc., are referenced; how documents are issued and dispatched; which documents have to be indexed, stored, and where; how technical information is maintained and up-dated; how telephone calls are noted; how unclear client instructions are resolved.

Section 2 Audits - Clauses 4.13, 4.16 describes procedures used for carrying out the internal quality audits of the system and how feedback is made effective. It should state how the audit is planned and who has to be notified; how the audit will be conducted; corrective actions needed and close-out of non-conformities; follow-up by senior management.

Section 3 Contract Review and Mobilisation - Clauses 4.3, 4.8, 4.10, 4.17. It should be noted that BS 5750 does not require pre-award activities such as marketing and tendering in the QMS, and although most companies will have integrated procedures for these items they should not be offered to quality systems assessors. This Section covers preparing programmes, method statements and work instructions for the Project Quality Plan. (*Note* documented instructions are only required under BS 5750 if the absence of such instructions will adversely affect quality). Other items include checking equipment as required under the Calibration Section; allocating resources; notifying the Health and Safety Executive; setting up communication with the client; assigning of trained personnel.

Section 4 Project Quality Plans - Clauses 4.2 4.3, 4.7, 4.8, 4.9 describes how the QMS is to be applied to a particular job, ranging from a standard one sheet plan for a Report on a few basic boreholes to a full manual for a complex investigation. *Standard* and *Special* Plans are described below and should be made available for second party assessment. Negotiations are likely to be required before an agreed Special plan is approved for the project QMS.

Section 5 Training and Development - Clause 4.17 describes the assessment of training, and training needs for the company and specific projects; recruitment and induction; how training at various levels is achieved and recorded; how and when personal jobs are reviewed;

Section 6 Purchasing - Clauses 4.5, 4.6. Few permanent materials are incorporated into SI works, so this section should concentrate on bought-in items which affect the quality of the work, eg core barrels. It describes how materials and equipment are referenced, purchased, received and recorded; quotations; acceptable suppliers and how assessed; requisitions and orders - details needed, personal limits; inspection/verification of compliance with order on receipt; rejection of non-conforming supplies and subcontractor data; supplier performance records.

Section 7 Sub-contracts - Clauses 4.5, 4.6 is similar to the procedures for purchasing - no differentiation is made in BS 5750, but a separate section is advisable in the present climate. The form of subcontract to be used should be cited, its relationship to the main Contract and presentation to the subcontractor.

Section 8 Control of Works - Clauses 4.8, 4.12, 4.13 describes how the site work, laboratory testing and reporting is to be monitored and controlled; it also covers routine items which would not be project specific, eg setting out methods, reference to standard specifications used by the company such as BS 5930; requests to the Engineer for approvals, concessions and variation orders; site storage of equipment and complying materials; self-checking procedures; remediation of defective work; checks for draft, interim and final reports.

Section 9 Calibration of Equipment - Clause 4.11 describes requirements for laboratory and field testing equipment (cross-referenced to the NAMAS accredited laboratory manual); designates the British Standards for methods, frequency and equipment for calibrations; identification and recording systems; corrective actions for non-conformances; monitoring environmental conditions; security seals on adjustable devices; storage.

Section 10 Records - Clause 4.15 these are the records of the Sections detailed above showing that the QMS has operated effectively. Ease of retrieval is vital.

Project Quality Plans - Clauses 4.2, 4.3, 4.7, 4.8, 4.9.
The Procedures Manual covers the functions which are common to most jobs carried out by SI contractors. However, each investigation can be said to be different in that each will have unique boreholes - number, depth, location, sampling etc - and therefore will need site-specific instructions to the personnel carrying out the work. This is the Project Quality Plan (PQP).

Standard PQP. This notes on a single sheet the details of the client, the client's brief, a short description of the work, any specific instructions for the driller, and any deviations from the company's standard technical procedures (duly authorised); the engineers to produce, check and approve the Report and the planned issue date. Specific schedules and method statements would be attached.

Special PQP. This would be in a manual format and follows the same headings as the 'Standard', but with additional items to meet the contract. These would include: issue control, distribution and amendment; project organisation chart and responsibilities of key members; references to relevant specifications, and statutory regulations; method of technical review; records; authorised personnel for checking/approving/issuing of data. In a System 2 contract the necessary 'Design Control Plan' as required by BS 5750 Part 1 (*Clause 4.4*) and the expanded and clarified acceptance criteria would be included in here, together with schedules, technical notes and method statements.

The Technical Manual will cover the routines for *Clause 4,7 - Product Identification* eg sample numbering; *Clauses 4.9, 4.11 - Inspection and Testing* eg lab manual; *Clause 4.14 - Packaging and Delivery* eg cores. *Clause 4.18 - Statistical Techniques* is not appropriate to SI contract management.

CONCLUSIONS
- all SI contractors should adopt a standard, BS 5750-based, QMS on the lines given above and progress rapidly to industry-wide self-certification
- a well-prepared QMS should not introduce additional duty of skill and care
- internal and second party assessments are far more appropriate to the way the industry will develop under quality management than BSI certification
- soils laboratories should be NAMAS accredited for testing to BS 1377
- a QMS cannot be a substitute for sound judgment when assessing the uncertainties/ variations inherent in SI
- under the current adversarial contract terms the benefits of quality management cannot be fully achieved as the contractor is not in full control of the end product
- as a main beneficiary of quality management, clients should help to lead the proposed changes.

In addition to greater control, the contractor has much to gain from the application of a QMS; notably, improvements in efficiency and market share, fewer frustrations, better margins and quicker response to the client - provided he continues to develop his system.

REFERENCES
1 British Standards Institution, BS 5750: 'Quality Systems'. London (1987).
2 Institution of Civil Engineers, 'Site Investigation in Construction' in 4 parts. London(1993)
3 Association of Geotechnical Specialists,'Quality Management in Geotechnical Engineering - a practical approach' AGS/G/1/90. London (1991).
4 Uff J F and Clayton C R I, 'Recommendations for the procurement of ground investigation' CIRIA Special Publication 45. London (1986).
5 Uff J F and Clayton C R I, 'Role and responsibility in site investigation', CIRIA Special Publication 73. London (1991).
6 Institution of Civil Engineers, 'New Engineering Contract'. London (1993).
7 Institution of Civil Engineers, 'Conditions of Contract for Ground Investigations'. London (1983).
8 Barber J N, 'Quality Management in Construction - Contractual Aspects', CIRIA Special Publication 84. London (1992).
9 British Standards Institution, BS 1377: 'Methods of Test for Soils for Civil Engineering Purposes'. London (1990).
10 Baden Hellard R, 'Total Quality in Construction Projects'. Thomas Telford, London (1994)
11 International Standards Organisation, Guide 2, 'General Terms and their Definitions Concerning Standardisation and Certification'. London (1980).
12 Building Employers Confederation, 'Quality Management for the Small Builder'. London (1991)

Site investigation for lime stabilisation of highway works

J PERRY[1], R A SNOWDON[2], and P E WILSON[3]

[1,2]   Transport Research Laboratory, Crowthorne, UK
[3]   Highways Agency, London, UK

INTRODUCTION

This Paper provides guidance on planning, conducting and directing site investigations for lime stabilisation in highway works. It considers the information required from the site investigation in order to design lime stabilised cappings and the best means of acquiring that information. Guidance is given on ground investigation strategy and laboratory testing. The main critical soil properties are discussed and their detection and evaluation described. Consideration is given to the chemical and geotechnical effects stabilisation may have on cohesive soil.

Different sites will have varying circumstances which will make the use of granular capping or stabilised capping more appropriate. Availability of granular material and the impact of importing this material is a consideration. On-site materials may not have properties suitable for stabilisation, in which cases granular material may be the only option. However, stabilisation may produce substantial benefits. These include the maximum use of on-site materials with less haulage off-site and reduced need for spoil tips, the saving of scarce resources and reduced impact of the construction on the surrounding environment.

The information presented is suitable for preliminary, main and supplementary ground investigations and covers site investigation for the purposes of designing highway capping using lime stabilisation only. It does not cover lime modification of bulk earthworks.

LIME STABILISATION TO FORM CAPPING

Capping is a higher strength layer placed and compacted on weak fills and cutting foundations. Material used in the capping must be of sufficient strength to provide a working platform for construction of the pavement layers and act as a structural layer in the longer term. This can be achieved by using either granular materials, of fine or coarse grading, or materials stabilised with lime or cement to the requirements of MCHW 1.

The lime stabilisation process involves the spreading of lime, either quicklime or hydrated lime, on deposited or intact cohesive material, and a subsequent process of pulverising and mixing followed by appropriate compaction to form the whole or a constituent layer of a capping. Lime stabilisation is a long-term effect, but initially the material will pass through the modification process. The capping design will rely on the stabilisation process but since a soil goes through the lime modification process first, it is necessary to describe the modification effects.

Lime Modification

The use and effects of lime in the modification of clay soils is different to the use and effects of soil stabilisation. When lime is mixed with cohesive material, the cohesive material is first modified before further chemical reactions occur which in most soils lead to stabilisation. Not all cohesive materials will stabilise but all will modify. The modification of a cohesive material using lime changes the soil properties in two ways if quicklime is used, or in one way if hydrated lime is used. Mixing quicklime with a wet soil immediately causes the lime to hydrate and an exothermic reaction occurs. The heat produced is sufficient to drive off some of the moisture within the soil as vapour and hence reduces the moisture content. The second effect is a reduction in plasticity as the clay particles flocculate. This reduction in plasticity also occurs with hydrated lime, or slaked lime as it is sometimes called, but since the lime is already hydrated, no exothermic reaction occurs. These effects are immediate and little affected by temperature providing the material is above freezing. They have been used as a mitigation measure on a number of water logged sites (Sherwood, 1992).

Lime stabilisation

In the longer term the lime reacts with the clay particles to produce cementitious products which then bind the soil together. This is termed stabilisation and is the principle behind the requirements given in Clause 615 MCHW 1 for capping. The site investigation will need to include an investigation strategy and testing regime to provide the information to encompass these requirements. The natural material which is to be stabilised is called Class 7E and once lime is added it becomes Class 9D. Stabilisation using lime is temperature dependent and must be carried out only between March and September and usually only when the shade temperature is above 7 °C (MCHW 1). Lime stabilisation requires the soil and stabiliser to be thoroughly pulverised and mixed on-site using mobile stabilising machines. The material may be stabilised in a single or multiple layer. A period of not less than 24 hours and not greater than 72 hours is normally allowed for the material to mellow. This period allows the lime to slake, if quicklime is used, provides time for the lime to migrate through the soil and, as a result of the plasticity changes, makes mixing of the material before final compaction easier. Stabilised cohesive materials will require final compaction according to a method specification (Method 7) and considerable amounts of water may need to be added as the soil is mixed, especially when using quicklime. The water should be introduced in the hood of the rotary mixer. It is essential that the moisture content is greater than the optimum moisture content otherwise inadequate compaction will occur. Material is not allowed to be deposited on the compacted layer nor construction plant allowed to traffic the layer until a specified bearing ratio (measured using the laboratory CBR test) has been achieved.

Sulphates, for example gypsum, although not affecting the reduction in plasticity, react with lime and cause swelling which can be detrimental to the pavement's strength and cause deformations of the road surface. The reactions are fully described in Sherwood (1993). The sulphates may be present within the soil already, be produced by the oxidation of sulphides for example pyrite, or be introduced by groundwater. The state of compaction has a significant effect on the amount of heave associated with sulphate and sulphide reactions. Adequate compaction, that is 5 per cent air voids or less (DMRB 4.1.1), for lime stabilised capping layers can help to reduce heave in both the short term, during the construction period, and long term. Organic materials may also prevent the stabilisation process occurring, the effect being dependent on the type and amount of organic materials (Sherwood, 1993).

PRELIMINARY SOURCES STUDY

The preliminary sources study (PSS) stage of the site investigation should ensure that lime stabilisation of capping is considered as an option from the very beginning of the SI and earthworks procedures. The PSS should include an assessment of local geology and geotechnical features associated with the route.

The list of all sources scrutinised for geotechnical data and other relevant general information should include those sources relevant to lime stabilisation. Such sources of information include the following.

(a)   Useful references - Sherwood (1993), British Lime Association (1990a and 1990b), British Aggregate and Construction Materials Institution (1988), Snedker and Temporal (1990), Bessey and Lea (1953), Cripps and Taylor (1986 and 1987), de Freitas (1981), Nixon (1978), Forster, Culshaw and Bell (1995) and British Geological Survey (BGS) handbooks, memoirs and special publications.

(b)   The British Geological Survey (BGS) regional geologist can give advice on whether clays, shales or mudstones are likely to contain sulphate or sulphide minerals.

(c)   Examination of previous site investigation reports where available either for the highway itself, adjacent highways or any adjacent construction project.

(d)   Records of any problems in the vicinity, not only with lime stabilisation, but also where heave of buildings or highways has occurred .

In carrying out walkovers, and investigating materials at exposures and in exploratory holes, recognition of sulphide and sulphate minerals is important. These minerals can be seen commonly in hand specimens using the naked eye or hand lens. Observations should also be made for evidence of organic material.

Most British clays will stabilise with lime but Dumbleton (1962) showed that the Weald Clay did not increase in strength with time as did other clays, such as London Clay and Lower Lias clay. Although the plasticity of the Weald Clay was reduced, the long-term stabilisation did not occur. This effect has been noted also by Heath (1992). Wood (1988) considered that the poor reaction with lime was due to the presence of kaolinite in the Weald Clay but Sherwood (1992) points out that kaolinite is present in most British clays including those that react well with lime. Other workers have also found that kaolinitic clays can stabilise well with lime (for example Anon, 1986).

GROUND INVESTIGATION

The purpose of the ground investigation is to provide information on which to judge, after consideration of the PSS, whether lime stabilisation is viable and to provide sufficient material to carry out testing to ascertain an adequate mix design. The strategy adopted to meet this purpose should be geared toward obtaining the maximum amount of representative data for the minimum cost and in a reasonable period of time.

The PSS will provide a broad and essential understanding of the site conditions. To plan the GI, these data and the probable location of cuttings and embankments should be used to establish the most efficient position of exploratory holes and depth of sampling.

Materials to be excavated for cuttings and used as acceptable material in embankments may have the potential for use as lime stabilised capping for embankments. Also the materials at or below formation level may be acceptable for lime stabilisation *in situ* as a capping layer at the base of cuttings.

The sampling and testing should reflect the intended use of the soil, as a capping material on fill or in cutting. In terms of lime stabilisation, sampling in cutting areas from formation to at least 1 m below is essential to include material which is most likely to be used in lime stabilisation. Less frequent sampling of the rest of the cutting area may only be possible as more material is involved and more mixing of material is likely. This has the advantage, however, of mixing characteristics detrimental to lime stabilisation, such as sulphur and organics, and diluting some of their effect. Much information on the feasibility of lime stabilisation can be obtained from the soil descriptions in the exploratory holes. The description with the naked eye and hand lens of soils in and from trial pits, trenches and boreholes, is the first step in the assessment.

The plasticity of some materials, for example Mercia Mudstone, varies depending on the degree of weathering and material properties (Chandler, 1969). The suitability for lime stabilisation can be related to the zoning or layering used to describe these variations. As clay particles are required for the stabilisation process, the suitability of weathered mudstone for lime stabilisation may well be governed by the degree of weathering.

Trial pits and trenches are particularly useful for obtaining information on the uniformity, extent and composition of soils. Deep trial pits, around 6 m deep, can provide a more accurate picture of the nature of the material in bulk compared to borehole observations. Deep trial pits give a more accurate profile of weathering, particularly in mudstones such as Mercia Mudstone and Lias Clay. The degree of weathering interpreted from boreholes can be less than from trial pits, due to disturbance during boring. Trial pits can also reveal the variability of the weathering profile in much more detail than borehole results. This has important implications for lime stabilisation in terms of identifying those less weathered materials that do not meet the plasticity requirements.

An awareness of the distribution and appearance of sulphate and sulphide minerals is important during the planning and execution of the GI. The presence of sulphide and sulphate minerals is indicative of likely heave problems which may occur after stabilisation has taken place. It should be emphasized that the description of the material from the core, or from or in the trial pit, is the key method for locating the presence of sulphide and sulphate minerals. The laboratory testing will give a precise figure at a particular location but will not give an overall assessment of the distribution of the sulphur minerals. Bessey and Lea (1953) and Sherwood (1957) noted that the sulphate contents of British soils are usually low in the surface layer and increase with depth. In the Oxford Clay, the upper 1 m shows a distinct lack of sulphates and an increasing amount of sulphate toward the base of the weathering zone as a result of leaching at about 2 m to 3 m, and then becoming only a small percentage at further depths. It is possible that a profile of sulphides in the same material would show a similar absence down to 1 m, but also an absence at 2 m to 3 m depth due to oxidation to form sulphates. However, below 3 m, sulphide contents can be significantly higher, although not uniformly distributed. The positioning of a trial pit or trench in the transition zone increases the likelihood of locating any sulphate minerals that may be present at the level where lime stabilisation may be used. Sulphide and sulphate minerals tend to occur in discontinuous nodules or finely disseminated making the macro-description crucial to obtaining an overview of the soil. There is little point conducting detailed laboratory tests on small samples taken from a non-sulphur bearing portion of a sample if large quantities of sulphate minerals were identified by eye and given in the exploratory hole description.

The oxidation of sulphide minerals, principally pyrite, and the associated formation of sulphuric acid and sulphates is time dependent (for example Sandover and Norbury, 1993). Therefore, laboratory tests for sulphates will tend to show higher values than occur *in situ* the longer the testing is delayed. Also the pH will decrease, that is become more acidic, the longer the testing is left. These changes will affect the interpretation of the acceptability of materials as higher total and water soluble sulphate contents will be recorded in the laboratory tests, whereas the undisturbed soil will have a lower sulphate content. In lime stabilisation, this can lead to an inaccurate assessment of material suitability. In order to reduce the effect of oxidation on samples to be used for sulphate testing, such samples should be kept in well sealed and full containers, thus limiting the amount of oxygen available for the oxidation process. In order to reduce the rate of oxidation, Mitchell (1986) recommends storing the samples at low temperatures (4° C).

Collection of information on groundwater is essential in the determination of groundwater regimes which affect the design of the earthworks. In particular, careful consideration needs to be given to ensuring the sulphate content of the groundwater is below an upper limit of 1.9 g/L of sulphate (expressed as SO_3). If a source of surface water can be identified which is likely to be used by the Main Works Contractor in lime stabilisation, tests will be needed to identify high sulphate contents and low pH values. Their effect can then be assessed in laboratory testing.

LABORATORY TESTS

Limits on the properties required for cohesive material before adding lime, and the limits once lime has been added, are essential to allow both the selection of suitable material and to give assurance of good long term performance of the pavement structure. Suggested laboratory tests to determine these limits for design purposes are given in Table 1 and a suggested test procedure is shown in Figure 1. Tests for suitability are those to determine whether or not the material is able to be stabilised and include tests such as initial consumption of lime and soaked California Bearing Ratio. Tests for acceptability are for determining whether the soil meets the requirements of the Classes in MCHW 1 Table 6/1 or for providing values for inclusion as limits in Appendix 6/1 of the Contract.

Soil Tests for Suitability and Acceptability of Material for Stabilisation (Class 7E).

Plasticity Index. It is necessary to ensure that there are enough clay minerals present in the untreated material to enable the soil to react with the lime and allow the pozzolanic reaction to occur. The lower limit of 10% for the Plasticity Index (I_p) in MCHW 1 Table 6/1 should ensure the suitability of the soil so that this reaction can take place.

Moisture Condition Value and Moisture Content. A calibration line for moisture content versus Moisture Condition Value (MCV) is needed to assess the suitability of the material for handling and trafficking. This is used to set the maximum moisture content, or minimum MCV, and is related to the bearing capacity of the soil for the construction plant and shear strength of the soil. An upper value of MCV or lower moisture content is superfluous as water could be added when mixing with lime and compacting although a value is commonly given.

Sulphates and Total Sulphur Content. In their natural state, sulphates occur as SO_4. The Building Research Establishment has revised its Digest on the effects of sulphates on concrete by changing from sulphates as SO_3 in BRE Digest 250 (Building Research Establishment, 1981) to sulphates as SO_4 in BRE Digest 363 (Building Research Establishment, 1991). However, in soil testing, sulphates are still normally expressed as SO_3. This is the situation in BS 1377 and is the convention used in this Paper. The test for the total sulphate content (also referred to as the acid soluble sulphate content) provides a measure of total amount of sulphate in the soil at the time of testing. The test for water soluble sulphate (BS 1377: Part 3 but amended by BSI News January 1991) gives a measure of the amount of soluble sulphate present in a soil in order to ascertain the potential for migration of sulphates. Sulphates may also be present in groundwater for which there is also a test in BS 1377. Sulphates (SO_3) can result from the oxidation of sulphides (S). It is therefore important that the sulphates converted from sulphides be measured as well. The test for total **sulphur** content is given in BS 1047 and provides a measure of the sulphates already in the soil plus those sulphates converted by oxidation from sulphides. The unit of measure in BS 1047 is percentage sulphur (S%) which should not be compared directly with the percentage sulphate

Material Property	Defined and tested in accordance with:	For suitability	For acceptability
Plastic limit	BS 1377: Part 2	x	
Liquid limit	BS 1377: Part 2	x	
Plasticity index	BS 1377: Part 2	x	
Particle size distribution	BS 1377: Part 2	x	
Organic matter	BS 1377: Part 3	x	x
Total sulphate content	BS 1377: Part 3	x	x
Total sulphur content	BS 1047	x	x
Initial consumption of lime	BS 1924: Part 2	x	
CBR	BS 1924: Part 2	x	x
Swelling	BS 1924: Part 2	x	
MCV for Class 7E	BS 1377: Part 4 (In Scotland, DMRB 4.1.4 SH7/83)		x
MCV for Class 9D	BS 1924: Part 2		x
Optimum moisture content for Class 9D (2.5 kg test)	BS 1924: Part 2		x
Frost susceptibility	BS 812: Part 124	x	

Table 1 Soil tests for suitability and acceptability

Plasticity Index and Grading - Within MCHW 1 requirements?

IF YES CONTINUE, IF NO REJECT

Establish Initial Consumption of Lime (ICL) - Is ICL established?

IF YES CONTINUE, IF NO REJECT

Total Sulphate Content / Total Sulphur Content / Organic Matter
CBR Tests :   (i)   mellow before compacting
(ii)   3 days curing
(iii)   followed by 4 days soaking
(iv)   then test
Swelling, monitor to day 28
Repeat tests at a range of moisture contents and lime additions from ICL in 0.5% stages
Determine design lime addition value
Is 7 day average CBR > 15% (no individual specimen less than 8%)?
Is average swelling ≤ 5 mm (no individual specimen more than 10 mm) and approaching an asymptotic value?

IF YES, THEN MATERIAL IS SUITABLE FOR STABILISATION
[subject to satisfactory water soluble sulphates in surrounding materials]
IF NO REJECT
ADDITIONAL LABORATORY TESTS REQUIRED:

Test for frost susceptibility
Establish MCV/mc calibration for Class 7E material
Establish MCV/mc calibration for Class 9D material at design lime addition
Determine OMC for Class 9D material at design lime addition
Water soluble sulphate test (BS1377:Part3 Amended in BSI News Jan 1991)

Fig 1 Testing for soil suitability

$(SO_3\%)$ in BS 1377 until the following formula has been used to convert the total sulphur percentage to a total sulphate percentage.

$$S\% \times 80/32 = SO_3\%$$

Once calculated, comparisons can be made on the same basis between the percentage of sulphate in the soil, and the percentage of sulphate in the soil plus that formed from the oxidation of sulphide.

Organic Matter. Testing for organic matter is required as it can interfere with the normal reaction between the lime and the soil by decreasing the soil's pH value.

Soil Tests for Suitability and Acceptability of Stabilised Material (Class 9D)

Initial Consumption of Lime. For adequate stabilisation with lime, sufficient lime needs to be added to allow the reaction with the clay to occur. The minimum value of available lime required to enable reaction with Class 7E material to be achieved must, therefore, be determined; this value is known as the initial consumption of lime (ICL). The ICL is a chemical test and does not dispense with the need to carry out strength determinations, as it does not establish whether the soil will react with lime to produce a substantial strength increase. Therefore, the ICL is used only to determine the minimum lime addition for strength testing.

California Bearing Ratio (CBR) and Lime Addition. The purpose of the CBR testing is to:
(a)   ascertain the lowest MCV, or highest moisture content, and minimum lime addition which produces a CBR value, based on laboratory CBR testing, adequate for carrying the next construction layer and the plant necessary to construct it;
(b)   ensure the structural integrity of the capping;
(c)   to find the lime content which produces a satisfactory CBR and degree of swelling.
Prior to testing it is recommended that the material samples are mellowed, in sealed containers to prevent carbonation, for a period of 24 to 72 hours to allow the lime to react with the soil and represent site procedures.

Sherwood (1992) reviewed the use of the CBR test with either soaked or unsoaked samples after a period for curing and proposed a test procedure which consists of preparing the test sample at the end of the mellowing period and then following the procedure given in BS 1924: Part 2 using the 2.5 kg rammer test. After allowing the test sample to cure in air for a 3 day period at 20 ± 2 °C, the sample is soaked for 4 days, at the same temperature, after which the CBR is measured. The test samples should be prepared with a range of moisture contents that enables the limit for acceptability to be determined. Dependent on the type of soil, this should cover moisture contents from about MCV 13 to MCV 8, that is from optimum moisture content through to wet of the plastic limit. It is necessary to establish the lower limit for the MCV, or equivalent moisture content, which gives a minimum CBR value (usually 15%) at 7 days. CBR tests should be carried out on a range of sample mixes with lime contents starting at ICL, and increasing in 0.5% steps, to establish the lime addition for the design that will produce an average CBR of greater than 15% at 7 days with no individual test specimen having a CBR of less than 8% (Wood, 1988). The number of test samples prepared should also allow CBR tests to be carried out at 28 days and the degree of heave to be monitored. The amount of heave that occurs during the soaking period is critical in determining the suitability of the Class 9D material. It is important, therefore, that measurements are recorded for up to 28 days to ensure that all swelling has ceased. This usually occurs by the fourteenth day with most materials. It is recommended that the average degree of swelling should be less than 5 mm (measured on the standard 127 mm high CBR mould sample), with no individual test specimen having a swell of more than 10 mm. If swelling is still occurring after 28 days but is still below this limit, some subjective assessment of changes in the rate of swell may be necessary: alternatively the swelling test period could be extended to 56 days provided that the rate of swell was declining.

Moisture Condition Values and Moisture Content. Lime stabilised soils behave in a very different manner to most naturally occurring soils and have completely different plasticity characteristics to the original unstabilised material; research has shown that the MCV is able to reflect these changes. Immediately following mixing with lime, the plastic limit and liquid limit change with time. Therefore, the timing of laboratory tests should be related to the likely time-scale of construction activities. In particular, the MCV/moisture content calibration should be made after the specified mellowing period, thus relating to the condition of the material at the anticipated time of compaction. If moisture content is likely to be used as a criterion for acceptability, as the alternative for MCV, then it is absolutely crucial that laboratory testing is carried out with due consideration to the effect of time on plasticity following the addition of lime to the Class 7E material. It is essential that the operator is aware that Class 9D material is likely to produce an MCV/moisture content calibration with 'wet' and 'dry' legs. Figure 2 shows a typical MCV calibration exhibiting this type of behaviour. The calibration leg which must be used is the 'wet' leg, which shows reducing MCVs with increasing moisture content values. This provides the correct range of MCV for use as the criterion for acceptability immediately prior to compaction.

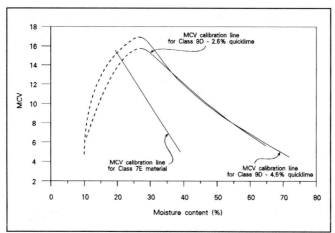

Fig 2 Typical MCV Calibrations for Class 7E & 9D materials

Laboratory Compaction. In this Paper, 'optimum moisture content' is the moisture content either at the maximum dry density or at 5 per cent air voids whichever is the wetter. Due to the rather flat dry density/moisture content curve that may result for some stabilised materials, some difficulty may occur in accurately defining the optimum moisture content. In such cases, it is recommended that the moisture content value at which the compaction curve crosses the 5 per cent air void line is taken as the optimum moisture content; this value will approximate to the moisture content in the field that should ensure an acceptable state of compaction to be achieved using Method 7 MCHW 1 Table 6/4.

For cohesive soils, only the 2.5 kg rammer method produces realistic results which relate to densities achieved on site. It is recommended that the laboratory samples are allowed to mellow before compaction for the same period as anticipated on site. This is particularly important if a moisture content value is to be used for the material acceptability criterion on site because the optimum moisture content of the mellowing soil will alter significantly with time. The actual period of time between the sample preparation and time of testing should be reported. This is a deviation from the BS 1924 test procedure which recommends testing immediately after mixing.

It must be appreciated how important it is to carry out the optimum moisture content testing at the same time as the MCV/moisture content calibration and at a similar time interval to that expected after mellowing on site. MCV is able to reflect the changes in plasticity and is less dependent on time of testing than optimum moisture content. Figure 3 illustrates this point for a stabilised heavy clay. Although the optimum moisture content has changed as the material becomes more granular in behaviour the MCV still stays around 13.5.

Frost Susceptibility. The test specified in sub-Clause 602.19 of MCHW 1 is an amended BS 812: Part 124 test which should be carried out on lime stabilised specimens of the materials proposed for lime stabilisation during the ground investigation. Further tests should also be carried out on specimens cured for at least 28 days after mixing with lime.

TEST INTERPRETATION

Initial Consumption of Lime. The establishment of an ICL will identify whether the Class 7E material is suitable for stabilisation and will also provide the starting value for the lime additions during the CBR tests. The ICL will depend upon the mineralogy of the unstabilised material, with typical ICLs ranging from 1.5% to 3.5%. If an ICL is not achieved then the Class 7E material is deemed to be unsuitable and must be rejected.

California Bearing Ratio. If the swell and strength criteria have not been met then the material is deemed to be unsuitable for lime stabilisation and should be rejected. From the CBR tests carried out over a range of available lime additions and moisture contents, the moisture content value, and hence the MCV, can be identified at which the minimum CBR strength requirement is achieved.

Lime Addition. The minimum amount of available lime required to achieve a CBR of greater than 15% will have been identified during the CBR tests. For the actual design mix value, it is recommended that an additional 0.5% is included to allow for variations in the available lime content of the bulk lime supplied to site, local variations in material mineralogy and inefficient site mixing.

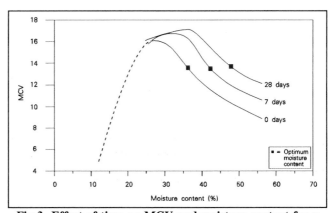

Fig 3 Effect of time on MCV and moisture content for a stabilised heavy clay

An overall minimum available lime content of 2.5% is often specified: this is a typical value, representative of materials tested to date. However, it must be replaced by any higher value of lime addition indicated by CBR testing. The amount of available lime varies with the type of lime manufactured. In quicklime the available lime content is about 90% to 95%, but in hydrated lime it is only about 60% of the total mass. The amount of lime added for the design will obviously have an influence on whether the process will be economical or not.

Moisture Condition Value and Moisture Content. For Class 7E material the limits on MCV, or moisture content, only need to reflect the limits on earthworking plant operation and material handling and hence only a lower limit is required. The value would typically be around MCV 7, or the equivalent moisture contents, although some stabilisation plant can cope with sites wetter than this. The lower and upper limiting values for the Class 9D material are based upon compaction requirements and the laboratory CBR. The lower limit would again be about MCV 7: the upper limit for compaction is discussed in the following section.

<u>Laboratory Compaction.</u> The optimum moisture content for the design mix should be used as the upper limit for the MCV, or the lower limit for moisture content, for acceptability of the Class 9D material. Any Class 9D material placed at a higher MCV, or lower moisture content, will probably have an air void content greater than 5% and will be highly susceptible to wetting up, swelling and strength loss in the longer term.

It has already been mentioned how important it is to carry out the optimum moisture content testing at the same time as the MCV/moisture content calibration and at a similar time interval to that expected after mellowing on site. In Figure 3 the optimum moisture content has changed as the stabilised heavy clay becomes more granular in behaviour but the MCV still stays around 13.5; consequently a specified limit of an MCV of 12.5 for a stabilised heavy clay would be wet of optimum moisture content.

Figure 4 (based on Parsons, 1992) shows that the MCV at optimum moisture content varies between about 12 and 14 for a number of stabilised materials ranging from lime stabilised glacial till and heavy clay to gravel-sand-clay. The MCVs for the different materials given in the figure are considered an absolute maximum for lime stabilisation and it is essential for compaction purposes to be at lower MCVs, that is to be 'wet' of optimum moisture content. The relation between MCV and optimum moisture content is the key to why the MCV is such a useful test for lime stabilisation; although the plasticity properties of a Class 9D material vary with time, the MCV remains relatively constant.

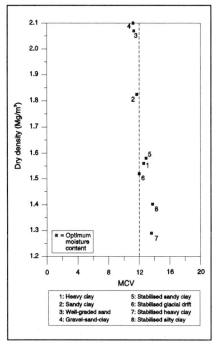

Fig 4 MCV/Dry density (2.5 kg rammer)/omc relations

1: Heavy clay	5: Stabilised sandy clay
2: Sandy clay	6: Stabilised glacial drift
3: Well-graded sand	7: Stabilised heavy clay
4: Gravel-sand-clay	8: Stabilised silty clay

<u>Sulphates and Total Sulphur Content.</u> Limits for acceptability of total sulphate content and total sulphur of Class 7E material will primarily depend upon the swell measured during the CBR tests. Values of total sulphate and total sulphur test, along with the amount of swelling measured in the soaked CBR tests, should be assessed to fix the limit for total sulphate and also to establish a limit for total sulphur. Selecting an upper limiting value for total sulphate content can be difficult. The upper limit for total sulphate content may be as high as 1% (HA 44; DMRB 4.1.1) (a higher limit is not recommended) and there is evidence that for some materials suggested values ranging from as low as 0.25% may cause swelling (Sherwood, 1992). Limits of total sulphate below 0.25% are likely to be too onerous for most soils. Information on limits on total sulphur is not readily available for stabilised soils. The testing regime described in this Paper will, however, provide sufficient information to set a limit on a site specific basis.

<u>Organic Matter.</u> An upper limit of 2% for acceptability of the Class 7E material is a useful guide, although there is some evidence to suggest that it is the type rather than the amount of organic matter which affects stabilisation (Sherwood, 1993). If the Class 7E material under investigation has an organic content greater than the 2% value, but has been successfully stabilised in terms of reaching acceptable CBR and swell values, then

consideration should be given to increasing the upper limit to accommodate the actual value measured.

CONCLUSIONS
It is recommended that the site investigation should:
(a)   include the early identification of sulphides and sulphates in the field;
(b)   include laboratory testing for sulphides and sulphates, if not found in sufficient amounts in the field to make lime stabilisation unfeasible;
(c)   emphasise achieving adequate compaction, and that the Moisture Condition Value (MCV) is the best means of doing this;
(d)   be guided by an upper MCV value of between 12 and 14 to ensure compaction wet of the optimum moisture content for all materials stabilised with lime;
(e)   control swelling of the capping by
        (i)   putting limits on swelling during extreme laboratory soaking tests;
        (ii)   advising that the total sulphate content should be below 1%;
        (iii)   investigating the potential production of sulphates from sulphides;
        (iv)   providing compaction to 5% or less air voids;
(f)   use organic matter limits based on laboratory test results for Initial Consumption of Lime and California Bearing Ratio;
(g)   emphasise that tests should be conducted at the same time and that these times are representative of site requirements. This leads to a contradiction with the test in British Standard BS 1924: 1990 for the optimum moisture content of stabilised material.

ACKNOWLEDGEMENTS
The work described in this Paper forms part of a Highways Agency, DOT, funded work programme conducted by the Transport Research Laboratory and is published by permission of Road Engineering and Environmental Division and the Chief Executive of TRL. Thanks are due to D J MacNeil and D Steele for their assistance, particularly in conducting the laboratory work.
Crown Copyright 1995. The views expressed in this Paper are not necessarily those of the Department of Transport.

REFERENCES
ANON (1986). Cementitious stabilizers in road construction. Draft TRH13. National Institute for Transport and Road Research, Council for Scientific and Industrial Research, South Africa.

BESSEY G E and LEA F M (1953). The distribution of sulphates in clay soils and ground waters. Proceedings of the Institution of Civil Engineers, 2, (1), 159-181.

BRITISH AGGREGATE AND CONSTRUCTION MATERIALS INSTITUTION (1988). 'Lime Stabilisation '88'. BACMI, London.

BRITISH LIME ASSOCIATION (1990a). Lime stabilisation manual. 2nd Edition.

BRITISH LIME ASSOCIATION (1990b). 'Lime Stabilisation '90'. BLA, London.

BRITISH STANDARDS INSTITUTION (1983). BS 1047 Air-cooled blastfurnace slag aggregate for use in construction. BSI London.

BRITISH STANDARDS INSTITUTION (1989). BS 812: Part 124 Testing Aggregates. Method for the determination of frost-heave. BSI London.

BRITISH STANDARDS INSTITUTION (1990a). BS 1377 Methods of test for soils for civil engineering purposes. BSI London.

BRITISH STANDARDS INSTITUTION (1990b). BS 1924 Stabilized materials for civil engineering purposes. BSI London.

BRITISH STANDARDS INSTITUTION (1991). Special announcement: BS 1377. BSI News. January, 1991. BSI London. 30.

BUILDING RESEARCH ESTABLISHMENT (1981). Concrete in sulphate-bearing soils and groundwaters. BRE Digest 250. Building Research Establishment, Watford.

BUILDING RESEARCH ESTABLISHMENT (1991). Sulphate and acid resistance of concrete in the ground. BRE Digest 363. Building Research Establishment, Watford.

CHANDLER R J (1969). The effect of weathering on the shear strength properties of Keuper Marl. Geotechnique, 19, (3), 321-334.

CRIPPS J C and TAYLOR R K (1986). Engineering characteristics of British over-consolidated clays and mudrocks: I Tertiary deposits. Engineering Geology, 22, 349-376.

CRIPPS J C and TAYLOR R K (1987). Engineering characteristics of British over-consolidated clays and mudrocks: II Mesozoic deposits. Engineering Geology, 23, 213-253.

DE FREITAS M H Editor (1981). Mudrocks of the United Kingdom. Quarterly Journal of Engineering Geology, 14, (4), 241-372.

DESIGN MANUAL FOR ROADS AND BRIDGES, HMSO.
HA 44 - Earthworks: design and preparation of contract documents (DMRB 4.1.1).
SH 7 - Specification for Road and Bridge Works: Soil suitability for earthworking - Use of the Moisture Condition Apparatus (DMRB 4.1.4).

DUMBLETON M J (1962). Investigations to assess the potentialities of lime for soil stabilization in the United Kingdom. Road Research Technical Paper No.64. HMSO, London.

FORSTER A, CULSHAW M G and BELL F G (1995). The regional distribution of sulphate in rocks and soils of Britain. In: Eddleston M, Walthall S, Culshaw M G and Cripps J C (eds). The engineering geology of construction. Engineering Geology Special Publication No. 11. Geological Society, London.

HEATH D C (1992). The application of lime and cement soil stabilization at BAA airports. Proceedings of the Institution of Civil Engineers (Transport), 95, February, 11-49.

MANUAL OF CONTRACT DOCUMENTS FOR HIGHWAY WORKS, HMSO.
Volume 1: Specification for highway works (December 1991): (MCHW 1).

MITCHELL J K (1986). Practical problems from surprising soil behaviour. Journal of the Geotechnical Engineering Division, ASCE Geotechnical Journal, 112.

NIXON P J (1978). Floor heave in buildings due to the use of pyritic shales as fill material. Chemistry and Industry, 4 March 1978, 160-164.

PARSONS A W (1992). Compaction of soils and granular materials: a review of research performed at the Transport Research Laboratory. HMSO.

SANDOVER B R and NORBURY D R (1993). On an occurrence of abnormal acidity in granular soils. Quarterly Journal of Engineering Geology, 26, (2), 149-153.

SHERWOOD P T (1957). The stabilization with cement of weathered and sulphate-bearing clays. Geotechnique, 7 (4) 179-191.

SHERWOOD P T (1992). Stabilized capping layers using either lime, or cement, or lime and cement. Contractor Report 151. Transport Research Laboratory, Crowthorne.

SHERWOOD P T (1993). Soil stabilization with cement and lime. Transport Research Laboratory State of the Art Review. HMSO.

WOOD C E J (1988). A specification for lime stabilization of subgrades. 'Lime stabilization '88', Symposium on Lime Stabilization. BACMI, London. 1-8.

Investigation and assessment of cohesive soils for lime stabilisation

L J BARBER
RSA Geotechnics, Stowmarket, UK (formerly of Babtie Geotechnical, Ipswich, UK).

INTRODUCTION

Recent well publicised failures of some highway capping layers formed from lime stabilised materials have led to considerable research into the suitability of various materials for lime stabilisation and into the criteria by which acceptability can be judged.

The recent state of the art review by Sherwood (Ref 1) on soil stabilisation with cement and lime includes an in depth account of the stabilisation process, methods of testing and specification of stabilised materials and the physical and chemical aspects affecting the strength of these materials.

The purpose of this paper is to outline an approach to the investigation of sites where lime stabilisation is being considered as an option for inclusion in road pavement layers. For the purpose of this paper lime stabilisation is considered in isolation. For highway schemes it is acknowledged that typically the investigation for lime stabilisation may be part of a larger general ground investigation of the site for the proposed highway and its associated structures.

LIME STABILISATION ACCEPTABILITY CRITERIA

Before a ground investigation can be planned it is necessary to know the criteria by which the material to be stabilised will be assessed. In the case of road schemes these criteria are specified in the Department of Transports Specification for Highway Works, (SHW) 7th Edition, Volume 1, Series 600. (Ref 2). Additional information is given in the DoT Departmental Advice Note HA 44/91 (Ref 3). The recommendations in the former document are only strictly applicable to DoT road schemes but form a good basis for any other road scheme. In these documents two material types are identified. Firstly, material suitable for lime stabilisation to form capping and secondly lime stabilised material.

Acceptability Criteria for Material to be Used as Lime Stabilised Capping

The raw material to be used in the lime stabilised capping layer should comply with the criteria given in Table 1, such material is defined as Class 7E material by the SHW.

Table 1. Acceptability Criteria for Class 7E material.

Class	General Description	Grading		Plasticity	Organics	Total Sulphates	Moisture Content or M.C.V.
7E	Cohesive	Sieve 75mm 28mm 0.063	%Passing 100 95-100 15-100	PI > 10%	< 2%	To be specified by design Engineer	To be specified by design Engineer

Appropriate values for total sulphate and moisture content/moisture condition value (MCV) have to be determined by laboratory testing.

Acceptability Criteria for Lime Stabilised Material
Once the Class 7 material is mixed with lime it is defined as Class 9D, lime stabilised cohesive material.

Clause 615 of the SHW requires that a minimum available lime content of 2.5% by dry weight is used. It also required that minimum and maximum MCV values are specified to control the moisture content of the lime-soil mix.

A minimum value of CBR is also required for the lime stabilised material.

SITE INVESTIGATION
To allow a confident decision to be made regarding the inclusion of lime stabilisation as an option in any road scheme it is imperative to have a good site investigation. This investigation should be built up around three key elements, the desk study, the ground investigation and the laboratory testing, the latter two elements culminating in the Factual Report. Each of these elements are the basis of any good site investigation, their particular reference to a lime stabilisation investigation is discussed in the following sections.

The Desk Study
To enable planning of the lime stabilisation ground investigation it is necessary to determine the general nature of the site, its underlying geology and whether the materials which are anticipated at the site are suitable for consideration as lime stabilised materials. It is the latter two of these that are most relevant to the lime stabilisation investigation.

Site Geology   A review of available geological literature of the site area will give a preliminary indication as to the strata likely to be encountered by the proposed road. This data can be refined into plans showing the anticipated materials down to and at road formation level in cuttings if the proposed vertical and horizontal alignments of the road are available.

Particular attention needs to be given to the clay lithologies anticipated in the area. To make lime stabilisation a viable option it is not only necessary to have clay present on site but also for this material to be of sufficient extent and regularity to allow efficient site operations during construction.

The ground investigation should be planned to optimise the use of trial pits and boreholes to recover the samples necessary for the laboratory testing stage.

<u>Anticipated Materials</u>   A review of the geological and geotechnical literature of the area should indicate the nature of the various lithologies present, with particular respect to lime stabilisation. Certain clay lithologies are known to have high levels of naturally occurring sulphate which may preclude their use for lime stabilisation.

Particular emphasis should be placed on reviewing the literature for examples of similar lithologies which have been lime stabilised elsewhere and of the success, or otherwise, of such operations. Attention should be paid to difficulties experienced not only with the lime stabilisation aspects of other schemes in similar lithologies but also to difficulties experienced in handling the materials during excavation and placement.

<u>Ground Investigation</u>
Generally the usual method of procurement for a ground investigation for road schemes is by competitive tendering. This requires the party letting the tender to prepare a Tender Document which details the required ground investigation and subsequent laboratory testing and reporting requirements. This document requires careful preparation to ensure that the investigation is appropriate for its purpose and that the estimates of quantities are reasonable. It is advised that a selective list of tenderers is drawn up based on their proven ability to perform the appropriate lime-soil mix design tests.

Two methods of exploratory hole are commonly used for ground investigation in soils for road schemes, these are cable percussion boreholes and trial pits. For lime stabilisation investigations the former are of limited use and the latter require a degree of refinement, both are discussed in the following sections.

<u>Trial Pits</u>:   Generally the type of trial pit employed for road investigation is between 3 to 4.5m in depth, excavated using a hydraulic excavator and is frequently not shored up to allow entry by logging staff. Typically six such pits could be excavated in a day. These pits would be used to recover samples of materials for laboratory testing and would give information of the nature of material encountered in cutting and at road formation level. Where cuttings are at greater depth than 4.5m boreholes would be necessary to determine formation level conditions.

For lime stabilisation purposes it is important that a detailed knowledge is built up of the materials encountered. This requires careful inspection and logging of the trial pits sides and can only be achieved by shoring up the pits to allow entry by logging staff. Experience from two investigations show that the economic and practical limits for such trial pits is in the order of 6.5m. Excavators with rubber tyres to enable them to travel rapidly and independently between trial pits by road and the practicality of transporting and installing the necessary shoring equipment, whilst retaining an adequate rate of progress, are the main constraints.

Experience has shown that the logging techniques used during the investigation are of great importance during subsequent periods of laboratory testing and interpretative reporting. The need for all staff involved with logging on site to be using the same logging criteria cannot be over emphasised. An engineering geological description of the materials encountered in the trial pit should be made. In addition to a description of the soil material it should include comments on fabric and chemical changes. General descriptive methods such as those recommended in Spink and Norbury (Ref 8) are appropriate. Particular importance should be placed on the need to describe the materials present and not merely to assign a prescribed weathering zone as a method of shorthand. Time spent at the start of the contract unifying the scheme to be used by all logging personnel will pay dividends at later stages. Experience has shown that in clays of the same lithology but having differing weathering zones, one zone may be suitable for lime stabilisation whilst another may not be. The consequence of incorrectly identifying weathering zones in such material could mean that the unsuitable material and suitable material may be mixed during sampling. The subsequent laboratory testing may indicate that the mixed material is appropriate for lime stabilisation, the error only coming to light during the construction period when the stabilisation process runs into difficulties.

At the ground investigation stage an assessment should be made on the practicality of distinguishing suitable and unsuitable materials during site works, when rapid assessment based on readily identifiable properties will be necessary during excavation works.

The level of sampling necessary is governed by the number of cohesive materials or material zones encountered. As an approximate guide a minimum of 500kg of each material or zone should be recovered during the whole investigation for lime stabilisation testing purposes. The more usual sampling philosophy of governing the number of samples taken in proportion to the extent of the material encountered should be avoided. It will become obvious from the laboratory test requirements that the mass of sample required for each suite of tests is constant. During trial pitting it is therefore necessary to take large numbers of samples from specific layers in individual trial pits to ensure that an adequate quantity of material is obtained for testing, this is particularly the case when relatively thin layers of material or materials having limited extent are encountered.

Another aspect that requires particular attention is the logging of any mineralisation present in trial pits. Sulphate bearing minerals are found in many clays and can have a significant effect on lime stabilisation, to the extent of precluding its use. If such deposits are evenly distributed throughout the clay laboratory testing will identify their presence. However if present in discrete concentrations, there is a potential that these materials may not be representatively sampled and therefore not identified during laboratory testing. Such concentrations may be mobilised during the stabilisation process having potentially serious consequences.

Cable Percussion Boreholes: Due to relatively small diameters of boreholes and the need for the large volumes of sample for lime stabilisation testing, boreholes are of limited use in providing sufficient samples. For the same reason they are also limited in their suitability for soil mass description. Their most significant use is to identify strata in areas where proposed cuttings are deeper than the depths practically attainable by trial pitting and to obtain samples for basic acceptability criteria. Generally the necessary samples can be obtained by careful location of pits in areas of the site where such materials are within the trial pitting depth mentioned above.

Laboratory Testing
Laboratory testing for lime stabilisation can be sub-divided into three phases, these are the acceptability criteria tests for 7E material, determination of the required lime content and the mix design testing. Ideally each phase would be run consecutively however if time constraints are tight then they can be run in parallel. The three phases form a suite of tests which must be undertaken on each type of material or material zone which is being considered for lime stabilisation.

Class 7E Acceptability Criteria Tests: These tests are performed to determine if the material is acceptable as a Class 7E material using the criteria stated in Table 1. The test, test method and approximate frequency of test is given in Table 2.

Table 2. Laboratory Testing for Class 7E Material

Test	Test Method	Minimum Test Frequency
Grading	BS1377:1990:Part 2 (Ref 6)	One test per material type per trial pit
Moisture Content	BS1377:1990:Part 2	One test per material type per metre per trial pit
Plasticity	BS1377:1990:Part 2	One test per material type/metre/trial pit
MCV/Moisture calibration	BS1377:1990:Part 4	Two tests per material type
Organic Content	BS1377:1990:Part 3	One test material type/metre/trial pit
Sulphate Content	BS1377:1990:Part 3	One test material type/metre/trial pit
* Compaction Test	BS1377:1990:Part 3	Two tests per material type
* Total Sulphur Content	BS1047:1983	One test material type/metre/trial pit

* Not required directly as acceptability tests.

Although not required as an acceptability test it is prudent to undertake Total Sulphur tests to determine the combined sulphate and sulphide content of the soil. Likewise compaction tests are desirable on the raw material to determine its behaviour with change in moisture content and to allow the appropriate Moisture Condition Values (MCV) to be specified.

From these tests it can be determined if the material is suitable for use as a raw material for lime stabilisation. The acceptability criteria for grading, plasticity and organic content are defined in Table 1. The acceptability criteria for total sulphate and MCV require interpretation.

As a preliminary guideline an upper bound total sulphate content of 0.25% is recommended by Sherwood (Ref 1.) The same author also suggests that the value of total sulphur should be considered and, again as a guideline, a value of 0.5% is recommended. Both these values should be treated with caution and soakage/swelling tests should be undertaken to determine if unacceptable swelling is being caused by the presence of sulphate minerals.

The purpose of the MCV limits at this stage according to HA44/91 (Ref 3) and the SHW Notes for Guidance (Ref 4) associated with the SHW are to control the moisture content so that the correct value of moisture is present when lime is mixed. If this was correct there would be little point in specifying an upper bound MCV (lower bound moisture content) as it is permissible to add water during the stabilisation process. The lower bound MCV (upper bound moisture content) would serve to limit the material from being too wet, however, the subsequent process of adding lime will cause a lowering of moisture content. If the relatively high moisture contents necessary to obtain a satisfactory lime-soil mix are applied to a Class 7E material then the 7E material would have very low MCV's and would probably be untrafficable by plant on site. It is considered by the author that generally the MCV limits for Class 7E material should be applied to define the range of MCV's over which the material is workable with typical site plant. To assist in this decision it is useful to undertake hand shear vanes on MCV samples to determine the approximate shear strength at each point on the MCV-moisture calibration curve.

Initial Lime Content Testing   It is necessary to determine the minimum lime content required to initiate the lime stabilisation process. This is normally carried out by the Initial Lime Consumption Value (ILCV) test defined in BS1924:Part 2:1990 (Ref 7). This will give the minimum lime content necessary for stabilisation. Whatever the result of the ILCV test the minimum lime content permitted on DoT Contracts is 2.5%. It may be found during the next phase of testing, the mix design phase, that higher levels of lime are required to obtain the required strength increases in the lime stabilised material.

The upper value of lime content is likely to be determined by economic constraints. A value of 5% is probably appropriate for this upper bound limit. It is therefore necessary to test the Class 7E material mixed with various contents of lime between 2.5% or the ILCV value, whichever is higher, and 5%, these tests are referred to as the mix design tests.

Mix Design Testing: The purpose of this testing is to determine the moisture content limits and the percentage lime necessary to achieve the required strength increase for a lime stabilised capping layer.

Many cohesive materials excavated from the ground and tested in the laboratory indicate reasonably good CBR values at their natural moisture content. These CBR values can be in excess of 15% which suggest that capping is not necessary. At their equilibrium moisture content however, their CBR values can drop considerably, often to values of less than 5%, thus capping layers become necessary. The objective of the lime stabilisation process is to obtain a long term strength in the stabilised capping layer equivalent to a CBR value >15%.

If the lime-soil mix moisture content is too high the necessary strength may not be achieved. If it is too low inadequate compaction is achieved and the stabilised layer becomes prone to heave when moisture infiltrates the soil structure. Figure 1 indicates a typical suite of tests necessary to determine the lime content and upper and lower bound moisture contents for a lime-soil mix.

Figure 1.   Typical Lime-Soil Mix Design Test Suite

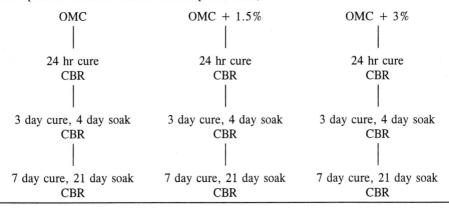

The compaction, MCV and CBR tests indicated in Figure 1 are carried out in accordance with BS1924:Part 2:1990. However it is recommended that the tests are modified to include a 24 hour mellowing period after mixing.

The method of sample selection for these tests is of considerable importance. Significant variation in laboratory test results could be anticipated due to natural variations in individual bulk samples. These variations could lead to difficulty in interpretation of laboratory test results. Experience has shown that the most effective way to minimise this risk is to mix a number of bulk samples of the same material forming one large relatively homogenous sample. This sample can be sub-divided for convenience and used as necessary.

It is also good practise to undertake two compaction and two MCV/moisture calibrations at each lime content, to safeguard against the potential for variations in individual test results.

<u>Microscope Examinations</u>
During the lime stabilisation process physical and chemical changes occur within the soil sample. One of the main areas of concern is the change of sulphate minerals into highly expansive forms such as ettringite. This secondary mineral, when fully developed, takes the form of needle like crystals up to a few millimetres in length. In relatively 'young' lime-soil mixes the crystals may not be so well developed. Examination through a binocular optical microscope is therefore recommended.

Work has been carried out using X-ray diffraction, X-ray fluorescence and scanning electron microscopy to identify the presence of deleterious material. These tests are of limited use. The techniques may reveal small amounts of ettringite in soil-lime samples, however, such small quantities can be below the resolution threshold for quantitative assessment. More seriously, these tests only examine very small samples, in the order of grammes, and therefore cannot be considered appropriate to the overall assessment for large scale earthworks. It is considered that such tests are better suited to back analyses of samples which swell to unacceptable levels, so that the swell mechanism can be identified.

Planning and Control of Laboratory Testing

Once the lime has been added to the soil sample it is extremely important that all mix design testing is carried out at the required period of time. Rapid changes in plasticity occur after lime addition and, as a result, the relationship between moisture content, MCV and dry density also changes. To obtain comparable test results it is therefore necessary to ensure that the testing laboratory is fully aware of the timescale effects and programming of mixing and testing periods should be carefully planned, taking into consideration weekends, holidays and other testing commitments. It is considered that the test timings should be accurate to within $\pm$ 2 hours of the stated period. Such a tight testing regime may seem excessive, it is, however, the only way in which the necessary degree of confidence in the results can be attained.

Assessment of Mix Design Test Results

This paper is not intended to give a detailed account of the interpretation of the laboratory mix design test programme, but suggests a general approach and highlights the aims of the assessment.

The need for sulphate testing has been discussed earlier. The remaining tests are all provided to determine the moisture content limits for lime and soil mixing and the methods by which these will be controlled on site.

Compaction Testing: This aspect of the mix design testing is critical. Incorrect specification of the minimum moisture content for compaction can lead to inadequate compaction and ultimately failure of the stabilised layer.

Figure 2 shows two compaction curves for a lime stabilised material. Curve B represents a test carried out some time after curve A, both curves are much 'flatter' than those which were obtained for the same material without lime. The lower bound moisture content should be selected by using the intersection of the compaction curve with the 5% air voids line. From Figure 2 it can be seen that if a compaction test is carried out at a later time, the compaction curve will 'shift', giving increases in optimum moisture content with time due to changes in plasticity during the stabilisation process. Clearly this would lead to problems in specification of a single moisture content in the Contract Document. These problems can be overcome by use of the MCV test as a method of moisture content control. Curve C indicates the compaction values for the same raw material before the addition of lime.

Moisture Condition Value: This test is designed to give a rapid assessment of moisture content and is widely used for moisture content control for earthworks in cohesive materials.

Work carried out by TRL and summarised in Parsons (Ref 5) indicates that the MCV obtained for various materials at optimum moisture content is relatively constant and is

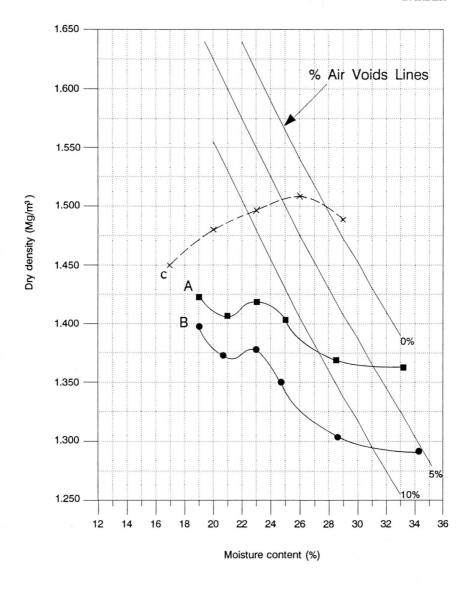

Figure 2. Typical Compaction Curve for Class 9D Material

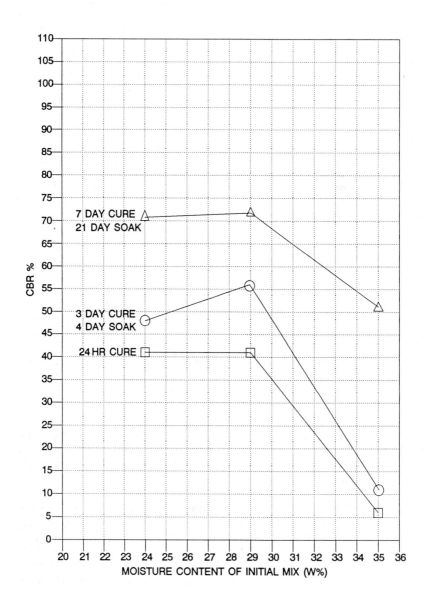

Figure 3. Plot of CBR Values Against Moisture Content for Varying Cure / Soak Periods

typically in the order of 12. With lime-soil mixes the position of the optimum moisture content and maximum dry density changes with time, as shown in Figure 2. The MCV obtained at this point remains reasonably constant. The MCV obtained at the intersection of the 5% air voids line with the compaction curve is also reasonably constant. Therefore, within the time periods expected for lime mixing and compaction on site, the MCV will give a good indication of the moisture content in relation to a given point on the compaction curve which is independent of time. For this reason, and due to the short time needed for this test, the MCV is the ideal control test for specifying the minimum moisture content for Class 9D material.

The Notes for Guidance to the SHW (Ref 4), and Advice Note HA 44/91, (Ref 3) suggest that the upper bound MCV value (lower bound moisture content) should be no greater than 12, although the latter document does state that higher values can be specified if ".... there are strong grounds for choosing a higher value". Recent experience has shown that with some materials values of MCV greater than 12 are appropriate for the upper bound MCV value. Such values should only be recommended if a high degree of confidence can be placed in the laboratory test results.

CBR Testing: The object of the CBR testing at various moisture contents and time periods is to determine the maximum lime-soil mix moisture content which will give a CBR > 15% This moisture content is used as an upper bound limit for material control on site. The 2.5 kg compaction method is used for sample preparation at various moisture contents. For a given moisture content the CBR values obtained after mixing the soil with lime generally increase over a period of days. As mix moisture content is increased, CBR values obtained at a fixed period of time decreases, as shown in Figure 3. The maximum moisture content should be limited to a value which allows the minimum CBR of 15% to be achieved within a suitable period of time. This period of time will depend on site operations but the value obtained from the three day cure, four day soak, test will generally be suitable.

CONCLUSIONS

The use of lime stabilisation of soils to provide capping layers has been hindered by some notable failures in this method. Analysis of both failures and successes has given a greater degree in confidence regarding the properties of materials which are important in the specification of lime stabilisation.

This paper has highlighted an approach to site investigation specifically designed to assess the suitability of cohesive materials on site for the lime stabilisation process. By adopting this type of investigation a decision can be made on whether lime stabilisation should be included as an option for capping layers in a road contract. This decision can be made on balanced engineering judgement based on a broad knowledge of the materials' behaviour with lime and may provide the client with a more economic or environmentally sensitive option to importing granular capping.

ACKNOWLEDGEMENTS

The Author gratefully acknowledges Babtie Geotechnical by whom he was employed during the period of the work described and preparation of this paper, and whose staff contributed the ideas which form the basis of this paper.

REFERENCES

1. SHERWOOD P.T., Soil Stabilisation with Cement and Lime, H.M.S.O., London, 1993.

2. DEPARTMENT OF TRANSPORT, Manual of Contract Documents for Highway works. Volume 1: Specification for Highway Works, H.M.S.O, August 1993.

3. DEPARTMENT OF TRANSPORT, Highways Safety and Traffic, Departmental Advice Note HA 44/91, D.O.T. 1991.

4. DEPARTMENT OF TRANSPORT, Manual of Contract Documents for Highway Works, Volume 2: Notes for Guidance on the Specification for Highway Works, H.M.S.O., 1993.

5. PARSONS A.W., Compaction of Soils and Granular Materials: A Review of Research at T.R.L., H.M.S.O., 1992

6. BRITISH STANDARDS INSTITUTION BS 1377 Methods of Test for Soils for Civil Engineering Purposes BSI London, 1990.

7. BRITISH STANDARDS INSTITUTION BS 1924 Stabilised Material for Civil Engineering Purposes. BSI London, 1990.

8. SPINK T.W., and NORBURY D.R. The Engineering Geological Description of Weak Rocks and Overconsolidated Soils. The Engineering Geology of Weak Rocks, Engineering Geology Special Publication No.8. A.A. Balkeema, Rotterdam 1993.

Spatial dependence in site investigation design

I. W. FARMER, G. C. XIAO AND P. G. CHALLINOR
Ian Farmer Associates, Newcastle upon Tyne, England

SYNOPSIS

It is important in site investigation to squeeze the last bit of information from costly field and laboratory data in order to enhance the quality and reliability of the resultant ground profiles and properties. It is also important that the design of the site investigation is the best possible one to facilitate this purpose.

Two simple applications of geostatistics are used to illustrate these premises. The first shows a method by which the quality of a simple variogram can be used to test the adequacy of borehole spacing to estimate the variation in thickness of a ground layer. The second shows how spatial correlation between data can be used to estimate characteristics (in this case RQD) of points or blocks, and how this resolution can be improved by incorporation of soft data.

INTRODUCTION

There is a need in site investigation for a more formal method of design than the reliance on experience, through the somewhat nebulous definition of a "geotechnical adviser", recommended by the Site Investigation Steering Group (1) of the Institution of Civil Engineers. To date this need has been met partly by the use of statistical techniques to assess the reliability of data and partly by probabilistic methods of "calculating" risk. The former is described in Webster and Oliver's (2) text on "Statistical methods in soil and land resource survey"; the latter in Whitman's (3) seminal Terzaghi lecture. In their second edition, Webster and Oliver extend their approach to include Matheron's (4) theory of regionalised variables as a basis for the calculation of the reliability of a ground investigation or the frequency of sampling. This approach has been used successfully in ground investigation by the authors (5,6), and Webster and Oliver quote several applications, both in soil distribution and contaminated ground investigation. Mackean and Rosenbaum (7) have applied the method to SPT data.

The theory and its associated empirical offshoot, kriging (8), which is the basis of geostatistics, has been used extensively in mineral exploration to estimate ore grades. Although some applications can be questioned (9,10), its validity is widely accepted. It relies on the principle of spatial dependence. This assumes that most properties of the earth's surface vary in space and that, consequently, values associated with points in space which are close together are more likely to be similar than those which are far apart. This is the basis of the variogram, which plots the change in semi-variance of a property against distance between data points. This can be used either in its original form or modified to satisfy the positive definite matrix of the kriging equation to estimate a property (usually grade) and its reliability for a block of earth in an area containing sampling points or boreholes. Its application to some site investigations is evident.

SPATIAL DEPENDENCE

The basis of the application of the theory of regionalized variables is the variogram or semi-variogram which is a device or hypothesis for describing the spatial variance of data. Stated most simply, the semi-variance function computes a measure of spatial correlation as a function of distance, by taking differences and squaring them:

$$\gamma(d) = \frac{1}{2} \times \frac{1}{n} \sum \left[z(x_i) - z(x_i + d) \right]^2$$

where n is the number of pairs of data points of location x at increasing lags, d, or in this case distances between boreholes, and $z(x_i)$ is the grade or property of interest.

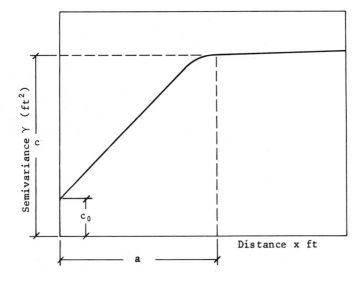

Figure 1. Semi-variogram (spherical model) defined by intercept Co, Sill C and range a. The range, which covers the sloping portion of the curve represents the distances between data points where there is spatial correlation.

An ideal semi-variogram plot is illustrated in Figure 1. It comprises a sloping portion from an ordinate intercept, which should be as small as possible, and which levels out parallel to the abscissa.

The classic semi-variogram has several defining characteristics. The slope portion represents separating distances or lags up to which there is spatial correlation. The steepness of the slope indicates the rate of change of variance with increasing separating distance. The variance increases with distance along the sloping portion until it reaches a maximum at which the curve levels out. This is the sill variance at which spatial dependence stops. Beyond this there is only regional variance. The separating distance at which the variogram reaches the sill is known as the range. Beyond the range the variance is unrelated to separating distance.

The ordinate intercept is known as the nugget variance. Since by definition the semi-variance at lag zero should be zero, this represents an error, usually attributed to spatially dependent variation over distances shorter than the shortest separation distance.

Variograms have simple forms. Complexity can usually be attributed to sampling variation. Simple functions capable of describing an ordinate intercept, a range of increasing variance and a sill may take one several forms including bounded linear, spherical and exponential. These can be fitted to the variogram by various methods to give an indication of goodness of fit and to provide the basis for a predictive approach. This is mainly achieved through kriging, where in its simplest form, the variogram is used to obtain an estimate of the actual value of a property at a point, or average value of a property over a block, from the unbiased weighted values of specific values of the property at sampling points. Such unbiased or ordinary kriging can be used to map a property, to estimate a property at a point or in a block or, through the relation between kriging weights and sampling configurations to design sampling schemes. An example would be to compute a graph similar in shape to a variogram plotting the kriging variance against sampling interval in order to determine the optimum sampling distance for an acceptable error.

GEOTECHNICAL APPLICATIONS

The kriging variance for a block is given by (see ref 3):

$$\sigma^2(B) = \sum_{i=1}^{n} \lambda_i \, \overline{\gamma}(x_i, B) + \mho - \overline{\gamma}(B, B) \tag{2}$$

where $\sum_{i=1}^{n} \lambda i = 1$ for ordinary kriging, is the sum of the weights applied to sampling points,

$\gamma(x_i, B)$ is the average semi-variance between the block and the i_{th} sampling point, $\mho$ is a Lagrange multiplier to achieve minimization and $\gamma(B, B)$ is the average semi-variance or value of the variogram within the block, equal to zero for a single point.

The average semi-variances in equation (2) are integrals of the semi-variogram. They can be computed numerically and the effect is to produce a smoothed graph of kriging variance against separation distance.

This can be used for several simple geotechnical applications, although care in interpreting block kriging will be required. The basic implications and applications are:

1:      The variance σ^2, or error can be estimated for a particular sampling interval within the sloping part of the curve. If an acceptable tolerance can be determined this will determine the sampling interval or borehole separation.

2:      The range at which spatial dependence occurs is determined by the sill variance. The distance equivalent to this spacing is the maximum sampling interval.

OPTIMUM INTERVAL

Ledvina et al (11) proposed a simple, if flawed, empirical approach to the optimum interval in site investigation. They suggested using an assessment of the quality of a semi-variogram plot to estimate this spacing, suggesting that the range distance represents "the practical spacing one must drill to adequately sample" the thickness distribution of a soil or rock layer. They illustrated this point by constructing semi-variograms based on an investigation to determine the thickness of a limestone layer. Initial exploration was over the full area of 35 square miles with average drill hole spacing averaging about 5000ft (Figure 2a). Secondary exploration was over about 6 square miles with average drill hole spacings of 1000ft (Figure 2b). The final drilling program was over about 0.02 square miles with drill hole spacings averaging 100 feet. From Figure 2c an optimum spacing of about 300ft is suggested from the best developed of the three semi-variograms.

This approach forms the basis of a very simple test of the adequacy of the drill hole spacing to determine the variation in thickness of one or more layers over a site area. The following guide lines may be suggested:

> 1. A minimum of 30 data points are needed for the results to be statistically meaningful. For 30 data pairs, a minium of 5 or 6 boreholes are needed. Ledvina et al's statement that: "Because of pairing, a small number of holes yields a substantial amount of information about spatial dependence... one gets a lot of mileage from a few drill holes," may be an over-simplification, but it is essentially correct.

> 2. Design a conventional site investigation using accepted borehole spacings and make a trial semi-variogram. This can be done quickly using either public domain software, such as the EPA Geoeas package (12) or an inexpensive package such as the University of Arizona Geobase software (13). The variogram should be examined and tested for continuity against the ideal.

> 3. If a semi-variogram form is found close to the ideal, then the spacing is good enough to adequately sample the thickness variation.

> 4. If the variogram shows poor continuity then either (a) select additional hole positions to obtain closer spacings or (b) select a smaller area within the larger area to explore in greater detail.

Arguably this approach is too simplistic for purists. The underlying assumption that for a spherical model variogram, the range of the variogram is the practical spacing one must drill to adequately sample, is not correct. This depends on the confidence level required on the estimate, and the spacing can be much less or greater than the range of the variogram. Strictly speaking the semi-variogram can only give the range of influence of samples or the

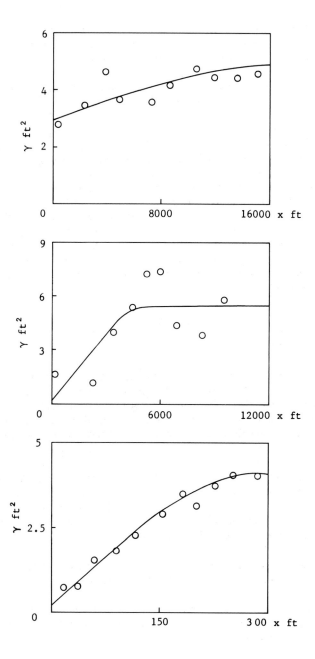

Figure 2. Semi-variograms (expotential model) of layer thickness for (a) area 40,000 x 50,000 ft; hole spacing 5,000 ft, (b) area 10,000 x 18,000 ft; hole spacing 1000 ft, (c) area 700 x 800 ft; hole spacing 100ft (after Ledvina et al, ref 11).

distance of sample correlation. Thus while 3 above may be an adequate test, 4 is more problematical. This will require particular attention if a smaller area is selected from a larger area, when questions of how the area is selected, what its size is in relation to the larger area, and how to determine the number and spacing of drill holes must be addressed.

During variogram modelling, there is an empirical rule which says that the computed variogram value at a lag distance greater than half the width of the field (in this case the area from which data are used) should be ignored, because these values are due to artifacts of the procedures. Thus the field should be sufficiently large in relation to the true range of the variogram in order to accurately determine the range of the variogram.

Arguably therefore in Figure 2b,c the computed ranges of the variograms of 4,000ft and 290ft are small in relation to the field widths of 10,000ft and 800ft respectively and could usefully be combined, raising questions about the validity of the methodology of selecting a smaller area for intensive exploration.

Nevertheless the approach is a valuable and useful method of evaluating site investigation programs, which provides a simple and practical assessment of spatial dependency. It can be used both for the simple case of layer thickness or for more complex distributions of geotechnical data.

PROPERTY DISTRIBUTION
A quite complex investigation of fracture parameters in predominantly schist and porphyry copper deposits (6) was carried out to determine data for slope design, blast design and heap size distribution at a copper mine. The purpose was to improve the predictions of rock fracture parameters in blocks of rock before mining. A total of 26 drill holes in an area of 3,200 by 2500ft at an average spacing of about 500ft and depth of 100ft formed the basis for data collection. One of the parameters used to describe the rock was RQD (others were % of +1 inch core and the longest pieces of core). RQD was calculated from 100 x 50ft lengths of core and was found to fit into two statistically distinct populations for schistose rocks and porphyry rocks. Semi-variograms were computed (Figure 3) for both types of rocks. They are reasonably defined and show a definite correlation of RQD in three dimensional space for the drill-hole spacings. They can, in other words be used for local estimations of a point or limited block in 3D space using its defined spatial dependence on nearby samples. Usually, the estimated value of a point or block is obtained as a weighted average of these nearby samples using ordinary kriging in such a way as to make the obtained estimate unbiased with a minimum of variance or estimation errors. For the data in Figure 3, Figure 4a plots the actual value of RQD against the value predicted by ordinary kriging. The fact that both rock types occurred as separate units in space was overcome by selective kriging so that parameters of each rock type were kriged using samples in the same rock unit. This made optimum use of all available hard information to produce estimates of RQD at particular locations in the rock mass. It is evident from Figure 4a that despite the spatial correlation of data, the scatter of data was substantial. Several ways of improving this were considered.

The first and most simple was to examine prior probability distributions of RQD obtained by measurement in various structural domains identified locally in the area under consideration. One of these was a triangular distribution histogram. The impact of incorporating this data is illustrated in Figure 4b, where the scatter is reduced significantly.

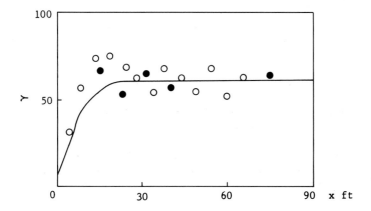

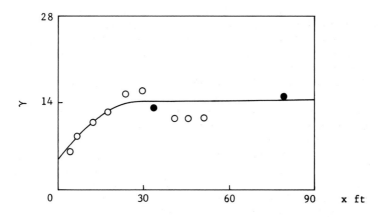

Figure 3. Semi-variograms (spherical model) of RQD for area 3,200 by 2,500ft; hole spacing 500 ft in (a) schist and (b) porphyry rocks. Note: the heavy points are those variogram values computed horizontally; the others are experimental values computed along the boreholes) after Kim et al, ref 6)

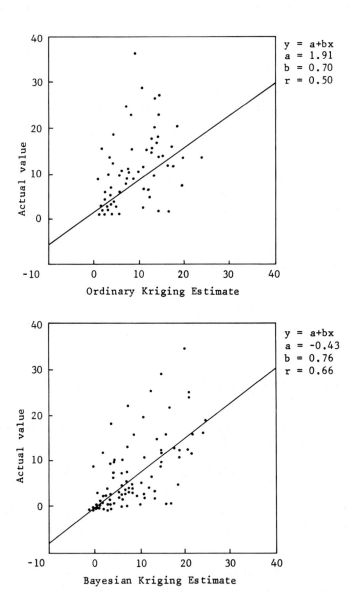

Figure 4. Scatter plots and regression equations for actual and predicted RQD (a) ordinary kriging and (b) soft kriging incorporating a single incorporating single triangular distribution (after Kim et al, ref 6).

The process is referred to by Journel (14) as soft or Bayesian Kriging and effectively incorporates soft data based on general statistics and histogram types derived from the actual (hard) data used for the semi-variograms. This data is coded as prior probability distributions, such as triangular and normal distributions. In the case of geotechnical data, where variability in sampling and testing techniques can often be incorporated into probability distributions this is a powerful and useful technique.

As well as this information it is possible to generate other types of soft data such as expert judgments - biased or unbiased - based on information generated by an expert field geologist or engineer. In the case of site investigations this can be obtained during preliminary investigations, from adjacent sites, from site inspections or from trial pits.

COMMENTS
There are several ways in which concepts of spatial dependency can be used to improve the design of site investigations. These do not involve the more complex applications of the subject. Two suggested approaches are:

1: Determine the optimum drill hole spacing or sampling interval for an acceptable variance or error, from a graph of kriging variance against sampling interval or from an estimate of the range of spatial correlation on a semi-variogram plot.

2: Improve the scatter of data in a semi-variogram plot for use in prediction of variance by incorporation of soft data based on general statistics and histogram types of the data used in the semi-variogram and by incorporation of expert information from preliminary site investigations or prior knowledge.

References
1. Site Investigation Steering Group, Site investigation in construction, Vols 1-4. Thomas Telford, London (1993)

2. Webster R. and Oliver M. A., Statistical methods in soil and land resource survey, 2nd Edn. Oxford UP, Oxford (1990)

3. Whitman R. V. Evaluating calculated risk in geotechnical engineering. Jl. of Geotechnical Engineering, ASCE, Vol 110, No 2, pp145-188 (1984)

4. Matheron G., The theory of regionalized variables and its application. Cahiers du Centre de Morphologie et Mathematique de Fontainebleau, no 5 (1971)

5. Kim Y. C., Farmer I. W. and Cervantes J. A., Application of Bayesian (soft) Kriging to underground rock characterization at San Manuel mine. Proc. 5th Annual Workshop, Mine Systems Design and Ground Control, VPISU, Blacksburg, Va, pp 213-225 (1987)

6. Kim Y. C., Cervantes J. A. and Farmer I. W., Predicting in-situ rock fracture parameters using soft kriging. Proc. 21st APCOM, Las Vegas, pp 237-252, Soc. Min. Engineers, Littleton, Co (1989)

7. Mackean R.A.N. and Rosenbaum M.S., Geostatistical characterization of SPT, Proc 6th

International IAEG Congress, Amsterdam, Vol 1, pp317-322, Balkema, Rotterdam (1990)

8. Krige D. G., Two dimensional weighted moving average trend surfaces for ore evaluation. Jl. S. African inst. Min. Metall, Vol 66, pp13-38 (1966)

9. Rendu J-M., Mining geostatistics - forty years passed. What lies ahead? Mining Engineering, Vol 46, pp557-558 (1994)

10. Merks J. W., Geostatistics or voodoo statistics? Engineering and Mining Journal, Vol 193, No 9, pp45-49

11. Ledvina C.T., Dowding C.H., Fowler S., Hunt G. and Nance R., Geostatistical guidance of exploration in roof control - how many drill holes are enough? Proc 5th Conference on Ground Control, pp14-30, S.I.U., Carbondale, Ill(1994).

12. Englund E. and Sparks A., GEO-EAS Users Guide, U.S. Environmental Protection Agency, EPA/600/4-88/033a, Las Vegas, NV (1988)

13. Kim Y.C., GEOBASE Users Guide, University of Arizona, Tucson AZ (1982).

14. Journel A.G., Constrained interpretation and soft Kriging. Proc. 19th APCOM, pp15-30, Soc. Min Engineers, Littleton, Co.(1986).

Overview of initial ground investigations for the Channel Tunnel high speed rail link

R C BECKWITH
Engineering Manager, Union Railways Limited (URL)
W J RANKIN
Geotechnical Manager, Secondment from Mott MacDonald to URL
I C BLIGHT
Resident Engineer, Secondment from Halcrow to URL
I HARRISON
Fieldwork Manager, Soil Mechanics Limited (SML)

1.0   INTRODUCTION
1.1   Scheme Description

The Channel Tunnel Rail Link will be a new 108 km (68 mile) railway between London and the Channel Tunnel near Folkestone. The announced route is shown on Figure 1.1.

The new railway has three main objectives:

. To provide capacity needed for growing international rail business via the Channel Tunnel, and reduce international journey times.
. To bring major improvements in capacity, journey time and quality for rail services between London and north and east Kent.
. To support economic regeneration and business development.

The Channel Tunnel Rail Link's London terminal will be in a refurbished and extended St Pancras station. There will be intermediate stations on the new railway in East Kent at Ashford, and in North Kent at Ebbsfleet, with another possible station at Stratford, East London. A connecting line has been designed to enable international trains to reach Waterloo International Terminal by leaving the Rail Link near Gravesend.

The Rail Link has been designed for safe operation of trains at speeds of up to 185 mph (300 km/h) in places. The Channel Tunnel Rail Link is designed to provide the capacity required for growing demand for direct passenger and be capable of accommodating freight train services using the Channel Tunnel. The Rail Link will also cut journey times saving 30 minutes using the Rail Link to Paris and Brussels.

1.2   Major Features

The overall programme for the works is shown in Figure 1.2 with construction starting in 1997 and opening scheduled for 2002. The percentage costs of the major elements of the scheme are shown in Figure 1.3, and average route costs per kilometre for various sections of the route are shown in Figure 1.4. This indicates the most costly

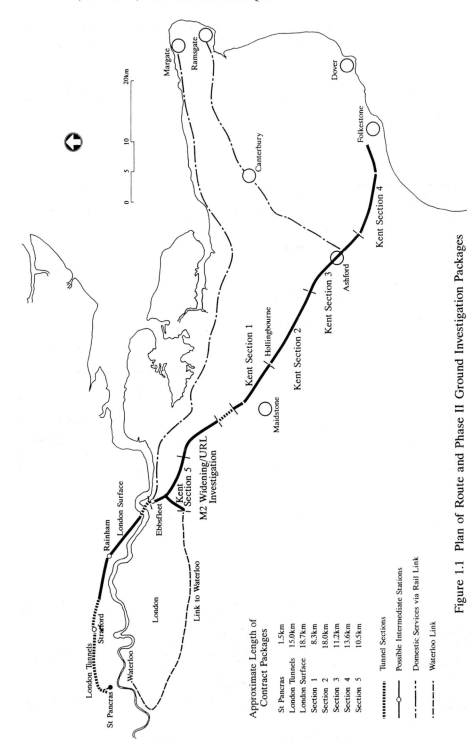

Figure 1.1 Plan of Route and Phase II Ground Investigation Packages

Approximate Length of
Contract Packages

St Pancras	1.5km
London Tunnels	15.0km
London Surface	18.7km
Section 1	8.3km
Section 2	18.0km
Section 3	11.2km
Section 4	13.6km
Section 5	10.5km

Tunnel Sections
Possible Intermediate Stations
Domestic Services via Rail Link
Waterloo Link

section to be around the Thames Crossing between Purfleet and Northfleet, it also highlights the cost difference between the London Tunnels, the Thames Tunnels and the surface sections.

The studies associated with the Channel Tunnel Rail Link (CTRL) have been partly funded by the European Union through the Transport Infrastructure and Trans European Networks Programmes.

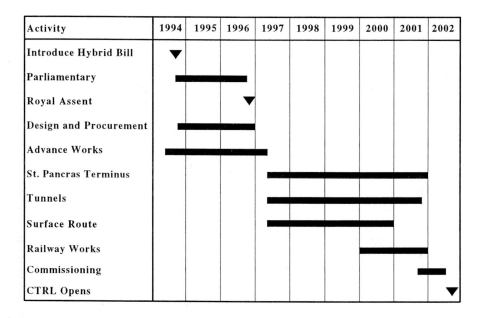

Figure 1.2 Overall Programme

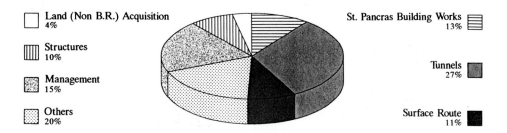

Figure 1.3 Proportion of Estimated Project Cost

121

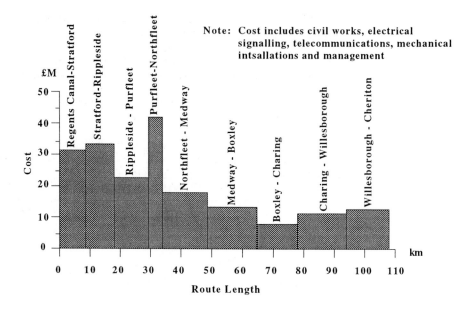

Figure 1.4 Average Route Costs Per km

2.0 GEOLOGICAL AND GEOTECHNICAL FRAMEWORK

2.1 Introduction

The combination of these aspects result in a variety of landform units which can be grouped together as shown on the geological plan in Figure 2.1 and are

.	The Thames Basin;
.	The North Downs;
.	The Weald.

Each of the aspects of the geological framework and the groups of landform units have been the subject of studies. A route wide desk study has been carried out on the geological materials and field studies of the landform units.

There is a large volume of existing data from wells, other boreholes, construction and national databases available to the project. Systems have been established to organise this data in an efficient manner for the project.

2.2 Regional Geology

The Thames Basin is dominated by a large synclinal structure in which the Lower London Tertiary deposits are encountered overlying the Upper Chalk. The nomenclature for the Tertiary strata has been considerably improved in recent years and the current terminology is summarised in Figure 2.2, Ellison, (1994). The distribution of the lower Tertiary strata across the London Basin is shown in Figure 2.3. This emphasises the continuity of both the Thanet Sand Formation and the Upnor Formation beneath the more variable Woolwich Formation.

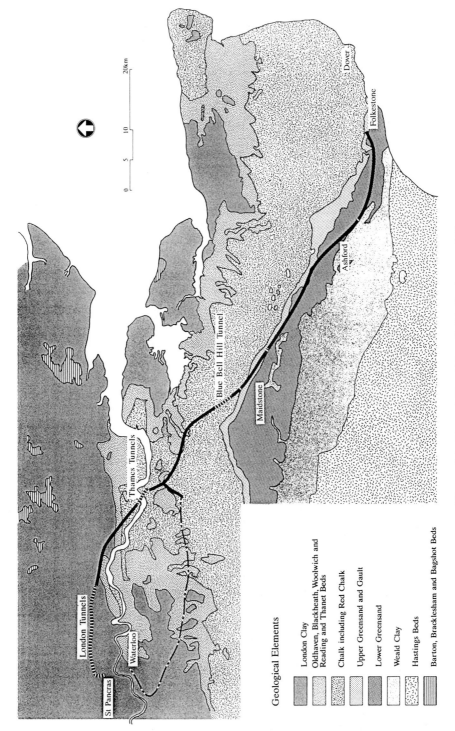

Figure 2.1 Plan of Simplified Solid Geology and Route

Geological Elements

London Clay

Oldhaven, Blackheath, Woolwich and
Reading and Thanet Beds

Chalk including Red Chalk

Upper Greensand and Gault

Lower Greensand

Weald Clay

Hastings Beds

Barton, Bracklesham and Bagshot Beds

Group	Formation	Informal Units/ Former Usage
Thames Group	London Clay Formation	
	Harwich Formation	"Swanscombe Member" "Tilehurst Member" "Harwick Member" Oldhaven Beds Blackheath Beds Hales Clay
Lambeth Group = Woolwich and Reading Beds	Reading Formation Woolwich Formation	"Reading Beds" "Woolwich Bed"
	Upnor Formation	"Woolwich Bottom Bed"
	Thanet Sand Formation	"Thanet Sands" "Thanet Beds"

Figure 2.2   Nomenclature for the Tertiary Deposits of The Thames Basin.
(After Ellison, 1994)

The North Downs are characterised by the chalk escarpment which runs from Blue Bell Hill to Dover. The modern nomenclature for the chalk, which has been developed by R Mortimore, 1986 is reproduced in Figure 2.4. This is intended to more closely reflect the engineering behaviour of the newly classified Divisions, however, the degree of weathering may well dominate engineering performance.

The route passes through the chalk escarpment in tunnel at Blue Bell Hill and then runs sub-parallel to the strike of the Lower Cretaceous strata which include the Gault Clay, the essentially granular Folkestone, Sandgate and Hythe Beds. The Atherfield and Weald Clays are encountered towards the southern end of the route.

Quaternary deposits occupy the Thames flood plain, the Medway Valley, the base of the chalk dry valleys and chalk hill tops. Head deposits are also distributed widely over the clays of the Weald. The key features of this sporadic mantle of Quaternary strata are as follows:

.   irregular contacts with the solid strata;
.   lithologically variable, both laterally and vertically over short distances;
.   inconsistent and in some areas relatively poorly mapped.

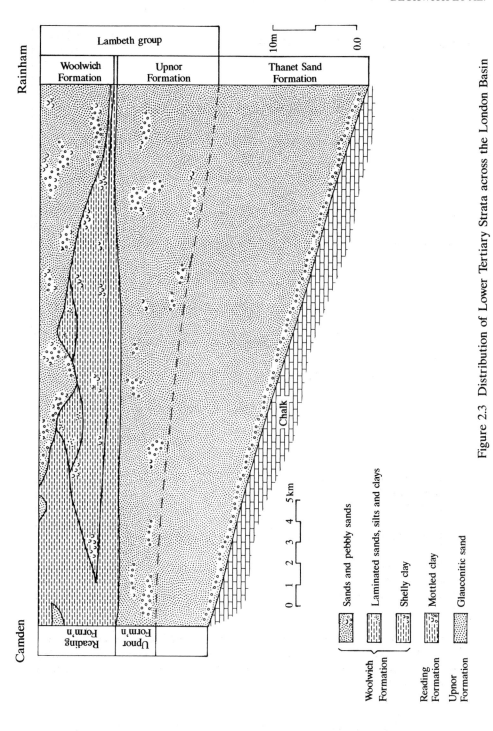

Figure 2.3 Distribution of Lower Tertiary Strata across the London Basin

2.3 Geomorphology

The landscapes of the Thames, Weald and North Downs were subject to considerable meso and micro-scale modification during the Quaternary. The main features that developed in this period include:

- periglacial shearing, cambering and valley bulging;
- formation of Head deposits;
- drainage evolution resulting in the formation of subsequent streams, cols and wind gaps;
- solution weathering of the chalk;
- coombe and coombe rock development;
- scour hollows and possible pingo development;
- cryoturbation and gelifluction;
- river terrace development;
- alluvial channel infilling;
- mass movement.

These features are specific to geological materials within certain structural settings, e.g. cambering on valley sides within mixed competent and incompetent strata.

2.4 Hydrogeology

There are three distinct hydrogeological areas; of the Thames Basin, the North Downs and the Weald.

Thames Basin: The Thames Basin Section comprises a minor superficial aquifer in the Quaternary deposits. A minor aquifer in the Tertiary deposits and a major aquifer in the chalk. Historically, the water levels within the Tertiaries are variable due to drawdown associated with groundwater supply boreholes into the underlying chalk in the City and East London. Also wells into the chalk to the east in the Dartford and Ebbsfleet areas for quarrying have resulted in relatively low groundwater levels.

North Downs: The North Downs Section comprises a superficial aquifer in the Quaternary deposits and a minor aquifer in the chalk. The chalk aquifer dominates the North Downs, but is significantly different to the Thames Basin in that it has little cover of Tertiary deposits and generally has a large unsaturated zone associated with a deep water table.

Weald: The Wealden Section comprises a superficial aquifer in the Quaternary deposits and a major aquifer in the Lower Greensand. The major aquifer comprises the complete sequence of the Folkestone, Sandgate and Hythe Beds. However, the Sandgate Beds comprise sand and mudstones and are classified under the NRA 1993 Groundwater Protection Policy as a minor aquifer as it offers limited hydraulic connection between the other strata.

2.5 Geotechnics

Surface Works: Many of the issues relating to the design and construction of the earthworks, foundations and earth retaining structures will be similar to those associated with highway design. However, high speed trains are particularly sensitive to variations in rail level due to both instantaneous and long term movements.

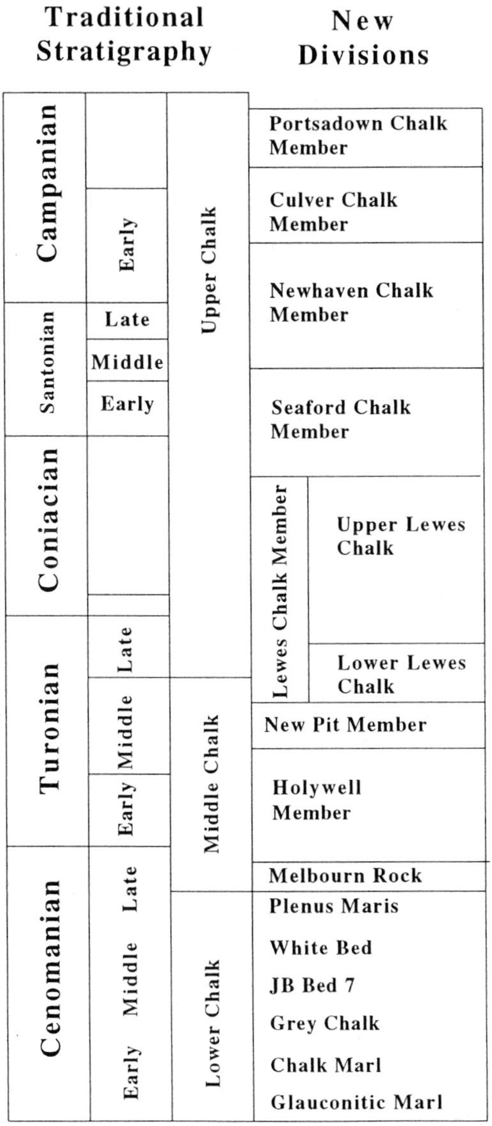

Traditional Stratigraphy			New Divisions	
Campanian		Upper Chalk	Portsadown Chalk Member	
	Early		Culver Chalk Member	
			Newhaven Chalk Member	
Santonian	Late			
	Middle			
	Early		Seaford Chalk Member	
Coniacian			Lewes Chalk Member	Upper Lewes Chalk
Turonian	Late			Lower Lewes Chalk
	Early Middle	Middle Chalk	New Pit Member	
			Holywell Member	
Cenomanian	Late		Melbourn Rock	
	Middle	Lower Chalk	Plenus Maris	
			White Bed	
			JB Bed 7	
			Grey Chalk	
	Early		Chalk Marl	
			Glauconitic Marl	

Figure 2.4   The Southern Province Chalk showing the new Divisions compared with the traditional stratigraphy. (After Mortimore, 1986)

Experience on existing high speed railways has shown that particular attention is required in forming embankments and at transitions with structures. The current requirements for transition zones is that the relative rotation (angular distortion) of the transitional length following opening to services should be less than 1:2000. To achieve the desired uniformity and stiffness, the track bed is made up of a number of layers which include ballast, sub-ballast plus a sand filter layer on a fine gravel subgrade. The total thickness of the track bed is typically no more than 1 m beneath base of sleeper. An assumption in the design to date is that over-consolidated clays will not be permitted in the track bed support embankment which shall comprise good quality granular or chalk materials. The desired track quality for railways is conventionally expressed as the standard deviation (SD) of the elevation of the rails over a given base length. The derived SD value for high speed railways is nominally $\pm$ 1 mm over 200 m, which represents a considerable challenge for the design, specification, construction and maintenance of the railway.

Tunnels: Many of the geotechnical issues associated with proposed tunnels relate to reliable identification of the ground and groundwater conditions along the alignment. Sufficient detail is required so that appropriate tunnel boring machines can be selected and means of minimising the incidence of geological features that may have an impact on the progress of the bored tunnels can be implemented in advance. Such features include:

.   the variation in levels of interfaces, especially between materials of contrasting characteristics; the presence of buried channels formed by a younger deposit eroding the surface of an older deposit;

.   the presence of hard strata or boulders within materials, such as limestone bands in the Woolwich and Reading Formation, flints in the chalk or at the base of the Thanet Sand Formation;

.   the permeability of the ground and the means by which to limit groundwater inflow during construction.

Man-made hazards that may have an impact upon the design and construction of the tunnels include the location of existing piles, wells and boreholes.

Similarly, an understanding of these conditions at shaft or cross passage locations, enlargements or sumps is as equally important. For example, the construction of cross passages may require hand excavation and hence appropriate ground treatment in potentially unstable soils.

Tunnel lining design will be influenced by considerations of the depth of tunnels, variations in ground water levels, durability, ground strength, stiffness and permeability as well as possible time dependent changes in ground stresses.

However, the selection of vertical tunnel alignment is dominated by factors other than geology, such as by the level of existing construction (piles, foundations, other tunnels and sewers), the ease and speed of construction, settlement, as well as complying with acceptable operating gradients and other fixed alignment criteria.

3.0 PHASED INVESTIGATIONS

3.1 *Main Geotechnical Issues*

A summary of the main geotechnical issues is given in Table 3.1. The tunnel lengths form key cost elements of the project, see Figure 1.3, as well as being the potentially high risk items in terms of potential influence of ground conditions upon construction progress and hence programme. Hence, these tunnel elements dominated much of the early planning of the investigations.

1. Tunnels (25 km in total length)		2. Earthworks (70 km in total length)	
St Pancras to Barking (15.2 km)	- Variable Tertiaries - Settlement - Length - Programme	Embankments over Soft Clay (13 km) of which: Thames Marshes (8 km) Stour Valley (4 km) Others (1 km)	- Settlement - Stability - Programme
Thames Crossing (2.8 km)	- Fissured Chalk - Permeability - Programme		
Bluebell Hill (3.2 km)	- Lower Chalk - Poor Chalk at Portals	Clay Cuttings (7 km) of which: Gault Clay (3.5 km)	- Stability - Landtake
Others (3.5 km)	- Variable Chalk - Portal Construction	(Other Structures 13 km)	

Table 3.1   Main Geotechnical Issues

The phased investigations have been focused at providing continuous information on ground conditions wherever possible, through high quality boring, drilling, sampling and testing. In addition, piezometers have been installed in most holes to allow groundwater monitoring to be undertaken over the period leading up to detailed design.

3.2 *Potentially Contaminated Land*

The project team were also required to include investigation of former industrial and waste disposal sites that could be potentially contaminated. Approximately 230 contaminated sites have been identified within 250 m of the route. Typical engineering concerns associated with such sites include:

- removal of unlicensed domestic landfill material from cuttings,
- construction through and next to gassing landfills,
- construction over and through unverified inert fill sites,
- foundation construction through asbestos fill,
- sub aqueous tunnelling within contaminated strata,
- removal of contaminated ground from beneath trackbed.

These sites present a range of potential contamination concerns. The certainty of URL's assessment of their risk depends to a considerable extent upon the quality and reliability of the available data. Consequently, sites which are well documented represent the least concern in the assessment even through they may be hazardous for route construction. Of the 230 significant sites, 146 were assessed to be of greater concern for the Hybrid Bill and the Environmental Statement because of a combination of uncertainty about the hazardous contaminants and uncertainty of the source data. All these sites have been the subject of further desk studies, including 1:2500 historical OS map reviews. Of these, 44 were recognised as bringing significant uncertainty to the assessment and were subject to more detailed desk studies and investigation.

3.3   Scope of Work

In August 1993, new impetus was given to the project's geotechnical activities with the establishment of a dedicated geotechnical management unit. Their initial brief was to assemble and collate historical data, establish a structured database, complete desk and development studies and carry out Phase I ground investigations to support the passage of the Parliamentary Bill. These studies were targeted at the top 15% of priority areas where there was perceived to be considerable uncertainty about ground conditions and hence a potential for risk in relation to construction. Detailed specifications for the ground investigations were developed whilst contractors were pre-qualified and contract documentation was issued in October/November 1993. The Phase I ground investigations were split into three to reflect the differing types of exploratory works and associated testing required for underground construction, surface works and potentially contaminated sites. Seventeen potentially contaminated sites were to be investigated at this stage.

By January 1994, the scope of the geotechnical works had been extended to include Phase II investigations, in order that approximately 50% of the ground information required for detailed design of the project be available by March 1995. Access to thirteen potentially contaminated sites was achieved in Phase I and a further twenty sites are to be investigated in Phase II. The locations of individual contracts Phase II is also shown on Figure 1.1. The corresponding organisation required to carry out this significant programme of works, valued in total at around £10 Million to date, is shown on Figure 3.1. The desk studies were completed at the end of 1993 and a summary desk study report issued to Tenderers in September 1994. A series of development studies were initiated in 1994, covering key geotechnical issues including topics such as:

- Construction over soft ground;
- Ancient landslips, shear surfaces, geomorphology;
- Chalk earthworks;
- Track Bed design;
- Seismicity;
- Groundwater conditions.

The overall programme for these activities within the context of other key activities during the early stages of the project are indicated in Figure 3.2. The main ground investigation activities undertaken in both Phase I and Phase II are indicated in Table 3.2. Thin wall sampling was carried out in the overconsolidated Tertiary clays and piston samples were obtained from the Alluvium and peats. In selected locations these high quality samples were taken continuously to provide a near continuous profile. The Phase I works comprised some 450 exploratory holes with a further 500 exploratory holes in Phase II.

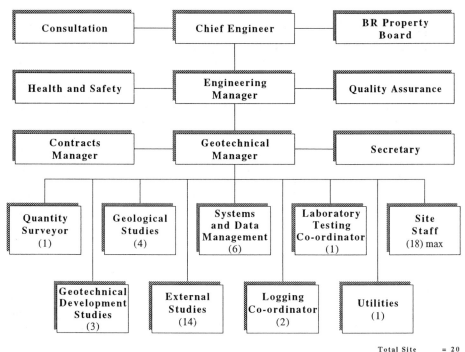

Total Site = 20
Total Office = 18
Figure 3.1 Geotechnical Management Unit Organisation Chart   Total External = 14

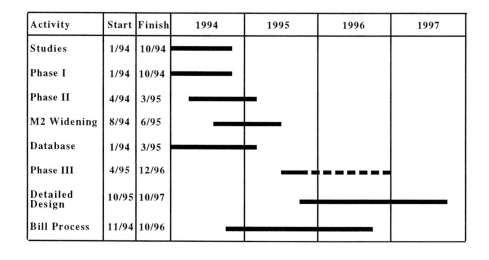

Activity	Start	Finish	1994	1995	1996	1997
Studies	1/94	10/94				
Phase I	1/94	10/94				
Phase II	4/94	3/95				
M2 Widening	8/94	6/95				
Database	1/94	3/95				
Phase III	4/95	12/96				
Detailed Design	10/95	10/97				
Bill Process	11/94	10/96				

Figure 3.2   Summary Geotechnical Programme

131

Exploratory Technique	Purpose
High Quality Rotary Drilling	Continuous stratigraphic profile, sampling
Light Percussive Drilling	Profiling and sampling
Deep Trial Pitting	Identification of weathering, solifluction, shear surfaces, sampling
Packer (single packers with transducers)	Permeability, grout take
Pumping Test	Permeability
Cone Test	Profiling, strength
Seismic Cone Test	Small strain moduli
Cone-Pressuremeter	Large strain moduli, horizontal effective stress
SBP/HPD/WRSBP Pressuremeter Test	In situ moduli, horizontal effective stress
Engineering Geophysics (a) Land	Solution features, depth to rockhead
(b) Marine	Depth to rockhead, geologic structure
Target: High Quality Reliable Data	

Table 3.2   Main Ground Investigation Activities

3.4   *Quality*

Union Railways Limited have demanded the highest quality of workmanship and professional skill since the inception of the project. The specification and controls set for the ground investigations have been aimed at maintaining these standards. Union Railways Limited's supervision on site has comprised staff from the consultants involved with the project and has been supplemented by recognised experts. In order to improve the consistency of logging across various contracts and differing contractors, a recognised national expert in logging was appointed to co-ordinate all engineering geological descriptions. Chalk core was logged in line with Spink and Norbury (1989) and CIRIA Project Report II (1994). In addition to this, the Tertiary strata logging was reviewed by Mr R Ellison of the British Geological Survey (BGS) and the chalk logging reviewed by Professor R Mortimore of Brighton University and Dr C Woods.

Similarly, during the Phase II surface works ground investigation the requirements for laboratory testing were co-ordinated through one person to assist standardisation of the scheduling of tests across staff from differing consultants and contracts. The Contractors have maintained a senior level of supervision on site, with all the core logging and with much of the factual reporting being prepared at the site offices. A requirement of the specification was to produce report quality borehole logs and in situ test results on site at the end of the fieldwork period using computerised techniques. The supervision was shared between the Resident Engineer and the Contractor with an approximately equal ratio of professional staff. This arrangement allowed each field activity to be supervised by a professional staff member. The specification for the works included the taking of high quality photographs of all core to provide an easy reference record for private sector Tenderers. The core is stored in a dedicated and modified warehouse in Kent. The site operations have been the subject of numerous safety, quality assurance and technical audits carried out by independent assessors. Some £6M has been expended on the direct costs of the Contractors for Phases I and II to date covering fieldwork, laboratory testing and factual reports.

4.0   KEY ASPECTS

This section presents some of the real practical difficulties faced in obtaining the data during fieldwork and provides suggestions as to means of improving techniques, as well as new data handling techniques developed for this project. The investigation for the tunnelled elements of the scheme have involved a wider range of more advanced exploratory techniques and some of these are considered here. For all the ground investigations, but particularly for the surface works there has been a detailed focus on soil logging and soil profiling, identification of weathering and existing shear surfaces, mapping of landform and surface features, aerial photographic interpretation as well as the establishment of realistic ground models. Some of these aspects are discussed in the companion paper to this conference by Waller and Phipps (in press).

4.1   *High Quality Drilling in the Tertiary Strata*

A fundamental requirement of the investigation was to obtain the best possible core, both in terms of quality and recovery, in the various 'strata' encountered. The investigation for the proposed London tunnels penetrated most of the Tertiary succession together with the underlying Seaford Chalk within a typical forty to fifty metre deep borehole. Therefore, the material drilled therefore ranged from a stiff clay, through dense sands to coarse gravels, to Chalk with numerous flint bands. Boreholes were typically advanced through any made ground and superficial deposits to around one metre into the underlying Tertiaries using cable percussive methods. Thereafter, the boreholes were generally extended to depth using wireline rotary drilling methods to produce 102 mm nominal diameter continuous cores.

The large diameter wireline drilling system offers the most suitable technique for achieving the objectives of high quality continuous core in such a wide range of materials. 'Geobor S' wireline drilling was first used for site investigation purposes on the Baghdad Metro ground investigation in 1982 - 1983 (Scarrow and Gosling, 1986) and subsequently introduced to the U.K. in 1987. Since that time the system has been used in a very wide range of materials and particularly in circumstances where conventional large diameter core drilling has failed to recover core of sufficient quality or quantity to allow a comprehensive appraisal of the materials encountered. The facility to operate a drilling system which cases the borehole as drilling progresses, effectively prevents caving and debris forming in the base of the borehole. When this is coupled with a low pressure flushing system this allows a large diameter core to be recovered with minimal disturbance and has proved to be a major step forward in geotechnical investigations.

Drilling induced core disturbance is minimised by the use of face discharge drill bits, low flush fluid pressures, and polymer drilling mud in conjunction with a triple tube system whereby a semi-rigid plastic lining tube is inserted within the inner core barrel. The lining tube also facilitates the removal of core from the core barrel and subsequent handling. A biodegradable polymer mud flush, GS550, was generally used though occasionally particular ground conditions necessitated the use of a different type of flush, see Table 4.1.

The wide range of ground conditions encountered can cause difficulties with drilling bit selection. It can be seen from Table 4.1 that the optimum drill bit varies according to the lithology.

Formation	Unit	Lithology	Ideal Bit Configuration	Drilling Problems	Solution
London Clay	London Clay	Stiff Clay	Diamond sawtooth	Softening/ core stretching in some areas	Water flush
Harwich Formation	Blackheath Beds	Sand (clay) Gravel	Diamond set ?	Gravels	Use Tungsten Carbide bit
Woolwich and Reading Beds	Upper	Clay/Sand	Diamond sawtooth	Loss of core due to flushing away matrix	Use Tungsten Carbide bit and change mud
	Lower Pebble Bed	Clay/Sand Gravel	Diamond sawtooth		
Upnor Formation		Silty Sand	Diamond set	None	
Thanet Sand	Thanet Sand	Silty Sand	Diamond set	None	
	Bullhead Beds	Gravel/ Cobbles	?	None	
Chalk			?	None	

Table 4.1   Table Showing Drilling Bit Selection

With the wireline system it is most desirable to complete the borehole with a single bit otherwise the whole string of casing has to be retracted. Although the drilling mud will help to maintain the stability of the borehole, it is probable on withdrawal of the casing that caving and possibly a large scale borehole collapse would occur. Whilst diamond bits would be preferable in some individual strata such as the London Clay or Thanet Sand, they were often rapidly rendered useless by the presence of gravels within the Harwich Formation, Pebble Beds within the Woolwich Formation, the Bullhead Bed at the base of the Thanet Sand and the numerous flint beds within the Chalk. For this reason a high quality face discharge tungsten carbide bit capable of drilling the whole succession was generally found to be the best option, albeit as noted above a compromise for some of the horizons.

To allow monitoring of core recovery, data from the daily journals was transferred to a simple spreadsheet and presented in graphical form, see Figure 4.1. As the work progressed useful recovery data was obtained in each stratum for particular areas leading to the definition of achievable norms. It was found that in certain specific areas the recovery in a particular stratum was consistently lower. This was independent of other factors that might otherwise be thought to be relevant such as the type of drilling bit, driller etc. It was possible to rapidly ascertain whether this was a departure from the acceptable range for the particular location and stratum. If low core recovery was not a function of the ground, and following routine checks, decisions concerning whether the drilling methods should be modified, or even if it was necessary to redrill a particular section of a borehole, were then quickly taken.

4.2   *Pressuremeter Testing*

General.   A total of over one hundred pressuremeter tests have been carried out in boreholes sunk along the proposed London, Thames and Blue Bell Hill tunnel sections. Over the length of the proposed London tunnels, dedicated pressuremeter boreholes were sunk at the planned locations of vent shafts and tunnel portals. The Cambridge self boring pressuremeter (SBP) was used in the London Tertiaries and the Hughes, high pressure dilatometer (HPD) was used in the underlying Seaford Chalk. Over the length of the proposed Thames tunnels, pressuremeter boreholes were sunk at three locations in the river and on both river banks. These tests were carried out using the high pressure dilatometer, with the majority of the tests being carried out in the Upper Lewes Chalk (Figure 2.4), which is towards the base of the traditional Upper Chalk. For the proposed Blue Bell Hill tunnels, pressuremeter boreholes were sunk at each of the planned tunnel portals and at one location between the portals. Tests using both the high pressure dilatometer and the weak rock self boring pressuremeter (WRSBP) developed at Newcastle University were carried out towards the base of the traditional Upper Chalk and throughout the traditional Middle Chalk.

Pressuremeter tests in the London Tertiaries.   Over thirty pressuremeter tests have been carried out using the self boring pressuremeter (SBP) within both the London Clay, the sands, silts and clays of the Woolwich and Reading Beds and the Thanet Sands. Experience has shown that pressuremeter test depths in these materials are best planned with reference to high quality, reliable geological information from adjacent boreholes, which for these investigations comprised continuously cored rotary boreholes sunk earlier during the investigations. It is particularly important for tests to avoid strata containing flint gravel, such as the Blackheath Beds, the Pebble Beds and some parts of the Upnor Formation, and also limestone beds, which occasionally occur within the Lower Mottled Clay. These materials tend to block the pressuremeter cutting shoe and can tear the pressuremeter protective lantern and membrane.

Drilling the self boring pressuremeter through the granular Tertiaries was particularly difficult with drilling times typically over one hour or more. High levels of control were required to adjust drilling variables such as the type of cutter, cutter shoe geometry and rotation speed, thrust pressure and flush pressure to achieve the optimum conditions. The results from a typical pressuremeter test in the Thanet Sand is shown in Figure 4.2.

This figure demonstrates the importance of 'holding periods' to minimise the creep effects for strain controlled tests. The specification required a delay period prior to unloading if creep effects became apparent. The delay period comprised the pressure being held constant until the rate of cavity strain reduced to zero. Without such a delay period the almost linear unload/reload loop shown in Figure 4.2 would tend to become more elliptical.

Pressuremeter Tests in chalk strata.   Over sixty high pressuremeter dilatometer tests and approximately ten weak rock self boring pressuremeter tests have been carried out in chalk along the proposed London, Thames and Blue Bell Hill tunnels. The most important factor dominating the method of testing chalk was the presence or otherwise of flints. Using the logs from adjacent rotary cored boreholes the presence of flints could be correlated, but their specific depth was often difficult to predict. For Chalk where the presence of flint was suspected, the HPD test rather than the WRSBP was adopted.

Given the presence of flints, the following methods were employed in an attempt to optimise the number of successful tests:

.   The use of wire-line casing to the top of the test pocket to prevent flint falling
        from the borehole wall into the test pocket.
.   The use of diamond, rather than tungsten, drilling bits with the aim of 'coring'
        the flints and to minimise flint shattering.
.   The drilling of longer pockets than theoretically required to provide a sump for
        flints that did fall from the test pocket wall.
.   At the completion of pressuremeter testing, the use of wire-line coring over the
        length of the test pockets, together with at least a half metre length of 100 mm
        diameter cored section between test pockets, in order to recover any loose
        flints that had fallen into the bottom of the test pockets.

Despite these precautions, a number of tests were unsuccessful due either to problems with flints or to the inherent weakness of the chalk. The more weathered, (Grades IV and V) chalk that typically underlies the Terrace Gravels in the vicinity of the Thames proved to be unsuitable for HPD pressuremeter testing, with only very small cavity stresses being measured for cavity strains at the limit of the pressuremeter's capacity.

Where the chalk is less weathered (Grade III or II), such as in the vicinity of the Thames at depth and in the chalk underlying the Thanet Sand in London, HPD tests have generally been more successful. A comparison of results is shown in Figure 4.3.

For chalk that did not contain flints, such as many of the units within the traditional Middle Chalk, the WRSBP was successfully employed. A result from a typical test using this equipment is shown in Figure 4.4.

The holding periods described above apply equally to the unload/reload loops for the HPD and the WRSBP tests in the Chalk. Figure 4.3 and Figure 4.4 show good unload/reload loops for this type of test.

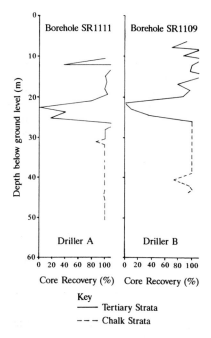

Figure 4.1 Identification of Zone of
Consistent Core Loss

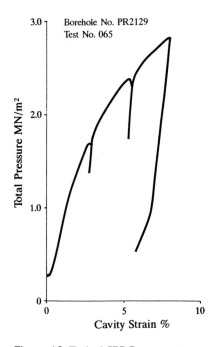

Figure 4.2 Typical SBP Pressuremeter
Test in Thanet Sand

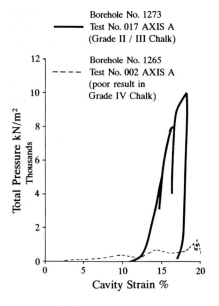

Figure 4.3 HPD Tests in Chalk

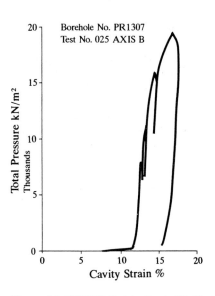

Figure 4.4 WRSBP Test in Grade III Chalk

137

4.3   *Advances in Packer Testing*

Some of the long standing problems associated with traditional pump-in packer testing using simple pneumatic packers, pressure measurement with analogue gauges at the surface and mechanical flowmeters have been alleviated by recent improvements to the equipment used on this investigation. The packer tests were carried out in the base of boreholes as drilling progressed. The equipment comprised a test section packer with a guard section packer located above it, see Figure 4.5.

By monitoring the pressure difference between the test section and the guard section packer it was possible to ascertain with certainty whether flow was taking place around the test section packer, a common phenomenon in more weathered and permeable material. This is difficult to detect with the simple early pneumatic packers used without a pressurised guard cell. In addition, the type of packer has been found to be important in achieving a high level of reliability. 'Petrometallic' packers proved to be reliable, robust and produce an effective seal in most conditions.

The test and guard section pressures were measured using electronic pressure transducers. The use of transducers allowed the pressure above and below the test section to be continuously and accurately monitored and provided immediate evidence of leakage past the test section packer. Formerly, pressure transducers which vented to atmosphere and recorded actual borehole pressures were used, but these proved difficult to change on site and were vulnerable to damage. Thus, transducers which record absolute pressure by reference to a vacuum in the instrument were adopted. These transducers proved more reliable than the direct reading instruments.

Electronic flowmeters, with a mechanical back-up in tandem were used because of their increased accuracy and reliability over a greater flow range than the mechanical units. A facility was provided using a large diameter riser to introduce high flow rates, of up to 25 litres per second. In addition to conventional single pump-in/permeability tests a number of pump-out tests were carried out using a small submersible pump located in a sealed riser pipe in communication with the test section, but located above the guard cell. Although these tests were primarily carried out for sampling purposes, some additional permeability data was obtained.

These advances in the packer equipment have provided a much better and more accurate test and allow a more realistic assessment of permeability. A summary of the results from one borehole in the Upper Chalk, Grade III/II is presented as Figure 4.6.

These show that the vast majority of tests display a pressure versus flow curve close to the classical shape. This is primarily due to the equipment adopted for the recording and maintenance of the pressure conditions occurring in the test section and the facility to control the tests more sensitively.

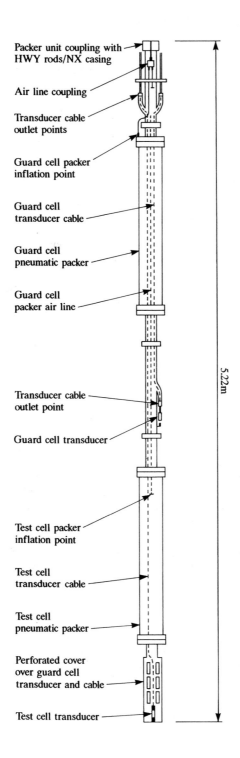

Packer unit coupling with HWY rods/NX casing

Air line coupling

Transducer cable outlet points

Guard cell packer inflation point

Guard cell transducer cable

Guard cell pneumatic packer

Guard cell packer air line

Transducer cable outlet point

Guard cell transducer

Test cell packer inflation point

Test cell transducer cable

Test cell pneumatic packer

Perforated cover over guard cell transducer and cable

Test cell transducer

5.22m

Figure 4.5 Packer Test Equipment with Pressure Transducers

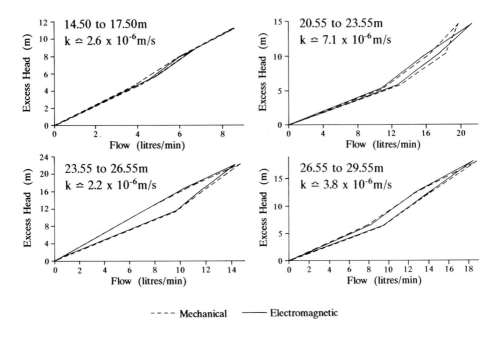

Figure 4.6 Parker Test Results from Borehole T1267 in Upper Chalk

4.4   *Data Handling*

Site Monitoring: Digital systems for on site monitoring were installed at each contract site office for the use of the Resident Engineers and their staff. The Contractor was also required as part of the specification to maintain computer systems on site and to guarantee standard formats for data interchange. The advantages of this approach proved numerous as follows:-

.   Reduction of paper records and number of copies
.   Rapid communication of information between contractors and head office
.   Cut, paste and compilation of data at head office
.   Instant analysis and decision making
.   Improved targeting of investigations with constantly updated geological information
.   More effective assessment of quality and performance
.   More effective assessment of quantities.
.   Effective cost and programme controls
.   Accurate checking of data

Digital Data Processing and Analysis: At the outset of the project a requirement for digital systems had been recognised to deal with the large quantity of information that would be produced. Three simple requirements were identified.

The system must allow;

1.   Rapid **assessment** of the data **available**   "You only see what you look for";
2.   Rapid **access** to the **right** data;
3.   The use of extensive **tools** for the **analysis** and **manipulation** of data.

In addition any solution must also satisfy the following criteria;

4.   Simplicity and a short learning curve
5.   Capability of dealing with large quantities
6.   Flexibility
7.   Familiarity
8.   Communicate via industry standards

Three solutions were chosen to address the identified requirements.

1.   MAPBASE is a simple DOS Geographical Information System for representing
      spacially referenced data on a base map of the UK. The power of MAPBASE
      lies in its graphics speed and ease of use. Currently six databases of
      information pertinent to geotechnics are attached to a map of the UK which
      also shows road and rail networks and urban boundaries at 1:625000 scale.
      These databases are shown as scalable icons and are:

      2500 boreholes
      1730 wells
      600 natural and artificial cavities
      250 landslips
      90 geo-morphological anomalies recorded during field mapping
      1800 shallow soils profiles for agricultural soil quality assessment

      The full UK map with all databases takes only 22 seconds to display on a
      standard 486 - 33 MHz PC and 45 seconds on a 386 sx. Since the graphics is
      in bitmap rather than vector format, the video memory requirement is also very
      low even for a large amount of information.
      Since each icon is referenced to a location and to a database there are powerful
      tools for locating data either spacially or depending on some user defined
      condition. This condition may relate to the source database of the icon or to
      some other database e.g. locate all boreholes in a defined area within 100m of
      a licensed abstraction well.

2.   CARDBOX, is a simple library database system, which was chosen to manage
      the referencing of the paper data.

3.   SID was chosen to manage the bulk of the geotechnical data. No single
      program was identified which could satisfy all the requirements of the Union
      Railways system specification. SID, however, did offer the flexibility to adapt
      to such a large project and a certain amount of continuity with the existing
      systems. The enhancement and customisation of the package benefited from
      the expertise of the existing personnel and has resulted in a powerful tool for
      analysis of data.

Historical Data: Historical data has been gathered from a wide range of sources including consultants, utility companies, local authorities, government institutions and British Rail. These arrive in a number of different paper and digital formats which demand a large amount of processing before they can be added to the databases. The data has been processed by Union Railways staff to provide only a synopsis of the geological profiles using stratum codes. Some level of confidence in the interpretation is indicated by the grade qualifiers in the data table of general information pertinent to the holes (location etc.). These are given in Appendix A1.

Paper copy data has been stored in a specially adapted geotechnical reference room within Union Railways. The paper data is stored in original complete report form reference by originator, title, date, etc and the borehole log information is also stored together with other reference data in geological/geographical sub-divisions of the route. These sub-divisions have been termed 'geosegments' and there are some 30 along the route length.

Phase 1 Contract Data: The digital results of Union Railways investigations are provided in AGS format with some additional fields specified to cover other investigation techniques. Submissions of data are provided at preliminary (on completion of fieldwork), draft (on completion of fieldwork and laboratory testing), pre-final (with corrections from clients review) and final (after outstanding corrections have been made).

The following checks are made on receipt of the data:

At preliminary stage
.   Import into spreadsheet as a "," delimited file to check formatting
.   Check headers and field names
.   Check presence of all exploratory locations
.   Return to contractor with notification of errors/request for clarification

At draft stage
.   Check corrections to data formatting
.   Check for illegal characters in fields, incorrect units etc.
.   Import into database and record any errors
.   Cross-check strata
.   Check exploratory locations against Resident Engineer's records
.   Draw long sections to check geological interpretation of stratum codes
.   Check random comparisons against paper records for inconsistencies
.   Check relational integrity of the database
.   Return to contractor with notification of errors/request for clarification

At Pre-final stage
.   Check corrections from draft data
.   Check that changes to draft paper reports have been carried forward to digital data
.   Check relational integrity of the database
.   Report maximum, minimum and mean values for fields
         Return to contractor with notification of errors/request for clarification

At Final Stage
.   Check corrections from pre-final data
.   Check relational integrity of the database
.   Inform contractor of errors or request clarification. Request new submission or agree to amend data in house.

SID provides automated tools for performing over 180 relational checks on the database structure and printing reports on data statistics. Future advances in SID are planned to include enhanced data statistics and reporting using filters to isolate impossible and implausable data values. Another powerful feature of the system is in-built QA tracking using automatically updated fields CHNG_BY and CHNG_DATE. These features enable up-dated data to be isolated from within complete standardised AGS (1992) submissions.

The data analysis tools in SID are powerful and have been customised to enable the user to pick geological sections from a map of London and Kent with the route overlain as well as produce over 160 graph types as standard including 3D surface and contour plots. Routines are also included for calculation of common derived parameters (liquidity index, equivalent undrained strengths from SPTs in clays etc.).

For advanced analysis of inter-related data sets exported data are manipulated using visualisation and numerical modelling tools.

5.0   CONCLUSIONS
The factual reports for Phase I fieldwork were largely complete by September 1994 with at least draft versions available to the Tenderers for inspection in a data room. Digital data and paper copies of the final factual reports will be distributed. The majority of Development Study reports are also available. Summary plans and sections are being prepared and progressively updated. Fieldwork for Phase II ground investigations is due to be complete at the end of November 1994 and factual reports are due to be complete in the period January to March 1995. This will achieve the objective of approximately 50% of geotechnical data being available prior to the start of detailed design such that geotechnics is now off the critical path within the project programme.

REFERENCES

Ellison, I.R. : Report to Union Railways (BGS 1994).
Mortimore, R.N. (1986) : 'Stratigraphy of the Upper Cretaceous White Chalk of Sussex'. *Proc. Geol. Ass. 97*, p97-139.
Spink, T.W. & Norbury, D. : 'The Engineering Geological Description of Chalk' *Proceeds of the International Symposium on Chalk* (Brighton 1989).
Lord, Twine & Yeow: CIRIA Report 11, Foundations in Chalk (1994).
Waller, A.M. & Phipps, P.J. : 'Terrain Systems Mapping and Geomorphological studies for the Channel Tunnel Rail Link' Submitted for the ICE Conference on *Advances in Site Investigation Practices* (30-31 March 1995).
Scarrow, J.A. & Gosling, R. : 'An example of Rotary Core Drilling in Soils' *Geological Society, Engineering Geology Special Publication No. 2* (1986).

Appendix A1

Grading of Borehole Data

The grade given to the record reflects the accuracy of the engineering geological descriptions as follows:

Grade	Description of Grade
A	Good lithological description and stratigraphical interpretation
B	Good lithological description and extrapolated startigraphy
C	Limited stratigraphic value but reasonable lithology
D	Poor quality record
E	Unvalidated electronic data

A grade is given to the record which reflects the accuracy of the reduced levels as follows:

Grade	Description of Grade
A	Accurate datum level
B	Corrected Liverpool datum
C	Suspect datum level e.g. from rail level
D	Estimated or interpolated datum level e.g. from contours or nearby spot heights
E	Unvalidated datum level

A standard has been applied througout all grading and interpretation to ensure consistency. These standards have utilised appropriate geological and geotechnical expertise from across the industry. A check of data interpretation is also undertaken by the database manager using the data analysis tools provided by SID. Reference to the high quality Phase 1 investigation results provides a useful constraint on the interpretations.

Historic maps in the investigation of land use and the identification of possible contamination

B. RIDEOUT and D. SMITH
Nabarro Nathanson

INTRODUCTION

There is nothing in the draft British Standard for site investigation DD 175 of 1988 or in the advisory documents of the Institution of Civil Engineers, or the Royal Institution of Chartered Surveyors or the Construction Industry Research and Information Association which deal with land use investigation beyond simple lists naming the archives in which research might be conducted. Whilst these eminent bodies agree that research of the archives is essential and highly cost effective none of the available material examines the short comings as well as the advantages of the respective archives.

Traditionally the "Desktop Survey" (which was the euphemism given to such archive research) was limited in scope, both because of a lack of understanding of the resources available and belief that errors and omissions within the "Desktop Survey" could be recovered by visiting site and taking samples for analysis. The ICE publication entitled "Without site investigation ground is a hazard" refers to research undertaken by NEDO and others over the last 25 years in both Civil Engineering and Building Projects which shows that the technical and financial risk which exists in ground related problems are responsible for more than 50% of all delays and cost overruns. In land which was previously developed many problems stem from (a) lack of knowledge of the archives on the part of researchers (b) inherent inaccuracies in the archive, which have not been properly appreciated, and (c) client's parsimony in authorising investigation.

A new and rapidly growing market exists for "Desktop Surveys" as the initial step in environmental investigations, land quality survey or environmental audits where there is no construction or development envisaged. Such reports are required by company management for background on health and safety programmes, risk assessment, financial planning, insurance and other purposes. Many banks and investment organisations, pension funds, property trusts and property investment companies require such reports prior to purchase or securing a property for loan purposes or on the occasion of portfolio valuation.

Whatever future changes in the law occur, it is unlikely that the financial markets will alter their requirements, now they have discovered their importance. Today no transaction, of any significance, takes place without investigation of the land use history of the site and the extent or otherwise to which the property may be contaminated.

"Desktop Surveys" usually concentrate upon Ordnance Survey maps and plans. It must be stressed that these are by no means the only documents to which reference should be made, however, their domination is such that this paper will concentrate on them.

ORDNANCE SURVEY MAPS

Maps produced by the Ordnance Survey are the basic guide to the changing British landscape from the beginning of the 19th century, showing it at a number of different scales.

1:2,500 - 25 inches to 1 mile

This is the essential scale for site investigation, 25 inch to 1 mile (1:2,500) which is the Ordnance Survey's basic scale for all cultivated areas. Although use of this scale rather than the 6-inch scale discussed below involves study of 16 times more sheets for any given area - the 25-inch sheets are very much easier and quicker to read than 6-inch due to reduced concentration of information into limited space. Researchers may find it difficult to requisition so many sheets even at intervals in archives and comparison is difficult when they are not bound together.

With the exception of urban areas, the 25-inch maps are the most detailed maps to be published for much of Great Britain. They aim to depict the permanent features of the man-made landscape at ground level as faithfully as possible without deliberately distorting the scale. Features are portrayed to scale, revealing their exact extent. The 25-inch landscape reveals not only industrial buildings but also such ground features as slag and soil heaps, open-cast workings, quarries, pits and so on. Many watermills are named and the use of many industrial premises defined. Descriptive names define the character or use of features which are otherwise unnamed, such as Allotment Gardens or Recreation Ground. Alternatively, distinctive names are noted, such as place names, with function defined for many industrial premises. However, names of buildings within large installations are omitted as are proprietary names unless they are required for identification. Mistakes were made, such as the false identification of iron ore calcines as lime kilns, on some first edition 25-inch maps. Military information was frequently omitted for strategic reasons on Ordnance Survey maps at all scales. Also 25-inch maps fail to distinguish between individual shafts within collieries, in the early editions in their representation of bell pits and other shallow workings and in areas of extensive lead mining activity. As on all maps, a place name may indicate an earlier industrial use, for example, "Tile Kiln Farm" and "Iron Mill Lane".

The first 25-inch survey was carried out between 1854 and 1895. However, it did not cover mountain, waste and uncultivated areas. The first full scale revision, excluding the Isle of Man and the Island of Lewis took place between 1891 and 1914. From 1882 the Ordnance Survey aimed to revise the 25-inch survey every 20 years so that the average age of sheets would be no more than ten years old. It maintained this schedule until World War I. The second revision, based on surveys started in 1904, was never completed. Certain urban developing areas were revised for Land Valuation Purposes in 1911-12 but not all were published. In 1918, a 40 year period was introduced for the revision of the most sparsely populated districts and in 1922 the 20 year revision period was confined to 6-inch sheets containing portions of urban administrative areas. In 1928, the system of revising at regular intervals was replaced completely by a selective system of continuous revision for areas of most rapid change. Thus, the planned revision programme, which had to be abandoned due to war and austerity, was replaced by selective revision. Consequently, groups of sheets were revised and printed independently of adjacent areas, especially in towns, although, frequently, blocks of sheets were revised together, covering an extensive area. Some 25-inch maps were revised between 1904 and about 1925; and some urban areas and their hinterlands were revised from about 1925 to 1944, with the later sheets in particular often being published only at 1:2,500 (25 inches to 1 mile) without being reduced to smaller scales.

After 1938 and during World War II, a Provisional Edition was produced with all available revisions. The first of a restructured 25-inch series on National Grid sheetlines appeared in 1948; the series being completed in 1983. From 1945, there was continuous revision of certain areas, determined by the amount of change.

The Ordnance Survey's 25-inch scale maps must be the essential tool for site investigation since they represent the Ordnance's basic scale of survey from which all smaller scale maps were derived. With very few exceptions, which will be explained, no small scale Ordnance Survey map can show any more or different information, only less than the 25-inch. From the inter-war period onwards many 25-inch surveys were published only at that scale and were not reduced to smaller scales. Furthermore, the system of continuous revision, in which individual sheets were revised according to the amount of change taking place, resulted often in the portrayal of numerous successive stages of development which were not recorded at the smaller scales. Certainly the 25-inch scale is the minimum scale necessary for study of densely developed urban areas in which site usage cannot be clearly portrayed at any lesser scale. The 6-inch maps [1:12,560] (surveyed before the introduction of the 25-inch survey) and [1:10,560] thereafter 1:10,000 maps show significantly less information than the 25-inch, with many industrial sites not being labelled or simply identified as "works" etc. Other difficulties of using 6-inch maps rather than 25-inch include problems in matching up labels with sites and cases where more than one label appears to apply to a particular site.

All maps, no matter what the scale, are inefficient in recording certain categories of site which tend not to be recorded in survey or not to be transposed to the finished map. These include scrap yards, garages, waste disposal sites, electricity sub-stations and those myriad "domestic" trades based on the home and small workshop, particularly of the late 18th and 19th centuries. It is the writers view that study of the 25-inch Ordnance Survey sheets should identify a high percentage (say 90%) of the potentially contaminated sites at the time of survey, but only at that time.

1:1,250 50 Inches to 1 Mile

In 1911, the Ordnance Survey published many plans at the scale of 50 inches to the mile (1:1,250) for the newly-formed Inland Revenue Land Valuation Department which were put on general sale. However, they were simply mechanical enlargements of the then 25-inch sheets. They show no more information than the 25-inch sheets. The same applies to 50-inch sheets produced at the request of local authorities and private individuals for land valuation and registration requirements from the same date. Since there is no additional information shown, it is pointless to search out these 50-inch sheets instead of the 25-inch sheets.

From 1946 a freshly surveyed 50-inch series began publication to cover all the "Major Towns" (Less than 6% of Great Britain at that time) although surveys were often extended to include complex industrial installations. The completion of the 1:1,250 50 inch series conventional mapping was overtaken by the advent of digital mapping, this being completed in 1991 with 56,073 plans published and a further 1,324 plans of coastal areas digitised but not published. Revision of 1:1,250 published sheets began in 1949 with continuous revision of areas experiencing significant change. Ordnance Survey planning allowed for at least one completely new edition of all 50-inch maps to be published every 25 years. However, in practice the continuous revision system resulted in many sheets being up-dated and published more frequently.

Although the 50 inch sheets are the largest of all current Ordnance Survey scales, it was still necessary, as in any map, for some details to be omitted, others generalised, and some features to be enlarged to a minimum size, and conventional signs to be used to codify much of the surface detail. However, these drawbacks do not reduce this scale's usefulness. From 1946, where they exist, the 50-inch sheets are the essential source for densely developed areas. Fortunately, to simplify study in some cases, the 1:1,250 was also published in a reduction at 25 inches to the mile between 1948 and 1973. These direct reductions without omission are readily recognised by the finer line work and lettering, the firm grid lines, and the compilation diagram where it has not been cropped from the sheet. Unfortunately, since the 25 inch coverage of 50 inch areas was very much subordinate in revision terms, not all editions of 25 inch sheets were published at 50 inches. Hence, only the 25 inch sheets offer a comprehensive record of site use over time,

1:10,560 6 inches to 1 mile

The 6-inch to the mile (1:10,560) is the second oldest Ordnance Survey series, the scale being adopted in 1840 for those areas of northern England and Scotland not yet covered by the 1-inch survey. However, when the 25-inch scale was introduced as the basic scale of survey for all cultivated areas from 1854, subsequent 6-inch maps were simply derived from 25-inch parent maps. All areas published at 25 inches, between 1854 and 1888 were also published at the reduced 6-inch scale.

The derived 6-inch maps omit certain elements, the numbers and acreage of different land parcels are not given; buildings in close town and village areas are blocked rather than shown as individual properties; certain boundaries are omitted in towns where they would be illegible; not all town streets are delineated to scale; and railways are portrayed conventionally rather than in plan.

Many features on the 6-inch series are depicted at their actual scale, but it was necessary to generalise many others, involving the selection and simplification of material to eliminate unwanted detail. This was the only way to achieve the required character and clarity of the maps. For the sake of clarity, some buildings are exaggerated from their true size and the distances between buildings and between other objects and buildings are increased. The generalisation of built-up areas involves, in particular, the omission of names of garages. Railways are not shown in scale plan and are generalised by the omission of linages, sidings, and also station buildings. In mining areas, the 6-inch maps fail to distinguish between the individual shafts and collieries. In overcrowded areas of the map, minor descriptive names are omitted, although the policy was to show as many as possible. Although many urban industrial sites are identified, the 6-inch maps may be less helpful than their 25-inch parent maps, particularly for densely developed areas. The compact nature of the 6-inch landscape makes it more difficult to read than the 25-inch, particularly in urban areas. It is the writers opinion that the 6-inch Ordnance Survey maps reveal some 10-15% less potentially contaminated sites than do the 25-inch maps, except in rural and semi-rural zones where it identifies most sites. The percentage falls to lower levels in dense industrial areas.

The 6-inch maps are original for Lancashire, Yorkshire, Fife, Kinross, Kircudbrightshire, Lewis, East and Mid Lothian, and Wigtownshire which were surveyed between 1842 and 1855, before the introduction of the 25-inch scale. They are also original for mountainous and moorland areas and coasts not covered at all by the 25-inch survey. The 6- inch scale is the largest scale to cover much of Scotland and Wales and is the largest to cover the whole country. Generally, the 6-inch is the only scale which covers land above the 1,000 foot contour line. Many mining areas were mapped only at 6 inches to the mile, creating maps which are of little use for the exact identification of mining sites. Pre-1854 6-inch maps, surveyed before the introduction of the 25-inch maps, show significantly less industrial information than the later post-1853 derived maps.

The 6-inch series was revised firstly between 1891 and 1914 as straightforward reductions of the 25-inch maps and, within the limits of the scale, are acceptable substitutes. The 25-inch partial revisions of 1904 to about 1925 were also published as a straight reduction at 6 inches to the mile.

In 1882, it was accepted that revision of the 6-inch maps should take place every 20 years and a new edition published. The aim was that no sheet should be more than 20 years old and the average age of a sheet should be ten years. Unfortunately, the First World War drastically interrupted this period system of revision and it became impracticable. In 1922, cyclic revision was confined to town areas, uncultivated areas were especially neglected.

In 1938 and 1939, 6-inch maps of urban areas were partially revised for air raid precaution purposes. Initially, these revised maps were published in a restricted edition for civil defence use, but later they were used for the 6-inch "Provisional Edition" which was published between about 1945 and 1950. Between 1944 and 1958, the 6-inch series was reconstructed on National Grid sheetlines with further partial revision. The newly-surveyed 25-inch sheets which appeared between 1948 and 1983 were reduced to the 6-inch scale. Since 1969, as part of the Ordnance Survey's programme of metrification, the 1:10,560 6 inch scale has been replaced by the 1:10,000. Unfortunately, on this and other new metric series, industrial sites are simply described as works or factories and many features identified on the previous imperial-scale Ordnance Survey maps are not shown. Where it is not possible to consult the original 25-inch map, study of the derived 6-inch map is the only alternative. However, it is essential to appreciate the 6-inch map's limitations.

1:63,360 - 1 inch to 1 mile

The only Ordnance Survey series surveyed prior to the 6 inch maps was the 1 inch to the mile (1:63,360) which appeared from 1801. After 1842, the 1-inch maps were simply derived, firstly, from the 6-inch surveys and, subsequently, from 1854 from the 25-inch surveys as they were completed.

The 1-inch maps published between 1801 and the appearance of those derived from 6 and 25 inch surveys are regarded as the least accurate of Ordnance Survey maps and have been widely criticised. However, they are superior to most other contemporary material of similar scale and were the most accurate produced to that time. Conventional signs locate mines, quarries, chalk and sand pits, lime kilns, canals, tramways, railways, docks, coal shafts, tide and powder mills, and so on. Unfortunately, features shown vary from sheet to sheet, reflecting the development of specifications for the maps and the increasing comprehensiveness of Ordnance Survey coverage. Even so, the 1-inch maps may be of limited use for the identification of sites in 'desktop' surveys where access to them can be arranged without difficulty. Although windmills are marked by a conventional sign and often named or function designated, watermills are frequently ignored. Industrial sites are only occasionally named. Obviously, the amount of detail relating to buildings and other features which can be shown in built-up areas is very restricted. Consequently, industry is never defined in developed areas. The 1-inch maps are most useful in identifying sites in rural areas between 1801 and the publication of the first 6 and 25-inch sheets of the same area. However, coverage is by no means comprehensive and the maps are difficult and tedious to analyse because so many sites are marked only by tiny conventional signs.

Revision of the 1-inch maps tended to be by piecemeal correction and amendment, particularly in respect of railways which were frequently added independently of any other revision ,as supplies of sheets ran out and demand dictated further printing. Revised sheets were not described as a new edition and the publication date was not altered. There may be many different versions of the early 1-inch sheets. In a few cases the whole sheet was revised and wholly re-engraved, providing a completely up-dated replacement, but mostly revision was partial, creating a landscape of variable date. Thus, the main problem of using the 1-inch maps is to determine the dates of original survey, revisions and printings of any particular sheet.

All 1-inch maps were revised between 1893 and 1898, and again between 1901 and 1912. A few sheets covering Berwickshire and East Kent were revised in 1909-10. The whole of Great Britain was again revised in 1912-31, and southern and eastern England only was revised between 1925 and 1939. With the exception of the Scottish Highlands and Islands, the series was revised between 1958 and 1971. It was completely republished at the metric 1:50,000 scale with a few areas revised one or more times. Revision for the 1:50,000 'First Series' in 1971-2 covered northern Wales and parts of the Midlands and the North-west only. Revision of the whole of Great Britain for the 'Second Series' or 'Landranger' series was carried out between 1971 and 1986. Since 1986 the entire series has been in a process of revision.

OTHER MAPS AND PLANS
Although the large-scale Ordnance Survey maps represent the basic source of archive information reference to them alone will leave gaps. Large-scale Ordnance Survey maps alone will not identify sites or usage which disappeared prior to the survey of the 6-inch and 25-inch sheets between 1842 and 1895. Nor do Ordnance Survey sheets record site use during periods between revisions of the maps. Site use could well have changed two or more times in the intervening period. Even the 25-inch scale cannot identify all sites in dense urban areas, particularly in the earlier surveys. If a comprehensive 'desktop' survey is intended, it is necessary to consult other types of maps in the hope of filling some of the probable gaps.

TOWN PLANS
The public health movement of the second half of the 19th century generated large-scale plans produced both by the Ordnance Survey and by private surveyors. The Ordnance Survey published plans of 59 English and 15 Scottish towns at the scale of 5 feet to the mile (1:1,056); produced plans of 26 English and two Welsh towns at the scale of 10 feet to the mile (1:528) in manuscript, 10 of which were subsequently resurveyed and published; published plans of eight English towns at 1:528; and published plans of 358 English, 26 Welsh and 55 Scottish towns at the scale of 10.56 feet to the mile (1:500).

Generally, only a single edition of most Ordnance Survey Town Plans was published. However, 25 plans were revised by the Ordnance Survey at national expense and reissued as a new edition and some underwent a process of ongoing un-noted piecemeal revision. Fifteen plans were revised by the Ordnance Survey at the expense of the town authorities between 1898 and 1908 at the 10.56 feet scale.

In other cases towns conducted their own revisionary surveys. Often revisions did not cover the whole of the area surveyed for the original plan and in some cases extended coverage by surveying newly-developed contiguous areas.

Ordnance Survey Town Plans note dates of survey, publication and revision on each sheet. Some provide the earliest large-scale coverage available, pre-dating the 25-inch plans and offering such additional information as the names of buildings and the use of commercial and industrial premises. With their often minute detail and ample space for annotation, these large-scale town plans are extremely useful. In all probability every industrial site in the area covered appears in these plans.

London was a special case. It was surveyed at its own expense by the Board of Ordnance, producing three sets of skeleton maps published by 1852 at scales of 6 and 12 inches and 5 feet to the mile. These maps were simply block plans showing only the outline of streets and bench marks. However, some areas were more fully surveyed at 10 feet to the mile, notably the riverside area upstream from the Houses of Parliament and the Kennington area. The skeleton plan provided a framework, trigonometrically correct and complete as far as street information was concerned, which could be used by other maps-makers as a foundation for the production of London maps of greater accuracy and detail then before.

Private map-makers surveyed and added topographical detail, most notably Edward Stanford whose 'Library Maps' to some extent filled the gap until the official detailed survey could be undertaken. The in-filling of the skeleton plans by the Ordnance Survey was surveyed between 1862 and 1872 and published at the 4 ft scale by 1876 on 326 sheets. Because of the expense of the 5-feet survey, London was never mapped at any larger scale. The series was completely revised between 1891 and 1895 and a new edition was published from 1893. The final sheet of '1894-1896 edition', or 'New Series' as it was called, appeared in 1898. A further revision took place between 1906 and 1907 extending coverage, for Land Registry purposes, to substantial areas of Kent, Middlesex and Surrey. Between 1906 and 1925 the Ordnance Survey revised large sections of the 5-feet map for the Land Registry and extended the survey into neighbouring built-up areas. However, only detail necessary for land registration was shown.

Despite the Ordnance Survey's impressive achievement in mapping so many towns at such large scales, the majority of towns were mapped by private surveyors commissioned by the local Board of Health to conform to the standards and instructions laid down by the central Board of Health. The Ordnance Survey was responsible for surveying only about 15% of towns covered under the 1848 Public Health Act due to its high tenders and charges. Thus, most large-scale town surveys prior to 1855 were conducted by private surveyors. Consequently, the resulting plans usually remained in manuscript in the hands of the town authorities, although some plans were eventually lithographed and made generally available.

Other useful maps generated by the public health movement are sometimes to be found in reports and enquiries concerning local conditions.

In addition to these officially produced and/or sponsored plans of towns, there is a vast number of privately produced plans dating from as early as the 16th century. Although quality and content are extremely variable, at best, large-scale plans, produced particularly in the 18th and 19th centuries, can be highly informative. Of course, privately produced town plans are snapshot views of the state of development and use at the time of survey. Whether produced for separate issue or inclusion in histories, topographies, atlases, directories, gazetteers, guidebooks and so on, such plans were rarely revised from fresh survey. Even when new editions of a plan were issued, with a new date of publication, only the most obvious changes were recorded.

The larger the scale of a plan, the more use it is likely to be. At smaller scales, only industry around the fringes of the towns tended to be recorded, and not always then depending on the plan's purpose. Even town plans inset on large-scale country maps produced between 1723 and 1827 may prove useful. Despite the fact that no privately produced, unofficial town plan ever provides a comprehensive record of industrial use, they may prove vital records for dense urban areas, making a significant contribution to the analysis of site use over time. Care needs to be taken in dating plans which appear in editions of books since the plan may be of significantly earlier date than the book edition.

FIRE INSURANCE PLANS

The most useful town cartographic records for checking site usage over time are the little known fire insurance plans. Although these were first produced by the fire insurance companies from the late 18th century, the most available and widely useful are those produced by the Goad Company from 1885. The central areas of the major towns and cities were surveyed by Goad within the next ten years. Goad's fire insurance plans relate particularly to warehousing and transport facilities, usually at the large scale of 1 inch to 40 feet (1:480). All sets of Goad's plans have a key plan at about 1 inch to 200 feet (1:2,400). Goad also produced selective plans of congested areas and of individual buildings and sites. These fire insurance plans offer the most comprehensive and detailed land use information for British urban centres from the late 19th century. Separate indexes list streets, buildings and firms.

Fire insurance plans were designed to give the underwriter, at a glance, all the information needed to determine the degree of fire risk and a fair premium. In addition to the nature of the structure itself and its use, it was essential to place it in a spatial context in order to understand the risks of fire developing in neighbouring premises and spreading to the insured property. Hence, in addition to much other information, Goad's plans precisely define the detailed use of individual buildings.

Insurance companies could not base risk assessment and premium calculation on out-of-date plans. Fire insurance plans were revised far more frequently than any other town plans. Goad's plans were never sold, being leased to subscribers on a long-term basis. On completion of a new survey, subscribers returned plans to Goad's for revision. Minor revisions, such as new ownership of a few plots, were made by pasting paper overlays or 'correction slips' on to the plans. However, major revisions might require preparation of a completely new sheet. Goad updated plans usually every five or six years, although larger, complex and more important centres might be re-surveyed every one or two years, creating a cycle of revision far more frequent than for any other type of map or plan. Goad's plans provide unrivalled data for increasingly complex areas badly recorded elsewhere. Where no accurate information is available of former usage and existing structures are not to be removed or samples collected for analysis upon a statistically acceptable grid fire insurance plans offer an unprecedented opportunity to learn the usage of every structure or division of a structure and those which adjoin.

CONCLUSION

The usefulness to the researcher of the different types of map or plan or other data will vary with the nature of the site and the historic period under consideration. For many, Ordnance Survey material will be sufficient and going beyond that material will be a rarity. The researcher must, however, always maintain an open mind and be prepared to use the most readily available material even if this means greater circumspection in its use.

Some of the statutory libraries have already noticed a significant increase in the numbers of researches involved in historic land use research. Some have expressed grave concern at the demands upon their resources. The British Museum Map Library is monitoring the position closely with a view to providing additional facilities. The more than 400 collections of historic documentation in county and local archives are under, perhaps, greatest threat in the longer term as there is no tradition of charging researchers the true economic cost of providing the research facility or in preparing copies of material from the archives in commercial volumes. Against an every tightening local authority purse string one wonders how much longer such facilities will be available.

Moderator's report on Session 1: planning, control, management and QA

N. A. TRENTER, Director, Sir William Halcrow & Partners Ltd

Introduction

It is just over fifty years ago that L F Cooling (1942) wrote in the introduction to his paper "Soil Mechanics and Site Exploration" that: "To the Civil Engineer there are few problems more likely to give rise to uncertainty than those relating to the design and construction of earthworks and foundations". He recognised that unless steps were taken to increase knowledge of ground and groundwater conditions at the site in question, then serious problems could occur during construction and incorrect assumptions could be made during design. It is likely that many, indeed most, practising engineers had realised the desirability of pre-knowledge of ground and groundwater conditions long before Cooling's paper, but his major contribution to the embryo British practice was to rationalise the requirements for information in a way which we now recognise as standard procedure for site investigations in this country.

Depth and extent of exploration were singled out by Cooling as a major element in site investigation and his approach was to combine theoretical considerations of extent of vertical and shear stresses beneath the foundation (noting the difference between their effects), with practical considerations such as the need to penetrate through a soft stratum to reach an underlying stronger one. Terzaghi (1939), in his James Forrest Lecture to the Institution of Civil Engineers, stated that: "The fundamental requirement for an adequate forecast of settlement is an intimate knowledge of the compressibility of all clay strata contained in the subsoil down to a depth equal to at least 1½ times the width of the area covered by a building". This good advice is maintained in UK site investigation practice to this day (eg Tomlinson 1986). As to the extent of the site investigation, Cooling noted that it was essential to investigate sufficiently to determine variation in strata thickness and properties, given that "differential movement of one part of the structure relative to another is an important consideration".

In the decades which followed Cooling's paper we can trace the development of the subject along two strands: investigatory technique and site investigation procedure. There is no doubt that investigatory technique gained enormously from progress made in Europe and the United States: The Standard Penetration Test was introduced into the UK from the USA in the early fifties (Tomlinson pers com) and the static cone penetration test from Holland at about that time (Harding 1949). Britain has some claim to the development of the vane test because whilst the idea was developed for site investigation use in Scandinavia (Carlson, 1948), according to Skempton (1949) the vane test was first suggested by the Building Research Station to the army in 1944 as a rapid method of estimating the bearing capacity of clays for tank manoeuvrability.

By the mid 1950s, a significant amount of British site investigation was being undertaken in London which proved a mixed blessing for UK practice because the London Clay presents very distinct problems in sampling and testing terms. More useful in these formative years would have been a mixed terrain of sands, clays, weak and hard rocks which would have provided an impetus to the use of rotary core drilling and in situ testing techniques. Cooling (1949) recognised the limitations of sampling procedures in British site investigation practice at the time, particularly sampling soft alluvial deposits and remarked upon continuous thin-walled sampling being developed in the United States by Hvorslev, permitting samples of large thickness of soft clays and silts to be recovered and inspected. Murdock (1949) emphasised the importance of trial pits in enabling small but important geological features to be identified. He observed that in conventional boreholes: "it was sometimes difficult to differentiate between isolated pockets and continuous strata, to decide between boulders and fissured rock when the fissures were filled with soft material and to detect thin layers of soil interleaved between harder material". The importance of small scale geological features on the performance of structures was, of course, underlined by Rowe in his Rankine lecture (1972).

Site investigation procedures including their planning and control, were well advanced by the end of the 1940s, as is evidenced by Harding (1949). His paper entitled "Site investigations including boring and other methods of sub-surface exploration" describes commonly used methods and equipment and the description of the tasks to be allocated to the boring foreman probably forms the basis of the specification for the preliminary or "driller's log" to this day. Harding's paper was published at a time when the first Code of Practice on Site Investigation had been issued in draft form for comment and was designed to amplify certain points made in it. A description is given of the common methods of boring and sampling together with rigs for land and over-water exploration and static cone and vane testing are mentioned.

Another major contribution to UK site investigation practice came in 1954 with Tomlinson's paper to the Institution of Civil Engineers on "Site exploration for maritime and river works". He placed site investigation in an historical perspective by relating how Telford investigated the foundations for the Glasgow Bridge by sinking a 72ft deep cylinder below the river bed and, from the information obtained, was able to foresee the possibility of a "blow" in a sheet pile cofferdam and made plans in advance against this risk. The emphasis of the paper was on over-water investigations with a detailed list of the preliminary data required for such works which would form an excellent basis for a specification today; he also recounted the use of staging and divers and procedures for setting out. Accounts were also given of the common investigation techniques available at that time, including the SPT with correlations between "N" value and unconfined compressive strength given for alluvial silts and clays.

The Papers

Given the concentration of earlier British site investigation resources in the capital, for comparative purposes it is useful to have the paper by **Pickles and Everton** to this Conference describing aspects of the site investigation for the Waterloo International Terminal (WIT). The terminal comprises a two storey structure with a basement which adjoins the existing BR station and is founded on a 170m by 60m reinforced concrete raft. Construction resulted in a net unloading of about $100kN/m^2$ which would have influenced the Bakerloo Line tunnels running beneath. Dimensions are not given but it would appear that the shallowest tunnel crown was some 7m below raft invert. Investigations were centred at five different locations, with rotary core drilling, percussion boring, thin wall 100mm diameter sampling, piezocone soundings, Marchetti Dilatometer and self boring pressuremeter profiles at each location. Further boreholes were sunk elsewhere to determine the fill and strata succession and there was probing using the flow-through sampler and the dynamic penetrometer to determine the Thames Gravel/London Clay interface. Additional piezocone and dilatometer profiles were made at set radial distances from the shallowest of the two running tunnels. Laboratory testing included "filter paper" suction tests, triaxial stress path "small strain" testing and stress controlled triaxial K_o determinations.

Beneath the fill, the London clay was proved by continuous coring to have significant sand and silt seams below about 15m depth at all five locations. Based on the results of filter paper suction tests from three of the five locations, Pickles and Everton plot mean effective stress "p'" v depth (their Figure 4) and show, as would be expected, that the mean effective stresses calculated from the results of tests made on rotary core samples are much lower than results obtained for thin wall samples retrieved from percussion boreholes. Mean effective stresses obtained from the thin wall samples are themselves significantly higher than those recorded for London Clay by Burland et al (1979), by over 100kN/m^2 at the surface, although reducing with depth. It would be interesting to know if all the suction test data available to the authors were used in plotting Figure 4 and, if not, on what criteria results were discarded.

Accurate measurement of the undrained shear strength of a stiff fissured clay like the London Clay is always of interest and the authors provide shear strength v depth plots for "c_u" obtained several different ways. Unconsolidated undrained tests on thin wall samples show scatter but, when allowance is made for the low results due to reduction in p' produced by sand and silt seams, the resulting strength v depth profile is similar to the one obtained using the results of triaxial tests consolidated to the estimated mean in situ stresses. The authors conclude that the latter method is the most appropriate for establishing a shear strength v depth profile. Shear strength values obtained from Marchetti dilatometer and self boring pressuremeter equipment were significantly higher than the corresponding triaxial test data. The authors' explanation for these differences would be helpful.

For three decades after Cooling's 1942 paper, geologists were to provide a valuable supporting role to the engineer in site investigation work, but in the UK it was not until the 1970s that they were to take the lead in developing new systems and procedures which were to make a major contribution to site investigation. Work reported by the Transportation and Road Research Laboratory and the Geological Society reports, including engineering geology in route planning design and construction (1972), the symposium on engineering geological mapping for planning design and construction in civil engineering (1979) and particularly the Working Party Report on land surface evaluation for engineering purposes (1982) were to yield an impetus for a new branch of the practice of which **Waller and Phipps'** paper to this Conference is a good example. The 108km Channel Tunnel Rail Link (CTRL) from Folkestone to St Pancras, London, is a linear structure covering a wide range of geomorphological and geological features with engineering structures ranging from tunnels to embankments and cuttings. Being a railway, the structures have to be engineered to significantly higher tolerances in settlement terms than would be the case with a highway. A £2.4M Phase I package of studies having been completed, the authors' strategy was to develop "Ground Models" both for future design and as an aid to planning a cost-effective £3.6M Phase II investigation.

Studies for all linear structures involve a synthesis of local data such as previous site investigations, well records, memoirs and similar, and the approach adopted by Waller and Phipps was to find a method for accurate extrapolation of these data, together with their storage and referencing. Fourteen "Terrain Systems" were identified (London

157

Basin Tertiaries, Tertiary Outliers, etc) on the basis of desk studies, augmented in some cases by field mapping and aerial photographic interpretation of colour stereographic pairs of "Type Sites". The route was also subdivided into a number of named "Geosegments" each having particular construction features and each belonging to one or more Terrain Units. This enabled 1:5000 scale plans and sections to be prepared upon which pertinent features within the Geosegments, such as hydrogeology and geo-hazards, were recorded.

Three dimensional ground models were prepared for each Terrain System on the basis of the Phase I data and the Phase II material will enable composite plans along the route centre-line at 1:5000 horizontal and 1:500 vertical scales to be produced, together with borehole logs and geo-hazard information. Moreover computer access to the more than 4000 relevant data items is available through databases, with information coded by Geosegment or by other location criteria. The work described by Waller and Phipps is a considerable response to the major challenge presented by the CTRL. No references were given in the paper, but it clearly draws upon the work of others. It would be interesting to know how much was spent on this part of the investigation, as a percentage of the total £6.0m apparently devoted to Phases I and II of the project. It would also be useful to know just how helpful the output will be to the engineer user, over and above the OS and BGS maps and memoirs which would have been the only material available to the early practitioners.

The contribution that site investigation can make to reducing total main works cost was well understood fifty years ago and it remains a major pre-occupation today. The publication "Inadequate Site Investigation" by the Institution of Civil Engineers (1993) highlighted deficient areas and underlined the need for a more effective approach to site investigation practice if the final or "out-turn" cost were to be kept within forecast. As a contribution to this challenge, **Winter, Mathieson and McMillan** in their paper to this Conference propose a knowledge based system for ground investigation in soil and rock for highway schemes, the other principal form of linear structure. The authors identify the objectives of a highway investigation as being the need to "provide information on conditions likely to be encountered during construction and to acquire data which can be used in design", to which one might possibly add the need to provide data for use during subsequent highway maintenance. They describe how deficiencies in site investigations are related not to shortcomings in the state-of-the-art, but to process planning and propose that an appropriately structured system would benefit.

Their system sub-divides a site investigation into preliminary, main and additional phases, becoming progressively less qualitative and more quantitative. The structure is illustrated in the authors' Figure 1, where it is puzzling to find that the additional ground investigation phase comprises no more than a review of all previous work; one would have thought some additional field or laboratory activity would have been necessary at this stage, which is presumably designed to accommodate route changes or to enhance knowledge in suspect areas. Procedures for the main investigation phase appear logical and appropriate but it would be useful if the authors could expand on their remark that "special studies are considered to be atypical of routine investigations", since it is by no means clear how, by their very definition, site investigations may ever be termed "routine" and, if so, why special studies are an inappropriate device. It would also be helpful if the authors would expand on their perception of the need or otherwise for instrumentation, such as piezometers, inclinometers and other such equipment in investigations for highways.

Knowledge based systems such as the one described by Winter et al could certainly not have been foreseen fifty years ago and the one the authors describe is clearly a major development. Whether or not it is a major advance will depend in part on the answer to the question: can any such system remove the requirement for training and experience on the part of the individual using it? The more qualified the individual the less need there is for the system; the less qualified the individual, the more comprehensive will the system need to be and the more relevant the information it has to provide. There is a fine line to be drawn and no doubt cost will have a major impact on the level of resources which can be directed to the system's exploitation before it can be gainfully employed by the industry.

Practical considerations in managing site investigation to ISO 9000 are addressed by **Ball and Whittlestone** in their paper to this Conference. They recognise that ISO 9000 is largely written for the manufacturing industry and, unlike the previously described paper, their contribution recognises that site investigations have to be site specific, even if the **methods** employed can often be routine. Commencing with an interpretation of quality management, they move on to outline their approach both to quality management and to site investigation.

The role of the client, consultant and contractor is defined and it is suggested that the consultant's role in site investigation is relatively easy to control "using some form of peer review". In contrast, the authors maintain that the contractor's product often depends on factors not under his control, in particular the ground itself being investigated. Here it would have been useful if the authors had stressed that the consultant's traditional role, to enforce the specification, benefits all parties to the contract and that technical audits by the consultant of the contractor are fundamental to modern site investigation management, and are as important as day-to-day supervision.

In the section headed "Contract Review" the authors correctly observe that the contract review is not simply a once-and-for-all operation, timed to occur between letting the tender and mobilisation on site, but is a procedure which progresses throughout the duration of the project. Moreover, the limitations of Quality Assurance are well described by Ball and Whittlestone in an example showing how logging samples to a specification other than BS5930 could satisfy quality procedures but could also yield a second rate borehole record.

After describing their management approach to non-conformances, Ball and Whittlestone propose "A Way Forward". This requires that "Quality" should begin with the client, that a contractor's technical expertise should be recognised, and that "teamwork is the new order of the day". Few would dispute the aims but more insight as to how they could be achieved in practical terms would have been useful. The most surprising omission in the authors' review is the absence of any significant discussion on health and safety matters, particularly the effects that implementation of CDM (Construction (Design and Management) regulations) will have on site investigation procedures in the field.

Another approach to site investigation management is put forward by **Woodward** who realises that "Quality", ie adherence to the specification is not achieved by chance and that a uniform Quality Management System (QMS) embodying "self-certification" should be adopted industry wide. Pointing out that the individual site investigation contractor has the staff, equipment and management procedures all under one roof, the author believes that site investigation contractors are well placed to adopt a "single responsibility approach". He notes that there are two methods of procurement under the ICE's suggested procedures (ICE 1993): System 1 where the engineer arranges the design of a site investigation and the supervision of a contractor for the fieldwork, testing and (normally) factual reporting, and System 2 where a single contractor undertakes the design and execution of the contract, including the (normally) interpretive report.

Although most site investigations are arranged under System 1, using the ICE conditions and specifications, Woodward considers that with an acceptable contract form such as that proposed by Uff and Clayton (1986, 1991) or the New Engineering Contract (ICE 1993), System 2 would offer distinct advantages, not least because of what he considers to be quicker response times to unexpected conditions and less document handling, Woodward follows Barber (1992) in proposing a clause for insertion into the ICE conditions (1983) specifically requiring that the contractor should operate a QMS complying with BS5750 including a quality manual and quality plan, with laboratory testing performed by a NAMAS accredited laboratory; the engineer would be required to audit the scheme as second party assessor.

The author recognises that surveillance is an area of difficulty under both Systems 1 and 2, but believes that System 2 is superior because intervention would only be necessary when the QMS was not being correctly operated. This is, of course, precisely the occasion when there would be intervention under System 1 arrangements, but Woodward appears to believe that System 2 is to be preferred because the contractor would be responsible for the "comprehensive design" of the investigation. As for the design of the investigation under System 1, Woodward asserts that System 1 can only be rationalised if the specification were re-written to eliminate approval clauses and fully to define acceptance criteria, which would be costly. Perhaps so, but there is no obvious difference between the contractor preparing the QMS for an investigation he designs, and the engineer writing the specification for an investigation which he designs. The cost is there in both cases, to be borne by the client ultimately.

It would be interesting to speculate on what the founding fathers of site investigation in this country, and indeed elsewhere, would think fifty years on, of the emphasis being placed on procedures rather than product. Quality Management was always of concern to the early workers and it is submitted that quality assurance has its part to play, but only a part, within the overall Quality Management framework. There should be at least as much emphasis in writing the correct specification as in documenting procedures for enforcing it.

After so much discussion on "Quality" in site investigations it is refreshing to turn to the paper by **Perry, Snowdon and Wilson** which actually describes how to do them or, more specifically in this case, how to undertake site investigations for lime stabilised capping in highway works. Following a brief review of the processes of lime modification and lime stabilisation and the different effects produced by quick and hydrated (slaked) lime, the authors discuss the requirements for a Preliminary Sources Study (PSS), the purpose of which is to ensure that lime stabilised capping is considered as an option for the main works from an early stage. The authors recognise that stabilised capping maximises use of on-site materials, thus saving haulage and tipping, and avoids further environmental pressures introduced by winning, transporting and placing granular materials, which are often the preferred alternative.

Guidance is given on investigation methods and emphasis is placed on the trial pit or trial trench as a tool permitting accurate portrayal of soil conditions, thus echoing Murdock's advice (loc cit) more than forty years ago. There is a significant discussion on the problems presented by sulphate and sulphide minerals because of the potential for heave which they pose. The distribution of sulphates for their materials is described as peaking at 2 or 3m depth, thereafter decreasing markedly; sulphides are suggested as being limited over the depth range to 2 or 3m, because of their oxidation to sulphates, but increasing in frequency below this depth. Thus soils excavated from depths greater than 3m, whilst containing little identifiable sulphate, could still present the opportunity for heave if they were used as lime stabilised capping, because of the potential for included pyrite to oxidise to sulphate at some stage during construction.

The authors note that the longer testing is delayed, the more opportunity there is for pyrite oxidation thus presenting, the authors suggest, a false picture of the amount of sulphates present in the soil at the time of sampling. This may be, but surely what is of interest to the engineer, when considering the lime stabilised capping option, is the total sulphate (sulphate plus potential sulphate from sulphide oxidation) at the time of capping construction. This is recognised in their discussion on laboratory testing where they recommend use of the test for total sulphur content as given in BS1047, with a conversion factor to bring the resultant sulphur content to sulphate content for comparison with the results of tests for total sulphate conducted to BS1377. There are two points which could perhaps be added: First, pyrite can occur finely disseminated and can only be recognised using electron microscopy; absence of pyrite in the hand specimen does not necessarily indicate absence of pyrite in the stratum. Second, when a cut is made, the soil below grade effectively commences a new weathering stage and attaining equilibrium could produce further chemical changes. As a consequence, sulphates could be brought into contact with the underside of the capping layer by moisture movements.

This most useful paper will greatly assist geotechnical practitioners in highway design as they grapple with costs and minimise pressures on the environment. It is a paper which would have been applauded by workers in the subject fifty years ago in that it recognises a need, produces clear recommendations and is prepared to relate investigation technique to construction procedures.

British site investigation has invested relatively little effort into statistical or probabilistic techniques. This was recognised recently by the Institution of Civil Engineers which held a Conference on Risk and Reliability in Civil Engineering in 1993. In their paper to this Conference, **Farmer, Xiao and Challinor** investigate spatial dependence in site investigation design, drawing upon two examples to illustrate their argument that spatial dependence (geostatistics) can both improve the quality of information obtained from site investigations and aid in planning field work in a more cost-effective way.

The first example is drawn from previously published work and attempts to demonstrate how the technique can be employed to determine the optimum borehole spacing. Factors such as geological structure and landforms are neglected and the problem is resolved by the use of an experimental semi-variogram (the authors' Figure 2). Figure 2(a) is not particularly convincing since a horizontal line would provide an almost equally good fit to the data and, whilst the curve is a much better fit for Figure 2(c), confidence limits are not given. The second example is also drawn from previously published work and attempts to show how a spherical semi-variogram may be used to assess Rock Quality Designation (RQD) values in two different rock types. No borehole records or core recovery data (RQD, TCR, AFS and similar) are given so it is not possible to judge the success of the technique in this case. It would be interesting to know why the authors selected an exponential variogram in their first example and a spherical variogram for the second; what criteria do they recommend? Their Figure 4 gives results of actual and predicted RQDs for "ordinary" and "soft" kriging techniques. Scatter is large and the regression lines shown would not appear correct on the basis of the data displayed.

There is undoubtedly room for geostatistical techniques in site investigations and there is no doubt that practice, at least in the UK, lags mining procedures. However, it is not clear that the authors' paper will have convinced many of the technique's advantages, at least on the basis of the examples employed.

Possible Future Developments

In the fifty years since Cooling's 1942 paper, the overall aims of the industry remain the same, although the industry itself has been transformed, due in part to the availability of relatively inexpensive and miniaturised control and record systems. These systems permitted developments in instrumentation and a variety of laboratory and in situ tests. At the same time, the growth in the number and complexity of analytical techniques and the availability of computers to service them has created specialist organisations, often attached to universities, and has put further pressure on the industry for still more accurate data.

Planning and control of site investigations are no longer in the hands of a few practitioners who made up with their experience for the deficiencies in the techniques available to them. Modern site investigation is a matrix of complementary and supplementary activities. Aids to securing better planning and control in such circumstances include:

- new developments in contractual relationships;

- electronic data transfer and computer databases;

- risk analysis techniques;

- quality management and various forms of accreditation.

Adoption of different forms of contract, such as the New Engineering Contract, is currently in vogue. Some may be more convenient for particular types of contract, such as design and build, as against the traditional consultant and contractor relationships, but ultimately it has to be recognised that entering into a contract can be a risky business and the different forms allocate risk in different ways. For example, removing "Clause 12" (unforeseen ground conditions in the ICE 5th Edition) from a set of contract documents does not make the problem of difficult ground go away; it simply shifts the risk, in this instance to the contractor. No doubt the contractor will in response cover the risk by increasing his rates, so it is doubtful if the employer gains financially in the long term.

Handling large amounts of field and laboratory data is both time consuming and expensive and the advent of the AGS electronic data system and the various forms of databases is a most significant step in controlling site investigations, with implications for both the accuracy and the presentational standard of the final report. Some schemes permit an audit trail and can therefore be manipulated with the Quality Assurance system for a particular job. Accreditation such as NAMAS (for laboratory testing) and the BDA (for drilling and boring) can only lead to improvements but the case for emphasis upon Quality Assurance as against Quality Control within the whole quality management scheme is perhaps more difficult to argue. Quality Assurance places the emphasis on procedure rather than upon content and arguably it is time to take stock of the role of "Quality" in site investigation and where the emphasis should be placed.

Whatever contractual arrangements are employed it is likely that site investigations in future, especially the larger projects, will be performed by a collection of organisations including perhaps a consultant, a contractor for general field work and testing, a specialist contractor (possibly a university) for other than routine testing, another specialist consultant for advanced analytical techniques, all under a project management team which could be from the employer's own organisation. In such a scheme lines of communication could become dangerously long and the ownership of risk blurred between the various parties. However, with strong management it will work. The traditional employer : consultant : contractor relationship will still survive, but will be called upon more and more to justify its existence.

REFERENCES

Barber, J N. Quality Management in Construction - Contractual Aspects, CIRIA Special Publication 84. London (1992)

Cooling, L F. Soil Mechanics and Site Exploration. J Instn Civ Engrs, Vol 17, No 5, pp32 to 61. March 1942

Cooling, L F. Discussion on Harding, J H B. J Instn Civ Engrs, Vol 32, No 6, pp 141 to 142. April 1949

Cooling, L F and Skempton, A W. A Laboratory Study of London Clay. J Instn Civ Engrs, Vol 17, No 3, pp 251 to 276. Jan 1942

Carlson, L. Determination in situ of the Shear Strength of Undisturbed Clay by means of a Rotating Auger. Proc 2nd Int Conf, Soil Mechanics, Vol 1, p265. 1948

Engineering Geological mapping for Planning, Design and Construction in Civil Engineering, Q J Eng Geology, Vol 12, No 3. 1979

Harding, H J B. Site investigations including Boring and other Methods of Sub-surface Exploration. J Instn, Civ Engrs, Vol 32, No 6, pp 111 to 137. April 1949

Institution of Civil Engineers, Conditions of Contract for Ground Investigations. London. 1983

Institution of Civil Engineers, New Engineering Contract. London. 1993

Murdock, L J. Discussion on Harding, H J B. J Instn Civ Engrs, Vol 32, No 6, pp 147 to 148. April 1949

Route Planning, Design and Construction. Q J Eng Geology, Vol 5, Nos 1 and 2. 1972

Rowe, P W. The Relevance of Soil Fabric to Site Investigation Practice. Geotechnique, Vol 22, No 2, pp 193 to 300. 1972

Skempton, A W. Discussion on Harding H J B. J Instn Civ Engrs, Vol 32, No 6, pp 151 to 155. April 1949

Terzaghi, K von. James Forrest Lecture, A New Chapter in Engineering Science. J Instn Civ Engrs, Vol 12, No 7, pp 105 to 142. May 1939

Tomlinson, M J. Site Exploration for Maritime and River Works. J Instn Civ Engrs, Vol 3, No 2, pp 225 to 272. June 1954

Tomlinson, M J. Foundation Design and Construction. Longman Scientific and Technical. 1986

Uff J F and Clayton C R I. Recommendations for the procurement of ground investigation. CIRIA Special Publication 45. London. 1986

Uff J F and Clayton C R I. Role and responsibility in site investigation. CIRIA Special Publication 73. London. 1991

Working Party Report on Land Surface Evaluation for Engineering Purposes. Q J Eng Geology, Vol 15, No 4, 1982

Discussion on Session 1: planning, control, management and QA

Discussion reviewer J. COUTTS
Furgro Ltd

Mr. S.J. Everton, Sir Alexander Gibb & Partners Ltd.

The Moderator noted that the undrained strengths and one dimensional compressibilities at Waterloo were not significantly different from those determined 40 years previously at the Shell Centre (Measor and Williams, 1962), suggesting little advancement in this period. The authors believe that perhaps we should not be expecting such increases or decreases in parameters - perhaps the real advance may be in the reliability we can place in the parameters, particularly with the "simple" but little understood undrained strength.

In fact the Shell Centre data formed an important part of the interpretation of the ground investigation data from Waterloo. We believe, however, that the Moderator has missed the main aim of the investigation. This was to attempt to accurately determine the expected movements of the underground tunnels beneath the raft. Therefore the advanced techniques were directed towards measuring the in situ stress state and the stiffness of the strata in a variety of loading and unloading cases. The techniques adopted allowed us to do this and provide a level of redundancy at the same time.

The Moderator has touched upon a few points of discussion in his report, and in Dr. Pickles' absence I wish to briefly address these.

The points raised in his written report were the following:

    I)   the depth of the London Underground tunnels below the foundation raft:

    ii)   the selection of suction measurements for interpretation purposes; and

    iii)   the higher undrained strengths measured by the self boring pressuremeter.

The depth of soil between the underside of the raft and the crown of the Bakerloo station tunnel is approximately 5.8 metres. The monitoring which has been performed to date has shown these tunnels to have remained within the expected and allowable range of movements, thereby justifying the scale of the ground investigation works. It is worth noting, however, that the ground investigation was performed for a much larger development than that finally constructed. The original scheme, which was to include a commercial development above the WIT, called for a double level basement for car parking. This would have put the foundation raft only approximately 2 metres above the Bakerloo tunnel.

With regard to the use of suction measurements to determine the in situ stress state, this is dealt with in more detail by Hight et al (1993). However, the selection of representative results is an important factor in the interpretation of suction measurements. At Waterloo, suctions were measured on both thin wall and rotary samples, but only the thin wall results were actually used in determining K_o. The thin wall sample tubes were rigorously examined to find any sign of damage or distortion, which was common and reduced the number of available samples for testing. Of the filter paper tests themselves, any set of three results from the same sample which showed large scatter was discarded, as was any which on completion was seen to contain sand or silt seams which had not been apparent on extrusion. Finally, results from samples which exhibited atypical stress paths during undrained shear were also discarded on the basis of being unrepresentative.

The undrained strengths from the self boring pressuremeter (SBP) were shown to be higher than those determined by triaxial tests, and also those from the empirical relationships with SPT "N" value and piezocone resistance. It is often questioned whether the triaxial results do give a "truer" representation of strength than the SBP. Notwithstanding this, from recent experience the ""apparent overread" is common in London Clay. The results presented in the paper are the "peak" strengths determined form the pressure-log volumetric strain data, due to the lift off pressures at Waterloo often being not easily discriminated; in addition the difference between the individual arms was large. Both of these phenomena are not particularly common in tests in London Clay, and may be the reason for unrepresentative results. This method of analysis is considered by Mair and Wood (1987) to be "unreliable" in any case. Other factors such as strain rate effects (Prapaharen et al,1989), end effects and local drainage are also generally considered to influence the results. These aspects are dealt with in more detail by Clarke and Sadeeq, and by Shuttle and Jefferies later in the Conference.

REFERENCES

HIGHT D.W., HIGGINS K.G., JARDINE R.J., POTTS D.M., PICKLES A.R., DE MOORE E.K., NYIRENDA Z.M. (1993) "*Predicted and measured tunnel distortions associated with construction of Waterloo International Terminal*"
Proceedings of Wroth Memorial Symposium, Oxford.

MAIR RJ, WOOD DM (1987) "*Pressuremeter testing: methods and interpretation*" Butterworths.

MEASOR EO, WILLLIAMS GMJ (1962) "*Features in the design and construction of the Shell Centre, London*"
Proc. I.C.E. 21 475-502

Mr. A.M. Waller, Union Railways

I should like to thank the Moderator for this considered comments on the paper and offer, in respect to the planning aspects of site investigations, some clarification of the points raised:-

- on the terms and techniques used.

- on the cost of the various studies; and

- how the information can be used in the engineering design and the construction process.

The 11km route alignment of the Channel Tunnel Rail Link (CTRL) follows existing transport corridors between St. Pancras, London and Cheriton, Folkestone, linked by sections of tunnels under the Thames and the North Downs

The chosen route traverses three distinct terrains, the Thames Basin, the North Downs and the northern limb of the Weald. These three areas differ in their geological materials, structural developments and their past and contemporary environmental processes.

These geological and geomorphological features are identified as making up 14 distinct landform assemblages or Terrain systems, (see Figure 3 of the paper).

The use of physical characteristics to describe the ground model of an area has been used at a variety of scales for a variety of applications, for instance, Cooke and Doonkamp 1990, Goudie 1981, Christian and Stewart 1952 and King 1987.

However, the main difference between many previous studies and that for the CTRL, is the linear area of study and that the route transects important terrain units over very short distances.

To establish which information is significant, each terrain unit was subdivided into terrain facets, though observation, reconnaissance mapping, geomorphological mapping or air photo interpretation. For each facet, a description is given of the near-surface block of ground, i.e. a ground model.

The development of a ground model for each of these facets involves the consideration and description of the model in terms which can be semi-quantified and assessed for accuracy and certainty, (even if this is only by judgement rather than some numerical technique).

These terms are:-

- the position and shape of the boundaries to the facets;
- the depth and variability in the strata or groundwater surfaces in the ground profile;
- lithological variations laterally and vertically within the main strata units;
- and the presence or absence of geohazards, e.g. landslides, deneholes and cambering.

The facets provide at the engineering scale of the High Speed Rail Link, a means to interpret and communicate a wide range of significant information. This has included most recently, the results of a recent phase of ground investigation which prove the ground models predicted by the facets in the majority of cases.

The resulting ground models provide a basis for prediction. They tie together a number of dispersed point data sets within a defined framework. The facets were used directly in the definition of objectives and strategy of the ground investigations.

The consideration of uncertainty in each part in this ground model formulation, (even if it is only in descriptive terms, can be communicated to the engineer and contractor as part of the assumptions for any future geotechnical modelling.

Costs

The costs of these geomorphological field and desk studies on the CTRL project was c.£37,500, or c 0.6% of the total geotechnical works. The total for all other specialist geological, hydrogeological and other specialist studies was £140,000. The total expenditure for all desk and field studies was £300,000. Total site investigation costs were of the order of £6m.

Conclusion

In conclusion, I should like to offer the following points to the discussion.

The Authors believe to successfully communicate information about the possible ground conditions, relevant information must be structured and described at a scale, and in a way to directly reflect it's accuracies and uncertainties, a method has been demonstrated in this paper.

This approach of terrain mapping has been applied so far in assisting the ground investigation design process with clearer objectives and strategy options for assessing ground conditions. The findings so far mostly agree with the ground model predictions.

As the project develops, the approach of terrain facet mapping, we believe, will offer a means to communicate the remaining uncertainties associated with the ground conditions, at the scale of the engineering works associated with the design and construction processes.

REFERENCES

CHRISTIAN C.S. and STEWART G.A, 1952, *Summary of General Report on survey of Katherine-Darwin Region* (CSIRO, Australia) Land Research Series I.

KING R.B, 1987, *Review of geomorphic description and classification in land resource surveys* in GARDINER V, (ed) *International Geomorphology* 1986 Part II, (Wiley, Chichester), p384-403.

COOKE R.U. and DOOKNKAMP J.C. 1990, *Geomorphology in Environmental Management*, (Clarendon Press, Oxford).

GOODIE A. 1981 (ed) *Geomorphological Techniques*, (George Allen and Unwin, London).

UNION RAILWAYS

TABLE OF GEOSEGMENTS, TERRAIN SYSTEMS AND TERRAIN UNITS

CHANNEL TUNNEL RAIL LINK		110 km in length		
TYPE of route division	*No.*	*STATISTICS*		
GEOSEGMENTS e.g. LCH (Based on geology, terrain and infrastructure)	29	Maximum length = 7.85 km (NLL - North London Line)	Minimum Length = 0.4 km (STP - St.. Pancras)	Average Length = 3.7 km
TERRAIN SYSTEMS e.g. VI - Chalk Dip Slopes (Based on geological materials, geological structure and environmental process)	14	Maximum number of times one terrain type is repeated = 6 (i.e. terrain system XI - Dissected Folkestone Beds Surface)	-	Each terrain type is repeated on average 3 times on the CTRL route.
TERRAIN UNITS e.g. XIA - Maidstone/Ashford Railway crossing to A20 crossing. (Specified extent of a terrain type traversed by the CTRL)	40	Maximum length = 11.43 km (IIB - Thames Floodplain)	Minimum length = c0.2 km (IXA - River Medway Floodplain)	Average Length - 2.7 km
	Terrain Units per geosegment	*No. of Geosegments*	*Examples*	
	>1	8	NAV - Nashenden Valley is included in unit VID	
	1	13	XIA - HOL - Hollingbourne	
	2	3	HAR - Harietsham includes units XIVA and IXB	
	3	3	MVC - Medway Valley Crossing includes units B,IXA and VC	
	4	1	NDC - North Downs Crossing includes VIO, VIE, VIHA and VIIIA	
		Total 29		
TERRAIN FACETS (Specified subdivision of a terrain unit traversed by the CTRL)	c.398 (for 70 km of the surface route of the CTRL in Kent)	Average Length = 0.45 km	Maximum Length = 1.0 km (XIB(XI)-Hythe Beds Footslopes)	Minimum Length = c0.05 km (IXA(V) - Sandgate Bed Valley)
	Number of Facets per Terrain System	*No. of Terrain Systems*	*Examples*	
	>5	8	XIII - Atherfield Clay Dissected Surface	
	5	2	X - Gault Clay Vale	
	6	3	VIII - Chalk Forelands	
	<6	1	VI - Chalk Dip Slopes	
		Total 14		

Mr Bill Rankin, Mott MacDonald

Preliminary Ground Investigations for the Channel Tunnel Rail Link

The route spans some 108 kilometres (approximately 68 miles) between St Pancras and Cheriton. The route runs to the north of the Thames through Stratford and Barking, crossing the river at Dartford and then crossing the North Downs before descending onto the Gault Clay and Lower Greensand deposits as it approaches the Channel Tunnel.

The project will cost around £2,500 million and will form a strategic link between the Channel Tunnel, which is now in use, and London thus integrating with the U.K. main line network. This is probably the only major project on the horizon in the U.K. at the present time.

The average construction route cost per kilometre may be summarised as follows:-

- London Tunnels   - in the order of £30 million per kilometre
- The Thames tunnels and approaches - around £40 million per kilometre
- Surface Sections   - around £14 million per kilometre

I joined the Union Railways' team in the Autumn of 1993 as Geotechnical Manager when outline design for sections of the route had been assigned to a number of Consultants and desk studies were underway. My task was to ensure the first two phases of ground investigation were carried out. The total cost of these two phases to date has been of the order of £10 million. The broad objective of Phases 1 and 2 was to achieve around half of all the ground investigations along the route complete by March 1995, with further site specific ground investigation to be carried out hand in hand with the detailed design in due course. Phase 3 is now ongoing and its extent is yet to be decided, as is the split between what will be commissioned by the Government and what is to be carried out by the Private Sector. The project is presently going through the Parliamentary Hybrid Bill process and the bidding Consortia wishing to design, build and operate the link submitted their bids on 14th March 1995. It is hoped that the project will have an owner by the end of this year.

The route contains 25 kilometres of tunnels at various locations, the main length being between St Pancras and Barking. The other key elements are 13 kilometres of embankment over soft ground, 8 kilometres of which are over the Thames Marshes. There are 7 kilometres of cuttings on the surface route, half of which are in the Gault Clay.

A significant element of the ground investigation was concerned with potentially contaminated land. There are some 150 sites on or close to the route which would influence the works design. The amount of information from desk studies varies from site to site. Of these 150 sites, 40 were identified as warranting investigation as part of the Phase 1 and 2 ground investigations. This probably represents the largest contaminated land investigation in the U.K. to date.

Our paper is intended to demonstrate the team effort that has existed between the Client, his Consultants and the ground investigation Contractors. The number of people involved on the consultancy aspects peaked at 40, distributed between the various Consultants and Union Railways staff. Half of the team were office based and half were site based supervising the £6 million of fieldwork performed in the last twelve months.

The office activities included geomorphological studies and the preparation of development study reports. These were presented in a single volume which draws together all of the information presently available on a particular topic. These included general earthworks, embankments over soft ground, seismicity and chalk earthworks. It is anticipated that this will give the designers, and the eventually successful Consortia, a flying start for their design.

The ground investigation was weighted in accordance with the design requirements of the route and the average cost as outlined above. The tunnels represent approximately 25% of the total Civils cost and the cost of the investigations for the tunnels, i.e. the London tunnels, the North Downs Tunnel and the Thames tunnels and approaches, received half (£3 million) of the ground investigation spend in the last 12 months.

The total percentage of ground investigation to Civils cost could eventually lie between 1.5%-2%. To date the spend has been between 0.5% to 0.6%

Mr. C.H. Adam, Fugro Scotland Limited

The moderator raised the question of the potential cost savings associated with the use of a jack-up platform compared with a floating barge.

The main cost items in any marine site investigation (SI), apart from possible the mobilisation charge, is often the delay due to adverse weather and sea conditions. The method of procurement of marine SI allows for the cost of weather/sea delays in a variety of ways. Unfortunately, the most common method of measurement is to quantify a fixed number of hours or days for the contractor to price. This procedure favours the cheapest marine spread which almost invariably is not the most appropriate for the prevailing site conditions.

An alternative method of measurement is to quote a lump sum for all weather/sea delays. This enables the contractor offering a more productive unit such as the jack-up platform with a dedicated marine drilling spread to submit a more competitive bid since the delays associated with using a jack-up platform are generally much lower and easier to estimate than a floating spread.

A further alternative at the tender evaluation stage is to apply different quantities of time to the different types of marine spread being offered.

To enable a lump sum to be quoted for weather/sea delays, my own company uses the great wealth of experience and in-house records from previous projects to estimate weather risk. Statistical data on weather and sea conditions can be obtained from various sources, but I believe that our own in-house data is more reliable.

In my view when all factors are fully evaluated, particularly weather delays, a nearshore marine investigation carried out using a jack-up platform is invariably more cost effective and provides better quality data compared with floating craft.

Dr. M.G. Winter, TRL, Scotland

The Moderator raises a number of points regarding our paper. These points are addressed in turn below.

First, he queries the structure of the additional ground investigation phase. He is quite correct to presume that further field or laboratory activity may be necessary to accommodate route changes or to enhance knowledge in suspect areas. Indeed, we state in our paper that further work may be necessary because of route alteration, changes in structures or their design, to remove uncertainties or to confirm a proposed course of action. What may not be so clear is that after reviewing previous work the user may then select any component from either the preliminary or main ground investigation phases as an activity within the additional ground investigation phase.

Second, he requests expansion on "special studies", a component within the main ground investigation phase. The description of investigations as "routine" is perhaps, unfortunate. We certainly do not imply, either by inclusion or omission, that special studies are inappropriate. In fact we state that this category has been included specifically to cater for circumstances or conditions which might be described as exceptional, or atypical. For example, we would include trial embankments and excavations in this category.

Third, he asks for comments on the need or otherwise for instrumentation, such as piezometers, inclinometers and other such equipment in investigations for highways. Clearly the use, or otherwise, of such equipment is inextricably linked to the ground under investigation, the prevailing ground water regime and the type of structure to which the investigation applies. In the foreword to Dunnicliff (1988), Peck (1988) states that "every" instrument installed on a project should be selected and placed to assist in answering a specific question".

Fourth, he raises a somewhat broader question: "can any system remove the requirement for training and experience on the part of the individual using it?" The simple answer is: "No!" I do not think that any system, whether described as knowledge-based or expert, can adequately substitute for training and experience. The system that we describe uses existing widely disseminated information. It presents this information in a unified and easily understood format and tailors it to the user's application. Applications include assisting in the design of ground investigations, checking the content of existing and proposed investigations and training. Such a system should be seen as a replacement for training and experience no more than should a program for slope stability analysis. Indeed, in my view, the description of systems as "expert" has dangers in itself, in particular in relation to the perception of the system by users.

Fifth, he asks about the level of staff operating the system. There are two levels within the system. The first level includes the basic organisation of a ground investigation and extends to giving general advice on the purpose of each phase and stage. This level is written with the client manager in view. The second level gives more detailed advice on the planning of investigations and assists the selection of appropriate methods and techniques. This level is written with the consultant manager in mind.

Finally, he asks how the system is integrated with contractual and management aspects of ground investigation. The system is designed to address technical issues and contractual and management issues are specifically excluded. This should enable the system to be used with any form of contract and under any management procedures.

In his summary The Moderator notes, quite correctly, that the removal of Clause 12 may not effect financial gain for the employer. Given that so much delay and extra cost on civil engineering projects is attributable to unforeseen ground conditions (Mathieson & Keir, 1978; Tyrrell et al.,1983; Anon,1988), perhaps we should consider that the employer's purpose in removing Clause 12 is not financial gain but relatively fixed out-turn costs.

REFERENCES

Anon (1988). *Faster building for commerce*. London: National Economic Development Office.

Dunnicliff, J (1888). *Geotechnical instrumentation for monitoring field performance*. J. Wiley & Sons, New York.

Matheson, GD and WG Keir (1978). *Site Investigation in Scotland*. TRRL Laboratory Report LR 828. Livingston: Transport Research Laboratory.

Peck, RB (1888). Foreword *Geotechnical instrumentation for monitoring field performance* by J. Dunnicliff. J. Wiley & Sons, New York.

Tyrrell, AP, LM Lake and AW Parsons (1983*). An investigation of the extra costs arising on highway contracts*. TRRL Supplementary Report SR 814*)*. Crowthorne, Transport Research Laboratory.

Dr. Peter Smart, Glasgow University, Civil Engineering Dept.

The initial cost of building the system was high; but that has been done. The user will have to face some additional costs in tuning the system to local requirements; but the system has been designed to make this as easy as possible. There will be a further cost in keeping the basis rule-set up-to-date.

One of the original ideas was to relieve a senior engineer of some of the tedium of training a junior assistant; that does not mean that the senior engineer is relieved of the responsibility of checking what the assistant is doing with the system.

Most firms have computer managers to whom the system will appear as just one more program.

There is an overlap between this type of expert system in general and managerial, contractual, and design systems; and this has to be taken into account in general.

Since the system runs on an ordinary PC, it ought to be possible to put it into a portable (laptop) computer and take it to a site office or in the car when using the observational method or making other field studies.

In some ways, the system presented is more of a computational program than an expert system. The principal product is a list of tests required; draft specifications of method can also be obtained.

Mr . T.G. Ball, Soil Mechanics Ltd.

Health and Safety regulations were not discussed in our paper for the simple reason that ISO 9000 does not include any specific requirements for health and safety management. In quality assurance terms the CDM regulations are simply additional legal requirements which must be addressed, just like any other standards, regulations or codes of practice. A corporate management system covers several subjects which are not included in ISO 9000; financial, environmental and corporate planning topics are also omitted form the standard. We take the pragmatic view that all these subjects need to be controlled, and, for simplicity, they are included within our documented quality management system even though they fall outside the scope of the standard and are not considered during accreditation. This is our decision and it is not governed by regulation.

The consultant's traditional role of enforcing the specification has been emphasised; this raises further questions. Whilst site investigation techniques are generally routine, ground conditions are unique to each site and poorly defined before the investigation. The investigative approach has to be site specific and is often modified as the investigation progresses; this should be reflected in the specification.

This month's Ground Engineering survey of site investigation contractors (Anon 1995) makes depressing reading, the main conclusion being that industry standards are unsatisfactory because of the forms of contract and specifications. The Civil Engineering industry continues to work in a confrontational contractual environment typified by enforcement rather than constructive problem solving. As a result there is a tendency for all parties to be dishonest about problems and blame is apportioned. This conflicts with the basic philosophy behind ISO 9000 where problems are resolved openly and teamwork is a prerequisite for progress. The Civil Engineering industry may well be improving its contractual attitude, however, we are not being sufficiently innovative.

Despite the oil industry's long history of confrontation recent contractual developments are dramatic and mirror similar progress in the Norwegian tunnelling industry; agreements are being built around the concepts of teamwork and trust (Beswick, 1995 and Barton et al, 1992). In the oil industry Operators, in practice a combination of Client and Consultant, are letting contracts which have incentives for speed, quality and cost, remove extraneous levels of supervision, and add contractual force to the principles of openness and co-operation. Technical performance is optimised by discussion and arguments are resolved by mutual agreement rather than in court. The Norwegian system is based on a similar level of trust and co-operation, prices are included in the tender for all likely methods, and the owner pays for the technically correct solution. Design and build and the New Engineering Contracts are steps in the right direction, but do not match these innovative solutions.

References:

ANON. 1995. *SI Psyche*. Ground Engineering March 1995, 22-27.

BARTON, N., GRIMSTAD, E., AAS, G., OPSAHL, O.A., BAKKEN, A., & JOHANSEN, E.D. 1992. *Norwegian method of tunnelling*. World Tunnelling August 1992, 324-330.

BESWICK, A.J. In press. *Drilling and associated aspects of CBM in Europe*, paper presented at the international conference: Planning for profit: Coal Bed Methane in the united Kingdom and Europe, March 1995

Mr. J.C. Woodward - Private Consultant

If the Industry wishes to see the proper application of quality management to SI contracting,- and I have believed this is essential since I was responsible for the first seminar on the subject in 1988, then it seems that we still need to get certain basis principles of QM understood:

Firstly, QM is entirely about the supplier producing acceptable end-product; the procedures are simply the means to that end, not an end in themselves.

Secondly, the specification and the QMS must therefore be two different but inter-related documents. The specification contains the standards which the Client wants the supplier to provide to produce an acceptable end product and the QMS contains the procedures which the supplier will use to produce the end product in accordance with the specification.

In other words the Client initiates quality, strict adherence by the supplier to a poor specification from the Client is likely to produce second rate results. This is why I do not follow the Moderator's argument on Mr. Ball's paper that a Quality Plan producing a borehole log to a specification different from the standard in BS5930 will produce a "second-rate log". This may or may not be so, but it cannot be the fault of the Quality Plan which has to be written to satisfy the Client's specification, whether BS5930 or something else. If it is not then there is a clear case of non-conformance on the part of the contract, and a breach of his contract.

Furthermore, the supplier cannot take it on himself to fill gaps in the Client's specification by anticipating in his QMS what the Client may accept. In most industries it is fundamental for fully detailed specs. to be addressed by well-designed QMSs. As a result there is no need for 'a consultant to enforce the specification' as the Moderator states, because the Client can be confident that the QM controls will show this is being done effectively by the supplier.

Our problem is that at present the SI contractor cannot be certain what is a non-conformance.

My third point in reply to the Moderator's reading of my paper is that it must be self-evident that there is a basic difference between a QMS written for a System 1 contract which requires the engineer to complete the specification as he goes along, and a System 2 single point responsibility under which the contractor warrants his results.

In System 1 the contractor's QMS will require him to stop work to await the engineer's decision on the non-specified matter; in System 2 the contractor himself has to provide the procedures to deal with geological uncertainties in order to produce an acceptable report.

It is only right that the Client should be able to check for non-conformance in any contract - and he does this in System 2 by using the established and effective surveillance techniques not by modifying the specification to accept/reject work in System 1.

The industry must therefore decide whether it genuinely wants to adopt QM in order to improve overall standards or muddle through on paying lip-service to costly half-baked QA applications. I have to say that the "founding fathers" mentioned by the Chairman were not plagued by inconsistent application of poorly drafted specifications as they were at the sharp end on site doing the work on a single responsibility basis.

However, I am pleased to report some positive progress. The BDA is considering QM seriously and is to examine if the principles of their successful drillers accreditation scheme can be developed to cover the self-assessment of a standard quality system for SI contractors.

Finally, in my paper, I have followed the usual legal position which rarely implies a duty of care in construction work designed and supervised by the Client's engineer.

This is now being changed by new case law as recently reported in Lindenberg v Canning (& others) in that regardless of what the engineer instructs the contractor to do, the contractor now appears to have a liability to check the correctness of those instructions if he is not to be found negligent.

This could have serious implications for SI contractors who, although following the engineer's instructions to the letter, omit to find a particularly significant piece of geology. Another very good reason for the contractor to be in control of the end product and be required to fully cover his liabilities by insurance as in Europe.

Mr. Ian Farmer - Ian Farmer Associates Limited

I find the comments of the Moderator a little surprising and confusing. Before I attempt to answer his specific queries, I'd like to restate our basic premise.

Spatial Dependency or geostatistics is based on an elegant mathematical explanation of the empirical observation that data points taken close to each other are more likely to have similar values than those taken further apart.

This is expressed as a semi-variogram which plots the change in seem-variance of a property against the distance between data points in space. Semi-variograms have simple forms and simple functions can be fitted to them. These can give an indication of goodness of fit and provide the basis for a predictive approach.

The simplest way to achieve this is to fit a smoothed variogram, based on unbiased weighted values of a property at sampling points. This can then be used to obtain an estimate of the value of the property at a point, or the average value of the property over a block.

Spatial dependency or Geostatistics are most viable where data is reliable, where there is plenty of it, and where the geology, or in the case of ore deposits, the controls on mineralisation, are understood. The core application is in evaluation of mineral deposits. The obvious geotechnical application is in contaminated land assessment.

In most site investigations, the applications are limited by two main factors, the shortage of data and its occasional lack of reliability. The paper sets out to examine how these might be improved, using the techniques of spatial dependency, in two simple and, I think innovative ways.

The first looks at how we might, if we have enough boreholes, closely enough spaced, to adequately sample a property - in this case the thickness of a layer, but it can be any property. Without going into detail, we can show that we need 5 to 6 boreholes to provide the minimum of 30 data pairs for the thing to be statistically meaningful, and the data need to fit closely enough to a variogram form in order to be shown to adequately sample the thickness variation. Going one stage further, if an acceptable error can be defined, we can estimate the optimum borehole spacing. This wasn't the aim of the example which was to illustrate rather than define.

As the Moderator suggests, a straight line could be fitted to Figure 2a which would mean having a pure nugget model. In such a case no correlation structure can be captured and no amount of consideration of geological structures and landforms will change this. The curve is a good fit for Figure 2c and demonstrates that the thickness variation has been adequately sampled.

The second example looks at how the variability of data from limited and possible unreliable sampling can be improved by incorporating soft data. In this case the soft data is a triangular distribution histogram of the hard data for RQD used for the semi-variogram. Incorporation of this data improves the correlation coefficients of the regression lines in Figure 4 from 0.50 to 0.66. The source of the soft data is made clear in the text. I hope this satisfies the Moderator's comments on borehole records. The regression lines in Figure 4 are correct, based on the scatters shown.

The choice between a spherical and exponential variogram form is based on how the correlation structure tapers off at the longer distance towards the range. A spherical variogram has a finite range of influence whereas an exponential model approaches the sill value asymptotically, in other words an infinite range. In the paper each variogram form is used appropriately.

One of the problems of geostatistics is the tendency to overelaboration, and there are numerous historical horror stories of fanciful, academic, computer models built on dubious data, ignoring both geology and common sense. If there are applications in site investigation, and I agree it is not proven, they must of necessity be simple. The two presented in the paper may arguably be simplistic, but they are certainly relevant.

Dr. John Perry - Transport Research Laboratory

Lime stabilisation can reduce construction costs, and reduce the impact of removing material off-site and associated cost of disposal. The Paper describes how to conduct a site investigation for this type of design. The Moderator has raised two points regarding the effect of sulphates which are considered in the site investigation. The first point recommends the inclusion of testing of material around the stabilised layer. This is in fact covered in the paper in Figure 1 and the associated text. The second comment regards the use of electron microscopy to identify disseminated pyrite. This point is open to discussion, but it would appear that given the identification of pyrite throughout the site investigation and the use of the heave test to show the effect of pyrite, the use of electron microscopy, which requires very small samples, is limited.

Mr. Liam Barber - R.S.A. Geotechnics

In response to the points raised by the Moderator, I would agree with the comments just made by Dr. Perry and would like to add the following. Sulphates and related minerals are problematic when present in a finely disseminated form, indeed at a meeting of the Engineering Group of the Geological Society some two years ago the observation was made that such minerals are at their most dangerous when they cannot be seen. The purpose of laboratory testing of FE material is to detect their presence. However, sulphate bearing minerals can also take the form of discrete pockets or may be present along fissure or fracture planes in soils and rocks. In this situation their detection by laboratory testing is more difficult. It is therefore necessary to pay close attention to the detailed logging of a trial pit side, this can only be achieved by entering shored trial pits.

With regard to electron microscopy these techniques have been used, however a sample in the order of grammes clearly cannot be representative of a much larger mass of soil and results must be treated with caution. Furthermore, levels of sulphates which may cause problems in lime stabilised soils may be below the threshold values of such techniques making quantitative assessment unreliable.

Mr. Kevin A. Corbett - Nabarro Nathanson Solicitors

Rideout and Smiths paper on historic maps is based on a major reference work the authors are finalising, publication of which is expected this Summer.

Nabarro Nathanson is one of the largest property law firms in Europe. It's Clients, local authorities, corporations, developers, financial institutions and investors demand very detailed advice on all land matters, including construction and environmental contamination.

My colleague, Barry Rideout, like our Moderator enjoys opening and reading old maps and this interest has maintained his commitment to the project. This paper and the to-be-published reference work provides the historical background to the various sources of information used in desk studies and thereby provides a useful checklist for practitioners.

It is interesting to see The Moderators NEDO slide showing the incidence of problems caused by unexpected underground obstructions. One would hope a very thorough desk investigation would limit or reduce such occurrences. However, I recall being called out to a site where ground works were in progress and a disused and unexpected air raid shelter was uncovered in the middle of the site.

In large developments some unexpected obstructions are clearly to be expected!

Mr. Philip Wilson - Highways Agency

Introduction

The moderators report for this session made reference to cost increases during construction. The National Audit Office (NAO) have identified unforeseen ground conditions as the major source of cost increases on highway construction contracts. (Reference 1). The NAO recognised that it is necessary to obtain a balance between the initial tender price and minimising cost increases during the contract. The balance is governed by the degree of risk for unforeseen ground conditions and the risk is mitigated by site investigation. The NAO recommended that expenditure on site investigation should be reviewed in relation to cost increases.

Research

A programme of research was undertaken to study the efficiency of site investigation practices. (Reference 2). The study attempted to determine a relationship between site investigation (SI) expenditure and total cost increases. Site investigation costs include the initial planning, preliminary studies and the cost of the physical ground investigation. The results are shown in Figure 1 and exhibit considerable scatter. A typical SI cost at 1.5% of tender indicates cost increases varying between 10% to over 45%.

A more detailed analysis of individual project records was undertaken to relate site investigation practice to cost increases due to geotechnical factors. The results are reproduced in Figure 2 and again show considerable variation.

The study was unable to establish any simple relationship between expenditure on site investigation and construction cost increases.

Conclusion

The theme of the conference is Advances in Site Investigation Practice.

The profession should learn from practice and practice what we learn.

References

1.   National Audit Office: 1992: *Contracting for Roads*. HMSO, London.

2.   Transport Research Laboratory: 1994: *Study of the Efficiency of Site Investigation Practices*, Project Report 60. TRL Crowthorne.

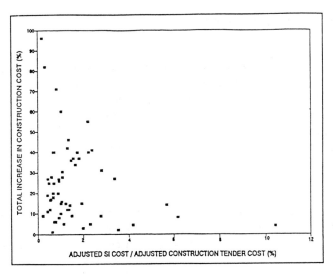

Figure 1 : Relation of total construction cost increase to SI expenditure

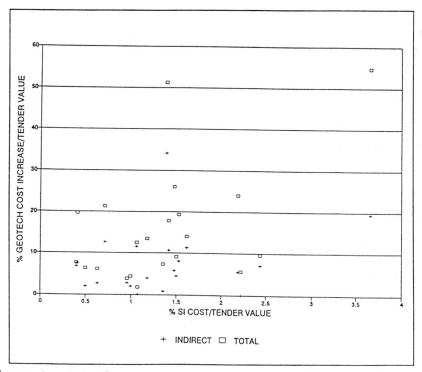

Figure 2 : Relation of geotechnical construction cost increases to SI expenditure

From paper to silicon chip — the growing art of computerised data management

R A NICHOLLS
A S PYCROFT
M J PALMER
Sir William Halcrow & Partners Ltd., Swindon, UK

J A FRAME
Gammon Construction Ltd., Hong Kong

INTRODUCTION

Modern investigation techniques can generate large volumes of data, "fast track" projects demand rapid assessment of information, often with design running concurrent with the investigation. Many analytical techniques require a large quantity of input data and generate similarly large outputs. These factors, combined with ready access to increasingly powerful computers, have resulted in a requirement for the digital transfer of information, and the computerised processing and management of the data.

The need for a common interchange format for the exchange of geotechnical data was recognised by the Association of Geotechnical Specialists (AGS), and in 1992 a specification for the electronic transfer of geotechnical data was published[1]. The Geotechnical Data Interchange Format (GDIF) produced by the AGS is intended as a neutral format to enable the electronic transfer of ground investigation data.

Changing the main medium of information exchange from paper to the silicon chip brings many benefits to the practising geotechnologist. However, establishing an effective electronic data transfer procedure and developing a geotechnical information management system is not without its difficulties. Recent experiences of processing and analysing large volumes of ground investigation data on major projects will be discussed. The paper highlights some of the areas which need to be addressed when establishing a computerised geotechnical data management system.

TERMINOLOGY

To avoid confusion the following terms are explained:

- **Field**   is the smallest named unit of stored data and contains a single value. For example the hole identification, easting, northing, sample depth, liquid limit and plastic limit are all fields.

- **Record**   is a collection of related **fields**. For example, all data associated with the location of one exploratory hole forms a record and all classification test results for one specimen would also form a record.

- **Group**   is a named collection of all occurrences of one type of **record**. For example, all hole location details would form one group and all classification test results would form another group.

- **Key Fields** are those fields required to uniquely identify a record within a data **group**, they are necessary to link groups together thus enabling data from one group to be combined with data from another. For example, to find all the classification tests within a defined area requires data in the classification test group to be combined with data in the hole location group. Without key fields all the location information would have to be stored with each classification test record.

Data are usually transferred from a data **supplier**, the ground investigation contractor, to a data **consumer**, the consulting engineer. The person examining, interpreting, analysing or in any way making use of the resulting data is referred to as the **user**.

CONSTRUCTING THE DATABASE

Prior to the introduction of the AGS GDIF data suppliers had to contend with preparing data in various non-standardised formats according to the requirements of individual consumers. The need to accommodate different formats led to inefficiency and an increased risk of data transfer errors. Introduction of the AGS GDIF has enabled advances to be made in the computerised management of site investigation data. However, the efficient transfer of data from supplier to consumer is only one part of a long and complex chain of activities which are required to create and operate a geotechnical database. As experience in the use of the AGS GDIF grows so the transfer of digital data will become more commonplace and more efficient.

The volume of digital data generated on medium to large projects can be quite substantial and consideration must be given to computer hardware and data storage requirements. Table 1 gives an indication of the requirements for recent projects.

Generation and Transfer of Digital Data

Site Investigation data are generated from many different sources, Figure 1. Ideally data transfer should take place on a digital to digital basis. That is to say the data capture, processing and final transfer all take place using computers thus eliminating errors associated with the manual input of data. This is feasible for many laboratory tests (e.g. triaxial, oedometer, shear box), and for some *in situ* tests (e.g. piezocones, pressuremeters). However, in many cases digital data capture is either not feasible or not practical and data capture becomes a manual process subject to human error.

Recent experience has shown that when suppliers are requested to supply information in AGS GDIF it is still fairly common for the data to be captured twice. Firstly to produce reports, printed borehole logs etc. using the suppliers' existing software and secondly data are re-typed for generating AGS GDIF files. There are substantial advantages to both the supplier

Project	Hardware Platform	Notes
Ground Investigation for a Large Transport Infrastructure Project	Server : INTEL 80486 DX2/50MHz, 8Mb RAM, 1.2Gb HDD, 14" VGA mono monitor. Novell Netware 386 v3.11 Analysis : INTEL 80486 DX2/66MHz, 16Mb RAM, 200Mb HDD, 21" SVGA colour monitor. Pre-Processing : INTEL 80486 DX/33MHz, 4Mb RAM, 200Mb HDD, 14" VGA colour monitor.	£2 million GI of 100km^2 site. Approx 320 boreholes, over 1300 piezocones, plus Mackintosh probes and trial pits, associated lab testing. 46Mb of AGS format data files received, of this 37Mb is piezocone data. 120Mb database files including extensive surface modelling.
Construction Monitoring of a Large Transport Infrastructure Project	Server : SUN SPARCstation 10GX model 51, 50MHz SuperSPARC processor, 64Mb RAM, 2 no. 1.05Gb SCSI HDD's, 20" colour monitor with graphics accelerator, OS Solaris 1.1 Network Nodes : 4 no. PC's INTEL 80486 DX2/66MHz, 16Mb RAM, 240Mb HDD, 1Mb VRAM, 14 & 20" colour monitors.	Expansion of the GI database to cater for instrumentation monitoring and analysis during earthworks construction. PC's running PC-NFS 5.0 and X-Terminal emulation software to enable DOS/UNIX networking.
Ground Investigation for a Motorway Widening Scheme	PC : INTEL 80486 25MHz, 4Mb RAM, 14" VGA colour monitor. Data storage on Novell network server.	£0.6 million GI for motorway widening scheme. 400 exploratory holes of which 173 boreholes, plus associated lab testing. 2.1Mb of AGS format data files received. Previous GI data combined from existing database. 11Mb total database files. 13Mb of figs etc.

TABLE 1   :   Hardware and Data Storage Requirements on Recent Major Projects.

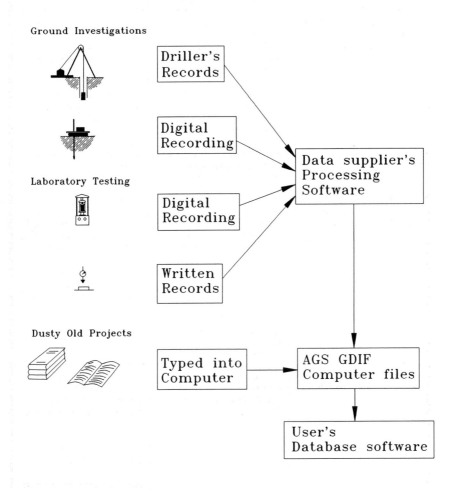

FIGURE 1 : Data Sources from Ground Investigation

and the consumer if the supplier uses software which generates printed reports from the AGS GDIF data, or which generates GDIF files and the required printed reports from the same set of raw data. The advantages are:

- no double handling of data, which leads to faster production and fewer errors;

- one data set, which simplifies updating and correction procedures;

- checking procedure is simplified, once the printed copy has been checked and the source data corrected then the contents of the GDIF files should also be correct.

Data suppliers need to establish rigorous data management and processing strategies. Because the syntax of the AGS GDIF has been defined in reasonably unambiguous terms it is relatively easy to generate files which satisfy its requirements. Difficulties arise in ensuring the file contents are correct. Suppliers' Quality Assurance (QA) systems should be geared towards ensuring filenames, hole identifications, sample numbers etc. are all recorded properly and that the correct information is entered into the appropriate fields. If the supplier cannot submit substantially correct data quickly then many of the advantages of electronic data transfer are negated, especially on fast track projects.

The importance of establishing a consistent syntax for key field values cannot be over emphasised. Without this it will not be possible to answer any but the most trivial of database queries. This needs forethought, planning and consultation with the end users before the site investigation works are in progress. On large projects where several investigation contracts are involved it is particularly important that a consistent system of hole identification and sample numbering is established by the data consumer. There is no reason why the GDIF could not be used as a two-way means of exchanging information. Data consumers could supply GDIF files with schedules of hole identifications and locations, sample numbers and depths, laboratory test requirements, etc. The data supplier could then return the completed electronic schedules, again as GDIF computer files.

When establishing a flow of electronic data from a ground investigation a proactive approach by the data consumer may have significant benefits. Consumers should be prepared to provide data suppliers with detailed guidelines, recommendations, examples and advice. This is particularly important when dealing with less experienced data suppliers.

Receiving and Checking Data
Whether information is captured digitally or manually and even if the majority of suppliers use software which generates both the conventional printed report along with the electronic data transfer files, the consumer must check the correctness and validity of the incoming data. It is very difficult for people to scan and check the contents (as opposed to the format) of a GDIF file. People find it much easier to check analogue format data.

On recent large projects computer software has been developed which automatically validates incoming GDIF files to ensure compliance with the AGS standard. Without a standardised file transfer format validation software would need to be adapted on a project by project basis. Computerised validation should be carried out by the consumer for at least the following reasons:

- there is no guarantee that all suppliers will have access to data preparation software;

- irrespective of software used there is always some manual input required, and therefore scope for human error;

- repetitive checks on large volumes of data are subject to human error;

- manual checking is time consuming.

Ideally validation of incoming data should be designed to check the following:

- syntax of data files;

- data type and range;

- formal data relationships, e.g. TCR $\geq$ SCR $\geq$ RQD;

- informal data relationships, e.g. low fracture index should be associated with high RQD;

- temporal relationships between values in a single field, e.g. should successive settlement gauge readings show heave.

It is relatively easy to develop software for validating the first two items, although setting of permissible ranges needs to be done with care. However, the last three items require substantially more programming effort and may require amendment on a project by project basis. In most cases it will be easier to carry out relational checks after data have been loaded into the chosen database software.

Having established a standardised transfer format and developed software for validating the compliance of incoming files the task of transferring the data from supplier to consumer becomes relatively straightforward. Nevertheless, consumers need to establish rigorous Quality Assurance procedures to ensure the integrity of the incoming data. A flow chart of a typical procedure for receiving GDIF files on disk is shown in Figure 2. Furthermore, the validation procedure forms a primary check on the data, whilst secondary checking of the correctness of the data needs to be carried out by the user.

The use of a computerised data management system allows a certain degree of automation in the explicit checking of input. Database software is available which allows the engineer to set acceptable ranges for data variables. If the computer encounters an out of range value whilst loading data a warning is issued and appropriate actions can be taken. In the days of paper the engineer transferring test results from the written report to a plot, or collating results into a table, would be carrying out an implicit check on the data. In the days of the silicon chip that level of implicit checking has been removed, since the physical process of

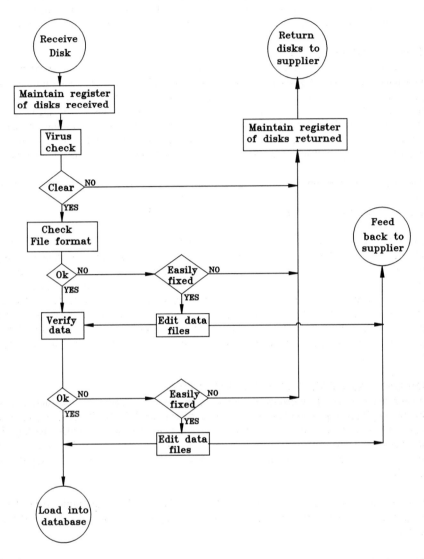

FIGURE 2 : Flow Chart Showing Quality Assurance Procedure for Receiving Data on Electronic Media

producing a plot or table is carried out by the software. It is therefore important that the engineer retains a 'feel' for the data under consideration. Ensuring that the information stored in the computer is correct remains the most important task in establishing a database.

Maintaining and Managing the Database

Rigorous procedures are required to ensure the integrity of data once they have been loaded into the database. This is particularly important on fast track projects where design work may be proceeding whilst results of a ground investigation are still being received. There is a need to keep track of when and by whom data were loaded into the database, or changes made to existing data. It is almost inevitable that some data will require re-submission and care is required to ensure that changes are reflected in any related data fields. For example, if a geological boundary is moved the classification of all affected samples should also change.

Problems are often encountered when diskettes are returned to suppliers for correction. It is not uncommon to find that previously corrected errors are re-introduced by subsequent submissions and consumers need to be alert to these possibilities.

Many of the problems of maintaining database integrity are difficult or tedious to incorporate into formalised procedures. The inclusion of automatic transaction tracking would be a major help, this feature is already incorporated into database management systems in other disciplines. In the meantime disciplined use by experienced personnel and good backup procedures are probably the best protection against serious damage to an established database. On multi-user computer systems access rights can be set according to users' ability, careful planning is required to establish such systems. On stand-alone PC's it can be very difficult to effectively protect against accidental changes to the data, particularly whilst a database is being constructed.

MANAGEMENT OF THE SITE INVESTIGATION

An integrated approach to scheduling of exploratory holes and database construction would enable the engineer to provide schedules to the contractor in AGS GDIF files. Implementing this system would force the consumer to think about the database structure and key field syntax before the site investigation commences. Inclusion of the Bill of Quantities into the database would enable final costs and measurements to be assessed more rapidly and efficiently.

Use of electronic data transfer methods implies a rapid availability of results from the site investigation. It therefore becomes possible on fast track projects to optimise the investigation strategy.

Knowledge of the location of available samples is essential for preparing laboratory test schedules and this information is readily available when there is access to a database. The structure of a database can be extended to accommodate information from test schedules that have been issued to the contractor. This information is readily extended to yield information on laboratory productivity and contract cost control.

INTERPRETATION AND USE OF THE DATA

The manual preparation of lists of specific data when writing interpretative reports is a very time consuming activity but it does, at least, ensure that data are checked as the lists are created. The disadvantage of manually building specific data lists is the reluctance to go back to the factual reports and build additional lists. There is a degree of inflexibility in a method that requires such time consuming activities as looking through nearly every page of factual reports several times to extract different parameters.

The introduction of computerised geotechnical databases and, in particular, the development of a comprehensive systematic manner of transferring all site investigation information has given the geotechnical engineer a new and powerful tool. The capture of all the site investigation data means that the interpretative report writer and subsequently the designer has the ability to assess all the information in different ways and can look for important trends and use the power of querying or filtering to assess whether parameters have significance and whether different parameters are related. Having data in a standardised retrievable form means that the newer tools of probabilistics and risk assessment can be used on any project and the geotechnical data fully integrated into project design. The interpretative report need no longer be for geotechnical engineers alone, selected contents can now be absorbed into the wider picture of project development and 'What if?' scenarios can be tested.

The output from an information management system must be tailored to suit the particular project and may well require the development of additional specific output formats, e.g. specialist parametric plots, tables etc. The requirements for a specific project need to be considered at an early stage to avoid the output of irrelevant correlations and plots, sufficient time should be allowed for the development and implementation of any project specific software.

To extract information from the geotechnical database the user has to be aware of the field names used and the general structure of the database. Customised menu systems can go some way to make it easier for the non-expert user to generate queries and extract information from the database. On all but the smallest investigations it is inevitable that data will need to be presented as sub-sets of the total database. To achieve this filters or queries will need to be set and some knowledge of the database structure is essential.

Once a sub-set of data has been extracted and presented as either graphical or tabular output it is essential that the filtering used to produce the output can be checked against the title or legend accompanying it. Data which have been graphed using software which is totally separate from the database software is particularly vulnerable to erroneous titling. On recent projects parametric plots have been produced which include QA audit trail information added by the software, see Figure 3. The QA trail includes such information as when and by whom the plot was created, from which database, filtering used to create the sub-set and how many data points were in and out of the plot range. In the example shown on Figure 3 the QA information shows that some data points exist outside the chosen range of the plot. Without the QA information this may not be immediately apparent.

Even though the QA information can be generated automatically some degree of checking is essential. Graphical data presentations are a powerful way of identifying incorrect values.

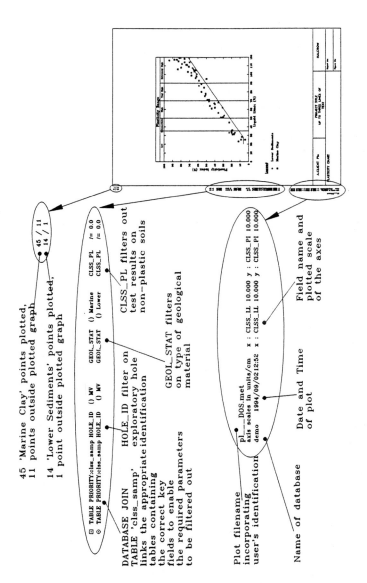

FIGURE 3 : Quality Assurance Audit Trail for Checking Database Output

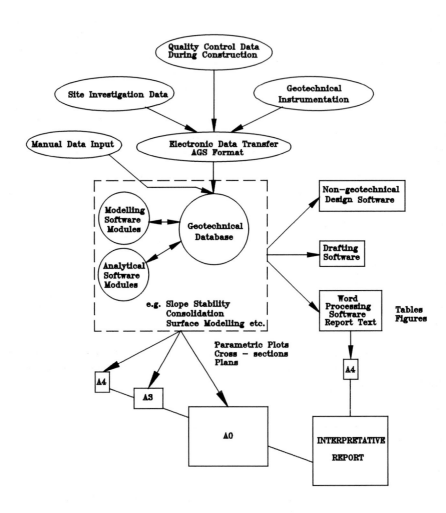

FIGURE 4 : Concept for an Integrated Geotechnical Information Management System

Any corrections or amendments made to the database need to be carried through to the source data, duly documented and passed to the data supplier if appropriate.

With the appropriate software parametric plots, cross-sections and plans can be produced rapidly and to a high presentation standard. This gives the engineer the ability to examine relationships within the data which would have been extremely time consuming using manual data preparation. For instance, parameters can be plotted against depth below ground level, elevation, or depth below a specified datum or geological boundary.

The ability to produce computer output needs to be used with some intelligent control and not used over zealously to make reports even larger. With the manual reporting system the minimum number of figures were prepared because of the effort and time required to tabulate additional information. With the advent of computerised systems there could well be a tendency to produce every conceivable figure. There is great benefit in looking at different parametric relationships but little benefit in their thoughtless reproduction. Having the ability to examine many options in a relatively short period of time can lead to optimal solutions, but cost efficiencies introduced by computerised data management can easily be lost if restraint is not exercised.

On large projects where the site investigations cover a wide area some form of surface modelling is likely to be required. The production of sections, contour plans and isopaches provides a powerful aid to interpretation, as well as forming an additional very valuable check on erroneous input. Generation of surface models from digital data is relatively quick and easy. However, experienced staff are required to ensure appropriate models are generated. Digital ground models can be interchanged with non-geotechnical software, for example to calculate volumes of proposed excavation. By exploiting the full potential for the interchange of digital data the geotechnical database can become part of a fully integrated geotechnical information management system. A conceptual representation of such a system is shown on Figure 4.

ACKNOWLEDGEMENT

The views expressed in this paper are those of the authors and do not necessarily reflect the views of their respective employers.

REFERENCES

1.   Anon. *Electronic Transfer of Geotechnical Data from Ground Investigations*. Pub. Association of Geotechnical Specialists, 1992.

A computer system for site investigation data management and interpretation

A.J.OLIVER AND D.G.TOLL
School of Engineering, University of Durham, Durham, UK

ABSTRACT

The use of a relational database management system (DBMS) for storing data from site investigations is discussed. The advantages of using a powerful DBMS to provide a central data store for a geotechnical company are outlined. A single central database with multi-user access (rather than individual stand-alone systems) ensures consistency of data and avoids redundancy. The database system described is also coupled to a knowledge-based system for assisting a geotechnical specialist with data interpretation. The combination of database and knowledge-based system technology can provide a powerful tool for data management and interpretation.

INTRODUCTION

Site investigation data storage is becoming increasingly computer based. Simple 'flat file' data storage, although widely used at present, is likely to become unsuitable as the quantity of data held by individual companies increases. There will be a need for a more efficient data management systems. This paper describes a site investigation database, GeoTec, implemented using a relational database model.

The database structure is based on the data exchange format proposed by the Association of Geotechnical Specialists (AGS, 1992). In addition, the work has taken the standard further by defining data types and lengths for the specific fields, an essential step in producing a relational database.

Additional features have been incorporated including multi-level storage of test data. This allows the more frequently accessed data to be available separately from the raw or semi-processed data. This increases the efficiency with which automated systems can extract the main measured properties, whilst allowing the raw data to be stored for manual examination. Another feature, unique to the database structure developed, is the ability to store the descriptions of layers of soil or rock in a structured form rather than as a simple text string. This allows the qualitative data stored in the description to be available in subsequent analysis routines.

GeoTec has been developed as the core of a knowledge based system to assist the geotechnical specialist with data interpretation. The system, called SIGMA (System for the Interpretation of Geotechnical Information), can assist a geotechnical specialist with data handling, assessment of design parameters and interpretation of ground conditions.

Advances in site investigation practice. Thomas Telford, London, 1996

DATA MANAGEMENT FOR SITE INVESTIGATION

As information technology becomes increasingly commonplace in the area of site investigation, the importance of ensuring the continuity and integrity of data increases. Electronic storage, manipulation and retrieval of data is only meaningful if the process is as least as productive and efficient, and hopefully significantly more so, than the respective manual transaction. As the use of electronic site investigation data gathering in the field and laboratory becomes more widespread, the requirements for computer-based storage and processing of this data become more valid. The quantity of data generated from even a medium sized site investigation is large and can involve a noticeable administrational overhead; for manual data handling the larger the investigation, the larger the overhead. Electronic data management becomes more effective as the size of the investigation increases.

This need for computer-based data handling has been recognised by several software companies who have successfully developed systems to mimic the manual process involved with the production of borehole logs and the presentation of test results. Some examples are: *Geodasy* (Howland, 1992), *gINT* (Staten & Caronna, 1992), *Keyhole* (Greenwood & Rothery, 1992), *SID-GDMS* (MZ Associates Ltd, Camarthen, UK) and *TECHBASE* (MINEsoft Ltd, Denver, Colorado, USA). The more sophisticated packages provide reporting facilities for presentation of borehole logs, laboratory test reports etc. as well as graphical displays of test results against depth (for example), cross-sections, contouring or fence diagrams.

These data management systems allow greater flexibility for the management of a project; more direct and immediate access to the data is provided and also a high level of data security. This security issue is not confined to the protection of sensitive or confidential data, but electronic data can be archived regularly by means of incremental backup systems, so preventing accidental loss or damage.

With the advent of the AGS Data Exchange Format (AGS, 1992;1994), the industry has a standard to adhere to for electronic transfer and hopefully this will lead to a more uniform approach, not only to data transfer but also to data management. Only within the framework of a national, or international, standard should the new generation of databases be designed, allowing the benefits (both economic and technical) of mass information storage and transfer to become apparent (Mott MacDonald *et al*, 1994). Whilst at first there may be a tendency to horde 'in-house' site investigation data, due to the initial capital outlay involved, in the long term the economic advantage of sharing the data must become apparent. If site investigation data has already been carried out on a particular location, it is both practically and economically foolish to duplicate the effort. Also, with the collection of large quantities of data, the potential for large scale analyses are increased, providing the industry as a whole with meaningful data.

Many companies are using personal computer (PC) based systems for managing their data. Some of these systems do not have the capability to handle multi-user access via networks (either local or wide-area). The use of a large scale commercial database management systems (DBMS) may be required to perform such a function. Many of these are still only feasible on a workstation platform, rather than on a PC, even with the rapid developments in PC technology. However, their use eases networking to peripheral machines, whether from within the same building or across the globe. All employees of an organisation can then have access to the same data source and as that source is continually updateable then everyone has access to the same level of data at the same time. Data verification need only occur once, reducing the duplication of effort and enforcing an organisation wide standard. In short, the potential offered for centralised data management significantly improves the operational efficiency of geotechnical companies, particularly when coupled with a system for data interpretation.

DATA INTERPRETATION SYSTEMS

Database systems of the type described above are limited in that they only provide data handling and presentation facilities, essentially replicating the manual data processing required to produce borehole logs, plots of the test data etc. The GeoTec database described here is also the core of a knowledge-based system called SIGMA. This system allows computer-aided interpretation of data, both for parameter assessment and interpretation of ground conditions (Toll *et al*, 1992; Toll, 1994). This combination of database technology and knowledge-based systems for data interpretation can provide a powerful tool for assisting the geotechnical specialist.

The interactions between different pieces of geotechnical software are shown in Figure 1. An interpretation system such as SIGMA can play a role in integrating these different packages. No other packages currently exist which can assist the geotechnical specialist with the interpretation process which must come between the 'raw' data, and conventional calculation packages (e.g. retaining wall or slope stability analysis packages).

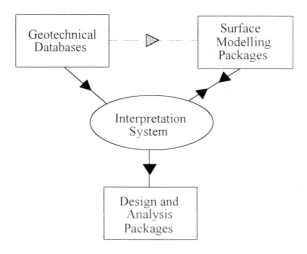

Figure 1. Links between software for geotechnical design.

Many database systems can achieve simple data checking by defining acceptable ranges for particular fields. However a knowledge-based data interpretation system can introduce a greater degree of 'intelligence' into the data checking. Use can be made of heuristic knowledge, such as correlations between parameters; knowledge of expected values for particular soil or rock types; knowledge of how measurements of the same parameter, determined by different tests, should compare. So, in addition to straightforward data validation, parameters can be cross-checked to ensure consistency between different pieces of data. As well as checking the numerical values, the system can also check that the soil or rock descriptions are compatible with the results of laboratory and field tests.

SIGMA goes beyond a database system in that it can assist a geotechnical specialist with the data processing required to assess design parameters. SIGMA is tightly coupled to the geotechnical database (Oliver, 1994) and has data handling facilities which allow the user to carry out basic database functions. Data in the AGS data exchange format is capable of being entered into the database. However, in addition to data manipulation the system can perform the preliminary data processing and present the results for consideration by the specialist.

ADVANTAGES OF RELATIONAL DATABASE SYSTEMS
The main advantage of using a database is that it provides centralised control of its operational data. By providing such control of the data, several other advantages can be obtained (Date, 1987).

1) **Redundancy can be reduced.**
   In non database systems, each application has its own private files. This can often lead to considerable redundancy in stored data. When similar data resides in several different files, the updating procedure can become clumsy and processor inefficient. This inefficient processing counteracts one of the technology's main advantages over other systems, namely high speed processing.

2) **Inconsistency can be avoided.**
   If a unique identifier is assigned to each record, no duplication of records will be allowed. This will eliminate the risk of having two inconsistent records for the same object. This occurs when, due to duplication, data in some file systems are not updated, leading to different values for the same data item.

3) **The data can be shared.**
   Different databases and applications can share the stored data without having to duplicate this information for each object.

4) **The data can be standardised.**
   By building a database the information stored and its format can be standardised. This is particularly desirable in cases where data is being exchanged between different systems and where it is necessary to have the information stored in a valid and complete form.

THE INGRES RELATIONAL DATABASE MANAGEMENT SYSTEM
The GeoTec database has been designed and implemented using the INGRES Relational Database Management System (RDBMS). INGRES provides multi-user access to a centralised data structure and is accessible via the industry standard SQL (structured query language), as well as various INGRES variant query languages (Relational Technology, 1990).

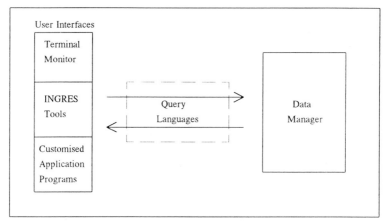

Figure 2. The architecture of INGRES

INGRES RDBMS consists of three main components; the data manager, the user interface and the query language (Figure 2).

The Data Manager. The data manager accepts the query language instructions and performs specified operations on the database. All basic INGRES tasks, such as data updates and retrievals, are performed by the data manager. However, the user never interacts directly with the data manager. Instead, the user must give instructions to INGRES through a User Interface

The Query Language. The query language passes instructions to the data manager from a user interface. There are many user interfaces that request different database tasks, as discussed below. Third party software can also communicate directly with the data manager using an embedded query language in the host software. Equally the host software can produce its own queries and pass them directly to the data manager.

The User Interface. A user interface enables the user to give instructions to the data manager. The terminal monitor, for example, allows the user to enter data and instructions for the data manager by directly entering query language statements. Other interfaces allow data manipulation by INGRES forms and menus. The user interface accepts instructions from the end user and forwards them to the data manager via a query language. A form based application subsystem frees the user from having to memorise the specific query syntax and provides a working environment that is suited to developing a database structure. Using third party software essentially bypasses the user interfaces provided by INGRES for accessing an existing database.

INGRES also provides network support via the INGRES STAR and INGRES NET subsystems. INGRES NET allows multiple INGRES sites to be connected regardless of the hardware platform, so enabling a PC in one location to access a workstation based INGRES database in another location. INGRES STAR allows for databases at several locations to be combined together as one database across a network, allowing data to be available for anyone who requires it and making data duplication unnecessary.

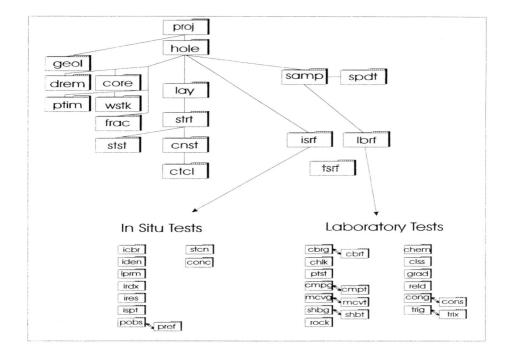

Figure 3. Schema for the GeoTec Database structure (see Figure 4 for legend)

THE DATA STRUCTURE OF GEOTEC

An outline schema for the GeoTec database is shown in Figure 3, where the boxes represent tables in a relational database structure. Each table stores data which represents a data group, the data group being a function, property or parameter of the site investigation. This structure produces an efficient structure for data retrieval and handling, necessary for the potential volume of data to be stored. A more detailed description of the database is given by Oliver (1994).

As far as possible the table names have been adopted to be compatible with the AGS headings (AGS, 1992). The top level table is the *proj* table which contains information on the location and date of the project and the parties involved.

A deviation from AGS is the use of a geology table *geol* which has been linked to the *proj* table through the *proj_id* key. The *geol* table allows storage of identified geological horizons which could exist at the site (rather than a layer description, as is the case for AGS). This stratigraphic information will generally be obtained from a desk study at the feasibility stage of the project before any ground investigation has been started (boreholes or trial pits). Therefore the information need not be related to specific holes but is attached at the project level. The information can be linked to specific layers identified at a later stage (during or after the ground investigation) using a *horz_no* field. Other tables for sources (desk study) information could be attached as separate tables to the *proj* table.

LEGEND				
proj	Project	drem	Depth Related Remarks
hole	Borehole / Trial Pit	core	Core
geol	Geology	frac	Fracture Detail
lay	Layer	ptim	Progress with Time
strt	Strata	wstk	Water Strike
stst	Stratum Structure	samp	Sample
cnst	Constituent	spdt	Specimin Detail
ctcl	Constituent Colour	isrf	In Situ Test Index
tsrf	Test Reference	lbrf	Laboratory Test Index

LEGEND - Laboratory Tests				
cbrg	California Bearing Ratio - General	cbrt	California Bearing Ratio - Detail
cmpg	Compaction Tests - General	cmpt	Compaction Tests - Detail
mcvg	Moisture Condition Value - General	mcvt	Moisture Condition Value - Detail
shbg	Shear Box Testing - General	shbt	Shear Box Testing - Detail
cong	Consolidation Tests - General	cons	Consolidation Tests - Detail
trig	Triaxial Tests - General	trix	Triaxial Tests - Detail
clss	Classification Tests	grad	Particle Size Distribution
reld	Relative Density	chem	Chemical Properties
chlk	Chalk Crushing Value	ptst	Laboratory Permeability Tests
rock	Rock Properties		

LEGEND - In Situ Tests				
icbr	In Situ California Bearing Ratio	iden	In Situ Density
iprm	In Situ Permeability	irdx	In Situ Redox
ires	In Situ Resistivity	ispt	Standard Penetration Test
stcn	Static Cone Penetration	conc	Cone Calibration
pobs	Piezometer	pref	Piezometer - Detail

Figure 4. Legend for GeoTec database structure

The term Hole has been adopted as the generic name for boreholes, trial pits or shafts (as per AGS). The *hole* table contains details of boreholes or trial pits such as location, date and method of boring. The details of the ground conditions observed at the hole are stored in the layer *(lay)* table. This contains depth and thickness information about the layers observed, and these can be linked to the appropriate geological horizon in the *geol* table.

Present in the *lay* table is a text field containing the layer description. Soil and rock descriptions can be long and complex. To allow more efficient access to this qualitative information a data structure has been devised in which the information contained within the description can be extracted and stored as separate fields within the database. The detailed implementation of this structured representation is described by Toll and Oliver (1994).

The structured representation of soil and rock descriptions consists of four tables: *strt* - stratum, *cnst* - constituent, *stst* - stratum structure and *ctcl* - colour. This is because descriptions of layers may contain more than one stratum (soil or rock), for example *SANDSTONE interbedded with SILTSTONE* or *CLAY with pockets of SAND*. In these examples two distinct strata are present within the layer, yet they cannot be distinguished as separate layers (a layer being defined by depth and thickness). Therefore the representation scheme allows for the possibility of multiple strata within a layer. This format can also deal with the case where a soil or rock changes significantly from the top to the base of a layer eg *Silty SAND becoming Clayey SAND*.

Strt represents the whole stratum eg *Silty sandy CLAY*, whereas *Constituent* indicates the constituents combining to make up the stratum eg *silt, sand, clay* etc. In the *Strt* table, information which relates to the whole stratum is stored such as *Moisture condition, Consistency* or *Weathering*. In the Constituent, *cnst*, table, information which relates to individual constituents is stored.

GEOTECHNICAL TEST DATA

Insitu or laboratory tests are linked to the *Hole* table as shown in Figure 3. Insitu tests can be identified directly from the depth at which they were carried out. Laboratory tests are carried out on samples. Information on samples is stored in the Sample, *samp*, and Sample detail, *spdt*, tables. Test results are identified by *proj_id* and *hole_id* but are not attached to a particular layer. This is because the test information will often be used in determining the layer boundaries.

Between the sample table and the test tables there are two additional tables, *isrf* - insitu test reference, and *lbrf*, laboratory test reference. These tables record which tests have been carried out, keyed on the investigation number. They act as an additional indexing mechanism to extract the specific test data and are automatically generated using simple SQL routines. The table *tsrf*, lightly shaded in Figure 3, contains reference data for all the different tests stored in the database. This table is shown without links to any other, as its contents may be accessed directly or by several of the tables in the data structure.

Laboratory test information can relate to different degrees of sub-division of samples. A sample is usually sub-divided in order to carry out different tests. For compatibility with AGS, this first sub-division of a sample is called a specimen. In some cases, the location of a specimen within a sample may be significant, for example, if the sample crosses a stratum or layer boundary. For this reason, a *Sample detail* table, *spdt*, is provided for identifying the location relative to the sample top. A comments field is also provided, for recording any peculiarities which are not true for the sample as a whole.

TRIG table (Level 1)

Project ID	Hole ID	Sample Ref.	Spec. Ref.	Cohesion	Angle of friction	Property code	Test code
7702	B5	12c	2	35		UPI	TRIQU
7702	B5	12c	2	4	27	EPI	TRIID

TRIX table (Level 2)

Project ID	Hole ID	Sample Ref.	Spec. Ref.	Test Ref.	Cell pressure	Max. deviator stress	Property code	Remarks
7702	B5	12c	2	12/2/1	100	275	UPI	B512c0-34
7702	B5	12c	2	12/2/2	100	277	UPI	B512c0-34
7702	B5	12c	2	12/2/3	100	279	UPI	B512c0-34

Table 2 - Example Test Tables showing multi level structure

For some tests (eg triaxial) the specimen is further divided into sub-specimens, each of which is tested in order to produce an interpreted result for the specimen as a whole (for the triaxial test the interpreted result would be c and ϕ). Alternatively a multi-stage procedure may be used in which the same specimen is tested under different conditions. These individual tests are identified by a Test reference, as per AGS, and the results obtained for the individual tests are also stored. This allows the individual results to be re-examined by the engineer, if there is some doubt about the interpretation. In the case of the triaxial test the individual results would be shear strengths measured at different cell pressures. Table 2 shows how this multi-level test storage operates - not all the fields in the data tables are shown to aid clear illustration.

There is also a further level of test data, which are the 'raw data' from which the results on specimens, or sub-specimens, were obtained. The triaxial test version of this would be the data points defining the stress-strain curve etc; for the moisture content test, it would consist of the weight of the sample, wet and dry and the tin weight. While it would be possible to define data structures for storing this level of detail, it is questionable whether this is worthwhile. Different companies will have differing views on what needs to be stored at this level; some might even question the need for this raw data to be included at all in a structured database. However, with the development of integrated data management systems, it would make sense for the data management system to have access to this raw data, particularly since the widespread use of laboratory data acquisition systems will mean that a large part of the data will already be stored in computer files.

To provide this facility, the database system developed can store pointers to files containing the raw data. These files do not need to be structured; they can be simple ASCII files, formatted Document files or Picture files. The _rem_ field in the appropriate test table has been allocated for this usage. If no raw data exists for ASCII storage, this field merely contains comments and remarks on the test, up to 250 characters. However, if raw data is available, this field stores the pointer and full path name of the relevant file or files.

The results from laboratory tests can therefore be stored at different levels. The top level is a result which applies to a specimen and is identified by Specimen Reference, _spdt_ref_. For the simple tests, such as moisture content, this is all that is required. The next level is a result which applies to a sub-specimen, and is identified by a test or stage number, _tesn_ (as per AGS). Pointers to files containing the raw data can be provided at either level. This provides the possibility of storing Level 2 data in an unstructured form. Doing this would mean that the Level 2 data would be available for reading by the engineer, but would not be suitable for access by the KBS.

DATA HANDLING MODULE
The data handling module is the user interface for entering data into the GeoTec database and for subsequently manipulating the data. The SIGMA knowledge-based system has been implemented using ProKappa, an object-oriented development environment. The INGRES database is mapped into ProKappa such that each database table (tuple) becomes a ProKappa object, the slots of which correspond to the fields (columns) within the table and the slot values are the data values contained in the database.

As well as providing the user interface, this module also performs the first level of data checking i.e. that the data entered is of the correct type to match the appropriate database field .

Since entering the soil descriptions in a structured form would be time consuming, the descriptions are simply entered by the user as text fields. A parser has been implemented for breaking down the text description into the component parts for entry to the database structure.

PARAMETER ASSESSMENT MODULE

The parameter assessment module in SIGMA provides data and knowledge to the geotechnical specialist in order to assist in the choice of a value for a particular parameter which will be used for design purposes. The parameter required may have been measured with a variety of different tests. On the other hand, it may be that no direct measurements have been made, and the parameter needs to be assessed from other information. SIGMA provides three levels of parameter assessment (Toll and Oliver, 1993):

1) From direct measurements of the parameter
2) From correlations with other test results
3) From the engineering description of the ground

If the parameter has been directly measured, the system extracts these data from the relevant data-tables. The results are presented separately for different test types, and knowledge about the reliability of the test to measure the parameter is also provided (This is knowledge contained in one of SIGMA's knowledge bases). The applicability of the test to measure the parameter in the type of ground of the chosen layer is also reported at this stage.

When the directly measured data are retrieved from the database, any other measured data are also extracted. A Parameter Correlation knowledge base can then be accessed to see if the parameter required can be obtained from correlations with these other data (Giolas, 1994). A wider scope of results are therefore given to the user from which to formulate a judgement. The system responds with a selection menu of correlations that are applicable. Each correlation may be executed in turn as required by the user. Results from the correlations can be compared with the directly measured results, providing a means of checking the validity of the measurements. In some cases the user might use the results from the correlations as well as, or even in preference to, the direct measurements in coming to a decision as to which value to choose for design.

As a final check, or in the event of there being no measurements which can be used, the parameter required can be assessed from the field description by accessing the Ground knowledge base. This will provide a broad range of typical values (Giolas, 1994). The more detailed the field description, the narrower and more precise will be the range of values.

INTERPRETATION OF GROUND CONDITIONS MODULE

The methodology used by SIGMA for interpreting ground conditions has been described by Vaptismas and Toll (1993). Observations of the ground conditions from boreholes are used to build up a model of the ground conditions across the site as a whole. The interpretation process is approached at two levels: Site-wide and Borehole-to-borehole. At the site-wide level, marker beds which stand out from the general ground conditions (and can therefore be more easily traced across the site) are examined. The continuity of these marker beds is then investigated at a number of different levels. Firstly links are established between possible marker beds in adjacent boreholes. These links are used to develop planar marker beds within groups of three boreholes (triangles). The compatibility between planar marker beds in adjacent triangles is then used to construct site-wide trends. A detailed examination of the ground conditions can then be made at the borehole-borehole level by considering all layers in pairs of adjacent boreholes. This leads to the generation of valid hypotheses which could explain the ground conditions between the two boreholes. The site-wide model is used to constrain the number of hypotheses generated at the borehole-to-borehole level. It also imposes an overall consistency onto the detailed consideration of the ground conditions, limiting local fluctuations which might otherwise be inferred incorrectly.

The SIGMA module for interpreting ground conditions is currently able to identify the similarity of two soils observed in adjacent boreholes, based on the soil descriptions. The qualitative terms are converted to quantitative representations which can then be compared using a Similarity Number (Toll *et al*, 1993). For each pair of boreholes a matrix of Similarity Numbers is produced indicating how similar each layer of the first borehole is with every layer of the second. By identifying the maximum Similarity Number between layers, a set of valid hypotheses is generated about the soil profile between the two boreholes, based on the principle that layers cannot 'cross over'. The hypotheses are ranked according to the number of layers participating and their average Similarity Number, and these can then be used to identify site-wide trends.

CONCLUSIONS
The efficient management of site investigation data, combined with computer systems which can assist with data processing and interpretation, could have a significant effect on the site investigation industry. Multi-user access to a central source of site investigation data ensures all users are utilising and making decisions with data that is consistent.

The combination of database and knowledge-based system technology can provide a powerful tool for the geotechnical specialist. A knowledge-based system for data interpretation can provide a means of linking database systems with other computer packages.

REFERENCES
AGS (1992) *Electronic transfer of geotechnical data from ground investigations*, Publ. AGS/1/92, Association of Geotechnical Specialists, Wokingham.

AGS (1994) *Electronic transfer of geotechnical data from ground investigations*, 2nd ed., Association of Geotechnical Specialists, Camberley, Surrey.

Date (1987) *An Introduction to Database Systems* (Vol.1, 4th ed.), Addison-Wesley, Reading Mass.

Giolas, A. (1994) *A knowledge-based system for estimating geotechnical design parameters*, PhD thesis, University of Durham (in preparation).

Greenwood J.R. and Rothery M.G.W. (1992) *Collection, Transfer, Storage and Presentation of Ground Investigation Data*, Proc. International Conference on Geotechnics and Computers, Paris, Presses de l'École Nationale de Ponts et Chaussées, Paris, pp 691-698.

Howland A.F. (1992) *Use of computers in the engineering geology of the Urban Renewal of London's Docklands,* Quarterly Journal of Engineering Geology, 25, 4, pp 257-267.

Mott MacDonald and Soil Mechanics Ltd (1994) *Study of the efficiency of site investigation practises,* Project Report 60, E063A/HG, Transport Research Laboratories, Berkshire.

Oliver, A.J. (1994) *A knowledge-based system for the interpretation of site investigation information,* PhD thesis, University of Durham.

Relational Technology (1990) *An Introduction to INGRES,* HECRC, ISG 245, Relational Technology Inc, California, USA.

Staten P.M. and Caronna S. (1992) *Geotechnical Data Management of a Major Highway Scheme in the UK,* Proc. International Conference on Geotechnics and Computers, Paris, Presses de l'École Nationale de Ponts et Chaussées, Paris, pp 741-748.

Toll, D.G. (1994) *Interpreting Site Investigation Data using a Knowledge Based System,* Proc. 13th Conference of Int. Soc. Soil Mechanics and Foundation Engineering, New Delhi, Balkema, Rotterdam, Vol.4, pp 1437-1440.

Toll, D.G. and Oliver A.J. (1993), *SIGMA: A System for Interpreting Geotechnical Information,* Proc. SERC Conf. on Informing Technologies for Construction, Civil Engineering and Transport, (eds. Powell J.A. & Day R.), Brunel University with SERC, London, pp 245-254.

Toll, D.G. and Oliver A.J. (1994) *Structuring Soil and Rock Descriptions for Storage in Geotechnical Databases,* Accepted for Geol. Soc. Special Publication on Geological Data Management.

Toll D.G., Moula M., Oliver A. and Vaptismas N. (1992), *A Knowledge Based System for Interpreting Site Investigation Information,* Proc. International Conference on Geotechnics and Computers, Paris, Presses de l'École Nationale de Ponts et Chaussées, Paris, pp 607-614.

Toll D.G., Vaptismas N. and Moula M. (1993) *A Methodology for Comparing Soils for use in Knowledge Based Systems,* Computer Systems in Engineering, 4, pp 317-324.

Vaptismas N. and D.G. Toll (1993) *Interpreting Borehole Information,* in Knowledge Based Systems for Civil and Structural Engineering (ed. B.H.V. Topping), Civil-Comp Press, Edinburgh, pp 153-159.

Management of spatial data for the heavy foundations of a blast furnace

PAUL NATHANAIL[1] and MICHAEL ROSENBAUM[2]
[1]CRBE, Nottingham Trent University, Burton Street, Nottingham NG1 4BU, UK
[2]Department of Geology, Imperial College of Science, Technology and Medicine, London, UK

ABSTRACT

A ground characterisation exercise was conducted in order to identify sites suitable for a blast furnace at the Redcar steelworks in northeast England. The purpose of this exercise, conducted some twenty years after the investigations were carried out and a decade after the steelworks had been constructed, was to develop techniques for managing spatial data using Geographical Information Systems (GIS) and geostatistics.

Borehole details, from the factual reports, were entered into a database using a relational structure with linked files. The borehole location and lithology files were keyed in using codes for geological material, weathering grade, strength, borehole type and verification status. Laboratory and field test data were input by scanning tables in the ground investigation reports and processing the scanned images using optical character recognition software.

A vector GIS was used to carry out point-and-query and spatial filtering operations both during the data validation phase and during the review of individual geological units.

The XBase programming language was used to execute graphing, GIS and geostatistics programs; thereby enabling repetitive tasks to be carried out by the computer. This both saved time and reduced errors and inconsistencies that could result from repeated manual execution of the same program. Some simple spatial operations could also be carried out entirely from within dBase, an example being the use of moving-window statistics employing rectangular windows aligned parallel to the axes of the coordinate system to study local stationarity.

GIS have been found to offer an effective tool for analysing, querying and presenting spatial site investigation information. Geostatistical kriging and simulation techniques offer powerful spatial modelling tools for creating visualisations of the spatial variability of parameters measured at points during a ground investigation.

INTRODUCTION

A ground characterisation exercise was conducted in order to identify sites suitable for a blast furnace at the Redcar steelworks in northeast England. The Redcar case history was chosen because it involved a relatively large site investigation with a significant, but not abundant, quantity of information, which is now no longer commercially sensitive. The purpose of this

Advances in site investigation practice. Thomas Telford, London, 1996

exercise, conducted some twenty years after the investigations were carried out and a decade after the steelworks had been constructed, was to develop techniques for managing spatial data using geographical information systems and geostatistics. The work was conducted in four phases:

a.   Compilation of a geotechnical database, scanning and digitising of paper maps.
b.   Geostatistical modelling (kriging and conditional simulation) of selected parameters.
c.   Integration of ground investigation information and geostatistical estimates in a GIS.
d.   Selection of suitable blast furnace sites.

Only the first two phase are described in detail here. Details on the third and fourth phase have been described elsewhere (Nathanail 1994; Nathanail & Rosenbaum 1994).

The heaviest structures in the new steelworks were to be the blast furnaces with a weight of 37,000 tons (for the structural tower and the hearth) and a tower height of 100 m above ground level (Jorden & Dobie 1976). The foundation slab was to be a 4 m thick square concrete slab of 38 m side length, weighing 18,000 tons, giving a total weight of 55,000 tons. Due to the weight of liquid metal, these furnaces have a very low tolerance for tilting (reflected in a limiting acceptable overturning moment of 110,000 ton-metres). Since they are being fed by conveyors and gravity chutes they must also not undergo excessive total settlement.

A series of site investigations was carried out during the early 1970s (Soil Mechanics Ltd 1972, 1973, 1974). A large number of boreholes were sunk for the whole site development, which covered an area of approximately 3 km by 1.5 km on the south bank of the Tees estuary.

SITE GEOLOGY

The regional geology, consists of Triassic and Jurassic bedrock overlain by fluvial, glacial, estuarine and aeolian superficial deposits (Institute of Geological Sciences 1883; British Geological Survey 1987). The Triassic succession beneath the site comprises Mercia Mudstone and Blue Anchor Formation of the Mercia Mudstone Group and Westbury and Lilstock Formations of the Penarth Group. The Jurassic sub-crop is restricted to the Redcar Mudstone Formation of the Lower Lias Clay.

The history of development at the site was explored by examining various editions of Ordnance Survey mapping. The Ordnance Survey 1:10560 sheets record the development of the Warrenby steelworks and the associated infilling of Bran Sands and 'The Marshes' (Ordnance Survey 1919 and 1930). Little change took place between 1930 and 1970 (Ordnance Survey 1970). The construction of the Redcar Steelworks, the subject of the present study, resulted in further reclamation of Bran Sands (Ordnance Survey 1988).

INTERACTION MATRIX FOR FOUNDATIONS

A generic matrix of potential interactions between geological history, environmental factors, engineering geology and engineering processes was established for foundations, using the principles described by Hudson (1992), to assist with the familiarisation of the ground investigation data and site circumstances (Figure 1). The leading diagonals of the matrix emphasise the environment (geological history, stress state and groundwater conditions), the mechanical properties of the ground (strength, deformability, permeability and durability) and engineering activity (foundation type and ground improvement) in response to the project objectives. During the creation and analysis of the database the generic rock-engineering

GEOLOGICAL HISTORY	residual stress; low seismic hazard	presence of ancient shear surfaces	strata of alternating properties	metastable mineralogy	development of fissures	age of groundwater	depth to suitable stratum	type and extent of unsuitable ground
determines tectonic evolution	IN SITU STRESS	strength is a function of stress state	differential stress state determines type of deformation	pressure solution	determines aperture of fissures	pore pressure influenced by total stress state	squeezing soils hinder cast in situ piles	high stress may require extra support
influences degree of fracturing	determines maximum stress that can be applied	STRENGTH	fractured/ failed material is more deformable	strong materials resist physical weathering	intact rock less permeable than fractured	failure can create new flow paths	influences type of foundation	weak material replaced or improved
influences response to changes in stress field	contrasts lead to local stress peaks	brittle materials fail suddenly	DEFORM-ABILITY	response to physical weathering	deformation can lead to closure of flow paths	consolidation results in water being removed	affects type and size of foundation	replace soft materials
durable materials form long term land masses	weathering affects how much stress can be maintained	weathered materials tend to be weaker	weathered materials deform in a ductile way	DURABILITY	solution can increase permeability	chemistry altered by solution of unstable minerals	exposed soil can wet and weather quickly	protect slakeable materials
rate of alteration	rate of dissipation of excess pore pressure	more pores/ fractures result in lower strength	voids are more compressible than solid material	rate of chemical weathering	PERME-ABILITY	Darcy's law $Q = KiA$	rapid flows wash away cement before it sets	drainage may be required
facilitates tectonic movement	reduces effective stress	saturated materials are weaker	water is less compressible than air	chemical weathering agent	hydraulic fracture or silting up of fissures	GROUND-WATER	chemistry influences cement used	quantity may control selected method
n/a	local redistribution of stress	improved by compaction	closure of fissures	excavation exposes soil to wetting	foundation loads can close fissures	affects flow path	FOUNDATION TYPE	determines need
n/a	local stress re-distribution	usually aim to increase strength	usually reduced	usually improved	reduced if fissures sealed	de-watering	cheaper foundation feasible	GROUND IMPROVEMENT

Figure 1   Generic rock - engineering interaction matrix for foundations

interactions (Figure 1) were replaced with those applicable to Redcar (Figure 2). Some interactions were found to be irrelevant to the circumstances at Redcar and the appropriate boxes indicated this.

The leading diagonals of the matrix in Figure 2 were selected to reflect the site environment, engineering geology and proposed construction. An extra parameter representing ecology could have been added and this would enable interactions causing an environmental impact to be highlighted. Alternatively, a separate matrix designed solely to assess the impact of the steelworks on the environment (rather than the impact of the environment on the steelworks) could be set up with appropriate leading diagonals, which could possibly include: air, water, soil, flora, fauna, geomorphology, anthropogenic activities, land use and steelworks development.

COMPILATION OF THE GEOTECHNICAL DATABASE

Details of 147 boreholes and four piston samples were entered into a database. The database was compiled from the factual site investigation reports using dBase IV; a relational structure with linked files was adopted. The borehole location and lithology files were keyed in using codes for geological material, weathering grade, strength, borehole type and verification status. Laboratory and field test data were input by scanning tables in the ground investigation reports and processing the scanned images using optical character recognition software.

Database Structure

Information was stored in seven tables The key fields used to link these tables were borehole number, lithology and sample depth. The LOCATION and LITHOLOGY files were linked by borehole number (BH_NO). The five tables of test results were linked to LOCATION and LITHOLOGY on borehole number and a condition box which ensured that the sample depth lay within the depth range of a single stratum. The expression in the dBase condition box was: depth2top < testdepth .AND. testdepth > depth2base. The relations were stored in dBase Query-By-Example (*.QBE) files. It was found, however, that it was both faster and easier to then store the QBE files as a DBF file, despite the extra storage overheads. The original tables were always treated as the master files and any amendments to the data were made in those tables.

Data Entry and Verification

Data were entered by keying-in either the full text or codes which were later expanded by a dBase program, and scanning followed by optical character recognition (OCR). Pick lists were used to constrain the range of allowable entries in fields for weathering grade, strength class, borehole type, geological unit and engineering material. A pick list is a .DBF file which is displayed in a window while entering data in a field and which contains permissible values for that field. This helps to ensure consistent terminology and rapid data entry.

A dBase program (THICKCHK.PRG) was written to calculate strata thicknesses from the depth to the top and bottom of a stratum and compare the calculated value with the keyed in value. Records with discrepancies were flagged and the original borehole logs checked. Most of the discrepancies were due to keying errors but a few were the result of errors in the original logs. In the latter case the depth information on the log has been assumed to be correct and the thickness value amended. The DEPTHCHK.PRG program was written to check that the depth to the base of one stratum was the same as the depth to the top of the underlying

GEOLOGICAL HISTORY	low residual stress & low seismic hazard	no strong rocks	deformable estuarine & lacustrine deposits	low durability gypsum deposited	alternating soils of low/ high permeability deposited	gypsum in Mercia Mst results in high SO_3^{2-}	unsuitable foundation level until rockhead	Bran sands need reclaiming
absence of major tectonic features	IN SITU STRESS	strength is a function of stress state	high OCR in Mercia Mst	pressure solution	determines aperture of fissures	pore water pressure influenced by total stress state	low stress environment	No high stress requiring extra support
low strength allowed Tees estuary to form	strong bands create stress shadows	STRENGTH	strong bands shield deformable bands	weak aeolian sands easily eroded	intact rock less permeable than fractured	failure can create new flow paths	weak shallow soils lead to use of deep piles	Compaction of loose fill to increase strength
mudstones were consolidated	concentrated in fresh bands of Mercia Mst	strong bands of Mercia Mst underlain by deformable bands	DEFORM-ABILITY	response to physical weathering	compaction of lacustrine deposits reduced vertical permeability		soft soils means deep piles needed	reclamation with thick layer of brittle slag
formed a low area	materials eroded rather than sustain high stress	weathered Mercia Mudstone is weaker	limestone bands weather less and are less deformable	DURABILITY	solution of gypsum can increase permeability	sulphate raised by solution of gypsum	piles avoid exposure of bedrock	n/a
chemical weathering of Mercia Mst	controls dissipation of excess pore pressure	more pores/ fractures result in lower strength	voids are more compressible than solid material	rate of chemical weathering	PERME-ABILITY	Darcy's law $Q = KIA$	Slag fill and aeolian sand likely to be water saturated	drainage of fill needed
absence gave rise to Mercia Mst weathering	reduces effective stress	saturated materials are weaker	water is less compressible than air	chemical weathering agent	solution of gypsum widens fissures	GROUND-WATER	sulphate-resistant cement may be needed	potential connection with sea may require coffer dam
n/a	local redistribution of stress	improved by compaction	closure of fissures	n/a	foundation loads close fissures	pile groups affect flow path	FOUNDATION TYPE	piles into rock - no improvement needed
n/a	fill increases vertical load	Marsh area improved	Fill spreads shallow loads	fill shields erodable soils from waves	reduced by layer of fill	Sea kept out of Bran Sand	More area available	GROUND IMPROVEMENT

Figure 2   Interaction matrix for blast furnace foundations at the Redcar steelworks

Figure 3   Multiple intersections of a unit

stratum; discrepancies were flagged and checked against the original logs. The grid reference of each borehole, as given on each log, was verified by plotting a site plan and comparing with the original drawings.

The ability to view the data spatially and to query individual points was important both during the data validation phase and during the review of individual geological units. Idrisi cannot query dBase files. A vector GIS which could read DBF files was, therefore, used to carry out point-and-query operations.

<u>dBase Programs</u>

DBase programs were written to:
a.   Link the various database tables (*.QBE).
b.   Verify the database tables.
c.   Calculate moving window statistics for any selected field.
d.   Write ASCII files for subsequent use in other programs.

Programs to link database tables were based on the output from dBase Query-By-Example module.

Moving window statistics were calculated using the WINDSTAT.PRG program (Nathanail 1994). The window can be square or rectangular and the amount of north-south and east-west overlap can be varied. The output, which is to an ASCII file, includes the minimum, maximum, mean and standard deviation of the desired fields. The sequence of output was such that it could be read into a raster GIS for further processing and analysis. The output was also used in studies of the relationship between mean and standard deviation to test for the presence of a proportional effect in the rockhead elevation data.

Some borehole logs indicated more than one intersection of a particular unit. Some of these intersections were contiguous and merely reflected variations within the unit. Others were due to an intervening stratum of a different unit. The contiguous intersections were identified and concatenated using the DUPLI.PRG program (Nathanail 1994). Remaining multiple intersections were examined manually. For example, in Figure 3 four intersections of sand (snd) are shown. The top three are contiguous and their thicknesses should be added together for a study of sand thickness. The fourth intersection of sand, below, is a band of estuarine deposits (ed) and would need to be excluded from the study of thickness of the sand above the estuarine deposits. It was important to ensure that the record sequence was in borehole number and depth-to-base order before running DUPLI otherwise it may not have been the deepest intersection of a stratum that was retained. To ensure this, the database was indexed on both borehole number (character field BH_No) and depth (numerical field DEPTHBASE), converted to a character string, using the following numerical expression: BH_NO + STR(DEPTHBASE).

SIMULATE.PRG (Nathanail 1994) executes the *sgsim.exe* module of GSLIB written by Deutsch and Journel (1992), reads the simulation into a DBF file and writes an Idrisi image (.IMG) file. An image documentation file (.DOC) for use within the Idrisi raster GIS is also written. Then *sgsim.exe* is executed to produce another simulation. This is repeated until all simulations have been created.

Tools for running DOS programs and reformatting data files were also produced using the dBase language. Programs were written to produce ASCII data files for use by GeoEAS, VarioC, Grapher and Idrisi.

SPATIAL DATA INSPECTION, ANALYSIS AND QUERY

The ability to view the data spatially and to query individual points was important both during the data validation phase and during the review of individual geological units. Therefore the Atlas GIS, which can read dBase DBF files, was used to carry out point-and-query and spatial filtering operations.

The dBase programming language was used to execute programs such as Grapher, Idrisi and GSLIB and enabled repetitive tasks to be carried out by the computer. This both saved time and reduced errors and inconsistencies that could result from repeated manual execution of the same program. Some simple spatial operations could also be carried out entirely from within dBase, an example being the use of moving-window statistics employing rectangular windows aligned parallel to the axes of the coordinate system to study local stationarity. Boreholes could also be selected, or excluded, on the basis of their grid reference. However, point-in-polygon selection ('cookie-cutting') with irregularly shaped polygons, while possible, was done more easily using other software! Programs were also written to produce ASCII data files for use by GeoEAS, Grapher and Idrisi.

SPATIAL SUMMARIES

Moving window statistics were calculated using the WINDSTAT.PRG program. The window can be square or rectangular and the amount of north-south and east-west overlap can be varied. The output includes the minimum, maximum, mean and standard deviation of the desired fields. The sequence of output was such that Idrisi image files of the window statistics could then be easily produced. The output was also used in studies of the relationship between mean and standard deviation to test for the presence of a proportional effect in the rockhead elevation data.

CONCLUSIONS

The dBase language has been found to be more powerful than DOS batch files and more flexible than programming languages such as FORTRAN. In particular the facility to access dBase data files directly rather than having to generate the rigorously formatted data files that FORTRAN requires saved a great deal of effort. Recourse to a GIS and specialist geostatistical software was necessary for spatial processing and interpolation.

The generic rock interaction matrix was found to be a useful tool for checking that all facets of the site circumstances had been considered. When using a raster GIS to model site conditions it is easy to keep on generating ever more layers and the building of the GIS model can become an end in itself! The interaction matrix helps one decide which layers are really necessary and how they should be combined in order to achieve the objectives of the project. In addition, the process of tailoring the interaction matrix to the situation at Redcar highlighted aspects that were not immediately obvious. For example, the geological history of the Redcar area interacts with the in situ stress regime to give rise to a low seismic hazard.

The advantage of the relational structure for geological characterisation in terms of reduced storage requirement is somewhat reduced by the need to generate secondary files for data transfer purposes. The decreasing financial cost of hard disks is in practice tending to negate this disadvantageous aspect of relational structures. However, the simplicity gained by storing, and therefore checking and correcting, 'master' information only once is still a real and significant benefit.

REFERENCES

BRITISH GEOLOGICAL SURVEY. 1987. *1:63,360 scale geological map sheet no. 33: Stockport (combined solid and drift)*. British Geological Survey, Keyworth.

DEUTSCH, C. V. & JOURNEL, A. G. 1992. *GSLIB: Geostatistical software library and user's guide*. Oxford University Press, New York.

HUDSON, J. A. 1992. *Rock engineering systems: Theory and practice*. Ellis Horwood, London.

JORDEN, E. E. & DOBIE, M. 1976. Tests on piles in Keuper Marl for the foundations of a blast furnace at Redcar. *Géotechnique*, **26**(1), 105-114.

INSTITUTE OF GEOLOGICAL SCIENCES. 1883. *1:63,360 scale geological map sheet no. 34: Guisborough (drift)*. British Geological Survey (Formerly Institute of Geological Sciences), Keyworth.

NATHANAIL, C. P. 1994. *Systematic modelling and analysis of digital data for slope and foundation engineering*. Phd Thesis, Imperial College, University of London (unpublished).

NATHANAIL, C. P. & ROSENBAUM, M. S. 1994. Selecting sites suitable for the heavy foundations of a blast furnace using a GIS, *Proceedings Seventh International Congress of the International Association of Engineering Geology*, Lisbon, September 1994 (in press).

ORDNANCE SURVEY 1919. *Yorkshire, north Riding, 1:10,560 map sheets VII and VIIA*, Revised edition, Ordnance Survey, Southampton.

ORDNANCE SURVEY 1930. *Yorkshire, north Riding, 1:10,560 map sheets VII and VIIA*, 2nd revised edition, Ordnance Survey, Southampton.

ORDNANCE SURVEY 1970. *1:63,360 scale map sheet 93*. Ordnance Survey, Southampton.

ORDNANCE SURVEY 1988. *1:50,000 scale map sheet 93: Middlesborough and Darlington*. Ordnance Survey, Southampton.

SOIL MECHANICS LTD. 1972. *Site investigation report No. 6039: Site investigation for a proposed major steelworks extension, Stage 2 development, Redcar, Yorkshire (in four volumes)*. Soil Mechanics Ltd, Doncaster (unpublished).

SOIL MECHANICS LTD. 1973. *Site investigation report No. 6260/1: Further site investigation for a proposed major steelworks extension, Stage II development, Redcar, Teeside*. Soil Mechanics Ltd, Doncaster (unpublished).

SOIL MECHANICS LTD. 1974. *Site investigation report No. 6369*. Soil Mechanics Ltd, Doncaster (unpublished).

ACKNOWLEDGEMENTS

The support of Wimpey Environmental during the research for this work is gratefully acknowledged by the principal author.

Management of geologic and instrumentation data for landslide remediation

LEE W. ABRAMSON and THOMAS S. LEE
Parsons Brinckerhoff Quade & Douglas, Inc., San Francisco, California, USA

ABSTRACT

Provo Canyon in Utah has a complex geologic history beginning as long as 300 to 500 million years ago. Continuing into the present, the Canyon has undergone active uplift and faulting, particularly along the Wasatch Front. Smaller scale erosion and landsliding have resulted in the present day topography. Widening of US-189 from two to four lanes within Provo Canyon has to account for a section of roadway through a major landslide area which has been moving for the last 50 years and probably longer. Slope movements are associated with the Manning Canyon Shale formation and are located along a 1.6-kilometer-long stretch of canyon beginning roughly 2.4 kilometers below Deer Creek Dam. This area is generally referred to as the Hoover Slides section of Provo Canyon. These active failures pose the most challenging geotechnically related issues along the existing roadway as well as the proposed widening.

Field work included borings, test pits, in situ testing, CPTs, downhole geophysical testing, and installation and monitoring of inclinometers, survey points and observation wells/piezometers. Additionally, geologic mapping was conducted, and seismic refraction methods were used to delineate soil/rock characteristics and boundaries.

The paper will discuss the collection and synthesis of exploration, laboratory testing and instrumentation data used to determine information such as probable failure plane(s), shear strength parameters, soil data, groundwater pressures, and inclinometer data necessary to analyze the landslide area and evaluate remediation methods. Various types of geotechnical computer applications were used to manage subsurface data collected from the exploration, laboratory testing and instrumentation programs. Relational data bases were required to store subsurface data. Procedures are detailed which allow the merging of geotechnical data with data from other disciplines using a project-wide Computer Aided Design and Drafting (CADD) system.

INTRODUCTION

Highway U.S. 189 through Provo Canyon, Utah (Figure 1) is being widened from 2 lanes to 4 lanes and realigned for an 80 kilometers per hour speed limit. The multimillion dollar 24-kilometer-long project is being designed and constructed in phases. The lower 12 kilometers have been constructed and are in use. The 3.2-kilometer-long middle section is called the "Narrows". This project includes two tunnels, 0.4 million cubic meters of rock excavation, and several thousand square meters of retaining walls. Construction on this section is

Advances in site investigation practice. Thomas Telford, London, 1996

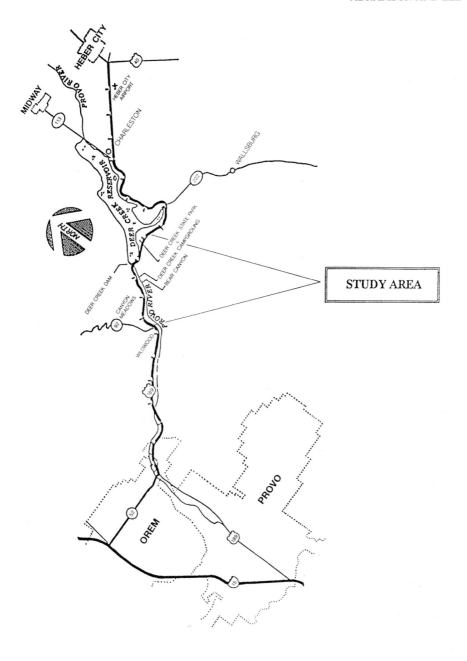

Figure 1 - Site Location Map

expected to start next summer. The upper 8-kilometer-long section starts from Wildwood to Deer Creek State Park. The widening of this section includes two large active landslide zones in the Canyon Meadows and the Horseshoe Bend areas of Provo Canyon. Both areas are collectively called the Hoover Slides area. The main challenge on this project is to develop methods for constructing the new road through the Hoover Slides area.

The Hoover Slides area has been active for over 50 years and actually consists of six active landslides. Of these six slides, at least three are significant slides measuring 150 to 210 meters long and 30 to 100 meters wide. In addition, slumps of a minor nature occur in cuts above the existing highway and in the embankment downslope from the highway.

The geotechnical data for this project includes 50 exploration boring logs and the associated laboratory, field, and geophysical testing results. To assist in managing the large amount of data collected, geotechnical computer applications were used to manage these subsurface data. Relational data bases were required to store subsurface data, and the steps required to create and manage these data are discussed. Procedures are detailed which allow the merging of geotechnical data with data from other disciplines using a project-wide Computer Aided Design and Drafting (CADD) system.

GEOLOGIC SETTING
The Hoover slides are located on the site of a large prehistoric landslide in the vicinity of Canyon Meadows. It is underlain by the Manning Canyon Shale formation which consists of black to brown shale with interbeddded slabby sandstone, thin beds of quartzite, and thin- to thick-bedded gray to black limestone. The shale weathers rapidly when exposed to the atmosphere and becomes highly plastic when wet.

The project area is heavily thrust faulted, resulting in intense fracturing of the rock. The Provo River has eroded through the upper plate of the thrust fault, creating the Sulfur Springs window as per the United States Geological Survey (USGS) 1964 and 1992 maps (Figure 2). The Manning Canyon shale is stratigraphically below the Oquirrh Formation limestone. In upper Provo Canyon, the Manning Canyon shale and Oquirrh Formation limestone are in contact along the thrust-fault.

DATA MANAGEMENT
Several computer programs were used for data management on this project. They fall into three major categories: (1) Data Bases; (2) Analytical Programs; and (3) Graphical Programs. To the extent possible, these programs have been linked to one common data base. Redundant data files were avoided. The data base programs were used to store the positioning and characteristics of geologic data developed from the borings, instrumentation, geologic mapping, geophysical testing, and joint surveys. These data bases were then used to perform stereonet and soil and rock slope stability analyses as well as to present the data and results graphically on plans and cross sections. The graphical (CADD) programs were used to generate three dimensional terrain and subsurface models with which evaluations of subsurface conditions (i.e. stratifications, shear zones, faulting, etc.) and slope stability could be modeled. As the design evolved, it was relatively simple to input various alternative alignments as changes were made and evaluated. This was particularly useful for evaluating potential impacts of new alignments on the landslide areas.

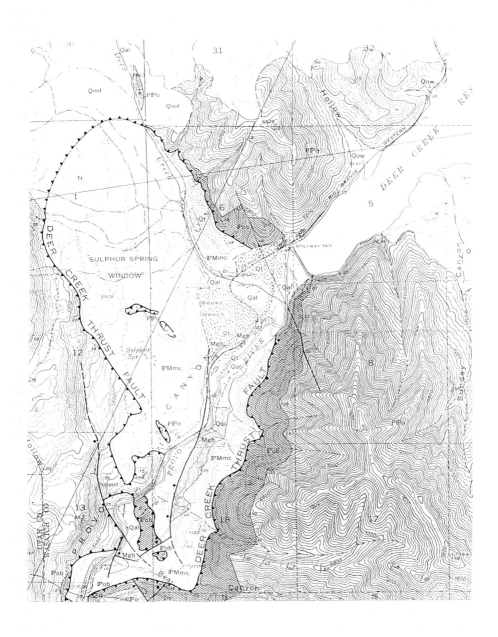

Figure 2 - Geologic Map

COLLECTION OF GEOLOGIC DATA

Geologic data have been collected by means of desk studies, surface studies and subsurface exploration. The desk studies consisted of reviewing maps such as USGS geological maps, landslide records, seismic records, aerial photographs, and published geologic reports and literature pertaining to the Hoover Slides area. Surface studies included site reconnaissance, geologic mapping, and joint surveys.

Desk Studies

Review of Maps and Records. USGS maps were reviewed in great length to identify geomorphologic forms and drainage patterns. These maps were extremely useful because they provided information on material types and geological structures such as thrust and normal faults. They served as a basis for developing the site-specific geologic maps of the Hoover Slides area.

Records of slope failures in the Hoover Slide areas are limited. The areas have been subject to instability, sloughing and raveling for at least the past fifty years.

Review of Publications and Literature. Geotechnical site investigations have been carried out during the past fifty years by the State Road Commission of Utah and later, the Utah Department of Transportation (UDOT) for construction of the highway between the Utah County Line and Heber City, Utah. Also, a feasibility study was made by the Bureau of Reclamation (BOR) in the early 1940's for the Salt Lake Aqueduct through Provo Canyon. The boring logs and laboratory testing results from these site investigations provided invaluable geotechnical data pertaining to the Hoover Slides area.

Review of Aerial Photographs. Aerial photographs dating back to the 1940's to the present time were reviewed. The Hoover Slide areas were thoroughly studied for signs of instability such that the limits of the Hoover Slide areas could preliminarily be delineated. Inferred fault lines and drainage patterns were mapped and then plotted on drawings by means of CADD.

Surface Studies

Site Reconnaissance. Site reconnaissance was performed to observe, recognize, and record surficial features which might affect the stability of slopes along the proposed US-189 alignment.

Geologic Mapping. The geology of the Upper Provo Canyon was originally mapped by Baker (1964) at a scale of 1:24,000 and later by Bryant (1992) at a more regional 1:125,000 scale. These U. S. Geological Survey maps served as a base for the project geologic map and regional geologic maps.

Geologic areas and contacts were digitized from the geologic maps through the use of CADD. The digitizing of the geologic maps was first started by setting up CADD files using UDOT CADD guidelines, which was subsequently followed by the translation of metric topography, planimeters, and the proposed alignments from Intergraph to GDS software. These new metric topography drawing and geologic maps serve as base drawings (Figure 3). With the geologic data stored in CADD and GDS files, any topography, planimetrics, geologic and alignment alternatives can be drawn and evaluated with ease on individual sheets, and any information such as boring locations, geologic data can be added or deleted easily.

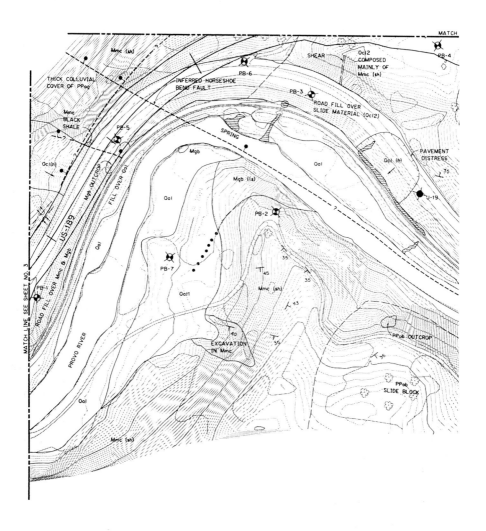

Figure 3 - CADD Plan Showing Geologic Features

Joint Survey. Twelve discontinuity transects (scan lines) were measured on bedrock outcrops along the existing US-189 alignment and on cuts near the Provo River. Generally, each transect spanned the entire length of an exposure to obtain as much data as possible. Several parameters were measured and classified in accordance with the method outlined by Hoek and Bray (1974). These parameters included: (1) rock types; (2) rock strength; (3) weathering; (4) structural types; (5) dip direction (or strike) and dip; (6) spacing; (7) persistence; (8) joint roughness coefficient (JRC); (9) aperture; (10) asperity height; (11) signs of seepage, and (12) types of joint infilling. The associated parameters for the joints in each transect are downloaded into a geologic data management file for stereonet and stability analyses.

Stereonet Analysis. Stereonets provide an invaluable tool for displaying spatial orientations of discontinuity trends. All twelve stereonets were plotted on Schmidt Equal Area Nets, each representing one of the twelve discontinuity transects (Figure 4). Because the number of data points on individual stereonets included up to 330 data points, the percentage contours were adjusted on each stereonet to best represent characteristic clusters of data points and to ignore the less significant data points. The contour clusters of data points were used to identify the characteristic joint sets for the transects. Representative attitudes of the characteristic discontinuity sets were taken from the center of the point cluster on the stereonets. The variability of joint set orientations was also evaluated from the stereonets based on the size and spatial orientation of the cluster.

Stability Analysis. With the major joint sets identified through stereonet analyses, stability analysis for each proposed rock cut slope along the new US-189 was performed by means of rock stability software programs for sliding, wedge, and toppling modes of failure.

Subsurface Exploration

Characterization of the landslides required the acquisition of specific subsurface information including borings and instrumentation.

Boring Exploration. In addition to the borings drilled several years ago, twenty four borings varying from depths of 60 to 150 feet were drilled by UDOT in the Canyon Meadows area between 1993 and 1994 (prior to commencement of design work). After design began, twenty-six deep borings of depths from 120 feet to 165 feet were drilled between May and September 1994 to determine the subsurface conditions at Horseshoe Bend and other areas in the vicinity of Canyon Meadows.

Instrumentation. Instrumentation being used to monitor the landslides include inclinometers and piezometers.

Piezometers Several open standpipe piezometers were installed by UDOT at the toe of the landslides in the Hoover Slides area in 1993. The water levels in these piezometers were periodically recorded. These data were plotted against rainfall data with time to see correlations between rainfall and height of water level (Figure 5).

Inclinometers Thirty inclinometers have so far been installed to monitor the ground movements in the Hoover Slides area. Of these thirty inclinometers, twenty three were installed in the Canyon Meadows area, and seven in the Horseshoe Bend area. These inclinometers have been periodically monitored since 1993. Readings are transferred from the field readings to computer data bases.

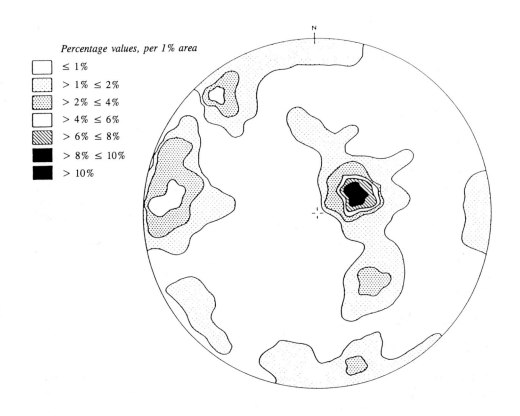

Percentage values, per 1% area

□ ≤ 1%

▨ > 1% ≤ 2%

▨ > 2% ≤ 4%

□ > 4% ≤ 6%

▨ > 6% ≤ 8%

■ > 8% ≤ 10%

■ > 10%

Figure 4 - Stereonet

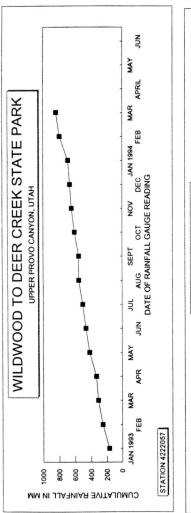
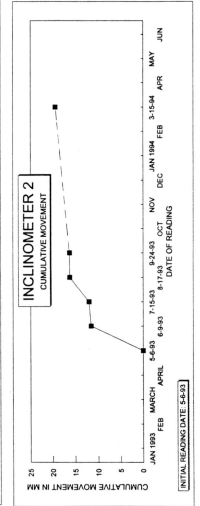

Figure 5 - Ground Movement Versus Rainfall Comparisons

Laboratory Testing. Selected soil samples and rock cores were tested for Atterberg Limit, unit weight/moisture content, slake durability, direct shear, unconfined compressive, grain size distribution and hydrometer, consolidation, and triaxial compressive tests. Clay samples retrieved from potential sliding zones were selected for X-ray diffraction analyses. Water chemistry tests were made on representative samples from selected boreholes and piezometers.

Geophysical Testing. Seismic refraction surveys have been carried out in selected areas within the Hoover Slides area to delineate the rock/soil interface, to determine the thickness of alluvial deposits below the Provo River, and to supplement the subsurface information between borings. Downhole geophysics logging was also performed on selected boreholes to confirm and refine the location of sliding plane.

SOLID MODELING

With the development of improved graphics software and hardware in recent years, it is possible to represent any object fully in three dimensions easily. These objects can be manipulated and viewed from different angles and at any scales. This technique can be used to present subsurface conditions in three dimensions and to interpret inclinometer readings in three dimensions. Figure 6 shows an example of a 3-D subsurface profile.

A specially-designed computer program was used to reduce and plot the inclinometer data. This information was then used to plot the displacements of multiple inclinometers in three dimensions using CADD to locate slippage planes and other anomalous features.

With the aid of the 3-D graphics, it is possible to determine the direction, inclination and thickness of potential sliding planes through the marriage and interpretation of boring log data and inclinometer readings. In addition, the 3-D graphics enable the designers to analyze the landslide areas and evaluate remedial works with a clearer perspective than would have been possible with 2-D drawings.

STUDY RESULTS

The data and analyses discussed above have been synthesized and used to develop models of behavior at the Hoover Slides area for use in alternative alignment studies. The sources and conclusions are described below.

Geological Models

Based on the geologic data synthesized, there are two distinct failure mechanisms occurring in the Hoover Slides area. The Horseshoe Bend area is acting as a rigid block sliding on a weak plane in the southern portion of the bend in the river. The northern portion of the bend in the river is a weak (semi-intact) rock mass sliding on a series of sliding planes at depths of 15 to 40 meters. The second area, Canyon Meadows, is a historic creeping landslide mass composed of very weak and weathered rock and soil which slide on a sheared/gouge zone of Manning Canyon shale at a depth of about 23 to 30 meters towards the alluvial deposits below the Provo River.

Verification of Geological Models

The inclinometer readings have indicated the ground movement is on the order of about 10 to 40 mm per year and is directed towards the alluvial deposits below the Provo River in the south to south east direction. The ground movement below the US-189 in the Canyon Meadows is deflected to the south to southeast direction as a result of a more competent

Legend

▦	PP_o	Oquirrh Limestone
▦	PM_mC	Manning Canyon Shale
▦	Mgb	Great Blue Limestone
▦	Landslide Debri	
▦	Alluvial Deposits	
☐	Fill	

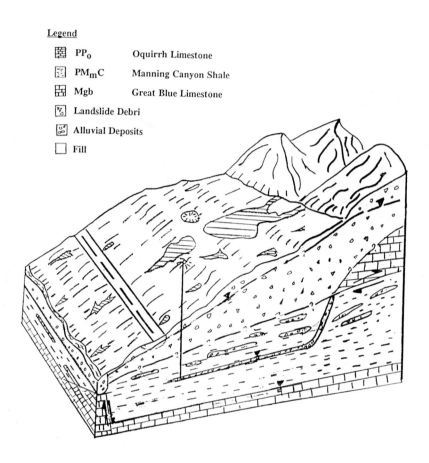

Figure 6 - Three Dimensional Block Diagram

limestone deposit acting as a buttress to movements of the slide materials. The movements occurred in deep-seated sliding planes as indicated by the inclinometer readings.

Results of X-ray diffraction analysis on samples taken from the sheared and gouge zones showed depletion of pyrite in the Manning Canyon shale (which formed sulfuric acid on hydration and oxidation of pyrite) and the presence of smectite clay minerals which have a very high swelling potential and which lose cohesion upon wetting. Direct shear tests indicated that the residual friction angles varying from 12° to 17°.

Stabilization Measures

With the geologic models established and verified, stabilization measures are being devised to mitigate the movements of the Hoover Slides area. Possible stabilization measures currently under study are large diameter drilled shafts, micropiles, tiebacks, soil nails, grouting, drainage measures consisting of horizontal drains and deep drainage trenches, or a combination of the above.

CONCLUSIONS

1. The use of computers is widespread and increasing to manage geologic data in a variety of applications. An example of this is presented in this paper.
2. The advance in computer technology has made it possible for geologic profiles and inclinometer readings to be plotted and presented in 3-D graphics. Through these 3-D graphics, locations of sliding planes and trends of movements of ground masses can be readily determined.
3. Computer systems allow information to be collected, verified, and checked more easily. Electronic gathering and transfer of data eliminate problems related to manual transfer of data.
4. Information and data entered into a computer can be easily transferred and analyzed electronically. There is less reliance on transfer of hard copies.
5. Various computer hardware and software can be used to match the project specific needs.

ACKNOWLEDGMENTS

The writers wish to thank their colleagues at the Utah Department of Transportation (UDOT), Centennial Engineering Inc. (CEI)., and Parsons Brinckerhoff Quade & Douglas, Inc. (PBQ&D) for their encouragement and support. However, any explicit and implicit statements made herein by the writers should not be inferred to be official or procedures of the UDOT, CEI, or PBQ&D.

REFERENCES

Baker, A. A., 1964, Geologic Map of the Aspen Grove Quadrangle, Utah: U. S. Geological Survey Map GQ-239, Scale 1:24,000.

Bryant, B., 1992, Geologic and Structure Maps of the Salt Lake City 1° x 2° Quadrangle, Utah and Wyoming: U.S. Geological Survey Map I-1997, Scale: 1:125,000.

Hoek, E. and Bray, J., 1974, *Rock Slope Engineering*, The Institution of Mining and Metallurgy, pp 358.

Investigation of London Underground earth structures

B T MCGINNITY
London Underground Limited
London
United Kingdom

D RUSSELL
Mott MacDonald
Croydon
United Kingdom

INTRODUCTION

The London Underground Limited (LUL) railway comprises some 400 route kilometres of track with just in excess of half of this forming the surface section of the system. Approximately 90% of the surface railway is more than 70 years old and many of the earth structures are showing increasing signs of deterioration and distress with a corresponding requirement for increased maintenance and the imposition of track speed restrictions. Occasional failures of earth structures occur with serious implications for railway operation and safety. The impact of failures on the railway was exemplified by the embankment slip and consequent closure of the Northern Line for two weeks between Burnt Oak and Colindale in January 1994.

Historically, there has been a lack of detailed knowledge of the nature and causes of the many earth structure difficulties facing LUL and remedial work has generally been carried out on a reactive basis following disruption to train services or to reduce excessive maintenance at particular locations. LUL are currently undertaking a programme of track replacement throughout the surface railway within the Track Replacement Project (TRP) and several lines (Central and Northern) are subject to modernisation schemes. Prior to this extensive capital expenditure on the surface railway, an earth structures strategic planning initiative was developed to obtain a comprehensive record of all LUL earth structure assets with their current condition and to reach a fundamental understanding of the nature of the difficulties associated with the surface railway earth structures and the means for their improvement.

This paper describes the recent major earth structures strategic planning initiative which has enabled the company to come to a better understanding of the nature, distribution and reasons for the continuing earth structure deformations and failures currently being experienced within the ageing surface railway.

EARTH STRUCTURES CONDITION SURVEY AND DATABASE

In the past LUL have not had a systematic, centralised formal means of recording either earth structure asset and condition information or routine maintenance works, although records of all major remedial works projects have been maintained. A core activity of the strategic planning initiative has been associated with the acquisition, storage and subsequent processing and interpretation of a comprehensive body of detailed information on the condition of LUL earth structures. A key element of this work has been the creation of a computerised relational database for the storage and manipulation of both historical and current earth structure asset and condition data.

In order to provide a framework for the acquisition and processing of condition data, an earth structure, vegetation and wildlife classification system was created and developed. The nature, geometry and current condition of LUL earth structures were then recorded in a walk over survey of the entire surface railway to clearly identify those factors and indicators most pertinent to earth structure stability.

Advances in site investigation practice. Thomas Telford, London, 1996

A) SURVEY (One side of track only)

Recorded by ☐   Date ☐   Time ☐

At Station   Y ☐

N ☐   Station ☐

From Station ☐   To Station ☐

BRS Code ☐   Chainage ☐

Line Direction ☐   Line Type ☐   Line Owner ☐

Weather: Fair ☐   Sunny ☐   Cloudy ☐   Drizzle ☐
Fog ☐   Rain ☐   Sleet ☐   Snow ☐

B) PHOTOGRAPHS

None ☐

Up Slope Time ☐   With Traffic Time ☐   Across Track Time ☐

Down Slope Time ☐   Against Traffic Time ☐

C) EARTHWORKS STRUCTURE TYPE

Accessible?   Yes ☐   No ☐

Cutting ☐   Embankment ☐   At Grade ☐   At a Structure ☐   Adjacent Depot or Sidings ☐

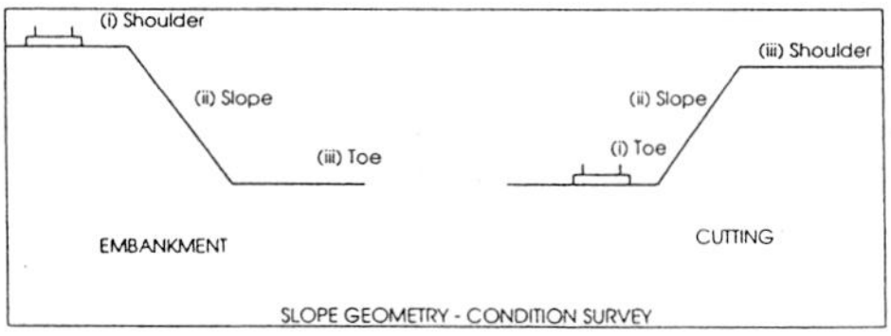

(i) Shoulder   (iii) Shoulder

(ii) Slope   (ii) Slope

(iii) Toe   (i) Toe

EMBANKMENT   CUTTING

SLOPE GEOMETRY - CONDITION SURVEY

D) CONDITION

No Evidence of Movement ☐

Movement Indicators

1) Lineside Services ☐   2) Slip Scars ☐   3) Slope Bulges ☐   4) Terracing ☐

5) Structure Distortion ☐   6) Dislocated Trees ☐   7) Toe Bulges ☐   8) Toe Debris ☐

9) Tension Cracks ☐   10) Dislocated Fence ☐   11) Cracked Roads ☐

Indicator Number ☐

Indicator Location - Section   (i) ☐   (ii) ☐   (iii) ☐

Offset from Running Rail ☐   Start (m) ☐   Finish (m) ☐
Offset from Breakpoint ☐   Start (m) ☐   Finish (m) ☐
Offset from Breakpoint ☐   Start (m) ☐   Finish (m) ☐

Indicator Number ☐

Indicator Location - Section   (i) ☐   (ii) ☐   (iii) ☐

Offset from Running Rail ☐   Start (m) ☐   Finish (m) ☐
Offset from Breakpoint ☐   Start (m) ☐   Finish (m) ☐
Offset from Breakpoint ☐   Start (m) ☐   Finish (m) ☐

Overall Slope Assessment

Movement Type: Creep ☐   Flow ☐   Subsidence ☐

Translational ☐   Rotational ☐   Unknown ☐

Depth of Movement: Shallow ☐   Deep ☐   Unknown ☐

Figure 1. Extract from Pro-Forma used in Earth Structure Survey

DATA MANAGEMENT

The earth structures condition classification and recording system was developed to meet criteria of completeness, consistency and user convenience (LUL 1994a). This was achieved by the creation of a set of standard tabular multiple choice data fields on a pro-forma for selection and completion by suitably geotechnically qualified and trained inspection teams (Figure 1). Information was collected under the following headings using either a paper based system, or hand held computers (Touch PC's) and on photographic film.

- Location
- Earth structure type
- Track condition
- Lineside services condition
- Grouting
- Geology

- Slope condition
- Slope geometry
- Retaining structures
- Earth structure drainage
- Water
- Vegetation and wildlife
- Adjacent property

In order to complement the information gathered from the condition survey, selected topographical surveys were conducted to provide accurate, co-ordinated and levelled earth structure cross-sectional information throughout the surface railway, in particular at sites showing evidence of earth structure instability.

A total of 453 Earth structure cross-sections were surveyed in sufficient detail to record all changes in gradient and in particular, any indications of instability. The Primary Control Network for the topographical surveying established using a Geographical Positioning System (GPS) comprised 27 stations forming a braced framework around London and has been named the "LUL grid" (LUL 1994b). The existing Ordnance Survey grid was thought incapable of providing the correct level of planimetric accuracy and consistency across the entire LUL system. Depending on the computation parameters, the determined accuracy of the network ranges between 0.6 parts per million (ppm) and 1.7ppm, meaning that the position of all points in the LUL grid are known to an accuracy of better than ±20mm.

The earth structures relational database was developed using Foxpro v2.0 (Dos) software for storing, managing and manipulating the data collected from the Condition Survey and other LUL historical records relating to construction drawings and previous ground investigation reports. The database currently comprises 26 linked datatables, contains some 600,000 entries and is some 9 megabytes in size. (Figure 2).

In order to facilitate the overall assessment of earthworks condition at individual locations and to permit the broad categorisation of slopes into either poor, marginal, serviceable or good condition and to determine priority areas for remedial action, a numerical scoring system was developed, whereby individual indicators of instability recorded during the condition survey were allocated severity ratings on a comparative basis and an overall condition rating calculated.

The allocation of severity ratings was based on qualitative application of geotechnical principles and a subjective appreciation of the relative importance of the different factors indicative of and contributing to earthworks instability.

The LUL Earth Structure Condition Ratings are as follows:

Category	Condition Rating
Poor	0% to ≤40%
Marginal	>40% to ≤65%
Serviceable	>65% to ≤80%
Good	>80% to 100%

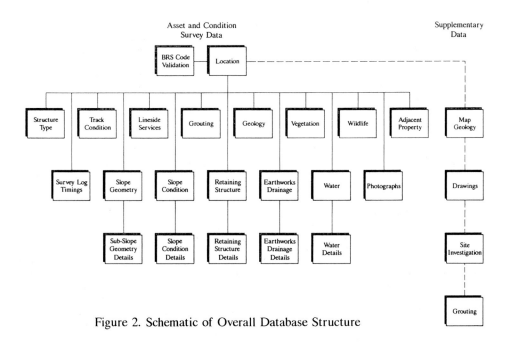

Figure 2. Schematic of Overall Database Structure

57% of LUL earth structures are classified as being in good or serviceable condition (110 lineside km), whilst 43% are classified as marginal or poor (84 lineside km) as shown in Figure 3. Some 16% (31 lineside km) of earth structures come within the poor category.

Summary earth structure condition data and ratings have been incorporated in the LUL Civil Engineering Asset Information System (CEAIS), this is based on the Intergraph/Informix Geographical Information System (GIS). The system stores geometric details referenced at various defined levels to Ordnance Survey digital mapping for easy viewing and retrieval. An example of the graphical output from this system is shown in Figure 4.

The GIS database will be updated by regular earth structure inspections to the LUL Standard (LUL 1994a).

The earth structures data and condition ratings have been analysed in order to determine the factors and mechanisms influencing the stability of LUL earth structures and to identify and rank the priority locations for investigation, assessment and remedial works enabling LUL to produce a strategic plan for future cost effective earth structures asset management (LUL 1994c).

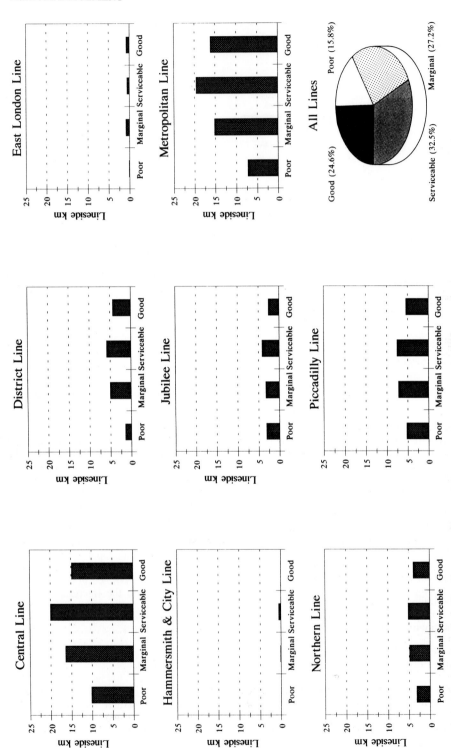

Figure 3. Summary Earth Structure Condition Ratings

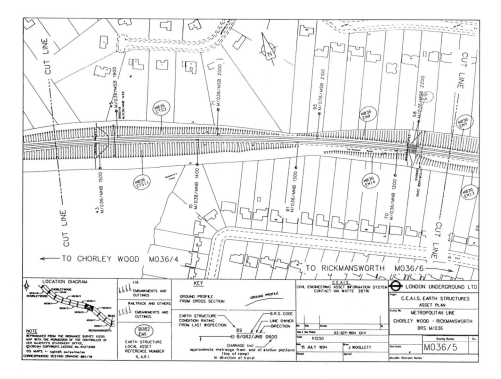

Figure 4. GIS Output for Typical Section of LUL Track showing Summary Earth Structure Condition Data and Ratings

VIDEO CAMERA TRIAL

In addition to the conventional condition surveys, a train mounted video camera recording (VCR) trial was undertaken over a section of LUL track on the Piccadilly Line with the primary aim of establishing the suitability and flexibility of using VCR as a technique for identifying changes in earth structure and vegetation condition in conjunction with regular conventional inspections. A basic provision of incorporating accurate locational information on a frame by frame basis on the video recording was developed. This could be linked to the earth structures database to enable LUL to maintain a time related easily retrievable condition information base. The trial however demonstrated that VCR is not currently a suitable means of monitoring changes in shape or form and hence stability of earth structures. A train mounted VCR system could however provide a record and comparative data of the condition of vegetation and lineside services. Future developments may also include computerised image enhancement and pattern recognition techniques.

TRACK RECORDING VEHICLE

The LUL track recording vehicle (TRV) is a modified train passenger car which is coupled to control cars at either end and provides a rapid and continuous means of regularly monitoring a series of pre-defined track quality parameters for the whole LUL network. The data acquisition devices on board the TRV record parameters relating to the vertical and lateral profiles of the two rails. Vehicle location and heading information are recorded by an on board tachometer and a gyroscope. Database and statistical post processing of TRV data captured on magnetic tapes is carried out and a range of reports produced. These comprise colour coded bar charts showing areas of satisfactory, adequate, bad and unacceptable track quality, a quality index trend report which gives the overall quality for selected track parameters within geographically defined areas, together with lists of sections on each route having the lowest quality.

Track displacement parameters recorded by the TRV are comprised of a number of component displacements, some of which relate principally to the geometry, construction details and condition of the track, whilst others have a greater possibility of being related to movement of the underlying earth structures. Track parameters thought to be most suitable for direct correlation with earth structure performance are the vertical measurements "left top", "right top", "dynamic crosslevel" and 2m and 10m track twists and the horizontal measurement of alignment deviation from recorded variations in track centre line. Left top and right top parameters measure the variations in surface level of the left and right rail surfaces. Dynamic crosslevel parameters record the variation in track cant under train loading. Track twist is a measure of the difference in rail levels over a specified base length. The 2m twist is measured across the TRV axle centres on a bogey and the 10m twist is measured over the car length between bogey centres. Fluctuations in track parameters are calculated from the continuous TRV data stream for different lengths of track depending on the type of analysis to be performed and the output report type. Requirements on the track scanning lengths are held on the route setting software programmed into the TRV before the start of each run, together with track parameter acceptance levels and other line specific information. The scanning length is typically 25m, although this can be reduced to 10m or some other appropriate length for sections with sharp curves or short transitions.

The regular transit of the TRV around the LUL railway approximately every seven weeks provides an opportunity for the early identification of embankment deformation. However the current data structures and software of the TRV information handling system do not permit the easy transfer of raw digital data for the analysis and processing of information other than for the purposes originally conceived. Modification of the TRV software and data handling system will be required to permit the fully automated analysis and interpretation of data for the specific purpose of detecting the occurrence and likely nature of earth structure defects.

The data taken prior to the Colindale Embankment failure in January 1994 (Figure 5) indicates the potential use of the TRV information as a predictive tool for embankment deformation based on the deterioration in track parameters.

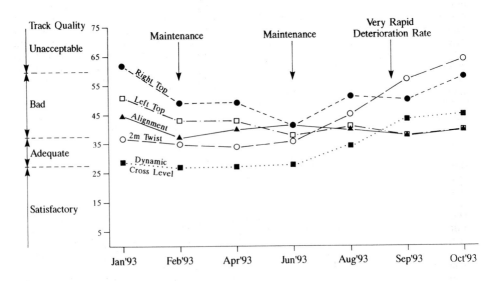

Figure 5. Colindale Embankment: Rapid Rate of Deterioration in TRV Parameters
Six Months Before Eventual Embankment Failure

INVESTIGATION OF LUL EMBANKMENTS

Although both embankments and cuttings along the surface railway contribute to the significant LUL earth structure maintenance requirements, it is the embankments which require proportionally much greater resource allocation. Some 29 lineside km of embankment have been identified during the Earth Structures Condition Survey as being in poor or marginal condition leading to continuing deformation and the need for frequent costly maintenance. Embankment failures also occur especially after prolonged wet periods and these can cause major disruption to LUL services and require the undertaking of expensive emergency repair work.

Prior to the recent investigation, very little information was available on the material used in the construction of the LUL embankments, the construction methods adopted, any difficulties experienced during construction or the contribution made to track instability by the ash which has been extensively used for maintenance and repair. In order to address this lack of information, a desk study comprising an investigation into the history of LUL embankment construction, subsequent remedial works and the engineering behaviour of ash was undertaken and this was followed by a preliminary ground investigation of a sample of nine LUL embankments. The embankments were as far as possible representative of the types and range of maintenance difficulties being encountered, the range of construction and remedial histories. They provided a reasonable spread both geologically and geographically over the surface railway system (Figure 6).

Design of the Embankment Investigation

The problems associated with working on an operating railway required a carefully designed ground investigation. Access to the railway track between the lineside cable runs is only usually available in Engineering Hours at night when the electric current is switched off. Occupation times vary from site to site with approximately 3 hours available on Sunday and weekday nights and 4 ½ hours on Saturday night.

These difficulties were mostly overcome by using trial pitting and easily portable exploratory sampling and probing techniques for the major part of the investigation on the track. These less traditional investigation techniques were calibrated against more formal borehole methods undertaken on the embankment shoulders in Traffic Hours (during the day when the trains are running and the current is on). Alternative techniques for investigating the track area, for example mounting a cone penetrometer rig on a railway wagon were considered, however operating difficulties and safety considerations made these options prohibitively expensive. The investigation on the track area in Engineering Hours was carried out in two stages. The first stage consisted of trial pitting through the railway ballast to the ash/clay interface. If the ash/clay interface was not reached at a depth of 1.5m a hand auger was used to establish the depth of the interface. Vertical plastic guide tubes were installed in the trial pits during backfilling to assist the second stage operations. The second phase consisted of window or dynamic sampling and dynamic probing through the guide tubes to the base of the embankment and into the underlying material.

Window or dynamic sampling is a method of ground investigation in which a hollow tube is driven into the ground using a hydraulically powered percussion hammer. The tube is then extracted so that the almost continuous soil sample can be logged through the window in the tube. Successively smaller tubes are used to reach depths for this investigation of up to 14m. The main advantages of the method are that the equipment is easily portable, it is much quicker than the more traditional cable tool boring, it provides a near continuous sample and instrumentation, e.g. piezometers can be installed. Disadvantages are that it produces lower quality samples than the traditional U100 and the technique does not allow standard penetration tests (SPT) to be carried out. There are also practical difficulties that can arise from the caving of loose granular materials and the reduced sample recovery in stiff clays. For these embankment investigations, principally through clay fill, the advantages of the technique were considered to far outweigh the disadvantages.

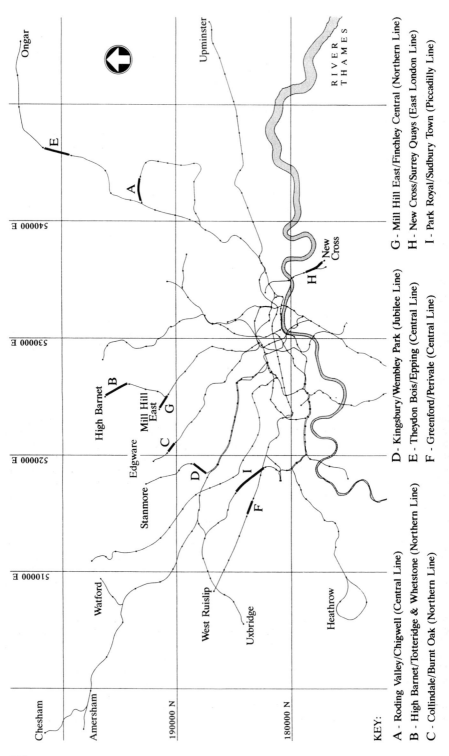

Figure 6. Location Plan for Embankment Investigations

KEY:

A - Roding Valley/Chigwell (Central Line)

B - High Barnet/Totteridge & Whetstone (Northern Line)

C - Collindale/Burnt Oak (Northern Line)

D - Kingsbury/Wembley Park (Jubilee Line)

E - Theydon Bois/Epping (Central Line)

F - Greenford/Perivale (Central Line)

G - Mill Hill East/Finchley Central (Northern Line)

H - New Cross/Surrey Quays (East London Line)

I - Park Royal/Sudbury Town (Piccadilly Line)

In the cohesive material from the window sampler, moisture content tests and pocket penetrometer strength tests were carried out every 0.25m. The pocket penetrometer and moisture content results together with index testing have economically provided a large body of information on the state of the clay fill. The pocket penetrometer results provide a convenient comparative guide to the variations and profiles of strength at different sites, although perhaps slightly overestimating the actual values of shear strength as conventionally measured in the laboratory. The presence of desiccated crusts and zones strengthened by pore-suctions can be detected by the moisture content, pocket penetrometer and odeometer tests. The dry crust has a high strength and low moisture content, but does not develop a high swelling pressure in the odeometer during confined saturation. Zones with high pore-suctions are also relatively strong, but are saturated and so have significantly higher moisture contents than the dried soil; they also show marked swelling pressures when submerged.

Dynamic probing using a standard 50kg hammer (BS 1377 1990: Part 9: Section 3.2) was used to provide quick semi-quantitative information on the comparative consistency of the embankment materials and sub-soils. However at several locations different materials, e.g. ash, clay fill and alluvium provided equally low resistances to probe penetration in which case it was difficult to distinguish the different materials from the probe results. A lighter hammer may have proved more sensitive to changes in materials.

A pulsed ground radar trial (PGR) undertaken on a high maintenance embankment between Colindale and Burnt Oak has shown that the technique when correlated with known features can identify some shallow sub-surface features and can be used to identify variations in the thickness of ballast to a depth of 500mm. However absorption of signals by the ash and more significantly by the clay resulted in only sporadic information being gained beneath the ballast and no useful information was obtained on materials below 2.5m depth. The non-destructive nature of the technique, the portability of the equipment, the low level of disruption by metallic objects and the speed of operation means that PGR would be convenient to employ in the railway environment. However at the present stage of development, it is not ready to replace the direct sampling and testing approach adopted in this investigation.

Results from the Embankment Investigation

From the desk study and embankment investigation, a picture of the original construction and past remedial works has been built up. The LUL surface railway was constructed at various times between 1856 and 1933, by modern standards the embankments were poorly constructed and have a long history of instability. The investigated embankments were found to comprise a central clay core, probably the spoil from adjacent cuttings, overlain by variable depths of ash (Figure 7). In all but two cases, the core comprised reworked London Clay. The ash was typically found as a 1 to 2m deep layer on the crest and as an irregular cover to the side slopes. Many investigated sections revealed localised deep zones of ash or granular fill, the result of remedial measures undertaken to repair previous slips.

The majority of the embankments were underlain by London Clay, however, many embankments crossed stream valleys and are therefore underlain at their highest sections by alluvial clays. A minority of other sub-grades were encountered, notably Terrace Gravels. Generally, the original topsoil and alluvial layers had not been removed or treated before embankment construction as would usually be the case today.

The embankment core would have been placed as excavated, in clods, often with minimal if any effort at compaction and would therefore initially have had a large amount of voids. The voids would have promoted significant settlement during and immediately after construction, particularly following heavy rain. During construction, these settlements would have been restored by the use of additional cohesive material. However, once construction of the section was complete other fill material would have been used to make up for the continuing settlement. As ash was readily available from the steam locomotives its use to make up settlement and as a ballast material appears to have been common. Following the demise of steam locomotives, power station ash has been used and is still used today to maintain the cess and embankment shoulders.

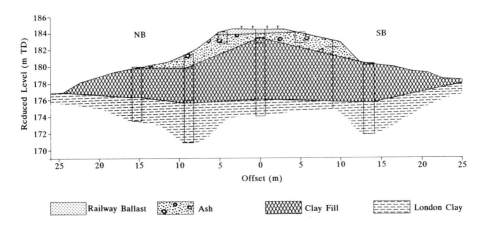

Figure 7 Typical LUL Embankment Cross Section

Settlement of any underlying alluvial clays would also have occurred in the first decade or so of operation. These settlements would generally have had to be made up as the alluvial layers were often adjacent to bridges which were founded below the compressible soils.

Failures of the poorly placed fills during construction would have been fairly common, particularly over the alluvial materials and where the side slopes were over-steepened during construction. Some failures may have been repaired by side tipping additional cohesive fill until a stable side slope was attained. Elsewhere stronger materials such as ash or granular fill appear to have been used to restore the original slope. Failures which were not fully excavated and replaced will have left internal shear surfaces of lower strength approaching residual values.

After construction, the upper clay fill generally would have swelled and reduced in strength and shallow post-construction failures may have been common in the early years before vegetation became established. The treatment of those failures would have been by infilling, possibly after some excavation of the slipped mass, with a readily available material. Ash again appears to have been widely used and more rarely, locally won granular fills.

Although the immediate embankment post-construction period has long since passed, on-going movements are still occurring and occasional failures still develop. Field tests showed vegetation to have a marked effect on the strength of embankment materials. On side slopes where there was no significant covering of ash but good plant cover, a desiccated clay crust comprising a stronger zone of drier material was generally formed. Beneath this crust and below ash on the slopes and crest, the embankment material was much weaker principally as a result of water infiltration due to the general lack of embankment and track drainage. The magnitude of strength difference is likely to vary seasonally, but the stabilising effect of mature vegetation and in particular that of the root systems, is likely to remain throughout the year. Figure 8 illustrates the wide range in undrained shear strength between the clay core and side slopes. Laboratory classification, strength and consolidation tests on the embankment clay core materials showed the moisture content of these materials to be higher than would be permitted in modern construction. Such material would today only be used if strict control on the compaction moisture content could be ensured. Evidence of the low residual shear strength of the clay was also found.

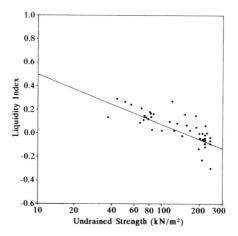

Figure 8. Liquidity Index against Undrained
Strength for typical Embankment Clay Fill

Laboratory testing of ash showed this to possess high shear strength and low density. Some crushing of the ash particles occurred during compaction tests but it is not thought that this degree of crushing would lead to significant track settlement where the ash is used as general fill. However, crushing may occur where the ash is acting as a capping to the weak underlying clay fill, resulting in inadequate support to the ballast.

Attempts have been made throughout the embankments' history to stop or reduce their deformations and evidence of these works was found at many of the investigated sites. Four previous types of remedial measures were identified, namely counterfort drains, toe drainage, toe walls and embankment grouting.

The effect of the embankment grouting operations undertaken over 28 lineside km in the 1970's does not appear to have produced as great or sustained an improvement as originally expected, unlike the reported experience of British Rail (Ayres 1994). The investigations revealed very little grout in the clay core. The grout was mainly found at the ash boundary with the clay fill or ballast. As the grout did not generally penetrate into the ash or clay fill, grouting could not resolve the overall problem of poor embankment performance.

The primary mechanisms for the observed long-term deformation relate to the and oversteepened embankment shoulders which are principally constructed of ash and the poor condition of the clay core. This results in both inadequate support to the track and low factors of safety, especially where potential slip surfaces pass along the scars produced by previous failures. The cyclic loading from trains significantly increases shear stresses in the slopes, resulting in overstressed embankment shoulders, progressive deformation, loss of support to the outside of the track and settlement of the outer rail relative to the inner one. The seasonal shrinking and swelling of the surface crust also promotes further ongoing downslope deformations of the oversteep slopes.

In this way the factor of safety of the embankment gradually reduces over time. Eventually, particularly following a sustained period of heavy rain that reduces pore suctions and raises porewater pressures, the factor of safety may reach unity and failure occurs. The above scenario implies that those embankments that are frequently maintained are most likely to fail. Permanent improvement of embankments requires that either further lateral support is provided or that train loadings are distributed deeper in the fill.

This investigation has provided a large amount of information on a representative sample of LUL embankments and adequately identifies the soil profiles and material properties such that conclusions as to the causes of deformation have been drawn. This has allowed the selection of priority sites for investigation, stability and risk assessment leading to a programme of detailed design and remedial work implementation to reduce embankment maintenance and prevent failure.

CONCLUSIONS
A condition survey of the LUL surface railway has provided a comprehensive record of the current stability of LUL earth structures. This survey results have been incorporated into a GIS/database for the fast, effective storage and manipulation of data.

Analysis of the condition survey data has identified the scope and nature of earth structures difficulties being experienced on the railway

The LUL condition rating system has categorised earth structures according to the weight of evidence for actual or potential instability thereby permitting the identification of areas at risk and facilitating the prioritisation of site investigation, slope stability assessment and remedial works at key locations. The use of train mounted video camera recording techniques and the track recording vehicle have been assessed for the monitoring of earth structure deformation.

Although both embankments and cuttings throughout the surface railway contribute to the significant LUL earth structures maintenance workload, it is the embankments which currently require proportionately much greater resource allocation. Desk studies and ground investigations have been undertaken at nine embankments distributed at representative locations throughout the surface railway to arrive at a better understanding of the materials and form of construction together with an appreciation of the dominant deformation and failure mechanisms.

These earth structures investigations will enable LUL to undertake cost effective earth structure asset management based on sound engineering principles and to undertake the phased investigation, assessment, design and implementation of appropriate remedial engineering works at high risk priority sites as the next stage of the strategic planning initiative.

ACKNOWLEDGEMENTS
The assistance of colleagues within London Underground Limited and Mott MacDonald in the development of this paper is hereby gratefully acknowledged. Any views expressed, however are the views of the authors' alone and cannot be taken as representing those of London Underground Limited or Mott MacDonald.

REFERENCES

Ayres, D J — Hydrofracture grouting of landslips in cohesive soils. Grouting in the Ground. Thomas Telford 1994

BS 1377: Part 9 — British Standard Methods of test for soils for Civil Engineering purposes. Part 9, In-situ tests: 1990

London Underground Limited (a) — Inspection of Earth Structures Standard, CED-SP-2035-A1, 1994

London Underground Limited (b) — Topographical Survey Standard, CED-ST-1101-A1, 1994

London Underground Limited (c) — Earth Structures Asset Development Plan 1994

Moderator's report on Session 2: data management

LEONARD THREADGOLD

Managing Director and Chief Engineer, Geotechnics Limited, Coventry, UK

1.0   INTRODUCTION

Papers in this session are concerned primarily with Information Technology and the ways in which it can relate to Geotechnical Data Acquisition, Presentation, Storage and Manipulation. Interaction between the data obtained and analysis is referred to but design or analytical software is not covered.

Developments in computer technology, particularly over the last decade, have rendered it accessible to even the smallest practitioners in geotechnics. The impact of computers was initially seen in relation to the analysis of problems of stress distribution, displacement and stability but it has become apparent that the ability of computers to store and manipulate large quantities of data will have a major, if not the major impact on the Site Investigation Industry.

Motivation for its application may be seen to have come from a number of sources:-

1.   From the Site Investigation Specialists, to aid preparation of reports for their clients.

2.   From Consultants, who wish to assimilate the data from site investigations into their systems to allow modelling of ground conditions and selection of design parameters.

3.   From Clients, who wish to store large quantities of data in their archives for future reference.

Their is indeed a further category which has not been a prime mover in this area, at least in the UK, but is likely to have a significant interest in it in the future, namely:-

4.   The Constructor who is required to tender for and build that which is designed.

The papers presented to this conference deal with the impact of computer systems on the second and third category of user although interaction between all elements is clearly vital to the successful application of such technology. It may be that trends within the industry may render such divisions less significant although the roles within The Team are likely to remain. Contributions from categories one and four are encouraged to redress the current imbalance of papers presented to the conference. Following a review of the papers presented the role of standard conventions for data transfer and adherence to them is addressed

2.0 REVIEW OF PAPERS
2.1 From Paper to Silicon Chip - The Growing Act of Computerised Data Management
2.1.1 Summary. The paper by *Nicholls et al* provides a valuable introduction to the components of the process of data acquisition, transfer, manipulation and management using computer technology. It highlights advantages, disadvantages, the importance of management and a "feel" for the data, as well as the dangers of plot proliferation. It is clearly based on experience from major projects and illustrates the importance of interaction between suppliers and users of the data. It is a useful guide to those who have embarked or wish to embark upon this process and illustrates the authors' perceptions and experience by using a table, flowcharts and an annotated plot.

The key to the implementation of the exchange of ground investigation data by electronic means as a common feature of investigations both now and in the future is recognised to be the establishment of a standardised format by the Association of Geotechnical Specialists (AGS, 1992). Further comment on the format is made later in this review.

Validation of data provided both in terms of compliance with the AGS format and by using relational criteria is particularly emphasised and on audit trial illustrated.

2.1.2 Comment. Guidance from consumers to suppliers is referred to although many suppliers believe that there is considerable scope for two way traffic in this respect. Later acquisition and transfer on a digital to digital basis is referred to as an ideal because of its elimination of human error. Unfortunately, it does not eliminate electronic errors or flaws which can often be more difficult to identify or correct than human ones. Furthermore, the scale of storage required by the digital to digital systems appears to be illustrated in their Table 1, where 80% of the data from a large infrastructure project was from one source, namely the piezo cone. Was the value of such data similarly proportional?

Concern is expressed at the availability of data giving rise to many plots but such plots are derived from the data and are not part of it and hence may be considered analogous to over-consumption rather than to over-production. "Raw" data should not be affected and it is vital that this remains so for any system.

2.2 A Computer System for Site Investigation Data Management and Interpretation
2.2.1 Summary. The paper by *Oliver and Toll* describes a computer system which they have developed for management and interpretation of site investigation data. It reviews the current situation in relation to computer-based systems for data handling and emphasises the benefit of the AGS format as a standard to be adhered to as a framework for the design of a new generation of data bases. They go on to identify the function of a knowledge-based system for computer-aided interpretation of data in terms of parameters and ground conditions using a knowledge of relationships between parameters. Such a knowledge-based system can be superimposed on a base of raw data. It further describes the benefits of multi-user access and provides examples of a relational database structure.

2.2.2 *Comment*. The work is clearly capable of considerable sophistication but the "rawness" of data which is required in some respects could go well beyond that envisaged when establishing the AGS format. This in turn places burdens on the supplier of the data and there is scope for considerable debate on how far this should go: this paper stimulates such a discussion. It is vital that all components of the process are not alienated from this vital contribution to the better use of geotechnical data. The example given of the incorporation of weights of wet and dry samples in moisture content determination into the database, even if this calculation has been derived using a computer programme, is going much to far. It will tend to unnecessarily load the database and delay and obscure the process of assimilation.

Other facilities within the software concern similarity between descriptions to allow various hypotheses of links between layers to be generated between boreholes and over the site as a whole. The computerisation of descriptions raises questions about what is factual, what is subjective judgement and what is interpretative. Here again, considerable care to ensure user flexibility and to allow their imagination to be applied in arriving at a model on which to base design is essential. The computer may merely give another opinion, rather than the definitive evaluation of the model.

A further concern is in the remoteness of the computer modeller from the soil with which the engineer has to deal. Unless the soil is described by the person modelling the ground conditions there is danger of a split between geotechnical engineer and computer specialist and of misinterpretations arising. There should be a geotechnical health warning on every computer and software package.

The foregoing comments should not be seen as a excuse for not using and developing such computer systems. Their potential is considerable and their development vital; such work is greatly to be encouraged.

2.3 Management of spatial data for the heavy foundations of a blast furnace
2.3.1 *Summary*. The paper by *Nathanail and Rosenbaum* reports on an exercise which was undertaken on data obtained some 20 years ago and recorded by conventional means. Its purpose was to develop techniques for managing spatial data using geographical information systems and geostatistics. It relates only to the compilation and modelling of parameters. The paper will be of considerable interest to those in possession of printed archives such as the Department of Transport in the UK and the Geotechnical Engineering Office in Hong Kong who may wish to make these data available in electronic format.

The data were entered in to a database using a relational structure with linked files. Codes were used to key in data such as weathering grade and borehole type and optical character recognition software was used to scan tables of data. Numerous programs were used to facilitate this process. The paper highlights the dangers of modelling becoming an end in itself.

2.3.2 *Comment.* The use to which the work was put has been reported elsewhere but it is regrettable for geotechnical engineers that at least a glimpse of the value of the application is not given in the paper. Perhaps this can be redressed in due course.

The geotechnical engineers reading this paper may find themselves wondering whether it is enlightening or confusing to refer to so many different proprietary programs to which they can only gain access through the authors and whether words such as "kriging" are a typing error or are symptomatic of an area into which they stray at their peril. It is important that such areas are accessible if they are to be an advance for Site Investigation practice.

2.4   Management of Geologic and Instrumentation Data for Landslide Remediation
2.4.1 *Summary.* The paper by *Abrahamson and Lee* deals with a specific site in America in which the data from desk studies, site investigation, instrumentation, survey and geological mapping were assembled using databases, analytical programs and graphical presentation software. These were used for stereonet plotting, stability analysis and three dimensional surface modelling. The site is in an area of landslip which poses a threat to the existing road and to its proposed widening and realignment. The data were analysed using CADD systems for modelling surface, sub-surface and inclinometer profiles. The analyses are currently being used to evaluate stabilisation measures. The paper highlights the use of 3-D graphics and existing software and hardware to aid the design. It is a useful case history of the collection and synthesis of data from a range of sources to aid design.

2.4.2 *Comment.* The impression given is that the data used is predominantly for graphical representation rather than for parameter reporting and selection to which much in the foregoing papers refer. Since the project is on-going it will be helpful if the authors could update us on the findings and the solutions adopted.

2.5   Investigation of London Underground earth structures
2.5.1 *Summary.* The paper by *McGinnity and Russell* relates to the investigation of embankments on the London Underground Limited (LUL) railway system. Movements and failure have created major problems and the paper describes the formulation of a classification system for earth structures and its integration with a topographical survey, particularly at problem sections, historical records of construction and ground investigation. Areas were identified primarily using a severity rating, based on quantitative and subjective judgement as well as measurement. From these data, areas for priority action and asset management are highlighted. The data have been incorporated into the LUL Civil Engineering Asset Information System (CEAIS) which is a Geographical Information System and can readily be updated and accessed. The paper goes on to highlight an interesting geotechnical problem which is a legacy of original construction processes for the embankments and deals with subsequent maintenance.

The use of a video camera recording technique was investigated although currently it is not considered suitable for monitored changes in shape and hence stability of earth structures. No doubt developments within the industry may render it more suitable in due course.

Records from a Track Recording Vehicle (TRV) has also been obtained and stored electronically but these data on track condition have not yet been integrated directly into the data storage and analysis system adopted for the remaining data. Records prior to a failure which occurred recently are shown as an example of its value in its present form.

2.5.2 *Comment.* The paper is of interest to those who have responsibility for maintenance of major infrastructures and utilities such as transport, water supply, and drainage and for river authorities. Major urban authorities would also have a significant interest. For example, Hong Kong has a large database for its slopes and covers those in private or public ownership. A significant feature of the latter is believed to be a factor which relates to the consequence of failure. It is perhaps implicit though not explicit in the paper and it would be interesting to learn whether this is so.

Presumably, when treated, the severity rating will change although scope to incorporate the results of monitoring of instruments installed to check on the effect of such works will be required so that evaluation and refinement of techniques of treatment would be facilitated.

3.0 DATA IN ELECTRONIC FORMAT
3.1 *Motivation*
The product of all investigations is information, as distinct from the other more tangible products of the construction industry. It is therefore vital that information is presented in a form which can readily be understood and can be accessed. This allows its use to be maximised and thus facilitates the development of designs which are safe and as economical as conditions permit. Increasingly, the short time scales within which designs have to be prepared either by consultants for conventional contracts of by consortia for Design and Build projects renders rapid assimilation essential. Computerised data compilation, transfer, manipulation and storage provides a powerful means of achieving this.

The potential value of compiling and transferring data from ground investigations in Electronic format was recognised by the Department of Transport in the late 1980's (Greenwood 1988) and pilot studies were undertaken. From these it became clear that unless all parties used identical software it would be necessary to link the proprietary software of producers, users and storers by the establishment of a standard interchange format.

Such a format was established by the Association of Geotechnical Specialists (AGS, 1992) by the publication of "Electronic Transfer of Geotechnical Data from Ground Investigations" which has become known the AGS Format. This has been subsequently updated (AGS, 1994). By the preparation of one interface program to allow transfer and receipt of data into and out of such a format all parties can communicate without knowledge of the other user's software, either at the time of initial acquisition or at some time in the future (Threadgold and Hutchison, 1992). As a consequence, software can be and has been written and developed without the inhibition that it may be usable only by individual sponsors rather than the industry at large.

3.2 *Benefits*
Computerisation of the data from site investigation provides benefit for producers and users at management, evaluation, design, presentation, construction and storage levels. This should lead to much better use of the data obtained within the time scales available.

- For the producer the borehole records and test results can be entered at preliminary and draft stages and easily edited as more data becomes available and correlations are undertaken, thus avoiding the need to re-type whole documents (Hutchison and Threadgold, 1992). Test schedule pro-formas, preliminary cross sections and quantities for costing and invoicing can be prepared from preliminary logs thus rendering the management function more efficient.

- For the designer, graphical presentations to aid interpretation can be greatly enhanced and plotting routines for study and analysis, using engineering judgement, statistical methods, or check correlations, are facilitated. This should lead to better and more economical design.

- For the client, the computerised data base should render their archives more accessible and hence valuable. Furthermore, the storage requirements for a given quantity of data are greatly reduced. These benefits also apply to regional or national archives held by publicly or privately owned bodies.

- The constructor having a computer system which can absorb the data in electronic format will be able to make much better use of the data made available to him than the current printed format allows. By setting up plotting routines for correlations which he considers to be important for both temporary and permanent works designs, he will have the potential to use all of the data in the limited time typically made available. This in turn should lead to more appropriate evaluations and tenders with benefits for him and his client.

3.3   Controls

The potential benefits which can be derived from the use of geotechnical data presented in electronic format are considerable. If the Site Investigation industry is to make use of this information "Super Highway", however, it is vital that those using it follow a "Highway Code", otherwise crashes are inevitable, long traffic jams will form and serious injuries will occur. The consequences of this are that much heavier traffic would return to the old network of "roads and lanes" and the use of older, slower and wasteful "modes of transport".

The "Highway Code" in this context is believed to be embodied in the AGS Format. This has been established by our industry for our industry and is specified in the recently published "Specification for Ground Investigation" (Site Investigation Steering Group 1993). Similar needs have been identified overseas and it has been adopted in Hong Kong and the USA as the basis for their systems. It seeks to facilitate transfer of most but not all of the data from ground investigations in a structured manner in accordance with a series of rules.

The format allows for extension to meet particular needs and for updating but if anarchy is to be avoided it is essential that those using it follow the rules and that they are not changed to meet the individual needs of the specifier.

4.0   CONCLUSIONS

It is evident that the use of computers and the transfer of data by electronic means will have an increasingly important impact on the Site Investigation industry. It is vital,. however, that the role of Geotechnical Engineers is not subordinated to that of the computer specialists and that their experience, skills and imagination are not impaired by pre-formed concepts or the obscure procedures of software designers. The freedom to correlate, manipulate and store data in an accessible manner requires the discipline of adherence to standards for data transfer. All sectors of the industry and its clients can benefit.

REFERENCES

AGS (1992). "Electronic transfer of geotechnical data from ground investigations". Publ. ASG/1/92 Issue 03/92 Association of Geotechnical Specialists, Wokingham.

AGS (1994). "Electronic transfer of geotechnical data from ground investigations". 2nd edition Publ. AGS/1/92. Issue 07/94 Association of Geotechnical Specialists, Camberley, Surrey.

Greenwood J.R. (1988), "Developments in Computerised Ground Investigation Data", Ground Engineering.

Threadgold L and Hutchison R.J, (1992). "The Electronic Transfer of Geotechnical Data from Ground Investigations", Geotechnique et Informatique, Colloque International, Paris, pp 749-756. Publ. Presses de l'ecole nationale des Ponts at Chaussées.

Hutchison R.J. and Threadgold L. (1992). "The Generation of Data in Electronic Format". Geotechnique et Informatique, Colloque International, Paris, pp 97-103. Publ. Presses de l'ecole nationale des Ponts at Chaussées.

Site Investigation Steering Group (1993) "Specification for Ground Investigation". Publ. Thomas Telford Services Limited.

Discussion on Session 2: data management

Discussion reviewer D. G. TOLL
School of Engineering, University of Durham, Durham

The discussion focused primarily on the use of computer-based data management systems for site investigation data. The experiences and needs of those involved at different stages of the data 'chain', from data producer to data user, were expressed. The Association of Geotechnical Specialists' data exchange format for electronic transfer of geotechnical data from ground investigations (AGS, 1994) was central to many of the contributions, reflecting the widening use of this standard format within the UK.

Rodney Hutchison (Exploration Associates Ltd) presented a data producer's view of data management. He pointed out the vast amount of work required to produce a report on a site investigation. There is a need to cross check and cross correlate virtually all the data, particularly between that obtained in the field and from the laboratory. The amount of work involved in this is often underestimated. With related data arriving from a number of sources (the field, from an in-house laboratory or an external testing organisation) at varying times throughout a project there is a need for a project specific data management system.

With the advent of powerful personal computers and modern database packages, it is now relatively straightforward to set up a computerised system which can provide an integrated data management and processing environment. Exploration Associates Ltd have such a system and computers are now an essential part of daily office practice. The motivation for the move to a computerised system is the need to reduce timescales, increase accuracy and above all reduce costs. Today, timescales are becoming even more important, particularly with design and build projects where readable, accurate data is required as the investigation proceeds.

As Exploration Associates' systems have developed they have seen significant improvements in efficiency and accuracy. Accuracy is improved through the reduction of data transcription and also by using cross checking routines. Much information in a report is the result of correlations between data and descriptions based on fixed "rules". These rules lend themselves to automated computer checking. However it is important that this remains only a check. Final decisions and amendments must remain with the geotechnical engineer. So much geotechnical work is based on engineering judgement which should not be clouded by over-reliance on computer output.

The Association of Geotechnical Specialists' (AGS) format is becoming more and more common as a means of communicating data between and within companies. This has brought reliability and consistency to data transfer in what was rapidly becoming anarchy. However, in some senses the AGS format has become a victim of its own success. Users are recognising the benefits and then asking for additional data to be included on an almost random basis. Users must recognise that such variations from the published format generally require a rewrite of the data producer's software. This results in:

◆ Loss of access by third parties unaware of variations
◆ Loss of reliability due to potentially unverified software
◆ Delays while software is amended
◆ Transcription errors from data re-entry
◆ Additional cost

The geotechnical profession must also remember that the concept of providing data on disc, particularly with the AGS format, is to provide summarised data only. The requirement "to provide all data" is inappropriate and largely impossible. In requesting data on disc, users must also be mindful of the problems of updating their own databases as updated discs are received. In particular, it is important to recognise that until the project is complete, the data are preliminary and subject to change. Data users must have systems in place to accommodate this.

Rodney Hutchison concluded with a plea to conform to the published AGS format. He recognised the need for the format to be continually modified. However, the forthcoming AGS User Group annual review meeting will provide a forum at which modifications can be agreed.

Quentin Leiper (Tarmac Construction Engineering Services) represented the view of the constructor. He responded to the question posed by Len Threadgold 'Do main contractors need site investigation data in electronic format?' To answer the question he described the process of tendering for the Medway Tunnel as a design-build contractor in 1991; a job which is now well on the way to completion. At that time a geotechnical engineer was tied up for three weeks drawing *one* cross section and plotting all the data required to design the casting basin retaining walls and cofferdam. In 1994, when tendering for the Cork Tunnel contract three weeks was again spent in transferring data from the site investigation report and plotting the data needed. However, by now Tarmac were using a spreadsheet package and a ground modelling program. Many more cross sections could be generated as well as isopachytes for each stratum. This greater amount of information allowed more efficient designs to be produced. There was more time to *engineer* the job and to assess (and hence reduce) the construction risk. Despite all the extra engineering input in data processing, Tarmac won the tender.

In the future, provision of data in electronic format would dramatically reduce the time needed to enter and process the data. The data would still need checking, but not as much checking. Therefore, if contractors were given data in electronic format they would be able to provide a better service. Therefore, Quentin Leiper's response to the question posed was that contractors would benefit from receiving data in electronic format, and this would result in a better service to their clients.

Mark Zytinski (MZ Associates) spoke on behalf of software developers. He raised the possibility of a further advance in data communications and suggested that the Internet was a very effective means of transferring data. Discs should be for storing data not as a means of transmission. North West Water were already making use of telephone modem links to site investigation contractors. Draft drillers logs could be used to generate three-dimensional block diagrams of a site while the investigation was underway. This allowed the possibility of modifying the investigation strategy as the data became available.

He also drew attention to difficulties in the implementation of the AGS format. Some companies were producing 'pseudo AGS format' which did not fully conform to the standard and contained errors. This was really a problem for the software developers rather than the users. The actual transfer mechanism should be transparent to the users. As a way of overcoming some difficulties he offered to provide free software which could scan AGS format discs and identify some of the common problems.

Lee Abramson (Parsons Brinckerhoff Quade & Douglas Inc, USA) discussed the data management for a highway project, with particular reference to graphical presentation of data. The route alignment inherited by the design team was defined by a federally and state approved environmental impact statement, arrived at through an environmental assessment process which had taken about 20 years to complete. Unfortunately this alignment included two bridges across a river with abutments founded on a moving landslide mass.

After analysing all the data and considering the costs of remaining with the alignment defined in the environmental study, the design team recommended moving the alignment above the unstable zone at significant cost savings to the project. A re-evaluation of the environmental study was required, but this had now been presented to the public and design for the new alignment was moving ahead.

However, Lee Abramson pointed out that in spite of the hours of processing and analysing the data using database systems and three-dimensional plots, the visual aid that was most convincing to the non-geotechnical engineers was an aerial photograph on which arrows indicating the rates and directions of movement at inclinometer locations were marked with coloured tape applied by hand. He also responded to a question posed by Len Threadgold as to whether all the field and laboratory data was entered into a database. The answer was no. At each major stage of the investigation the team had wanted to do something else with the database system, but they had found that it could not be done with the present data structure. He noted the use of the AGS format in the UK and felt that this was a step in the right direction.

In terms of the efficacy of using computer-based data processing he felt that it had speeded up the process, but that when each task was completed, it simply identified more problems which had to be tackled. There is the possibility of the data managing us (rather than us managing the data), the ground is what it is and we have to react to it.

Brian McGinnity (London Underground Ltd) described a different type of data management - that of storing earthwork condition data. London Underground have 200km of earth structures (above ground) of which about 90% is over 70 years old. The database stores records of visual inspections of earth structures. This is used to calculate an overall slope condition rating based on geotechnical principles to permit a broad categorisation into "good", "serviceable", "marginal" and "poor". Priority areas for site investigation and slope stability assessment are identified from the database. Results from these are then put into a Quantified Risk Assessment of the slope to determine the likelihood and consequences of an earthworks failure and to allow prioritisation of stabilisation works.

Following the initial condition survey of all London Underground Ltd earth structures in 1993, regular inspections were now being undertaken to ensure a record of current condition was maintained in the database. It is planned that all earth structures would be inspected at a maximum 10 yearly interval. Earth structures in a "poor" condition were typically inspected at yearly intervals. For critical earth structures a special regime of daily or weekly inspection would be introduced. Following the completion of stabilisation works the earth structure would be inspected and the database updated.

He noted that instrumentation data was not currently held in the Earth Structures Condition Database. Instrumentation monitoring data was managed as part of the site investigation process. He suggested that significant benefits would accrue if the AGS format was expanded to include field monitored data. Len Threadgold said that edition 2.0 of the AGS format now included horizontal and vertical movement records as well as piezometer observations.

Syd Pycroft (Sir William Halcrow & Partners Ltd) discussed the advantages of digital to digital data transfer. He noted that on a recent large project the data supplier had entered information into the computer twice; once to provide the written record (the report) and a second time to produce AGS format data interchange files. This situation obviously increases the possibilities for data entry errors. He endorsed Rodney Hutchison's comments on the desirability of entering data once only into an integrated system which fulfils all the data processing requirements.

The supply of data should be seen as a chain of activities from data source to data users. Electronic errors can propagate down the chain and each activity should not be seen in isolation. It was also important that data consumers had a clear view of the end use of the data and establish precisely which data are required. Otherwise unnecessarily large volumes of data are generated. However, on large projects or with particular investigation techniques (e.g. piezocone) it is inevitable that large volumes of data will be generated.

The concern had been raised by Len Threadgold that the ability to provide data in a digital format could dominate the selection of a geotechnical specialist rather than their geotechnical quality. Syd Pycroft expressed the opinion that the transfer of data by digital means should become the norm rather than the exception. Provision of digital data should provide the geotechnical engineer with improved access to the data and hence enable more effective use of the data.

Tim Spink (Mott MacDonald) spoke as a receiver of data. He suggested that the AGS data interchange format had proved to be of considerable value and marked a significant step forward for the UK geotechnical industry. Prior to AGS, of the order of 50% of the time involved in producing a geotechnical report was spent in checking the data, hand plotting or keying the data into spreadsheets and graphing packages; only 50% of the time would be spent analysing, understanding and interpreting the data. This situation should have improved dramatically since the introduction of the AGS format. However, it was found that significant amounts of time still had to be spent on checking and correcting the digital data received from site investigation contractors before it could be entered into the database and used for producing plots and cross sections.

There are a number of different database programs in use by data receivers. The one thing they have in common is that they are unforgiving of incorrectly formatted or internally inconsistent data. For example, a borehole referred to as 'BH1A' in one part of the data must not be called '1A' elsewhere. Such problems were common in the data received by Mott MacDonald. Therefore, before the digital data can be used a rigorous series of manual and automated checks were required. These consisted of:

◆ Is the data in the correct AGS format?
◆ Is the data internally consistent?
◆ Is the data complete?
◆ Does the digital data agree with the paper report?
◆ Is the data factually correct?
◆ Is the data mathematically correct?
◆ Have the test results been correctly analysed and reported?

The only way for data producers to achieve data consistency would be for each company to have a unified database system that covers all aspects of their work from borehole logs, through field testing, laboratory scheduling and sample tracking, to laboratory testing. The same software that produces the paper report must be used to produce digital data. Some UK site investigation contractors are closer to this ideal than others, and it shows in the quality of the reports and digital data they produce.

Tim Spink agreed that the site investigation industry has come a long way in the three years since the introduction of the first edition of the AGS format. However, further development of the data producers' software systems were still needed.

Dr David Toll (University of Durham) suggested that the industry should be looking to the future when computer-aided geotechnical design would require much closer linkages (or at least forms of data interchange) between software packages than currently exist. Geotechnical database systems were now common, and ground modelling packages (such as geographical information systems) were likely to become more widely used in the geotechnical profession. In addition, knowledge-based systems could be used to assist geotechnical specialists with the process of data interpretation. A system had been developed at Durham University which could provide the link between a geotechnical database and conventional calculation packages (e.g. slope stability or retaining wall analyses).

The AGS format had been very successful in facilitating the transfer of factual ground investigation data from the data producer to the data receiver. However, the current format could not allow the interchange of *interpreted* geotechnical data. Increasingly there would be a need for a standard format for storing interpreted data which would allow access by a range of different software packages. It would also allow access to interpreted data by a number of different users within the receiving organisation. Possibly, in the future, there could be exchange of interpreted data between organisations. However, the legal implications as to who is responsible for carrying out the interpretation is likely to militate against that.

Len Threadgold noted that the AGS working party had specifically decided not to become involved with interpreted data. However, with a move to design-build contracts he could see that interchange of interpreted data might become necessary.

Dr David Cox (University of Westminster) discussed the use of videogrammetry as an important recorder of information. He described how a video camera could be attached to the optics of a theodolite or level in order to record the cross-hair image and also angles and distances. It provided a simple and quick survey technique which could monitor movements to better than 1mm. It had the capability of generating millions of frames in a day. A difficulty for digital data storage was that each frame required about 2 MBytes. However, conversion to digital storage was not always necessary and video data could be considered as an alternate technique. For example, the precise x,y,z co-ordinates of almost every brick in a large structure had been recorded on a single video tape, for long term monitoring purposes.

Video cameras had been used to record information in other applications: a video penetrometer, video triaxial tests and video oedometers. Video oedometers could be used to observe consolidation behaviour which took place in less than a second.

Ulf Bergdahl (Swedish Geotechnical Institute) described a database developed for the 17km Øresund link to be constructed between Copenhagen in Denmark and Malmö in Sweden. The site investigation for the project included about 50 cored boreholes, more than 1000 laboratory tests of different kinds and at least 50 km of seismic profiles.

To be able to handle all these data and make use of them in the future a database called the Geomodel was created. The data was entered into the database after a quality assurance process. Copies of the Geomodel were sent out to the pre-qualified contractors for their use during tendering. As it is a design and construct project it might be expected that the Geomodel will be expanded to contain another 500 to 1000 cored and drilled boreholes and 1000-3000 laboratory tests.

He noted that in creating such a major database it is important to agree in advance a mutual nomenclature and presentation format for the data. This is particularly difficult in a project which involves institutions in more than one country, as customs differ from country to country.

Chris Eldred (Sir Alexander Gibb & Partners) stated that with modern data logging systems large quantities of data could be gathered and stored. Sir Alexander Gibb & Partners have been associated with many projects where large volumes of data had been gathered and the biggest challenge was visualising the data in a manner which was useful to engineers and geologists.

On one project, the Client who had access to thousands of pounds worth of state-of-the-art software and hardware, resorted to constructing a cardboard tube representation of the geology which meant more to himself and the audience than numerous computer generated models. The challenge is therefore for software developers to produce packages that meet these needs.

CONCLUSIONS

It was clear from the discussion that the Association of Geotechnical Specialists' (AGS, 1994) data exchange format for electronic transfer of geotechnical data from ground investigations was a valuable step forward in data management. There was strong support for the format from all involved, from data producer to data receiver/user, and also software developers. Contractors tendering for work would also benefit from receiving data in digital form and therefore be able to produce a better service for their clients.

Care was still needed in the use of the AGS format. A number of speakers identified the problems of 'pseudo AGS' data, which did not fully conform to the standard, and errors or inconsistencies in data entry. Integrated data management systems which eliminated multiple data entry could help to reduce such difficulties. Modifications to the standard format should not be undertaken on an ad hoc basis.

It was also clear that visualisation is a key element in data processing. Graphical presentation software was vital for assessing large quantities of data. However, this had to be considered in context and sometimes simpler, non-computer-based presentation tools could convey a concept perfectly adequately.

In the longer term, the possibility of developing standard formats for interpreted site investigation data would need consideration. In addition, other forms of data (not only ground investigation data) needed to be stored. Records of surveys were one example, including visual inspections and techniques such as videogrammetry. Videogrammetry, in particular, would produce large quantities of data, if it were to be digitised for storage.

Developing the means to exchange data had produced major advantages for the site investigation industry in the UK. Exchanging data in a different form had also been successful for a project in Sweden and Denmark. However, developing means of exchange across national boundaries would be more difficult, as different countries use different nomenclatures and presentation formats for geotechnical data.

REFERENCE

AGS (1994) *Electronic transfer of geotechnical data from ground investigations,* 2nd ed., Association of Geotechnical Specialists, Camberley, Surrey.

Dynamic (window) sampling — a review

C S ECCLES AND R P REDFORD
Soil Mechanics Limited, Wokingham, Berkshire, UK

INTRODUCTION

In recent years the use of various dynamic sampling techniques for ground investigation have become increasingly widespread due to the cost effectiveness, flexibility and speed of operation of the system. The need for such lightweight sampling systems have been driven by the increase in the number of ground investigations for house settlement problems and also by the requirement for shallow reconnaissance sampling of contaminated sites over relatively large areas. In addition, there has been an increase in the number of sites where access is poor or operating space limited and dynamic sampling has been found to be a valuable technique for many of these locations. The Nordmeyer type 'window' sampler is increasingly becoming the most widely used of the dynamic samplers in the UK.

This paper reviews the Nordmeyer type window sampling technique together with the advantages and disadvantages of dynamic sampling, the limits of its capability in terms of depth of investigation, geology and the quality of sample retained. Case histories from four diverse sites are presented to illustrate its operation. A model specification for dynamic sampling is also included to give some assistance to geotechnical specialists wishing to specify this technique who may not be familiar with it.

THE EQUIPMENT AND THE SAMPLING PROCESS

The dynamic 'window' sampler is a light weight sampling unit which uses specially constructed sampling tubes driven into the ground by a hydraulically powered or petrol driven high frequency percussion hammer. The sample tubes are then extracted with either hand operated or hydraulic jacks (see Plates 1 to 3).

Typically the sample tubes are constructed of hardened steel in 1 m, 2 m and 3 m lengths and incorporate a 'window' slot along their length. The tubes range in diameter from a nominal 80 mm diameter to the smallest diameter of a nominal 38 mm, the standard Nordmeyer sizes are shown on Table 1. A heavy duty cutting shoe is screwed onto the base of the sample tube and this can incorporate a basket spring to assist with sample retention. A clear plastic inner liner may be available for use with some types of sampling systems. Also, a closed tube may be used for sampling below the water table. The sample tubes are carried to depth on one metre driving rods with EW threads or similar construction. The rods are adapted for driving at the hammer with a special connector rod with integral anvil.

The percussion hammer may be hydraulically powered (Stanley type) from separate power pack units located at or some metres from the sampling position or be a self contained petrol driven unit (typically a Wacker BHF 30s). The rods and sample tubes are extracted using a hollow stem jack with a ball cone clamp. Two jacking systems can be used: hand operated, or hydraulically activated using the same power pack as the hammer. The hydraulic system used by Soil Mechanics Limited is based on the flow through sampler described by Martin and Suckling (1984).

Advances in site investigation practice. Thomas Telford, London, 1996

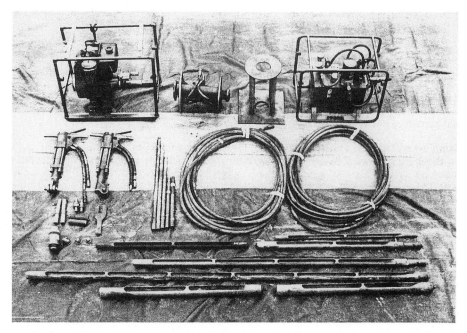

Plate 1 : Hydraulically Powered Dynamic Sampling Equipment

Plate 2 : Driving an 80 mm Diameter Sample Tube

The sampling system operates dry with no flushing medium. Initially a metre long tube is driven with the hammer connected directly onto the tube using the driving adaptor. The operator will generally work from a small elevated working platform to provide a comfortable operating position. Once the tube is driven its length, the hammer is disconnected and a rod added to permit the hollow stem jack to be used to extract the tube. The jack grips the rod about 0.75 m above ground level reacting against a specially designed lifting table. Once extracted the soil sample retained within the tube can be described through the window slot by a geologist or experienced technician (see Plate 4). The sample may be photographed and subsampled using a spatula to dig out the soil. Once described and subsampled the tube is then cleaned out for reuse.

Sampling at depth is achieved by repeatedly driven sample tubes which are carried deeper on the driving rods. Usually the diameter of the sample tube is progressively reduced to minimise side friction and ease returning the tube to the base of the hole. Where surface soils are in a loose state it may be possible to drive in a short length (0.75-1.50 m) of plastic casing to stabilise the top of the hole. However, the system essentially relies on rapid progressive sampling in an uncased hole. Dependent on the geology and density or stiffness of the soils, sampling to depths of 7 m to 15 m may be achieved. Generally the more powerful hydraulically driven systems will achieve a greater depth of penetration.

The soil sample recovered is highly disturbed by the driving process. The sample tubes have a high area ratio (see Table 1), two to four times that of the standard U100 sampler. However, where the soil conditions are reasonably favourable, a near continuous soil profile may be recovered. Nevertheless, care should be taken to recognise areas of redriving, where the sample tube has been driven through materials that have collapsed from higher up the sides of the uncased hole. In cohesive soils it is possible to take U38 tube samples from the hole using the driving rods with a suitable sliding hammer and at shallow depths to use a hand vane with extension rods. The sampling process should not be delayed greatly due to the inherent instability of the uncased hole.

The sample holes may be utilised for the installation of standpipes, gas monitoring tubes or other instrumentation, within the restrictions of the limited borehole diameters.

| Sample tube | Sample Shoe | | Area |
Outside Diameter (mm)	Outside Diameter (mm)	Inside Diameter (mm)	Ratio (%)
38	40	26	137
50	54	38	102
60	64	46	94
80	84	66	62
SML 100 mm piston sample			8
U38			19
U100			25 to 30
SPT split spoon			112

Table 1 : Comparison of Area Ratios of Samples

Plate 3 : Jacking Rods And Sample Tube Out

Plate 4 : Description And Taking Disturbed Samples

THE ADVANTAGES AND DISADVANTAGES

The initial impetus for the growth in the use of dynamic sampling techniques in the UK arose from the expansion in the late 1980's of small scale investigations of house settlements, coupled with the general increase in the number of environmental investigations. The window sampler compared with the traditional cable tool methods offered a relatively cheap and highly manoeuvrable shallow sampling system which suited the requirements of this type of investigation. The equipment is light enough to be manhandled to almost any sampling location. With the heavier rated hydraulically driven units, it is often possible to leave the power packs remote from the sampling location and run out hydraulic hoses for distances up to 60-70 metres. Thus the equipment has a significant advantage in allowing shallow sampling in areas of very poor or restricted access. In this respect the economics of the system compared to cable tool methods are very compelling. The light weight nature of the equipment also means that ground damage or disturbance resulting from the sampling operations can be minimised. This makes the system especially attractive for use in gardens, landscaped areas, car parks etc, where the costs of unavoidable damage may add considerably to the cost of an investigation.

Rapidity of operation is another attraction with this equipment. An average progress rate of 40 to 50 m of sampling per day compares very favourably with cable tool methods where daily progress would be in the order of 10 to 15 m per day. The labour resources employed by both methods are similar and thus the meterage costs for window sampling are potentially lower. However, the cost of equipment breakages are relatively high, reducing the cost advantage. Also the progress rates that can be achieved below 6 to 7 m depth reduce rapidly with increasing depth.

As discussed earlier the sampler recovers a totally disturbed sample which can only be regarded as Class 3 samples as defined by BS 5930 : 1981 and are therefore only generally suitable for classification and moisture content testing. However, under reasonably favourable conditions a full profile of the soils penetrated is retained in the sample tubes. This includes relatively thin horizons which may not easily be recognised in a cable tool borehole where only about 30% of the volume of the borehole is sampled. The soil profile displayed in the window tube may be subsampled for classification tests but the lack of undisturbed material for laboratory strength tests is a major disadvantage. In cohesive soils it may be possible to recover U38 samples between the sampling runs. Consideration may also be given to using the sampler as a profiling tool between standard cable tool boreholes in which U100 or piston samples are recovered.

Chandler et al (1992) have used filter paper suction tests (Chandler and Gutierrez, 1986) to assess clay desiccation around low-rise buildings. Soil suction profiles were found to be a better measure of desiccation than moisture content profiles. Therefore filter paper tests would be a valuable method for testing soils in conjunction with window sampling. However filter paper suction tests are normally carried out on undisturbed samples and thus standard window sampling methods do not recover samples suitable for this test. However, Crilly and Chandler (1993) stated that "it may be possible to obtain reasonable measurements of soil suction for the purposes of determining relative degrees of desiccation from disturbed soil samples", indicating that soil recovered from the dynamic sampler should be suitable for filter paper tests. It is therefore desirable that a system of dynamic sampling with a plastic liner is developed to overcome this disadvantage. In Germany, a plastic liner system has been used with dynamic sampling equipment.

The use of the sampler is also limited by the soil type. Dynamic sampling has been successfully carried out in a variety of soil types including: soft alluvial clays, very stiff overconsolidated clays, sands, sandy gravels, chalk and made ground. Coarse grained soils or coarse fill materials cannot successfully be sampled and a sample hole may be blocked by a single piece of coarse gravel or a cobble. This is mitigated by the relative ease and speed of moving the equipment to a new location to recommence the aborted hole. Due to the dynamic nature of driving the sample tubes, coarse gravel is often broken into a number of pieces. This enables holes to be progressed through gravel when there is only a limited proportion of a coarse fraction. The relative density of the soils to be penetrated is also a controlling factor with the use of this equipment. The hole is uncased and may thus collapse in loose granular deposits. Experience has shown this not to be a great problem above the water table. In very dense soils the depth of sampling will be restricted.

Sampling below the water table in all types of soil presents difficulties although a closed tube sampler may replace the window sample tube. Sampling in granular soils becomes almost impossible.

A tabulation of the advantages and disadvantages of the window sampling is presented as Table 2 for ease of reference. The following Case Histories illustrate the range of applications of the technique in ground investigations.

ADVANTAGES		DISADVANTAGES	
1.	Lightweight, highly manoeuvrable equipment which can be manhandled to almost any sampling location.	1.	Limited depth of sampling, ie 10 to 15 m.
2.	Rapid operation, 40 to 50 m of sampling per day.	2.	Limited by geology, ie no good in gravels, coarse fill materials.
3.	Relatively cheap.	3.	The hole is uncased, therefore loose ground can limit sampling.
4.	Possible full recovery of soil profile.	4.	No undisturbed samples.
5.	The operation of equipment causes minimal disruption and ground damage.	5.	Sampling below the ground water table is problematic to impossible.
		6.	The operation of sampling equipment is relatively noisy.

Table 2 : Advantages and Disadvantages of Window Sampling

CASE HISTORIES

Site A

This site is located to the south west of Stratford in London and is presently occupied by a supermarket with garage and an extensive car park. The site area was previously occupied by a number of potentially contaminative industries including a chemical works and an asbestos works. Ground conditions comprise tarmac overlying 3 to 5 m of Made Ground and Alluvium over London Clay. In three days a total of 22 dynamic sample holes were put down to depths of up to 6 m to take samples for chemical testing and to confirm the location of buried foundations below the car park. No additional plant was required to break out the tarmac and pavement sub-base as the hydraulic hammer and a chisel from the dynamic sample equipment was used as a breaker. To prevent cross-contamination between the various holes and samples, each window sample tube was cleaned using a jet wash prior to use. Gas monitoring wells with a geotextile filter wrap were installed at eight of the positions. At this site, dynamic sampling was a rapid method that minimised the disturbance caused by carrying out the survey and minimised the reinstatement required on completion. A "conventional" survey comprising trial pits and cable tool boreholes would have:

      i)   caused more disruption to the public using the car parks
      ii)   required more reinstatement
      iii)   cost more and taken longer

Site B
Approximately 250 dynamic sample holes were put down over a period of ten weeks at nine sites around London as part of a preliminary investigation into the stability of railway embankments for London Underground Limited.

A typical downward succession from the top of an embankment comprised track ballast (up to 1.5 m thick) over a variable thickness of granular material (typically gravelly sand of ash and clinker) overlying cohesive embankment fill before natural ground was encountered. A typical borehole log is presented in Fig 1. At the sites access was usually very poor with holes being located either between or adjacent to the railway tracks or on the steep overgrown embankment slopes. At many locations plant and materials could only be moved onto position during short night time track possessions. Work adjacent to the tracks was limited to night shifts which typically allowed a 5 hour working period. On most night shifts only one hole could be completed, the maximum depth achieved was 15 m. Temporary plastic casing was required to support the loose near surface granular deposits. Each sample tube was photographed immediately after recovery, described and sub-samples taken for moisture content and index testing. Pocket penetrometer tests were also carried out. The moisture content and penetrometer results for the specimen log presented in Fig 1 are plotted against depth in Fig 2.

The relative costs of various investigation methods carried out on this project are indicated in Table 3. The costs calculated are based on all work being carried out during a day shift on the embankment slopes. It can be seen that the cost of dynamic sampling is nearly twice that of dynamic probing, however, far more information is gained from dynamic sampling. Cable tool boring on the embankment slopes cost significantly more.

Set Up and Carry Out 10 m deep investigation	Relative Cost
Dynamic sample	1.0
Dynamic probe	0.6
Cable tool borehole	6.0
1.2 m deep hand dug trial pit	0.8

Note: The above relative costs do not take into account any mobilisation, sampling or testing costs

Table 3 : Approximate relative cost of various exploratory methods deployed at Site B

⌬ Soil Mechanics					BOREHOLE No. SPECIMEN				
					Sheet 1 of 2				

Equipment & Methods	Location No. 7848
Dynamic sampling in 80mm diameter to 1.30m, 60mm diameter to 3.40m, 50mm diameter to 6.10m and 38mm diameter to 9.20m. Hydraulic hammer and jacks, EW rods.	Location INVESTIGATION OF EARTH STRUCTURES STAGE II

Carried out for	Ground Level	Coordinates	Date
London Underground Limited	118.72 m A.O.D.	Chainage 319m / Off Set 11.2m N	13.08.93

Description	Reduced Level	Legend	Depth (Thick)	Samples/Tests					Field Records
				Depth	Sample		Test		
					Type	No.			
Dark grey silty fine to coarse SAND with some angular to subangular fine and medium ash and slag gravel. (EMBANKMENT FILL)	118.72		(1.25)	0.00 - 1.20	D	1			
				0.00 - 1.30	DS				212 seconds 1300mm recovery Hole collapsed to 1.00m
GROUT recovered as sandy angular gravel.	117.47 117.32		1.25 (0.15) 1.40	1.30 - 3.40	DS				382 seconds 1760mm recovery Hole collapsed to 1.50m
				1.50 - 2.00	D	2			
Stiff brown mottled orange brown slightly fine sandy CLAY with rare angular to subangular fine ot coarse flint gravel. (EMBANKMENT FILL)			(2.96)						
				3.40 - 6.10	DS				416 seconds 1980mm recovery Hole collapsed to 5.50m
Firm dark grey mottled grey brown slightly fine sandy organic CLAY with rare fine flint gravel. (ALLUVIUM)	114.36 114.07		4.36 (0.29) 4.65	4.68 - 5.08	D	3			
Firm orange brown fine to coarse sandy CLAY. (ALLUVIUM)			(1.40)						
	112.67		6.05	6.10 - 9.20	DS				242 seconds 2220mm recovery
Yellow brown slightly silty very fine to coarse sandy angular to subrounded fine and medium flint GRAVEL. (TERRACE GRAVEL)			(3.15 pen)	7.10 - 8.20	D	4			
BOREHOLE ENDS AT 9.20 m.	109.52		9.20						

Remarks	Logged by
1. 75mm diameter plastic casing installed to 1.25m.	MC
2. Backfilled with bentonite pellets.	Scale
3. See sheet 2 for moisture content and pocket penetrometer results.	1:50

Notes:
Materials are described in accordance with Appendices. For explanation of symbols and abbreviations see Fig. 1. (c) Soil Mechanics
All depths and reduced levels in metres. Thicknesses given in brackets in depth column.   29.07.94/09.07   (Ver 4.1.28)

Fig. 41

FIG 1 : SITE B - SPECIMEN BOREHOLE LOG

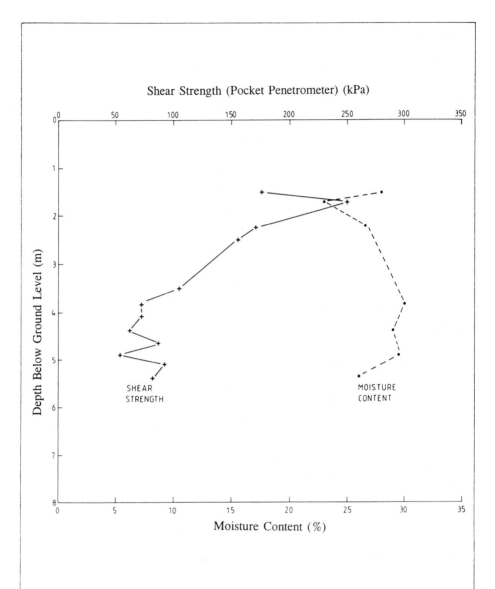

FIG 2 : SITE B - SHEAR STRENGTH AND MOISTURE CONTENT AGAINST
DEPTH BELOW GROUND LEVEL FOR SPECIMEN BOREHOLE

Site C
Soil Mechanics Ltd has carried out dynamic sampling on embankments and cuttings on a number of motorways including M3, M4, M6 and M25. Dynamic sampling has been carried out in cuttings as steep as 1:1.5 (vertical:horizontal) without the use of temporary scaffold platforms or benching which would be required by most other investigation methods. Typical daily production rates have been five or six holes to 7 m depth but this was reduced significantly at some locations due to difficult access or short working days due to motorway restrictions.

Many dynamic sample holes were primarily put down for the installation of instrumentation in slopes showing signs of instability. Research has shown that shallow slope failures are a significant problem on motorways (Perry, 1989). When investigating and monitoring these failures dynamic sampling has been found to be a highly economic method of installing arrays of instruments at relatively shallow depths down these slopes. Both pneumatic and standpipe piezometers have been installed together with slip indicators.

Site D
The conventional method of investigating ground conditions for houses and other low-rise structures which have been damaged due to subsidence or heave on clay sites has been to carry out a combination of trial pits (possibly with hand augering) and/or cable tool boreholes. This work often causes much disruption to residential properties and the location of boreholes is often severely restricted due to access and other constraints. However, by carrying out a combination of hand dug trial pits to investigate the nature of the existing foundations and carrying out a number of dynamic sample holes, more information can usually be obtained in a shorter time causing less disruption than with conventional field techniques. At site D a bungalow constructed in the early 1930's was showing signs of distress. In one day a team of two technicians and a geologist carried out the site work comprising two hand dug pits to 1.2 m and three dynamic sample holes to 7 m as indicated in Fig 3. Ground conditions encountered were a thin superficial layer of reworked London Clay over weathered London Clay. The trial pits revealed that the foundations were inadequate only extending to 0.55 m below ground level. The Atterberg limits and moisture content results are presented for one of the dynamic sample holes in Fig 4. This shows that below about 2.00 m the moisture content of the London Clay is relatively uniform with depth and is approximately equal to the plastic limit. Above 2.00 m depth the moisture contents are significantly lower than the plastic limit. Between about 4.0 and 6.0 m the moisture contents are slightly lower than the plastic limit. The results indicate that desiccation has occurred near the surface at this site and that shrinkage of the London Clay which is of high to very high plasticity is probably the cause of the foundation movement. The reduced moisture contents between 4.0 and 6.0 m may be due to the invasion of roots causing desiccation.

CONCLUSIONS
Dynamic (window) sampling is a technique which has only been used in the UK for three to four years. It is now being increasingly used as many geotechnical specialists have found dynamic sampling to be a relatively rapid and economic method of carrying out ground investigation to depths of up to 15 m in a wide variety of soil types.

Dynamic sampling is not included in the soon to be published revised version of BS 5930, therefore in this paper the authors have attempted to give some guidance to engineers considering adopting this technique. A specification for dynamic sampling was not included in the recently published 'Specification for Ground Investigation' (SISG, 1993). To give assistance to geotechnical specialists wishing to carry out dynamic sampling a proposed model specification is appended to this paper.

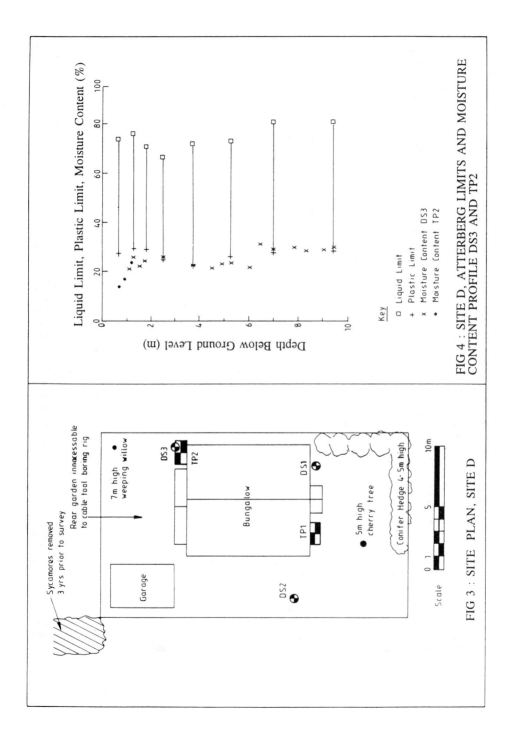

FIG 4 : SITE D, ATTERBERG LIMITS AND MOISTURE
CONTENT PROFILE DS3 AND TP2

FIG 3 : SITE PLAN, SITE D

ACKNOWLEDGEMENTS
Acknowledgement is given to the Directors of Soil Mechanics Limited for their support in publishing this paper and to London Underground Limited for their permission to present data from Site B.

REFERENCES

BRITISH STANDARDS INSTITUTION. 1981. BS5930. Code of practice for site investigation.

CHANDLER, R. J. & C. I. GUTIERREZ. 1986. The filter paper method of suction measurement. Geotechnique Vol 36, No 2, pp 265-268.

CHANDLER, R. J. & M. S. CRILLY AND G. MONTGOMERY-SMITH. 1992. A low cost method of assessing clay desiccation for low-rise buildings. Proc of Inst Civil Engineers, Civil Engineering, Vol 92, No 2, May, pp 82-89.

CRILLY M. S. & R. J. CHANDLER. 1993. A method of determining the state of desiccation in clay soils. BRE Information Paper IP4/93. Building Research Establishment.

MARTIN J. H. & A. C. SUCKLING. 1984. Geotechnical investigations for Mount Pleasant Airfield, Falkland Islands, using the Flow Through Sampler. Ground Engineering, November 1984.

PERRY J. 1989. A survey of slope condition on motorway earthworks in England and Wales. Transport and Road Research Report 199. Department of Transport.

SITE INVESTIGATION STEARING GROUP (SISG). 1993. Specification for Ground Investigation. Site Investigation in Construction, Part 3. Thomas Telford, London.

APPENDIX - MODEL SPECIFICATION

The equipment shall be capable of sampling to depths up to 15 m below ground level in suitable ground and shall utilise a high frequency percussion hammer to drive the sampling tubes into the ground.

On reaching the required depth, the driving hammer shall be removed and the sampling probe jacked out.

Each sample tube shall have a large 'window' down one side to allow visual examination (and photographing if required) of the soil sample recovered prior to the removal of any disturbed samples.

Sample tubes of 1, 2 and 3 m lengths of nominal 80, 60, 50 and 38 mm diameters shall be available. Rods and joints between rods and to the sample tube shall be robust so as to resist the forces imposed during driving and extraction without damage.

Successive sample tubes shall generally be driven in a sequence of decreasing diameters until the specified depth is reached. A virtually continuous recovery of the ground profile shall be produced for immediate visual examination (and photographing if required). Provision shall be made for taking disturbed samples for subsequent laboratory testing and carrying out pocket penetrometer tests as required.

Deep rotary cored boreholes in soils using wireline drilling

P HEPTON
Soil Mechanics Limited, Doncaster, South Yorkshire, UK

INTRODUCTION

During early 1994 Soil Mechanics Limited carried out a site investigation for Nuclear Electric plc as part of the ongoing works for the proposed Sizewell C Power Station in Suffolk. The work included rotary cored boreholes put down to over 100m depth through shelly sands of the Crag deposits overlying clay and sand of the Lower London Tertiaries, and penetrating the underlying Chalk bedrock. The purpose of the cored boreholes was to obtain a complete stratigraphic profile and obtain high quality core samples, suitable for subsequent small strain triaxial and dynamic laboratory testing. Some boreholes were logged by wireline geophysical methods and permanent plastic liners were also installed for cross hole seismic logging.

The borehole construction techniques are discussed in detail, with particular reference to the drilling muds used to aid core recovery in the weak formations and provide interim stability of boreholes left uncased during geologging and installation of permanent grouted liners. The use of a semi-rigid plastic liner in the inner core barrel to assist in core handling is described, together with core sample selection, preservation and storage for subsequent transport and testing off site.

GROUND CONDITIONS

The site is located on the Suffolk coast close to the village of Leiston. Ground conditions are summarise in Table 1.

Material	Typical depth (m) to base of stratum
Made ground and/or beach deposits overlying peat and soft organic clay	10 (where present)
Shelly sand of the Crag Deposits	45
London Clay	60
Sand and clay of the Lower London Tertiaries	85
Chalk (Upper)	

Table 1: Summary of Ground Conditions

Ground level is generally less than +5m OD with groundwater level close to the surface, although artesian heads of up to 5m were encountered in the Chalk.

CORE DRILLING EQUIPMENT
The Core Drilling System

The boreholes were commenced by cable tool boring to penetrate the made ground and alluvial deposits, and to install temporary 250mm diameter steel casing for drilling mud containment and recirculation. The boreholes were extended by rotary core drilling using lorry mounted top drive Dando 250 and 220 drilling rigs, each with a 6m stroke. A small amount of work was carried out with a lorry mounted Edeco H40 conventional drive unit (see Plates 1 and 2).

Rotary core drilling was carried out using the Geobor S wireline drilling system. The equipment consists of hollow steel drill pipe (nominally 140mm OD and 128mm ID) with a coring bit fitted to the bottom length which acts as the outer tube to the core barrel. The nominal 1.5m long inner core barrel, complete with the core lifter case, spring and semi-rigid plastic liner, is run down inside the drill pipe and fixed to the outer barrel by a latching mechanism. After a typical 1.5m long core run, the inner barrel assembly, complete with core, is recovered to the surface using an overshot device run on a wire from a high speed winch. The overshot connects to the top of the inner barrel, unlatching it from the outer barrel, and allowing the whole inner barrel assembly to be withdrawn up the drill pipe.

The advantages of wireline coring over conventional coring (using drill rods and core barrels) are that separate temporary lining casing for borehole stabilisation is not required, the drill string is more rigid, and interruption to the coring sequence by the removal of the core barrel is minimal. This minimises the disturbance to the borehole base, walls and cores recovered and allows rapid sub-sampling of cores to minimise the potential for softening effects of drilling muds. In addition, the thin 3mm annulus between the drill pipe and borehole wall requires significantly lower volumes of flush fluid than for conventional drilling to achieve adequate up-hole velocity for removal of cuttings from the hole. As a consequence lower flush velocity at the bit is achieved which reduces core and borehole wall erosion. The main disadvantage of the system is that the coring bit cannot be examined or replaced without withdrawing the drill string, a time consuming operation which can also cause damage to the borehole wall.

Core Drilling Bits

Since introducing the Geobor S wireline system and its predecessor, the SK6L system, to the UK in the mid 1980's, it has been SML's experience that the highest quality core is obtained with diamond core bits. These have a smoother cutting action and produce less disturbance in weak and laminated material than the more aggressive tungsten carbide set bits. The bits used for all strata were face discharge seven step surface set diamond bits with 10 to 15 stones per carat and 65 carats per bit. The bits were set to cut a nominally 146mm diameter hole and 102mm diameter core.

The use of expensive diamond bits in some horizons, particularly chalk strata with frequent flints, may result in excessive bit wear and damage and prove to be uneconomic. In such cases cheaper tungsten carbide toothed core bits are frequently used to reduce cost while generally still providing an acceptable but slightly lower quality core. At this site, the flint content of the Upper Chalk was generally low, allowing the use of diamond bits at low overall cost with bits producing up to 190m of core with minimal wear.

Plates 1 and 2: General View of Lorry Mounted Drilling Rig and Associated Equipment

Core Drilling Techniques

The core drilling process was carried out by lowering the inner core barrel, complete with semi-rigid plastic liner, to the base of the drill string and adding a 1.5m length of wireline casing at the surface. The drill string was returned to the base of the borehole, having been withdrawn some 300mm at the end of the previous run, while flushing and then rotated to commence coring. Bit pressure, speed of rotation, and drilling mud flow rate were controlled by the foreman driller using the optimum settings of these variables to provide high core quality and maximum recovery. Unfortunately there is little or no published optimum data in this regard and therefore much is dependent on the skill and experience of the drill rig operator. This was recognised at contract tender stage by the Client who required that a Drilling Supervisor should be on site during the drilling programme to provide specific expertise, in addition to that of the rig operators, to achieve both the core quality and recovery specified and the objectives of the boreholes in total.

The contract specification followed the normal practice of requiring core runs of less than 90% recovery to be followed by two short runs. It was found in the Crag deposits that generally good recovery was obtained for 0.75m long runs, while relatively less percentage core was recovered for 1.5m long core runs. The reasons for this are not clear, however, it may be that after a certain length of core has been retained in the core barrel the increased penetration force required to overcome the frictional resistance between the core and the inner surface of the liner causes compression and thus disturbance to the soil in the region of the bit kerf thus leading to consequent erosion and loss of material.

DRILLING MUD

Mud Constituents and Mix

Particular attention was given to the design of the flushing medium to maximise core recovery, reduce torque on the drill string, control artesian water flows and ensure stability of the borehole. The latter was especially important where the wireline casing had to be withdrawn to allow geologging of the borehole and the installation of permanent lining casings.

The drilling mud serves three major functions in addition to the bit cooling, lubrication and cleaning properties required of any flushing medium:-

(1)   The increased viscosity of the mud, compared to that of water, facilitates transportation of the core cuttings at a lower up-hole velocity. The lower flow rates reduce the jetting effect at the bit kerf and thus potential for erosion of core and the borehole wall.

(2)   The mud permeates the formation at the borehole wall and develops a gel-like filter cake which adds stability to the borehole wall.

(3)   The specific gravity of the mud can be raised by the addition of weighting agents to increase the mud pressure in the borehole above that of the groundwater. This prevents subartesian and artesian groundwater flowing into the borehole which can result in borehole wall abrasion and possible "piping" of granular materials.

The mud constituents were chosen to satisfy these three criteria using target Marsh Funnel and density parameters based on previous drilling work on the site as a starting point. The constituents chosen also had to be tolerant to saline conditions. For the majority of boreholes the mud was based on an organic polymer, to provide gel strength and act as the main agent for filter cake development. However some polymer muds have only limited capacity for carrying a weighting agent, therefore, an attapulgite clay-based salt water viscosifier was added to maintain the weighting agent (Barytes) in suspension as well as assisting in the development of the filter cake.

For boreholes in which piezometers were to be installed, clay based drilling muds were not considered appropriate and for these boreholes a high quality polymer mud, which also acts as a salt water viscosifier, weighted with Barytes was used. The polymer was broken down prior to the installation of piezometers by adding bleach to the recirculated mud. The use of this drilling mud was restricted to piezometer boreholes as the cost of the high quality polymer was prohibitive compared to the lower grade polymer/attapulgite clay combination mud.

Mud Control

Target parameters were set for the drilling muds at the outset of the works by specifying operational Marsh Funnel viscosities and mud weights. For boreholes to approximately 50m, viscosities of between 50 and 55 seconds and mud weights of 8.7 to 8.9lbs per US gallon (SG of 1.04 to 1.07) were used. Where boreholes were extended to 100m and were likely to encounter artesian groundwater conditions in the chalk, the viscosity was increased to a maximum of 80 seconds to allow for weighting up to 9.4lbs per US gallon (SG of 1.13).

During drilling the mud was monitored by the Drilling Supervisor. Addition of mud constituents was required to maintain the drilling fluid properties which were continually changing due to:

(1)   The increase of mud volume in the system as the borehole volume increased with depth.

(2)   The addition of natural materials to the drilling mud arising from the strata penetrated.

(3)   The influence of the ground and groundwater chemistry on the chemical and physical properties of the mud.

Mud Handling

All the mud constituents were supplied as 25 kg bagged solids, and were mixed into potable water in the drilling tanks via a hopper with a venturi nozzle (see Plate 3). Three shallow tanks were used with a total capacity of approximately 2500 litres. The mud return from the borehole was fed via an outlet fitted to the temporary casing into the first tank then carried by gravity flow through the second and third tanks. Recirculation of the mud between the second and third tanks was carried out by pumping to prevent the mud gelling and settling to the bottom of the tanks. The recirculation flow rate was set at a compromise between achieving adequate shearing (hence lowered viscosity) to allow the cuttings to settle out of the fluid, and preventing resuspension of the cuttings by turbulent flow within the tanks. Recirculation in the first tank was not carried out as this was the first part of cuttings collection and removal. The mud was pumped from the third tank to the drill string using a pump rated at a maximum flow of 180 litres per minute at low pressure, although considerably less flow than this was normally used.

Plate 3: View of Drilling Mud Tanks

Plate 4: Core Transportation Box

CORE HANDLING
Core liner
The inner core barrel was lined with a semi-rigid plastic liner to act as a sample tube and provide a triple tube barrel configuration. This reduced disturbance to the core during extrusion and handling by providing sufficient rigidity for a full 1.5m length of core to be handled by one person without risk of damage. On retrieval of the inner core barrel, the liner was pulled out from the inner barrel by hand, without vibration, cut to length if necessary (for example after a short run) and the open ends capped. The liner was then marked with the depths at the beginning and end of the core run. The liner used was pre-printed with an up-direction arrow to prevent mislabelling of the top and bottom of the sample.

The liner can be supplied either opaque or transparent. The latter potentially allows the core to be viewed before removing the liner, for selection of sub-samples or assessment of stratum boundaries, for example. However, in practice the inside is frequently smeared by drilling mud rendering it effectively opaque. The liner is manufactured to a nominal outside diameter of 110mm to fit the internal diameter of the inner barrel (111mm), with standard wall thicknesses of 1.5mm (107mm ID) or 3mm (104mm ID) being available depending on the degree of sample restraint and rigidity required. Core bits are set to cut nominally 102mm diameter core and consequently core recovered using standard thickness liner will have either a 1 or 2.5mm clear annulus. While a tighter fit may be desirable in terms of preventing disturbance to the core, adequate clearance is required to allow the core to pass freely up the inner barrel, including if required, allowance for potential swelling of some cohesive soils.

At the onset of the investigation a liner of nominal 3mm wall thickness was specified with the actual product supplied having an internal diameter of 104.2mm. However, during the initial boreholes it was observed that cores recovered from the Crag deposits were clearly cutting undersize compared to the liner and concern was expressed that sub-samples taken from the core would not be adequately restrained to prevent disturbance during handling. To reduce this effect a nominal 4mm wall thickness liner with a smaller internal diameter was manufactured. Subsequently this was used for coring in the Crag deposits, with the 3mm wall option used for core drilling in other deposits. Protection to the core with the enlarged thickness was found to be greatly improved with no reduction in core recovery.

Sampling, preservation and storage
The project required the core to be initially examined and sub-sampled (if required) within one hour of it being recovered to the surface. To achieve this, the core was transferred to the core store after each 1.5m run inside polystyrene lined transportation boxes to reduce handling disturbance and thermal shock (see Plate 4). On arrival at the store, the core was received by the logging geologists and, where required, samples of Crag deposits were cut directly from the core still in its liner using a bench mounted diamond tipped circular saw (see Plate 5). The samples were taken approximately two thirds from the top of the core run, where the effects of any drilling disturbance were expected to be minimal, and examined at each end to confirm their suitability for retention. Acceptable samples were fitted with push-on plastic end caps, taped on to prevent moisture loss. Samples of granular material from the Lower London Tertiaries were also preserved in a similar manner. The samples were stored in the site cool store within polystyrene lined sample boxes to protect them from thermal and physical shock and maintain them in an upright position (see Plate 6). The cool store was kept at a temperature of between 8 and 12° Celsius.

Plate 5: Bench Mounted Saw

Plate 6: Sample Storage Box

The remaining sample tube was cut longitudinally along each side using an angle grinder mounted on a sliding bed (see Plate 7). Where suitable the core was split, one half being photographed (see Plate 8) and the other logged, before putting the two halves together, taping up the sides and fitting end caps. The cores were kept in 1.5m long 2 channel boxes within the core store, which was maintained at a temperature of between 5 to 25° Celsius. Daily records of the core store, cool store and ambient maximum and minimum temperatures were kept.

Specimens from the London Clay and cohesive horizons of the Lower London Tertiaries were taken after the liner had been removed. Approximately 350mm of core was then cut from the lower part of the run, using the bench saw described above. Any drilling fluid on the core surface was removed by dabbing with absorbent cloth and the samples then wrapped in three layers of wax and cling film. A low melting point wax was used, kept just liquid in a shallow tray on an electric thermostatically controlled hot plate. Each core sample was first rolled in the wax (approximately 20mm deep), each end dipped in, and then allowed to cool. Waxed cling film was then rolled around the sample with small squares of waxed cling film placed over each end. The process was repeated twice more and the samples placed in the cool store as described above.

BOREHOLE INSTALLATIONS
Standpipe piezometers were installed in a number of the boreholes through the wireline drill string by conventional means. However, in three boreholes it was required to grout in 101mm internal diameter plastic casing to approximately 103m depth for crosshole geophysical surveying. As this casing would not fit through the wireline core bit, it was necessary to withdraw the full length of wireline drill pipe, leaving about 100m of unsupported hole. The stability of the borehole was maintained by mixing a mud to the maximum weight and viscosity and pumping this into the borehole to displace the thinner drilling mud. The drill pipe was withdrawn while adding further mud to compensate for the displacement volume of the drill string and maintain excess mud pressure above the groundwater pressure within the borehole. Where geophysical logging of the borehole was required prior to permanent casing installation the mud level was maintained whilst logging was carried out.

The mud mix ensured that the boreholes remained stable while uncased for up to 12 hours. Following this, PVC casing with a one-way valve fitted to the bottom was installed, using mud pumped inside to counteract the buoyancy. It was found that below about 30m the pressure outside the casing was sufficient to force mud through the casing joints, allowing it to sink in a controlled manner to within about 2m of the base of the borehole. Mud was then pumped inside the casing, through the non-return valve and up the outside to clear any cavings pushed ahead of the casing and to ensure a clear pathway before grouting commenced.

The first casing was to be grouted in place by attaching a pressure cap to the top of the casing, pumping grout down the casing and through the one-way valve at the bottom until grout was observed returning to the surface between the borehole wall and the permanent liner. The liquid grout inside the casing was then to be removed by flushing with clean water. Using this method it was found that the borehole could only be partially grouted as part way through the operation the pressures required to lift the grout and mud column exceeded the bursting pressure of the cap to the permanent liner.

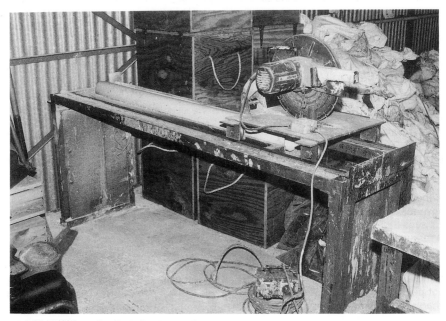

Plate 7: Longitudinal Cutting of Core Liner

Plate 8: Example Photographic Record of Core

Following the experience with the first installation a review was carried out of the procedures used. It was concluded that the time taken to grout the installation was excessive which resulted in the grout reaching a gel strength requiring pressures in excess of the capacity of the system for it to flow. A revised method statement was prepared which included the following:

(1)   The time lost in filling the casing with grout could be considerably reduced by using a one-way valve threaded to fit conventional drill rods so that grout could be pumped down the rods directly through the valve and up outside the casing.

(2)   Two air driven grout mixing and pumping units were employed so that grouting could be carried out as a continuous process with the grout remaining mobile at all times.

(3)   A slightly less viscous grout was used by increasing the water:cement ratio to 1.5 (by weight), the specified cement:bentonite proportions remaining at 3:1.

It should be noted that although the specification required grout trials to be carried out in advance of the grouting operation these can only provide guidance to the performance of the full scale grouting operation. The pregrouting trials included preparing a range of mixes at different water:cement ratios, measuring their Marsh Funnel viscosities and subsequently assessing their strengths. After the initial problems with the production grouting, the different mixes were reassessed by comparison with the viscosities of drilling mud with a view to establishing a lower bound water:cement ratio that could just be pumped. However, it was not possible to account for the increasing gel strength with time nor changes in the grout characteristics due to mixing with the drilling mud and loss of water to the formation.

CONCLUSION

The large diameter wireline drilling system with suitably designed diamond drilling tools and drilling muds produced core recoveries summarised in Table 2.

Material	Range of recovery (average per borehole)	Mean recovery from all boreholes
Crag Deposits	42% to 79%	65%
London Clay	92% to 99%	95%
Lower London Tertiaries		
(Granular)	67% to 96%	85%
(Cohesive)	93% to 98%	95%
Chalk	76% to 98%	95%

Table 2: Summary of Core Recovery Percentages

It should be noted that the apparently low recovery of chalk in one borehole resulted from the loss of core from a single run due to a flint obstruction where the total length of chalk cored was less than 10m.

Core quality of the cohesive soils, in conjunction with careful sample preparation as soon as possible after recovery, generally provided suitable high quality specimens for small strain triaxial and dynamic laboratory testing. Selected samples of the granular materials were also considered worthy of such sophisticated testing, although the test results are not yet available to allow definitive assessment of the sample quality.

In addition to the works described above, a 400mm diameter water well was successfully constructed to 52m depth in the Crag Deposits for a pumping test, using a conventional tractor-mounted site investigation drilling rig together with a mud system based on the experience of the core drilling programme described. This experience further demonstrates the virtues of the use of designed drilling mud on site investigation projects.

ACKNOWLEDGEMENTS
Thanks are due to Nuclear Electric Plc and the Directors of Soil Mechanics Limited for permission to publish this paper and the support of colleagues within Soil Mechanics, in particular Mr Stephen Tomlinson for input during the drafting.

Sample disturbance in rotary core tube sampling of softrock

TATSUOKA,F., KOHATA,Y.,
Institute of industrial science, University of Tokyo, Tokyo, Japan
TSUBOUCHI,T.
Institute of Technology, Tokyu Construction Co. Ltd., Sagamihara, Japan
MURATA,K.
Tokyo Soil Research Co. Ltd., Tokyo, Japan
OCHI,K.
Institute of Technology, Tokyu Construction Co. Ltd., Sagamihara, Japan
WANG,L.
Graduate Student, University of Tokyo, Tokyo, Japan

INTRODUCTION

In a number of case records of large Civil Engineering structures constructed in and on sedimentary softrock having a compressive strength of 3 – 10 MPa in Japan, the major principal strain ε_1 in the ground is about 0.2 % or less (Tatsuoka and Kohata, 1994). For sedimentary soft mudstones, as ε_1 at the peak stress state is around 0.5 % or less, with a sufficient margin to the failure of ground, these in–ground strains should be small as described above. These structures are foundations for giant suspension bridges; the world–longest 3,910 m–long Akashi Strait Bridge and 800 m–long Rainbow Bridge, high–rise buildings, ground excavations for a 50 m–deep shaft described in this paper and very large shafts for 80,000 and 200,000 kl under–ground LNG tanks and tunnels at deep places. Their full–scale field behaviour was simulated successfully well by numerical analyses using the stiffness values determined based on the elastic moduli obtained from field seismic surveys while accounting for the strain– and pressure–dependency of stiffness evaluated by triaxial compression (TC) tests (Ochi et al., 1993, 1994, Tatsuoka and Kohata, 1994). Therefore, the elastic deformation characteristics and the stiffness at those small strains of sedimentary softrocks are important design parameters.

These numerical analyses assumed negligible effects of cracks, joints and faults in the field and essentially elastic (recoverable and strain rate–independent) deformation characteristics at very small strains of softrocks (Tatsuoka and Shibuya, 1992, Tatsuoka and Kohata, 1994). In fact, in all the cases described above and others, the elastic shear modulus $G_f = \rho \cdot V_S^2$ from the shear wave velocity V_S measured for a field mass is very similar to the average of the maximum shear moduli G_{max} obtained from a number of TC tests using high–quality core samples retrieved from that mass. The analysis procedure is also based on the fact that the stress–strain relation is not very non–linear at strains less than about 0.1 %.

Yet, the issue of the disturbance of softrock samples retrieved from bore holes by rotary core tube sampling is not well understood. **Fig. 1** shows a typical triple–tube sampler. This sampler is a sort of open sampler, which has a metal outer tube with a metal bit at the bottom, a metal inner tube with a shoe at the bottom end, through which cored material is pushed in, and a PVC lining tube placed inside the inner tube to contain cored material. The triple core pack tube sampler uses a thin plastic sheet in place of a PVC lining tube. The inner tube is separated from the rotation of the outer tube through a bearing, but no measures is taken to prevent its rotation if it occurs. Research into this issue is very limited, partly because only in a few cases, both 1) rotary core tube sampling from bore holes and 2) block sampling or direct coring from the excavated ground have been performed at the same site. Moreover, the conventional triaxial testing method is inadequate to obtain accurate stress–strain relations at strains smaller than 0.1 %, which are essential for evaluating the quality of a given core sample. Because of the sample disturbance problem, one may consider that plate loading and pressuremeter tests in the field are superior over laboratory stress–strain tests, even when stress and strain measurements in the laboratory tests are very sensitive and accurate. However, these field tests are not able to obtain a <u>continuous</u> stress–strain relation for a range of strain from less than 0.001 % to that at the peak state <u>under well–controlled conditions</u> of drainage, strain rate or so. Moreover, the interpretation of the field test results are usually not straightforward due to uncertainties in the boundary conditions and possible errors in the displacement measurements. Indeed, laboratory and field tests are supplementary to each other, and both are important for the purpose of site characterization.

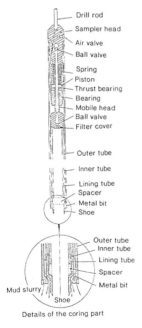

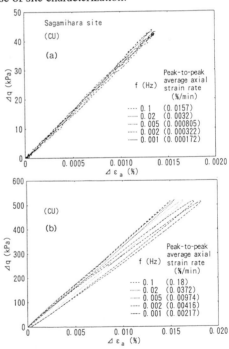

Fig. 1 Triple–tube sampler
(JSSMFE, 1982)

Fig. 2 Results of cyclic triaxial tests
using a DC sample of sedimentary soft mudstone
(see Fig. 2.60 of Tatsuoka and Kohata, 1994)

In view of the above, the following two schemes of research were conducted: 1) A comparison of the elastic Young's moduli, pre–peak stress–strain relations and compressive strengths evaluated by TC tests using core samples obtained by rotary core tube sampling and those obtained by block sampling or direct coring. 2) Compilation of pairs of the elastic shear modulus $G_f=\rho\cdot V_S^2$ from a field seismic survey and the initial shear modulus G_{max} from a triaxial test obtained for sedimentary softrocks together with artificial softrocks (cement–mixed sands and clays), obtained in relation to a number of construction projects. Their data were analysed on the same basis considering their very similar deformation and strength

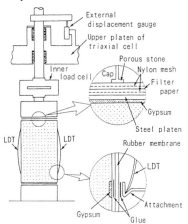

Fig. 3 Triaxial testing system

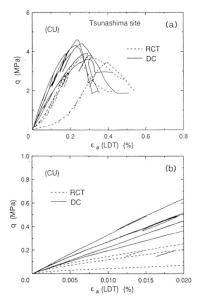

Fig. 5 Comparison of $q(=\sigma_a-\sigma_r)$ – local ε_1 relation, CU TC tests using RCT and DC samples

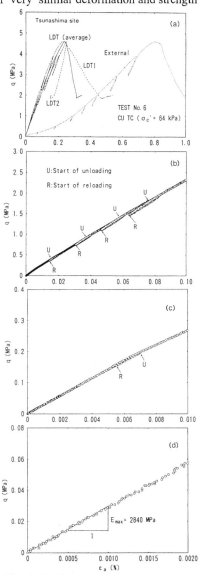

Fig. 4 Typical CU TC test result, DC sample of sedimentary soft mudstone

characteristics.

ELASTIC BEHAVIOUR AT VERY SMALL STRAINS

Fig. 2 shows the result of consolidated undrained cyclic triaxial tests applying symmetric sinusoidal cyclic deviator stresses to a specimen consolidated isotropically to the field effective overburden pressure σ_v'(in–situ)= 0.47 MPa. The specimen (5 cm in diameter x 15 cm high) of sedimentary soft mudstone was obtained by direct coring at Sagamihara Test Site (explained later). The compressive strength of the sample from a consolidated undrained (CU) TC test performed after these cyclic loading tests was 6.8 MPa. The cyclic loading frequency **f** was changed for a range from 0.001 Hz to 0.1 Hz with a single amplitude axial strain $(\varepsilon_a)_{SA}$ equal to either about 0.0007 % or about 0.008 %. At constant **f** and $(\varepsilon_a)_{SA}$, eleven cycles were applied, during which any noticeable change in the stress–strain relation was not observed. The axial strain and stress were measured locally by using a pair of Local Deformation Transducers (LDTs; Goto et al., 1991) and a sensitive load cell set inside the triaxial cell (**Fig. 3**; Tatsuoka et al., 1994). For a comparison purpose, these relationships have been plotted shifting the recorded ones in the vertical and horizontal directions so that the bottom ends of the curves be located at the common origin. For $(\varepsilon_a)_{SA}$= 0.0007 %, the relations are very linear and the effects of **f** (i.e., strain rate) on the relations is negligible, while for $(\varepsilon_a)_{SA}$= 0.008 %, some effects of strain rate can be seen. This result suggests that in seismic surveys, is which the strain level is of the order of 0.001 % or less, the deformation of sedimentary softrock is essentially linear elastic.

Fig. 4 shows the result of a typical CU TC test using a sedimentary soft mudstone sample retrieved by direct coring at Tsunashima site (explained below). The specimen (5.5 cm in diameter x 15 cm high) was isotropically consolidated to σ_v'(in–situ)= 64 kPa and the external axial strain rate was 0.01 %/min. In Fig. 4a, the two readings of a pair of LDTs and its average are shown, while in Figs. 4b – 4d, only individual data points of the average are shown. The difference between the local strains and those measured externally with a gauge set outside the triaxial cell (Fig. 4a) is due primarily to the effects of bedding error at the top and bottom ends of specimen resulting from the extra compression of a drainage layer and a thin disturbed zone formed during end trimming. In Figs. 4b–4d, high linearity and recoverability at strains less than about 0.01 % may be seen. The latter feature can be noted from the fact that in Fig. 4c, the stress–strain relations during an unload/reload cycle applied during otherwise monotonic loading overlap almost perfectly with the primary stress–strain relation. The initial Young's modulus E_{max} was defined at strains less than 0.001 % (Fig. 4d).

TSUNASHIMA SITE (SEDIMENTARY SOFT MUDSTONE)

At Tsunashima in Kanagawa Prefecture, just west of Tokyo, from an about 1.5 million years old sedimentary soft mudstone (Kazusa Group), core samples of 7.0 cm in diameter were obtained using a triple–tube sampler prior to the ground excavation for a high–rise building. The drilling was at 50 rpm using mud slurry. The 7.0 cm–in–diameter and 17 cm–high TC test specimens were prepared. These samples will herein be called RCT samples. Samples of 5.5 cm in diameter were obtained from the excavated ground by direct coring using a well-fixed core barrel having a diamond bit rotating at 900 rpm using water as drilling liquid. Detrimental rocking motion of the core barrel during coring was unlikely. These samples will herein be called DC samples.

A series of CD and CU TC tests were performed on RCT and DC samples reconsolidated isotropically to σ_c' equal to 0.5, 1, 2, and 4 times σ_v'(in–situ) prior to excavation (**Fig. 5**). The effects of σ_c'/σ_v'(in–situ) were found negligible. The S–shape seen for the stress–strain curves of the RCT samples cannot be seen for those of the DC samples. The RCT samples exhibited generally lower maximum Young's moduli E_{max}, lower compressive strengths q_{max} and larger axial strains ε_{1f} at the peak stress state when compared with those of the DC samples (**Fig. 6**). The average of the E_{max} values from the CU TC tests using the DC samples is very close to the field value $E_f = 2(1+\nu) \cdot \rho \cdot V_S^2$ ("the average E_{max}"/$E_f = 1.023$) (Fig. 6b), whereas the E_{max} values of the RCT samples are much smaller. It is seen from Fig. 6c that the ε_{1f} values are largely overestimated when based on external axial strains. **Fig. 7** shows the relationships between the "tangent Young's modulus $E_{tan} = dq/d\varepsilon_a$"/$E_{max}$ and the shear stress level q/q_{max} for the RCT and DC samples. For some range of q/q_{max}, the maximum values of E_{tan}/E_{max} for some RCT samples are larger than those of the DC samples, which is not due to larger E_{tan} values of the RCT samples, but due to the smaller E_{max} values. The difference in E_{tan}/E_{max} (and E_{tan}) between the RCT and DC samples is the largest when q/q_{max} is around 0.1. **Fig. 8** compares ε_{1f}, q_{max}, E_{max} and "E_{tan} at $q/q_{max} = 0.1$". The difference between the RCT and DC samples is larger in the order of "E_{tan} at $q/q_{max} = 0.1$", E_{max}, q_{max} and ε_{1f}. Smaller values of "E_{tan} at $q/q_{max} = 0.1$" of the RCT samples correspond to the S–shaped stress–strain curves (Fig. 5).

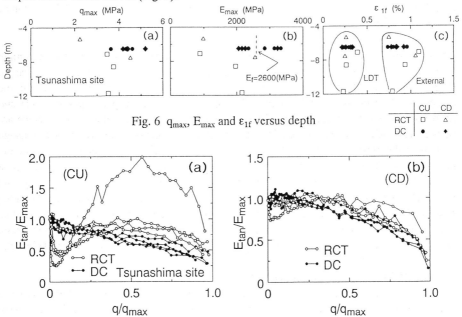

Fig. 6 q_{max}, E_{max} and ε_{1f} versus depth

Fig. 7 E_{tan}/E_{max} (based on local strains) – q/q_{max} relations of RCT and DC samples

SAGAMIHARA TEST SITE (SEDIMENTARY SOFT MUDSTONE)

In an about 2 million years old sedimentary soft mudstone (Kazusa Group) at Sagamihara City, west of Tokyo, a 50 m–deep shaft and a series of short tunnels having different cross–

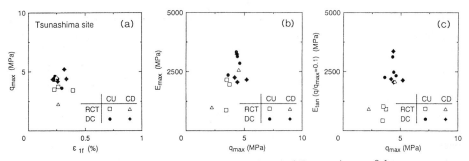

Fig. 8 Comparison among ε_{1f}, q_{lmax}, E_{max} and E_{tan} at $q/q_{max} = 0.1$

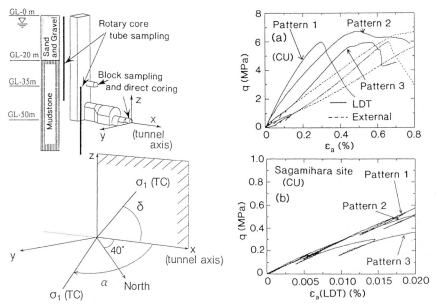

Fig. 9 Sagamihara Test Site
and sampling methods

Fig. 10 $q - \varepsilon_a$ relations,
CU TC tests using RCT samples

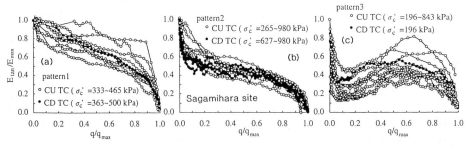

Fig. 11 E_{tan}/E_{max} (based on local strains) – q/q_{max} relations, CU TC tests using RCT samples

sections were excavated for research purposes (**Fig. 9**). The field test data and full–scale behaviour and their numerical analyses were reported by Ochi et al. (1993, 1994) and Tsubouchi et al. (1994). The following methods were used to obtain specimens for TC tests: 1) Rotary core tube (RCT) sampling from the ground surface to a depth of 45 m using a triple–tube sampler with a core diameter of 11.6 cm. 5.5 cm–in–diameter and 15 cm–high specimens were re–cored from them. 2) RCT sampling from a depth of 35 m to a depth of 85 m using a triple core pack tube sampler with a core diameter 5.5 cm. 5.5 cm--in–diameter and 15 cm–high specimens were prepared from them. 3) Block sampling (BS) using an electric chain saw in a test adit at depth of 35 m and in a tunnel at a depth of 48 m. 5 cm–diameter samples were re–cored in the laboratory by using a diamond core barrel from these blocks. The axis of the samples obtained by the methods 1), 2) and 3) was vertical in–situ. 4) Direct coring (DC) using a diamond core barrel in the tunnel with a core diameter of 20 cm. Coring was performed first at the tunnel end face at a distance of 30 m from the tunnel entrance in various directions in vertical planes at the angles δ of the sample axis direction to the in–situ horizontal direction equal to 90, 67.5, 45, 22.5 and 0 deg. In the second step, coring was performed in a horizontal plane at the adjacent side wall of the tunnel, in which the angle α of the sample axis direction were 45 and 90 deg. The diamond core barrel was firmly fixed against rocking motion. Specimens with a diameter of 5 cm and a height of 15 cm were re–cored from the central part of each 20 cm–in–diameter core sample. The inherent anisotropy of this softrock is not significant (Tatsuoka and Kohata, 1994).

The TC test results of RCT samples were reported elsewhere (Tatsuoka et al., 1993, Kim et al., 1994). By comparing the TC test results of RCT samples with those of BS and DC samples, it was revealed that some of the RCT samples had been noticeably disturbed (Tsubouchi et al., 1994). **Fig. 10** shows the results of typical CU TC tests using three RCT samples isotropically consolidated to σ_v'(in–situ)= 0.35, 0.63 and 0.65MPa. The classification into the three patterns was made based on the E_{tan}/E_{max}–q/q_{max} curves (**Fig. 11**) (Tsubouchi et al., 1994). The sample disturbance is considered to be larger in the order of the patterns 3, 2 and 1, since BS and DC samples exhibited only the pattern 1 behaviour (**Fig. 12**). In **Fig. 13**, the values of q_{max}, E_{max} and ε_{1f} are plotted against depth. **Fig. 14** compares ε_{1f}, q_{max}, E_{max}, and "E_{tan} at $q/q_{max}= 0.1$ and q_{max}". Again, the effects of sample disturbance for the RCT samples are generally larger in the order of "E_{tan} at $q/q_{max} =0.1$", E_{max}, q_{max} and ε_{1f}, while the difference between the RCT and DC samples is smaller than in Tsunashima Site case. In particular, the values of E_{max} of the RCT and DC samples are very similar to each other, and they are similar to the field value Ef (Fig. 13b). The E_{50} values from unconfined compression tests (measuring axial strains only externally) using vertical DC samples with a core diameter of 5 cm obtained during the excavation of the shaft are also shown in Fig. 13b. The sample disturbance cannot be analysed based on such quantities.

For Tsunashima and Sagamihara cases, the behaviour of the RCT samples different from that of the DC samples should be due to sample disturbance resulting from rotation and rocking motion of the inner tube during rotary coring. Although the inner tube is designed not to rotate together with the outer tube, it can rotate when in contact with the outer tube. In that case, the cored sample is twisted, which may result into cracking in the direction inclined from the axial direction of the sampler. Although the bottom of the outer tube is fixed against lateral movement being fixed against the ground, the other parts of the outer tube are not, and

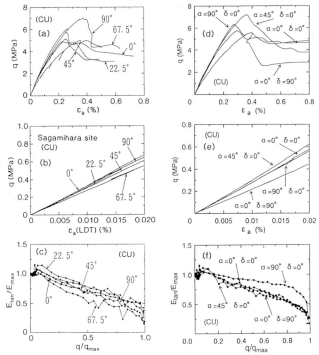

Fig. 12 q –ε_a relations and E_{tan}/E_{max} (based on local strains)
– q/q_{max} relations, CU TC tests using DC samples

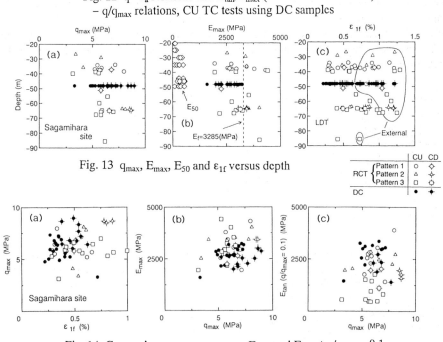

Fig. 13 q_{max}, E_{max}, E_{50} and ε_{1f} versus depth

Fig. 14 Comparison among ε_{1f}, q_{max}, E_{max} and E_{tan} at q/q_{max} = 0.1

the axis of rotation and the sampler axis may not be co-axial. Therefore, rocking motion of the sampler is quite possible, particularly when drill rod and bore hole are long and they are not straight. Such rocking motion may result into lateral transverse cracking in a cored sample of softrock. This mechanism of sample disturbance is different from that in fixed-piston thin-wall tube sampling. Note also that the RCT samples used for the TC tests were obtained by selecting continuous parts not including visible cracks, while the other parts in each sampler tube included many cracks to an extent much higher than that for the cracks observed on the wall of the shaft. This fact also indicates that most of these cracks were made during rotary coring.

Sensitive and accurate local strain measurements made possible such a detailed analysis of sample disturbance shown above. When based on external axial strains, however, the difference in the stress-strain relations between disturbed and high-quality samples becomes much less discernible (see Fig. 10a). This would be one of the reasons why the sample disturbance problem in rotary core tube sampling of softrock has been over-looked.

TRANS-TOKYO BAY HIGHWAY PROJECT (CEMENT-MIXED SOIL)

At Kawasaki Man-Made Island (**Fig. 15**) in Trans-Tokyo Bay Highway (see Fig. 2.8 of Tatsuoka and Kohata, 1994), a large cylinder with an inner diameter of 88 m, a thickness of 13 m and a height of 30 m was constructed between the outer and inner steel jacket structures by placing underwater slurry of cement-mixed sandy soil with a volume of 118,000 m^3 by means of Tremmie pipes. Within this cylinder, subsequently a huge diaphragm wall was constructed to a level of −114 m below the sea level. About three months after the placement of the slurry, undisturbed samples were retrieved using a triple core pack tube sampler with a core diameter of 6.5 cm from three bore holes, in which PS logging was performed. TC specimen of 5 cm in diameter x 10 cm high were prepared by trimming. The CU TC tests described herein are those performed using specimens consolidated isotropically to σ_v'(in-situ).

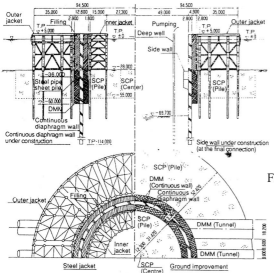

Fig. 15 Kawasaki Man-Made Island, Trans-Tokyo Bay Highway Project (Uchida et al., 1993) (left)

Fig. 16 compares the values of E_{max} from the CU and CD TC tests and CU cyclic triaxial tests and the field values E_f in one of the bore holes. Generally they are not very dissimilar, but some E_{max} values are particularly smaller than the E_f values. A summary of pairs of E_{max} and E_f are shown in **Fig. 17** (together with other data). In general, samples with E_{max} values noticeably lower than the E_f values exhibited q_{max} values lower than those of the samples which showed E_{max} values similar to the E_f values, while exhibiting S–shaped stress–strain relations. We consider that this behaviour is due primarily to sample disturbance.

To support the main structure of Kawasaki Man–Made Island, the existing soft clay deposit with a volume of 132,000 m^3 was improved by mixing in–situ with cement slurry (Deep Mixing Method method, Fig. 15). About 2.5 years after the ground improvement, during the excavation of the ground inside the diaphragm wall, undisturbed samples were retrieved from exposed ground by direct coring using a diamond core barrel with a core diameter of 10 cm, from which TC specimens of 5 cm in diameter x 10 cm high were re–cored. The compressive strength by CU TC tests using block samples was 6.7 – 7.6 MPa. **Fig. 18** shows the result of a typical CU TC test using a DC sample, which is very smooth unlike those of some disturbed RCT samples of cement–mixed soils.

All the pairs of E_f and E_{max} for the fills made of cement–mixed sands and the existing soft clay deposits improved by DMM are summarized in Fig. 17. These ground improvement methods are described in Uchida et al. (1993). For the DC samples, the ratio E_f/E_{max} is on average close to 1.0, while for many of the RCT samples, the ratio is larger than 1.0, which is likely due to sample disturbance. **Fig. 19** is the summary of pairs of E_{max} and q_{max} values, which include some CU TC data of the samples made by pouring in air cement–mixed sand in the slurry condition into a mold with inner dimensions of 5 cm in diameter and 10 cm high. The ratios E_{max}/q_{max} for most of the data lie within a relatively narrow band. The reasons for some scatter include: 1) the curing time spans over a very wide range of time, whereas the ratio E_{max}/q_{max} decreases with curing time (Tatsuoka and Shibuya, 1992), and 2) the ratio E_{max}/q_{max} for some data of the low–strength type DMM is particularly low. Based on the results of sedimentary soft mudstone shown above, the latter is likely due to that the effects of sample disturbance was larger on the E_{max} values than on the q_{max} values.

SUMMARY

A summary of pairs of the G_{max} value from consolidated triaxial tests using RCT, BS and DC samples and the G_f value from a field seismic survey (mostly the down–hole PS logging) for sedimentary softrocks and cement–mixed soils is presented in **Fig. 20**. To obtain the shear moduli G_{max} from the values of E_{max}, Poisson's ratios of 0.4 and 0.2 were assumed for the undrained and drained tests. Each data point represents the average value for several similar TC tests (up to 18 tests). The following trends may be noted:

1) All the values of G_{max} evaluated based on external axial strains are considerably lower than each corresponding value of G_f due to the large effects of bedding error, irrespectively of the sampling method.

2) The values of G_{max} based on local axial strains are much closer to G_f.

3) When based on local axial strains, the BS and DC samples exhibit G_{max} values very close to the G_f values.

4) When based on local axial strains, some data for the RCT samples exhibited G_{max} values

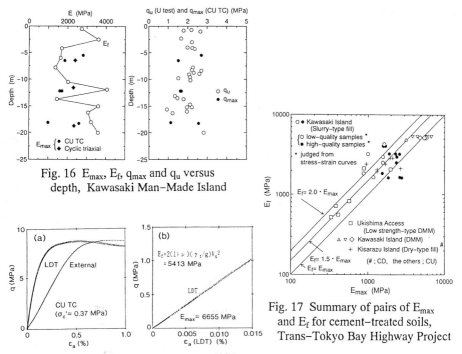

Fig. 16 E_{max}, E_f, q_{max} and q_u versus depth, Kawasaki Man–Made Island

Fig. 17 Summary of pairs of E_{max} and E_f for cement–treated soils, Trans–Tokyo Bay Highway Project

Fig. 18 Typical CU TC test result, DC sample from a soft clay deposit improved by DMM, Trans–Tokyo Bay Highway Project

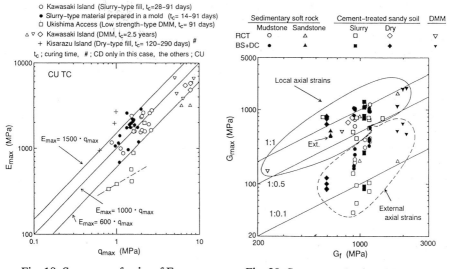

Fig. 19 Summary of pairs of E_{max} and q_{max} for cement–treated soils, Trans–Tokyo Bay Highway Project

Fig. 20 Summary of pairs of G_{max} and G_f for sedimentary softrocks and cement–mixed soils

close to G_f (e.g., the case of Sagamihara Site data). But in many other cases, the values of G_{max} are noticeably smaller than G_f.

CONCLUSIONS

1) The conventional rotary core tube sampling method may be in-adequate to retrieve sufficiently high-quality core samples of softrock. The sample disturbance in RCT sampling may decrease substantially by preventing the rotation and rocking motion of the inner tube by using some measures.

2) The effects of sample disturbance are usually much more serious on the stiffness at small strains than the compressive strength. Therefore, the effects of sample disturbance may not be noted in the results of unconfined compression tests and triaxial tests measuring axial strains only externally.

3) The elastic stiffness from relevant TC tests using high-quality samples of softrock is in general very close to the field value from seismic surveys.

REFERENCES

Goto,S. Tatsuoka,F., Shibuya,S., Kim,Y.S. and Sato,T. (1991). A simple gauge for local small strain measurements in the laboratory, Soils and Foundations, 31-1, pp.169-180.

JSSMFE (1982). Site Investigation Method (second ed.), Japanese Society of Soil Mechanics and Foundation Engineering.

Kim,Y.S., Tatsuoka,F. and Ochi,K. (1994). Deformation characteristics at small strains of sedimentary soft rocks by triaxial compression tests, Geotechnique, 44-3, pp.461-478.

Ochi,K., Tatsuoka,F. and Tsubouchi,T. (1993). Stiffness of sedimentary softrock from in-situ and laboratory tests and field behaviour, Geotechnical Engineering of Hard Soils-Soft Rocks, Balkema, Vol.1, pp.707-714.

Ochi,K., Tsubouchi,T. and Tatsuoka,F. (1994). Deformation characteristics of sedimentary softrock evaluated by full-scale excavation, Proc. Int. Symp. on Prefailure Deformation Characteristics of Geomaterials, IS-Hokkaido '94, Balkema, Vol.1.

Tatsuoka, F. and Shibuya,S. (1992) Deformation characteristics of soils and rocks from field and laboratory tests, Proc. 9th Asian Regional Conf. on SMFE, Vol.2, pp.101-170.

Tatsuoka,F., Kohata,Y., Mizumoto,K., Kim,Y.-S., Ochi,K. and Shi,D.(1993). Measuring small strain stiffness of softrocks, Geotechnical Engineering of Hard Soils-Soft Rocks, Balkema, Vol.1, pp.809-816.

Tatsuoka,F., Sato,T., Park,C.S., Kim,Y.S., Mukabi,J.N. and Kohata,Y. (1994) Measurements of elastic properties of geomaterials in laboratory compression tests, GTJ, ASTM, 17-1, pp.80-94.

Tatsuoka,F. and Kohata,Y. (1994). Stiffness of hard soils and soft rocks in engineering applications, Keynote Lecture, Proc. Int. Symp. on Prefailure Deformation Characteristics of Geomaterials, IS-Hokkaido '94, Balkema, Vol.2.

Tsubouchi,T., Ochi,K. and Tatsuoka,F. (1994). Non-linear FEM analyses of pressuremeter tests in a sedimentary soft rock, Proc. Int. Symp. on Prefailure Deformation Characteristics of Geomaterials, IS-Hokkaido '94, Balkema, Vol.1.

Uchida,K., Shioi,Y., Hirukawa,T. and Tatsuoka,F. (1993). The Trans-TokyoHighway Project - A huge project currently under construction, Proc. of Int. Conf. on Transportation Facilities through Difficult Terrain (Wu and Barrett eds.), Balkema, pp.57-87.

Rock and soil description and classification — a view from industry

DAVID NORBURY & DICK GOSLING
Soil Mechanics Limited, Wokingham, UK

ABSTRACT

It may seem inappropriate to have a paper on the, supposedly, well understood subject of description and classification of soils and rocks in a conference addressing recent advances in site investigation practice. It is the authors contention that this subject is neither as well understood nor as standardised as many think, a situation which recent work on Standards has been addressing.

The subject of description and classification has advanced a long way from the early days of the sciences of soil and rock mechanics. In UK practice this was most recently formalised in BS 5930:1981. This Code advanced the subject substantially, but is generally acknowledged to contain significant omissions, errors and ambiguities. The problems in the current Code will be discussed, and suggested revisions to that Code presented, to indicate the way forward.

The paper outlines the historical background, to put current practice into context, and the principles of description and classification of soil and rock for engineering purposes. The process by which the degree of interpretation increases from an exposure, to samples, to strata descriptions, to interpretative reports will be described. Some of the more common problems in current description and classification are discussed briefly.

INTRODUCTION

Geological description of natural materials is a prerequisite to understanding the ground, but such descriptions are often of limited value in engineering. Description of soils and rocks for engineering purposes requires the reporting of relevant information that affects features such as strength, permeability and deformability; it is these parameters that control the response of the ground to the engineering works. The need for a geological description of the materials as well as a description of physical properties must be recognised and accommodated. In addition, it is becoming more commonly recognised that features of the ground that may affect its response to environmental influences are also very important.

Systematic description and classification is necessary for communication between geologists and engineers who may be employed on a project in a variety of roles. It is usual for the design engineer, and even more common for the construction contractor, never to have seen the materials at the site and therefore to rely totally on the descriptions presented to them. Description, and particularly classification, acts as a shorthand to allow general experience in materials of similar geological or engineering characteristics to be invoked, and for judgements to be made as to the engineering models which can reasonably be applied to the ground. It is often the case that a clear, and preferably concise, description by an experienced practitioner will be more valuable than whole suites of test results.

HISTORICAL BACKGROUND

The rush of work on airfields in the 1930's and 1940's (Casagrande, 1947) led to the Unified Soil Classification System (USCS), and many of our current practices have evolved from there. Casagrande's approach adopted factual particle size and plasticity classification based on a categorised experience of how soils would actually behave. Early similar work on rocks in the 1950's and 1960's was summarised in Deere (1963).

The first attempt in the UK at unifying the approach to description was the publication in 1957 of CP 2001. This Code said little about soil description, and even less about rocks. However, the basic precept laid down was that soils were to be described in terms of their likely engineering behaviour.

Through the 1960's, description was based on CP 2001 but, as the industry expanded, each company or individual tended to evolve their own particular system only loosely based on the Code. In the 1970's and 1980's this situation was addressed by various published guidelines. First to appear were the Working Parties of the Engineering Group of the Geological Society with their "Rock Cores" and "Maps and Plans" reports (1970 and 1972 respectively). Later there were reports by the Engineering Group (Rock Masses, 1977) and by ISRM (1978, 1981) and IAEG (1981). These documents mostly concentrated on rocks.

The flurry of published guidance culminated in 1981 with BS 5930, a Code of Practice which had been some 15 years in the writing in parallel with the other documents. In some respects the various publications are not compatible with each other, which is hardly surprising given the lack of precedent for the recommendations. However, it is BS 5930 under which most soil and rock description has ostensibly been carried out in the UK, and in a number of other English speaking parts of the world, since 1981. Three years later, the Engineering Group's Guildford Conference Proceedings on the Code were mostly critical, particularly with regard to Section 8: Description of soils and rocks.

The main problems with the 1981 Code can be summarised as follows (Norbury et al, 1986). For soils, the Code laid down the 35% fines content as the approximate boundary between fine soils and coarse soils, thus abandoning the precept of description based on likely behaviour. The Code also introduced the British Soils Classification System (BSCS) which quantified many of the boundaries used in soil description and is similar to the USCS. For rocks, the Code introduced a weathering classification system derived from the other publications which separated the description of weathering of the rock material from the weathering of the rock mass; this system is often not workable in UK or international practice. The Code provided virtually no guidance on the description of rock discontinuities, regrettably, because it is the discontinuities that are so important in assessing the behaviour of a rock mass.

BS 5930 is actually a Code of Practice. In the words of British Standards Institution such a document details professional knowledge and practice; it is written as guidance only and is not intended to provide objective criteria by which compliance may be judged. BSI state that, unlike a British Standard, a BS Code of Practice normally would not be called up directly in a job specification. However, BS 5930 is frequently invoked within a contract specification in respect to soil and rock description. This is not acceptable; major changes introduced in the Code conflicted with previous practice and have given rise to problems. This practice also has contractual implications which have led to unnecessary constraint and caution in the application of the recommendations.

As a result of this practice, the specialist practitioners have each attempted to achieve a reasonable, and contractually safe, working compromise within the spirit of the Code. The Code could thus be considered to have failed in that different companies interpret and apply the Code differently. The ten year review of BS 5930, currently underway, will hopefully remove the existing errors, omissions and ambiguities.

The 1981 Code has provided a framework around which a higher standard and more uniform approach to description has started to develop in this country, and in other areas where British Standards are used. There is also movement afoot to spread this uniformity of approach more widely. UK practice is amongst the better and more consistent in the world, and UK delegates have been closely involved in drafting International and European Codes. The revisions within the BS are essentially consistent with these other documents.

CLASSIFICATION OR DESCRIPTION

Reference has been made above to purely factual description of soils, ie their **classification**. This sort of approach is epitomised in the BSCS, and in the USCS, but is not the approach generally favoured for **description**, the main reason being the level of information lost if description is curtailed. The parameters used to classify soils are primarily the particle size and the plasticity; these tests require the soil to be totally remoulded. The classification approach therefore omits any reference to soil fabric which can assume considerable importance in the behaviour of soils (see for example Rowe, 1972). It is therefore only appropriate in the authors' view to use such a classification approach when the engineering works will themselves totally remould the soil, such as in earthworks.

Furthermore, reliance on classification tests to determine engineering properties requires an unrealistic level of understanding about soil behaviour. We cannot, for example, generalise as to what percentage of clay has a significant effect on strength, as this depends on a large number of factors including the particular clay mineralogy and the overall shape of the grading curve. Further, the critical percentage of clay will be different for a similar level of significance in permeability.

It is for these reasons that the authors do not favour rigorous adherence to the 35% cut off between fine and coarse soils in the 1981 Code. Child (1986) has made arguments for this cut off to be anywhere between 5 and 50%, and different countries in Europe define a boundary somewhere between 25 and 40 %. The recommendation should be that practitioners are encouraged to make a description, and to use classification tests as back up and calibration. In certain cases a combination of a complete description together with a factual classification may provide the best of both worlds.

However, a difficulty arises in that the logging practitioners need to be experienced. Commercial pressures often dictate that logging is carried out by the youngest (and therefore cheapest) members of our profession who have generally been taught geological description rather than engineering description at university or college. Even where suitably trained and experienced staff carry out descriptions, checking engineers have been known to discount these and place inappropriate levels of reliance on laboratory classification tests.

THE PROCESS OF DESCRIPTION AND CLASSIFICATION IN PRACTICE

Description and classification are fundamental actions and are part of a progression. The description provides information on the material and mass characteristics, is of immediate use and does not need test results; this is also the stage at which the **art** of description should be exercised in order to convey the maximum possible amount of information sensibly to the reader. This requires experience to decide what is relevant or significant. A borehole sample is described, initially and usually in simple terms by the driller, and then in fuller detail by the engineer or geologist.

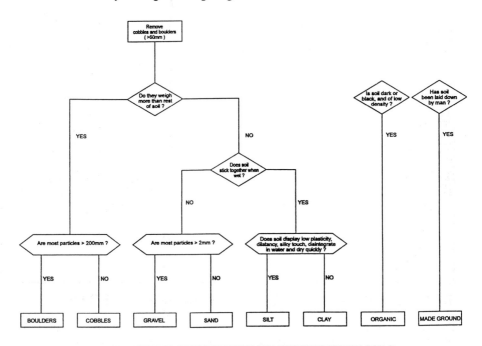

FIG 1 : GENERAL IDENTIFICATION AND DESCRIPTION OF SOILS

The sample is classified, initially in simple terms, so that decisions can be made on what testing is appropriate. This level of classification is illustrated on Fig 1. Full classification requires test results and is based on defined criteria, with each category identifying certain engineering characteristics for use in design; however classification is usually based on laboratory tests carried out on remoulded samples and so reflects material properties only. The basic principles of soil classification are shown on Table 1. An example of a specific earthworks materials classification (Dept of Transport, 1991) is given on Table 2. Elsewhere in Europe, for example, a classification based on grading alone is used in Sweden (Swedish Geotechnical Society, 1981) and a "universal" classification similar in principal to the USCS is used in Germany (DIN 18196, 1988).

The geotechnical practitioner is faced with the everyday problem of describing a sample, core or exposure; the paramount need is to record the observations in a manner that is accurate, factual and complete, and in terms that any other practitioner will understand without ambiguity. That a description should be factual is a common area of misunderstanding. The basic description made at the sample bench, in the trial pit or at the quarry face should normally be complete without involving interpretation. The probability of this being achieved is clearly greater at the quarry face than with samples from a cable tool boring. Even at this level there could be an element of selection in deciding what is relevant; for example whether the piece of 'foreign' gravel is naturally placed or unnaturally placed. It is not always easy to discard even those artefacts placed there by the boring tools or sample handling procedures.

			CLASSIFICATION INTO GROUPS OF SIMILAR BEHAVIOUR			FURTHER SUBDIVISION AS APPROPRIATE BY:
WET SOIL DOES NOT STICK TOGETHER	VERY COARSE	most particles > 200 mm	B	Bx		Requires special consideration
				Kb, Bk		
		most particles > 60 mm	K			
	COARSE	most particles > 2 mm	G	Ks, Kg	Ksg	PARTICLE SIZE
				Gk, Sk	Gks	SHAPE OF GRADING CURVE
						Relative density
				Gs, Sg	Gsl, Sgl	Permeability
		most particles > 0.06 mm	S	Gl, Gt; Sl, St	Gst, Sgt	(Mineralogy)
				Sp		(Particle shape)
WET SOIL STICKS TOGETHER	FINE	low plasticity dilatant	L	Ls	Lsg	PLASTICITY
					Lst	Water content
				Lt, Tl		Strength, sensitivity
		plastic non−dilatant	T		Tsg	Compressibility, stiffness
				Lp, Tp		(Clay mineralogy)
DARK COLOUR LOW DENSITY	ORGANIC		P	Ps, Pl, Pt		Requires special consideration
NOT NATURALLY DEPOSITED	MADE GROUND		A		Man made materials	Requires special consideration
				Ax	Relaid natural soils	As for natural soils

KEY TO SYMBOLS

Cases requiring special consideration should be classified according to national or project requirements

Soil Type :	Principal component	Secondary or tertiary component
Boulders	B	b
Cobbles	K	k
Gravel	G	g
Sand	S	s
Silt	L	l
Clay	T	t
Organic	P	p
Made Ground	A	−
	−	x

G and S can be subdivided into fine (f), medium (m) or coarse (c)

any combination of components

TABLE 1   PRINCIPLES OF CLASSIFICATION OF SOILS

That samples from borings can often be described completely and factually may be of little help to the reader who wants to know the condition of the ground in situ. Some interpretation·may therefore be necessary to achieve an appropriate description for further use. This interpretation may be to identify the likely in situ condition or merely to highlight the difficulty. After all, the opportunity to examine the fresh samples in sequence, and make an interpretation thereof, is short; the original logger's view is thus very significant (cf comment above on experience).

The degree of interpretation included in a log should however be kept to a reasonable minimum. If the quality of sample or exposure is such that it is not readily possible to describe the in situ condition, the description will need to be qualified with terms such as "possibly" or "probably". The presence of such terms in a description should flag the need for further investigation, perhaps using a different approach, so as to achieve the objective of a complete and accurate description.

The preparation of the borehole, pit or face log will require compilation of several descriptions to provide the overall stratum descriptions. This compilation will necessarily

CLASS	DESCRIPTION	TYPICAL USE	GRADING	UNIFORMITY	PLASTICITY	WATER CONTENT	STRENGTH	MCV	CHEMICAL	OTHER
										PARAMETERS USED IN CLASSIFICATION (see Note 1)
1A	Well graded granular	General fill	*	*				*		
1B	Uniform granular	General fill	*	*				*		
1C	Coarse granular	General fill	*	*						
2A	Wet cohesive	General fill	*		PL	*	P		*	
2B	Dry cohesive	General fill	*		PL		Ur	*	*	
2C	Stony cohesive	General fill	*		PL	*	Ur	*	*	
2D	Silty cohesive	General fill	*			*	Ur	*	*	
3	Chalk	General fill	*			*		*		Porosity
4	Various	Landscape fill	*					*		
5	Topsoil	Topsoiling	*							
6A	Selected well graded granular	Below water	*	*	PI			*		
6B	Selected coarse granular	Starter layer	*	*	PI		P			
6C,D	Selected uniform granular	Starter layer	*		PI	*	P	*		
6E	Selected granular	Capping by stabilisation	*		LL,PI			*	*	
6F	Selected granular	Capping	*				P	*	*	
6G	Selected granular	Gabion filling	*				P	*	*	
6H	Selected granular	Drainage layer	*		PI		P	*	*	
6I	Selected well graded granular	Reinforced earth fill	*	*			E	*	*	Friction/adhesion
6J	Selected uniform granular	Reinforced earth fill	*	*			E	*	*	Friction/adhesion
6K,L	Selected granular	Bedding, buried steel	*		PI		P	*	*	
6M	Selected granular	Surround, buried steel	*		PI		P	*	*	Permeability
6N	Selected well graded granular	Fill to structures	*	*			U,E,P	*	*	
6P	Selected granular	Fill to structures	*			*	U,E,P	*	*	Permeability
7A	Selected cohesive	Fill to structures	*		LL,PI	*	U,E	*	*	
7B	Conditioned PFA	Fill to structures					U,E	*	*	Density, Permeability
7C	Selected wet cohesive	Fill to reinforced earth	*		LL,PI	*	E	*	*	Friction/adhesion
7D	Selected stony cohesive	Fill to reinforced earth	*		LL,PI	*	E	*	*	Friction/adhesion
7E	Selected cohesive	Lime stabilisation	*		PI			*	*	
7F	Selected silty cohesive	Cement stabilisation	*		LL,PI	*		*	*	

NOTES

1 Guidelines for some parameters are provided; those parameters required for any soils available to a project are defined as and when required.

2 Plasticity – PL = Plastic Limit, LL = Liquid Limit, PI = Plasticity Index.

3 Strength – Ur = undrained shear strength of remoulded soils, U = undrained shear strength, E = Effective shear strength, P = Particle strength.

4 Chemical includes pH, sulphate content, chloride content, organic content, resistivity, redox potential, microbial index activity.

5 MCV (Moisture Condition Value) widely used in the UK and related to the compactability and strength of soils.

TABLE 2 – SUMMARY OF PART OF THE UNITED KINGDOM CLASSIFICATION FOR EARTHWORKS
Manual of Contract Documents for Highway Works. Volume 1: Specification for Highway Works (December 1991): HMSO (MCHW 1).

require a further degree of interpretation to provide a concise and meaningful description. In the easiest cases this interpretation is largely editorial, but more fundamental problems are far from uncommon. Information will usually be deleted, but a blanket deletion of minor constituents will probably be misleading. Similarly, excessive wordage on ranges of characteristics will successfully obscure the important features.

The stratum description, and indeed the whole of the log within which it appears, is intended to be "as objective as possible a record of the ground at the borehole or face location before the ground was subjected to loss and disturbance by the drilling or sampling process, and thus necessarily includes an element of interpretation" (extract from the "ten year" revision to BS 5930, Section 7). The confusion about the status of a borehole log arises because, for convenience, the log appears in the "factual" report even though it includes interpretation.

Finally an overall statement of the ground conditions is presented in the report which involves substantial interpretation. Within this interpretation is the final and comprehensive classification of the ground; this can be at several levels and developed from the basic field classification outlined on Table 1 and Fig 1. At its simplest the classification may merely be that the materials are soft clays, or dense sands, or weak rocks. Alternatively the classification may be geological, so that the groupings are of alluvial soils, glacial moraines or coal measure rocks as appropriate. A third option, and that most commonly used, is for a classification based on the engineering requirements of the project and built around the understanding of the geological succession. The materials would thus be classified into geologically based groupings of similar engineering characteristics for purposes of assessment and prediction of ground conditions pertinent to the earthworks, foundations, tunnelling conditions, aquifers or whatever is appropriate. It is important to note here the analogy to the fundamental precept in CP 2001, namely that soils should be described in accordance with their likely engineering behaviour. Abandonment of this precept is both illogical and inadvisable if correct interpretation of the overall ground behaviour around the engineering works is to be made.

THE FUTURE AS REPRESENTED WITHIN THE REVISED CODE

The proposed revisions arising from the ten year review of BS 5930 have been published for public comment. In particular in the context of this paper, there are significant revisions to Section 8 on the description and classification of soils and rocks. To summarise the main proposed alterations, the revised Table 6 "Field identification and description of soils" is reproduced for information. The amendments to the description of soils which are summarised on Table 6 are:

- relaxation of the 35% boundary between coarse and fine soils
- reinstatement of compact term for density of silts
- removal of the subdivisions to the clay consistency scale and an amendment of terms above 150 kPa
- provision for soils to run into rocks in descriptive terms (it is intended that the word order for soil and rock descriptions be unified so that, for example, very stiff clay / very weak mudstone can be easily used)
- secondary constituent classifiers provided for before and after principal soil type
- clarification of application of secondary constituent terms when more than three size fractions present
- exclusivity of clay and silt terms made unambiguous

SOIL GROUP

Coarse soils (over about 65% sand and gravel sizes) — Very coarse soils; Coarse soils

Fine soils (over about 35% silt and clay sizes)

Organic soils

Density/Compactness/Strength — Field Test

Term	Visual / Field Test	Borehole SPT N-Value
Loose	By inspection of voids and particle packing	
Dense		
Loose	Can be excavated with a spade. 50mm wooden peg easily driven	Very Loose
		Loose 4
Dense	Requires pick for excavation. 50mm wooden peg hard to drive	Medium Dense 10
		Dense 30
		Very Dense 50
Slightly cemented	Visual examination, pick removes soil in lumps which can be abraded	
Uncompact	Easily moulded or crushed in the fingers	Cu
Compact	Can be moulded or crushed by strong pressure in the fingers	
Very soft	Finger easily pushed in up to 25mm	20
Soft	Finger pushed in up to 10mm	40
Firm	Thumb makes impression easily	75
Stiff	Can be indented slightly by thumb	150
Very stiff	Can be indented by thumb nail	300
Hard (or very weak mudrock)	Can be scratched by thumb nail. See 41.2.2	
Fibrous	Fibres already compressed together	
Spongy	Very compressible and open structure	
Plastic	Can be moulded in hand and smears fingers	

Discontinuities

Scale of spacing of discontinuities — Term	Mean spacing mm
Very widely	over 2000
Widely	2000 to 600
Medium	600 to 200
Closely	200 to 60
Very closely	60 to 20
Extremely closely	under 20

Fissured — Breaks into blocks along unpolished discontinuities
Sheared — Breaks into blocks along polished discontinuities

Plant remains — Fibrous; recognisable and retains some strength
Pseudo-fibrous — Plant remains recognisable, strength lost
Amorphous — Recognisable plant remains absent

Spacing terms also used for distance between partings, isolated beds or laminae, desiccation cracks, rootlets etc.

Bedding

Scale of bedding thickness — Term	Mean thickness mm
Very thickly bedded	over 2000
Thickly bedded	2000 to 600
Medium bedded	600 to 200
Thinly bedded	200 to 60
Very thinly bedded	60 to 20
Thickly laminated	20 to 6
Thinly laminated	under 6

Inter-bedded — Alternating layers of different types. Prequalified by thickness term if in equal proportions. Otherwise thickness of and spacing between subordinate layers defined.
Inter-laminated

Transported mixtures:
Slightly organic clay or silt — grey
Organic clay or silt — dark grey
Very organic clay or silt — black
PEAT — Accumulated in situ, black

Colour

Red, Orange, Yellow, Brown, Green, Blue, White, Cream, Grey, Black etc.
Supplemented as necessary with: Light, Dark, Mottled, and Reddish, Orangish, Yellowish, Brownish etc.

Composite soil types (mixtures of basic soil types, see 41.3.2.4)

For mixtures involving very coarse soils / coarse soils:
Term before principal soil type — Slightly (sandy▲) / — / (sandy▲) / Very (sandy▲) — gravelly and/or sandy
SAND AND GRAVEL ▲

Approximate % secondary: with a little (sand▲) <5; with some (sand▲) 5 to 20; with much (sand▲) >20; SAND AND GRAVEL about 50

For fine soils:
Term before principal soil type — slightly (sandy△) / — / (sandy△) / Very (sandy△) — gravelly and/or sandy
Approximate % secondary: with some <35; with much 35/65; 65; >65

Minor Constituents

% defined on a site or material specific basis or subjective:
Shell fragments, pockets of peat, gypsum crystals, flint gravel, fragments of brick, rootlets, plastic bags etc.

Calcareous, shelly, glauconitic, micaceous etc.
Slightly (calcareous ●) / — (calcareous ●) / Very (calcareous ●)

Particle Shape

Angular, Sub-angular, Sub-rounded, Rounded, Flat, Elongate

Particle Size (mm)

Principal Soil Type	Particle Size (mm)	
BOULDERS	200	
COBBLES	60	
GRAVEL	coarse	20
	medium	6
	fine	2
SAND	coarse	0.6
	medium	0.2
	fine	0.06
SILT	coarse	0.02
	medium	0.006
	fine	0.002
CLAY		

Visual Identification

BOULDERS / COBBLES — Only seen complete in pits or exposures. Often difficult to recover whole from boreholes.

GRAVEL — Easily visible to naked eye; particle shape can be described; grading can be described.

SAND — Visible to naked eye; no cohesion when dry; grading can be described.

SILT — Only coarse silt visible with hand lens; exhibits little plasticity and marked dilatancy; slightly granular or silky to the touch. Disintegrates in water; lumps dry quickly; possesses cohesion but can be powdered easily between fingers.

CLAY/SILT — Intermediate in behaviour between clay and silt. Slightly dilatant.

CLAY — Dry lumps can be broken but not powdered between the fingers; they also disintegrate under water but more slowly than silt; smooth to the touch; exhibits plasticity but no dilatancy; sticks to the fingers and dries slowly; shrinks appreciably on drying usually showing cracks.

Composite soil types (see 41.3.2.4) — as for minor constituents; NOTES:
□ Or described as coarse soil depending on mass behaviour
■ Or described as fine soil depending on mass behaviour
● Minor constituent type
▲△ coarse or fine soil type assessed excluding cobbles and boulders

Colour:
grey — Contains finely divided or discrete particles of organic matter, often with distinctive smell, may oxidise rapidly.
as mineral / dark grey — Describe as for inorganic soils using terminology above.
black — Predominantly plant remains, usually dark brown or black in colour, distinctive smell, low bulk density. Can contain disseminated or discrete mineral soils.

Stratum Name

(RECENT DEPOSITS), (ALLUVIUM), (WEATHERED BRACKLESHAM CLAY), (LIAS CLAY), (EMBANKMENT FILL), (TOPSOIL), (MADE GROUND OR GLACIAL DEPOSITS ?) etc.

Examples of composite soil types, indicating preferred order of description:

Loose brown very sandy subangular fine to coarse GRAVEL with small pockets (up to 30mm) of clay. (TERRACE GRAVELS)

Medium dense light brown clayey/fine SAND with some fine gravel. (GLACIAL DEPOSITS)

Stiff very closely fissured orange mottled brown sandy CLAY with a little rounded quartzite gravel. (REWORKED WEATHERED LONDON CLAY)

Firm thinly laminated grey CLAY with closely spaced thick laminae of sand. (ALLUVIUM)

Plastic brown clayey amorphous PEAT. (RECENT DEPOSITS)

TABLE 6 FIELD IDENTIFICATION AND DESCRIPTION OF SOILS

- description of organic soils clarified
- BSCS deleted entirely.

The main changes in the description of rocks are the provision of guidance on the description of discontinuities, and revision of the approach to the description and classification of rock weathering following the work of the Working Party of the Engineering Group of the Geological Society (1995). Some of these amendments are discussed in more detail below where appropriate. The BS 5930 draft for public discussion does not contain all these amendments; it was considered more important to begin the consultation period rather than finally polish every section of the revisions.

It is hoped that the revised Code achieves the aim of fewer ambiguities and omissions than the 1981 version, and that it provides a standard for the UK geotechnical industry to move forward with over the next 10 years. Lest anybody be left with the wrong impression, the authors would emphasise that this paper and previous criticisms have concentrated only on the areas of difficulty with the 1981 Code. As it was drafted from scratch, it provided a major step forward in UK practice, and we all owe the authors a debt of gratitude. This Code, and the British practice it encapsulates, is still, we believe, at the forefront of practice in Europe and in numerous other parts of the world. The authors of this paper hope that the comments made here and incorporated in the revised Code will retain the momentum and lead already established.

SOME COMMON PROBLEMS IN DESCRIPTION AND CLASSIFICATION
There are some areas in description of soils and rocks which, in the authors' view, create problems on an almost daily basis. This is often due to varying interpretations of and confusion about what the Code actually says. A selection of these are discussed below.

i)   Is it ever possible to describe very coarse soils properly, given the need for a representative sample and the limitations of most sampling techniques?
This is one of the most problematical and common contractual arguments, particularly in tunnelling and piling. The main difficulty is that our common investigation techniques are unable to recover very coarse particles; even a 200 mm diameter shell has great difficulty in recovering particles larger than about 60 mm. Special care is required to ensure that all relevant information is presented together on the log: difficulties with casing, chiselling records and the presence of broken angular fragments should be brought together with a comment as to a possible or probable boulder. The presence of even one or two boulders in the boreholes is likely to be greatly magnified as a problem in the actual construction works. Trial pits are only better in theory: reference to Fig 10 in Part 2 of BS 1377:1990 shows that the minimum size of representative sample required from a soil with a maximum particle size of 200 mm is 1000 kg; samples of this size are not common in site investigations. The revised Code emphasises these difficulties and comments that the defined terminology may often be inappropriate.

ii)   How is the boundary between coarse and fine soil defined and identified in practice, particularly bearing in mind the fundamental precept in CP 2001 that soils should be described in terms of their engineering behaviour?
The revised Code addresses this matter as follows. "... where a soil (omitting any boulders or cobbles) contains about 35% or more of fine material, it is described as a fine soil (CLAY or SILT dependent on its plasticity). With less than about 35% of fine material, it is described as a coarse soil (SAND or GRAVEL dependent on its particle size grading). The 35% boundary between fine and coarse soils is approximate being

dependent, primarily, on the plasticity of the fine fraction and the grading of the coarse fraction. Although the 35% limit is used in other countries and is often reasonably appropriate, soils with the boundary as low as 15% are not unknown. All soils should be described in terms of their likely engineering behaviour, the descriptions being supplemented with and checked against laboratory tests as required." This does not and cannot address the problem that arises when considering the same soil from the point of view of, for example, foundations as against dewatering; the relevant boundary is not the same in percentage terms. However, this is a question that can only be addressed by ensuring that the appropriate field and laboratory tests are carried out.

iii)   What is the definition of the clay - silt boundary in engineering terms?
The 1981 Code created problems in this matter. It was intended that the terms silt and clay should be mutually exclusive, but this was not specifically stated and conflicting examples were provided. The revised Code sets out to remove any ambiguity in this regard as follows: "most fine soils are mixtures of clay and silt size particles; these can include silt size aggregates of clay minerals and clay size particles such as quartz. The distinction between clay and silt is often taken to be the A-line on the plasticity chart, with clays plotting above and silts below. The reliability of the A-line in this regard is poor as might be expected (Child, 1986; Smart, 1986); the effects of clay mineralogy and organic content are also significant. Fine soil shall be described as either a SILT or a CLAY depending on the plastic properties; these terms are to be mutually exclusive and terms such as silty CLAY are thus unnecessary and are not to be used. The field distinction between CLAY and SILT should be made using hand tests. Where these hand tests are genuinely indecisive or ambiguous, the hybrid term CLAY/SILT may be used which may flag the need for additional plasticity determinations in the laboratory to provide further information."

iv)   How are organic soils defined, described and identified?
This is not a very common problem, but potentially significant confusion arises and the revised Code addresses this problem. Small quantities of dispersed organic matter can have a marked effect on plasticity and hence the engineering properties; increasing quantities of organic matter heighten these effects. Soils with organic contents up to about 30% by weight and moisture contents up to about 250% will behave largely as mineral soils, albeit with different parameters (Hobbs 1986, 1987). Such materials are generally transported and would not be described as peat which strictly accumulates in situ in a mire. This morphological distinction may be difficult to define, eg within a fluvial sequence, and distinction based on engineering behaviour should be made

v)   How should soils with more than two size fractions be described?
There is widespread confusion about this question, which is common as most soils consist of more than one size fraction. The terms silt and clay are mutually exclusive in secondary constituents as well; which term is used depends on the plasticity of the fine fraction. It is therefore not possible to have a silty clayey SAND; it is either silty or clayey. The terms sandy and gravelly may both be used, in which case the percentages are assessed separately. Thus a "slightly sandy CLAY with a little quartz gravel" contains up to 35% sand and up to 35% gravel. Care should be taken to avoid soils with combined fractions of more than 100%.

In the authors experience the greatest difficulties occur in describing soils where the full range of fine and coarse particle sizes are present, the classic case being a "boulder clay". While the proportion of gravel in the soil can be readily assessed visually the same is not true of the sand content. The sand is often intimately bound up with the silt and clay size particles forming the matrix in which the gravel size clasts are set. The application of the about 35% distinction between a fine and coarse soil can be virtually impossible in these circumstances; arguably it would also be erroneous. In terms of soil behaviour the question is surely whether the material is clast supported or matrix supported.

vi)   What is the boundary between rock and soil?
The frequency of conferences on weak rocks suggests that this is a difficult question, and it certainly has not yet been resolved. Most standards avoid addressing this issue. In the revision of the Code, the practice suggested by Spink and Norbury (1993) has been followed; namely that the upper limit of very stiff is set at a shear strength of 300 kPa, so that the term very weak extends between unconfined compressive strengths of 600 and 1250 kPa, see Table 6.

vii)   Do you consider the weathering terminology in BS 5930 satisfactory and useable?
A Working Party Report of the Engineering Group of the Geological Society has recently been published on this topic (Anon, 1995). A review of weathering classifications over the last 35 years shows how weathering terms for mass and material have been interchanged and why this is an area of confusion. The conclusion drawn in the Report is that it is not possible to compile or operate a single weathering classification scheme that will encompass all rocks in all climates. In addition, it is recognised that many engineering investigations see insufficient of the rock mass or the weathering profile for classification to be possible. It is therefore recommended that the first and most important step is to ensure that a full and complete description of the effects of weathering is made. The incomplete proposals in the draft Code will be replaced by the Working Party recommendations.

Following the description, a range of approaches to the classification of material and mass are proposed, covering stronger and weaker rocks respectively. There is a close analogy here between soils and rocks. A clear preference is stated to **describe** the rock followed, if required and appropriate, by a **classification** of the weathering.

The features to be described factually that are due to weathering include: discolouration; changes in fracture state; reduction of strength; presence, character and extent of weathering products. Guidance on the description of these features is provided and it is emphasised that their extent and character should be quantified wherever possible. This will require the use of "standard" descriptors as well as "non-standard" English descriptors. A preference is noted for recording actual measurements of extent rather than relying on the standard descriptors, such as closely spaced or thinly bedded. This sort of more flexible approach should be adopted more widely throughout the descriptive art; it provides opportunity for more information to be sensibly communicated. Over reliance on codified words will eventually kill off this ability to communicate.

viii)   Are you sure of the definition of solid core that you use in measuring RQD?
The 1981 Code provides an ambiguous definition of solid core, as a result of which several other indices have been proposed, and there is no standard approach in the UK. The revised Code provides a definition based on axial length rather than the right cylinder definition favoured by some. The illogicality of this latter definition is discussed by Spink and Norbury (1990), who show how the two approaches can give very different measured indices. The definition in the revised Code is as follows: "ratio of solid core

recovered to length of core run; solid core has a full diameter but not necessarily a full circumference (ISRM, 1978; Deere and Deere, 1988)." The preference for this approach is based on the better repeatability of measurements, and the American intention to include the axial definition in ASTM procedures.

CONCLUSIONS
The reader's attention has been drawn to the preference for full and complete description, and for this to be followed by classification only when appropriate and relevant. Problems in the everyday description of soils and rocks remain, sometimes due to misunderstanding of the engineering problem, sometimes to inexperience of the logger and therefore inability to exercise the art of description, and sometimes due to the constraints placed by using the Code as a specification. The spirit of the Code should allow experienced practitioners, who log every day, to achieve the required objective of readable, relevant and meaningful descriptions. Imposition of the Code limits the freedom of expression which is so important in this regard. The 1981 Code represents a milestone in UK practice on which we are now trying to build; it is hoped that the revised Code will maintain the impetus into the next century.

ACKNOWLEDGEMENTS
The authors wish to thank the Directors of Soil Mechanics Limited for permission and encouragement to publish this paper. Thanks are also due to our colleagues who have provided so much background in the philosophy in our approach; particular thanks are due to Geoff Child and Tim Spink, our erstwhile colleagues in Soil Mechanics.

REFERENCES
BS 5930 : 1981 : Code of practice for site investigations. BSI.

BS 1377 : 1990 : Methods of test for soils for civil engineering purposes. BSI.

CP 2001 : 1957 : British Standard Code of Practice: Site investigation. BSI.

Casagrande A : 1947 : Classification and identification of soils. Proc Amer Soc Civil Engineers. Vol 73, pp 783 - 810.

Child G H : 1986 : 'Soil Descriptions - Quo Vadis?' Geol Soc Eng Geol Special Publication No 2. Proc 20th Regional Meeting Eng Grp, Guildford

Deere D U : 1963 : Technical description of rock cores for engineering purposes. Rock Mech Eng Geol. Vol 1, pp 18 - 22.

Deere D U & Deere D W : 1988 : 'The Rock Quality Designation (RQD) Index in Practice'. Rock Classification Systems for Engineering Purposes. ASTM STP 984.

Department of Transport : 1991 : 'Specification for Highway Works.' Volume 1 of the Manual of Contract Documents for Highway Works. HMSO, London.

Deutsche Norm : 4022 : Part 1 : 1987 : 'Classification & description of soil and rock. Borehole logging of rock and soil not involving continuous core sample recovery'. DIN

Deutsche Norm : 18196 : 1988 : 'Soil Classification for civil engineering purposes'. DIN

Geological Society, Engineering Group Working Party : 1970 : 'The logging of rock cores for engineering purposes.' QJEG, Vol 3, pp 1 - 24.

Geological Society, Engineering Group Working Party : 1972 : 'The preparation of maps in terms of engineering geology.' QJEG, Vol 5, pp 295 - 382.

Geological Society, Engineering Group Working Party : 1977 : 'The description of rock masses for engineering purposes'. QJEG, Vol 10, pp 355 - 388

Geological Society, Engineering Group Working Party : 1995 : Description and classification of weathered rocks for engineering purposes. QJEG, Vol 28.

Hobbs N B : 1986 : 'Mire morphology and the properties and behaviour of some British and foreign peats. QJEG, Vol 19, pp 7-80

Hobbs N B : 1987 : 'A note of classification of peat'. Géotechnique 37, No 3, pp 405-407

IAEG : 1981: Rock and soil description and classification for engineering geological mapping. Bull IAEG, Vol 24, pp 235 - 274.

ISRM : 1978 : 'Suggested Methods for the Quantitative description of Discontinuities in Rock Masses'. Int J Rock Mech Min Sci & Geomech Abstr, Vol 15, pp 319-368

ISRM : 1981 : 'Basic Geotechnical Description of Rock Masses'. Int J Rock Mech Min Sci & Geomech Abstr, Vol 18, pp 85-110

Norbury D R, Child G H & Spink T W : 1986 : 'A Critical Review of Section 8 (BS 5930) - Soil and Rock Description'. Geol Soc Eng Geol Special Publication No 2, Proc 20th Regional Meeting Eng Grp, Guildford

Rowe P W : 1972 : The relevance of soil fabric to site investigation practice. Geotechnique, Vol 22, pp 195 - 300.

Smart P : 1986 : 'Classification by Texture and Plasticity'. Geol Soc Eng Geol Special Publication No 2, Proc 20th Regional Meeting Eng Grp, Guildford

Spink T W & D R Norbury : 1990: Discussion in International Chalk Symposium, Brighton. pp 205 - 207. Thomas Telford.

Spink T W & Norbury D R : 1993 : 'The engineering geological description of weak rocks and overconsolidated soils'. Eng Grp Geol Soc Special Publication No 8, Proc, 26th Annual Conference, Leeds. Balkema.

Swedish Geotechnical Society : D8 : 81 : 1981 : 'Soil classification and identification'. (SGF Laboratory Manual, Part 2) Byggforskringsrådet

A new core orientation device

I P WEBBER, BEng, MSc, MICE FGS
Technical Director, W A Fairhurst and Partners, Glasgow, Scotland.
D GOWANS,
Managing Director, Drillcorp Limited, Seaham, Co. Durham, England.

ABSTRACT

As part of the initial investigation for the Highways Agency of proposed improvements to the A167, south of Durham, it was necessary to collect discontinuity data in areas of proposed rock cutting, remote from rock exposures.

A trial was therefore carried out to compare available techniques for measuring discontinuity data from boreholes. Comparison of results obtained from impression packer, CCTV, local exposures and the core orientator system are presented.

The core orientator system relies on a pendulum which on an inclined borehole under gravitational force locates the lowest point of the core. The system is described fully within the paper.

INTRODUCTION

It has long been recognised that there is a need for a system of recording discontinuity data from boreholes (Hoek and Bray, 1974, Burwell and Nesbitt, 1954). A number of techniques have been reported in the literature (Gamon, 1986). A new core orientation system has recently been manufactured by Archway Engineering. The system relies on a pendulum attachment fixed to the head of the core barrel which indicates the lowest position of an inclined borehole.

As part of the initial investigation of the proposed A167 improvement, south of Durham, it was decided to carry out a trial of the core orientator system to compare it with other available techniques. For comparison purposes discontinuity data was obtained from the same borehole using CCTV. Impression packer tests were carried out in an adjacent borehole, and information was also obtained from a nearby rock exposure.

This paper describes the work carried out and the techniques used. The results of the investigation are presented for comparative purposes.

Advances in site investigation practice. Thomas Telford, London, 1996

INVESTIGATION

The proposed realignment of the A167, south of Durham, will involve the excavation of a number of rock cuttings in both Magnesian Limestone and Middle Coal Measures. Rock exposures within the route corridor are limited to small outcrops, predominantly within railway cuttings and valley sides, remote from the proposed main cuttings. In order to obtain more information on rock discontinuities, orientated rock coring and insitu testing, in the form of CCTV and impression packer tests, were carried out during the preliminary investigation. In one area, adjacent to a valley side rock exposure, all of the above techniques were adopted to allow comparisons to be made with data obtained from discontinuity mapping of the rock exposure. It was intended that the results of the comparisons would indicate which borehole technique could be best adopted in the main investigation.

The equipment used in the investigation is described below.

Core Orientator

The core orientator system incorporates a pendulum which moves under gravitational force while drilling to indicate the lowest position at an inclined borehole. The system is designed to operate in boreholes with a minimum inclination of 5° from the vertical. The system depends on maintaining a fixed rotational relationship between the inner tube of the corebarrel and the orientation device containing the pendulum. This is achieved by rigidly fixing the orientation device to a modified spindle in the corebarrel head.

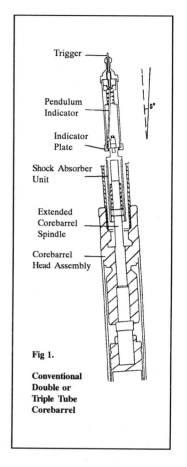

Trigger

Pendulum Indicator

5°

Indicator Plate

Shock Absorber Unit

Extended Corebarrel Spindle

Corebarrel Head Assembly

Fig 1.

Conventional Double or Triple Tube Corebarrel

Once the core run is complete an overshot trigger is lowered to activate the core orientator. The overshot device requires the use of internally flush drill rods of sufficiently large internal diameter (Figure 1). The overshot device latches onto the corebarrel assembly, triggering the pendulum by pushing it downwards against the action of a spring. The lowest position of the inner tube in the inclined borehole is then indicated by the point of the pendulum which emerges through one of 72 small holes on the indicator plate. The pendulum position and that of the inner tube is then fixed. The corebarrel can then be removed.

Prior to removal from the corebarrel the rock core was marked (by drilling a small hole) to identify the lowest point of the core. From this the orientation of the core was determined by correcting the low point of the cored hole with the orientation and dip of the drilling mast.

The drilling work carried out incorporated the use of coreliner tube, on which the alignment of the core was then marked. The core was then logged by an engineering geologist on site. The results of the data obtained are reproduced in the form of a contoured pole plot, as shown on Figure 2. An equal area stereonet was used to plot the data.

CCTV

CCTV surveys included two surveys of a particular borehole. The first survey provided an axial view down the length of the borehole, whilst the second provided a lateral view of the strata by means of an angled mirror mounted ahead of the camera lens. The camera was equipped with a compass during the axial survey in order to assess discontinuity orientations. The orientations were assessed by an engineering geologist on site. The contoured pole plot of results is shown on Figure 3.

Impression Packer

Impression Packers were carried out over four 1.5m sections of an adjacent borehole. These were logged on site by an engineering geologist. The contoured pole plot of results is shown on Figure 4

RESULTS

Discontinuity data was obtained from a local exposure of medium grained, thickly cross bedded, sandstone within the Middle Coal Measures. This was used to identify the structural domains present. The contoured pole plot for the rock exposure is shown on Figure 5 and this shows four principle discontinuity sets.

The data from the rock exposure was taken as being representative of the discontinuities present in the local rock mass and used as a reference against which the data from the insitu techniques could be compared

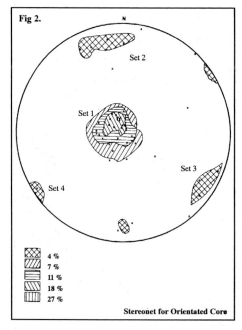

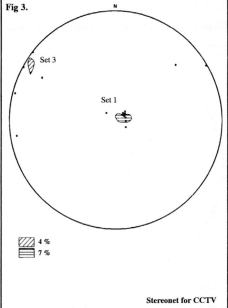

Table 1

Discontinuity Set	Type	Exposure Azimuth/Dip	CCTV Azimuth/Dip	Impression Packer Azimuth/Dip	Core Orientator Azimuth/Dip
1.	Bedding	125/23	262/07	068/05	162/10
2.	Joint	355/88		336/70	157/74
3.	Joint	300/82	123/84		300/87
4.	Joint	254/78			

All the methods used for the investigation provided some data on the discontinuities intercepted in the boreholes.

The bedding planes, Discontinuity Set 1 were identified by all techniques. The three borehole techniques all show a near horizontal bedding. The variations in the dip and azimuth of the bedding is likely to reflect the cross bedded nature of the sandstone, as opposed to bringing into question the accuracy of any of the techniques employed.

Discontinuity Set 2 was identified by both the core orientation device and the impression packer, although the orientated core gave a better correlation with the rock exposure.

Discontinuity Set 3 was identified by both the core orientation device and the CCTV, although the orientated core again gave better correlation with the rock exposure mapping.

Discontinuity Set 4 was identified by the core orientation device only.

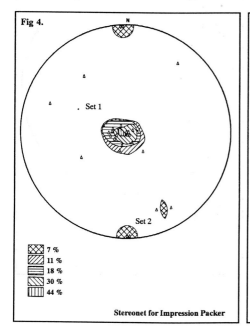

Fig 4.

Set 1

Set 2

▨	7 %
▨	11 %
▤	18 %
▧	30 %
▥	44 %

Stereonet for Impression Packer

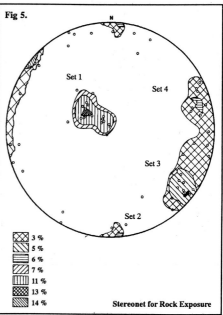

Fig 5.

Set 1

Set 4

Set 3

Set 2

▨	3 %
▨	5 %
▤	6 %
▧	7 %
▥	11 %
▨	13 %
▨	14 %

Stereonet for Rock Exposure

DISCUSSION

The trial demonstrated that information on rock discontinuities could be obtained by all of the borehole techniques used. However, a number of limitations were apparent from the trial and from previous experience of the use of the techniques.

CCTV

The CCTV camera recorded some discontinuity data that correlated with Discontinuity Sets 1 and 3. However, due to the lack of discontinuity data recorded by this method it was not possible to carry out any significant contouring of the plotted poles to discontinuities.

It is possible that the near vertical nature of Discontinuity Sets 2 and 4 may have meant that they were not encountered in the particular borehole in which the CCTV survey was carried out. However, this theory is somewhat negated by the fact that Discontinuity Set 2 was recorded by the Core Orientator device when it was used in the same borehole. Elsewhere on the site, and in other investigations, the use of borehole CCTV was abortive due to the suspended solids within groundwater encountered in the borehole. In general, the borehole CCTV has been found to be a good additional source of discontinuity information wherever it has been successful in obtaining any information. The quality of the results are highly dependent on the skill and diligence of the operator. The current specifications for borehole CCTV (e.g. Department of Transport 1987) do not emphasise the level of detail or accuracy required in the reporting of CCTV surveys. Where rock discontinuity data is required for design, the specification requires to be augmented accordingly.

Impression Packer

The impression packer provided some results for Discontinuity Set 2, although an apparent difference in dip and dip direction of 18° and 19° was respectively recorded when compared with the rock mapping. The near vertical Discontinuity Sets 3 and 4 were not recorded, and this may be due to them not being encountered in the borehole or that sufficient data was not collected. A strong correlation was found between the data representing Discontinuity Set 1, as recorded by the Impression Packer and Rock Exposure Mapping. The limitations of the impression packer have been observed on other sites, particularly with respect to identifying infilled joints. The method of orientation of the impression packer is particularly critical to the accuracy of the results. Further work on the orientation of impression packers is warranted.

Core Orientator

The core orientation device recorded discontinuity data representative of all the Discontinuity Sets described by the Rock Exposure Mapping. The strongest correlation was between the Discontinuity Set 1 data obtained by each method. However, as before, the near vertical nature of Discontinuity Sets 2, 3 and 4 will have precluded these discontinuities from being encountered in the individual boreholes in any great numbers.

The accuracy of recording of the orientation of the core is one of the limitations of the core orientator. The pendulum and preset holes creates a maximum accuracy of ± 5°. The subsequent transfer of this information to the core is the other significant potential for inaccuracies. Recent modifications to the core orientator have replaced the pendulum mechanism with a ball bearing which is locked in place prior to removal of the core barrel. A number of alignment systems have been trialled, including a lazer alignment kit. As with other techniques the final alignment is down to the operators diligence.

CONCLUSIONS

This paper describes a trial carried out of different techniques for obtaining rock discontinuity data from boreholes. The trial demonstrated that CCTV, impression packers and core the orientation device can all provide meaningful data. The core orientator system has been shown to be a useful additional tool in obtaining discontinuity data, and has few limitations on the conditions under which it will provide meaningful data.

ACKNOWLEDGEMENTS

This paper is published by permission of the Highways Agency. The views expressed are those of the authors and should not be attributed to the Highways Agency.

The authors are grateful of the contribution and support of their colleagues in the preparation of this paper.

REFERENCES

Burwell E.B. & Nesbitt R.H. - The NX Borehole Camera. Transactions American Inst. Mining Engineers, 1954.

Department of Transport - Specification and Method of Measurement for Ground Investigation August 1987.

Gamon T.I. - A Comparison between the Core Orienter and the Borehole Impression Device, Geological Society, Engineering Geology Special Publication No. 2 ,1986.

Hoek E. & Bray J.W. Rock Slope Engineering, The Institution of Mining and Metallurgy, London. 1974

The use of down-hole CCTV for collection of quantitative discontinuity data

P. McMILLAN[1] , A.R. BLAIR[1] and I.M. NETTLETON[2].
[1] Transport Research Laboratory Scotland, Craigshill West, Livingston, West Lothian EH54 5DU.
[2] Dept. of Civil Engineering, University of Newcastle Upon Tyne.

SYNOPSIS: Rock excavation design requires discontinuity orientation and spacing data representative of the rock mass in the area of the proposed excavation. Ground investigation (GI) in rock has traditionally relied on rock outcrop mapping to provide such data. Where outcrop is poor or absent, subsurface investigation methods have been employed with varying degrees of success.

Down-hole methods for obtaining quantitative discontinuity data are becoming common. Sophisticated down-hole geophysical methods, developed for the oil industry, have been used to obtain quantitative discontinuity data but are expensive to use and are therefore rarely used in engineering GI. Down-hole CCTV methods have previously been used to extract qualitative or semi-quantitative discontinuity information. The research and results described in this paper demonstrate that representative, accurate, quantitative discontinuity orientation data can be obtained using down-hole CCTV.

INTRODUCTION

Designing excavations in hard rock for engineering or extractive purposes requires accurate, quantitative discontinuity orientation and spacing data recovered from the area of the proposed excavation (Matheson 1983). This data is traditionally collected from rock exposures by either judicious or scanline mapping, or both (Matheson 1983; Priest & Hudson 1981). Such an approach becomes difficult or impossible if rock exposure is poor or absent. Additionally, if the excavation is in a structurally complex rock mass the discontinuity geometry at outcrops may not be representative of the geometry at depth within the proposed excavation. In such circumstances subsurface investigation is required.

Techniques of recovering quantitative discontinuity data from boreholes have been in use for some time. However most of these methods have been developed for use in the oil or mineral exploration industries and many are expensive and impractical for use in conventional engineering GI. Down-hole CCTV methods have been in use for over 10 years in the UK for the recovery of qualitative and semi-quantitative discontinuity data (Gunning 1992) using relatively simple equipment. However, recovery of accurate, quantitative data from down-hole CCTV surveys requires more sophisticated equipment and systematic operation. In particular, accurate camera orientation control and measurement are required and reliable interpretation procedures must be established and followed. If these are done then orientation data can be recovered from discontinuity traces on the borehole wall.

Advances in site investigation practice. Thomas Telford, London, 1996

This paper describes the down-hole CCTV operating and interpretation systems developed by TRL Scotland. The results of laboratory studies and field trials comparing down-hole CCTV surveys with conventional methods of obtaining discontinuity orientation data are presented.

Down-hole CCTV

General

A basic down-hole CCTV system comprises a camera, a monitor for viewing the camera image, a VCR for recording the image and a camera control unit for controlling camera focus, iris, and lighting. Many of the systems in use in the UK are adaptations of pipe and sewer inspection systems.

This basic CCTV equipment can be used only for obtaining qualitative or at best semi-quantitative orientation data from a borehole. If down-hole CCTV is to be used for collecting quantitative discontinuity data then more complex equipment is required.

Down-Hole CCTV for Quantitative Data Collection

The minimum requirements of a CCTV system for collecting quantitative discontinuity data are as follows:-

- Side view camera capable of rotating to scan at least a 180° arc of the hole wall.
- Camera mount to position the camera axis coincident with the hole axis.
- Orientation control unit to provide dip, azimuth and axial orientation of camera.
- Depth measurement unit.
- Camera control unit
- A means of measuring the rotation of the side view camera.

In addition it is useful to have a forward view camera as part of the assembly to inspect borehole conditions ahead of the equipment and evaluate the risk of entrapment.

The system used in this research comprised the following components:-

- Two Rees Instruments R93 cameras and control units, one fitted with a forward view head the other with a side view head.
- Two monitors; one each for forward and side view images.
- Two VCRs for recording the images.
- Magnetic orientation sonde.
- Mounting skid for cameras and sonde.
- Depth counter.
- Portable computer.
- Winch and tripod.
- Generator.
- Cables and guide rods.

The side view camera head provides a view of the borehole wall via a 45 degree mirror which can be rotated through 360 degrees. However when the camera is mounted in the skid the borehole wall can only be viewed through 210 degrees as the remainder is obscured by the skid. The 45 degree rotating mirror is divided, in a vertical plane, by a thin steel plate intended to reduce internal reflections. When the image from the camera is viewed on a monitor this plate appears as a dark, vertical line which is thinner toward the top of the hole (Figures 2, 3 & 4). The mirror plate trace is important to interpretation of CCTV scans as it provides a direct measure of mirror rotation, a guide to the up hole direction and a reference for measurement of apparent dip and azimuth from the discontinuity trace.

The magnetic orientation sonde provides an indication of the side view camera orientation. The sonde gives a constant readout of the dip, azimuth and axial rotation (Magnetic Tool Face, MTF, value) of the camera unit. These orientation data are simultaneously displayed on the side view monitor and recorded on video tape. All three orientation data sets are required to determine the apparent dip and azimuth of discontinuities relative to a fixed datum line of known orientation. These apparent dips and azimuths can then converted to true dips (relative to the horizontal) and azimuths (relative to magnetic north).

OPERATION and INTERPRETATION PROCEDURE

The procedures described in this section have been developed at TRL Scotland and, with the exception of drilling and borehole preparation, were used throughout this study.

Drilling and Borehole Preparation

Down-hole CCTV surveys can be carried out in cored holes or open holes drilled using any technique so long as the hole is stable and the walls are not obscured by casing, debris or flush fluid.

To ensure the maximum amount of data is recovered from down-hole CCTV surveys it is necessary to correctly orientate boreholes with respect to discontinuities present in the rock mass. If there is outcrop present the main discontinuity sets should be established and the borehole(s) drilled to intersect these sets at as large an angle as possible. If no outcrop is present the discontinuity pattern is unknown and to ensure representative results a minimum of three holes should be drilled as near orthogonal as possible.

Successful, quantitative, down-hole CCTV surveys can be conducted in either dry or water-filled holes. For best results in either case the borehole wall should be clean and stable. If the hole is full of water then measures should be taken to ensure that this water is clear enough to give a side view image of the borehole wall. This is normally done by circulating water in the hole until the returns are clear and then allowing a period of 24 hours for any fine sediment to settle.

Operation

Effective operation of a down-hole CCTV system requires experience and an understanding of the data collection, interpretation and end use requirements. Ideally the operator should be an engineering geologist who will also be involved in interpretation of the results. If this is not possible then an engineering geologist should be present during the field work to provide guidance for the operator.

Boreholes are normally scanned from the top down using down-hole CCTV. The procedure described below was used for investigating discontinuity orientation and spacing patterns.

The system is set up so that orientation sonde and depth measurements are displayed on the side view camera monitor and recorded with the side view images. Depth measurements are recorded to the centre of the side view camera image.

Whenever possible the initial axial orientation of the down-hole CCTV assembly is maintained constant during the survey. The survey then follows an imaginary reference line of constant orientation which simplifies interpretation. If deviation occurs the sonde orientation data can be used to correct any measurements relative to the initial imaginary reference line.

When a discontinuity or suspected discontinuity trace is encountered with the side view camera, forward progress of the system is halted. The trace is followed by manoeuvring the skid and rotating the mirror on the side view camera until both the cusp and dip limb (Figure 1) of the discontinuity are located. If the apparent dip of a discontinuity plane is large (>80°), the discontinuity trace forms a very elongate ellipse. Following such traces can be difficult as it involves large movements of the skid. If the trace is truncated, impersistent or markedly non planar and it is not possible to locate both the cusp and dip limb then the orientation cannot be determined.

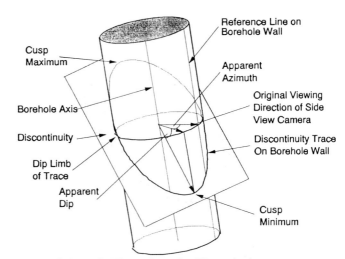

Figure 1 Schematic Illustration of a Discontinuity Intersecting a Borehole Wall

Where areas of closely spaced, non parallel, discontinuities are encountered in a borehole wall it can be very difficult to follow a specific trace in the field. In these circumstances it is important to ensure that the entire fractured section of borehole wall is scanned systematically over the full range possible. If this is done then it may be possible to identify and orientate at least some of the discontinuities during careful interpretation of the scans in the laboratory.

Interpretation

Interpretation is carried out by playing back the side view camera video on a large, high quality monitor. The procedure can be split into three stages as described below.

Stage 1, Recording Factual Data: When a discontinuity trace is encountered the following information is noted from the on-screen displays:-

      a. Sonde dip and azimuth.
      b. Sonde MTF Value.
      c. Intercept depth.

315

The uncorrected apparent azimuth of a discontinuity trace is measured in a plane perpendicular to the hole axis relative to the initial viewing position of the side view camera (see Figure 1). This is determined as follows:-

a. The video is played until the discontinuity trace cusp is located (this is sketched). The nature of the cusp, maximum or minimum (Figure 2), is noted. If the cusp is a maximum, 180° is used as the reference for determining apparent azimuth. If the cusp is a minimum, 360° is used as the reference for determining the apparent azimuth.

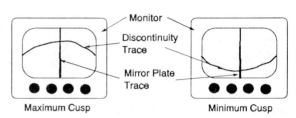

Figure 2 View of Discontinuity Trace and Mirror Plate Trace on Side View Monitor

b. The video is continued until the mirror plate trace is perpendicular to the tangent to the cusp apex.

c. The angle through which the mirror plate trace has rotated is measured (Figure 3) and the direction of rotation, clockwise or anti-clockwise, noted.

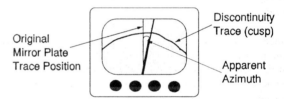

Figure 3 Measurement of Apparent Azimuth From Side View Monitor Image

d. If the rotation is clockwise the measured angle is added to the reference. If the rotation is anti-clockwise the measured angle is subtracted from the reference.

The apparent dip of the discontinuity trace is measured relative to a plane perpendicular to the hole axis (see Figure 1). This is determined as follows:-

a. The position of the mirror plate line, when perpendicular to the cusp tangent, is marked on the monitor screen with adhesive tape.

b. The video is played or rewound, depending on the original sequence of recording, until the mirror plate trace has rotated through 90 degrees. The dip limb of the discontinuity trace is now on screen.

c. The angle between the adhesive tape and discontinuity trace is measured (Figure 4).

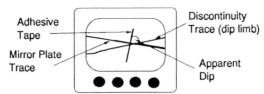

**Figure 4 Measurement of Apparent Dip From
Side View Monitor Image**

Stage 2, Correcting Factual Data for Axial Rotation of the Camera: All discontinuity spacing and apparent orientation data recorded on a CCTV scan must be referenced to the imaginary reference line, of fixed, known orientation on the borehole wall. The position of the CCTV equipment when viewing this line is recorded by the initial MTF value. If the CCTV equipment undergoes axial rotation during a survey measurements will be made relative to a line of changing orientation. However, axial rotation of the cameras and sonde is recorded by the MTF value and data can be converted with reference to the original MTF value. The MTF recorded at a given discontinuity is compared with the original value and the amount and direction of camera rotation is calculated. The apparent discontinuity azimuth is then corrected by either adding (clockwise rotation) or subtracting (anticlockwise rotation) the amount of axial rotation.

Discontinuity spacing can be corrected for axial rotation by reviewing the video tapes and noting the intercept depth to the discontinuities when rotation of the side view camera has compensated for axial rotation of the CCTV equipment.

Stage 3, Calculating True Orientation and Principle Spacing: The discontinuity orientation data measured from CCTV videos are apparent dips and azimuths relative to a reference line on the borehole wall. To be useful in design these must be converted to true dip, relative to the horizontal, and azimuth, relative to magnetic north. Three methods for correcting apparent dip and azimuth of discontinuity traces to true dip and azimuth were developed by Lau (1983). One of these methods, the spherical trigonometry method, was modified slightly to adapt it for use with the results obtained from CCTV surveys. The modified method has been programmed in Turbo Pascal to allow rapid processing of orientation data.

Discontinuity spacing data as derived from scanlines or CCTV surveys is of limited engineering use. Of much greater use is the Principle Spacing which is defined as the spacing between two adjacent discontinuities belonging to the same set measured along the normal to the mean orientation of the set (Wang et al. 1991). The principle spacing is determined by simple calculation from the orientation and spacing of discontinuities recorded on a scanline or CCTV scan of known orientation.

VERIFICATION STUDIES

Laboratory Evaluation

Experimental Procedure: Laboratory evaluation involved using a perspex tube as a borehole and perspex sheets as discontinuities. Slots were cut into the perspex tube and the sheets of perspex were fitted to the tube. The tube was mounted on a bench in a modified drill orientation device (Matheson 1984) that allowed the dip of the tube to be varied and measured accurately. The azimuth of the tube could be varied by rotating the bench and the orientations of the perspex sheets could be varied by rotating the tube about its axis. This allowed a range of discontinuity orientations intersecting a range of borehole orientations to be studied.

At each set-up position the tube was surveyed with the CCTV equipment. The traces of the perspex sheets on the tube wall were recorded on video and the true dip and azimuth of the sheets were measured using a Clar compass. Apparent dip and azimuth of the sheets were determined from analysis of the videos as described above. True dip and azimuth of the sheets were calculated from the CCTV data using the software developed for this purpose.

Results: The results of the laboratory evaluation are shown graphically in Figure 5. These show the very good correlation between dip and azimuth measurements made with the Clar compass and the CCTV equipment. The results clearly illustrate the potential of the system for recovering quantitative discontinuity orientation data.

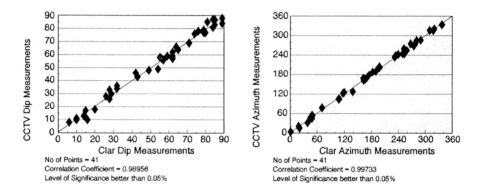

Figure 5 Laboratory Comparison of Clar Compass and CCTV Discontinuity Dip and Azimuth Measurements

Field Evaluation - Trial Rock Excavation
During the research into verifying and improving the methods of obtaining down-hole CCTV techniques for ground investigation the opportunity arose to make use of a trial rock excavation for a major road project.

The proposed excavation was a small box cut, formed in two lifts both 5m high, which was being carried out to test the performance of presplit blasting in massive dolerite prior to it being used on the main road contract. The design angle of the cutting lift faces varied from 55 to 75 degrees. The programme for the trial was extremely tight and any research had to be carried out without causing disturbance or delay.

Field Work: Forming the trial excavation required drilling and charging a large number of presplit blast holes. The presplit blast holes were left open between drilling and charging. It was possible to survey 20 of these and two additional holes, drilled behind the line of the presplit, with down-hole CCTV. The additional holes were optimally orientated with respect to the discontinuity sets present within the dolerite rock mass. As a result of the tight schedule for the trial it was not possible to clean any of the holes prior to survey. The down-hole CCTV surveys were therefore being tested under very unfavourable conditions.

Once the rock faces were exposed a proportion of the presplit blast hole half barrels remained on the finished faces. Some of these corresponded to holes that had been surveyed with the CCTV. Scanline discontinuity mapping was carried out down each of these half barrels. In addition to mapping the half barrels, five other optimally orientated scanlines and judicious mapping were carried out on the excavation faces.

Discontinuity data were recovered from videos of the presplit blast holes using the methods described earlier in this paper. The results of the CCTV surveys were compared with the discontinuity data recovered from the traditional scanline and judicious mapping techniques.

Results: Direct comparison of individual discontinuity measurements from the CCTV scans and presplit scanlines proved very difficult. There was overburden or fill at the top many of the presplit holes when they were drilled and the CCTV depth measurements were referenced to ground level. On excavation this overburden or fill was removed so it was impossible to reference the scanlines to the same datum. In addition the top of many of the presplit holes could not be located after excavation due to overbreak at the top of the cutting lifts. It was therefore not possible to reliably cross reference the depth of individual discontinuities from CCTV and scanline surveys.

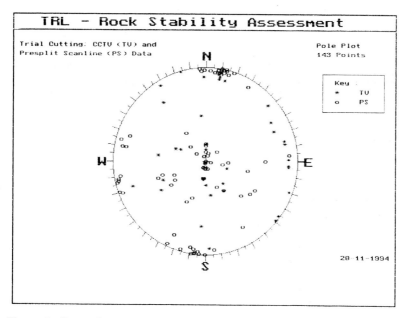

Figure 6 Comparison of CCTV and Presplit Scanline Discontinuity Data

Combined discontinuity orientation data from CCTV surveys were, however, compared with those from presplit scanlines, orientated scanlines and judicious mapping on stereographic projections (Figures 6, 7 and 8 respectively) plotted using the ROCKSTAB program (Matheson 1991). In all cases the discontinuity data recorded from the CCTV surveys follow the same general pattern as those from the other investigation techniques.

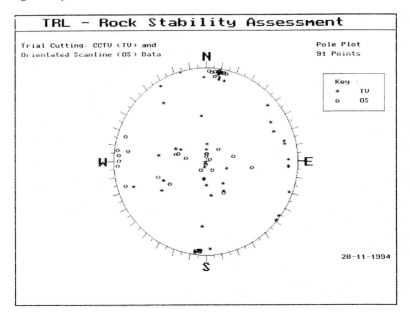

Figure 7 Comparison of CCTV and Orientated Scanline Discontinuity Data

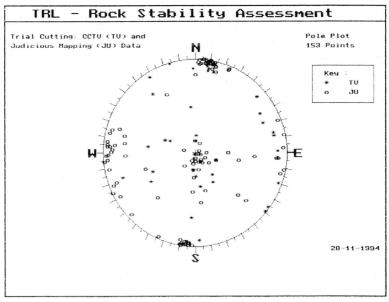

Figure 8 Comparison of CCTV and Judicious Mapping Discontinuity Data

The mean orientations of discontinuity sets were determined using a procedure in the DIPS (Hoek & Diederichs 1989) software. The results of these mean orientation determinations are given in Table 1. Once again there is a good correlation between all of the data sets.

Table 1 Comparison of Discontinuity Set Mean Orientations

Discontinuity Sets	Mean Orientations							
	Judicious Mapping		Orientated Scanlines		Presplit Scanlines		Down-hole CCTV	
	Dip	Azm	Dip	Azm	Dip	Azm	Dip	Azm
Set 1 (see note 1)	09	324	03	064	07	026	08	357
Set 2	88	188	88	187	87	189	84	187
Set 3 (see note 2)	84	093	82	090	83	093	84	267

Note 1: The azimuths of joints with very low dip angles are characteristically variable, even when they belong to one set. As a result the mean orientations for set 1 joints give apparently poor correlations. However figures 6,7 and 8 show that set 1 joints from all data sources plot in the same general area of the stereonets.

Note 2: Set 3 discontinuities are close to vertical and individual discontinuities from the set occur either side of vertical with azimuths of approximately 090 and 270. The mean for the CCTV data falls on the opposite side of the vertical to those for the other data sources.

An attempt was made to correlate the principle spacing data derived from CCTV surveys with that recorded from the presplit scanlines and orientated scanlines. However, due to the poor quality of CCTV surveys obtained from the unprepared presplit blast holes there were insufficient CCTV spacing measurements to carry out a meaningful comparison. Rock flour and other debris completely obscured 39% of the total length of the boreholes surveyed and some of the remaining 61% was partly obscured. As a result an unknown number of discontinuity traces on the borehole walls were not identified during the CCTV surveys. Because of the missing discontinuities the spacing data derived from these CCTV scans is likely to be misleading and is not reproduced here. This problem emphasises the need for good borehole preparation prior to CCTV survey, particularly when accurate discontinuity spacing data is required.

DISCUSSION

Experience in applying down-hole CCTV surveys has shown that the quality of data recovered from the surveys is generally improved if engineering geological expertise is at hand during the survey.

Down-hole CCTV is not without its limitations. In particular survey results are dependant upon the cleanliness of the borehole wall and fluid in the hole. If the hole wall is smeared or it is not possible to achieve clear water in the hole then alternative down-hole methods must be considered. The interpretation procedures described in this paper for determining apparent dip and azimuth from CCTV survey videos rely on manual measurements made from a video monitor. This is time consuming. However, the possibility of automating these measurements using video capture software and direct camera mirror rotation readout is being explored. It is therefore likely that the speed and accuracy of CCTV interpretation will improve.

The principle limitation of discontinuity data recovered from CCTV and other down-hole surveys is the lack of persistence data. It is not possible to obtain a measure of discontinuity persistence from down-hole survey techniques. It is therefore very difficult to evaluate the relative importance of discontinuities encountered in a borehole.

The accuracy of CCTV discontinuity spacing data could not be demonstrated during this research due to lack of time to clean the presplit blast holes for the trial excavation prior to survey. However experience elsewhere with correctly cleaned holes, has shown that it is possible to recover a complete record of the borehole wall along with all discontinuity traces on it. Determination of spacing and principle spacing from such high quality scans is a relatively simple matter. In addition the discontinuity spacing data recovered from CCTV scans is likely to be more accurate than that recovered from borehole cores as the possibility of significant induced fracturing is largely eliminated.

CONCLUSIONS
The results of laboratory and field trials have demonstrated that discontinuity orientation data recovered using down-hole CCTV can be comparable with that recovered using conventional outcrop mapping techniques. Such data could therefore be used for rock excavation design purposes.

Correct borehole orientation and preparation are critical to obtaining quality, representative discontinuity orientation and spacing data from down-hole CCTV surveys. If clean borehole walls and fluid cannot be achieved then alternative techniques should be considered.

The usual application of down-hole CCTV discontinuity surveys is to supplement surface investigations and to improve confidence in the extrapolation of discontinuity orientation and spacing patterns to depth. However, this research has shown that the system can be used to recover discontinuity orientation data, comparable with that from surface techniques, in areas with no exposure.

ACKNOWLEDGEMENTS
The authors wish to thank Prof. G.D. Matheson, Director of TRL Scotland and Mr C. McCaul, formerly TRL Scotland and now retired, for guidance and support on this work.

This research has been conducted at TRL Scotland as part of an ongoing programme of research for the Scottish Office Industry Department, funded by the Chief Scientist Unit of the Department of Transport. The paper is published by permission of the Director of Roads, Scottish Office and the Chief Executive, TRL.

REFERENCES
Gunning A.P. CCTV Borehole Surveying and its Application to Rock Engineering. in Hudson J.A. ed. "Rock Characterization" ISRM Symposium: Eurock '92, Thomas Telford, London. p 174 - 178.

Hoek E. and Diederichs M. 1989. "DIPS Version 2.2, Users Manual - Advanced Version" University of Toronto, Dept. of Civil Engineering, Rock Engineering Group.

Lau J.S.O. 1983. The Determination of True Orientations of Fractures in Rock Cores. Can. Geotech. J., Vol. 20, p221 - 227.

Matheson G.D. 1983. Rock Stability Assessment in Preliminary Site Investigations. Dept. of the Environment, Dept. of Transport, Transport and Road Research Laboratory LR 1039.

Matheson G.D. 1984. A Device for Measuring Drill Rod and Drill Hole Orientations. Dept. of the Environment, Dept. of Transport, Transport and Road Research Laboratory SR 817.

Matheson G.D. 1991. The use of Field Data in the Design of Rock Slopes. in Roegiers J-C. ed. "Rock Mechanics as a Multidisciplinary Science" Proceedings of the 32nd US Symposium on Rock Mechanics. AA Balkema, Rotterdam. p 1083 - 1094.

Murdoch J. and Barnes J.A. 1981. Statistical Tables for Science, Engineering, Management and Business Studies, 2nd Edition Revised and Expanded. Publ. Macmillan London.

Priest S.D. and Hudson J.A. 1981. Estimation of Discontinuity Spacing and Trace Length Using Scanline Surveys. Int. J. Rock Mech. Min. Sci. & Geomech. Abstr., Vol. 18, p 183-197.

Wang H., Latham J-P. and Poole A.B. 1990. In-Situ Block Size Assessment from Discontinuity Spacing Data. 6th Congress of the International Association of Engineering Geology, p 117 - 127, AA Balkema, Rotterdam.

Advances in survey techniques for contaminated land

DR P C ROBERY*, DR W L BARRETT# AND MRS S M HERBERT**
*TBV Stanger, Elstree, #Tarmac Quarry Products Ltd, **TBV Science, Birmingham, UK

INTRODUCTION

Contaminated sites present hazards, which can cause harm to susceptible targets such as humans, surface or groundwater bodies, or buildings. The risks presented by such hazards, assessed as the probability that the harm will occur, can present significant technical, financial and liability constraints to the operation of a business or a planned development. Human health and safety can also be adversely affected and damage caused to the environment. The uncertainty associated with the possibility of risks from contamination, can be equally damaging.

It is important that the nature and extent of any risks are evaluated and, if necessary, appropriate measures designed and implemented to reduce the risks. This requires "hazard-pathway-target" scenarios, which are plausible for the specific site conditions, to be identified and characterised through investigation. Accordingly, site investigation is critically important in the effective risk management of contaminated land[1,2].

Investigations carried out at sites which are suspected of being contaminated, appear to break the first and most important unwritten rule, namely "don't touch anything until you know what's there". But without investigation, it is not possible to establish whether any potential risks exist. Therefore, investigations will always be undertaken under conditions of uncertainty. The solution is to undertake all investigations with the utmost care and to ensure that proper planning is carried out and appropriate techniques always used.

Deficiencies in the site investigation can lead to errors of judgement and can affect the technical sufficiency, cost and duration of works subsequently undertaken, irrespective of whether they are remedial measures for contamination or for construction works[3]. Furthermore, poorly informed and executed site surveys may cause environmental damage and/or expose investigation personnel and the public to health risks.

This paper briefly reviews the strategy and planning for investigations into contaminated or potentially contaminated land. Both non-intrusive and intrusive survey techniques are described, which will provide maximum information cost effectively and with minimum risk. Case histories illustrate applications for some techniques described. Recent guidance documents by CIRIA[2] and the ICE [1] provide more detailed information on available techniques and their suitability for use on contaminated land projects.

PLANNING AN INVESTIGATION STRATEGY

As with all investigations, proper planning and implementation is the key to success. An agreed plan significantly increases the likelihood of obtaining a clear and thorough

Advances in site investigation practice. Thomas Telford, London, 1996

understanding of the site characteristics in the most financially and technically effective manner possible. This requires consideration of key factors including:

* objectives
* scope and content
* health and safety measures
* timing
* phasing
* environmental protection

Management factors, such as the need for legal approvals and operational matters such as access, resource requirements etc., will also need to be addressed.

Defining the general purpose of a survey of contaminated land, such as the scale and extent of the contamination, is straightforward. However, to enable a risk assessment approach to be made of the significance of contamination, even for a qualitative assessment, it is vital that the investigation deals with the factors that affect the behaviour of the contaminants. The investigation must be designed therefore, to address[1,2]:

* contamination   - nature, extent and distribution,
* geology   - physical characteristics of the ground and contaminated media,
* hydrology   - sensitivity of potential water targets and likely transport and
                       fate of contaminants in the water environment,
* pathways / targets - fundamental to deciding whether there is a risk of harm.

Site investigation is frequently required for other purposes, concurrent with determining whether contamination exists. Planning and designing an investigation for several purposes is clearly sensible in cost and duration terms, but this must not compromise the objectives of the work: for example, locations of exploratory holes for geotechnical purposes may not match the locations required for contamination purposes. Satisfying geotechnical objectives at the expense of contamination objectives invariably leads to inadequate data for the risk assessment.

INITIAL ENQUIRIES

Investigations into contamination should be programmed to allocate as much time as possible for a proper phasing of activities. Phasing is essential to maximise the amount of information obtained and to advise on health and safety precautions for personnel. The first phase should always be a desk study[4,1,2], completed and appraised before any site reconnaissance or fieldwork is carried out. This ensures that personnel are not exposed to unnecessary health risks. In some circumstances, an exploratory investigation stage involving preliminary sampling, non-intrusive survey techniques or monitoring, may be appropriate before either the main, or the detailed investigation stage is carried out.

The potential significance of health and safety issues, especially for investigation personnel, is a key factor that differentiates contamination investigations from those for other purposes. The British Drilling Association (BDA) recognised this in 1992, publishing guidelines for safe investigation of landfills and contaminated land, which were subsequently reproduced by the Site Investigation Steering Group (SISG)[5]. The guidelines propose three categories for potentially contaminated land, namely Green, Yellow and Red, in increasing order of potential for risks to drilling personnel.

Review of desk study information should enable a site to be categorised and allow appropriate precautions to be adopted during the fieldwork. However, experience over the last few years has shown that often insufficient information is available to allow the investigation contractor to have a satisfactory level of confidence in the categorisation. Accordingly, there is a trend to assume that sites are Red, unless there is definitive information to prove otherwise[6]. This can have serious financial implications: for example, the cost of setting up full decontamination and safety measures for a Red site can exceed £40,000.

There is increasing pressure on both consultants and contractors to obtain maximum quality information under frequently severe financial constraints, while not exposing their personnel to health risks, the environment to damage or themselves to litigation. This reinforces the need for more careful selection of survey techniques and for the greater use of existing techniques other than conventional boring, drilling or trial pitting. Those designing and planning an investigation into contaminated or potentially contaminated land, must ensure that the latest specialist expertise is incorporated into the process and that maximum use is made of the variety of survey techniques available today.

NON-DESTRUCTIVE TECHNIQUES

Commonly, site records will be poor or nonexistent, leading to automatic classification of the land as Red, as described above. Proceeding with a slow, tentative investigation, will have implications for both the costs of the works and delays to project schedules.

Depending on the circumstances, information can be obtained by wholly nondestructive means to supplement inadeqate site records. Two classes of technique have been used successfully on contaminated sites, including landfills:

* Aerial surveys, for the rapid identification of suspect areas,
* Geophysical investigation, providing a detailed interrogation of the sub-surface.

It is important to recognise the merits and limitations of the different techniques, so that the correct method can be chosen for a given situation.

Aerial Surveys

Infra-red thermography is becoming an accepted tool in the armoury of techniques suited to the monitoring and investigation of landfill sites[7,8] and other sources of contamination. The technique has a proven record for gas detection, as well as identifying leachate escape and underground fires. Once suspect areas have been identified, follow-up ground based surveys to investigate the cause of the problem, can be targetted more effectively. The result is a significant cost saving.

The technique can be applied to many other situations where heating or cooling is taking place. These include monitoring the environment around a site or factory, or locating warmer water discharging into streams, rivers and the sea, which may be carrying pollutants. A distinction must be made between infra-red imaging and infra-red photography. Only the former detects surface warming effects, whereas infra-red (false colour) photography detects changes in materials that are not apparent to the naked eye such as distressed vegetation in times of drought.

326

Often, the most suitable method of examination is by helicopter, rather than a fixed wing aircraft. This is for a variety of reasons, including:

* rapid transit between sites,
* manoeuvrability and the ability to stop over features of interest,
* operation at low flying levels for viewing three-dimensional features,
* operation in most weather conditions,
* the client can join the survey to view sites at close quarters,
* the helicopter can fly from a small local base.

Some localised problems can be investigated using thermographic equipment on a lorry-mounted hoist.

Geophysical Methods

There are a variety of techniques that can be used to provide information on the sub-surface condition and content of the ground before resorting to excavation or boring. Selection of the most appropriate technique is important, as each has individual applications, benefits and limitations. Advice is needed to decide what can and cannot be done with each method for the particular site conditions and features to be identified.

The techniques can be used in situations where it is suspected that there are buried objects or other features. One example would be where buried drums of waste may be present; another situation may be the exact positioning of underground voids or mine shafts. Table 1 gives a summary of the different geophysical techniques, their common applications and general limitations. This must be treated as an outline guide only. Factors that can influence the choice of survey technique include the type of soil over the site, the depth of the water table, whether areas have been capped by a reinforced concrete raft and whether features such as mine shafts have been backfilled.

It is normal to use more than one geophysical technique to investigate a particular site. The use of multiple methods not only enables ambiguous results to be clarified but also helps differentiate between the types of targets; some materials will not respond in the same way to different survey methods.

All techniques are based on establishing a close-centred survey grid over the site using fixed lines or optical survey stations. The spacing of the grid lines is decided by experience, based on the possible size of the targets to be located, the ground conditions and the consequences arising from missing a target object in the particular circumstance.

Data is usually recorded in digital form, such that it can be easily processed back at the laboratory. Unusually high or low values may be due to the presence of a target beneath ground level. To differentiate between background levels and buried features, the site data are usually processed using contour-plotting computer software. Detailed exploration in areas giving anomalous readings, enables the data to be calibrated against actual features, thereby confirming the size and nature of the targets being detected.

PHYSICAL EXPLORATION

Once all available information has been gathered, physical exploration can begin. In contaminated land investigations, care in the selection of exploratory techniques for

Table 1   Geophysical Methods

METHOD	APPLICATIONS	LIMITATIONS
Impulse Radar	Specially-developed impulse radar can be used to profile sub-surface geological, hydrological and geophysical features. The rapid-scan technique can locate suspect features and delineate their size, enabling considerable reductions to be made in the time and number of exploratory excavations made on a site.	Radar scans form precise sections through the surface, resulting in features to the side of the scan line being missed. Certain ground conditions such as clay soils can limit the depth of penetration. Scanning through reinforced concrete slabs can be a problem.
Magnetic Prospecting **- profiling** **- gradiometry**	Measures the earth's magnetic field at specific grid points, which is affected by changes in the magnetic susceptibility of the underlying ground, due to magnetic objects etc.	Accurate surveying may require topographical information if the terrain is especially varied. Infrequently, magnetic storms due to solar activity can completely stop work for a particular day. The technique can be strongly affected by industrial 'noise', such as overhead power lines, metal fences etc.
Gravity	By measuring the earth's gravity field, anomalies due to the presence of buried objects or voids in the ground can be detected.	The technique is slow and requires accurate level data for each grid point. It will detect nonmagnetic features.
Seismic Techniques **- refraction** **- high-resolution reflection**	Accurate information on a variety of features such as depth to bedrock, stratigraphy, buried channels etc.	Can involve relatively lengthy fieldwork times in comparison with other techniques.
Electromagnetic Techniques - various such as conductivity profiling and EM-sounding	Various, including fluid quality and buried artifacts.	Can be subject to industrial 'noise'
Electrical Techniques - various such as Self Potential and Vertical Electrical Sounding	Various - can be used in conjunction with EM and other techniques to differentiate between buried objects and features.	Can be slower in the field than EM techniques.
Borehole Geophysics	Various 'insitu' measurements such as fluid flow movements, formation types and thicknesses etc.	Requires physical intrusion prior to monitoring.

borehole formation is vital, because of the potential for contamination of clean ground and groundwater. Precautions may be needed in the following areas[2]:

*   use of appropriate casing kept coincident with the base of the borehole throughout formation,
*   use of permanent sealing (e.g. bentonite) to stop movement of contaminants and leachate,
*   measures to prevent extraneous matter falling into the borehole (e.g. use of casing and sealable/lockable covers),
*   avoidance of the use of water-assisted drilling, where the potential for dilution of contamination is significant,
*   avoidance of the use of foam flush, because of the potential for introducing organic substances into the borehole,
*   use of filters on compressed air for air-flush drilling, when monitoring for organic contaminants, to avoid contamination of samples with oil from the compressor,
*   the Health and Safety implications (i.e. the discharge of contaminated water and cuttings) during air-flush drilling in highly contaminated ground and the displacement of volatile compounds when using air-flush techniques.

Pre-determination of a site category is essential if the correct safety equipment, investigation methods and rig specifications are to be employed on any particular job. Insistence upon and policing of compliance with the BDA/SISG guidelines by the Health and Safety Inspectorate would help to raise safety standards of both equipment and techniques.

It must be accepted that often, the categorisation of land may remain indeterminate, despite desk studies and nondestructive surveys. Thus, before proceding with the detailed investigation, there can be benefits in obtaining further, accurate data cost-effectively, using systems such as described below.

Geoprobe System
A significant advance in the field of exploratory investigation and safe categorisation of contaminated land is provided by the geoprobe/geolab system. It consists essentially of a percussive drill with attachments for the sampling of gases, liquids and solids and a self-contained mobile laboratory for on-site testing. Using this system, sealed samples are obtained, avoiding the risks of potential contaminants being liberated at the surface. The American-designed geoprobe is limited in its depth capability, with 5 to 10 m being the normal depth of penetration. Despite this, the system provides significant safety improvements over other intrusive techniques and offers a cost-effective and advantageous method for categorisation of sites and providing preliminary contamination data.

Although it is recommended that this method be initially used for potentially contaminated sites to safely establish the site category (Green, Yellow or Red), detailed investigation works will continue to rely upon conventional drilling, excavation and sampling methods.

Conventional Drilling Techniques
Significant advances in the methods for safely investigating and monitoring contaminated land have developed over the last 10 years. The advances have arisen through the systematic evaluation of established drilling techniques, rather than the discovery of

revolutionary new methods. Increasing awareness of the potential hazards coupled with the threats of litigation and the new legislative measures, have provided impetus to the need to find improved techniques.

Particular recent developments have stemmed from experiences gained from the investigation of landfill and other contaminated fill sites compared with that of natural strata. Certain methods of drilling and installation of monitoring equipment have been found more suitable than others. Table 2 summarises the advantages and disadvantages of the methods, and is supplemented by the text below. Whilst some of the comments relate spcifically to use in landfill situations, it is considered that they will also be of value for other contaminated land sites.

Cable percussive boring (Shell and Auger). This is the most common UK method for site investigation, and is still widely used for contamination surveys. The technique is relatively inexpensive and the rig is highly mobile but is slow in dealing with obstacles in the ground and there is a high risk of cross contamination down the hole, particularly when water is encountered. In addition the casing may cause compression and sealing of the sides of the hole while dropping the shell is likely to generate sparks when it strikes on stone or metal in the hole. Large quantities of arisings are produced, with consequential risks to health and safety, the potential for surface contamination and the need for safe disposal. In some cases, the disadvantages inherent in this technique outweigh the advantages and consideration is needed as to whether it is appropriate.

150 & 200 mm Continuous Flight Augers. A moderately powerful rig, often mounted on a truck, tractor or low ground pressure vehicle, is best suited to operate this method. No water or air flush is needed, reducing secondary contamination, and the auger flights carry displaced materials to the surface. Although relatively cheap and quick, the method is incapable of taking discrete samples and its main use is in giving an estimate of the ground conditions, for example establishing the depth of landfill or other fill material. Other problems with this method include:

* poor ability to overcome obstacles in the ground,
* propensity to get stuck,
* severe disruption to the ground next to the holes (in landfill),
* high chance of deviation off-course,
* health risk arising from cleaning augers by hand,
* severe vibration and wear and tear on the rig,
* limited diameter holes of somewhat suspect stability,

Penetration deviation and sticking problems can be reduced by developing custom-built auger bits equipped with the tungsten carbide pick teeth.

Hollow Stem Augers (250 mm). The system is similar to continuous flight augers but the outside diameter is larger to incorporate a hollow tube through which sampling, testing and placement of borehole instrumentation can be achieved. No air or water is needed and samples with low risk of contamination can be obtained. However, disadvantages include:

* a powerful, very high torque rig is needed, which can suffer heavy wear

Table 2   Comparison of Methods for Investigation and Monitoring Landfill and Contaminated Sites.

A. Within Fill Material

TECHNIQUE	MAIN ADVANTAGES	MAIN DISADVANTAGES	COST/m (Approx)
Shell and Auger	Relatively cheap, highly mobile, reasonable samples in cased sections	Requires water, serious cross contamination especially from water, risk of sparks, splashing of leachate., large amounts of waste brought to the surface. Difficulty with obstacles, causes disturbance of ground adjacent to hole.	£12.00
150 & 200 mm continuous flight augers	Relatively cheap, rapid penetration. Needs no air or water for flushing.	Poor ability to overcome obstacles, deviation of holes, small diameter holes, severe vibration, hard on rigs, serious cross contamination, disturbance around holes, collapses common, large amounts of sample produced. Cleaning difficult. Performance can be improved with special bits.	£15.00
250 mm hollow stem augers	Good stable straight holes. Good samples. Little danger of cross-contamination, ease of installation of hole equipment. No air or water needed. More acceptable hole size guaranteed stability.	Costly, slow progress, poor obstacle passing ability. Flights easily damaged and costly to refurbish, tendency to leak at joints, large high torque rig needed. Very punishing on rigs.	£25.00
300 mm continuous flight augers	Reasonable penetration rates but poor obstacle performance. Less hole deviation. No air or water needed. Appropriate diameter holes	Relatively expensive, large rigs needed. Holes tend to collapse, poor obstacle performance, very large amount of material brought to the surface . Cleaning problems, very punishing on rigs.	£20.00
200 - 400 mm single tube barrels	Excellent holes, minimal danger of cross contamination. Little deviation, excellent obstacle penetrating ability, minimal ground disturbance, large precise and manageable samples, modest size rigs.	Relatively slow and costly, difficulty in extracting samples from the barrel. Frequent re-tipping of barrel needed but not overly costly. Penetration poor in natural ground.	£25.00

Table 2   Comparison of Methods for Investigation and Monitoring Landfill and Contaminated Sites (Cont...)

B. In Natural Strata

TECHNIQUE	MAIN ADVANTAGES	MAIN DISADVANTAGES	COST/m (Approx)
250 mm hollow stem augers (sand and gravel)	Used with pilot bit or detachable cap, guarantees good straight hole, making installation of in-hole equipment easy. Minimises risk of contamination. No air or water needed.	Costly, slow progress and high wear and tear rates. Poor at coping with obstacles, auger liable to damage and costly refurbishment. Large rig needed and technique punishing on rig. Danger of leakage at joints.	£15.00
Soft rock formations open hole and if necessary casing advancer.	Appropriate rock rollers used with air flush. Rapid progress, not overly expensive and can give good holes. If collapsing occurs casing can be seen-in with casing advancers.	Compressor hire is costly. Air flush may propel contaminants into the air. Irregularities in hole sides can cause pipe installation problems.	£10 - £15*
Hard rock-dth hammer and casing advancer if necessary.	Air flush hammer holes lined if necessary using casing advancer system. Rapid progress often good quality straight holes	As for open holes	£10 - £20*

* Drilling costs only, no casing or reaming included.

* the ability of the method to overcome obstacles is limited,
* compaction and/or disturbance of the material next to the hole can be severe,
* deviation of the holes is quite common,
* wear and tear on the auger flights is heavy and expensive to repair,
* progress is slow.

300 mm Continuous Flight Augers. This method has been used on landfills in both continuous flight mode and with a telescopic Kelly bar behind a single flight and bit. As anticipated a satisfactory larger diameter hole can be achieved at a fast rate and lower cost than with the hollow stem augers. Disadvantages include:

* very poor ability to overcome obstacles in the ground,
* a high likelihood of deviation in the hole,
* very large amounts of material are brought to surface,
* health and safety risks from handling and cleaning the augers,
* a large and powerful rig is required, which will suffer heavy use.

Large diameter single tube barrels (250 - 400 mm). Continuing research directed particularly at the aspects of health and safety, penetration and vibration damage to the rigs, led to the development of and trials with large diameter coring barrels. Experience to date suggests that this method is often the most suitable technique for landfill/filled sites. Rigs of moderate power and torque rating (but with sufficient retracting power) are satisfactory for this technique that has many other advantages:

* obstacles are easily overcome, with cores having been obtained through old engine blocks and pieces of concrete,
* the technique is effective below the water table (but care is needed if cross contamination is of potential concern),
* in landfills no air or water flush is needed,
* the holes suffer little deviation and excellent, precisely-located samples can be obtained,
* cross contamination within soils/fills is virtually eliminated,
* the coring method is very "rig friendly", because the barrel is rotated on low speed and there is little vibration,
* R & M costs are much reduced on the rigs when using this method,
* the waste is safely contained within the barrel and can be discharged in a controlled manner,
* the technique has great depth capabilities (40m in landfill has been proved),
* the capital outlay is low since the barrel can be driven on existing rods of 100 mm diameter or more.

Inevitably there are also a few relatively minor disadvantages, including:

* slow progress, especially in sub-tip strata,
* difficulty in extruding compressed waste from the barrel,
* relatively frequent refurbishment is needed for the cutting end of the barrel.

Borehole and Monitoring Installations in Natural Strata

Construction of deep monitoring wells poses particular problems. Over the years monitoring holes have been drilled and cased to depths of over 140m and as a result of considerable experimentation, favourable techniques have been developed for sand and gravel, soft rock and hard rock (Table 2B). Historically, small diameter (19mm ID) pipes were used but 50mm diameter pipes are normally now the appropriate minimum.

Sand and Gravel Areas. Where exploratory holes of only limited depth are required, hollow stem augers of 250 mm diameter, with retractable pilot bit, have been found effective, drilled to the required depth before withdrawing the pilot bit and putting the casing down the tube. On completion of the installation the augers are withdrawn carefully and an appropriate security cap is fitted.

Soft Rock. In soft rocks, ground stability can pose problems, especially for deeper holes. Open hole drilling methods can be used successfully to install monitoring pipes and any fabric wraps or packing in the hole. If the ground shows signs of collapsing, a casing advancer system can be employed, following which the monitoring pipes are installed and the casing recovered.

Hard Rock. In hard rock conditions down the hole percussive hammers with air flush are used to drill holes up to 250 mm in diameter. Such holes will often be stable, enabling the monitoring pipes and any specified packing to be installed directly. If problems are encountered, these holes also can be reamed out and cased using a casing advancer system. As with the soft rock installations the casing can subsequently be recovered for reuse.

CASE HISTORIES

Landfill Site - West Midlands

A densely developed West Midlands Borough had the task of monitoring over forty completed landfill sites for methane gas activity and risk to adjacent properties. Approximately half of the sites were believed either to be inert or to have no remaining biodegradable matter. However, records were of insufficient detail to prove this. It was estimated that to install a sufficient number of boreholes and associated equipment would cost in the region of £10,000 per site, with a commitment to monitor periodically the estimated 1,000 boreholes. This presented an impossible task, both in financial terms and given limited staff resources.

An aerial thermographic survey was commissioned to investigate all the sites. A major part of the brief was to prioritise the sites into high and low risk categories, with the aim that where there was little evidence of gassing activity, a reduced scale ground sampling and monitoring exercise would be sufficient for final verification. The helicopter-based survey was completed in two flights of one hour's duration each. The results largely confirmed the expectations of the client that eighteen sites were no longer a risk with respect to methane. When the cost of both the aerial survey and supporting ground based investigation needed for corroboration, were compared with the costs of a conventional survey, it was estimated that a saving of around £135,000 was made.

Suspected Contaminated Site - Essex

A developer in London had acquired 10 acres of a "brownfield" site for housing development. The site was known to have been used for several industrial uses and that industrial wastes such as drums of cyanide had been stored in some areas. Although the vendor had given assurances that all drums of waste had been removed, concern had been expressed that buried metallic tanks or "capsules" may remain beneath the ground surface. If these were damaged by excavators, this would exacerbate the problem.

Investigation was undertaken using a Portable Proton Total Field Magnetometer. This instrument measures the total magnetic field of the Earth at a fixed point above ground level and is influenced by the presence of metallic objects such as drums, which interfere constructively or destructively with the Earth's general field value. The survey found steady and repeatable readings from the Proton Magnetometer on the site despite its history of industrial use and the amount of metallic debris on the ground surface and reinforcement within existing concrete ground slabs. The investigation also found several anomalous areas, which were delineated for closer inspection. The survey confirmed that the site was free from any drums of waste, giving comfort to both the developer and subsequent house buyers.

Underground Fire - South Wales

This site was an old, closed landfill with a small industrial estate developed on top, but known to have a fire burning at depth. All attempts at extinguishing the fire had previously failed and there was uncertainty as to the exact current location and extent of the fire, which was a cause for concern in view of the businesses on the site.

The thermographic survey easily pinpointed the seat of the fire and its extent. There was some indication that limited migration had occurred towards one industrial unit over the period since the fire was first reported. This element of the total investigation was completed within approximately five minutes. It was recommended that subsequent investigation should be carried out using a thermographic camera fixed to a lorry-mounted hoist, providing the most cost-effective method of periodically monitoring the progress of this fire until an effective long-term solution had been found.

CONCLUSIONS

The investigation and monitoring of existing sites of potential contamination can be both difficult and hazardous. In an attempt to improve safety standards the SISG and BDA have issued guidance on the safe drilling of landfills and contaminated land. Three categories of contaminated sites have been suggested, based on the anticipated level of hazard, with appropriate safety equipment and procedures proposed for each. Despite this, the problem of reliably establishing site categories (especially on older sites) is very real. Without good site records, the only completely safe course is to assume all sites are in the Red category until investigations prove them to be of a lesser hazard. This assumption requires a substantial advance investment in safety equipment and facilities.

By using a combination of non-destructive techniques and special sampling methods, confidence can be gained economically as to the true site condition. Rapid scan aerial thermographic surveys can locate heat sources such as escaping methane. Geophysical methods can supplement site data by locating buried objects, strata and fluid flow measurements. Finally, suspect areas can be investigated by the geoprobe system, allowing sealed samples of liquids, solids, and gases to be safely obtained and tested in an on-site laboratory. Using this approach, the category of site can be safely established ahead of detailed examination by conventional investigation methods.

Once the site has been properly classified, more detailed sampling and testing can be undertaken more cost-effectively and safely. Involvement over many years with the investigation and monitoring of contaminated land, especially landfill sites, has enabled many existing investigation techniques to be evaluated and refined. Now, preferred and acceptably safe drilling and boring methods have been evolved for different situations.

ACKNOWLEDGMENTS
The authors acknowledge the assistance given by their colleagues in the provision of information and formulation of views for this paper. However, the views expressed are entirely those of the authors and do not necessarily represent those of their employers.

REFERENCES
1.   Harris, M R and Herbert, S M (1994) Contaminated land: investigation, assessment and remediation. ICE Design and practice guide. Thomas Telford (London).
2.   Harris, M R, Herbert, S M and Smith M A (1994) The remedial treatment of contaminated land. CIRIA Special Publication SP101 - 112. CIRIA (London).
3.   Site Investigation Steering Group (1993) Without investigation, ground is a hazard. Site investigation in construction, Volume 1. Thomas Telford, (London).
4.   British Standards Institution (1988) Draft for development code of practice The identification of potentially contaminated land and its investigations. DD 175. BSI.
5.   Site Investigation Steering Group (1993) Guidance on the safe investigation by drilling of landfills and contaminated land. Site investigation in construction, Volume 4. Thomas Telford, (London).
6.   Herbert, S M and Barrett, W (1994) Drilling Techniques for a BDA Red site. Paper to British Drilling Association seminar on Drilling on Contaminated Land, 23 March 1994, Stoneleigh, Warwickshire.
7.   Titman, D.J., Aerial Thermal Imaging of Landfill Sites, Polluted + Marginal Land '92, Brunel University.
8.   Titman, D.J., Aerial thermographic Surveys of Landfill Sites, Discharge Your Obligations 1993, Kenilworth.

Moderator's report on Session 3: drilling, boring, sampling and description

DR D. W. HIGHT
Geotechnical Consulting Group, London, England.

INTRODUCTION

The brief for the Moderators was defined by the Organising Committee as follows: "to summarise the papers in the session, highlighting the salient points, extrapolating key information and promoting subsequent debate". The Moderator's Report was to be "based on the points in the papers and their own knowledge (including papers in recent conferences)". Session III is to consider "exploratory drilling and boring, sampling techniques with special reference to granular soils and weak rocks, and the description and classification of the ground".

This Report has been written with this brief and with the title of the Conference, "*Advances in Site Investigation Practice*", in mind. It draws on recent literature and on papers allocated to other Sessions in order to provide a more complete coverage of the topics and to identify developments and trends. Papers presented to this Conference and cited in this Report are shown in italics.

The Report has been written in two parts: Part I - Drilling, Boring and Sampling; Part II - Description of the Ground. To reflect the Moderator's particular interests two threads run through Part I of the Report, namely:

- the effects of sampling in different materials, using the techniques under discussion, and

- assessment of the level of sample disturbance, using comparisons of G_{max} from in situ and laboratory measurements; this serves as a link between the topics discussed here and in Sessions V and VI.

Before beginning the review, a brief background is provided to sample disturbance and to its assessment.

PART I: DRILLING, BORING AND SAMPLING

BACKGROUND

Sample Disturbance

A sample taken from the ground and brought to the laboratory for a mechanical test is subject to a variety of strains and is relieved of the stresses to which it was subject in situ. The sequence of strains and of stress relief depends on the method of boring and sampling.

In clays, the sampling process is generally undrained (constant volume) and a mean effective stress is retained in the sample by capillarity (soil suction). Referring to Hight (1993), Vaughan et al (1993) and Tavenas and Leroueil (1987), the effects of sampling are to cause:

- a change in mean effective stress, p'; in soft clays p' is reduced; in stiff overconsolidated plastic clays p' is increased by tube sampling

- water content redistribution post sampling - from the periphery to the centre in tube samples of soft clay and from the centre to the periphery in tube samples of stiff plastic clay; rotary coring with water-based drilling muds leads to an overall increase in water content and reduction in p' in all materials

- mechanical damage to the clay structure which is manifest as a shrinking of its bounding surface (or lowering of its failure line); this damage may result from undrained shear strains during sampling or by volumetric and shear strains during subsequent swelling; in tube sampling of hard brittle clays, fractures may propagate from the cutting edge

- movement on existing fissures.

In sands, the sampling process is drained and the suction that can be sustained after stress relief is severely limited. The effects of sampling are to cause:

- a change in volume (void ratio)

- mechanical damage to any cementing (soil structure) as a result of shear and volumetric strains

- a major reduction in p'

- changes in contact distributions.

In weak rocks, sampled by rotary coring, there is a reduction in p' as a result of a low B value and of subsequent swelling. Samples taken from substantial depths may be damaged by the initial isotropic relief of stress (Holt et al, 1993), as well as by subsequent swelling. Other damage, including the formation of cracks, may result from torques transmitted to the core, from rocking of the core barrel (*Tatsuoka et al*), from erosion, and from strains imposed during extrusion from the core barrel. Failure to sustain a suction across discontinuities will allow them to open and displace during handling; mismatch of joint surfaces as a result of stress relief can have an important effect on small strain stiffness measured on samples of weak rock (Cook, 1992).

Assessing Level of Sample Disturbance. Various methods for assessing the level of sample disturbance (sample quality) have been proposed and these were reviewed by Hight (1993). Reference will be made to two methods here:

- the comparison of initial mean effective stress in samples, p_i', and in situ, p_o'

- the comparison of in situ and laboratory measurements on samples of shear wave velocity, v_s (or small-strain shear modulus, G_{max}).

The basis for adopting comparisons of v_s in the laboratory and field as a measure of mechanical disturbance is explained in Figure 1 for the case of sands. In sand, v_s depends on stress state, void ratio (Fig. 1(a)), degree of cementing (Fig. 1(b)), age (Fig. 1(c)), and contact distribution. Figure 1(c) illustrates how a disturbance, in this case a twist, such as might be applied during rotary coring, changes v_s; the effects of aging are removed, and, since the void ratio actually reduces, the reduction in v_s must be related to a change in contact distribution. A similar argument applies to clays in which the development of structure at constant stress is associated with an increase in v_s, while destructuring during sampling results in a reduction in v_s. In weak rocks, v_s will depend inter alia on the distribution and state of discontinuities and on the degree of cementing; Figure 2 illustrates how removal of cementing (destructuring), in this case by weathering, results in a progressive reduction in v_s.

For the comparisons to be valid, the laboratory samples must be representative, in terms of fabric, discontinuities, etc., and must be restored to their in situ stress state, particularly since v_s depends on the complete stress state. Measurements of v_s should also be made with shear wave propagation in the same direction as in the field, with the same plane of polarisation and at a similar frequency.

Shibuya, Tanaka and Mitachi provide a valuable summary of Japanese experience with comparisons of G_{max} from field measurements (seismic survey - using downhole, crosshole or seismic cone) and laboratory measurements on retrieved samples (refer to their Table 1 and Figure 1). It should be noted that, in most cases, samples were isotropically consolidated to a stress equal to the in-situ effective overburden pressure, σ'_{vo}. Lefebvre et al (1994) have shown that in undisturbed samples of soft structured clay this procedure can lead to an overestimate of G_{max}; *Shibuya et al* present data (see Figure 3(c)) which support this.

ADVANCES IN THE PRACTICE OF TUBE SAMPLING

Disturbed Dynamic Sampling

Eccles and Redford provide a useful background to dynamic sampling techniques. They ascribe the growing popularity of dynamic sampling to:

- its being cost-effective and rapid (40-50m of sampling can be achieved in one day, although the rate obviously reduces with increasing depth of investigation); damage, and, therefore, reinstatement costs are minimal

- its flexibility, being lightweight and man-handleable, and, therefore, suitable for sampling at locations with poor access and limited operating space

- the increasing need for shallow reconnaissance over large areas in connection with studies of contamination and house settlement.

Eccles and Redford describe in detail the Nordmeyer type window sampler. It comprises 1m, 2m and 3m long hardened steel sampling tubes with a 'window' slot along their length.

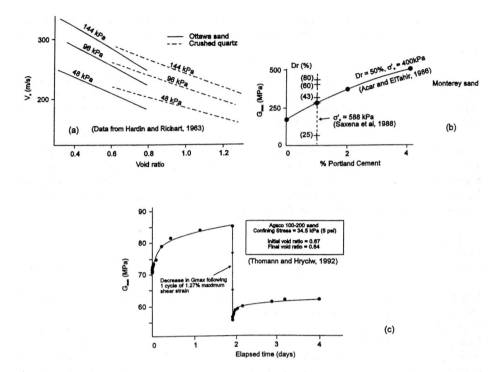

Figure 1. Background to the use of v_s (or G_{max}) for assessing sample disturbance in sand
    (a) Dependency of v_s on void ratio
    (b) Dependency of G_{max} on degree of cementing
    (c) Sensitivity of G_{max} to age and disturbance

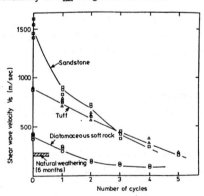

Figure 2. Effect of destructuring by weathering cycles on v_s in weak rocks (Nishi et al, 1989)

The tubes, which have internal diameters of 38, 50, 60 and 80mm, are driven by a high frequency percussion hammer. The sample is described through the window and extruded in the field to allow re-use of the tubes.

Dynamic 'window' sampling produces a continuous profile of disturbed soil over depths of 7m to 15m, depending on soil density and stiffness. It relies on rapid progressive sampling in an uncased hole and, therefore, involves driving through cave-ins. The sampling holes may be used for the installation of standpipes, gas monitoring tubes, etc.

Eccles and Redford present four case histories illustrating the successful use of dynamic window sampling in a wide range of materials (difficulties arise in coarse gravels and when cobbles are present). Additional information on their second case is given by *McGinnity and Russell*. The use of another dynamic sampling tool, the flow-through sampler (Martin and Suckling, 1984) is described by *Pickles and Everton*.

Undisturbed Tube Sampling
Developments in undisturbed tube sampling are sparse and none have been reported to this Conference. However, the trends discussed below have been identified from the literature.

Soft Clays. In soft clays, there appears to be growing awareness of the shortcomings of conventional fixed piston sampling, at least in low plasticity structured clays. Specialised large diameter samplers, such as the Laval sampler (La Rochelle et al, 1981) and the Sherbrooke Sampler (Lefebvre and Poulin, 1979) are being used increasingly on major projects and for establishing benchmarks for the sampling of clays around the world. For example, the Laval sampler has been used in France (Magnan et al, 1994), in Japan (Oka et al, 1992), in the UK (Hight et al, 1992a), in Sweden, USA (National Test Site at Amherst) and Pisa (Leroueil, pers. comm.); the Sherbrooke Sampler has been used in Norway (Lacasse et al, 1985), in the UK and USA.

In the Bothkennar Characterisation Study in Scotland (Hight et al, 1992b), it was shown that, for this material, conventional UK sampling practice for soft clays produced samples that were inferior to Sherbrooke and Laval samples and were unsuitable for determining the yield characteristics of the clay.

Several examples are now available of the assessment of sample disturbance in soft clays using comparisons of field and laboratory measurements of v_s or G_{max} Three examples are shown in Figure 3:

(a)   shows disturbance in piston samples of Bothkennar Clay leading to a reduction in v_s compared to in situ values (destructuring by piston sampling was demonstrated independently as shrinking of the soil's bounding surface, Hight et al (1992b));

b)   shows good agreement between v_s values for a Sherbrooke sample of St Alban clay at 4.0m, but differences in v_s values at 7.9m, where the clay is of very low plasticity and more susceptible to disturbance;

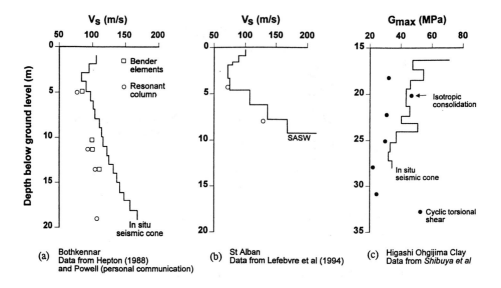

Figure 3. Assessing sample disturbance through comparisons of v_s (or G_{max}) from in situ and
laboratory measurements in soft clay.

(c)   shows disturbance in samples of Japanese clay taken with thin-wall fixed piston
samplers and is taken from *Shibuya et al's* paper to this Conference (note the higher
value of G_{max} obtained in the isotropically consolidated sample and its coincidence
with the in situ value).

<u>Stiff Plastic Clays</u>. The use of pushed thin-wall tube sampling in stiff clays was promoted
in the UK by the BRE but appears to have been taken up by the Site Investigation industry
only reluctantly. Its use in London Clay at the Queen Elizabeth II Conference Centre was
described by Burland and Kalra (1986); its use in London Clay at the Waterloo International
Terminal was described by Hight et al (1993) and more detail is given by *Pickles and
Everton*. It has been argued that the technique has little to offer over conventional driven
tube sampling, since differences in strength between samples taken using the two methods
are small.

In the Moderator's view, the benefits of thin-wall sampling in stiff plastic clays are:

•   to reduce the thickness of the highly distorted zone around the sample periphery and
so reduce the amount by which the mean effective stress increases as a result of this
distortion (Vaughan et al, 1993)

•   to reduce the shear strains imposed and the mechanical damage that results.

The two effects may compensate for one another in unconsolidated undrained (UU) triaxial
compression tests, as illustrated in Figure 4. The thin-wall samples show a lower mean

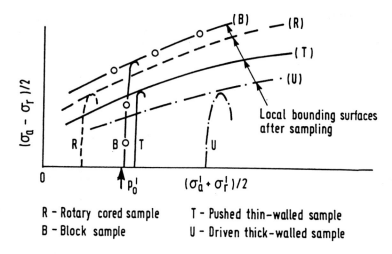

R – Rotary cored sample   T – Pushed thin-walled sample
B – Block sample   U – Driven thick-walled sample

Figure 4. Effects of different types of sampling in stiff plastic clay

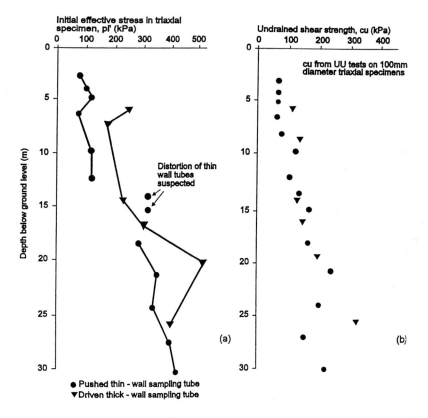

● Pushed thin - wall sampling tube
▼ Driven thick - wall sampling tube

Figure 5. Comparison of thin-wall and thick-wall tube sampling in stiff London Clay
(a) initial effective stress, p_i' (b) undrained shear strength, c_u

effective stress than the thick-wall samples (both are shown as exceeding the in situ p_o') but effective stress paths for the thin-wall samples reach a higher effective stress failure envelope, reflecting the reduced mechanical damage. The two types of sample have a similar undrained strength. Evidence to support this postulate is presented in Figure 5; part (a) shows higher values of p_i' in driven thick-wall tube samples than in pushed thin-wall tube samples taken alternately in the same borehole; part (b) shows little difference in c_u between the two types of sample.

Techniques used for thin-wall sampling in practice have been described by Harrison (1991). Difficulties are sometimes encountered with tubes distorting or buckling, particularly if the base of the borehole is not flat; distortion of the thin-wall tubes is thought to have affected two of the data points in Figure 5(a). In brittle clays having an undrained strength in excess of 200 kPa, it appears that major damage occurs to the sample, as a result of fractures propagating from the cutting edge.

Assessment of sample disturbance in stiff plastic clays has been made on the basis of measurements of p_i' and the location of the bounding surface (Figure 4). An attempt has also been made to assess disturbance in such clays by comparing field and laboratory measurements of G_{max}. The comparison is shown for thin-wall tube samples of London Clay in Figure 6. The higher values measured in the laboratory samples may be the result of differences in frequency and wavelength between the in-situ and laboratory methods of measurement or a lack of representativeness of the sample. The comparison serves to illustrate the care needed in applying this approach to assessing sample quality.

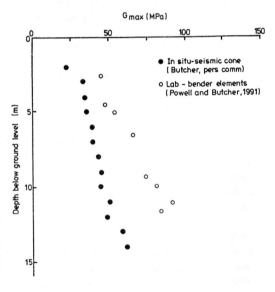

Figure 6. Assessing sample disturbance through comparisons of G_{max} from in situ and laboratory measurements in stiff London Clay.

Sands. Since the introduction of the Modified Bishop sampler by Hanzawa et al (1980) there appear to have been few developments in the tube sampling of sands. One development of which the Moderator is aware concerns the use of the Laval Sampler in sands (Leroueil, pers. comm.). The advantages of using large diameter samplers follows from:

- a reduced risk of plugging during penetration

- the possibility of obtaining a central core in which changes in void ratio are acceptably small.

Volume changes caused by the penetration of samplers, model piles, etc., into sand have been studied widely. A very clear illustration of the volume changes and their radial extent was presented by Yoshimi et al (1978) and is shown in Figure 7.

Dense sands dilate on tube sampling, loose samples contract. This is illustrated convincingly in a comparison of field and laboratory measurement of G_{max} made by Yasuda and Yamaguchi (1985) in reclaimed sands (see Figure 8), which were presumably unstructured and in which changes in G_{max} reflect changes in void ratio (see also Figure 1(a)).

ADVANCES IN THE PRACTICE OF ROTARY CORING
The idea of obtaining samples of stiff clay by using rotary coring equipment developed for rocks is not new. Ward et al (1959) describe rotary coring in London Clay; Samuels (1975) used these techniques in investigations for the Ely-Ouse tunnel in Gault Clay.

However, several important developments in rotary coring techniques have taken place in the last 15 to 20 years. These include the introduction of:

- large diameter triple tube wireline drilling systems

- UPVC liners in a triple tube core barrel configuration

- new drilling fluids, including biodegradable polymer muds

- specifications for core handling, sealing and storage.

In his paper to this conference *Hepton* describes the current state-of-the-art in rotary coring in geotechnical investigations and its application at Sizewell 'C'. Use of these techniques is also described in the papers by *Pickles and Everton*, *Maddison et al*, *Johnson et al*, *Davis et al*.

Wireline Coring
Wireline core drilling systems were introduced into the UK for geotechnical investigations in the mid-1980s. This followed their successful use in 1982/83 on the Baghdad Metro Project, where 102mm diameter cores of sands and gravels and of lightly overconsolidated clay were successfully recovered using the Craelius SK6L system (Scarrow and Gosling, 1986). The principles of wireline coring are outlined by *Hepton*.

The advantages of wireline coring over conventional drill rods and core barrels are summarised in Table 1. With regard to sample disturbance, the increased speed of retrieval, reducing the time for destructuring by swelling, is an important advantage in clays and mudrocks. The systems currently in use -Geobor S and its predecessor SK6L - produce a 102mm diameter core and 146mm diameter borehole.

Core-liners
Semi-rigid plastic liners were introduced in 1986 as the inner barrel of a triple tube configuration. These have eliminated the need for extrusion of the core, since the liner can be cut along diametrically opposite sides, and have facilitated core handling; disturbance to cores has, thereby, been reduced. The fit of the liner into the barrel and of the core into the

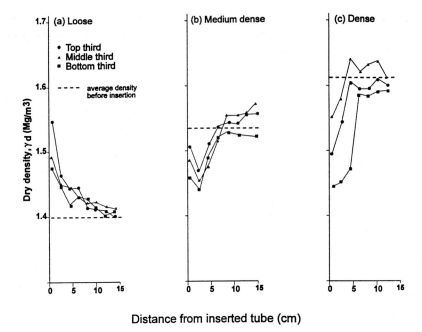

Figure 7. An example of density changes in sand due to tube penetration (data from Yoshimi et al, 1978)

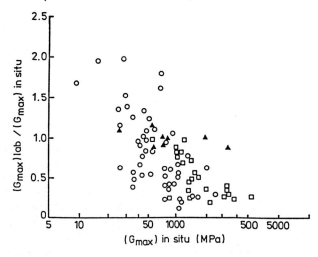

○ Reclaimed sand   - Thin wall tube samples   (Yasuda and Yamaguchi, 1985)
□ Pleistocene sand   - Triple tube sampling

▲ Sand + sandy gravel   In situ freezing + coring   (Tokimatsu and Oh-hara, 1990)

Figure 8. Assessing sample disturbance through comparisons of G_{max} from in situ and laboratory measurements in sand

liner is critical. *Hepton* describes a clearance between the core and liner to allow for potential swelling in some cohesive soils; in the Moderator's view, clearances should be an absolute minimum to avoid damage by swelling.

Table 1. Advantages of wireline core drilling systems over conventional
          drill rods and core barrels

•   Casing is not required for borehole stabilisation
      -      caving of borehole walls is largely eliminated
 	
 •   Required volume and velocity of flushing fluid is reduced
      -      potential for erosion of core and borehole wall is reduced
 	
 •   Removal of drill rods to recover core barrel is avoided
      -      time for swelling of core and of base of hole is reduced
 	
 •   Rigidity of drill string is increased
      - risk of rocking of core barrel is reduced

Drilling Muds

Drilling muds are selected and proportioned to ensure lubrication and cooling of the bit, stability of the borehole wall and base, and removal of cuttings. An important parameter concerns the viscosity of the mud which determines the up-hole velocity required to remove cuttings and, therefore, the risk of erosion to the core and borehole wall. The range of muds now available has increased and includes environmentally acceptable biodegradable polymer muds.

Sensitivity of core quality to drilling mud and local groundwater chemistry is illustrated in Figure 9. The two cores of London Clay were taken using the same triple tube wireline drilling system; the core in Figure 9(a) was obtained using a synthetic polymer mud, which had been used successfully in London Clay in East London; the core in Figure 9(b) was obtained after switching to a natural gum-based mud (Epps, pers. comm.). Figure 9(b) illustrates the quality of core that can now be obtained.

Handling and Sealing of Rotary Cores

Specifications developed by the Geotechnical Consulting Group have attempted to build on the improvements in sample quality achieved by the innovations in rotary coring by aiming to minimise further the risk of damage to the cores as a result of swelling. It is required that the cores of stiff or hard clay or weak rock are brought to a site hut immediately they are retrieved on the wireline and the core in its liner is removed from the core barrel. There, the sample liner is cut along diametrically opposite sides, without scoring the sample, and drilling fluids are removed from the core surface. The length of sample that is selected for laboratory testing is sealed using a procedure developed in Eastern Canada (La Rochelle et al, 1986) and successfully employed in the Bothkennar Characterisation Study (Hight et al, 1992b). This involves applying a coat of low melting-point wax to the core, before wrapping the core in Clingfilm which has been dipped in the same wax; the sequence of wax coating and Clingfilm wrap is repeated three times. For protection the sealed core is replaced in its plastic liner, the two halves of which are taped together.

Figure 9. Rotary core samples of London Clay: effect of drilling mud
(a) synthetic polymer mud   (b) natural gum-based mud

In combination, these developments in rotary coring techniques have led to increased core recovery and improved core quality. Full recovery can generally be achieved in uncemented sands and silts, in overconsolidated clays and in weak rocks. A summary of the formations which have been successfully investigated in the UK, together with example projects and references, is presented in Table 2.

It is, perhaps, surprising that with all these developments the recording of drilling parameters (rotation speed, rate of penetration, thrust on drillstring, torque on drillstring, pressure in drilling fluid) is still not routinely carried out. The advantages of so doing are described by Bécue et al (1993) for the case of coring in carbonate sands.

Rotary Coring in Stiff and Hard Clays
The advantages of rotary coring over tube sampling of stiff and hard clays relate to both the description of the ground (see Part II) and to the measurement of mechanical properties. Provided that destructuring by swelling can be avoided, then considerably less disturbance is imposed by rotary coring than by tube sampling. This was not evident in the past when comparisons were made in terms only of total stresses between undrained strengths measured in UU tests on rotary cored and tube samples; the rotary cored samples were found to have lower undrained strengths. A similar picture emerges even when the current state-of-the-art in rotary coring is applied in London Clay, refer Figure 10(b).

As explained above, tube sampling of stiff plastic clays results in an increase in mean effective stress and in shrinking of the soil's bounding surface (mechanical disturbance). It is obvious that during rotary coring, the sample has access to water in the drilling fluid, so

Table 2. UK Formations Successfully Investigated Using Large Diameter Triple Tube Wireline Drilling Systems (prepared with the assistance of Soil Mechanics Ltd and Foundation and Exploration Services).

Formation	Project	Reference
Norwich Crag	Sizewell 'C'	1, 2
Barton Beds	Pennington STW	-
London Clay Woolwich and Reading Beds	Sizewell 'C' Waterloo International Terminal CrossRail Jubilee Line Extension (JLE) Channel Tunnel Rail Link (CTRL)	1, 2 3, 4 5 - -
Thanet Sands	Third Blackwall Tunnel CrossRail JLE, CTRL	6 7 -
Chalk	JLE, CTRL Port of Felixstowe	- -
Folkestone Beds Sandgate Beds	Folkestone Wastewater Treatment	-
Hythe Beds Atherfield Beds Weald Clay	CTRL	-
Hastings Beds	Dungeness	-
Kimmeridge Clay	Wissington Silos	9
Oxford Clay	RAF Molesworth	9
Kellaway Beds	RAF Fylingdales	-
Lias Upper Lower	Hinkley 'C' Batheaston Fulbeck	1 11 11
Mercia Mudstone	Second Severn Crossing	8
Upper and Lower Coal Measures	Telford Blaydon Bridge, Newcastle	- -
Drift, etc. Compacted Clays and Mudstones Chalky Boulder Clay Glacial Till	Carsington Dam RAF Molesworth Blaydon Bridge, Newcastle	10 9 -

1 *Davis et al*; 2 *Hepton*; 3 Hight et al (1993); 4 *Pickles and Everton*; 5 Lehane et al (1995); 6 Eccles (pers. comm.); 7 *Johnson et al*; 8 *Maddison et al*; 9 Newman (pers. comm.); 10 Rocke (1993); 11 Eldred (pers. comm.).

that it can swell and its mean effective stress can reduce. Figure 10(a) compares the initial mean effective stress, p_i', in rotary cored samples and tube samples of London Clay taken at Waterloo International Terminal (Hight et al, 1993; *Pickles and Everton*). The rotary cored samples were taken either using triple tube wireline techniques, with polymer mud, or with conventional drill rods and core barrels, using foam flush. There is a clear hierarchy in the values of p_i', with the tube samples having the highest p_i' and the cores samples, taken with mud flush, having the lowest. The difference in p_i' between the samples contributes to the differences in strength evident in Figure 10(b).

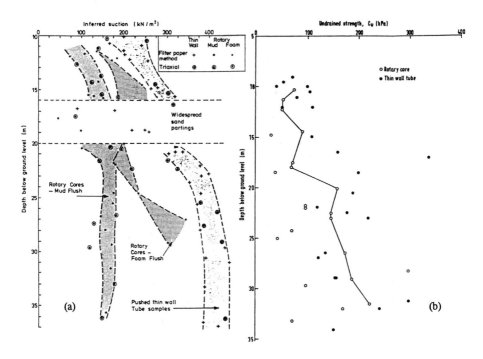

Figure 10. Comparison of rotary core and thin-wall tube sampling of London Clay
(a) initial effective stress   (b) undrained shear strength, c_u

When the data from the UU tests is examined in terms of effective stresses it is apparent that, although the initial mean effective stress is lower in the rotary cored samples, the effective stress paths in these samples rise to a higher failure envelope (bounding surface) than the tube samples. Clearly, mechanical disturbance is less in the rotary cored samples (see Figure 4).

The greater disturbance caused by tube sampling is even more evident when comparing effective stress data from rotary cored and driven tube samples of hard Upper Mottled Clay of the Woolwich and Reading Beds. Figure 11(a) shows effective stress paths from UU and CIU triaxial compression tests on rotary cored samples; these indicate a peak failure line approximately corresponding to c' = 200 kPa, ϕ' = 27° and a post-rupture failure line of c' = 0, ϕ' = 22°. Effective stress paths from similar tests on driven tube samples of the same deposit (Fig. 11(b)) climb only to the post-rupture failure line, indicating major disturbance.

This level of disturbance associated with tube sampling of stiff and hard clays has only become apparent with the availability of high quality rotary cored samples of these materials.

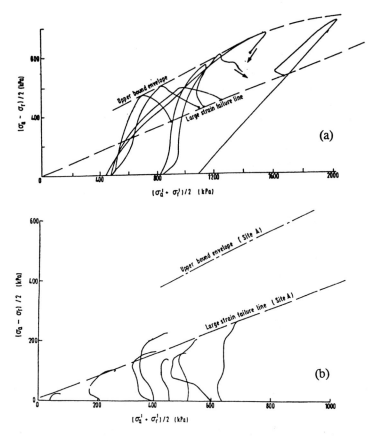

Figure 11. Comparison of effective stress failure envelopes for Upper Mottled Clay of Woolwich and Reading Beds
(a) rotary cored samples (Site A)   (b) driven thick-wall tube samples (Site E)
(from Hight and Jardine, 1993)

Rotary Coring in Sands

Without Freezing. Since the pioneering work on the Baghdad Metro (Scarrow and Gosling, 1986), the wireline rotary coring techniques described above have been used to sample a number of sand formations in the UK (see Table 2). Three investigations involving rotary coring of sands are referred to in this Conference: *Davis et al* and *Hepton* present information on sampling of the Norwich Crag in the 1988 and 1994 investigations at Sizewell

'C'; *Johnson et al* present properties measured on rotary cored samples of the Thanet Sands taken during the CrossRail investigation.

It is obviously essential to establish the level of disturbance in samples taken this way. Inspection of the cores at Sizewell revealed that sample quality was variable and possibly related to the degree of cementing. The same appears to be the case for the Thanet sands, based on the descriptions given by *Johnson et al* who suggest that "relatively undisturbed samples" could be obtained where the samples were weakly interlocked and of high relative density.

Unfortunately, comparisons of G_{max} values from measurements on laboratory samples and in situ are not available for either site at present. However, the relatively soft stiffness characteristics measured in triaxial compression tests on isotropically consolidated undisturbed samples of the Thanet Sand point to loosening of this initially dense sand.

Japanese experience with rotary coring in unfrozen sands points to important disturbance. Yoshimi et al (1989) provide the following comparison of relative densities for samples taken by rotary coring, both with and without in situ freezing - coring after in situ freezing is thought to provide a reliable measure of in situ density.

Relative Density (%)	
In Situ	After Sampling
55	78
78	83
87	72

In situ and laboratory measurements of G_{max} on rotary cored samples are compared in Figure 8 and illustrate the damage that can occur. In recent sands, such a change can be understood on the basis of the test data shown in Figure 1(c) which illustrates the effect of an applied twist on G_{max}.

With Freezing. It would appear that truly undisturbed samples of sand can only be obtained if the in situ sand structure is fixed in some way before sampling. The most common method of doing this is by freezing the ground in situ before carving out large diameter samples by rotary coring (Yoshimi et al, 1978). Although no papers submitted to this Conference describe the technique of sampling frozen sand, and the Moderator is not aware of its use in the UK, the method is gaining in popularity around the world, continuing to be used in Japan, as summarised by Yoshimi et al (1994), and recently introduced to Canada (Konrad, 1990; Davila et al, 1992) and to Italy (O'Neill, pers. comm.).

The success in obtaining undisturbed samples in this way is illustrated in Figure 8, which compares in situ and laboratory measurements of G_{max} on samples cored after freezing and shows the ratio to be unity for a range of in-situ G_{max}.

Rotary Coring in Weak Rocks
Three papers provide details on rotary coring in weak rocks. *Maddison et al* describe the investigation for the Second Severn Crossing in which wireline rotary coring was used to obtain 300mm and 100mm diameter cores of Triassic mudrocks. The Authors point out that there was "little experience with large diameter rotary core drilling in the UK and Europe"

prior to the investigation. Cores of 300mm diameter were considered to be more representative than 60mm or 100mm diameter cores and could be used in large laboratory direct shear testing; they were regarded as the maximum diameter that could be practically handled. *Maddison et al* point out that the 300mm diameter cores were of better quality than smaller cores. However, it is noted that an inner plastic liner was not used and this affected core recovery and core handling practice.

An interesting example of the improvement in core quality that results when using larger diameter coring equipment and a triple tube, rather than double tube, configuration is presented in Figure 12. Fifity-four mm core taken from a double tube barrel is compared with an annular overcore taken at the same location with a triple tube barrel producing 112mm core. Several of the discontinuities and erosion evident in the 54mm core do not exist in the overcored material. The example is from the Trias Mudstones at the Second Severn Crossing; the 54mm core was taken when forming a pocket for a high pressure dilatometer test.

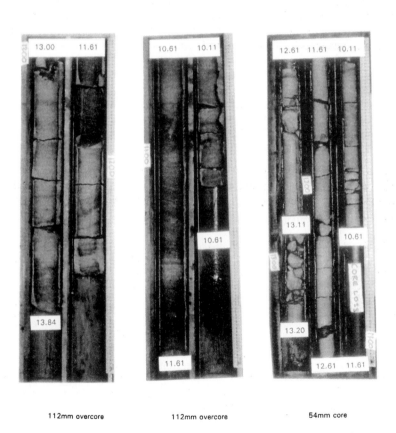

112mm overcore   112mm overcore   54mm core

Figure 12. Comparison of 54mm core (double tube) and 112mm overcore (triple tube) of Mercia Mudstone (photographs by courtesy of Laing - GTM)

In their paper to this Conference, *Tatsuoka et al* describe evidence for disturbance in samples of soft rock, taken using triple tube rotary coring techniques. They compare data on small-strain stiffness, measured using transducers mounted over a central gauge length of specimens, prepared from samples taken by rotary coring (RCT), direct coring (DC) and by carving as blocks. Direct coring refers to samples taken by coring into the rock face from galleries or adits, involving, therefore, only a short length through which the torque is transmitted and ensuring that core is taken without imparting any rocking motion. A key comparison is presented by *Tatsuoka et al* in their Figure 5. The Authors ascribe the differences in small-strain stiffness and shape of stress-strain curves to the disturbance in the rotary cored specimens resulting from rocking and torque imparted to the inner tube. They point out that these differences have only become apparent because of the use of local displacement measurements, being swamped by bedding errors, etc., in external measurements. These findings are consistent with the views of Cook (1992), quoted earlier in this Report, where misalignment of fissures may be occurring as a result of the twist applied to the core, as well as due to stress relief. Presumably the risk of rocking motion is reduced with wireline drilling because of the increased rigidity of the drillstring (see Table 1).

Tatsuoka et al also report close agreement between G_{max} measured in situ in seismic surveys and G_{max} found in triaxial compression tests on high quality DC and block samples. For this finding to confirm a lack of disturbance requires the samples to be representative in terms of fabric and stress state. The finding is in direct contrast to that of *Gordon et al* who report G_{max} values measured on cored samples of chalk that are well in excess of in situ values. However, *Gordon et al* tested specimens of 38mm diameter by 76mm high, which, presumably, did not contain a representative set of discontinuities, while *Tatsuoka et al* tested 70mm diameter by 170mm high specimens, with a correspondingly longer gauge length, which, presumably, were representative. The depth of the samples and the ambient stress level are also of relevance in such a comparison, since the significance of discontinuities reduces as they close under higher confining stress (Cook, 1992).

PART II: DESCRIPTION OF THE GROUND

The importance of an accurate and complete description of the ground cannot be overstated. So often engineering problems arise because of details of the soils or rock's fabric (e.g. a surface of weakness or a seam of high permeability) which were not noted or the importance of which were not appreciated. Several techniques may contribute to building up a picture of the ground at both a macro- and micro-scale; some are described in papers to this Conference: *Norbury and Gosling, Paul et al, Maddison et al, McMillan et al* and *Webber and Gowans*.

ADVANCES IN SOIL OR ROCK DESCRIPTION AND CLASSIFICATION

Information on the ground's state is communicated to the engineer via soil or rock descriptions and classifications. *Norbury and Gosling* discuss the principles involved in description and classification, their history and the problems with BS5930: 1981; they also provide a valuable background to the proposed changes to Section 8 of that document.

Norbury and Gosling make a number of important distinctions in their paper:

- between *classification* and *description*: *classification* is a factual description based on tests on remoulded soil - there is no reference to the all-important fabric; *description* provides information on the material and mass characteristics - it is of immediate use and does not depend on test results. They advocate description first with the use of classification tests for back-up and calibration

- between description for *geology* and *engineering*: for engineering purposes description requires the reporting of information that affects strength, permeability, deformability and response to environmental influences

- between *factual* and *interpreted* information: they suggest that interpreted information should be kept to a useful minimum. In the Moderator's view, factual and interpreted information can easily be separated on a borehole log and we should make maximum use of the expertise of the engineer/geologist making the descriptions. As a minimum, we should expect guidance on which fabric features have been induced by boring and sampling; why core might have been lost; what depositional and post-depositional processes could have given rise to features in the core.

To eliminate the perceived shortcomings to the present code (BS5930: 1981), *Norbury and Gosling* describe how the proposed alterations will re-introduce descriptions based on likely engineering behaviour and will introduce guidance on the description of rock discontinuities. They describe some common problems in soil description and pose a number of questions which should promote discussion.

The hallmarks of good descriptions are concinity and consistency. Even with these traits, the Moderator has to admit to a short concentration span when it comes to reading soil and rock descriptions and prefers to rely on personal inspection and photographic records. Sadly the quality of core photographs available in many site investigation reports does not match that shown in Figure 13 (from Ellison, 1991). The photograph is presented, not only to illustrate the high quality of core photograph that one should expect, but to serve as a reminder of one of the major advantages of rotary core ("total volume") sampling: namely, the provision of a continuous profile which allows the lithology and fabric of the deposits to be examined in detail. Because of the advantages of wireline drilling (see Table 1), cave-ins are less likely and damage to the core is reduced, leading to better quality samples and easier interpretation of ground conditions.

BENEFITS OF MACRO- AND MICRO-FABRIC STUDIES

Detailed studies of macro- and micro-fabric rarely find a place in site investigations despite advances in techniques and in the quantification of data. Space does not permit a review of the engineering benefits but reference may be made, for example, to the following:

- Paul and Barras (1994), whose studies of the Bothkennar clay have identified the different forms of cementing (polysaccharide cements, welded junctions, clay bridges) and have helped to explain the unusual combination of high friction angle and high compressibility in the material.

Figure 13.   Example of rotary core photograph of Woolwich and Reading Beds
(Ellison, 1991)

- *Paul et al*, who illustrate the use of X-ray densimetry in the description of sediments
  and evaluation of geological processes to which they have been subjected.

X-ray photographs of samples and cores, and other scanning techniques, can demonstrate the
presence of laminae, discontinuities, inclusions, etc., before extruding samples, or selecting
specimens for testing.   It is surprising that the techniques are not more widely used,
particularly in evaluating the effects of coring in weak rocks.

An appreciation of fabric is essential to interpreting the response to sampling and forecasting
in situ behaviour.

ADVANCES IN ESTABLISHING DISCONTINUITY DATA IN ROCKS

The rôle of discontinuities and the need to establish their orientation and spacing is obviously
critical in hard rocks, in which, for example, they dominate behaviour in excavations.   Two
papers are presented to this Conference, in which techniques for determining discontinuity
data are discussed.

Webber and Gowans present a comparison of available techniques for measuring discontinuity
data in rocks: impression packer, CCTV, local rock exposures and inspection of cores, the
orientation of which have been established.   They describe in detail a core orientator, which
is fixed to the head of the core barrel and relies on a pendulum to indicate to $\pm 5°$ the lowest
position of the core in inclined boreholes, having a minimum inclination of $5°$ from the
vertical.   The system depends on maintaining a fixed rotational relationship between the inner
tube of the core-barrel and the orientator.   The orientation is found by correlating the lowest
point of the cored hole with the orientation and dip of the drilling mast.

Webber and Gowans conclude that each technique provides some information. Problems caused by suspended solids were encountered with CCTV and the information on discontinuities was not complete. Impression packers did not identify infilled joints. The oriented core gave a more complete correlation with discontinuity data based on the local exposure, presumably because discontinuities tended to open with full stress relief of the core.

This experience reinforces the need for the integrated use of different techniques to build up a more complete picture, as in all aspects of site investigation.

McMillan, Blair and Nettleton describe the use of down-hole CCTV to obtain quantitative discontinuity data, to supplement rock outcrop mapping or, as a substitute when outcrops do not occur. They describe the equipment, its operation and their interpretation procedure. By recording the camera's depth, rotation and orientation as a discontinuity trace in the borehole wall is followed, the apparent dip and azimuth of the discontinuity can be established. This is then corrected to true dip relative to horizontal and true azimuth relative to magnetic north.

A field evaluation of the techniques by *McMillan et al* emphasised the need for a stable borehole with clean walls, not obscured by debris or flush fluid, and clean borehole fluid. A capability for inclined boreholes is required to allow the main discontinuity sets to be intersected at as large an angle as possible.

As with any down-hole survey technique, there is no information on the likely persistence of the discontinuities and it is difficult to evaluate their importance. However, it is almost certain that the discontinuities in the borehole wall are natural rather than having been induced by drilling, which remains the problem in using borehole core for evaluating discontinuities (refer Figure 12).

TOPICS FOR DISCUSSION

Based on this review, the following topics are suggested for discussion:

- Sample disturbance
  - its effect on sands and weak rocks
  - its assessment using v_s/G_{max}
  - its reduction

- Dynamic sampling
  - methods and experience

- Rotary coring
  - when can it be justified in sands and stiff clays?
  - when is 300mm diameter coring in weak rocks justified?
  - when is freezing of sands justified?
  - why do we not record drilling parameters?

- Description of soils and rocks
  - how much interpretation should we expect?
  - benefits of microfabric studies

- Discontinuity data
  - methods of establishing

- Are we matching the sophistication of our testing to the quality of samples?

ACKNOWLEDGEMENTS
The Author is pleased to acknowledge the assistance given during the preparation of this Report by Peter Gee and Chris Eccles of Soil Mechanics Ltd, Roger Epps of Foundation and Exploration Services and Tom Power of Laing Technology Group.

REFERENCES

ACAR, Y. B. and EL-TAHIR, E-T.A. (1986). Low strain dynamic properties of artificially cemented sand. ASCE, JGE, 112, 11, 1001-1015.

BÉCUE, J. P., PUECH, A. and LHUILLIER, B. (1993). Recording of drilling parameters: a complementary tool for improving geotechnical investigation in carbonate formations. Proc. Int. Symp. on Geotechnical Engineering of Hard Soils-Soft Rocks, Athens, 283-289.

BURLAND, J. B. and KALRA, J. C. (1986). Queen Elizabeth II Conference Centre: geotechnical aspects. Proc. Instn Civ. Engrs, Part 1, 80, 1479-1503.

COOK, N. G. W. (1992). Natural joints in rock: mechanical, hydraulic and seismic behaviour and properties under normal stress. Jaeger Memorial Dedication Lecture. Int. J. Rock Mech. Min. Sci. & Geomech. Abstr., 29, 3, 198-223.

DAVILA, R. S., SEGO, D. C. and ROBERTSON, P. K. (1992). Undisturbed sampling of sandy soils by freezing. Proc. 45th Can. Geo. Conf., Toronto, Paper 13A-1-13A-10.

ELLISON, R. A. (1991). Lithostratigraphy of the Woolwich and Reading Beds along the proposed Jubilee Line Extension, south-east London. British Geological Survey Technical Report WA/91/5C.

HANZAWA, H., MATSUDA, E. and HIROSE, M. (1980). Evaluation of sample quality of sandy soil obtained by the modified Bishop sampler. Soils and Foundations, 20, 3, 17-31.

HARDIN, B. O. and RICHART, F. E. (1963). Elastic wave velocities in granular soils. ASCE, JSMFD, 89, SM1, 33-65.

HARRISON, I. R. (1991). A pushed thin wall tube sampling system for stiff clays. Ground Engineering, April, 30-34.

HEPTON, P. (1988). Shear wave velocity measurements during penetration testing. Proc. Conf. 'Penetration Testing in the UK', Telford, London, 275-278.

HIGHT, D. W. (1993). A review of sampling effects in clays and sands. Proc. SUT Conf: Offshore Site Investigation and Foundation Behaviour, 115-146.

HIGHT, D. W. and JARDINE, R. J. (1993). Small-strain stiffness and strength characteristics of hard London tertiary clays. Proc. Int. Conf. on Geotechnical Engineering of Hard Soils-Soft Rocks, Athens, 533-552.

HIGHT, D. W., BOND, A. J. and LEGGE, J. D. (1992a). Characterisation of the Bothkennar Clay: an overview. Géotechnique, 42, 2, 303-347.

HIGHT, D. W., BOESE, R., BUTCHER, A. P., CLAYTON, C. R. I. and SMITH, P. R. (1992b). Disturbance of the Bothkennar Clay prior to laboratory testing. Géotechnique, 42, 2, 199-217.

HIGHT, D. W., PICKLES, A. R., DE MOOR, E. K., HIGGINS, K. G., JARDINE, R. J., POTTS, D. M. and NYIRENDA, Z. M. (1993). Predicted and measured tunnel distortions associated with construction of Waterloo International Terminal. Proc. Wroth Memorial Symp., Oxford, 317-338.

HOLT, R. M., KENTER, C. J., UNANDER, T. E. and SANTARELLI, F. J. (1993). Unloading effects on mechanical properties of a very weak artificial sandstone: applications to coring. Proc. Int. Symp. on Geotechnical Engineering of Hard Soils-Soft Rocks, Athens, 1609-1614.

KONRAD, J-M. (1990). Sampling of saturated and unsaturated sands by freezing. ASTM Geotechnical Testing Journal, 13, 2, 88-96.

LACASSE, S., BERRE, T. and LEFEBVRE, G. (1985). Block sampling of sensitive clays. Proc. 11th ICSMFE, San Francisco, 2, 887-892.

LA ROCHELLE, P., SARRAILH, J., TAVENAS, F., ROY, M. and LEROUEIL, S. (1981). Causes of sampling disturbance and design of a new sampler for sensitive soils. Can. Geotech. J., 18, 1, 52-66.

LA ROCHELLE, P., LEROUEIL, S. and TAVENAS, F. (1986). A technique for long-term storage of clay samples. Can. Geotech. J., 23, 602-605.

LEFEBVRE, G. and POULIN, C. (1979). A new method of sampling in sensitive clay. Can. Geotech. J., 16, 1, 226-233.

LEFEBVRE, G., LEBOEUF, D., RAHHAL, M. E., LACROIX, A., WARDE, J. and STOKOE, K. H. (1994). Laboratory and field determinations of small-strain shear modulus for a structured Champlain clay. Can. Geotech. J., 31, 61-70.

LEHANE, B. M., PAUL, T. S., CHAPMAN, T. J. P. and JOHNSON, J. G. A. (1995). The apparent variability of the Woolwich and Reading Beds. To be published in Proc. XI ECSMFE, Copenhagen.

MAGNAN, J-P, KHEMISSA, M. and JOSSEAUME, H. (1994). Influence du prelevement sur le comportement des argiles. Proc. XIII ICSMFE, New Delhi, 1, 317-320.

MARTIN, J. H. and SUCKLING, A. C. (1984). Geotechnical investigations for Mount Pleasant Airfield, Falkland Islands, using the Flow-Through Sampler. Ground Engineering, November.

NISHI, K., ISHIGURO, T. and KUDO, K. (1989). Dynamic properties of weathered sedimentary soft rocks. Soils and Foundations, 29, 3, 67-82.

OKA, F., HASHIMOTO, T., NAGAYA, J. AMEMIYA, M., OHYAMA, H., FURUKAWA, S. and ANDO, Y. (1992). Application of Laval type large diameter sampler to soft clay. Proc. Symp. on Soil Sampling, JSSMFE, 35-38 (in Japanese).

PAUL, M. A. and BARRAS, B. F. (1994). Soil microstructure and cementation in the Bothkennar clay. Bothkennar Research Report, BRR-12/94.

ROCKE, G. (1993). Investigation of the failure of Carsington Dam. Géotechnique, 43, 1, 175-180.

SAMUELS, S. G. (1975). Some properties of the Gault Clay from the Ely-Ouse Essex water tunnel. Géotechnique, 25, 2, 239-264.

SAXENA, S. K., AVRAMIDIS, A. and REDDY, K. R. (1988). Dynamic moduli and damping ratios for cemented sands at low strains. Can. Geo. J., 25, 2, 353-368.

SCARROW, J. A. and GOSLING, R. C. (1986). An example of rotary core drilling in soils. Geol. Soc., Eng. Geol. Special Publn No. 2, Site Investigation Practice: Assessing BS5930, 357-363.

TAVENAS, F. and LEROUEIL, S. (1987). Laboratory and in situ stress-strain-time behaviour of soft clays: a state-of-the-art. Int. Symp. on Geotech. Eng. of Soft Soils, Mexico City, Vol. 2, 1-46.

THOMANN, T. G. AND HRYCIW, R. D. (1992). Stiffness and strength changes in cohesionless soils due to disturbance. Can. Geotech. J., 29, 853-861.

TOKIMATSU, K. and OHARA, J. (1990). Soil sampling by freezing. Tsuchi-to-Kiso, JSSMFE, 38, 11, 61-68.

VAUGHAN, P. R., CHANDLER, R. J., APTED, J. P., MAGUIRE, W. M. and SANDRONI, S. S. (1993). Sampling disturbance with particular reference to its effect on stiff clays. Proc. Wroth Memorial Symp., Oxford, 685-708.

WARD, W. H., SAMUELS, S. G. and BUTLER, M. E. (1959). Further studies of the properties of London Clay. Géotechnique, 9, 2, 33-58.

YASUDA, S. AND YAMAGUCHI, I. (1985). Dynamic shear modulus obtained in the laboratory and in situ. Proc. Symp. on Evaluation of Deformation and Strength of Sandy Grounds. JSSMFE, 115-118. (In Japanese.)

YOSHIMI, Y., HATANAKA, M. and OH-OKA, H. (1978). Undisturbed sampling of saturated sands by freezing. Soils and Foundations, 11, 3, 59-73.

YOSHIMI, Y., TOKIMATSU, K. and HOSAKA, Y. (1989). Evaluation of liquefaction resistance of clean sands based on high-quality undisturbed samples. Soils and Foundations, 29, 1, 93-104.

YOSHIMI, Y., TOKIMATSU, K. and OHARA, J. (1994). In situ liquefaction resistance of clean sands over a wide density range. Géotechnique, 44, 3, 479-494.

Discussion on Session 3: drilling, boring, sampling and description

The organizers and publishers very much regret that a verified Discussion Review could not be produced for this Session. Although this item presents the essence of the proceedings some contributions may not have been included.

A speaker from the floor described the use of triple tube coring technology in the UK developed in the late 1980s, based on overseas experience and the specification in 1986 by Sir Alexander Gibb and Partners of large diameter (100 mm core) triple tube core barrels at Fulbeck for UK Nirex Ltd. The main advantages of the system are to maximize core retrieval and enhance core presentation by minimizing disturbance.

Experience indicates that virtually all types of soil can be cored successfully using this system, including glacial tills, man-made fills and sands and gravels, as well as stiff cohesive clays and weak rocks. The factors which have the greatest impact on the overall quality of samples are the qualifications and experience of the driller, selection of the drilling fluid, rotating speeds and core barrel set-up and maintenance. Polymer is the preferred flush fluid, although, in certain instances, foam, water and air have all been used successfully. While there are no firm rules for rotational speeds, experience suggests that the highest possible rpm should be used commensurate with stability of the string and flush velocities.

Core barrel set up and maintenance is usually given insufficient attention by contractors and consultants. The dimensional difference between the diameter of the throat of the bit and the internal diameter of the liner is one of the key factors influencing the quality of the undisturbed samples. For general coring in competent strata, a 2 mm thick liner and 2 -3 mm annular clearance is advisable. The gap between the bottom of the core catcher and the back of the bit should also be minimized to direct the flush to the face of the bit and so as not to erode the core as it enters the barrel. In 1986 much effort was expended to optimize the system. In the intervening years this seems to have been overlooked and examples of annular clearances of 10 mm and above are now too common.

Another contributor described the use of 300 mm diameter cored holes on the Second Severn Crossing project. Large diameter rotary coring was selected to give better quality core, especially in weaker strata such as mudrocks, and to enable large samples to be obtained for laboratory testing. Coring at 300 mm diameter was chosen as the largest diameter core which was practical to handle. No core liner was used and the core was extruded into a steel trough. In retrospect, use of core liner and end caps might have improved ease of handling and transportation of the core to the laboratory intact. Laboratory test data from 300 mm and 100 mm diameter cores were compared, showing little difference in the data. However, the use of 300 mm diameter cores allowed very much better core recovery in the weaker rocks, which is important since otherwise it may not be clear whether low results from laboratory tests reflect weak strata or poor quality recovery.

One contribution provided further information on the Second Severn Crossing project, focusing on the implications of site investigation data for construction. The shear strength of the mudstone was a crucial parameter for a number of construction elements (including the size of the caisson used), and the value to be adopted would therefore have a critical influence on costs. Coring at 300 mm diameter was introduced with three main aims:

(a) to provide a large mass sample of rock for inspection
(b) to provide large discontinuity samples for testing in 300 mm shear boxes
(c) to provide a large diameter sample of the rock below locations of the borehole plate bearing tests.

The first aim reflected the expectation that it would be very difficult to undertake trial pitting in the river bed at low tide, because of time restrictions and the isolated, distant location of the exposed rock platforms. In fact, this concern proved unfounded as three trial pits were eventually excavated.

The 300 mm core provided only a few discontinuities for testing and the majority of the shear strength data eventually used was obtained from tests on discontinuity samples on 100 mm core. One 300 mm core was affected over virtually its entire length by a vertical joint, a situation which was thought unlikely to occur with the smaller diameter core in this rock.

Problems initially arose with handling the core, including extrusion from the barrel. However, once these were resolved, the large diameter core provided very good information on fracture spacing and the condition of the joints, to over 15 m depth.

A new sample for use in stiff clays and weak rocks with the objective of obtaining high quality samples was described by one speaker. This sampler has a number of particular features:

(a) the inner tube does not rotate when the outer tube rotates, enabling the inner tube to retain the sample
(b) overburden pressure is maintained
(c) the inner sleeve moves with the sample to reduce friction.

The sampler, called the Planet Sampler, can be used in a wider range of soil types and, because it creates less disturbance, provides more reliable results than earlier samplers. General details are shown in Fig 1.

The need to comply with the Construction (Design and Management) Regulations 1994 and to balance cost savings against health and safety risks was highlighted by a contributor. In the light of this, the question of specifying site investigation techniques was raised, for example for the use of 300 mm diameter core, or for window sampling, where the short and long term risk to the health of the operator of the jack hammer needs to be considered. The 300 mm diameter core was referred to as the maximum diameter which could 'practically' be handled. The observed difficulties in handling encountered by the drillers suggest that perhaps the Manual Handling Regulations would prevent this size being used in common practice.

Parts of Planet Sampler

Operation diagram

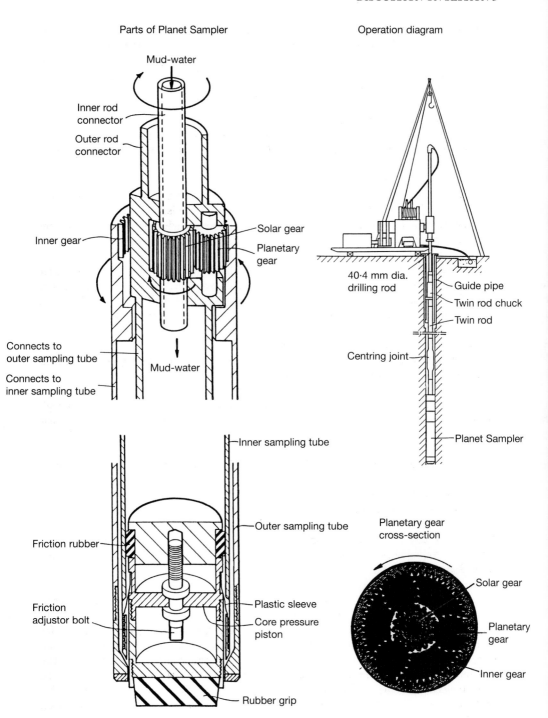

Mud-water

Inner rod connector

Outer rod connector

Inner gear

Solar gear

Planetary gear

Connects to outer sampling tube

Connects to inner sampling tube

Mud-water

40·4 mm dia. drilling rod

Guide pipe

Twin rod chuck

Twin rod

Centring joint

Planet Sampler

Inner sampling tube

Friction rubber

Outer sampling tube

Friction adjustor bolt

Plastic sleeve

Core pressure piston

Rubber grip

Planetary gear cross-section

Solar gear

Planetary gear

Inner gear

A speaker replied that experience from the Second Severn Crossing project indicated difficulties in handling 300 mm core. However, for future projects consideration could be given to using mechanical lifting gear.

A contributor described a case study involving in situ ground freezing of a loose sandy deposit, to provide a high quality undisturbed sample and to establish whether any ageing effects occur with this technique. In situ and laboratory tests were performed on intact and reconstituted samples for comparative purposes, with the intact samples having been obtained both using in situ freezing and from tube samplers. The results suggest that in situ freezing is a promising technique for obtaining undisturbed samples for laboratory tests. Ageing effects on small-strain stiffness were not found to be significant in a loose sandy deposit of 20 years of age.

The CANLEX project was described, which aims to characterize the behaviour of loose sands (hydraulically placed natural sands and naturally deposited sands). Also to assess liquefaction potential through modelling and conducting a major liquefaction event to evaluate characterization and modelling. The project also includes the development of cost-effective techniques for obtaining undisturbed samples of sands by ground freezing, and subsequent laboratory testing. This is a three year collaborative project with input from industry, engineering consultants and the University of Alberta.

The results to date suggest that the ground can be frozen selectively at target depths (e.g. at 7 m depth). Temperature sensors are used to determine when the freezing achieves the set objectives, and then the sample is cored. An example of ground freezing at a target depth of 27–37 m was described, and a comparison of the effectiveness of core sampling using ground freezing against information obtained from slimline geophysical logging, SPTs and several other methods was made. Core samples were initially obtained using a 100 mm diameter core barrel and a dry coring technique, although this has recently been upgraded to 200 mm diameter which allows horizontally oriented samples to be taken from the core. The project also includes a literature study on the effectiveness of freezing using small strain tests including shear wave velocity tests.

A speaker expressed interest in determining whether there is any UK experience in ground freezing techniques for obtaining undisturbed samples as he believed that the technique has been used in Italy.

One speaker suggested that when the technique has been used in the UK, it has been found to be very expensive and hence unsuitable for routine testing. The researchers were requested to advise on the costs involved.

A speaker replied that an estimated cost is £25,000 which, he said, is very expensive for one sample, but can be cost-effective for many samples. It was thought unlikely that the technique will be used routinely, but it could be very valuable for major, million-dollar decision projects where liquefaction potential is significant. The CANLEX project is also evaluating conventional sampling techniques.

One speaker asked whether any difficulty arises in reconstituting sands at the void ratios involved.

The researchers replied that this was not a problem that they had encountered.

The effect in terms of volume expansion was queried, and the researchers replied that it is important to have a slow open freezing system.

A contributor raised a point with Mr David Norbury on the comments made in his paper on geological descriptions. Having spent the last three years restraining the descriptive art aspect of soil and rock logging and realising that BS5930 and ISRM standards are not perfect, he was instrumental in the development of a project specific logging manual, which is used by all parties involved with the logging of soil and rock on the Nirex Geological Investigations project at Sellafield. This necessitates the use of standard terminology, to allow descriptions to be correlated and meet the overall QA requirements of the project. The requirement for basic standardized terminology does not preclude the addition of further information. As the contributor aptly stated, we are all aware that two geologists and one core sample can result in three descriptions! He stressed the fact that there will always be the need for some standardization in order to allow effective communication between the interested parties.

A speaker from the floor emphasized the importance of communication and made a plea for contractors to be provided with Interpretative Reports.

The value of macro- and microfabric studies in soil description using an example from the EPSRC Bothkennar test site was described by one contributor.

Detailed studies of both macro- and microfabric can greatly benefit site investigations. When integrated with other geological and geomorphological data, they provide information from which soil behaviour may be better understood. Generalized, engineering facies (including weathering facies) can be erected and geological models can be developed to explain both geotechnical profiles and the spatial distribution of facies. These models may in turn be related to more complex systems based, for example, on geostatistical concepts or on GIS classifications.

The application of macrofabric-based facies to the classification and stratigraphic interpretation of the clay soil at the Bothkennar test site has been described in detail elsewhere (Paul, Peacock & Wood, 1992; Paul, Peacock & Barras, 1995). This scheme relied on the identification of fabric features at a scale of around 1–10 mm using visual observation, conventional and x-ray photography and x-ray densimetry (cf. Paul, Sills, Talbot & Barras, Session 6, Figs 5–7). Radiocarbon dating has established that deposition occurred rapidly between 5000 and 3000 BP, during which period the local water depth decreased from about 20 m to inter-tidal, and that the succession of facies records changes in geological processes that occurred during this fall. These changes are reflected in the profiles of several geotechnical parameters, notably grain size, water content and void index (Paul, Barras & Peacock, 1994a) and the facies succession allows a local engineering stratigraphy to be constructed for the site as a whole (Paul, Barras & Peacock, 1994b). Thus simple observations of soil fabric, in conjunction with other data, can be used to construct models of considerable complexity and predictive power.

Fig. 1. Hierarchy of fabric and facies scales

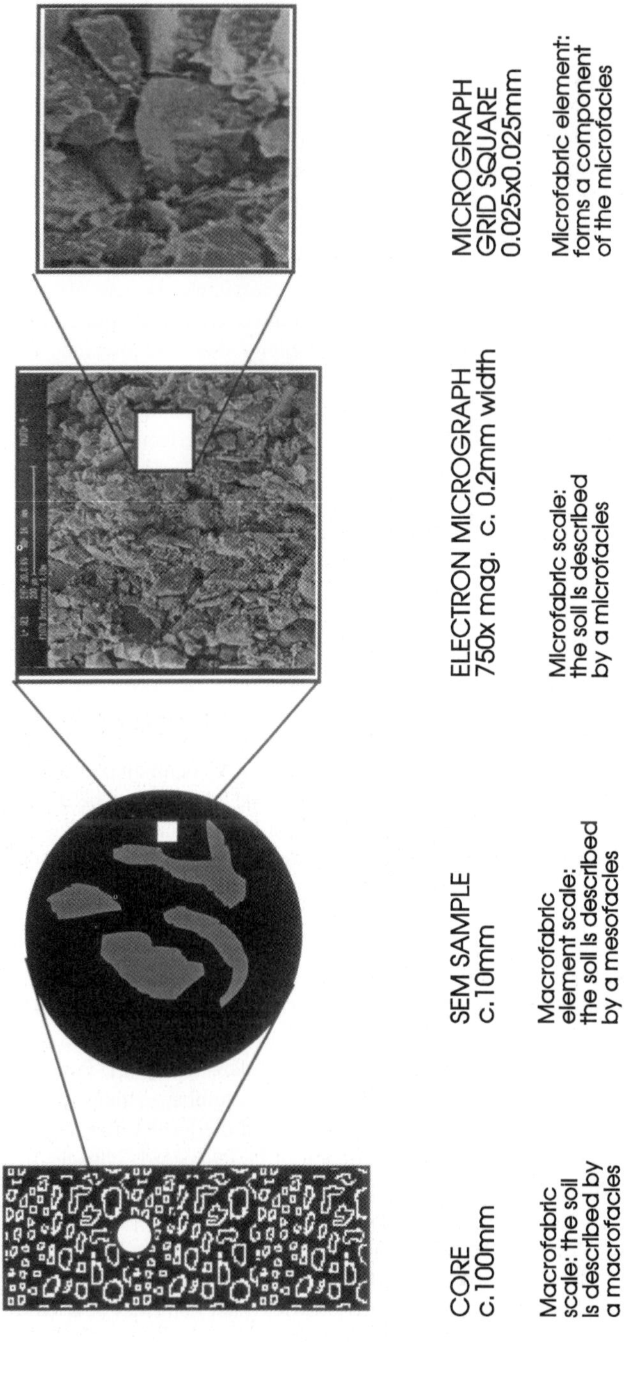

MICROGRAPH
GRID SQUARE
0.025x0.025mm

Microfabric element:
forms a component
of the microfacies

ELECTRON MICROGRAPH
750x mag. c. 0.2mm width

Microfabric scale:
the soil is described
by a microfacies

SEM SAMPLE
c.10mm

Macrofabric
element scale:
the soil is described
by a mesofacies

CORE
c.100mm

Macrofabric
scale: the soil
is described by
a macrofacies

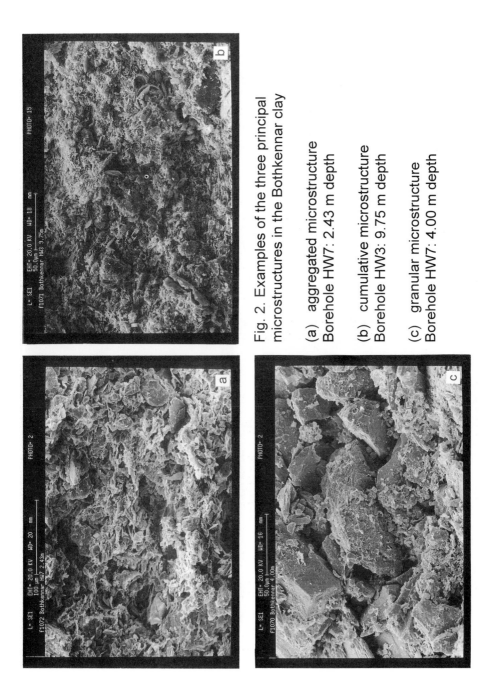

Fig. 2. Examples of the three principal microstructures in the Bothkennar clay

(a) aggregated microstructure
Borehole HW7: 2.43 m depth

(b) cumulative microstructure
Borehole HW3: 9.75 m depth

(c) granular microstructure
Borehole HW7: 4.00 m depth

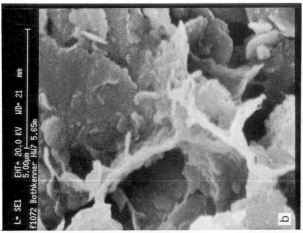

Fig. 3. Examples of cementation
(a) clay bridge between silt grains
Borehole HW7: 5.79 m
(b) welded junction in clay framework
Borehole HW7: 5.65 m
(c) biogenic structure with mucal
cement
Borehole HW3: 7.21 m

Fig. 4. Backscatter images showing
the principal microstructures

AGGREGATED

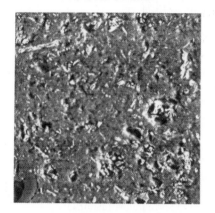

HW3 7.21m MOTTLED FACIES   Type III
1cm below upper boundary
Magnification x500

GRANULAR

HW7 2.15m BEDDED FACIES
Contact non-erosive, transition
Magnification x500

CUMULATE

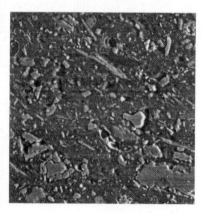

HW3 9.78m MOTTLED FACIES
Type II - Sparse
Magnification x500

Figure 5

Backscatter Images showing
the principal microstructures

In parallel with these macrofabric investigations, a detailed study of soil microfabric under the electron microscope has been carried out (Paul & Barras, 1994) using both conventional secondary electron imaging and backscatter imaging. Fig. 1 shows how these microfabric observations are related in hierarchical terms to the larger scale macrofacies. In the Bothkennar clay several distinctive microfabric elements have been identified which can be classified statistically into just three principal types of microfacies (Fig. 2). The work has also shown the presence of significant inter-particle cements (Fig. 3), principally biogenic mucus and bridges of clay particles. The backscatter images (Fig. 4) have allowed quantitative measurements of the soil fabric to be made (based on image processing techniques, cf. Tovey & Hounslow, 1995). These have shown that at several scales there are statistical differences between the facies in the anisotropy of the microfabric. Further work is in progress to quantify the total amount and size distribution of both void space and mineral types (using x-ray mapping). At this time this appears to be a most promising development.

The microfacies types and occurrence of cement can both be related to specific geological processes, which are probably of general importance in all alluvial clay soils. These include the formation of biogenic pellets in the water column, which subsequently accumulate to form the clay sediment itself, the reworking of the sediment by bottom-dwelling organisms during breaks in sedimentation, the stabilization and hardening of the sediment surface also during these breaks and the local winnowing of fines to leave thin silt laminae.

The relative proportion of the three microfacies types and the extent of cementation in any given sample varies with the macrofacies from which the sample was drawn and so soil behaviour at the microscopic sale is hence intimately related to its macrofacies classification. This detailed study of the microstructure has also shed light on the reasons for several curious geotechnical properties of the Bothkennar clay, not normally found in combination (cf. Hight, Moderator's report, Session 3). These include the high ($37°$) friction angle, which is probably the result of the angularity of the silt particles; the high yield stress ratio (around 1.6, rising to 2.5+ in the inter-tidal deposits) which is due in part to cementing by mucal polysaccharides; the high compressibility, which is the result of the open clay boxwork; and the loss of plasticity upon drying which is again the result of organic geochemicals attached to individual clay particles.

In summary, the use of detailed fabric descriptions, combined with other geological data, should not be overlooked in any site investigation. The benefits may include better understanding and correlation of in situ tests, better understanding of mechanical behaviour and unusual tests results and the basis on which to erect soundly based, quantitative ground models in which lateral and vertical variation can be properly included and understood and from which numerical predictions of increased reliability can be made.

References

Paul M. A., Peacock J. D and Wood B. F., 1992. The engineering geology of the Carse Clay at the National Soft Clay Research Site. Bothkennar. *Géotechnique*, **42**, 183–198.

Paul M. A. and Barras B. F., 1994. *Microstructure and cementation in the Bothkennar clay.* Bothkennar Research Report BRR 12/94, Heriot-Watt University, Edinburgh.

Paul M. A., Barras B. F. and Peacock J. D., 1994a. *Depositional history of the Bothkennar clay*. Bothkennar Research Report BRR 9/94, Heriot-Watt University, Edinburgh.

Paul M. A., Barras B. F. and Peacock J. D. 1994b. *A revised stratigraphy for the Bothkennar clay*. Bothkennar Research Report BRR 10/94: Heriot-Watt University, Edinburgh.

Paul M. A., Peacock J. D. and Barras B. F., 1995. Flandrian stratigraphy and sedimentation in the Bothkennar-Grangemouth area. *Quaternary Newsletter*, **75**, 22–35.

Tovey N. K. and Hounslow M. W., 1995. Quantitative micro-porosity and orientation analysis in soils and sediments. *J. Geol. Soc.*, **152**, 119–129.

Development of automated dilatometer and comparison with cone penetration tests at the University of Adelaide, Australia

W. S. KAGGWA, M. B. JAKSA and R. K. JHA
Department of Civil and Environmental Engineering, The University of Adelaide,
Adelaide, Australia

ABSTRACT
The cone penetration and flat dilatometer tests continue to enjoy widespread use as reliable in situ test methods. When used together, enough data are obtained for the design of footings in most soil conditions. While the cone penetrometer has undergone extensive development, aimed at increasing the speed of testing, analysis and interpretation of the results, the flat dilatometer test continues to require manual operation and recording of data. This paper describes the developments in automated data recording and processing of the cone penetration and flat dilatometer tests at the University of Adelaide over the past decade or so. Results obtained in two types of soils are presented.

INTRODUCTION
The use of in situ tests in site investigations for foundation design, either independently or in conjunction with laboratory tests, is very common. However, the choice of the type of in situ test depends largely on the complexity and cost of the construction project and knowledge of, or lack of, information about the soil conditions at the proposed site. The availability of experienced site investigation personnel, with the in situ test device, also plays a part in the final choice.

Of the many available in situ test devices, the electric cone penetration test (CPT) is one of the most widely used devices in cohesive soils and extensive experience has been gained since it was developed almost thirty years ago. The use of the dilatometer test (DMT), on the other hand, has steadily increased in cohesionless soils where it offers less variability than that of the standard penetration test (SPT). However, there is a lot of concern over the empiricism employed in estimating design soil parameters. The main advantages of both the CPT and DMT are that they cause minimal disturbance to soil and provide relatively simple means of determining variations in the soil profile, shear strength and stiffness parameters.

The continued acceptability of the CPT is partly due to the ease of testing and partly due to the availability of quick data recording and retrieval for processing. Data acquisition systems have developed from the pen and chart recorders, when the CPT was first introduced, to current systems where data can be printed in the form of tables or charts, with suggested soil types and design parameters.

Advances in site investigation practice. Thomas Telford, London, 1996

DESIGN CRITERIA FOR DATA ACQUISITION SYSTEMS

The cone penetrometer

A number of guidelines and standards are available, such as ISOPT-1 (De Beer et al., 1988), ASTM D3441 (1986), AS 1289.F5.1 (Standards Association of Australia, 1977), which set minimum requirements for the accuracy of CPT measurements. There appears to be no exclusive specifications regarding all aspects for the CPT. The following aspects are desirable for data acquisition of the CPT.

De Beer et al. (1988), in ISOPT-1, specify that the precision of measurement of q_c and f_s should be greater than 5% of the measured value and 1% of the maximum value of the resistance in the layer under consideration, and that there should be a facility to view the data throughout the test, as well as store the data on tape (or disk). With regard to sampling rate, Jamiolkowski et al. (1985) recommend sampling within the range of 0.1 to 1 second, whilst Lunne et al. (1986) suggest a sample interval of 1 second. For the standard penetration of 20 mm/s, these rates are equivalent to depth intervals of between 2 mm and 20 mm.

With regard to depth measurements, an accuracy of 0.25% was reported by de Ruiter (1981). In addition to the precision of the data acquisition system, it is desirable that all components of the equipment should be robust enough to withstand the rigours of field testing in a variety of weather conditions.

In designing the University of Adelaide data acquisition system for cone penetration, one aim was to obtain data for spatial variability analyses (see Jaksa et al., 1993) and a measurement spacing of 5 mm was targeted.

The flat dilatometer

The flat dilatometer was developed by Silvano Marchetti in the early 1970's. The ASTM published a suggested method for performing the flat dilatometer test (Schmertmann, 1986), which consists of (i) forcing the dilatometer blade vertically into the soil to a desired depth, measuring the thrust required to accomplish this penetration, and (ii) expanding the circular steel membrane by gas pressure. The operator measures and records the pressure required to expand the membrane to deflections of 0.05 mm and 1.1 mm.

Recording of data from the flat dilatometer test is essentially manual, and relies on reading the gas pressure on a gauge to within 0.2 bar (20 kPa), at three preset positions of membrane deflection. Readings of the lift-off pressure, pressure at 1.1 mm membrane deflection, and the pressure at recovery, are manually recorded. These readings are then corrected for membrane stiffness and the expansion itself. In addition to operator measurement errors, it should be remembered that these two readings form the basis of subsequent analyses.

At the University of Adelaide, a data acquisition system has been developed to measure and record the membrane deflection and the gas pressure during pressurisation, as well as the resistance during blade penetration (blade tip resistance). The depth measuring system is also used to measure the depth below the ground surface.

THE UNIVERSITY OF ADELAIDE DATA ACQUISITION SYSTEMS

Instrumentation of cone penetrometer

The data acquisition system consists of five components. These include the electric cone penetrometer, the depth box, the alarm button, the microprocessor interface and the micro-computer. The arrangement described by de Ruiter (1971) for the cone penetrometer is followed. The depth box was developed to measure accurately the depth of the cone penetrometer. The depth box consists of a thin metallic cable, one end of which is attached to the hydraulic ram that drives the cone penetrometer into the ground, and the other end is wound around a metal drum. A shaft encoder is connected to the shaft of the drum and, as the drum rotates, a pulse is transmitted every one-five-hundredth of a revolution. The metal drum has an external diameter of 100 mm. Therefore, each pulse corresponds to 0.62832 mm of travel of the cone penetrometer. A light torsional spring is also attached to the metal drum to enable the cable to retract into the depth box as the cone penetrometer is driven into the ground. By recording the phase of the pulse, and counting the number pulses, the depth and direction of movement of the cone penetrometer are accurately determined.

The alarm button is connected to the depth box and located near the drilling rig operator. Its primary function is to communicate to the microprocessor interface when to start sampling. The operator presses the button when the cone penetration test is ready to commence. Also, a warning alarm sounds when the drilling head has been raised by more than 200 mm, (either to connect additional drilling rods or when the cone penetrometer is being withdrawn from the ground at the completion of testing) indicating that sampling has ceased. The alarm continues to sound, informing the operator that data acquisition will not continue until the button is depressed.

The functions of the microprocessor interface include multiplexing, signal conditioning and amplification, analogue to digital conversion and data storage and transfer. The microprocessor has been set to measure cone tip resistance q_c , in the range of 0 to 15 MPa and sleeve friction, f_s, in the range of 0 to 500 kPa. The microprocessor interface is also fitted with a cone amplification switch, which, when activated, decreases the ranges by a factor of 5 so that the maximum value of q_c = 3 MPa, thereby increasing the resolution, which is useful in soft soil deposits. Measurements of q_c , f_s , and depth are stored and transmitted by the microprocessor interface and are scaled, displayed and saved onto floppy disk on an IBM compatible micro-computer. A portable personal computer is currently used.

A detailed description of the data acquisition system is given by Jaksa and Kaggwa (1994).

Instrumentation of flat dilatometer

Modifications to the standard flat dilatometer were undertaken to at least achieve the same level of precision of measurement as obtained in the CPT. Also, empirical correlations rely on values of tip resistance obtained from a cone sounding adjacent to the dilatometer test. To overcome inaccuracies associated with operator error and soil variability, three key modifications were made to the standard flat dilatometer equipment. These included (i) installation of a load cell between the dilatometer blade and the drilling rod, (ii) connection of a pressure transducer to the pressure line, and (iii) redesigning of the measuring system for membrane deflection. The first and second modifications were relatively straightforward. The third modification involved installation of a spring to which were attached two strain gauges. At one end, a pin rests on top of the spring while at the other end the spring pushes against the dilatometer membrane. Figure 1 shows the standard and modified measurement systems.

The modifications necessitated attention to the use of the gas cable for wiring of electrical leads from the strain gauges of the load cell, and the deflection measuring spring in the dilatometer blade. Because of limitations of the internal diameter of the cable, half bridge strain gauges were used throughout, and a special adaptor manufactured to provide low resistance contacts for connection to the microprocessor interface.

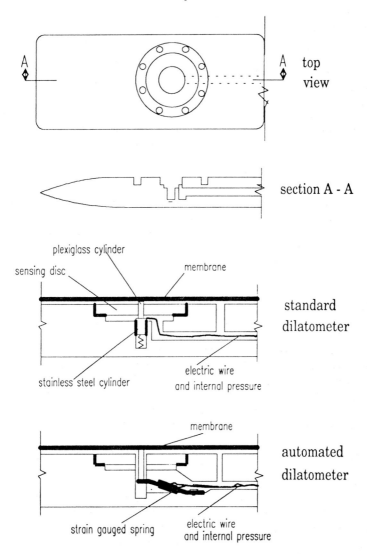

Figure 1: Details of deflection measurement systems of the standard and automated dilatometer tests

Additional calibrations were necessary to determine (i) the stiffness of the dilatometer membrane during expansion and rebound, (ii) the relation between spring deflection and membrane deflection, and (iii) the load cell strain gauges and pressure transducer. Two sets of calibrations were performed, the first, carried out in the instrumentation laboratory, was to check and calibrate the strain gauges and transducers, and the second, using a calibration chamber, for determining the membrane stiffness.

Computer software

A suite of five computer programs was written primarily to optimise the time taken to process and present the recorded information. One program reads the cone penetrometer load cells, depth box and alarm button, temporarily stores these data in RAM, and at various stages, transfers the data to the micro-computer. The other programs (i) allow interaction between the operator of the CPT and the data acquisition system, (ii) allow visual inspection of cone penetration data on the screen both in the field and at a later date, (iii) enable the user to process the CPT data stored on disk and to prepare and print report quality plots and tables.

The software developed for the cone penetration test was modified to permit measurement of dilatometer tip resistance, q_D, and depth during penetration, as well as pressure and membrane deflection during pressurisation of the dilatometer.

EXAMPLES OF RESULTS OBTAINED FROM RESEARCH PROJECTS

The automated cone penetration test has been used on a number of research projects such as the study of variability of Keswick Clay, that underlies the City of Adelaide (Jaksa et al., 1993). Results from this research project are presented to illustrate some of the features of the data acquisition system. Results of tests in calcareous sand, using the modified dilatometer and cone penetration tests, are also presented.

An extensive investigation into soil variability was conducted in 1992 in Keswick Clay. These results are presented in Figures 2 to 4. Figure 2 shows the display on the screen during field testing from *CPTest*, and Figure 3 shows the same data during inspection after the test. After analysis of the data, the same data are plotted using *CPTPlot* (see Figure 4), ready for inclusion in a report.

Dilatometer testing

Another series of tests were conducted in a uniform deposit of sand containing between 20% and 25% carbonate content in the form of biogenic matter. Figure 5 shows the layout of the in situ tests used. These tests were primarily aimed at comparing the results of standard dilatometer tests and those obtained using the automated dilatometer test. The depth interval of the DMT tests was 200 mm. The drilling rig is a modified 3-tonne truck with anchors to provide stability and additional capacity up to 5 tonnes. Care was taken to arrange the sequence of testing so as to provide comparisons among the different types of tests, to conduct the DMT tests prior to the cone penetration test, and to orient the DMT membrane away from the adjacent test position.

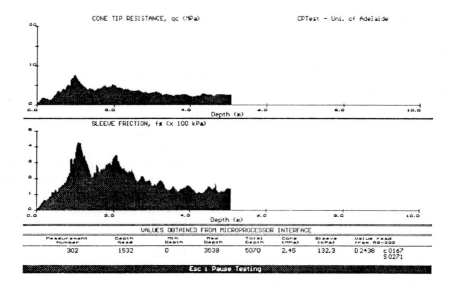

Figure 2: An example of the *CPTest* screen, as seen in the field

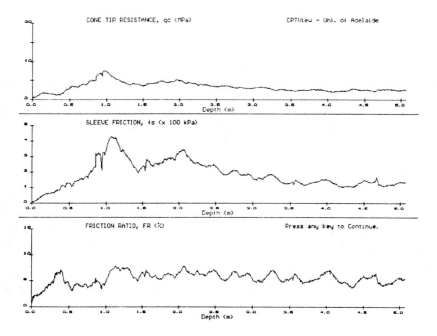

Figure 3: An example of a screen plot produced by *CPTView*

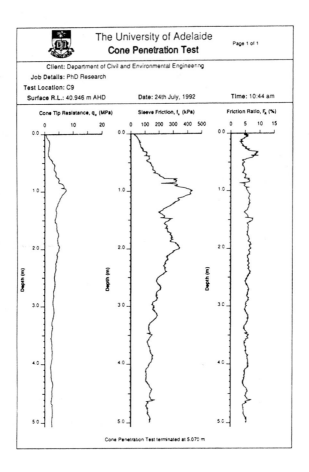

Figure 4: An example of a graphical plot produced by *CPTPlot*, for report presentation

These results are shown in Figures 6 to 8. Figure 6 shows the soil profile obtained from the cone penetration test whereas Figure 7 shows the profile obtained from the standard and automated flat dilatometer tests. It can be seen that the variations with depth of tip resistance, q_c, in Figure 6 are similar to the variations of dilatometer pressures, p_0 and p_1, in Figure 7.

Figure 8 shows the variation of dilatometer tip resistance, q_D, with depth, and pressure versus membrane deflection at one of the stages of the test. It can be seen that there was little variation in tip resistance in the silt/sand layer of the soil profile. The pressure rises to the lift-off pressure of 100 kPa (comparable to p_0 in Figure 7) and increases non-linearly as the membrane deflection increases. The pressure corresponding to 1.1 mm is comparable to that obtained from the standard dilatometer (see Figure 7). The figure allows a closer examination of the deflection of the membrane, during the loading and unloading stages of the test.

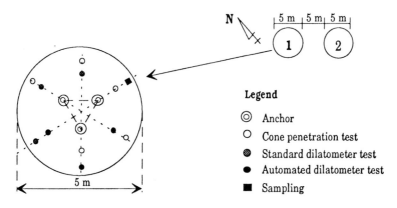

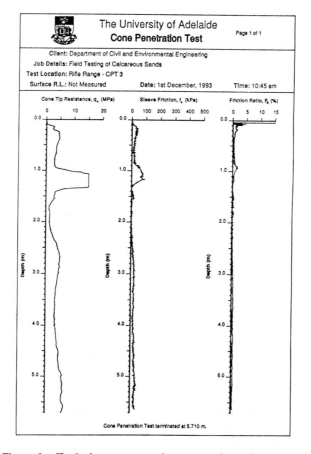

Figure 5: Layout of in situ tests showing the two locations and arrangement of tests at each location

Figure 6: Typical cone penetration test results at the test site

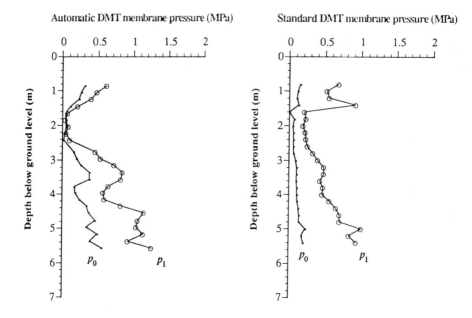

Figure 7:   Profiles of dilatometer corrected pressures p_0 and p_1

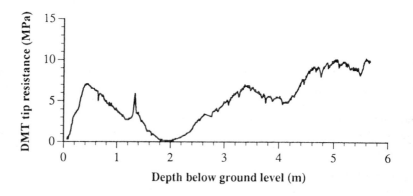

Figure 8:   Variation of dilatometer tip resistance obtained from automated dilatometer

Analysis of automated dilatometer test data

To properly analyse the pressure versus membrane deflection curve obtained from the automated dilatometer, the non-linear relation between output voltage and deflection, as well as the stiffness of the membrane have been allowed for. These corrections are applied to the raw data, to obtain the pressure applied to the soil, and this pressure is then used to determine the soil parameters. A typical pressure versus membrane deflection curve in calcareous sand is shown in Figure 9. It should be realised that the effects of cavity expansion, during blade penetration, have not been allowed for here, and more detailed analysis of the results is proceeding.

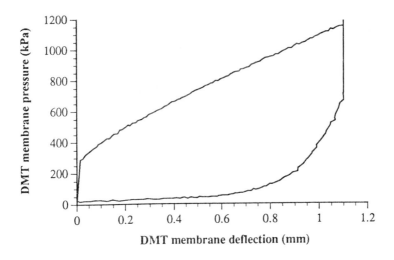

Figure 9:   Typical pressure versus membrane deflection curve obtained from
the automated dilatometer

SUMMARY
The development of automated data acquisition systems for in situ test devices will continue to gain momentum, especially those involving robust equipment. Of necessity, the hardware must be robust, and the software should allow for both field monitoring of the test as well as reduction of the data to report quality format. This way, the need for field operators to have ready access to test results, and reduction in time for data presentation will be achieved. The University of Adelaide data acquisition systems go some way in meeting these objectives.

ACKNOWLEDGEMENTS
The authors wish to acknowledge the technical support of the staff of the Department of Civil and Environmental Engineering, University of Adelaide, in particular Bruce Lucas and Tad Sawosko, and Dr. J. N. Kay, now of the University of Hong Kong, who initiated the project of automation of field testing. Funds for developing the data acquisition systems have been provided by the University of Adelaide and the Australian Research Council. This financial assistance is gratefully acknowledged.

REFERENCES
American Society for Testing Materials (ASTM) (1986), Standard method for deep, quasi-static, cone and friction-cone penetration tests of soil (D3441). *Annual Book of Standards,* Vol. 04.08, ASTM, Philadelphia, pp. 552 - 559.

De Beer, E.E., Goelen, E., Heynen, W.J. and Joustra, K. (1988), Cone penetration test (CPT): International Reference Test Procedure. In *Penetration Testing, Proc. of the First Int. Symposium on Penetration Testing* (ISOPT-1), de Ruiter (ed.), Orlando, Florida, A.A. Balkema, Rotterdam, pp. 27 - 51.

De Ruiter, J. (1971), Electric penetrometer for site investigation, *J. of Soil Mechanics and Foundations Division,* ASCE, Vol. 97, SM2, pp. 457 - 472.

De Ruiter, J. (1981), Current penetrometer practice. In *Cone Penetration Testing and Experience,* Geotechnical Engineering Division, ASCE, St. Louis, Missouri, pp. 1 - 48.

Jaksa, M.B., Kaggwa, W.S. and Brooker, P.I. (1993), Geostatistical modelling of the undrained shear strength of a stiff, overconsolidated, clay. *Proc. of Conference of Probabilistic Methods in Geotechnical Engineering,* Canberra, pp. 185 - 194.

Jaksa, M.B. and Kaggwa, W.S. (1994), A micro-computer based data acquisition system for the cone penetration test, Research Report No. R116, Dept. of Civil and Environmental Engineering, University of Adelaide, May, 31p.

Jamiolkowski, M, Ladd, C.C., Germaine, J. and Lancellotta, R. (1985), New developments in field and laboratory testing of soils. *Proc. 11th Int. Conf. on Soil Mech. and Foundation Engg.,* Vol. 1, San Francisco, pp. 57 - 154.

Lunne, T., Eidesmoen, T., Gillespie, D. and Howland, J.D. (1986), Laboratory and field evaluation of cone penetrometers. In *Use of Insitu Tests in Geotechnical Engineering,* ASCE, GSP No. 6, Blacksburg, Va., pp. 714 - 729.

Schmertmann, J.H. (1986), Suggested method for performing the flat dilatometer test, *Geotechnical Testing Journal,* GT-JODJ, ASTM, Vol. 9, No. 2, pp. 93 - 101.

Standards Association of Australia (1977), Determination of the static cone penetration resistance of a soil - Field test using a cone or a friction cone penetrometer. In *Methods for testing soils for engineering purposes,* AS 1289.F5.1, Sydney.

Dynamic probing and its use in clay soils

A. P. BUTCHER, K. McELMEEL and J. J. M. POWELL
Building Research Establishment, Watford, Herts, UK

INTRODUCTION

Dynamic Probing has long been established as an exploratory prospecting tool in many countries throughout the world. It is probably the oldest and simplest soil test. It basically consists of percussing a metal tip into the ground using a drop weight of fixed mass and travel. Its use as an investigative tool for site investigation prior to construction has been limited to a few countries in Europe, the rest of the world following the United States of America in the use of the Standard Penetration Test (SPT) for this purpose. The Building Research Establishment (BRE) has carried out a programme of research to evaluate the potential of dynamic probing as an economical, repeatable and operator insensitive test for profiling and assessing soil properties.

This paper describes the dynamic probing test procedure, equipment specifications and the treatment of results. Dynamic probing was carried out at 10 clay soil test bed sites, 6 in the UK and 4 in Norway, to cover a reasonably wide range of undrained shear strengths and plasticities. Correlations between dynamic probing data and undrained shear strength as well as the results from SPT and static cone penetrometer tests have been obtained and, potentially, will allow wider application of dynamic probing to ground investigation.

TEST PROCEDURE AND EQUIPMENT SPECIFICATION.

The dynamic probing test consists of driving a point into the ground, via an anvil and extension rods, with successive blows of a free fall hammer. The number of blows required to drive the point each successive 10cm (N_{10}) is recorded, so creating a record of blows/10cm with depth of penetration of the point. When driving is halted to add a further extension rod (every metre) the torque required to rotate the extension rods that are already in the ground is measured to assess the friction on the rods from the soil. The free fall hammer may be raised by hand (only practical with lightweight hammers) or by a device incorporating a motor driven continuous chain with an automatic latch and release arrangement for the hammer. The automatic latch arrangement avoids operator influence on the test which is inevitable with hand operated or manual latch type equipment. Typical of the equipment readily available is the motorised dynamic probing rig used by BRE which is shown in Figure 1.

Standards for test equipment and procedure

A draft standard on dynamic probing was written in Germany in 1974 (DIN 4094 part 1) to set out equipment specifications and in 1980 DIN 4094 part 2 outlined uses and comparisons with other penetration tests in some granular soils. Under the auspices of the International Society for Soil Mechanics and Foundation Engineering (ISSMFE) a technical committee was set up to review and report on all penetration testing in soils. The work of this committee was then extended to set out Recommended Test Procedures (RTP) for each penetration test examined including equipment specifications and tolerances. The RTP for dynamic probing was formulated in 1977 and finalised and reported in 1989 (ISSMFE 1989) but does not include the interpretation and use of the results.

The specifications for various configurations of equipment, as given in the RTP, is reproduced in Table 1. The RTP includes a standard range of blows/0.1m penetration of 3-50 for light (DPL), medium(DPM) and heavy (DPH) and of 5-100 blows/0.2m penetration for superheavy (DPSH). Specified tolerances for wear of cones, dimensions of rods and cones and the mass and height of fall of hammers is also included. In general DPL, M and H have the same hammer drop height but with different hammer weights and DPH and M use the same extension rods. Dynamic Probing Super Heavy (DPSH), as can be seen from

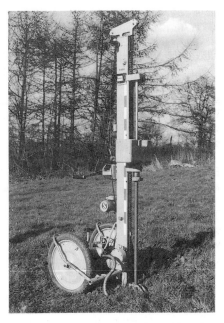

Figure 1: Motorised dynamic probing rig

Factor	Test Specification				
	DPL	DPM15	DPM	DPH	DPSH
Hammer mass, kg	10 ±0.1	30 ±0.3	30 ±0.3	50 ±0.5	63.5 ±0.5
Height of fall, m	0.5 ±0.01	0.5 ±0.01	0.5 ±0.01	0.5 ±0.01	0.75 ±0.02
Mass of anvil + guide rod (max), kg	6	18	18	18	30
Rebound (max), %	50	50	50	50	50
Rod length, m	1 ±0.1%	1 ±0.1%	1 ±0.1%	1 ±0.1%	1 ±0.1%
Mass of rod (max), kg	3	6	6	6	8
Rod eccentricity (max),mm	0.2	0.2	0.2	0.2	0.2
Rod OD,mm	22 ±0.2	32 ±0.2	32 ±0.2	32 ±0.2	32 ±0.3
Rod ID,mm	6 ±0.2	9 ±0.2	9 ±0.2	9 ±0.2	9 ±0.2
Cone apex angle, deg.	90	90	90	90	90
Cone area (nominal), A cm^2	10	15	10	15	20
Cone dia. new, mm	35.7 ±0.3	43.7 ±0.3	35.7 ±0.3	43.7 ±0.3	50.5 ±0.5
Cone dia. worn (min) mm	34	42	34	42	49
Mantle length, mm	35.7 ±1	43.7 ±1	35.7 ±1	43.7 ±1	50.5 ±2
No. of blows per x cm penetration	N_{10} :10	N_{10} :10	N_{10} :10	N_{10} :10	N_{20} :20
Standard range of blows	3 - 50	3 - 50	3 - 50	3 - 50	5 - 100
Specific work per blow (Mgh/A) kJ/m	50	98	150	167	238

Table 1: Details of dynamic probing test specifications

the specification in Table 1, is designed to closely resemble the Standard Penetration Test (SPT).

Also included in the test specifications in Table 1 is the configuration DPM15, a variation used by BRE, which is essentially a DPM test but with a 15 sq cm cone. The specific work per blow of DPM is close to that of DPH and 3 times that of DPL. The DPM15 has a specific work per blow value only twice that of DPL and fits better between the DPL and DPH; it also has a greater clearance between the cone and the rod diameter than DPM which can help reduce the build up of friction between the soil and the extension rods. The significance of this will be discussed later in the paper.

BS 1377:1990, part 9 only includes the heavier weight DPH and DPSH dynamic probing with a test procedure close to that recommended by the RTP.

Table 2 sets out the dynamic probing test specifications available in the UK and the relevant standards for equipment and test procedure. Also included is the specification for DPM15 which uses equipment specified in the DIN 4094 part 1 standard but is not a recognised configuration; this will be discussed again later.

	Test Specification				
	DPL	DPM	DPM15	DPH	DPSH
Equipment	DIN 4094/1	DIN 4094/1	DIN 4094/1	DIN 4094/1 BS1377:1990	BS1377:1990
Procedure	DIN 4094/2	DIN 4094/2	-	DIN 4094/2 BS1377:1990	DIN 4094/2 BS1377:1990

Table 2: Standards for dynamic penetration tests available in the UK.

Treatment of results

The results from probings are usually presented as blows/10cm penetration (N_{10}) for DPL, DPM, DPM15 and DPH and blows/20cm (N_{20}) for DPSH against depth as a straight field record and should be within the standard range of values stated in the RTP (see Table 1). Also included with the field results will be the torque readings, as specified in BS1377:1990 and the RTP, taken to assess rod friction when each extension rod is added. The effect of rod friction on the N_{10} values can be significant but no mention of an allowance for this effect is made in the standards or the RTP. The effect of rod friction will be examined later in the paper. The N_{10} values can be interpreted, according to the RTP, to give the unit point resistance (r_d) or the dynamic point resistance (q_d) all against depth of point using the following formulae:

$$r_d = \frac{Mgh}{Ae}$$

$$q_d = \frac{M}{(M+M')}r_d$$

Where:

r_d and q_d are resistance values in Pa

M   is the mass of the hammer in kg

g   is the acceleration due to gravity in m/sec^2.

h   is the height of fall of the hammer in m

A   is the area at the base of the cone in m^2

e   is the average penetration in m per blow ($0.1/N_{10}$ from DPL, DPM15, DPM, and DPH, and $0.2/N_{20}$ from DPSH)

N_{10} is the number of blows per 10cm

M' is the total mass of the extension rods, the anvil and the guiding rods in kg

The value of r_d is an assessment of the driving work done in penetrating the ground and further calculation, to produce q_d values, modifies the r_d value to take account of the inertia of the driving rods and hammer after impact with the anvil. The calculation of r_d includes the different hammer weights, the height of fall and the different point sizes. The different sizes and numbers of extension rods is included in the calculation of q_d, and so should allow comparison of different equipment configurations.

TEST BED SITES AND THEIR SOIL PROPERTIES

In this study dynamic probings have been carried out on a total of ten well documented test bed sites having known soil properties and extensive data bases of *in situ* soil testing data.

In order to obtain data on a wide range of clay soils with different soil properties the six UK test bed sites were complimented with four sites in Norway as part of collaborative work with the Norwegian Geotechnical Institute (NGI) and the University of Trondheim.

The principal soil properties of the ten sites are given in Table 3 including references for the data from each site. The soils tested ranged from stiff heavily overconsolidated London and Gault clay, through stiff glacial tills to soft silty and marine clays. Plasticity Indices range from 17 to 55% and sensitivities up to 8.

Four of the UK sites were used for comparisons of the different configurations of dynamic probing, the BRE sites of Brent and Canons Park (London Clay), Cowden (Glacial Till) and the EPSRC test site at Bothkennar (soft silty clay).

RESULTS

Repeatability

Initial investigations by BRE considered the repeatability of dynamic probing in clay soils. For all the configurations listed in the RTP it was found that good repeatability could be obtained for any one configuration. The motor driven equipment is free of operator influence but the hand operated DPL needs care to get repeatable results. Figure 2 shows five DPM15 profiles from the Canons Park site two of which included the use of bentonite slurry to reduce rod friction. Down to 5m depth, where low torque values were recorded in all the tests, the N_{10} values from each test are very similar. The high values of N_{10} between 1 and 2.5m depth represent a gravel layer which includes stones larger than the cone tip which would obviously destroy any repeatability. It can be seen in Figure 2 that rod friction builds up at different rates in the tests without bentonite which makes the repeatability look, at first glance less encouraging, but as will be discussed below this was far from the case.

Site	Depth m	Soil Description	Density Mg/m³	w %	w_p %	I_p %	c_u kPa	S_t	Reference
Canons Park	3	Firm to stiff brown silty clay	1.95	26	30	46	70	1.2	Uglow & Powell
	4	Stiff brown silty	1.95	26	30	52	75	1.2	(1988)
	5	fissured clay	2	30	30	48	120	1.2	
	6		2	28	28	35	110	1.2	
	7	Stiff brown silty fissured	2	28	28	42	100	1.2	
	8	with sand laminations	2	29	28	42	100	1.2	
	9 - 10	Stiff blue London clay	2	29	29	42	120	1.2	
Brent	2	Stiff finely fissured	1.95	31	26	56	46	1.2	Powell et al (1988)
	3	brown weathered	1.95	30	23	54	52	1.2	
	4	clay	1.95	30	26	52	62	1.2	
	5		1.95	29	27	52	69	1.2	
	6 - 8		1.93	30	29	52	85	1.2	
	9	Stiff Highly fissured	1.98	29	27	52	90	1.2	
	10	grey-blue, unweathered	2	29	27	50	96	1.2	
	11	clay	2	30	24	45	100	1.2	
	12		2	29	23	45	108	1.2	
	13		2	29	29	45	112	1.2	
	14		2	29	29	45	120	1.2	
Madingley	2	Firm intact grey-green	1.88	31	31	55	90	1.0	Butcher & Lord
	3	mottled, weathered	1.88	31	31	55	90	1.0	(1993)
	4	silty clay	1.9	31	31	55	110	1.0	
	5	Stiff grey fissured,	1.93	31	31	55	140	1.0	
	6	weathered silty clay	1.94	31	31	55	140	1.0	
	7-10	Very stiff dark grey	1.95	30	30	50	140	1.0	
	11 - 13	closely fissured	1.95	30	30	50	150	1.0	
	14	silty clay	1.95	29	29	50	160	1.0	
Cowden	2	Stiff brown weathered	2.2	17	22	20	160	1.0	Marsland & Powell
	3	stony clay till with	2.2	17	19	20	140	1.0	(1985)
	4	some fissuring	2.2	17	19	20	110	1.0	
	5-6	Stiff dark grey-brown	2.2	17	21	20	105	1.0	
	7 - 10	unweathered stony clay till	2.2	17	16	20	110	1.0	
BRS	3 - 4	Stiff mottled brown-grey	2.2	18	18	28	200	1.0	Marsland & Powell
	4 - 5	clay till with some fissuring	2.1	19	19	28	200	1.0	(1989)
	7 - 8	Chalky highly fissured clay	2.05	20	20	25	150	1.0	
Bothkennar	2 - 5	Soft to firm fissured black organic very silty clay	1.68	56	28	36	22	4.0	Powell et al (1988)
								4.0	
	6 - 8	Soft intact dark grey-black very organic silty clay	1.586	66	30	40	28	4.0	
								4.0	
	9 - 12	Soft slightly fissured black organic silty clay	1.626	62	28	48	35	4.0	
	13	Soft slightly fissured black	1.666	60	30	45	40	4.0	
	14	silty clay	1.68	52	28	48	45	4.0	
Museumpark	5	Silty clay	1.753	40	25	20	25	6.0	Dyvik (1985)
	7	Plastic clay	1.753	52	29	20	28	6.0	
	8 - 10		1.753	52	30	30	30	8.0	
	11		1.753	51	26	25	34	7.0	
	12	Lean clay	1.75	45	20	25	34	4.0	
	13		1.855	40	20	20	26	5.0	
	14		1.855	32	20	15	28	3.3	
Lierstranda	5 - 6	Soft clay	1.83	40	25	20	19	4.0	Masood et al
	7		1.83	40		22	25		(1990)
	8 - 9		1.83	40	24	23	30	4.0	
	10 - 11		1.88	40	24	25	36	4.0	
	12	Soft silty clay	1.88	39		24	36		
	13 - 15		1.88	38	22	22	27	4.0	
Valoya	4	Stiff Homogeneous clay	1.845	40	31	17	70	6.6	Sandven (1990)
	5		1.85	42	37	15		4.7	
	6 - 7		1.845	42	30	23	60	4.1	
Glava	2	Soft silty clay	1.825	45		17	28	8.6	Sandven (1990)
	3 - 5		1.876	35	23	15	35	5.8 - 7.7	
	6 - 12		1.937	32	22	14	38	4.8 - 8.1	

Table 3: Soil properties of the test sites

Rod friction and torque corrections

Friction on the extension rods can affect the values of N_{10} recorded. Experience has shown that torque readings in excess of 200Nm generally mean the driving rods are at some point forced off line and further driving would permanently bend the affected driving rods which would then not comply with the RTP. For this reason tests were usually terminated when a torque reading reached 120Nm.

The RTP mentions the use of drilling mud to reduce friction on the extension rods. To incorporate this into the tests a special extension rod, with holes through the rod wall, was used immediately above the cone and the drilling mud, in the form of a bentonite slurry, was introduced down the centre of the hollow extension rods. For stiff clay sites bentonite was poured into the rods and was sucked out or forced out by gravity into the annular space behind the probe tip during driving. The level of bentonite was topped up after the addition of an extension rod. This method did not work for soft clay sites so a special pressure system was used to maintain a pressure, at just above the mean *in situ* stress, in the bentonite slurry so forcing it into the space behind the probe tip.

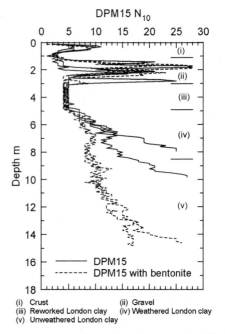

Figure 2: DPM15 N_{10} depth profiles from Canons Park

(i)   Crust   (ii)   Gravel
(iii) Reworked London clay   (iv) Weathered London clay
(v)   Unweathered London clay

Torque measurements were taken during all the tests, including those where bentonite drilling mud was used. Examples of the effect of the bentonite on recorded N_{10} values for the DPM15 configuration tests at Canons Park are shown in Figure 2.

The measured torque values were plotted against the average blow count per 0.1m penetration (N_{10}) over the preceding 1m of driving. A torque correction was calculated using that part of the profile where the penetration and torque values were constant. This gave the correction as the torque equivalent to 1 blow of the driving hammer. By dividing the measured torque by the torque correction the number of blows needed to overcome the rod friction can be calculated and then subtracted from the recorded N_{10} value. The torque corrections obtained at the stiff clay sites, Canons Park, Cowden and Madingley, are given in Table 4 and show remarkable similarity for similar equipment configurations.

Torque corrections are harder to establish at the soft clay sites where it is difficult to getprofiles without significant torque for clear comparison. The use of the pressurised bentonite mud reduces the measured torque to zero or low levels however, but then gives N_{10} values below that recommended in the RTP.

The torque measurements made during probing were only taken at the addition of an extension rod, that is at 1m depth intervals. An estimate of the torque at other depths, that

	Torque corrections in Nm/blow		
	Soil Type		
Dynamic probing configuration	Gault clay	London clay	Glacial till
DPH	11.1	12.0	12.2
DPM15	6.8	7.0	8.3
DPL	-	3.8	4.0

Table 4: Corrections to blow counts for torque at Cowden, Canons Park and Madingley.

is between torque measurements, was obtained by linear interpolation between torque measurements. Experience with correcting for rod friction allowed the discrimination between torque effects and soil properties. For instance, if the blow count immediately after a torque measurement was lower than the previous blow count it was probably due to the action of taking the torque measurement, especially if the next blow count was back to the pre torque measurement value. In this case the torque correction would not be applied to the low blow count but the interpolation would use the next, that is higher, blow count. This technique enabled a more realistic correction for torque in each profile. Figure 3 shows the data from Figure 2 corrected for torque using the appropriate correction factor from Table 4 and the techniques mentioned above. Apart from the values in the gravel layer between 1.2 and 3.2m depth the corrected N_{10} profiles show good repeatability.

Comparison of configurations

The available configurations of dynamic probing equipment, with their different specific work per blow, should allow soils of different dynamic point resistance to be investigated. However, the inherent variability of the soil and the need for correlations with other soil test data requires that reproducible resistance values be obtained from probings of different configurations. Tests were carried out at three sites having different soil properties, London clay, glacial till and soft alluvial clay, to compare the results from the different available dynamic probing configurations.

Comparison in London clay

Comparisons of different configurations of dynamic probing equipment in London clay were carried out at the Brent site. Figure 4a gives typical profiles of blow counts (N_{10}), corrected for rod friction, obtained using

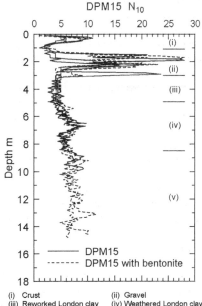

(i) Crust (ii) Gravel
(iii) Reworked London clay (iv) Weathered London clay
(v) Unweathered London clay

Figure 3: DPM15 N_{10} depth profiles from Canons Park corrected for rod friction.

DPL, DPM15, and DPH configurations. The DPL profile gave the greatest detail with less from the DPM15 profile and least from the DPH. The DPH N_{10} values down to 5m depth and the DPM15 N_{10} values down to 3m depth were at or below the RTP minimum value of 3.

In Figure 4b the data from 3 configurations from Brent are shown as profiles of r_d. The DPM15 and DPH data are very similar with better resolution being evident from the lighter weight configuration. The DPL profile, however, while giving better detail than the DPM15, gave r_d values up to 50% higher than DPM15.

Figure 4c presents the same profiles as in Figures 4a and 4b but in terms of q_d. The overall profiles are now very similar for all configurations; however, it is noticeable that the q_d values from the lighter weight tests again gave the most detail and the DPM15 tended to form the lower bound to the q_d profiles.

In the London clay the values of q_d obtained were found to be independent of the dynamic probing configuration and demonstated good profiling ability by clearly showing the different clay layers in the profile which match those given in Table 3. The r_d profiles also showed the different clay layers but the values varied according to the equipment. However the lighter weight equipment generally gave better sensitivity though a probing could be halted more easily by obstructions.

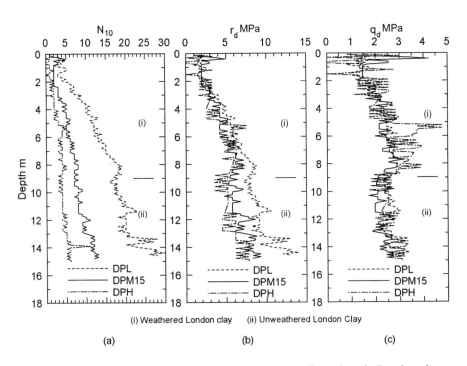

Figure 4: Comparison of dynamic probing equipment configurations in London clay:
a) N_{10} vs depth, b) r_d vs depth, c) q_d vs depth

Comparison in a Glacial Till

The comparison of results from different configurations of dynamic probing tests in glacial till was made at the BRE Cowden test bed site. Figure 5a shows the blow count depth profiles from DPL, DPM15, DPH and DPSH and again illustrates the greater detail picked up by the lighter weight tests with the heavier DPH and DPSH giving profiles of blow counts close to the RTP minimum values (ie 3 for the DPH and 5/20cm or 2.5/10cm for DPSH). The DPL, DPM15, and DPH tests used bentonite to reduce extension rod friction whereas the DPSH did not and a build up of rod friction has increased the blow counts below 6m.

Figure 5b presents the same profiles as Figure 5a in terms of r_d. The r_d profiles are very similar in shape but the heavier types (DPH and DPSH) gave higher r_d values but again with less detail and highlighted their lower resolution. Figure 5c presents the calculated q_d profiles of each configuration. The q_d values from DPH and DPM15 gave very similar profiles whereas the DPL q_d values were up to 1.5 MPa lower than the DPH and DPM15 (at 3.5m to 4m) but generally compare well with the heavier configurations but forming the lower bound of the q_d profiles. This may be due to the stony nature of the till at Cowden and the relative sizes of the cones used in the different configurations or due to the inherent variability of glacial tills. The DPSH gave slightly higher values though the coarseness of the resolution made interpretation difficult.

In general at Cowden the heavier weight equipment is seen to give poor resolution of values with the lighter weight equipment giving more detail. Only with the q_d values did the different configurations give comparable values. The values of r_d varied with the configuration

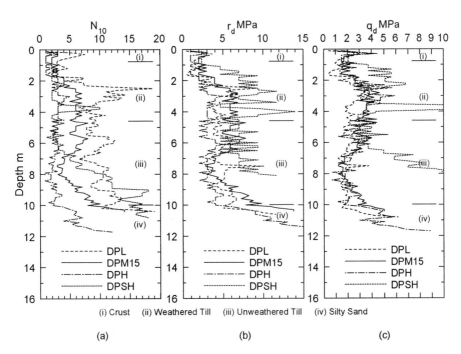

Figure 5: Comparison of dynamic probing equipment configurations in Glacial till:
a) N_{10} vs depth, b) r_d vs depth, c) q_d vs depth

of equipment used. Again, as at Brent, the q_d profiles demonstrate the profiling ability of dynamic probing showing a change of profile from the crust to the weathered till to the unweathered till as given in Table 3.

Comparison in soft alluvial silty clay

Four configurations of dynamic probing were used in the soft alluvial clay at Bothkennar. The blow counts were generally in the 1 to 3 range for all the configurations which is below the minimum value of 3 in the RTP. These data endorse the application of a minimum number of blow counts because similar blow counts were obtained from all the test configurations until the build up of torque increased the recorded numbers. Not surprisingly the calculation of q_d revealed very low values which by that fact were similar but did not cover the same ranges of values for each configuration.

CORRELATION WITH UNDRAINED SHEAR STRENGTH

Based on the outcome of the work on the comparison of configurations, q_d was chosen as the most valid parameter for correlation with peak undrained shear strength. Data from the ten test bed sites, five each of stiff and soft clay, were used in the correlation with q_d values from dynamic probing tests using bentonite, to reduce extension rod friction, and, where necessary, with corrections for torque. The shear strengths used were either from small scale *in situ* tests or laboratory triaxial tests and ranged from 18 to 200kPa across the ten sites. Figure 6 shows dynamic point resistance (q_d kPa) against undrained shear strength (c_u kPa). It is immediately evident that the data groups separately for stiff and soft clays. The correlations are for stiff clays:

$$c_u = (q_d/22)$$

and for soft clays:

$$c_u = (q_d/170)+20.$$

It can be seen in Table 3 that the soft clays all had sensitivities greater than 4 which, it was felt, might have had a significant effect on the results. An empirical correction for the data by dividing the dynamic point resistance (q_d) by the sensitivity (S_t) produces the relationship in Figure 7 and gives a consistent relationship for all the clay sites tested of $c_u = 0.045(q_d/S_t) +10$.

CORRELATION WITH OTHER *IN SITU* TESTS IN CLAY SOILS

Correlations of dynamic probing data, corrected for torque, have been made with the Standard Penetration Test (SPT) and static cone resistance q_t (q_c corrected for pore water pressure effects).

Correlation of Dynamic probing with Standard Penetration Test (SPT)

The current study has enabled a correlation between dynamic

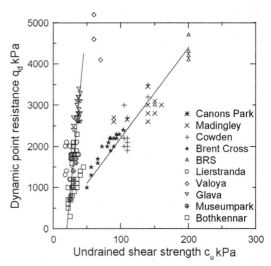

Figure 6: Dynamic point resistance vs undrained shear strength

probing and the SPT in stiff clays. The dynamic probing N_{10} values were obtained from DPH tests, made adjacent to the SPT boreholes, using bentonite to reduce the extension rod friction and were corrected for torque before being used in the correlations. The SPT values were obtained using the split spoon sampler as per BS 1377:1990 without a cone.

In order to compare the tests the DPH dynamic probing blow counts were averaged over the depth increment that closely matched those used for the SPT (excluding the seating drive). The averaged blow count was then correlated with the SPT 'N' values and plotted in Figure 8 which shows acorrelation, based on limited data, of 'N' = $8N_{10}$ - 6.

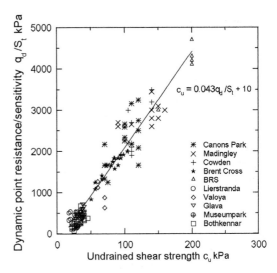

Figure 7: Dynamic point resistance/Sensitivity vs undrained shear strength

Correlation of dynamic probing with static cone penetrometer (CPT)
Data from ten test bed sites have been used to correlate dynamic probing q_d with static cone resistance q_t. The stiff clay data are shown in Figure 9 which includes a $q_d = q_t$ line which fits the data reasonably well. The soft clay data are shown in Figure 10 which includes the correlation line of $q_t = 0.24q_d + 0.14$. The soft clay data form a reasonably tight banding close to the correlation line with the Glava data having the biggest spread. As with the correlation with undrained shear strength the factor which determines whether the clay fits the stiff clay or soft clay correlation appears to be the

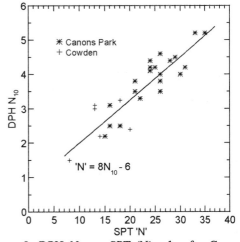

Figure 8: DPH N_{10} vs SPT 'N' value for Canons Park and Cowden

sensitivity. The sensitive Valoya clay has the same dynamic cone resistance as the glacial till site at the BRS site, for example, but has a much lower static cone resistance.

CONCLUSIONS AND RECOMMENDATIONS
The following conclusions and recommendations have been drawn from the forgoing:

1. Dynamic probing has a good profiling capability, especially with the lighter weight

equipment configurations and the use of bentonite to reduce extension rod friction.

The equipment specifications in the RTP are generally too heavy for the soft clays and work is required to develop further the lighter weight specifications from the existing equipment e.g. DPH but with a 10kg hammer or DPL but with a 15sq cm cone. In many cases the DPH configuration is too heavy for stiff clays and the DPM15 configuration is recommended. With its lower specific work per blow and greater rod to cone diameter ratio than the DPM, the DPM15 configuration was found to have better sensitivity yet still cope with stiff clays without exceeding the RTP blow count maximum.

2. Dynamic probing gave very reproducible values of N_{10}, r_d and q_d, with very similar values of q_d being obtained from different configurations of equipment in the same clay soil profile.

The use of q_d has the potential to allow the configuration of equipment to be varied down a profile as the blow counts fall too low (reduce the hammer mass) or rise too high (increase the hammer mass). To accommodate this procedure equipment with modular hammer weights is recommended to allow rapid changes to configuration, during a probing if necessary, to ensure the best sensitivity and avoiding too high, or too low, an energy input.

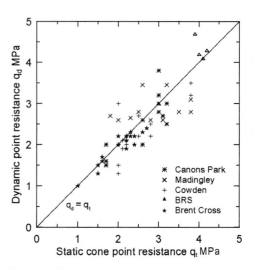

Figure 9: Dynamic point resistance vs static cone point resistance for stiff clays

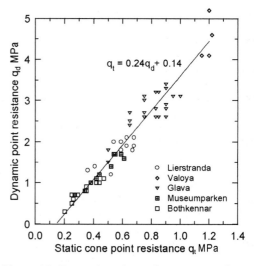

Figure 10: Dynamic point resistance vs static cone point resistance for soft clays

4. Two separate relationships have been found for c_u from q_d for soft and stiff clays. An empirical relationship, based on the data available, has been proposed between undrained shear strength and q_d/S_t for all the clays.

5. A correlation has been established between SPT `N' value and DPH N_{10} values for the stiff

overconsolidated clays at Cowden and Canons Park.

6. Values of q_d correlate closely with CPT q_t values but differ between stiff and soft clays. This opens up the possibility of using correlations for soil properties developed for static cone penetrometers.

ACKNOWLEDGEMENTS
The authors would like to thank the Norwegian Geotechnical Institute, in particular T.Lunne, and Trondheim University, in particular R. Sandven, for the use of their test sites and access to the soil property data from the test sites.

© Crown Copyright

REFERENCES
BS 1377:1990. Methods of testing soils for civil engineering purposes. British Standards Institution, London.

Butcher, A.P. & Lord, J.A. (1993). The engineering properties of the gault clay in and around Cambridge UK. *Geotechnical Engineering of Hard Soils-Soft Rocks.* Ed Anagnostopoulos, Schlosser, Kalteziotis and Frank. Vol 1, 405-416. Balkema, Rotterdam.

DIN 4094 part 1. 1974 Dynamic and Static penetrometers, Dimensions of apparatus and method of operation. Deutsches Institut für Normung e.V. Berlin.

DIN 4094 part 2. 1980 Dynamic and Static penetrometers, Application and Evaluation. Deutsches Institut für Normung e.V. Berlin.

Dyvik. R. 1985. In situ Gmax versus depth profiles for four Norwegian test sites. NGI Report No 40014-10.

ISSMFE . Report of technical committee on penetration testing of soils - TC 16 with Reference Test Procedures. Swedish Geotechnical Institute, Information 7. 1989.

Marsland A. and Powell J.J.M. 1985. Field and Laboratory investigations of the clay tills at the Building Research Establishment test site at Cowden Holderness. *Proc. Int. Conf. on Construction in Glacial Tills,* Edinburgh.

Marsland A. and Powell J.J.M. 1989. Field and Laboratory investigations of the clay tills at the test bed site at the Building Research Establishment, Garston Hertfordshire. *Quaternary Engineering Geology,* Geological Society Engineering Geology Special Publication No7. pp 229-238.

Masood, T., Mitchell, J.K., Vaslestad, J., Lunne, T. and Mokkelbost, K.H. 1990. Testing with lateral stress cone, special dilatometer and stepped blade at three sites in Drammen. NGI Report No 521600-1.

Powell, J.J.M., Quarterman, R.S.T. & Lunne,T. (1988). Interpretation and use of piezocones in UK clays. *Proc. Conf. on Penetration Testing in the UK,* Birmingham. Thomas Telford.

Sandven R. 1990. *Strength and deformation properties of fine grained soils obtained from piezocone tests.* Theses submitted in partial fulfilment for the degree of Doctor of Engineering at the University of Trondheim, Norway.

Uglow, I.M. and Powell, J.J.M. 1988. Application of Dilatometer for pile design - Dilatometer tests at five UK sites. BRE Note 22/88

The development of the seismic cone penetration test and its use in geotechnical engineering

P. A. JACOBS[1] and A. P. BUTCHER[2]
[1]Fugro Ltd, Hemel Hempstead, England [2]Building Research Establishment, Watford

ABSTRACT

The use of in situ testing in soils is proven, in most situations, to be the most accurate method of testing to determine a soil's properties. Whereas the testing of samples of soil recovered from any form of drilling method will cause some degree of disturbance to the sample and one of the soil parameters most affected by sample disturbance is stiffness. The seismic cone penetration test (SCPT) has become a commonly used test to determine profiles of in situ low strain shear modulus (G_0) for soils. The test has been used as a more economic alternative to other geophysical seismic testing techniques such as crosshole borehole seismic testing. The equipment has developed from a standard electric cone fitted with a single array of triaxial geophones to a dual array piezo friction cone complete with highly accurate inclinometers. The G_0 data determined from this equipment has, until recently, been used primarily in the analysis of the behaviour of soil when it is subjected to dynamic loading. The close relationship between G_0 from in situ tests and soil stiffness back figured from instrumented foundations (Burland 1989) has shown the potential use for seismic tests, such as the SCPT, in the analysis of geotechnical problems. The SCPT can also be used as a powerful benchmark test to compare stiffness data from other sources such as small strain laboratory tests and surface seismic testing.

INTRODUCTION

The seismic cone penetration test is a development of the quasi static cone penetration test (CPT) which is widely used to investigate soil stratification at sites. The same basic equipment used in the CPT method is utilised in the SCPT procedure. With the addition of the hammer trigger, recording system and the seismic cone penetrometer (Figure 1). The seismic cone penetrometer was originally developed about a decade ago (Robertson et. al. 1986) and comprises a standard cone penetrometer fitted with an array of small geophones.

These geophones are mounted in the three orthogonal planes (X , Y , Z) and fit inside the cone body which has a diameter of typically 44 mm. As for the standard CPT method the seismic cone is pushed into the ground using a hydraulic jacking system mounted in a ballasted vehicle which has the capacity to apply 30 tonnes of thrust. During the penetration phase of the test measurements can be made of cone end resistance, local sleeve friction and pore water pressure. This data is used to provide a profile of the type and strength of soil being penetrated, together with more detail on layering from the measured pore water pressure. The inclination of the cone is also measured during testing to ensure that the cone penetrates vertically into the ground.

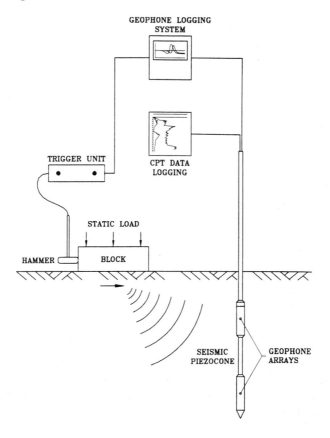

Figure 1   SCPT Method

At the required depth where seism'c measurements are to be made the penetration of the cone is stopped. As for the downhole borehole test method the seismic cone is the receiver and is also referred to as the downhole SCPT. The seismic source is typically a beam clamped to the ground by the cone penetrometer vehicle which is struck by a hammer horizontally in the direction of the long axis of the beam. This type of source has been found to produce mainly horizontally polarised shear waves with very low energy compression waves. This seismic source together with the cone's horizontally mounted geophones results in clear recorded shear wave signal arrivals. The strength of the shear wave pulse received at the seismic cone reduces as the depth of penetration increases. When the signal strength or background noise prevents accurate data interpretation then signal stacking is employed. Typically stacking is necessary when the seismic cone is more than twenty five metres below ground level. Signal stacking is a routinely used technique in surface seismic testing, where more than one seismic event is used and the received signals are stacked one on another to enhance the signal received by the geophones. The time for the shear wave to travel from the ground surface to the cone is measured using a seismograph or computer controlled data logging system.

EQUIPMENT DEVELOPMENTS
Single Array Seismic Cone Test - Incremental Test
The seismic cone, as originally developed had a single set of geophones. In order to improve the definition of the shear wave arrival two shear waves are generated at the surface for each test depth where each signal has a reverse polarity of the other. This is achieved by striking opposite ends of the clamped beam. A typical result of this type of single array testing is shown on Figure 2. Here the two signals of opposite polarity have been plotted together clearly showing the first arrivals, and peak amplitudes. The cone is then advanced to the next test depth and the test procedure is repeated. In order to determine the travel time between test depths the shear wave velocity (V_s) is computed from the difference in the shear wave travel times to each test depth. The travel paths of the shear waves are assumed to be straight and are corrected for the horizontal offset between shear wave source and cone string axis and the incremental depth between tests. This type of testing with a cone fitted with a single set of geophones is often called the incremental SCPT.

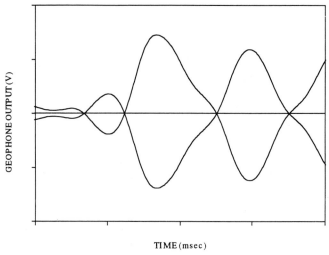

Figure 2   Typical Shear Wave Reversal Seismic Test Result

Dual Array Seismic Cone Test
To reduce the time taken to perform a series of SCPT tests in the field and the reliability and accuracy of the data interpretation a seismic cone with two arrays of geophones was developed. The two arrays of geophones are mounted a fixed distance apart vertically, typically 1m (Figure 1). This equipment set up gives both total and differential shear wave travel times from a single hammer significantly reducing the time taken to complete a series of tests.

The dual array SCPT has a number of advantages over the incremental SCPT system when interpreting shear wave velocities. The major advantages are:

(1)   measurements of arrival time are independent of trigger accuracy
(2)   measurements are unaffected by variable strength seismic energy source
(3)   accuracy of depth control is not critical to the accuracy of results.

The dual array seismic cone requires 15 data channels to be recorded which presents technical problems of using standard cone testing equipment because of the limited number of cables which can fit inside the cone testing rods. This has been overcome by using micro switches fitted inside the body of the dual array seismic cone which switch logging cable from cone channels to seismic array channels as required.

<u>Signal Logging</u>
The signals from the seismic cone are normally recorded on either a 12/24 channel seismograph or an 8 channel computer controlled A/D card. The A/D card system offers a number of benefits over the seismograph:-

     (1)   cheaper cost of system
     (2)   immediate post test filtering of signal noise
     (3)   automated calculation of shear wave velocity (dual array cone data)

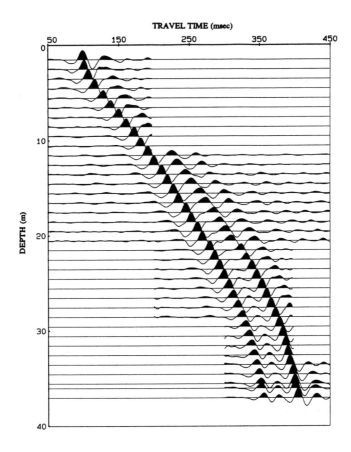

Figure 3   Travel Time Profile from Dual Array Seismic Cone

A set of results using the latest dual array seismic piezocone is presented on figure 3. The tests were carried out from a jack up pontoon platform in fifteen metres of water. Tests were performed at 2 metre intervals with a 1 metre geophone array spacing and the shear waves were generated using a hydraulically powered underwater hammer. It can been seen from the results that the accuracy of system produces a continuous profile of travel times even though the tests were made at two metre centres.

MEASUREMENT OF SOIL STIFFNESS
General
The dynamic shear modulus, also called the zero strain modulus, G_O is calculated using elastic theory which relates bulk density of soil, ρ_b and V_s to G_O as follows:

$$G_O \;=\; \rho_b \,.(V_s)^2 \qquad\qquad .... \quad (1)$$

The need to investigate the low strain stiffness of soils has come from the large discrepancies between the predicted ground movement under foundations, based on conventional laboratory tests, and the observed ground movements under the installed foundation. Soil stiffness back figured from monitored foundations on clay soils have been shown to be of the order of five to ten times higher than those measured in the laboratory.

At the foundation working load most of the ground surrounding the structures experiences a shear strain of less than 0.1%. St. John (1975) and Stevens et al. (1977) have shown that the undrained soil stiffness from conventional laboratory tests (triaxial compression) is an order of magnitude lower than the soil stiffness back analysed from measurements of foundation movement (Figure 4).

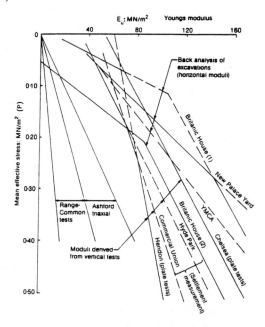

Figure 4   After St. John (1975)

London Clay at Blackwall

Soil stiffness of London Clay has been gathered from a comprehensive site investigation for the A102 Third Blackwall Crossing in London. These stiffnesses have been determined from a variety of in situ and laboratory tests.

A typical result of the one of the friction CPTs performed at the Blackwall Crossing site is presented in Figure 5 as profiles of cone end resistance (qc) and friction ratio (Fr).
Figure 5 shows the London Clay to be quite variable in strength and grading identified by the cone parameters qc and Fr respectively. This material has been found in adjacent boreholes to have a number a sandy layers. This is confirmed by the variability in qc and Fr where the weaker clay layers (lower qc) have an associated higher Fr values. This may clearly be seen in Figure 5 at depths of 13 to 14 metres. This variability in strength and grading will result in some variability of measured stiffness.

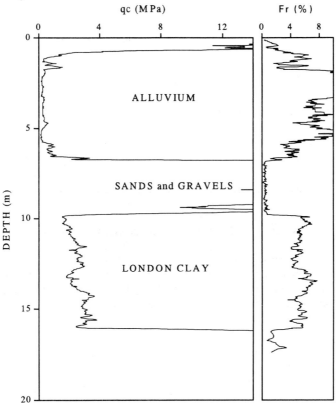

Figure 5   CPT Data - Blackwall Crossing - North Site

In Situ Measurements Of Shear Wave Velocity

At the Blackwall site shear wave velocity was measured in situ using the crosshole borehole test (CBT) and SCPTs test methods. The shear wave velocity results from these different test methods are shown in Figure 6. It can be seen that the results from the SCPT work agree well with the CBT data carried out in the London Clay at Blackwall.

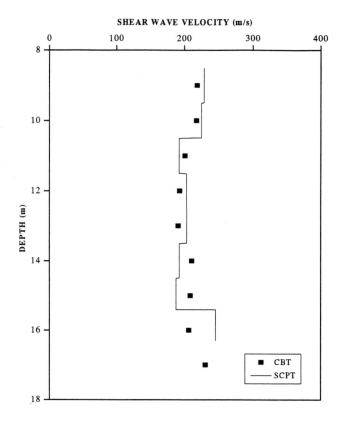

Figure 6   Seismic Data - London Clay (Blackwall)

Robertson et al. (1986) have presented results of CBT and SCPT testing performed at six test sites which all illustrated good agreement between the two techniques in both normally consolidated sand and clay deposits. Also Baldi et al. (1988) have illustrated the good correlation between results from these two test techniques for a sand site.

Laboratory Measurement Of Stiffness
Resonant Column Testing   The soils from the Blackwall site were tested in the Hardin oscillator apparatus which is a "fixed free" torsional Resonant Column (RC) machine. The RC test was employed on Blackwall London Clay samples to determine the shear modulus of samples over a range of strain levels between 10^{-5} to 10^{-1} per cent shear strain. The material damping (D) of the soil is also determined from the test.

The maximum or plateau shear modulus results from the RC tests are presented in Figure 7 with the calculated shear modulus values from the SCPT and CBT. The results illustrate that the soil stiffness from the SCPTs at the Blackwall site are similar to the plateau values presented for the RC tests. This shows a close relationship between the dynamic soil stiffness from in situ seismic testing and laboratory measurements in the resonant column apparatus as has also been shown by Barwise (1994) to hold for a large range of strains.

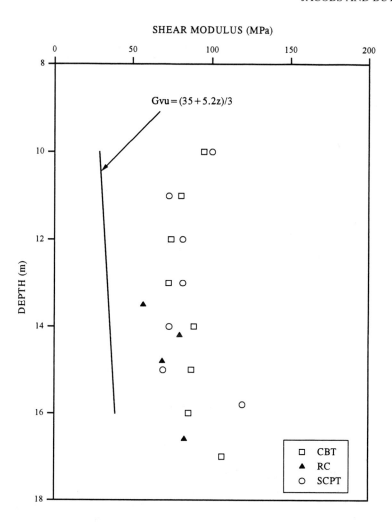

Figure 7   In Situ and Laboratory Soil Stiffness (Blackwall)

The shear modulus profile derived from Owsianka (1988) for vertical loading (undrained) used as a model for predicting foundation movements on London Clay is given by:

$$Gvu = (35 + 5.2\,z)/3 \qquad\qquad (2)$$

assuming a Poisson's ratio of 0.5 and

where

z = depth below ground level in metres
units of Gvu are MPa

403

The Owsianka profile is shown on Figure 7 with the laboratory stiffnesses and compared to the dynamic stiffness profile is some 2½ times stiffer than the derived ground stiffness profile described in Equation 2 and this ratio appears to be constant with depth for the London Clay.

Triaxial Testing with Local Strain Measurement   The measurement of sample strain locally on the sample is used to avoid the problems of 'end-effects' in the triaxial apparatus. The measurement of local strain in the apparatus used for the Blackwall samples was performed by using the Hall effect gauges (HEG).

The stiffness of the London Clay from the instrumented triaxial tests are in the form of undrained secant modulus (Eu) at axial strains of 10^{-4} (0.01%) and 10^{-3} (0.1%) these are presented as Eu $_{0.01}$ and Eu $_{0.1}$ in Figure 8.

It can be seen from Figure 8 that the stiffnesses at the smaller strain level are significantly greater than those at the higher level of strain. Although the maximum value of Eu measured as G from the RC tests are very close to the Eu $_{0.01}$ data. The data shows some scatter and the trend is an increase of stiffness with depth.

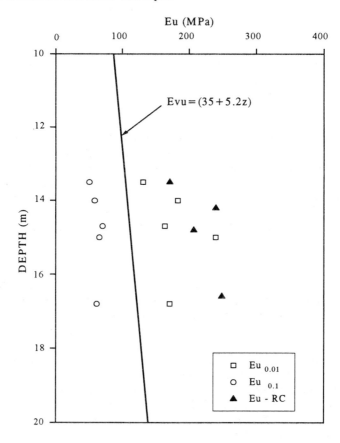

Figure 8   Soil Stiffness Data for London Clay (Blackwall)

The good agreement between Eu $_{0.1}$ and Owsianka's profile would confirm the reported work of Costa Filho and Vaughn (1980) which showed that the majority of the ground beneath a foundation on London Clay is experiencing an average shear strain of 0.1%.

DISCUSSION

The SCPT has become an economic and routinely available in situ test technique for determining the very low strain stiffness of soils. The equipment has developed sufficiently that the geotechnical engineer can qualitatively assess the results produced by the test. The test results are being used for the dynamic foundation design of structures as well as the benchmark for data from other in situ and laboratory stiffness determination tests. More recently the results from tests carried out with a constant energy source are being used to determine in situ soil damping.

The Owsianka stiffness profile was found by Clayton et al. (1991) to be conservative in predicting ground movements when compared to their back figured soil stiffness from the Grand Buildings site in London. The back figured stiffnesses in London Clay presented by Clayton et al. for a number of foundations in the centre of London illustrate that there is a wide range of stiffness profiles. The upper bound of this range is almost twice as stiff as Owsianka's profile which is approaching the stiffnesses measured in the dynamic laboratory and seismic field tests.

The profile of stiffness from in situ and laboratory tests on the London Clay at Blackwall are found to be 2½ times stiffer than the back figured stiffness profile from measured foundation movements on London Clay. This ratio was found to be almost constant with depth in the London Clay.

The results of the two in situ seismic methods (CBT and SCPT) give very similar profiles of dynamic or zero strain shear modulus, G_0 as shown previously. This is an interesting result because the London Clay is extremely anisotropic, and would be characterised by the horizontal stiffness being greater than the vertical stiffness. For instance Clayton et al. used a value for horizontal undrained stiffness, Ehu equal to 1.6 Evu to model the ground movements at the Grand Buildings. It would be expected that because the shear waves generated by the two test methods are perpendicular to each other in both orientation and propagation they would produce significantly different stiffness profiles in an anisotropic soil such as reported by Butcher and Powell (1995).

The RC test applies torsional oscillations to the sample in the horizontal plane which causes strain in both a vertical and horizontal plane. The results of the RC tests also lies within the data from the in situ dynamic tests. This close correlation between in situ and laboratory low strain testing has been shown to apply to overconsolidated clay by Barwise (1994).

The stiffness measured on London Clay samples from Blackwall in the RC give similar stiffnesses to those measured in the instrumented triaxial system at an axial strain of 0.01% (Eu$_{0.01}$). From the examination of the results of shear modulus from RC tests for increasing strain levels it is observed that for increasing strain the shear modulus only starts to reduce from a near constant value of G_0 at a shear strain of approximately 0.005%. This behaviour explains the close agreement in stiffness from the RC with Eu$_{0.01}$.

REFERENCES

Baldi, G., Battaglio, M., Bruzzi, D., Jamiolkowski, M. and Superbo, S. (1988). Seismic cone in Po River sand. *Proceedings*, Penetration Testing, International Symposium on Penetration Testing. Orlando, Fla., vol. 1, 643-650.

Barwise, A.M. (1994). An Investigation Into The Effect of Strain Amplitude on the Shear Modulus and the Effect of Dispersion on the Primary Wave Velocity, for a Clay and a Sand Site. MSc Thesis, University of Wales, Bangor.

Burland, J.B. (1989). "Small is beautiful"-the stiffness of soils at small strains. Canadian Geotechnical Journal. 26, 499-516.

Clayton, C.R.I,. Edwards, A., and Webb, M.J. (1991). Displacements in London Clay during construction. *Proceedings*, Tenth European Conference on Soil Mechanics and Foundation Engineering, vol. 2, Firenze, Italy, Balkema, Rotterdam, 791-796.

Costa Filho, L.F. and Vaughn, P.R. (1980). *Discussion on* A computer model for the analysis of ground movements in London Clay. *Géotechnique*, **30**, No. 3, 336-339.

Jardine, R.J., Symes, M.J. and Burland, J.B. (1984). The measurement of soil stiffness in the triaxial apparatus *Géotechnique*, **34**, No. 3, 323-340.

Owsianka, A. (1988). *Discussion on* Burland & Kalra (1986). *Proceedings* of the Institution of Civil Engineers, Part 1, 84, 111-114.

Robertson, P.K., Campanella, R.G., Gillespie, D. and Rice, A. (1986). Seismic CPT to measure in situ shear wave velocity. ASCE, Journal of Geotechnical Engineering, vol. 112, No. 8, 791-803.

St. John, H.D. (1975). Field and theoretical studies of the behaviour of ground around deep excavations in London Clay. PhD thesis, University of Cambridge, United Kingdom.

Stevens, A., Corbett, B.O. and Steele, A.J. (1977). Barbican Arts Centre: the design and construction of the substructure. Structural Engineer, No. 55, 473-485.

Applications of penetration tests for geo-environmental purposes

P.K. ROBERTSON [1], T. LUNNE [2] and J. POWELL [3]
[1] Professor of Civil Engineering, University of Alberta, Edmonton, Canada
[2] Division Head, Norwegian Geotechnical Institute, Oslo, Norway
[3] Building Research Establishment, Garston, UK.

ABSTRACT

The application of penetration tests for geo-environmental purposes has grown rapidly in recent years. The Cone Penetration Test (CPT) has become recognized as a valuable in-situ testing technique because of its speed, reliability, cost effectiveness and excellent soil profiling capabilities. In recent years additional sensors have been added to the CPT equipment to enhance and expand its capabilities for environmental purposes. These additional sensors include, electrical resistivity/conductivity and permittivity measurements, pH and ion detectors as well as special fiber optic devices for fluorescence measurements. Special penetrometer sampling probes have also been developed to obtain liquid and/or vapor samples as well as soil samples. These sample probes are installed with CPT equipment and vary from simple inexpensive devices for bulk water samples to sophisticated probes for very high quality water and/or vapor samples from specific zones. These sample zones are often identified using the continuous profiles from CPT equipment with special sensors. A summary of recent advances in CPT and penetrometer equipment related to geo-environmental applications is presented and examples are shown to illustrate these applications.

INTRODUCTION

Site characterization has been a major part of geotechnical engineering practice and various drilling, sampling and in-situ testing techniques are now well established. In recent years there has been a steady increase in geo-environmental engineering projects where geotechnical engineering has been combined with environmental concerns. Many of these projects involve some form of contaminant in the ground. These contaminants can take the form of vapors, liquids and solids. Hence, in recent years there has been a change in site characterization techniques to accommodate these environmental issues related to contaminants. Drilling techniques have been modified to account for possible contaminated ground. However, drilling techniques generally produce considerable disturbance to the materials surrounding the drill hole, which can have a significant effect on subsequent sample quality. With increasing application of data quality management, drilling and sampling techniques are becoming less acceptable. Also, drilling and sampling methods produce cuttings of the material removed from the drill hole. If these cuttings are contaminated they may require special handling and disposal methods. In many states of the USA there are now regulations that require all drill cuttings removed from geo-environmental site investigations to be disposed or stored in an acceptable manner. This can increase the cost of a day of drilling by as much as US $1,000. Hence, there have been clear incentives to develop techniques that do not produce cuttings from the subsurface.

The most rapidly developing site characterization techniques for geo-environmental purposes involve direct push technology, i.e. penetration tests. The direct push devices generate essentially no cuttings, produce little disturbance and reduce contact between field personnel and contaminants, since the penetrometer push rods can be decontaminated during retrieval.

A variety of penetrometer tests exist for both geotechnical and geo-environmental investigations. In general these tests can be divided into three main categories; logging, specific and combined tests. The most popular logging test for geotechnical investigations in soil is the Cone Penetration Test (CPT). The CPT provides a continuous profile of measurements, the test is rapid, repeatable, reliable and cost effective. Specific tests include the field vane test and the pressuremeter test since these measure specific soil parameters and are often carried out in locations identified by the logging test. Combined tests typically combine the features of logging and specific tests into one test, examples of which are the seismic CPT and the cone pressuremeter.

The objective of this paper is to summarize penetrometer technology applied to geo-environmental site characterization.

OBJECTIVES FOR GEO-ENVIRONMENTAL SITE CHARACTERIZATION
The main objectives for a geotechnical site investigation are to determine the following;
1. Nature and sequence of the subsurface strata (geologic regime)
2. Ground water conditions (hydrogeologic regime)
3. Physical and mechanical properties of the subsurface strata.

For geo-environmental site investigations where contaminants are possible, the above objectives have the additional requirement to determine;
4. Distribution and composition of the contaminants.

To complicate matters the contaminants can exist in vapor, liquid and solid forms. The above investigation should be carried out in sufficient detail as required by the project. For geotechnical projects this is usually a function of the proposed structure and the associated risks. The geotechnical engineer is often in control of the risk process and hence, selection of the required site investigation detail. For geo-environmental projects the extent of detail required for the determination of the distribution and composition of the contaminants maybe controlled by various regulatory agencies, for which the engineer may have little control. With the rapid improvements in measurement technology, the in-situ detection limits required by some agencies for certain contaminants are decreasing at an alarming rate.

For geotechnical investigations the information is often obtained at one instance in time and projections are made to predict changes in ground conditions due to such factors as seasonal rainfall. For major projects where the observational method may be applied critical parameters such as deformations can be monitored to evaluate the changing mechanical conditions. For geo-environmental projects where potential contaminants are identified long term monitoring and sampling maybe required for both design and either remediation or containment. Hence, the objectives for geo-environmental site characterization can be quite different from those for a more traditional geotechnical site characterization.

CPT TECHNOLOGY FOR SITE CHARACTERIZATION
The Cone Penetration Test (CPT) has become an important in-situ test for the characterization of soils where penetration is possible. Penetration can be difficult through cemented materials and materials with large particle sizes. The CPT provides excellent near continuous profiles of soil type and detailed stratigraphy. If pore pressures are measured (CPTU), improved stratigraphic detail can be obtained as well as important additional information on equilibrium ground water conditions, consolidation characteristics and hydraulic conductivity. Empirical and semi-theoretical correlation's are available to estimate a full range of mechanical properties (Roberston and Campanella, 1983). The CPTU measures the mechanical response of the ground or material to the penetration process through cone penetration resistance (q_c), sleeve friction (f_s) and the pore liquid pressure (u). If a solid contaminant has mechanical properties significantly different from those of the surrounding soil then the CPT can identify the presence of the material. However, the CPT can not identify the chemical composition of the contaminants. Hence, sensors have been developed that can be added to cone penetrometers in an attempt to identify certain contaminants. The following section describes some of these recent developments.

The measurement of equilibrium pore pressures can be an important part of an investigation to evaluate the direction of ground water flow and vertical pressure head distribution and hence the hydrogeologic regime. Most cone penetrometers that measure pore pressures contain high capacity pressure transducers because penetration pore pressures can be very large in soft soils. To improve the measurement of the equilibrium pore pressure some cone penetrometers included a low pressure transducer connected to the outside of the probe via a control valve so that equilibrium pore pressures can be measured to a very high degree of accuracy (e.g. 30 mm head of water). This can minimize the number of possible permanent monitoring wells to measure ground water flow regimes. However, care is needed when penetrating soft fine grained soils since in can take a considerable time to dissipate the high penetration pore pressures. The use of small diameter probes (e.g. 1 cm^2) can be advantageous for speeding up dissipation time in fine grained soil in cases where knowledge on in-situ pore pressure is vital.

GEO-ENVIRONMENTAL PENETROMETER LOGGING DEVICES

The CPT and CPTU are excellent logging devices that provide near continuous profiles of mechanical parameters. Sensors have been added to cone penetrometers to enhance their application for geo-environmental site characterization. One of the earliest sensors included to a cone penetrometer was a temperature sensor. Initially temperature sensors were used to aid in either calibration corrections or to locate zones of different ground temperature, such as frozen soil. Recently temperature sensors have been used to aid in the identification of contaminants that generate heat due to biological and/or chemical activity.

The next major sensor that has been added to the CPT is for the measurement of electrical resistivity or conductivity. The conductivity is the inverse of resistivity, with the following as a useful guide;

Conductivity (μS/cm) = 10,000 $\div$ Resistivity (Ω-m)

The measurement of electrical properties was first developed to evaluate in-situ density of sands (Kroezen, 1981) but more recently it has been used to evaluate contaminated soils (Horsnell, 1988; Campanella and Weemees, 1989; Woeller et al., 1991, Strutynsky et al., 1991). The rationale for making electrical measurements is that in many circumstances, the electrical properties of the soil will be changed by the presence of contaminants. Therefore, by measuring soil resistivity, the lateral and vertical extent of soil contamination can be evaluated. Unsaturated soils and saturated soils with many non-aqueous-phase-liquid (NAPL) compounds exhibit very high electrical resistivity (low conductivity). Dissolved inorganic compounds, such as those contained in brines and landfill leachates, significantly decrease soil resistivity.

The resistivity CPT works on the principle that the measured voltage drop across two electrodes in the soil, at a given excitation current, is proportional to the electrical resistivity of the soil. The resistivity electrodes are typically steel rings from 5 mm to 15 mm wide that are set apart by distances that vary from 10 mm to 150 mm. The larger the spacing the greater the depth of penetration for the electrical field into the surrounding soil. Some probes have several electrode spacings so that lateral penetration varies. Some devices use small circular electrodes mounted around the circumference of the probe. Figure 1 shows a typical resistivity cone penetrometer with two ring electrodes (Woeller et al., 1991). The electrodes are designed to be reasonably wear resistant and have a high electrical conductivity. A non-conducting plastic or other material is used as the insulator separating the electrodes. The resistivity measurements are typically made by applying a sinusoidal current across the electrodes and measuring the resultant potential difference between the electrodes. The current can be regulated by a downhole microprocessor that adjusts the current when the resistivity changes appreciably to ensure a linear response. This enables resistivity measurements between 1 and 250 ohm-m to be made with an accuracy of +/- 0.2 ohm-m. A 1000Hz source is typically used to avoid polarization of the electrodes. Electrical resistance is not a material property but a function of the electrode spacing and size. To convert from resistance to resistivity, which is a material property, a laboratory calibration of the probe geometry is necessary. The resistivity of soil is for the most part influenced by the resistivity of the pore liquid, which in turn is a measure of the pore liquid chemical composition. Hydrocarbons are non-conductive and will therefore exhibit high resistivity. The resistivity CPT has been used

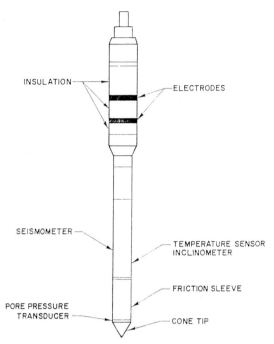

Figure 1.   Example of two ring electrode resistivity cone penetrometer (After Woeller et al., 1991).

successfully in acidic ground conditions. The main disadvantage with electrical measurements is that the bulk resistivity or conductivity is not directly controlled by the chemical properties of the surrounding material. The measurements are strongly influenced by the background soil and pore liquid.   Hence, it is important to obtain measurements of the background uncontaminated ground for comparison. For relatively uniform soil conditions it is possible to develop local site specific correlation's between the bulk resistivity and selected contaminants (Campanella et al., 1994). The primary advantage of the resistivity CPT is that it provides continuous profiles of bulk resistivity along with the full CPT data in a rapid cost effective manner. The profiles of resistivity measurements can then be used to identify potentially critical zones where detailed sampling and/or monitoring can be carried out. An example of a resistivity CPTU profile is shown in Figure 2. This profile was obtained at a site where the main contaminant was creosote from a timber treatment plant (Campanella et al., 1994). The measured bulk resistivity values are larger compared to the background values in zones with the contaminant. The free product was verified by monitoring well sampling.

The electrical measurements discussed above relate to resistivity. However, electrical measurements can also be made to measure the dielectric constant of the material surrounding a penetrometer. The resistivity is somewhat insensitive to contaminant type whereas, the dielectric constant can be very sensitive to contaminant type. The dielectric constant is frequency dependent. However, above about 50 MHz the dielectric constant is essentially constant. Delft Geotechnics have developed a High-frequency-impedance-measuring (HIM) probe to measure both dielectric constant and conductivity of soil samples as a function of frequency and hence, detect the presence of contaminants. The probe is a cone penetrometer based instrument with a retractable cone. The cone is pushed into the ground in the closed position and at the required depth the tip is stopped while the outer conductor is allowed to move on, so that a soil sample fills the space. The sample chamber contains a central pin which acts as an antenna and the rim of the sample chamber acts as a receiver, as shown on Figure 3. A high frequency electromagnetic field pulse is generated at the ground surface and

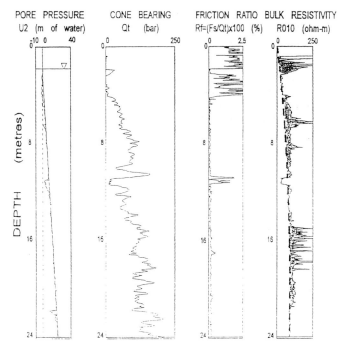

Figure 2.   Example resistivity CPTU sounding at an organic contamination by heavy oils
(After Campanella et al., 1994).

1 = Coax cable
2 = Isolator
3 = Retractable rim receive
4 = Soil sample
5 = Central antenna

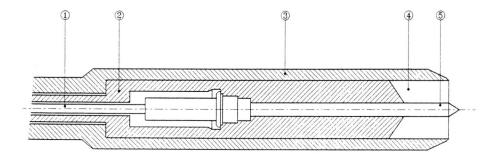

Figure 3.   Cross Section of HIM-probe (After Stienstra et al., 1994).

transmitted through a coaxial cable to the sample in the HIM probe. The resulting dielectric constant and conductivity are determined as a function of frequency (10-500MHz). After the measurement the sample is pushed out of the inner cylinder re-establishing the cone shape. At the next depth the procedure is repeated. Accuracy of the HIM probe is 5% for both the dielectric constant and conductivity, however, the issue of cross-contamination may be a problem. Contaminants such as light and dense NAPL zones can be detected by the low values of dielectric constant and conductivity in relation to the background surrounding water saturated soil. The location of the contaminant in the sample cup showed some effect on the measured values (Stienstra and van Deen, 1994).

The acidity of a material can be measured using pH sensors mounted either inside a cone or on the surface of the probe. The major disadvantage of sensors mounted inside the cone is that a sample of pore liquid must be drawn into the measuring cell and then expelled. This process can be difficult in low permeability soils such as clay and the cell and sensor must be cleaned after each reading. More details about sampling techniques will be given in a later section. Sensors mounted on the outside of the cone have the advantage of direct exposure to the surrounding material, however, abrasion and damage to the sensor can be a major problem. Several CPT manufactures and operators have placed a pH sensor in a small recess a short distance behind the friction sleeve. This recess is designed to produce a small vortex for pore liquid to enter during cone penetration. The sensor is then well protected from abrasion and damage from the surrounding solid material and can measure continuous variations of the pH of the pore liquid. The pH measurements are sensitive to temperature changes and generally a temperature sensor is mounted adjacent to the pH sensor to allow automatic correction for temperature effects. The continuous measurement of pH can be a useful guide for detecting certain contaminants with significantly different pH values from that of the background soil. The pH sensor was successfully used to differentiate between acidic tar material and waste drilling mud at an old disposal site in California. The temperature profile was also elevated over the acid tar depth interval and confirmed that chemical reactions were occurring in the acid materials. At this site the resistivity profiles could not delineate between the acid tar material and the drilling muds (Bratton, 1994).

The oxygen exchange capacity of a material can be measured with a sensor for redox potential. A CPT probe (Chemoprobe) described by Olie et al. (1992) carries out the measurements of redox potential, pH and conductivity. These three parameters are major variables of chemical equilibrium for inorganic substances. The sensors are mounted inside the cone and a sample is drawn into a measuring cell located a short distance behind the cone tip. The measurements are made under a nitrogen atmosphere to reduce the exchange of atmospheric oxygen with dissolved gases from ground water. A slight excess pressure of nitrogen is used during penetration of the probe to stop the flow of liquid into the 15 ml measuring cell. The nitrogen is supplied from a gas cylinder at the surface. The stainless steel porous filter is cleaned by demineralised water, which is also used to clean the sensors and check the calibration. The water is pushed out by an excess nitrogen pressure. A pressure sensor is also included to monitor the flow of liquid into the measuring cell and to estimate the hydraulic conductivity of the surrounding soil. Olie et al. (1992) showed that the measurement of redox potential, pH and electrical conductivity enabled the monitoring of in-situ sanitation, designed to dissolve a floating layer of versatic acid with infiltration of hydroxide solution beneath a storage tank at a petrochemical plant in the Netherlands. A similar probe (Chemicone) was described by Woeller et al. (1991) and is shown in Figure 4. Further details of devices that sample the surrounding pore liquid or vapor are given in the next section.

Gamma and neutron sources and sensors have been added to cone penetrometers in the past in an effort to measure in-situ density and moisture content (e.g. Mitchell et al., 1988; Sully et al., 1988). However, these devices have not become popular because they contained active radiation sources which can present significant problems if the probes should become lost in the ground and require expensive recovery. Recently there has been a trend toward adding passive gamma-ray sensors in an effort to detect radioactive contaminants (Marton et al., 1988). A variety of different passive sensors are available, the selection of the appropriate sensor is based on efficiency, range of gamma-ray energies expected, temperature dependence

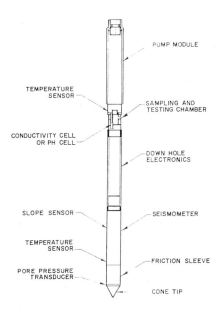

Figure 4.   Example of a sampling cone penetrometer-chemi-cone (After Woeller et al., 1991)

of the sensor and sensor ruggedness. An example of a radiation detection cone is shown in Figure 5. The application of gamma-ray sensors is clearly limited to environments where specific gamma-ray emitting contaminants are possible.

The most recent sensors to be added to the cone penetrometer for environmental applications are those that involve laser induced fluorescence (LIF). Hydrocarbons are one of the most common ground contaminants and most hydrocarbons, because of the poly-aromatic constituents, produce fluorescence when irradiated with various forms of light. LIF technology applied to the environmental field is relatively new with the first published work conducted by Hirshfield et al. (1984) and Chudyk et al. (1985). The initial work and much of the recent work has centered around the development of field portable LIF systems. The U.S. Army Engineer Waterways Experiment Station (WES) under the sponsorship of the U.S. Army Environmental Center (AEC) has developed a Site Characterization and Analysis Penetrometer System (SCAPS) for investigating and screening sites for ground contamination using LIF (Lieberman et al., 1991). SCAPS incorporates existing cone penetrometer technology with LIF. A 6.35 mm diameter sapphire window is mounted flush on the side of the cone approximately 60 cm above the tip. This window provides a view port for the fibre optic based LIF system. A pulsed nitrogen laser light (337 nm wavelength) is sent down to the window over a 400 micron diameter 60 m long silica optic fibre. Fluorescence generated in the surrounding material is carried back to the surface by a second fibre where it is dispersed using a spectrograph and the intensity quantified with a time-grated, one-dimensional photo diode array. Readout of a fluorescence emission spectrum requires approximately 16 milli-seconds. A microcomputer based data acquisition and processing system controls the fluorometer system, acquires and stores sensor data once a second, and plots the data in real-time as vertical profiles on a CRT display. A schematic diagram of the SCAPS probe is shown in Figure 6. Initial field trials at petroleum-oil-lubricant (POL) contaminated sites have been promising although results have been qualitative since calibration for specific contaminates is difficult.

Research has shown that common fuel contaminants such as heating oil, jet fuels, gasoline and diesel fuel marine exhibit strong fluorescence signatures, with the degree of fluorescence

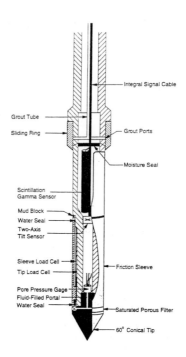

Figure 5. Example of a radiation detection cone (After Applied Research Associates, Inc., 1994).

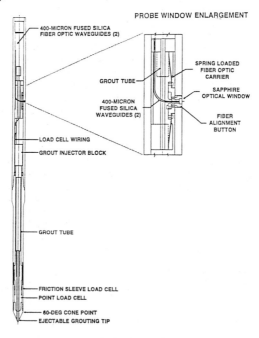

Figure 6. Schematic of SCAPS probe (After Liebermann et al., 1991).

depending on the excitation wavelength (Gillispie and St. Germain, 1993; Chudyk et al., 1985). However, common chemical contaminants such as chlorinated hydrocarbons (e.g. TCE, PCE) do not fluoresce and are not suitable for the fluorescence technique. The intensity of fluorescence is a function of excitation wavelength and recent efforts have been made to develop tunable laser fluorimeters, i.e. systems that can vary the wavelength of the laser light source (Bratton et al., 1993), and hence, detect a greater range of contaminants. More information about the contaminants can be obtained if the complete wavelength-time-intensity matrix is recorded, although this measurement takes a little longer to perform and a pause in the penetration is required. Fluorescence research to date has concentrated on the aqueous phase and very little work has been carried out to evaluating the LIF characteristics of contaminated soils (Apitz et al., 1992). The intensity of the LIF signal is strongly dependent on soil type, with sands having a stronger signal for a given concentration than clayey soils. Bratton et al. (1993) suggest that LIF research in soils is in the infancy stage and even standard laboratory procedures for evaluating LIF response in soil materials have not yet been developed. The calibration and detection limits of LIF in soils are complicated by soil type, soil grain size effects, natural organic compounds (humic acid) and the influence of time on the contaminant degradation. There are also problems of low signal to noise ratios for the LIF systems. Currently, field correlation's of LIF intensity to contaminate concentration are preferred since laboratory calibrations are still uncertain (Bratton et al., 1993). Concerns also exist over the long term durability and maintenance of the fiber optic cable.

Van Ree and Olie (1993) have also developed a fluorescence CPT probe (Hydrocarbon probe). The probe is 55 mm in diameter and contains a UV light source as well the fluorescence detection system. No fiber optic cables are required since the complete sensing system is located in the probe. During penetration measurements are made by illuminating the material surrounding the probe with a small mercury lamp to produce the UV light source placed behind a clear window. The fluorescence emitted by the hydrocarbons is detected in the probe by a small photomultiplicator tube. The signal is conducted through the electrical CPT cable to a data processing system at the ground surface. A detection limit of 50 mg/kg dry weight for light NAPL is claimed. The intensity of the radiation emitted by the contaminant is an indication of the concentration of the product in the soil, however, specific calibration is required. The detection system can also handle other wavelengths for identification of other contaminants by using filters to control the excitation wavelength. Preliminary results at a demonstration site where NAPL layers of domestic fuel oil were present show excellent results (Olie et al., 1993). An example profile from the hydrocarbon probe is shown in Figure 7. These results show excellent baseline stability and clear sharp peaks in a series of alternating NAPL layers. Van Ree and Olie (1993) also investigated the effects of smearing and the displacement of the measured contaminant due to the penetration process. The limited results indicate that the soil effectively cleans the window on the surface of the cone and that the displacement (smearing) of the detected layer can be as much as 5 cm.

GEO-ENVIRONMENTAL PENETROMETER SAMPLING DEVICES
The cone penetrometers described in the above section are primarily screening devices that log the ground profile for geotechnical and chemical measurements. Based on these measurements it is often possible to identify potentially critical zones or regions that may require more selective testing to measure or monitor specific contaminants. Sampling and monitoring wells are usually installed for this purpose and penetrometer technology has been used to develop a complete range of short and long term sampling probes. Sampling probes have been developed to sample either vapors, liquids or solids. The following section describes some of the main developments in this area.

Most of the available vapor and liquid samplers have some common features. Almost all the samplers are pushed to the desired depth based on adjacent CPT profiles. Sometimes the sampler is pushed down the same hole as the CPT, and since the sampler is of a larger diameter (typically 50 mm) than the CPT contact can be maintained with the surrounding material. Generally, the push rods are pulled back to expose a filter to the surrounding material. This avoids contamination of the filter before reaching the required depth. Selection of the appropriate filter material is based on the type of sample (gas or liquid) and the expected

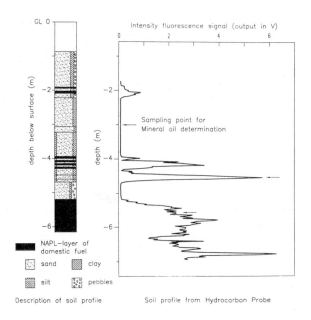

Figure 7.   Example profile from hydrocarbon probe (After Olie et al., 1994).

contaminant type. A sample is pulled into the sampler using either a vacuum or the natural in-situ fluid pressure. The sampler can then be either withdrawn to the surface for sample retrieval and sampler decontamination or the sample can be taken to the surface via a tube or by wireline and the process repeated at a greater depth. The following describes some typical samplers.

Liquid Samplers
The most common discrete depth in-situ water samplers are the Hydropunch and the BAT Enviroprobe. The Hydropunch and its variations is a simple sample tool that is pushed to the desired depth and the push rods withdrawn to expose the filter screen. The filter screens can be made from a range of materials although stainless steel is the most common. Screened intervals can vary in length from 100 mm to 1500 mm depending on ground conditions, required sample depth, contaminant type and hydraulic conductivity of surrounding material. A small diameter bailer is lowered through the hollow push rods and body of the sampler to collect a liquid sample. A peristaltic pump can also be used to pump larger volumes of non-volatile liquids to the surface. This type of push-in liquid sampler is very common, simple to use and can produce large samples. However, there is little control over the sampling process and liquid samples are often turbid (i.e. contain suspended solids), especially when using a coarse screen. Modified versions of the Hydropunch concept have been developed to install long term monitoring wells. These include innovative techniques to seal and grout the sampler into the ground after the sampler is pushed to the required depth.

The BAT Enviroprobe was developed by Torstensson (1984) and consists of three basic components:
1.   A sealed filter tip with retractable sleeve attached to the push rods.
2.   Evacuated and sterilized glass sample vials, enclosed in a housing and lowered to the filter tip via a wireline system.
3.   A disposable, double-ended hypodermic needle which makes a hydraulic connection with the pore fluid by puncturing the self-sealing flexible septum in the filter tip.
The filling rate of the sample vials can be monitored using a pore pressure transducer attached to the vial. The monitoring shows when the fluid infiltration is complete, assuring that

pressure inside the vial is equal to the in-situ fluid pressure. These measurements can be used to estimate the hydraulic conductivity of the surrounding material. Various modified versions of this concept are now in use for ground fluid sampling. The system is most applicable for retrieving fluid samples where limited volumes are sufficient (<150 ml). Small filter size allows for discrete sampling intervals, but can create longer sampling times in less permeable materials. The protective sleeve and fine filter screen produces fluid samples with low turbidity. However, sometimes the needle can become blocked with fine material passing through the filter resulting in incomplete filling of the vial.

Modified versions of the above samplers have been developed to provide a mixture of the two techniques. Samplers exist that allow large volume (>1200 ml) samples to be taken under a back pressure of argon or nitrogen gas to avoid volatilization. The measurement of pressure with time can be used to monitor the sampling process.

Vapor Samplers

Vapor (gas) samples can be obtained in a manner similar to that described above for liquid samples. However, special care is required to purge the sample tubes and store the samples. Typically, the sampler is pushed to the required depth, the filter element exposed and a vacuum applied to draw a vapor sample to the surface. The volume of gas can be monitored using special monitoring equipment and typical sample containers are Tevlar bags, gas tight syringes and glass or steel sampling vessels. Special disposable plastic tubing is used to draw the sample to the surface.

Vapor sampling modules have also been added to cone penetrometers to allow sampling during a CPT. The sampling module is typically located a short distance behind the cone. Samples can be taken during short pauses in the penetration. To minimize cross-contamination for subsequent samples within one vertical sounding requires either a two line sampling design for purging of the vapor collection lines between samples or a single line system with nitrogen or argon gas purging. A small positive internal gas pressure can stop the inflow of gases during the penetration process.

Solid Samplers

A variety of push-in soil (solid) samplers are now available. Most are based on designs similar to the Gouda or MOSTAP soil samplers. The samplers are pushed to the required depth in a closed position. The Gouda type samplers have an inner cone tip that is retracted to the locked position leaving a hollow sampler with 31 mm diameter stainless steel or brass sample tubes. The hollow sampler is then pushed to collect a sample. The filled sampler and push rods are then retrieved to the ground surface. The MOSTAP type samplers contain a wire to fix the position of the inner cone tip before pushing to obtain a sample. Modifications have also been made to include a wireline system so that solid samples can be retrieved at multiple depths rather than retrieving and re deploying the sampler and rods at each interval.

SEALING AND DECONTAMINATION PROCEDURES

The hole produced by the penetrometer requires sealing usually with a special grout. The grouting can be carried out either after the push rods are removed or during the removal process. Typically the grouting is carried out after rod removal using special grout push rods with disposable tips. The hole is grouted from the bottom up as the grout rod is pulled from the ground. The grout rods generally follow the previous penetrometer hole since the rods follow the path of least resistance. However, some operators use penetrometers that allow grouting during penetrometer retrieval. The grout is usually delivered to the penetrometer through a small diameter grout tube pre threaded through the push rods. The grout can exit the penetrometer through either a sacrificial cone tip or through ports above the cone located on the friction reducer. A special high torque pump is used to provide the grout.

To clean the push rods and equipment after a sounding all downhole equipment can be pulled through a rod-washing, decontamination chamber. The chamber can be mounted at the base of the hydraulic thrust cylinders. The push rods can be steam cleaned as they pass through the chamber, before handling by field personnel. Rubber wipers at the entry and exit of the chamber control water leakage. The waste water is collected for later disposal. Many of the special truck mounted penetrometer systems also include stainless steel sinks for decontamination of samplers and penetrometers.

FUTURE TRENDS
The potential cost savings of penetrometer based geo-environmental site investigations has fostered considerable research expenditures into additional sensors. Considerable research is underway on improvements to the fluorescence techniques. Research is also underway to investigate new sensors that could be incorporated into cone penetrometers, such as, Raman spectroscopy, fiber optic chemical sensors, laser-induced-breakdown-spectroscopy (LIBS), time-domain-reflectometry (TDR), ground penetrating radar and integrated optoelectronic chemical sensors.

Raman spectroscopy involves a powerful laser light which is focused onto the reverse side of a glass or quartz slide which has been coated with a species-specific chemical. When a contaminant vapor is in contact with the coated surface of the sensor the reflected light can be interpreted using spectral analyses of the shift in wavelength. The shift is a characteristic of the contaminant, although the shift is generally very small. Improved levels of detection can be made by pressurizing the vapor and a large range of contaminants can be identified using different chemical coatings. Fiber optics can be used to carry the laser and reflected light to the processing data acquisition system at the ground surface. Laser-induced-breakdown-spectroscopy (LIBS) can be used to identify hazardous metal compounds. The laser light must be focused directly onto the material to form a short lived plasma which emits light that can be collected using fiber optics and analyzed. Time-domain-reflectometry (TDR) sensors are under development to measure soil moisture content, which can be an important parameter in unsaturated soils.

Ground penetrating radar (GPR) technology for surface measurements has developed considerable in recent years and systems are under development that will incorporate the technique into penetrometer probes. GPR responds to changes in dielectric constant of the material which can be sensitive to contaminant type. However, the dielectric constant of soils varies over a wide range and is strongly influenced by the presence of water. The depth of penetration is also controlled by soil type and excitation frequency. Penetration of the GPR can be very limited in saturated clayey soils.

One of the most interesting areas of current research is on the development of integrated optoelectric chemical sensors. They contain a small diode laser which is focused on a chemically selective overlay. These sensors can be small and in modular form and can be inexpensive.

A major problem with the development and application of chemical sensors relates to the interaction of the sensor with the contaminant. Most of the sensors require that the contaminant (usually in vapor or liquid form) be pulled into the probe so that the chemicals can interact with the sensor. This produces problems related to the cleaning of the sensor, measuring cell and filter element to avoid cross-contamination, as well as the time required for this process. Little research has been carried out to evaluate these problems and the issues related to; the interaction of the contaminant and soil, the interaction of the contaminant and measuring device, the contaminate state and contaminate mixtures.

SUMMARY
To better characterize potentially hazardous sites, improved investigation devices and methods are being developed which use cone penetrometers. The CPT gathers high quality in-situ geotechnical information in a rapid cost effective manner. Electrical and chemical sensors have been adapted to cone penetrometer probes to enable mapping of subsurface contamination in sufficient detail to reduce the need for more costly invasive subsurface sampling and monitoring points. Traditional drilling and sampling methods when compared to CPT methods have high waste management costs from handling and disposal of

contaminated materials. Also, CPT methods minimize exposure of field personnel to hazardous environments.

Significant developments have been made in recent years to improve geo-environmental site characterization using penetrometer technology. Sensors have been added to the cone penetrometer to enhance the logging capabilities for both mechanical and chemical measurements.

Bratton and Higgins (1992) describe a synergistic approach to 3-D site characterization by utilizing a combination of surface geophysical technology and direct-push (i.e. penetration test) technology. By taking advantage of the synergism between the two a significant improvement can be achieved in the interpretation of the geophysical data.

REFERENCES

Apitz, S.E., Theriault, G.A., and Lieberman, S.H., 1992. Optimization of the Optical Characteristics of a Fibre-Optic Guided Laser Fluorescence Technique for the In-Situ Evaluation of Fuels in Soils. SPIE, The International Society for Optical Engineering, Proceedings Vol. 1637, Environmental Process and Treatment Technologies (T. Vo-Dinh, Ed.). OE/LASE '92, Los Angeles, California, January 1992.

Bratton, J., 1994. Personal communication.

Bratton, J.L. and Higgins, C.J., 1992. A Synergistic Approach to Three-Dimensional Site Characterization. Proceedings Federal Environmental Restoration Conference and Exhibition, Hazardous Materials Control Resources Inst., Greenbelt Md., pp. 278-285.

Bratton, W.L., Shinn II, J.D., Bratton, J.L., 1993. Air Force Site Characterization and Analysis Penetrometer System for Fuel-Contaminated Sites. Third International Symposium on Field Screening Methods for Hazardous Wastes and Toxic Chemicals, Las Vegas.

Campanella, R.G., Wesmees, I.A., 1989. Development and Use of an Electrical Resistivity Cone for Groundwater Contamination Studies. Proceedings 42 Canadian Geotechnical Conference, Winnipeg, pp. 1-11.

Campanella, R.G., Davies, M.P., Boyd, T.J., and Everard, J.L., 1994. Geoenvironmental Subsurface Site Characterization Using In-Situ Soil Testing Methods. First International Congres on Environmental Geotechniques, Edmonton, Alberta.

Chudyk, W.A., Carrabba, M.M. and Kenny, J.E., 1985. Remote Detection of Groundwater Contaminants Using Far-Ultraviolet Laser-Induced Fluorescence. Analytical Chemistry, 57, 1237, American Chemical Society, 1985.

Stienstra, P. and Van Deen, J.K., 1994. Field data collection techniques - unconventional sounding and sampling methods. Proceedings of the 20 - year Jubilee Symposium of the Ingeokring 1994, Delf, The Netherlands, Balkema, Rotterdam, pp. 41-56.

Gillispie, G.D., and St. Germain, R.W. In-Situ Tunable Laser Fluorescence Analysis of Hydrocarbons. North Dakota State University, Dept. of Chemistry, Fargo, ND.

Hirshfield, T., Deaton, T., Milanovich, F., Klainer, S.M., and Fitzsimmons, C., 1984. The Feasibility of Using Fiber Optics for Monitoring Groundwater Contaminants. Project Summary, Environmental Monitoring Systems Laboratory, U.S.E.P.A., January, 1984.

Hornsnell, M.R., 1988. The Use of Cone Penetration Testing To Obtain Environmental Data. Penetration Testing in the U.K., Institution of Civil Engineers, pp. 289-295.

Kermabon, A., Gehin, C., and Blavier, P., 1969. A Deep Sea Electrical Sediments. Geophysics, Vol. 34, No. 4, pp. 554-571, August.

Kroezen, M., 1981. Measurement of In Situ Density in Sandy Silty Soil. Canadian Geotechnical Society Newletter, Sept., Vol. 18, No. 4., p13.

Marton, R., Taylor, L., and Wilson, K., 1988. Development of an In-Situ Subsurface Radioactivity Detection System - the Radcone. Proceedings of Waste Management '88, Univeristy of Arizona, Tucson, February 28 - March 3, 1988.

Mitchell, J.K., 1988. New Developments in Penetration Tests and Equipment, in Vol. 1, Penetration Testing, 1988. Proceedings of the First International Symposium on Penetration Testing/ISOPT-1, Orlando, Florida, March 1988., ed. J. DeRuiter, A.A. Balkema, Rotterdam/Brookfield, 1988.

Olie, J.J., Van Ree, C.C.D.F., and Bremmer, C., 1992. In-Situ Measurement by Chemoprobe of Groundwater from In-Situ Sanitation of Versatic Acid Spill, Geotechnique 42, No. 1, pp.13-21.

Olie, J.J., Meijer, J.C. and Visser, W. 1993. Status Report on In-situ Detection of NAPL-Layers of Petroleum Products with Oil Prospecting Probe Mark 1.

Robertson, P.K., 1990. Soil Classification Using the Cone Penetration Test. Canadian Geotechnical Journal, Vol. 27, No., 1, pp. 151-158.

Robertson, P.K., and Campanella, R.G., 1983. Interpretation of Cone Penetration Tests. Canadian Geotechnical Journal, Vol. 20, No. 4.

Sully, J.P., Echezuria, H.J., 1988. In-Situ Density Measurement With Nuclear Cone Penetrometer, in Vol. 2, Penetration Testing 1988. Proceedings of the First International Symposium on Penetration Testing/ISOPT-1, Orlando, Florida, March 1988, ed. J. DeRuiter, A.A. Balkema, Rotterdam/Brookfield, 1988.

St. Germain, R.W., and Gillispie, G.D., 1992. In-Situ Tunable Laser Fluorescence Analysis of Hydrocarbons. Environmental Process and Treatment Technologies (T. Vo-Dinh, Ed), SPIE 1637, pp.159-171.

Strutyusky, A.I., Sandiford, R.E., and Cavaliere, D., 1991. Use of Piezometric Core Penetration Testing with Electrical Conductivity Measurements for the Detection of Hydrocarbon Contamination in Saturated Granular Soils, Current Practices in Groundwater and Vadose Zone Investigations, ASTM STP, 1118.

Torstensson, B.A., 1984. A new System for Groundwater Monitoring. Ground Water Monitoring Review, Vol. 4, No. 4, pp. 131-138.

Van Ree, C.C.D.F. and Olie, J.J., 1993. The development of in-situ measurement techniques by means of CPT - equipment in the Netherlands. Proceedings Third International Symposium Field Screening Methods for Hazardous Wastes and Toxic Chemicals, February 24-26, 1993, Las Vegas, Nevada, pp. 296-303.

Vlasblom, A., 1977. Density Measurements in situ. LGM-Mededelingen, XVIII (4), pp. 69-70.

Woeller, D.J. Weemees, I., Kohan, M., Jolly, G., and Robertson, P.K., 1991. Penetration Testing for Groundwater Contaminants. Proceedings of Geotechnical Engineering Congress, ASCE, Boulder, Colorado, pg. 76-83.

The multifunctional Envirocone® test system

D. A. O'NEILL[1], G. BALDI[1] and A. DELLA TORRE[2]
[1]ISMES S.p.A., Bergamo, Italy; [2]Geotec srl, Brussaporto (BG), Italy

ABSTRACT

The multifunctional Envirocone® test system (METS), developed by ISMES for rapid, comprehensive in situ characterization of contaminated sites, enables sampling of pore water and soil gas up to depths of about 50m. The heart of METS is the 4.4cm-diameter Envirocone, the key element in the hydraulic connection between the subsurface strata of interest and the physico-chemical analytical instruments and/or sample receptacles in the penetrometer cabin at ground surface. A continuous flow of ground water sample is tagged in real-time with five indicator parameters: temperature, pH, redox potential, dissolved oxygen and electrical conductivity. Similarly, soil gas can be readily monitored for dissolved oxygen, total volatile organics, combustible gas, H_2S, CO_2, etc. Another METS module permits sampling of soil solids. Environmental data are complemented by the simultaneous acquisiton of tip resistance, pore pressure and pore fluid temperature during Envirocone penetration and the option of conducting constant head permeability tests. Thus, geotechnical site characterization, ultimately required should the site be earmarked for construction, but also necessary for the rational design and implementation of clean-up and/or containment efforts, is facilitated. Such penetrometric systems also avoids various difficulties associated with conventional contaminated site characterization such as the generation of potentially polluted effluent from borehole drilling, purging and sampling of wells, etc. A field program was conducted at a site with both cohesionless and cohesive strata to evaluate physical capabilities of the METS such as sampling rate, maximum achievable depth, and repeatability of physico-chemical parameters. Envirocone penetration parameters (q_c and u) compare well with conventional piezocone penetration test results. Hydraulic conductivity measured with the Envirocone is consistently two orders of magnitude less than estimated in situ values due to penetration effects but appears to mirror trends. Three case studies using METS are presented: ground water quality evaluation in an agricultural zone; rapid evaluation of vadose zone contamination in the vicinity of a landfill; and evaluation of the extent of underground migration of tar sludge from coal-paraffin processing waste pond. System improvements currently underway are mentioned including use of in-line solid phase microextraction methods for component specific ground water analysis and various mechanical aspects associated with the Envirocone.

INTRODUCTION

Subsurface characterization with respect to both contaminants and geotechnical parameters is a critical step towards the rational treatment of contaminated sites, best taken after having effected an exhaustive historical review of the site. The design and successful implementation of containment and/or clean-up efforts depend on the accuracy and completeness of the information garnered from such an investigation program. The exploratory technique discussed herein also serves well in the monitoring and/or verification of the effectiveness of reclamation activity.

Current ground water investigation techniques at depths in excess of a couple of meters commonly involve borehole drilling with screened portions at the elevations of interest. A great variety of well-sampling techniques are in use with varying degrees of success [1], and these are being increasingly refined as more data become available relative to the effect of the type of pumping system, purge volumes and sampling rates, etc. on the chemical component concentrations measured [2,3]. Recent research has demonstrated that well design, construction, performance and purging can cause a variability of up to an order of magnitude in the accuracy of the analytic results for metals and organic compounds during purging of up to ten well volumes [4]. Furthermore, the effort, time and cost associated with the drilling process, the disposal of potentially contaminated drilling fluid and unwanted recovered soil, and well development, etc., are warranted only for at least semi-permanent monitoring networks at a previously characterized site. The negative aspects of drilling are most taxing on reconnaisance investigative work, tending to encourage coarse sample location grids thus increasing the risk of missing potential hot spots even with use of geostatistical methods. Essential stratigraphic and geotechnical information must then be obtained by additional in situ or laboratory testing, frequently requiring additional drilling.

Soil gas survey techniques, useful also for delineating hydrocarbon-contaminated groundwater [5], commonly involve installation of less complicated steel probes or wellpoints via manual or van-mounted driving systems to a fixed depth usually less than two meters.

Penetrometeric methods for characterizing and/or sampling contaminated strata represent an attractive alternative to these approaches given enhanced flexibility for horizontal and vertical sampling positions, more data, and not insignificant cost savings. Several concepts have been implemented:
- pumping or bailing from the surface to a penetrometer-installed wellpoint (e.g., Geoprobe [6]);
- sampling from passive elements in the penetrometer tip to obtain specimens which are then retrieved for analysis by extracting the cone (e.g., the Tortensson BAT Enviroprobe or MK [7]; the Hydropunch™ [8]); and
- outfitting a penetrometer near the tip with physico-chemical or other sensors (e.g., the Delft Geotechnics' chemoprobe [9]; the U.S. Waterways Experiment Station laser induced flourescence probe [10]; the electrical resistivity cone [11]).

Holistically addressing the issues relative to contaminated site characterization and sampling, ISMES S.p.A., of Bergamo, Italy has, instead, developed the in situ multifunctional Envirocone® test system (METS) which implements a piezocone element as a conduit for sending liquid and gas samples to the surface, as a tip resistance (q_c) and pore pressure (u) measuring instrument and as a soil solids sampler. METS facilitates the rapid, quantitative delineation of subsurface contamination plumes and the estimation of geotechnical parameters, both areally and with depth (up to 50m). Calibration of the device and case studies of use on three sites are presented herein.

DESCRIPTION OF METS
The design of the Envirocone test system incorporates the conventional piezocone penetrometer concept in a sampling instrument that creates a conduit between the water or

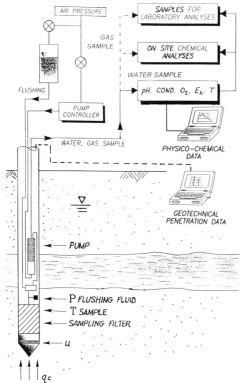

Fig.1 - METS: the multifunctional Evirocone® test system.

Fig. 2 - The 4.4cm-diam. Envirocone® with filter rings (left) and sintered stainless steel filter collar.

soil-gas sample and analytic devices at ground surface (Fig.1). The Envirocone element (Fig.2) consists of a 43.7mm-diameter, stainless steel cylindrical module, about 1.5m long, capped by a conical, 60deg-apex tip with a projected area of 15cm². The pore pressure filter is located immediately behind the conical tip. Instrumentation located inside the stainless steel housing includes a load cell, two pressure transducers, and a thermoresistor that measure q_c, u, sampling/flushing pressure (P) and temperature (T), respectively.

The 75 mm-long filter zone through which water and gas samples are drawn is located approximately 10 to 50mm above the pore pressure filter, to allow space for the silicone oil deairing procedure without inundating the filter zone. Two filter types have been successfully employed: stainless steel rings with 20μm to 200μm-deep channels and 50μm sintered steel collars (Fig.2). The most recent version of the Envirocone has a 100mm filtered length to enhance the sampling rate and is equipped with inclinometers and an additional pressure transducer to monitor pump and flushing fluid pressure, to avoid possible hydraulic fracturing and consequent risk of cross contamination during flushing and to better evaluate the applied head difference for permeability measurements given the significant friction head losses.

The pump module, mounted immediately behind the piezocone tip, consists of a positive displacement, polytetraflouroethylene (PTFE) accordian pump activated by up to 6 to 8 bars of air pressure cycled by a synchronized, double-output pump controller. The pump is oriented with the opening pointing upward so as not to trap hydrocarbon product. Positive displacement pumps are preferred over peristaltic and suction pumps as well as over bailers or gas driven systems, not only for maintaining maximum chemical integrity but also because they provide a continuous stream of sample [2]. A bladder pump has also been employed but is not treated herein.

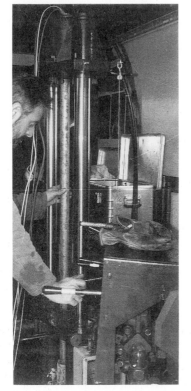

Fig.3 - Hydraulic penetration system.

Sampling and flushing are effected through two flexible PTFE lines (ID=2 mm) that wend their way to the surface through the penetration rods. PTFE was chosen as pump and sample line material as it sorbs lesser amounts of hydrocarbons from and leaches lesser amounts of metals and organics into acqueous specimens as compared to polypropylene, polyethylene, polyvinyl chloride (PVC) and silicone [2, 12].

A truck-mounted hydraulic system is used to penetrate the Envirocone to the desired depth. The cabin of the approximately 20t vehicle also houses the air/water pressure flushing system, the pump controller, the flow cell, a gas sampling vacuum pump, a gas bag sampling box, field portable analytic devices, and a fume hood. As with typical piezocone penetrometer systems, penetration parameters are calculated in real time and recorded by computer. The penetrometer vehicle is outfitted with a steam cleaning device to decontaminate the rods during extrusion.

For ground water investigations, the sample is pumped through the flow cell (Fig.4) outfitted with a series of electronic sensors for measuring pH (an Ag-AgCl cell in KCl gel), reduction potential (E_h), dissolved oxygen (O_2) (polarographic Ag tube and Pt wire anode), electrical conductivity and temperature. This real-time characterizing or "tagging" of the fluid sample flow with physico-chemical indicator parameters is stored on a portable computer at a rate of up to 2 readings/sec. For soil-gas sampling in the vadose zone, a vacuum is either applied directly to the Envirocone filter or to the bag sampling box. Field portable instruments such as a combustible gas indicator (CGI), photoionization devices for total organic vapor analyses, and Dräger tubes can be put on-line or in contact with bag samples to perform

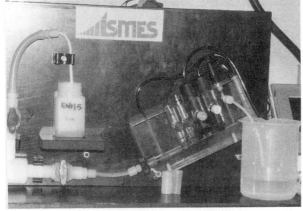

Fig.4 - Flow cell manifold with sensors and bypass valve.

measurements. A portable gas chromatograph is foreseen for use in the near future which will greatly enhance the component specificity of the system. A soil sampler module can be mounted on the Envirocone for retrieving disturbed, 45 mm-diameter, approximately 500mm-long specimens in either rigid PVC (rigid PVC sorbs less organics than PTFE [12]) or stainless steel liners, depending on the type of contaminant, for eventual identification and analytic testing.

In situ hydraulic conductivity can be evaluated with the Envirocone in two ways: (1) performance of constant head tests (permeable soils) and (2) interpretation of pore pressure dissipation curves according to consolidation concepts (cohesive soils) [13].

Among the advantages of accessing contaminated strata with a system such as METS are:
- continuous stratigraphic evaluation of the subsurface profile based on q_c and u simultaneously with the sampling process, ensuring that the desired contaminated strata has been tapped, along with providing data to evaluate strength, stiffness and permeability;
- precise evaluation of in situ ground water pressure for identification of acquifers and evaluation of hydraulic gradients;
- sampling of highly representative pore water and soil gas at different elevations without extracting the probe through use of the flushing system;
- immediate physico-chemical tagging of fluid samples, acquired on computer, permitting immediate assessment of representation and providing a digital sampling history;
- access to the flow of fluid samples in a relatively controlled environment enabling analytical field screening of samples which can radically reduce the risk of sample handling errors and ensure maximum sample integrity ;
- flexibility relative to its analytical capabilities in that expected contaminants of a given project can be monitored by simply positioning the appropriate field portable equipment for chosen indicator parameters on-line with the sample outlet;
- maximizing of sample integrity with highly representative water and gas samples travelling through a PTFE line directly from their point of origin to the measurement sensor or sample bottle/bag. Thus, for example, it is likely that volatile organic compound (VOC) measurements on Envirocone-retrieved specimens will be highly representative of actual conditions. Soil samples come in direct contact only with either PVC or stainless steel sampler casing; and
- relatively rapid definition of contaminant plumes given the investigative flexibility enabled by the four-wheel-drive, truck-mounted system, permitting effective well location where long term monitoring is necessary with consequent considerable cost savings.

As a penetrometer-based system, one drawback is its restriction to use only in penetratable soils such as fine gravels, sands, silts and clays. Encountering compacted or cemented strata, dense gravelly soils, glacial tills and miscellaneous fills can result in bending of the Envirocone or the penetration rods, especially when encountered at relatively shallow depths. It is also possible that such strata require penetration pressures in excess of the approximately 60 MPa capacity of the Envirocone. Typical methods of skirting this problem include pre-penetration with an uninstrumented penetrometer and pre-drilling through the hostile layer(s).

GROUNDWATER SAMPLING WITH METS
Calibration program
This calibration of METS focussed primarily on its ground water sampling capabilities. One phase of the study was executed with the Envirocone placed in a water-filled tube to physically calibrate the system and quantifying such aspects as the effects of ground water pressure and

425

pump cycle on the output of two prototype accordian pumps (versions 1 and 2). To investigate the mixed interface preceding a fluid sample, a buffer solution of known characteristics was introduced at a precise time. Simulating worst-case conditions, 120 m of tubing was attached to the Envirocone module during these tests, resulting in a dead volume of 365 ml. Prior to testing, the Envirocone circuit was flushed for several hours with 0.1N nitric acid followed by several liters of distilled water.

A second phase of the calibration program involved penetration and ground water sampling at the relatively uncontaminated stratified San Prospero site in the Po River valley. The behavior of the pump and different types of sample filters was investigated in different soil types and at different depths. Only the accordian pump version 1 was employed and 120 m of tubing was used between the Envirocone and the flow cell. Prior to introduction into the ground, METS was decontaminated by circulating, in succession, phosphoric acid (5% by vol.), distilled water, 0.1N nitric acid and distilled water through the lines. Indicator tags of samples retrieved via the Envirocone were compared with those from reference wells.

System characteristics

Water sampling flow rates merit some attention. On the one hand, in cases where eventual chemical analyses involve extraction processes requiring sample volumes on the order of 1000ml, hours of pumping may be implied if pump rates are low; on the other hand, should VOC's and gas sensitive parameters be involved, pump rates should not exceed about 100ml/min so as to maintain chemical integrity [14].

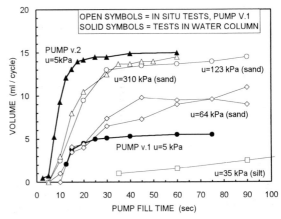

Fig.5 - *Effect of hydrostatic pressure on pump cycle and output.*

The accordian pump achieves its maximum volume output and minimum fill time as a function of increasing ground water head. As illustrated in Fig.5, the minimum fill time for pump version 1 is on the order of 20 sec when flow is uninhibited (i.e., no soil). This flow rate was not restricted in the presences of either sintered steel or steel ring filters. The output per cycle of pump version 1 in the field, also shown in Fig.5, is greatly affected by the hydrostatic pressure and the soil type. Field measurements with pump vers.1 indicate that the maximum output is probably reached when the tip is about 8 m below the water table. Optimum fill time in the sand layer at d=10 m is about 45 sec; other values ranged from 30 sec in the deeper sand layers to greater than 90 sec in the shallow silt layers. An additional 30 sec is required to empty the accordian.

Under a hydrostatic pressure of 5kPa, pump version 2, not yet tested in situ, has a greater output than vers.1 due to its longer body and the application of a force that actively extends the accordian. The imparting of suction to the filter zone by this active extension is expected to enhance the rate of sample acquisition at a given hydrostatic head, effectively tripling pump output while reducing cycle duration. Additional tests indicate that pump vers.2 achieved its maximum output under a 2m head while pump vers.1 only reached half its maximum at this pressure. The housing for the longer pump vers.2 has been redesigned such that a potential maximum of about 25 ml/cycle can be sent to the surface by the Envirocone.

The results of the flow cell sample arrival calibration are presented in Fig.6. Conductivity is shown to stabilize after about 185 ml (1150ml less 600ml less dead volume) of fluid has flowed past the sensors following first indications of sample arrival. E_h, on the other hand stabilized only after approximately 2000ml of flow. Both O_2 and E_h sensors are very sensitive to flow: during a pump cycle, these values decrease immediately upon cessation of flow. Thus, the upper boundary of the measurements represent actual sample values. It is believed that a large part of the 185ml "mixed" volume prior to achieving stabilized pH and conductivity values is attributable to the dead volume of about 7ml in the accordian in the empty position. This volume is reduced by 90% in the most recent Envirocone prototype.

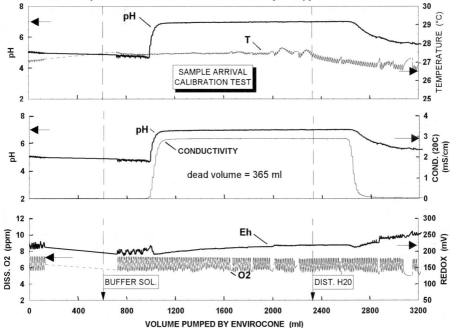

Fig.6 - Sample arrival calibration: distilled water (pH=5) and buffer solution (pH=7).

Such calibrations serve to identify the most appropriate tagging or indicator parameters for different pollutants. For example, for VOC sampling, both conductivity and dissolved O_2 have been recommended [1]. Other research, however, has demonstrated that pH, conductivity and temperature are not sufficiently sensitive to evaluate whether or not all stagnant well water has been removed [3].

Metals potentially leachable from stainless steel (Fe, Mn, Zn, Ni, Va, Cd, Cr, and Cu) were analyzed spectrophotometrically for a sample of deionized water (pH=5) retrieved by the Envirocone in the laboratory and compared with the results for a pristine sample of the same water untouched by the probe. These results indicate a slight contribution due to contact with the Envirocone of Fe and Zn (0.1ppm and 0.25ppm, respectively) and negligible amounts of the other metals.

Site characterization and ground water quality investigation

A ground water quality program was effected in the fluvially-deposited Po River sand. The soil profile comprises an interstratified clay and sand stratum underlain by a 1m thick sandy silt layer which overlies more than 35m dense to medium dense uniform sand, as demonstrated by the Envirocone penetration resistance and pore pressure measurements presented in Fig.7, which indicate good agreement with the conventional CPTU test. In this soil profile a maximum depth of 35m was achieved with the 15cm²-Envirocone while a maximum of 52m was reached with the 10cm²-piezocone. Differences indicated below a d=20m are associated with variations with the elevation of the clay strata (low q_c) at the site. Penetration pore pressures are also well-mirrored. Temperature data are those obtained during penetration and thus exhibit frictional warming. The cooling indicated at a d=16m was registered during an overnight pause and thus better represents the the ambient temperature of the fluid at that depth. Unless assisted by pumping, reaching equilibrium temperatures can take several hours.

Prior to advancing to a given sampling depth, the sampling lines and filter zone are flushed with 100ml to 400ml of deionized water. Part of this procedure can be dedicated to the performance of constant head permeability tests from which the hydraulic conductivity, k, can be calculated as k (m/sec) = Q/Fh where Q (m³/sec) is the flow monitored either in the air/water interface tank or, in the case of clay, utilizing a flow pump to minimize equilibrium time; F is the Hvorslev shape factor ($=(2\pi L/\ln[L/d+(1+(L/d)^2)^{0.5}]$) which is about equal to other estimates of this factor for L/d≈2 [15]; and h is equal to the difference in pressure head taking into account the significant friction losses in the small bore flushing line. Some of the scatter in the constant head test results in Fig.7 can be attributed to inaccuracy in the

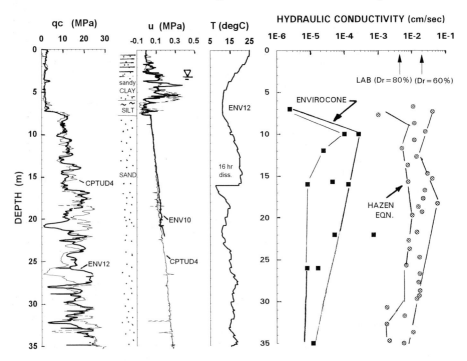

Fig.7 - Geotechnical characterization at San Prospero with the Envirocone®.

calculation of the friction head loss - a problem avoided in the newest Envirocone which is equipped with a flushing fluid pressure transducer.

As seen in Fig.7, hydraulic conductivity measured in the silt layer was about 2×10^{-6}cm/sec. In the sand stratum, k ranged between about 2×10^{-4}cm/sec at d=10m to about 1×10^{-5}cm/sec at d=35m. The results of a large number of grain size analyses on sand sampled from various depths indicate the stratum to be composed of a uniform sand ($c_u \approx 4$, $D_{10} \approx 0.2$mm) with typically 4% to 7% fine particles. Permeability estimated according to Hazen's equation ($k(cm/sec)=D_{10}(mm)^2$) is consistently about 100 times greater than the Envirocone values. Values of k measured with the Envirocone in the sand are also about two orders of magnitude less than the vertical permeability measured in a rigid wall laboratory permeameter on a specimen of Ticino sand with 5% fines added to simulate in situ material, compacted to D_r=60% (approximately the in situ density). Compacting the specimen to D_r=80%, however, reduced the difference to only about 1.5 orders of magnitude. Thus, local densification due to penetration as well as possible particle crushing are likely explanations of the significantly lower measured k. It is encouraging to note that, in any case, the slight trend of decreasing k from Hazen's equation seems to be mirrored by the Envirocone measurements. Lutenegger et al. (1991) report that the permeability measured in clay with a push-in piezometer is about one tenth that measured with a pre-drilled piezometer. Additional comparisons in different soils between push-in devices and less disturbed self-boring or pre-drilled piezometers would provide an empirical basis which could render penetration device k measurements useful.

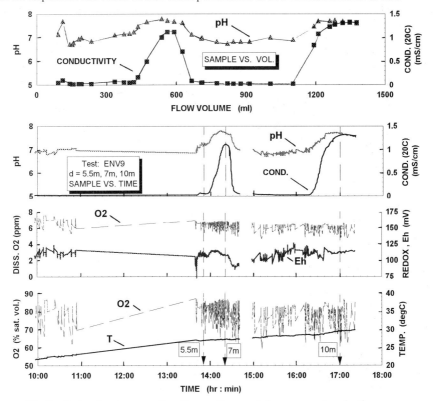

Fig.8 - Ground water sampling flow cell records from succesive depths.

Critical to the flushing process which decontaminates the pump, the sampling line and the filter zone is the double-output of the pump controller that automatically synchronizes pressurizing of the flushing fluid and emptying of the pump. With such a flushing system, a sample can be "pushed" to the surface with a flow of distilled water. To capitalize on this feature it is necessary to maintain a running record of cumulative flow. In Fig.8, flow cell records are marked for the arrival of samples from depths of 5.5m (sandy clay), 7m (silt) and 10m (sand). Given the low permeability and low hydrostatic pressure in the clay layer, pump output was less than 1 ml/min and only 35ml was pumped at this level. The Envirocone tip was advanced to a depth of 7m without flushing, and 200ml were pumped at an increased sampling rate. The samples from both 5.5m and 7m were then "pushed" to the surface by pumping deionized water.

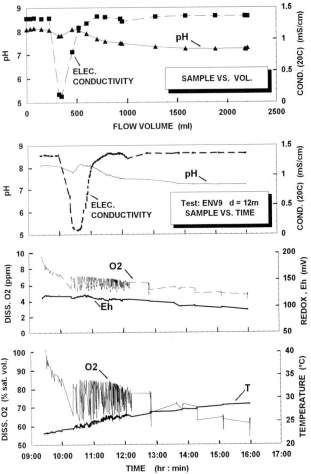

Fig.9 - Flow cell record for sample from d=12m, S. Prospero.

From the calibration in Fig.6, it is evident that about 185ml to 200ml of sample flow is required to achieve stabilized values. Thus the flow cell data for 5.5m are not representative while those for 7m are only just valid. On the other hand, the measurement for d=10m can be considered representative since a volume in excess of 650ml was pumped, 285ml of which passed through the flow cell. A typical flow cell record is presented in Fig.9 wherein the arrival of about 100ml of distilled water flushed through the line (as identified by the reduction in conductivity to about 0.1mS/cm), is followed by a flow of almost 2 liters of sample. Conductivity stabilized first, followed by pH then, approximately 500ml later, by O_2 and E_h.

Stabilized flow cell measurements selected from flow records from five Envirocone soundings and summarized in the upper part of Fig.10 together with the results from three different reference wells (RW), sampled for comparison with Envirocone data. The wells were screened over 1m lengths and were sampled with a positive displacement bladder pump following purging of about three to five well (casing) volumes. RW samples were tested in the flow cell on the same day as Envirocone-retrieved samples were tested at the same depth: RW at 11.4m

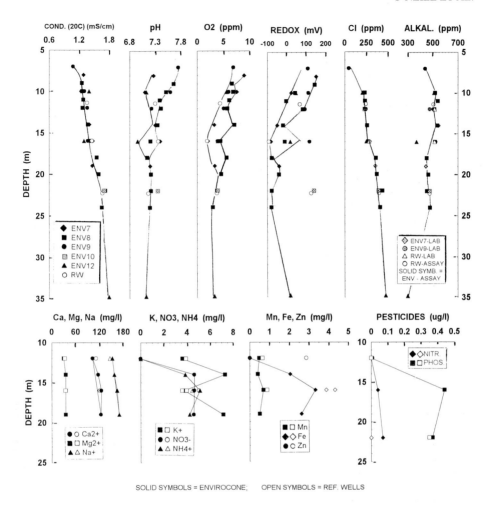

Fig. 10 - Ground water quality at San Prospero based on Envirocone® and well samples.

should be compared with test ENV9, RW at 16m with ENV7 and RW at 22.4m with ENV10.

Electrical conductivities from the five Envirocone tests (ENV7 conducted in winter, ENV8-ENV12 in late spring/summer) attest to measurement repeatability, as do other ionic, assay-measured quantities (chloride and alkalinity). Notoriously more instable parameters: O_2, E_h and pH, on the other hand, indicate scatter quite possibly attributable to the substantial unrelated pumping activity at d=12m and 22m in the nearby RW during tests ENV8 through ENV12. In all cases, comparison between Envirocone and RW samples is excellent, particularly considering the E_h value at d=22m, probably anomalously high due to the aforementioned pumping activity. It is worth noting also that the large number of low-cost, assay kit analyses conducted in situ for Cl and alkalinity on both Envirocone and RW samples compared well to laboratory analyses. Temperature values are not presented as they effectively represent atmospheric conditions due to the approximately 100m of tubing at ground surface through which the sample flowed.

Laboratory analyses were also conducted for metals and ion balance on Envirocone and RW specimens, the results of which are presented in the lower section of Fig.10. Envirocone specimens were drawn from the sample line prior to flowing past the sensors in the flow cell to avoid contamination. The wells consistently indicated approximately 30% greater manganese levels, 25% greater iron levels and about 100 times greater zinc levels. Such contributions could conceivably be attributed to well casing or well drilling operations. Neither specimens had detectable levels of Cu, Cd, Ni, Cr, V or Mo. Ion balance measurements from Envirocone samples agreed well with RW samples (including undetected sulphate, carbonate and floride). These data together with the alkalinity measurement indicate that the hardness of the water is partially permanent as indicated by the chloride measurements.

Low O_2 content at depths greater than about 14m, indicating between about 20% to 50% of saturated O_2 values, strongly suggest the presence of some type of organic pollutant. Given the agricultural activity in the area, analyses were performed for pesticides. Pesticides were not detectable at a depth of 12m in accordance with the greater O_2 levels. For samples from depths of 16m and 22m, however, levels of phosphatic pesticides close to and slightly exceeding permissible drinking water limits were detected. In addition to the borderline pesticide levels, excess levels of Fe, Mn and Cl^- and hardness suggest that, untreated, the water below about 12m is not potable.

SOIL GAS SAMPLING
Landfill soil gas measurements
Following the explosion of an isolated residential structure in the vicinity of a municipal solid waste landfill in northern Italy, an investigation was effected in an attempt to ascertain whether or not the landfill was the source of the combustible gas. A preliminary gas survey to evaluate the possibility of biogas migration was performed with the Envirocone along the periphery of the landfill and in the direction of the explosive incident which occurred near Envirocone test E21. The results of one transect of the survey are presented in Fig.11 including soil gas analytical results and the soil profile as inferred from five of the penetration tests. The stratigraphy at the site included approximately 6m of silty sand overlying cemented layer above the potential conducting gravelly sand strata. It was frequently necessary to employ an uninstrumented cone to punch through the cemented layer.

Typical municipal landfill gas in the methagenic or anaerobic phase is composed of about 50% methane and 40% CO_2. These values were confirmed for this landfill upon analysis of the gas vent. Analyses from E3 and E5 indicate some escape of gas from the landfill while E22 already begins to indicate significant dilution. The availability of penetration resistance values permits disregarding the relatively clean air ($O_2 \approx 20\%$ and low CO_2 an CH_4 measurements) indicated by sounding E9 since this soil gas was sampled above the suspected biogas transmitting layer.

Levels of CH_4 in excess of 70% were detected in tests E20 and E21, approximately 200m distant from the landfill near the location of the incident. Such levels are substantially greater than those typically associated with landfill-generated biogas and with those measured at the landfill. Natural gas, on the other hand, is typically composed of between 55% and 90% CH_4. This information would suggest that the landfill is unlikely to be the source of combustible gas

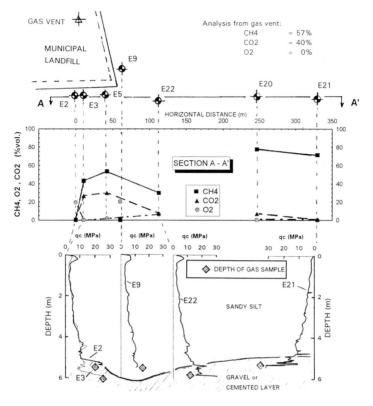

Fig.11 - Soil gas survey in the vicinity of a municipal landfill.

and a leaking natural gasline could be the culprit, thus providing rational redirection of the investigation.

Soil gas measurements at an abandoned coal mining site

Among the environmental scars left by coal mining and processing activity during the century between the 1830's and the 1930's in a 5 km² area located in east central Germany is a man-made pond formed in a sand quarry depression in which the waste sludge generated by the distillation of lignite to obtain paraffin was impounded. Currently, the area is misleadingly bucolic with the algea-covered pond perched on a small hill, ringed by light woods and surrounded to the south by a relatively level meadow and gentle, rounded slopes in the other directions. The meadow of the tar-filled pond, presently in agricultural use, served as the venue of a METS investigation program delineated in the site plan presented in Fig.12, as part of a larger survey effected in the company of a mobile analytical laboratory group. Two sources of contamination from the pond include breaching of the earth walls by the rainwater trapped over the impermeable tar sludge and escape of the viscous sludge itself from the impoundment into the surrounding area. The focus of the METS program was ultimately the latter.

Eight soundings ranging in depth between 3m and 7m were performed in three days. The soil profile inferred from the penetration test results and shallow soil sampling consisted of a silty fine sand interstratified with silty sand and gravel layers overlying an impenetratable cemented quartzitic layer. Due to the presence of the tar layer and the cemented stratum and the fact that all tests were executed above the water table, sampling was effected with the uninstrumented 38mm diameter gas cone. Two gas sampling well points were also installed with the penetrometer system. The areal extent and vertical location of the sludge within the soil profile as evaluated visually from the penetrometer tests is presented in Section B-B' (Fig.12).

A preliminary survey effected in this area involving conventional, manually-installed steel probes to a depth of approximately 70cm produced no measurable amounts of monoaromatic hydrocarbons. However, signficant values of benzene, toluene, xylene and phenol (Fig.13) were found when employing METS. The filter zone at the cone tip was penetrated into the stratum to be sampled, and high quality gas samples were obtained directly into tedlar gas bags at a rate of 4 liters/min by applying a vacuum to the sampling box. (Only gas sampling at

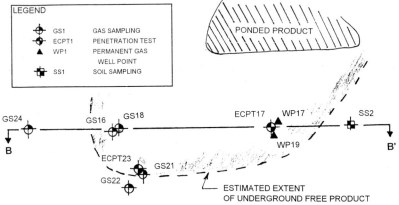

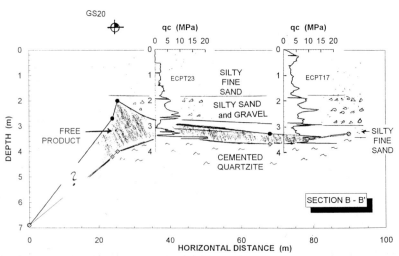

Fig.12 - Subsurface exploration plan for abandoned lignite processing disposal pond.

GS17 and GS16 involved positioning of the filter zone in a pre-penetrated stratum.) Evaluation of BTX concentrations were effected both on site with Dräger tubes as well as at the mobile laboratory equipped with a gas chromatograph (GC) located about 2 km off site.

Notwithstanding their paucity and preliminary nature, the data presented in Fig.13 suggest some interesting aspects associated with gas sampling. That most immediately notable is the vertical gradient of hydrocarbon concentrations in the meter above the tar layer, indicative of the chemical flux from the free product source to the atmosphere, apparently steepest for the semivolatile phenol. Also, in the 40cm stratum of finer grained material (lower q_c values) staining of soil seemed to indicate some product migration possibly due to capillary rise of contaminated surface water. O_2 concentration, simply and rapidly measured in this case with a CGI, served as an excellent indicator of the presence of measurable quantities of BTX, when less than the atmospheric concentration of 20.9%.

Only recently has attention been paid to limiting the vertical extent of filtered zones to avoid "well-averaging" which can easily mask actual plume concentrations given the commonly steep vertical concentration gradients [4]. A possible example of "well-averaging" are the elevated O_2 results from GS16 and GS17 as these data were obtained in holes penetrated to depths greater than sampling depth immediately prior to gas sampling with consequent poor sealing with the soil as well as possible effects of aeration of the sampling zone. O_2 concentrations measured in the wellpoints may also demonstrate such an effect in that although both points were installed at 3m depth, they act on different size filter zones (WP19 was backfilled with 10cm of gravel and WP17 with about 110cm). In Fig.13, the wellpoints are plotted about midheight of their respective filter zones. Purge time, equal for the two wellpoint samplings notwithstanding the diverse well volumes, could also have resulted in lower volatile gas (and greater O_2) concentrations for WP17 in comparison with WP19 [17].

The greater concentrations of toluene and xylene measured in the field in comparison to those evaluated by GC in the mobile laboratory could be attributed to the volatility of the compounds

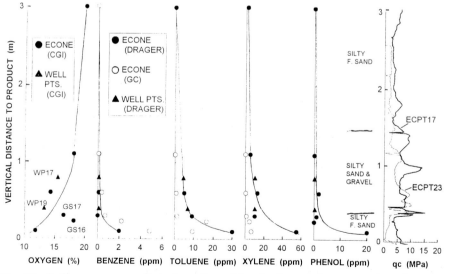

Fig. 13 - Soil gas survey results in the soil overlying the free product-contaminated stratum.

435

and consequent loss due to additional handling in the relatively warmer lab air. Dräger tube measurements for benzene were tested on a 20ppm benzene source and are considered reliable to within 15%. Greater GC benzene concentrations could be due to sampling sequence.

The fact that the conventional, shallow survey did not result in measurable quantities of BTX could have been due to poor sealing between the steel tube and the soil, insufficient purging and/or distances greater than 1m above the tar. In any case, a system such as METS is flexible enough and capable of sampling sufficiently high quality samples to overcome initial subsurface unknowns. Only a small amount of data were obtained in this very short test program but enough to illustrate the potential of METS as an investigative tool.

SYSTEM IMPROVEMENTS
In addition to the other mechanical modifications indicated in the relevant sections, a flow volume measurement sensor to facilitate the recording of correlations between physico-chemical indicators and volume sampled is being incorporated into the flow cell.

A concerted effort is being focused on improving the component specific capabilities of the MET system. The key component is a field portable GC with both photo and flame ionization detectors. Given the applicability of VOC's as a general purpose ground water contaminant indicator [18], and the relative ease of GC testing, a solid phase microextraction module is under design for on-line analysis of volatile and semi-volatile compounds in liquid samples with the GC.

CONCLUSIONS
The Envirocone test system serves as an effective tool for characterizing subsurface conditions at contaminated sites with respect to both geotechnical and environmental parameters, particulary when very little data is available. Soil type and strength of interested strata can be estimated together with the direction and magnitude of the areal hydraulic gradient. Hydraulic conductivity measured with the Envirocone in sands exhibits the effects of penetration and is thus about two orders of magnitudes less than that estimated as actual in situ values.

The representativity of pore fluid and soil gas samples can be quantified on site. Contaminant plume delineation is facilitated by the flexibility of areal and vertical sampling points. This flexibility together with the relatively small filter zone also enables rapid evaluation of vertical gradient variations that might otherwise mask actual plume form.

The time required to extract ground water samples from soil is an important element to be borne in mind when designing a site investigation program. In sandy soils, it is estimated that pore fluid can be sampled at about 25 ml/min. In finer grained soils, sampling rates of about 5ml/min are expected. Gas samples, on the other hand, can be obtained at a rate of up to about 4 l/min in damp, fine sandy soils with up to about 40% fine particles. In more permeable soils, gas sampling rates should be limited to respect sample integrity.

Analytic results for Envirocone ground water samples correspond very well with those from conventional wells after purging of about three well volumes. Well casing appears to contribute to metal concentrations such as zinc and iron. Envirocone gas samples tend to be more representative than those of conventional methods due to the limited filter zone and well volume avoiding difficulties associated with "well-averaging" and stringent purging requirements.

REFERENCES

[1] Barcelona, M., H. Wehrmann, M. Varljen (1994) "Reproducible well-purging procedures and VOC stabilization criteria for ground water sampling", Ground Water, Vol.32, n.1, pp12-22.

[2] Barcelona, M., J. Helfrich and E. Garske (1988), "Verification of sampling methods and selection of materials for ground water contamination studies", Ground Water Contamination - Field Methods, ASTM STP963, pp221-231.

[3] Maltby, V. and J. Unwin (1991), "A field investigation of ground water well puring techniques", Current Practices in Ground Water and Vadose Zone Investigations, ASTM STP 1118, pp281-299.

[4] Barcelona, M. and J. Helfrich (1991), "Realistic expectations for groundwater investigations in the 1990's", Current Practices in Ground Water and Vadose Zone Investigations, ASTM STP 1118, pp3-23.

[5] Marrin, D. and G. Thompson (1984), "Remote detection of volatile organic contaminants in ground water via shallow gas samplin", Proceedings Petroleum Hydrocarbons and Organic Chemicals in Groundwater, National Water Well Association.

[6] Geoprobe, Salina, KS, USA, internal newsletter publication.

[7] Rad, N.S., S. Sollie, T. Lunne, and B.A. Tortensson (1988), "A new offshore soil investigation tool for measuring the in situ coefficient of permeability and sampling pore water and gas", BOSS Conference Proceedings, pp409-417.

[8] Edge, R. and K. Cordry (1989), "The hydropunch: an in situ sampling tool for collecting ground water from unconsolidated sediments", Ground Water Monitoring Review, Vol.IX, (3), pp177-183.

[9] Olie, J., C. Van Ree and C. Bremmer (1992), "In situ measurement by chemoprobe of groundwater from in situ sanitation of versatic acid spill", Geotechnique, 42, n.1, pp13-21.

[10] Applied Research Associates, So. Royalton, VT, USA, technical literature.

[11] Campanella, R. and I. Weemees (1990), "Development and use of an electrical resistivity cone for groundwater contamination studies", Canadian Geotechnical Journal, 27, pp557-567.

[12] Parker, L. (1991), "Suggested guidelines for the use of PTFE, PVC and stainless steel in samplers and well casings", Current Practices in Ground Water and Vadose Zone Investigations, ASTM STP 1118, pp217-229.

[13] Baligh, M., Levadoux, J., "Consolidation after undrained piezocone penetration, II: interpretation", Journal of Geotechnical Engineering, Vol.112, No.7, (1986), pp727-745.

[14] Barcelona, M., and J. Gibb (1988), "Development of effective ground water sampling protocols", Ground Water Contamination - Field Methods, ASTM STP963, pp17-26.

[15] Tavenas, F., M. Diene and S. Leroueil (1990), "Analysis of the in situ constant-head permeability test in clays", Canadian Geotechnical Journal, Vol.27, pp305-314.

[16] Lutenegger, A. and D. Degroot (1991), "Measurement of hydraulic conductivity in clay using push-in piezometers", Current Practices in Ground Water and Vadose Zone Investigations, ASTM STP 1118, pp362-376.

[17] Karably, L. and K. Babcock (1990), "Effects of environmental variables on soil gas surveys", Hazarous Materials Control Research Institute Monograph Series: Sampling and Monitoring, Vol. 1, pp32-35.

[18] Pastor and Frich (1991), "Considerations in selecting indicator parameters for the statistical evaluation of ground water quality", Current Practices in Ground Water and Vadose Zone Investigations, ASTM STP 1118, pp411-426.

Use of piezocone tests in non-textbook materials

TOM LUNNE[1], JOHN POWELL[2], PETER ROBERTSON[3]
[1]Division Head, Norwegian Geotechnical Institute, Oslo, NORWAY
[2]Division Head, Building Research Establishment, Garston, ENGLAND
[3]Professor of Civil Engineering, University of Alberta, Edmonton, CANADA

ABSTRACT
Cone penetration tests (CPT) and more recently piezocone tests (CPTU) have been used in geotechnical practice for a number of years. However, most of the geotechnical literature and available interpretation methods for the test results are for 'text book' soils, i.e. pure clay and pure sand. There is an increasing tendency now to also use CPT and CPTU more and more in other soils and materials. It is, therefore, important to know how to interpret the test results in the most meaningful way in these other materials.
This paper considers practical examples of the use of CPT and CPTUs in a range of non-textbook materials including:
- intermediate, partly drained soils in the silt range
- peat/organic soils
- underconsolidated clays
- soft rock including chalk
- mine tailings

INTRODUCTION
Cone penetration tests (CPT) and more recently piezocone tests (CPTU) have been used in geotechnical practice for a number of years, and the tendency is for increased use in many countries all over the world. One of the main advantages of the CPT, and especially the CPTU, is the excellent profiling capability. However, the use of the data for identification of penetrated material (e.g. Robertson, 1990) and for deriving soil parameters for foundation design is also very important. Over the last 10-20 years significant developments have taken place regarding the interpretation of CPT/CPTU data in terms of soil parameters for pure clays (undrained conditions) (see f.inst. Robertson and Campanella, 1983 and Powell et al., 1988) and for sands (drained conditions) (see f.inst. Baldi et al., 1986; Lunne, 1991). Engineers are daily interpreting the results in these "text book" soil types. However, for soils or other materials that do not fall into the clay or sand categories, it is much more difficult for practising engineers to know how to use the test results in a meaningful way. In addition the CPT/CPTU is used increasingly in man-made (e.g. mine tailings, engineered fills, puddle clay core) and other "unusual" material (e.g. peat, chalk and silts) where it is also important to identify material type and to derive soil parameters for foundation design or engineering works. The purpose of this paper is to show examples of CPT/CPTU results in 'non-textbook' materials and how the results may be used in a meaningful way. Due to space limitations only a few selected material types will be covered in some detail whereas other material types will only be listed with key references.

CPT/CPTU PARAMETERS TO BE USED IN INTERPRETATION
The measured parameters from a cone penetration test (CPT) are: q_c = cone resistance and f_s = sleeve friction. In a piezocone, CPTU, pore pressure, u, is measured in addition. Nowadays the most usual filter position for measuring u is immediately behind the conical part (u_2) as will be used in this paper. Pore pressure may also be measured on the cone face (u_1) and occasionally at the upper end of the friction sleeve (u_3). It is important that the measurement location is given with all CPTU results. It is now standard practice to correct

q_c for pore pressure effects according to the following formula:

q_t = corrected cone resistance = q_c + (1-a)u

where a = area ratio over cone (e.g. see Lunne et al., 1986).

If both u_2 and u_3 are measured a similar correction can be applied to the sleeve friction, to obtain a corrected value f_t. However, in practice this is seldom possible. For interpretation of CPT/CPTU results the following derived parameters are used:

$$R_f = \text{friction ratio} = \frac{f_s}{q_t} \times 100\% \; ; \quad B_q = \text{pore pressure ratio} = \frac{u_2 - u_o}{q_t - \sigma_{vo}}$$

where u_o = in situ pore water pressure and σ_{vo} = total vertical overburden stress.

INTERMEDIATE SOILS

In this paper intermediate soils are taken as those which according to their grain size distribution range between pure clay and pure sand. Examples are clayey sands, sandy silts, silts and possibly silty clay. Interpretation methods valid for sand or clay may not be applicable for intermediate soils since penetration in this type of material can be partially drained. Bugno and McNeilan (1984) suggested that undrained response will occur for the standard cone (10 cm^2) and penetration rate (2 cm/sec) if the permeability of the soil is less than 10^{-7} to 10^{-6} cm/sec. Soils having permeabilities between 10^{-6} and 10^{-3} cm/sec, which is supposed to include most silts, are believed to behave in a partially drained manner.

Based on a comprehensive study of the soil investigation data from the North Sea site Gullfaks C, Hight et al., (1994), confirmed that the clay content is very important for the penetration behaviour in clayey sands. Figure 1 shows the results of a typical CPTU at the Gullfaks C site, with q_c (uncorrected) and pore water pressure measured on the cone (u_1) vs depth. Penetration through clayey sands is marked by large variations in both q_c and u which Hight et al. argued was due to variations in clay content which are also shown in Figure 1.

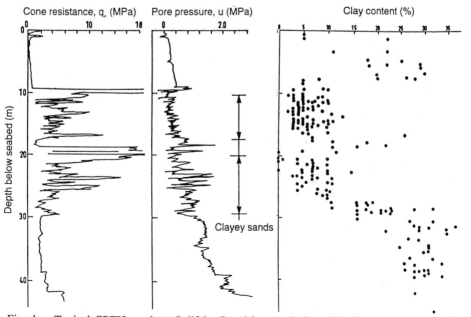

Fig. 1   Typical CPTU result at Gullfaks C and its correlation with clay content
(after Hight et al., 1994)

439

Hight et al. found that the relation in Figure 2 could be used to assess the in situ variation in clay content and the drainage conditions during penetration.

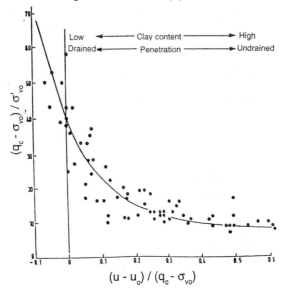

Fig. 2 $\Delta u_l/(q_c-\sigma_{vo})$ vs $(q_c - \sigma_{vo})/\sigma_{vo}$' from CPTU records at Gullfaks C (after Hight et al., 1994)

For the interpretation of CPT data in intermediate soils it is also important to identify the drainage conditions expected both in the design problem and during the cone penetration test. If the design problem will involve the undrained shear strength and cone penetration is also undrained, the CPT data can be interpreted in a manner similar to a clay. On the other hand, if the design problem will involve drained loading and the cone penetration is also drained, the CPT data can be interpreted in a manner similar to a sand. Interpretation is more difficult if the design problem is expected to involve drained loading but the cone penetration process is undrained or partially drained. In this case effective stress strength parameters are needed for design. Senneset et al. (1988) concluded that effective stress strength parameters can be estimated from CPTU results in a manner similar to that proposed for clay. Senneset et al. (1988) presented typical values of attraction (a) and friction (tanϕ') for silts that may be used for reference, see Table 1. Attraction is defined as cohesion, c, divided by tanϕ':

Table 1   Typical values of attraction (a) and friction (tanϕ')

SOIL	Effective stress shear strength parameters		
	a, kPa	tanϕ'	ϕ'$^\circ$
Silt, soft	0-5	0.50-0.60	27-31
Silt, medium	5-15	0.55-0.65	29-33
Silt, stiff	15-30	0.60-0.70	31-35

Undrained shear strength (s$_u$)
The cone factor, N_{KT} = $(q_t - \sigma_{vo})/s_u$ was found by Senneset et al. to range from 15-30 when CAUC (consolidated anisotropically, undrained test sheared in compression) triaxial tests

were used for reference s_u. However, the relevancy of using undrained shear strength in silts may be debateable. Senneset et al. (1982) argued that s_u-values obtained by such correlations are particularly questionable for soils generating small pore pressures, especially corresponding to $B_q < 0.4$. Such conditions may be representative for materials coarser than about clayey silts.

Constrained modulus

Under conditions where one dimensional loading applies (such as in a thin layer underneath a large platform or in an oedometer test), settlements can be computed using a constrained modulus, M_o, which can be expressed as (Senneset et al., 1988):

$$M_o = m \sqrt{\sigma'_{vo} \cdot \sigma_a}$$

m = dimensionless modulus number
σ_{vo}' = effective overburden pressure
σ_a = reference stress, taken as 100 KPa
$\Delta\sigma_v$ = additional vertical effective stress

The study by Senneset et al. (1988) of silty soils indicated that a linear correlation between the corrected cone resistance, q_t and the tangent modulus, M_o, gave the best fit, see Figure 3. As can be seen from the figure, there is a tendency for M_o/q_t to increase with increasing q_t, at least for $q_t < 5$ MPa which is the actual range of q_t in most silty soils.

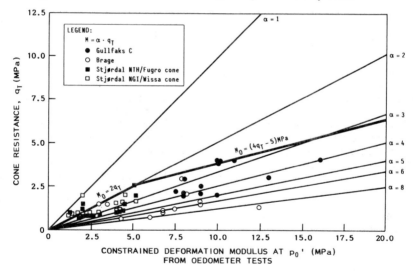

Fig. 3   Constrained modulus (M_o) vs cone resistance, q_t for silty soils
(after Senneset et al., 1988)

Based on the data in Figure 3 a reasonably conservative estimate of M_o can be summarized as follows:

for $q_t < 2.5$ MPa M_o = $2q_t$ MPa
for 2.5 MPa < q_t < 5 MPa M_o = $(4q_t-5)$MPa

The constrained modulus for the stress range $\sigma_{vo}' + \Delta\sigma_v'$ may be estimated from the following formula:

441

$$M = M_o \sqrt{\frac{\sigma'_{vo} + \Delta\sigma'_v/2}{\sigma'_{vo}}}$$

Small strain shear modulus, G_o

Figure 4 shows a correlation proposed by Gillespie (1990) between G_o/q_c and q_c/σ'_{vo}. Silts were found to fall between sands and clays with typical G_o/q_c values of between 20 and 40. The small strain shear modulus can be measured directly using the seismic CPT (Campanella et al., 1986)

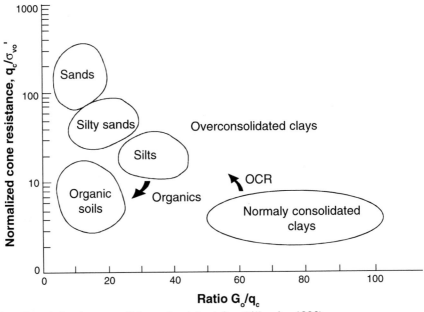

Fig. 4   Correlation between C_o/q_c and q_c/σ'_{vo} (after Gillespie, 1990)

Coefficient of consolidation

To get a first estimate of coefficient of consolidation the methods used for clay (see Robertson et al., 1992) may be used.

However, it should be born in mind that the theoretical initial pore pressure distribution used in the models is developed on the assumption of ideal undrained response during penetration. This assumptions may be questionable in silty soils with $B_q < 0.4$ and partial drainage considerations are of great importance in understanding and interpreting dissipation data in silty soils.

If the cone penetration is considered to be partially drained it may be possible to change the rate of penetration to ensure either a drained or undrained penetration. This procedure may be feasible in silt deposits where a reasonable change in rate of penetration may produce the required change in drainage. However, since permeability varies by orders of magnitude it may require changes in the rate of penetration by orders of magnitude. Such large variations in the rate of penetration are often difficult to achieve.

PEAT

As discussed by Landva (1986) peatland soils may be subdivided into different groups: peats, peaty organic soils, organic soils and soils with organic content. Peats are derived from plants and are therefore very fibrous materials. As the degree of humification increases, the peat becomes less and less fibrous until it is transformed into an amorphous mass without any discernible structure. Any testing in peat should be related to a classification system. Vos (1982) showed examples of CPT profiles from Holland which included peat layers. One example is included in Figure 5. Vos reported that in Holland a friction ratio larger than 5 normally identifies peat. Figure 6 shows two examples of a piezocone test from the coast of Germany which identified a peat layer that was later confirmed by sampling in a nearby borehole. As can be seen from Figure 6 the peat layer can be identified by a very high friction ratio combined with negative pore pressure in one case and positive pore pressures in the other case. In the one with negative pore pressures some sand was mixed in the peat. Experience in the Vancouver area of British Columbia has shown that fibrous peat has high friction ratio and low penetration pore pressure. However, amorphous peat has high friction ratio and high pore pressures. Also the rate of pore pressure dissipation is slower in the amorphous peat (Gillespie, 1994, private communication). In conclusion peat can generally be identified with a high friction ratio, local correlations should be established for use of pore pressure ratio to assess type of peat.

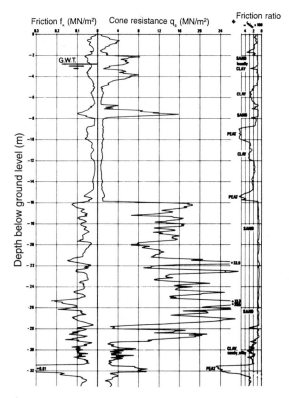

Fig. 5   Example of CPT profile from Holland with peat layers (after Vos, 1982)

UNDERCONSOLIDATED CLAY

Underconsolidated soils are soils which are not fully consolidated under their present overburden stress, this means that the pore water pressures in the soil are higher than the corresponding hydrostatic pressure (i.e. depth below water table multiplied by the average unit weight of pore water). The reason for a soil to be in this state can be that excess pore pressures have not had sufficient time to dissipate since the soil deposition, the placement of a fill or other aspects related to its geological history. Frequently the main problem is to determine whether the in situ pore pressures are higher than hydrostatic or not. This can be found by dissipation tests, but in clay it may take a prohibitively long time to read static pore pressure conditions. This is specially so for offshore investigations.

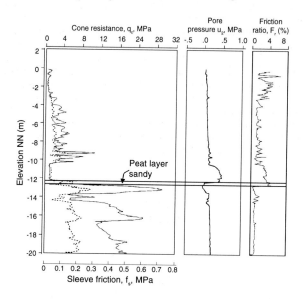

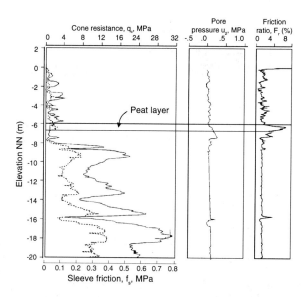

Fig. 6   Example of CPTU profiles from German coast with peat layers

Tanaka and Sakagami (1989) reported results of piezocone tests in underconsolidated soft marine clay in Osaka Bay, Japan. The clay was underconsolidated because of placement of a fill, and the excess pore pressures caused by this fill had not had time to dissipate. Tests were carried out in both normally (before placement of fill) and underconsolidated clay. Tanaka and Sakagami showed that by plotting the results in terms of Δu (= u_2 - u_o) vs q_t - σ_{vo} the underconsolidated clay could very clearly be identified as shown in Figure 7. When $\Delta u > 3/4$ (q_t - σ_{vo}) the clay was underconsolidated. Note that u_o is here the corresponding hydrostatic pore pressure. By using reference values of the in situ excess pore pressure (measured in piezometers and by dissipation tests) Tanaka and Sakagami developed a procedure for assessing this using the measured penetration pore pressure. It is necessary to have the corresponding (q_t - σ_{vo}) and Δu values in n.c. clay, with the same classification characteristics to use this procedure. It must be mentioned that the correlation obtained by Tanaka and Sakagami (1989) is valid for Osaka clay only. The relation between Δu and q_t-σ_{vo} that shows if the clay is underconsolidated or not will vary from clay to clay. An additional approach to assess if a clay is underconsolidated or not is to compute undrained shear strength, s_u, from a CPT vs depth (see f.inst. Powell et al, 1988). This s_u profile should then be compared to a corresponding normally consolidated s_u profile. If the CPTU calculated s_u profile is lower than the corresponding normally consolidated s_u profile this is an indicator that the clay may be underconsolidated.

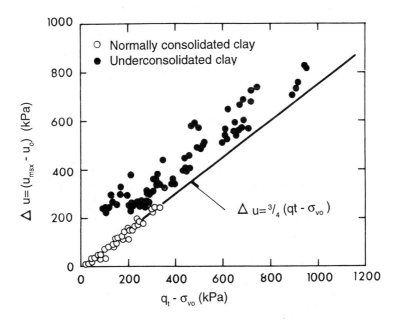

Fig. 7   Possible identification of underconsolidated clay layers
(after Tanaka and Sakagami, 1989)

SOFT ROCK INCLUDING CHALK

Power (1982) presented a comprehensive study on the use of CPT in chalk. Based on testing in various types of chalk he proposed the classification of chalk grades according to q_c and f_s values as shown in Table 2. An example of a CPT in chalk and the resulting Chalk Grade description is given in Figure 8. The sharp peaks in the profile are thought to not only be the results of flints (which frequently occur in chalk), but rather the manner in which penetration resistance builds up, then is followed by grain crushing and/or closure of fissures, together with the possible effects of variability in density, degree of cementation and jointing and fissuring. For one site (Mundford) Power (1982) also developed a tentative correlation between Young's Modulus and q_c.

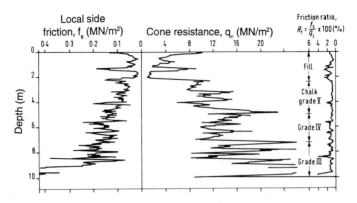

Fig. 8   Typical cone penetration test profile in Middle chalk at Mandford, Norfolk
(after Power, 1982)

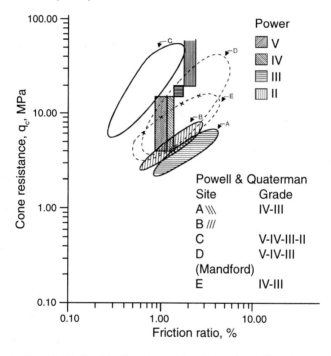

Fig. 9   Tentative chalk classification chart (after Powell and Quarterman, 1994)

More recently Powell and Quarterman (1994) have examined the use of the CPT on a
number of chalk sites. They found that the trend in Power's work of increasing values of
both q_c and f_s with improving chalk grade (i.e. V → II) was valid, but that the actual ranges
for any given grade varied from site to site. Figure 9 summarizes their findings and also
shows Power's correlations. They also considered the potential influence of the chalk density
on the CPT results and whilst site A had a lower density than the others and formed the

lower bound of the data, no significant difference was evident amongst the other site data. It should be remembered that chalk is a very variable material where the state of fracturing and fissuring and the nature of infill material is very important. Thus care should be exercised in extrapolating correlations from one area and chalk type to another. Ideally, specific correlations need to be developed for each area. Also to facilitate comparisons between sites, it is essential that CPT results should be accompanied by careful descriptions of the chalk.

Table 2   Chalk grades related to CPT values (after Power, 1982).

Grade	Brief description	q_c (MPa)	$R_f(\%)$
VI	Extremely soft structureless chalk containing small lumps of intact chalk.	< 5	-
V	Structureless remoulded chalk containing lumps of intact chalk	5-15	0.75-1.0
IV	Rubberly partly-weathered chalk with bedding and jointing. Joints 10 to 60 mm apart, open to 20 mm, and often infilled with soft remoulded chalk and fragments.	5-15	1.0-1.25
III	Rubberly to blocky unweathered chalk. Joints 60 to 200 mm apart, open to 3 mm, and sometimes infilled with fragments.	15-20	1.25-1.50
II	Blocky medium-hard chalk. Joints more than 200 mm apart and closed.	> 20	1.5-2.0
I	As for Grade II, but hard and brittle.	No penetration	

MINE TAILINGS

Stability of tailing dams is a major problem confronting many geotechnical engineers. These deposits are by their nature usually very layered and the ability of the piezocone to detect layering is particularly useful. The properties of mine tailings will vary with the ore being mined, the type of grind to produce the tailing, the solids concentration during tailings disposal, the method of contraction dam and the location on the dam.

To illustrate the use of the CPTU in tailings, the results form the Zelazny Most in Poland will be used as an example. The Zelazny Most tailings dam has been under construction since 1977 and used to store the flotation wastes from the mechanical processing plants of all Polish copper mines. The waste is continuously transported hydraulically and deposited in the reservoir. Because of its size (are 12.3 km², actual height of dam - 40 m, planned 100 m) thorough investigations have been carried out to check its stability. Piezocone testing is an essential part of the investigations. Figure 10 shows the results of two typical CPTU profiles, one on the so called beach of the reservoir and one in the pond. The material in the pond is very uniform with depth while the material on the beach is very layered. Mlynarek et al. (1994) divided the material into 5 groups according to the grain size as shown on Table 3:

Table 3   The classification of post-flotation sediments according to grain size
(after Mlynarek et al., 1994)

Sediment group	Material type	Sand content (%)	Silt content (%)	Clay content (%)
I	Fine sand	>10	<10	<2
II	Silty sand	70-90	10-3	<2
III	Sandy silt	50-70	30-50	<2
IV	Silt	<70	>30	2-10
V	Silty clay	<40	>50	>10

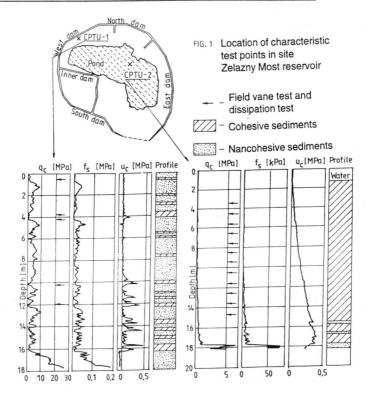

FIG. 1   Location of characteristic test points in site Zelazny Most reservoir

— — Field vane test and dissipation test

▨ – Cohesive sediments

▨ – Nancohesive sediments

Fig. 10   Results of typical CPTU profiles on the beach and in the pond of the Zelazny Most
tailings dam (after Mlynarek et al., 1994)

Mlynarek et al. (1994) found that CPTU results could be used for classifying the deposits
according to the above sediment groups. The chart shown in Figure 11 was found to work
well for the first three sediment groups, while for groups IV and V it is more difficult to get
a reliable definition using the chart. By using a B_q vs q_t - σ_{vo} plot as shown in Figure 11 it
is, however, possible to make a more clear distinction between groups IV and V. The
correlations in Figure 11 can also be used to estimate the clay content of the material. The
authors agree with Mlynarek et al. (1994) that once the material group has been defined
using CPTU results, then site specific correlations to shear strength and deformation
characteristics can be developed in a similar manner as for "normal" soils.

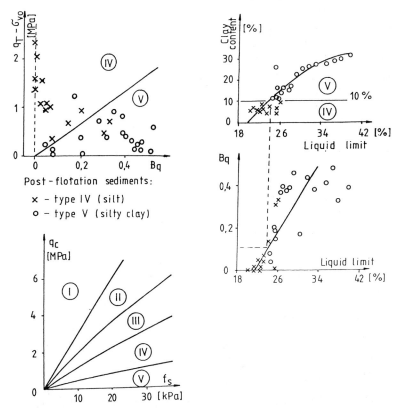

Fig. 11 Identification of mine tailings material type (from Mlynarek et al. 1994)

OTHER MATERIAL TYPES

The previous sections have given examples of CPT/CPTU results and their interpretation in several different material types that may be encountered in civil engineering practice.

In other parts of the world experience exist in other material types. To provide a quick reference for engineers who may come across some of the materials, the table below has been prepared as a guide to further reading.

Table 4 Selected references on CPT/CPTU results in various material types

Material type	Selected references
Calcareous soils	Ebelhar et al., 1988
Cemented sands	Rad and Tumay (1988)
Residual soils	Ajayi and Balogun (1988)
Slurry wall	Manassero (1984)
Permafrost and ice	Ladanyi et al. (1991)
Snow	Schaap and Føhn (1986)
Gas hydrates	Tzirita et al. (1991)

SUMMARY AND CONCLUSIONS
The CPTU is an excellent profiling tool in a wide range of material, both natural and man made. With accumulated experience it is possible to identify material type using CPTU parameters. Even if most theories and interpretation methods are applicable usually to pure sands (drained conditions) or pure clay (undrained conditions) the principles can also be used for other materials. However, it is necessary to understand the drainage conditions during cone penetration and to make the necessary modifications to the interpretation methods. In some cases completely new material specific interpretation methods need to be worked out. Several examples are elaborated in this paper.

ACKNOWLEDGEMENT

The authors would like to acknowledge Dr. Joe Keaveny, NGI and Dr. Fernando Danziger, research fellow at NGI, for their thorough review of the paper.

REFERENCES

Ajayi, L.A. and L.A. Balogun (1988). Penetration testing in tropical lateritic and residual soils - Nigerian experience. Proceedings of the First International Symposium on Penetration Testing, ISOPT-1/Orlando/20-24 March 1988, pp. 315-328.

Baldi, G., R. Bellotti, V. Ghionna, M. Jamiolkowski and E. Pasqualini (1986). Interpretation of CPT's and CPTU's; 2nd part: drained penetration of sands. Proceedings of the Fourth International Geotechnical Seminar, Singapore, pp. 143-156.

Bugno, W.T. and T.W McNeilan, (1984). Cone penetration test results in offshore California silts. Strength testing of marine sediments; Laboratory and in situ measurements, a symposium sponsored by ASTM Committee D-18 on Soil and Rock. San Diego 1984, ASTM Special technical publication, 883, pp. 55-71.

Campanella, R.G., P.K. Robertson and D. Gillespie (1986) Seismic cone penetration test. ASCE Spec. Conf. In Situ 86, Use of In Situ Tests in Geotechnical Engineering, Blacksburg, Virginia, USA, pp. 116-130.

Ebelhar, R.J., A.G. Young and G.P. Stieben, (1988). Cone penetrometer and conductor pullout tests in carbonate soils offshore Africa. Engineering for Calcareous Sediments, Proceedings of the International Conference on Calcareous sediments, Perth, Vol 1., AA Balkema, pp. 155-163.

Gillespie, D.G. (1990). Evaluating Velocity and Pore Pressure Data from the Cone Penetration Test. Ph.D. thesis. Department of Civil Engineering, University of British Columbia, Vancouver, B.C.

Hight, D.W., V.N. Georgiannou and C.J. Ford (1994). Characterization of clayey sands. Proceedings, Behaviour of Offshore Structures, BOSS '94, Boston, July, 1994. Vol. II, pp. 321-340.

Ladanyi, B., T. Lunne and W. Winsor (1991). Experience with the performance of load controlled cone penetrating tests in permafrost soils. Proc. 44th Canadian Geotechnical Conference, Calgary, September 1991.

Landva, A. (1986). In Situ Testing of Peat. ASCE Speciality Conf., Blacksburg, Virginia. In Situ 86', Proceedings pp. 191-205.

Lunne, T., Eidsmoen, T., Gillespie, D. and Howland, J.D. (1986). Laboratory and Field Evaluation of Cone Penetrometers, In Situ '86, Specialty Conf., ASCE, Blacksburg, Virginia. Proceedings, pp. 714-729.

Manassero, M. (1994). Hydraulic conductivity assessment of slurry wall using piezocone test. Journal of Geotechnical Engineering, Vol. 120, No. 10, pp. 1725-1746.

Mlynarek, Z., W. Tschuschke and T. Lunne (1994). Techniques for examining parameters of post flotation sediments accumulated in the pond. 3rd International Conference on Polluted and Marginal Land, Uxbridge, UK. Proc. pp.17-23.

Powell, J.J.M., R.S.T. Quarterman and T. Lunne (1988). Interpretation and use of the piezocone test in UK clays. Proc. Geotechnology conference on Penetration Testing in the UK, Birmingham, July 1988. Proc., pp. 151-156.

Powell, J.J.M. and R.S.T. Quarterman (1994). A reappraisal of CPT testing in chalk. BRE Report No. G/G p/9412.

Power, P.T. (1982). The use of electric cone penetrometer in the determination of the engineering properties of chalk. ESOPT, Amsterdam, Proc. Vol. 2, pp. 769-774.

Rad, N.S., and Tumay, M.T., (1988). Effect of cementation on cone penetration resistance of sand. NGI Publication No. 171, Norwegian Geotechnical Institute, Oslo, Norway.

Robertson, P.K. and R.G. Campanella (1983). Interpretation of cone penetration tests Part II: Clay. Canadian Geotechnical Journal, Vol. 20, No. 4, pp. 734-745.

Robertson, P.K. (1990). Soil classification Using the Cone Penetration Test. Canadian Geotechnical Journal, Vo. 27, No. 1, pp. 151-158.

Robertson, P.K., J.P. Sully, D.J. Woeller, T. Lunne and D.G. Gillespie (1992). Estimating coefficient of consolidation from piezocone tests. Canadian Geotechnical Journal, Vol. 29, No. 4, pp. 551-557.

Schaap, L. and P. Føhn (1987). Cone penetration testing in snow. Canadian Geotechnical Journal, Vol. 24, pp. 335-341.

Senneset, K., N. Janbu and G. Svanø (1982). Strength and deformation parameters from cone penetration tests. 2nd European Symposium on Penetration Testing. Amsterdam. Proc. Vol. 2, pp. 863-870.

Senneset, K., R. Sandven, T. Lunne, T. By, T. Amundsen (1988). Piezocone tests in silty soils. International Symposium on Penetration Testing ISOPT-1. Orlando, Florida, USA. Proc., Vol. 2, pp. 955-966.

Tanaka, Y. and T. Sakagami (1989). Piezocone testing in underconsolidated clay. Canadian Geotechnical Journal, Vol. 26, No. 4, pp. 563-567.

Tzirita, A., P. Jeanjan, J.-L. Briand and W.A. Dunlap (1991). Detection of gas hydrates by in situ testing. Offshore Technology Conference, Houston, May 1991. Proceedings Paper No. 6536, pp. 325-334.

Vos.J. de (1982). The practical use of the CPT in soil profiling. ESOPT II, Proc. Vol. 2, pp. 933-939.

Cone penetration testing in volcanic soil deposits

KOJIRO TAKESUE[1], HIKARI SASAO[1], and YORIO MAKIHARA[2]
1) Kajima Technical Research Institute, Tokyo, Japan
2) Tokyo Soil Research Co., Ltd., Tokyo, Japan

ABSTRACT
Deposits of a volcanic soil, known locally as "Shirasu", are widely distributed in southern Kyushu, Japan. The unique physical properties of Shirasu affect its mechanical properties and lead to geotechnical difficulties. The purpose of this study was to examine the applicability of the Cone Penetration Test with pore pressure measurement (CPTU) in Shirasu deposits. CPTU data for Shirasu deposits was compared with data obtained from conventional tests including a vertical loading test on a bored cast-in-place pile and the Standard Penetration Test (SPT). The main conclusions are as follows: 1) Shirasu manifests a SPT-CPTU relation that differs from such relations for other soils. 2) Soil classification charts by Robertson (1990) are applicable even for Shirasu deposits. 3) A simple method using CPTU data is proposed for estimating the mechanical behavior of a pile, and the validity of the method is proved for Shirasu deposits.

1. INTRODUCTION

Deposits of a volcanic soil, known locally as "Shirasu", are widely distributed in Kagoshima prefecture in southern Kyushu, Japan. Shirasu is mainly composed of vesiculate and easily crushable volcanic glass grains of angular shape, whose specific gravity ranges lower than that of ordinary quartz sands. A detailed study on the geological, physical and mechanical properties of Shirasu has been done by Haruyama (1973). The unique physical properties of Shirasu affect its mechanical properties and lead to geotechnical engineering difficulties. The Standard Penetration Test (SPT) is one of the most commonly used in-situ tests in Japan. However, the conventional SPT-data-based design correlations, which are constructed mainly from ordinary soil data, do not apply well to Shirasu deposits. The strength of Shirasu estimated from SPT N-values generally tends to be lower than that obtained from other reliable tests (e.g., Hatanaka et al. 1985; Makihara et al. 1994). One factor affecting this could be the high crushability of Shirasu grains. Because of this crushability, static cone penetration tests are presumed to be preferable to dynamic penetration tests such as SPT, especially for evaluating the static mechanical properties of Shirasu.

In recent years in Japan, the Cone Penetration Test with pore pressure measurement (CPTU) has become widely used, and the advantages of pore pressure and friction measurements have been reported. Most of the studies, however, have been limited to soils other than Shirasu. The purpose of this study was to examine the applicability of CPTU to Shirasu deposits. CPTU soundings with a maximum penetration depth of about 70 m were conducted at a site on thick Shirasu deposits in the coastal area of Kagoshima. CPTU data was compared with the data obtained from conventional in-situ

Advances in site investigation practice. Thomas Telford, London, 1996

tests conducted at the same site (including a vertical loading test on a bored cast-in-place pile as well as the SPT and sampling), and from laboratory tests. SPT-CPTU relationships, a comparison of CPTU soil classification results with the results by observation and laboratory tests, and the effects of penetration rate on CPTU data are presented in the first half of this paper. A simple method for evaluating load-settlement curves of piles using CPTU data as well as relationships between ultimate pile skin friction and CPTU sleeve friction are presented in the second.

2. RESEARCH SITE

The distribution area of Shirasu and the location of the research site are shown in Fig. 1. A summary of the soil profile based on SPT and sampling results at a typical point of the site is shown in Fig. 2. From the surface down, there is a fill layer, an alluvium and then a diluvium with boundary depths of about 7 m and 48 m. The texture of the alluvium gradually changes from sand to silt; the alluvium could be subdivided into upper As_1 (sand), middle As_2 (silty sand) and lower Ac (silt) layers. The upper As_1 layer is dominantly composed of ordinary quartz sand, while the middle As_2 and the lower Ac layers are dominantly composed of resedimentation Shirasu. This is marked by the distribution of soil particle density; the soil particle density in the As_1 layer ranges higher than 2.5 g/cm^3, while

Fig. 1. Distribution Area of Shirasu and Location of Research Site

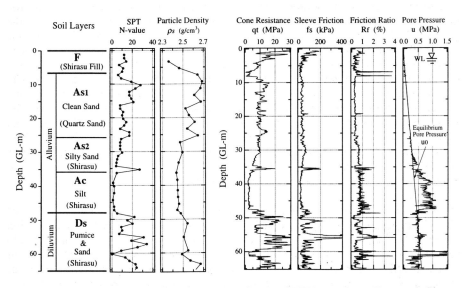

Fig. 2. Soil Profile at Research Site   Fig. 3. CPTU Results at Research Site

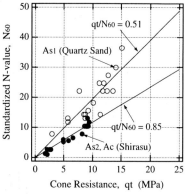

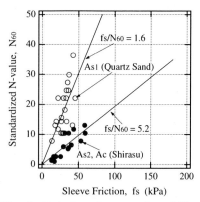

Fig. 4. Relations between q_t and N_{60} in Alluvium

Fig. 5. Relations between f_s and N_{60} in Alluvium

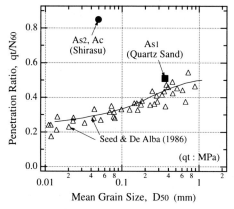

Fig. 6. Variation of q_t/N_{60} Ratio with Mean Grain Size

that in the As_2 and Ac layers ranges lower than 2.5 g/cm³. The Ds layer is a diluvial Shirasu layer mainly composed of pumice and sand with pumice. It contains a thin silt layer near a depth of 60 m, where consequently there is a local reduction in N-value.

Fig. 3 shows the results of CPTU at the same point where the boring investigation shown in Fig. 2 was conducted. In the CPTU for this study, a cone penetrometer as specified in the "International Reference Test Procedure for Cone Penetration Test (CPT)" (ISSMFE, 1989) was used with a penetration rate of 2 cm/s. The distribution of the cone resistance, q_t, in Fig. 3 generally corresponds to the distribution of SPT N-values in Fig. 2.

3. SPT-CPTU RELATIONSHIPS
Fig. 4 shows the relationships between cone resistance, q_t, in the alluvium with the standardized SPT N-value, N_{60}. N_{60} is an N-value corrected in terms of an effective energy of 60% of free fall energy and equals 1.3 times the N-value measured with the mechanical trip device that is widely used in Japan (Seed et al. 1985). Fig. 4 also shows the separate regression results: the regression line for the As_1 layer dominated by quartz sand and that for the As_2 and Ac layers dominated by Shirasu. The slope of the

regression line for the As2 and Ac layers clearly differs from that for the As1 layer. Fig. 5 shows relationships between sleeve friction, fs, and N_{60}, along with the separate regression results, as in Fig. 4. The layers with different components give different slopes of regression lines in Fig. 5, as in Fig. 4.

It is generally recognized that the qt/N_{60} ratio depends on the grain size of the soil, and that qt/N_{60} becomes higher as the mean grain size, D_{50}, becomes larger (Robertson et al. 1983; Seed and De Alba 1986). Fig. 6 contrasts the regression results shown in Fig. 4 with the relations ascertained by Seed and De Alba (1986). In Fig. 6, the regression results in Fig. 4 are plotted against the average D_{50} values for the respective soils. Although the relation between D_{50} and qt/N_{60} for the As1 layer dominated by quartz sand agrees fairly well with the relations found by Seed et al., the ratio for layers As2 and Ac principally composed of Shirasu is about three times greater than that determined by Seed et al. for the same D_{50}. This result indicates that the ratio of static penetration resistance to dynamic penetration resistance for Shirasu is higher than that for ordinary soils. One of the major causes of this phenomenon could be the high crushability of Shirasu, which is composed chiefly of angular and vesiculate volcanic glass grains. The crushability of Shirasu could be more conspicuous when it is exposed to dynamic penetration force.

4. SOIL CLASSIFICATION

Several charts have been proposed for evaluating soil type from CPTU data. The necessity for preliminary examination before applying these charts to a local soil is generally accepted. The soil classification charts proposed by Robertson (1990) are found to be capable of yielding comparatively good results for ordinary soils in Japan. To eliminate the effect of survey depth on the classification results, the charts by Robertson employ indices (normalized cone resistance, Qt, normalized friction ratio,

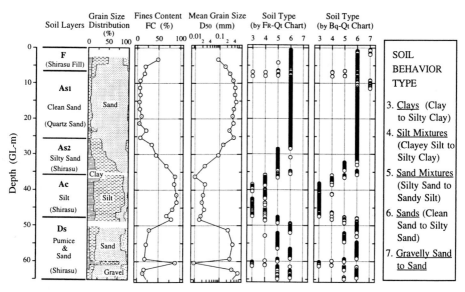

Fig. 7. Comparison of Sampling Test Results and Soil Classification Results from CPTU Data Using Charts Proposed by Robertson (1990)

F_R, and pore pressure ratio, B_q) that are obtained by normalizing the measured values in terms of overburden stress and equilibrium pore pressure. There are two types of charts by Robertson (1990): one uses indices Q_t and F_R and the other uses indices Q_t and B_q.

The results of soil classification using the two charts by Robertson and the CPTU data shown in Fig. 3 are illustrated in Fig. 7, together with the sampling and laboratory test results for the same point. The results of soil classification according to the charts are indicated by classification numbers that gradually increase as the soil grains become coarser, in accordance with the original charts. In this case the classification numbers start with 3 (clays) and end with 7 (gravely sand to sand). Both results of soil classification using the two charts show distribution patterns that match well with the results of grain size analysis of soils: grain size distribution, mean grain size D_{50}, and fines content FC. Approximate soil classification results can be obtained from CPTU data even in Shirasu deposits using the charts by Robertson. The charts, however, have no segment for pumice and the pumice in the Ds layer tends to be classified as a sand or sand mixture in the charts. Moreover, the silt in layer Ac tends to be classified as a clay, especially in the B_q-Q_t chart. In utilizing the charts by Robertson for the Shirasu deposits these features should be noted.

5. PENETRATION RATE EFFECT

The penetration rate, which is one of few controllable factors affecting CPTU data, is generally specified as 1–2 cm/s as in the American Standard (ASTM, D3441-86). The effects of penetration rate in layers As2 (silty sand) and Ac (silt), both of which are mainly composed of Shirasu, are summarized in Fig. 8. Here each value for the vertical axis is an average of the measured values in the same four-meter range in each layer. Data on clayey silt from a study by Campanella et al (1983)

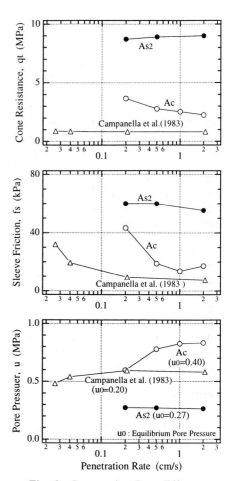

Fig. 8. Penetration Rate Effects

Table 1. Soil Properties

	Shirasu		Campanella et al. (1983)
	As2 (Silty Sand)	Ac (Silt)	(Clayey Silt)
Sand (%)	61	9	10
Silt (%)	33	77	70
Clay (%)	6	14	20
FC (%)	39	91	90
LL (%)	NP	NP	38
PI (%)	NP	NP	15

PI = Plasticity index.
LL = Liquid limit.
NP = Nonplastic.

are also plotted in Fig. 8 for reference. Table 1 shows soil properties of the layers for which data is plotted in Fig. 8. Fig. 8 indicates that variation in penetration rate within the standard range (1–2 cm/s) had relatively little effect on the values measured in any layer. In layer As_2, where the fines content, FC, is low, the pore pressure, u, almost equals equilibrium pore pressure, u_0, and values for cone penetration resistance, q_t, and sleeve friction, f_s, remain constant for the penetration rate range between 0.2 and 2 cm/s. In contrast, in layer Ac, where FC is high, penetration rates between 1 and 2 cm/s produce excess pore pressure (u-u_0) of around 0.4 MPa, and a penetration rate of about 0.5 cm/s or lower tends to cause the excess pore pressure to decrease, and correspondingly q_t and f_s to increase. For the clayey silt examined by Campanella et al., which had a FC comparable to that of layer Ac, a trend similar to that noticed for layer Ac is observed for f_s, but variation in f_s starts at a penetration rate of around 0.05 cm/s: one order of magnitude lower than the rate at which variation was found to begin in layer Ac. Excess pore pressure, being easily influenced by change in penetration rate, indicates that penetration rate also has a considerable effect on values other than u, especially f_s, for that soil. For a further discussion of the topic, the concepts of effective stress interpretation should be needed, as suggested by Campanella et al (1983). However, a qualitative knowledge of the penetration rate effect on f_s could help in understanding the relation between f_s and skin friction resistance of a pile, as discussed later.

6. ESTIMATION OF PILE SKIN FRICTION

In the coastal area of Kagoshima, where a thick layer of alluvial Shirasu has been deposited, many structures are supported by friction piles because the firm bearing layer for the piles is deep. The CPTU sleeve friction, f_s, is a direct measure of friction between the in-situ soil and the sleeve of the cone penetrometer. The relations between the f_s value and the frictional resistance of a pile are discussed, and a method of estimating a load-settlement curve at the pile top using the CPTU results are proposed.

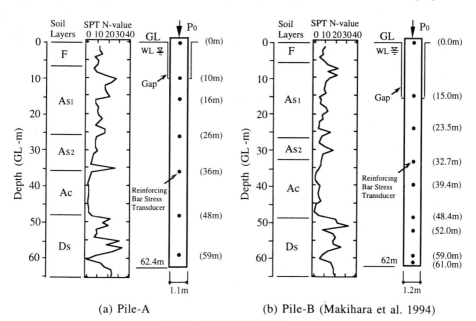

(a) Pile-A   (b) Pile-B (Makihara et al. 1994)

Fig. 9. Test Pile Profiles

6.1 CPTU Sleeve Friction and Ultimate Pile Skin Friction

Fig. 9 shows the profiles of vertically loaded test piles. The test on Pile-A was conducted at the same point where the soil investigation data shown in Fig. 2 and Fig. 3 was obtained. The test on Pile-B, whose data is quoted from a paper by Makihara et al (1994) for comparison, was conducted at a site with a soil profile similar to that at Pile-A. Both were bored cast-in-place piles installed by the reverse-circulation boring method.

Fig. 10 shows the relation between CPTU sleeve friction, f_s, and the ultimate skin friction, τ_p, of Pile-A for sandy soil; Fig. 11 shows that for clayey soil. Fig. 10 and Fig. 11 include data obtained by Shibata et al. (1987) on bored cast-in-place piles from four locations in Osaka, for reference. The relation $\tau_p=f_s$ holds for both sandy layers As1 (mainly quartz sand) and As2 (mainly Shirasu). The data on sandy soil from Shibata et al., though dispersed, can also be approximated on average by the relation $\tau_p=f_s$. For clayey soil layer Ac, τ_p is higher than f_s, forming the relation $\tau_p=3f_s$. The data of Shibata et al. for clayey soil, though somewhat dispersed, indicates a general relation $\tau_p>f_s$. One of the prime reasons τ_p is larger than f_s in clayey soil could be the difference between the settlement rate of the pile in the loading test and the penetration rate of the penetrometer in CPTU. The settlement rate of Pile-A in the loading test was less than 0.004 cm/s, or less than one-five hundredth the penetration rate in CPTU. The fact that the $\tau_p>f_s$ relation occurs for layer Ac can be qualitatively accounted for by referring to the relation between cone penetration rate and f_s in Fig. 8. Just as the effect of cone penetration rate on f_s depends on soil properties, so could the relation between f_s and τ_p. Differences in soil properties among soils classified as sandy soil or clayey soil could be one cause of the dispersed distribution of the f_s-τ_p relations reported by Shibata et al.

In practical design work the ultimate skin friction, τ_p, of a pile in clayey soil is often estimated from the undrained shear strength, c_u, of the soil. Fig. 12 shows the relation between f_s and the undrained shear strength, c_u, of undisturbed samples taken from four different depths in layer Ac. Plotted distinctively from one another are values for c_u obtained as one-half the unconfined compression strength and values for c_u obtained from the triaxial UU-test, along with regression lines passing through the origin for each set of values. Values for c_u obtained from the triaxial UU-test tend to be about 1.5

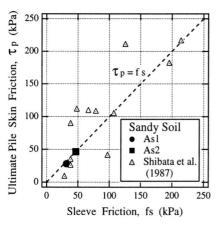

Fig. 10. Relation between f_s and τ_p
(Sandy Soil)

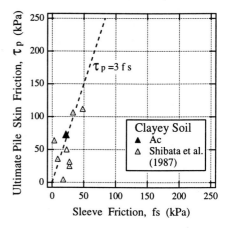

Fig. 11. Relation between f_s and τ_p
(Clayey Soil)

times larger than values obtained from the unconfined compression test. However, comparing Fig. 12 and Fig. 11 reveals that the c_u obtained from the unconfined compression test is closer to the τ_p in layer Ac.

6.2 Estimation of Load-Settlement Curves

Fig. 13 shows the analytical model of the load-transfer method proposed by Seed et al.(1955). In the model, a pile that is divided into multiple elastic elements resists the applied load with skin-friction springs, Kf_i, attached to each element and a tip-resistance spring, Kt, attached to the tip element. By giving these springs appropriate properties, the model generates pile behavior such as a load-settlement curve and a load distribution curve. Some studies have been made on methods for theoretically determining skin friction spring properties for the model using parameters obtained from laboratory soil tests. However, a method that needs laboratory test results is not suitable for Shirasu deposits that are mainly composed of sandy soil with difficulties in taking undisturbed samples. Therefore, with practicality in view, a simple method of evaluating friction spring properties using CPTU results is studied in this paper.

Fig. 14 shows a model of the friction spring property, the relation between pile skin friction, τ, and pile displacement, δ. The curve between the origin O and the yielding point A (δ_y, τ_p) is a hyperbolic curve that represents the shearing resistance of the soil surrounding the pile. Between points A and B the skin friction is constant, representing

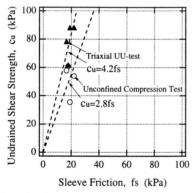

Fig. 12. Relations between f_s and c_u in Ac Layer

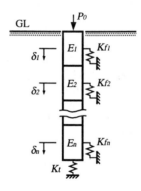

Fig. 13. Model for Load-Transfer Method

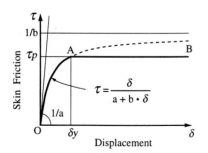

Fig. 14. Model for δ-τ Relation

Table 2. Parameters for δ-τ Model

Soil Type	α	β	δ_y (cm)
Sandy Soil	1	1.3	1
Clayey Soil	3		2

the pile slipping against the soil next to the pile. To fix the model, three out of four parameters (a, b, δ_y and τ_p) must be determined. Here, the coordinates of yielding point A (δ_y, τ_p) and the asymptotic value of the hyperbolic curve, 1/b, are determined by the following methods:

(1) τ_p and 1/b are set using the following formulae:

$$\tau_p = \alpha \cdot f_s$$
$$1/b = \beta \cdot \tau_p = \alpha \cdot \beta \cdot f_s$$

where

    α, β: parameters dependent on soil properties, pile type and pile installation method.

    f_s: CPTU sleeve friction

(2) δ_y is set in consideration of soil properties, pile type and pile installation method.

Fig. 15 shows the relationship between skin friction, τ, and displacement, δ, for Pile-A in alluvial layers, contrasting measured values with modeled curves. Table 2 shows the parameters used for forming the models. These parameters were determined by considering the relation between f_s and τ_p shown in Fig. 10 and Fig. 11, so the modeled curves would approximate the measured values; the modeled curves in Fig. 15 generally do. Moreover, applying the parameters listed in Table 2 to the model of the Ac layer produces two relations: $\tau_p = 3 \cdot f_s$ and $1/b = 3.9 \cdot f_s$. It is noteworthy that the two relations approximate the relations shown in Fig. 12 between c_u obtained from the unconfined compression test and f_s ($c_u = 2.8 \cdot f_s$) and between c_u obtained from the triaxial UU-test and f_s ($c_u = 4.2 \cdot f_s$), respectively.

Fig. 16 illustrates the relations between δ and τ in layer Ds for Pile-A and Pile-B. In the curve for Pile-B, skin friction peaks when the pile is displaced about 1 cm, as is the case for the curves for sandy layers in Fig. 15, while the curve for Pile-A skin friction tends to increase even after pile displacement has approached 8 cm. In vertical loading tests of bored cast-in-place piles, such a δ-τ relation is occasionally seen for coarse-grained soil layers such as those composed of gravely sand. The reason for this could be related the phenomenon that the particle structure of coarse-grained soil layers is prone to loosening as a result of reduction of confining pressure due to boring.

Layer Ds, composed mainly of pumice (maximum diameter about 10 cm), is a coarse-grained layer which should be easily loosened by boring. The fact that 20 hours were spent from the end of boring up to the start of concrete placement for Pile-A, four times

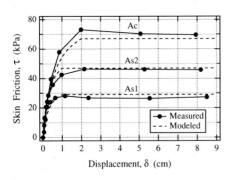

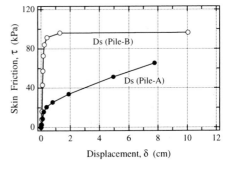

Fig. 15. Measured and Modeled δ-τ Curves in Alluvial Layers

Fig. 16. δ-τ Curves in Ds Layer

as long as was spent for Pile-B, indicates a high probability that loosening of the soil in layer Ds took place in the case of Pile-A. Differences in installation work such as this seem to have caused differences in the form of δ-τ relations for Pile-A and Pile-B. Since, as mentioned above, the δ-τ relation depends not only on soil properties but also on the conditions of the pile installation work, sufficient account should be taken of this fact in formulating a model of the δ-τ relation.

Here a model of the relation is formed using the values shown in Table 2 as a basis and incorporating the effect of soil loosening by varying only δ_y (due to a lack of sufficient knowledge about the matter).

Fig. 17 shows pile top load-settlement curves for Pile-A, contrasting a measured curve with curves calculated by the load transfer method. The calculations were performed by dividing the pile into about two-meter-long elements and using a δ-τ model determined for each pile element by CPTU data of the corresponding range. In Case 1 the values in Table 2 were used in determining the skin-friction spring property; in Case 2 the calculation was done assuming that δ_y in layer Ds was 10 cm to allow for loosening of the soil. Classifications of the sandy soil and clayey soil were made using the chart by Robertson. The measured relation between displacement and resistance at the tip was applied directly to the tip resistance spring property. The results for Case 1 agree with the results of measurements for loads up to about 7 MN, but for higher loads calculated values for settlement at the pile top are too small because the effect of loosened soil in layer Ds is neglected. The results for Case 2, in which the effect of loosened soil in layer Ds is taken into account, are closer to the measured values than those for Case 1, but the settlement values are still small when the load rises above 9 MN. Fig. 18 shows the load distribution curves for Case 2. Each calculated curve fits well with the measured one, especially when the applied load is less than 9 MN. Though Case 2 reflects the effect of loosened soil in layer Ds, only δ_y of the three parameters in the δ-τ model takes account of this effect. Further study on the parameters are necessary to make the calculated values approximate the measured values more closely.

Fig. 19 shows pile top load-settlement curves for Pile-B. Since CPTU data at Pile-B was not available, in the calculation, each δ-τ model was determined by applying the f_s value estimated based on the f_s value at Pile-A considering the ratio of SPT N-value at Pile-A to that at Pile-B for each layer. The values in Table 2 were used in the

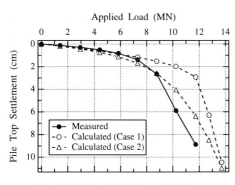

Fig. 17. Load-Settlement Curves
at Pile Top (Pile-A)

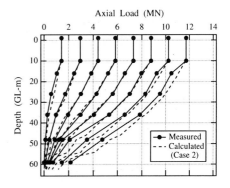

Fig. 18. Load Distribution Curves
(Pile-A, Case 2)

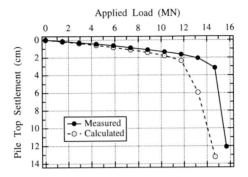

Fig. 19. Load-Settlement Curves at Pile Top (Pile-B)

calculation. Although the calculated values agree with the measurements for loads up to 12 MN, they exaggerate pile settlement for greater loads. Estimating f_s in the manner described above may be one cause of the disagreement between the calculated and measured results.

A simple method using CPTU data for estimating the bearing mechanism of piles that mainly depend on frictional resistance was proposed and examined. The validity of the method was proved to some degree, though it is based on limited data. In using the method, parameters must be determined in consideration of pile type and pile installation method. A study on these parameters is needed hereafter by accumulating data on various types of piles, installation methods and soil properties.

7. CONCLUSIONS

Applicability of CPTU in Shirasu deposits was studied. A simple method for estimating the mechanical behavior of a pile on the basis of CPTU data was also proposed and studied for Shirasu deposits. The conclusions drawn from this study are as follows:

1) The ratio, q_t/N_{60}, of cone resistance to standardized SPT N-value in alluvial Shirasu deposits is three times that presented by Seed et al. for soils of the same mean grain size.
2) Approximate soil classification results can be obtained from CPTU data even in Shirasu deposits using the charts proposed by Robertson.
3) Although the effects of penetration rate on CPTU data depend on soil properties, the effects are small enough to ignore, even in Shirasu deposits, when the rate is within the standard range of 1 to 2 cm/s.
4) The relation between CPTU sleeve friction and ultimate pile skin friction depends on soil properties. One reason for this could relate the difference between CPTU penetration rate and pile settlement rate.
5) The displacement-skin friction relation of a pile can be estimated based on CPTU data; the mechanical behavior of a pile in Shirasu deposits can be estimated by the proposed method.

NOTATION

$\quad$ a, b = Parameters in hyperbolic equation.

$\quad B_q$ = Pore pressure ratio

$\quad c_u$ = Undrained shear strength.

$\quad D_{50}$ = Mean grain size.

FC = Fines content.
F_R = Normalized friction ratio.
f_s = Sleeve friction.
LL = Liquid limit.
N_{60} = Standardized SPT N-value.
PI = Plasticity index.
Q_t = Normalized cone resistance.
q_t = Cone resistance (strictly total cone resistance for CPTU).
R_f = Friction ratio (f_s/q_t).
u = Pore pressure.
u_0 = Equilibrium pore water pressure.
α, β = Parameters relating sleeve friction to displacement-skin friction model.
δ = Pile displacement.
δ_y = Yield pile displacement.
ρ_s = Soil particle density.
τ = Pile skin friction.
τ_p = Ultimate pile skin friction.

REFERENCES

1) Campanella, R. G., Robertson, P. K. and Gillespie, D. (1983): "Cone penetration testing in deltaic soils," Canadian Geotechnical Journal, Vol. 20, No. 1, pp.23-35.
2) Haruyama, M. (1973): "Geological, physical and mechanical properties of Shirasu and its engineering classification," Soil and Foundations, Vol. 13, No. 3, pp. 45-60.
3) Hatanaka, M., Sugimoto, M. and Suzuki, Y. (1985): "Liquefaction resistance of two alluvial volcanic soils sampled by in situ freezing," Soil and Foundations, Vol. 25, No. 3, pp. 49-63.
4) ISSMFE (1989): "International reference test procedure for cone penetration test (CPT)," Report of the ISSMFE technical committee on penetration testing of soils - TC16 with Reference Test Procedures CPT-SPT-DP-WST, pp. 6-16.
5) Makihara, Y., Enami, A., Kurokatsuhara, G., Kumagai, Y., Sakimoto, H. and Yoshimori, M. (1994): "Vertical load test results of long friction pile in Shirasu deposits," Summaries of Technical Papers of Annual Meeting, AIJ, pp. 1361-1362 (in Japanese).
6) Robertson, P. K., Campanella, R. G. and Wightman, A., (1983) : "SPT-CPT correlations," Journal of Geotechnical Engineering, ASCE, Vol. 109, No. 11, pp. 1449-1459.
7) Robertson, P. K. (1990): "Soil classification using the cone penetration test," Canadian Geotechnical Journal, Vol. 27, pp. 151-158.
8) Seed, H. B. and Reese, L. C. (1955): "The action of soft clay along friction piles," Proc. ASCE, Vol. 81, Paper No. 842, pp. 1-28.
9) Seed, H. B., Tokimatsu, K., Harder, L. F. and Chung, R.M. (1985): "Influence of SPT procedures in soil liquefaction resistance evaluations," Journal of Geotechnical Engineering, ASCE, Vol. 111, No. 12, pp. 1425-1445.
10) Seed, H. B. and De Alba, P. (1986): "Use of SPT and CPT tests for evaluating the liquefaction resistance of sands," Proceedings of In Situ '86, ASCE, pp. 281-302.
11) Shibata, T., Yashima, A., Emi, S. and Horikoshi, K. (1987): "The estimation of the shaft friction of bored piles," Proc. 32th Soil Eng. Symposium on Foundations without using Piles, JSSMFE, pp. 7-10 (in Japanese).

Moderator's report on Session 4(a): penetrometers

P.K. ROBERTSON
Professor of Civil Engineering, University of Alberta, Edmonton, Canada

INTRODUCTION

This paper represents the Moderators report for the first part of Session IV dealing with Penetrometer testing. Session IV deals with the topic of In-situ testing. The first part of Session IV is devoted to Penetrometer Testing. A total of seven papers were received for this part of the Session; most involve the Cone Penetration Test (CPT) with one also describing developments to the Dilatometer Test (DMT) and one describing work with various dynamic probes. It is clear from these papers that the CPT is now one of the most popular forms of penetrometer test. In many areas of the world the Standard Penetration Test (SPT) is still the most common in-situ test for geotechnical investigations. However, the CPT is becoming more common due primarily to advantages in cost, reliability, continuous results and repeatability. The lack of papers to this Conference on the SPT is some indication of the limits to this form of penetration testing.

In the areas of off-shore engineering and geo-environmental applications the CPT has clearly become the most dominate in-situ test. For geo-environmental applications this has resulted partly from regulations in some areas of the world that control the disposal of waste cuttings from drilling operations. The high cost of disposal has increased the cost of drilling substantially and resulted in a rapid expansion of direct push technology using various forms of penetrometers. The direct push devices generate essentially no cuttings, produce little disturbance and reduce contact between field personnel and contaminates, since the penetrometer push rods can be decontaminated during retrieval. The CPT has therefore become the main vehicle for new direct push devices with a vast array of new sensors added to the cone penetrometer.

This moderators report will concentrate on the main areas addressed in the papers to this Session. They are; equipment development, software, geo-environmental applications and applications in unusual or non-textbook materials.

EQUIPMENT DEVELOPMENTS

The paper by Kaggwa, Jaksa and Jha describes research work carried out at the University of Adelaide, Australia on the development of an automated dilatometer (DMT) and then comparison of the DMT with the CPT results. The paper describes a data collection system for both the CPT and DMT. Microcomputer data acquisition systems are now common for in-situ testing and this paper describes a system developed for research at the University of Adelaide. The main part of the paper describes the instrumentation of the flat dilatometer to measure the push force during penetration, the pressure behind the expanding membrane and the pressure expansion response of the membrane. The resulting DMT is very similar to that described by Campanella and Robertson (1991). The paper briefly describes the problems with calibration corrections to derive the DMT pressure-expansion curve due to the highly non-linear response of the stiff constrained dilatometer membrane. The paper also presents a typical profile of DMT results, an example of the DMT penetration resistance profile and a typical pressure-expansion curve for the DMT membrane at site consisting of uniform sand with between 20 to 25% carbonate content. There is a similarity between the DMT membrane expansion curve presented and that observed for typical self-boring pressuremeter tests in sand. The authors are

Advances in site investigation practice. Thomas Telford, London, 1996

encouraged to compare their results with those presented by Campanella and Robertson (1991).

Jacobs and Butcher describe developments using the seismic CPT (SCPT). This paper describes a SCPT system with two sets of geophones for improved accuracy and reliability. This is not a new concept since the early work on the SCPT used two sets of geophones (Robertson et el., 1986). This early work showed that the single array of geophones gave very similar results when compared to a two array system. However, experience over the last 10 years has shown that a two array system can be faster and provide better resolution of shear wave velocity, especially at depths greater than 20 m. Jacobs and Butcher illustrate this with an excellent profile of shear wave velocity down to a depth of 38 m using the SCPT over water. The SCPT has the advantages of providing independent measurements of CPT data as well as soil stiffness profiles. An advantage of the SCPT that is not well recognized is its potential to identify unusual soils, as will be discussed in a later section.

The paper by Butcher, McElmeel and Powell describes work carried out at the Building Research Establishment (BRE) using various Dynamic Probes. The main conclusions from this work is to confirm the importance of rod friction and standardization of hammer, anvil and penetrometer equipment. The need to remove the effects of rod friction are clearly shown in several profiles. The use of bentonite mud to reduce rod friction should become a standard. In many parts of the world dynamic probing tests are not well standardized and the importance of rod friction is not always recognized. For smaller projects and where ground conditions are relatively uniform, the dynamic probing tests, when carried out with care, can be very useful for basic stratigraphic profiling. However, the results can be misleading in complex and interbedded profiles and care should be used in the interpretation and application of the results. Like most dynamic testing methods the interpretation of results can be subject to many variables.

SOFTWARE

The paper by Kaggwa, Jaksa and Jha describes software developed to collect and present CPT and DMT data. In recent years with the rapid development of fast microcomputers many software packages have been developed to collect, process, interpret and present penetrometer test data. Most of these developments have been made for the CPT and many have been made by the specialty manufacturers and contractors supplying the service of cone penetration testing. The software has become a useful value added component for the contractors. Figure 1 shows an example of a modern CPT plotting software package. This profile is produced immediately after completion of the CPT sounding. The plot can show the measured parameters of cone resistance (Q_c), sleeve friction (F_s) and penetration pore pressures (U) versus depth as well as the calculated values of friction ratio (Rf). The software also allows the corrected cone resistance (Q_t) to be plotted, as shown in Figure 1. Included in the plot are profiles showing the interpreted soil behaviour type (SBT) based on the charts proposed by Robertson and Campanella (1988) and Robertson (1990). The software provides a visual representation of the soil type using colour bars. Each soil type has a different colour and length of bar. Colour monitors and printers can be provided in the CPT vehicle to enhance the presentation. Alongside the coloured soil profile the software produces an automatic description of the soil profile using standard soil type descriptions. The software automatically tracks the layers and provides break lines where appropriate. The level of detail can be adjusted based on a minimum layer thickness criteria. Software is also available that can produce CPT profiles to include interpreted parameters, such as, undrained shear strength, friction angle, relative density and equivalent SPT values.

Software packages that can produce profiles similar to that shown in Figure 1 and interpretation profiles are now becoming more common. It is now possible for engineers or field personnel to have complete profiles of CPT results including interpretation of soil behaviour type and estimated geotechnical parameters in the field immediately after completion of a CPT sounding. Some CPT vehicles now come with laser printers, cellular phones and fax machines to allow rapid transfer of results to remote offices for evaluation and analyses. This has allowed greater flexibility in site investigation supervision and control. A senior project engineer can now review the results from a remote location and provide rapid advise regarding future sounding locations and additional testing. This has become very useful on geo-environmental projects where groundwater and/or soil sampling locations need to be adjusted based on the interpretation of soil type and stratigraphy from adjacent CPT soundings.

Software developments have also been made in the area of geotechnical analyses using CPT data. Software packages are now available to estimate pile capacity and load settlement response, shallow foundation settlement and soil liquefaction analyses using CPT data. These programs make use of the near continuous CPT data to provide, for example, profiles of pile capacity with depth and liquefaction resistance with depth. Data is taken directly from the CPT files and design profiles produced. This can significantly enhance the speed of design and allow the engineer greater time to carryout sensitivity analyses.

It is clear that developments in software will continue and that these developments will further enhance the flexibility of site investigations and the interpretation and application of in-situ test results.

GEO-ENVIRONMENTAL APPLICATIONS

The paper by O'Neil, Baldi and Della Torre describes the Multifunctional Envirocone Test System (METS) developed by ISMES in Italy. This probe is a 15 cm^2 CPT with a sampling element located above the cone to sample pore water and/or soil gas. The probe has a small accordian, positive displacement pump to push water or gas samples to the surface for chemical analyses. Details are given regarding probe design and calibration as well as several case histories. Preliminary results look promising and good comparison is shown between the METS and conventional methods. The only concern with the system appears to be the long time periods required to obtain relevant samples. In some cases it maybe more productive to try various devices to probe the subsurface and sample the relevant layers for contaminants. Experience in the USA has shown that, in some cases, less expensive sampling probes are used in combination with CPT and sometimes, surface geophysics to obtain the required geo-environmental information. Clearly, there is a need for a balance between cost, speed of information and reliability of data.

The paper by Robertson, Lunne and Powell provides an overview of the many developments in the area of penetration testing for geo-environmental purposes. Brief descriptions are provided for penetrometers that include sensors such as; pore pressure, inclination, electrical resistivity (conductivity), dielectric constant, pH, redox potential, natural gamma radiation, laser induced fluorescence (LIF) using fibre optics and UV fluorescence. The paper also describes the recent developments in sampling devices using direct push technology. Samplers have been developed to obtain liquid, gas and solid samples. Sealing and decontamination techniques are also briefly described. The paper also provides insight into future trends in the area of penetration testing for geo-environmental purposes and clearly illustrates the intense activity in this new area.

For penetration testing in geo-environmental applications there are some interesting topics that may stimulate discussion. The first relates to the different regulations controlling geotechnical and geo-environmental problems. For geotechnical projects the extent and detail of the site investigation is usually a function of the proposed structure and the associated risks. The geotechnical engineer is often in reasonable control of the risk process and hence, selection of the required site investigation detail. For geo-environmental projects the extent and detail required for the determination of the distribution and composition of contaminants maybe controlled by various regulatory agencies, for which the engineer may have little control. With the rapid improvements in measurement technology, the in-situ detection limits required by some agencies for certain contaminants are decreasing at an alarming rate. Hence, engineers are often asked to aim for a moving target in geo-environmental investigations. There is a need for engineers to play a more important role is the selection and definition of the regulations that govern geo-environmental work.

The hole produced by the penetrometer requires sealing usually with a special grout. The grouting can be carried out either after the push rods are removed or during the removal process. Typically the grouting is carried out after rod removal using special grout push rods with disposable tips. The hole is grouted from the bottom up as the grout rod is pulled from the ground. The grout rods generally follow the previous penetrometer hole since the rods follow the path of least resistance. However, some operators use penetrometers that allow grouting during penetrometer retrieval. The grout is usually delivered to the penetrometer through a small diameter grout tube pre threaded through the push rods. The grout can exit the penetrometer through either a sacrificial cone tip or through ports above the cone located on the friction reducer. Regulations will increasingly control this sealing process. It will be important that engineers become more involved with the evaluation of various sealing methods since regulations may make the process increasingly more complex and in some cases unnecessary. Engineers need to maintain some level of control over the standards required for site investigation.

A major problem with the development and application of chemical sensors for penetrometers relates to the interaction of the sensor with the contaminant. Most of the sensors require that the contaminant (usually in vapor or liquid form) be pulled into the probe so that the chemicals can interact with the sensors. In the case of the METS described by O'Neil et al., the sample is taken all the way to the surface. This interaction between sample and probe can produce problems related to the cleaning of the sensor, measuring cell, sampling tubes and filter element to avoid cross-contamination, as well as the time required for this process. Little research has been carried out to evaluate these problems and the issues related to; the interaction of the contaminant and soil, the interaction of the contaminant and measuring device, the contaminate state and contaminate mixtures. Clearly there is much research required in this area.

PENETROMETER TESTING IN 'UNUSUAL' OR NON-TEXTBOOK MATERIALS

The papers by Lunne, Powell and Robertson and Takesue, Sasao and Makihara present CPT data in 'unusual' materials. Lunne et al. show results from intermediate soils, such as silts, organic soils, such as peat, underconsolidated clays, soft rock, such as chalk and mine tailings. They also provide references to papers that describe CPT results in materials such as, calcareous soils, cemented sands, residual soils, permafrost and ice, snow, gas hydrates and slurry materials used for slurry walls. Lunne et al. correctly stress the link between the drainage conditions during cone penetration and that expected in the design problem. If the cone penetration process is undrained, the CPT data can be interpreted in a manner similar to that in a clay. However, if the cone penetration process is undrained or partly drained and the design problem is expected to involve drained loading, then the interpretation is more difficult.

Takesue et al. present some interesting results of CPTU and SPT as well as laboratory tests and pile load tests in a volcanic soil found in southern Kyushu, Japan. The unique physical properties of this volcanic soil affect its mechanical properties and lead to some geotechnical difficulties. This volcanic soil is mainly composed of vesiculate and easily crushable volcanic glass grains of angular shape, whose specific gravity is lower than that of quartz sands. Hence, this soil is highly crushable and compressible. Results are presented from CPTU to illustrate the effect of rate of penetration on the volcanic silt. As the rate of penetration is decreased from the standard 2 cm/sec the penetration moves from an undrained process to a drained process at a rate of about 0.1 cm/sec. This change in drainage has a larger effect on the sleeve friction than on the cone resistance, since the sleeve friction is directly controlled by effective stresses. The cone resistance is a total stress measurement and hence, responds to both changes in effective stresses and pore pressures. This rate effect also influences the relationship between the CPT and SPT, since the measured SPT N value is low partly due to the higher rates during the dynamic penetration process. The crushability of the soil also influences the CPT and SPT differently, since the CPT measures only tip resistance, whereas the SPT is a summation of tip resistance and friction along the inside and outside of the SPT sampler. Some crushable soils have slightly smaller friction along penetrometers, although this does not appear to be a major factor for these volcanic soil. This is further reflected by the fact that the soil behaviour charts developed by Robertson (1990) provided consistent estimates of soil type in these volcanic soils.

Takesue et al. provide additional data related to the application of CPT results for pile design. They applied the measured CPT sleeve friction values to estimate the unit shaft resistance for cast-in-place bored piles. The values they suggest are consistent with previous published values for silica sands. This again supports the conclusion that the frictional properties of this volcanic sand is similar to that of other sands. However, the high crushability of the volcanic sand produces somewhat lower CPT penetration resistance.

An observation that can be made from the papers by Lunne et al. and Takesue et al. is the importance to be able to identify unusual soil types from in-situ testing. Current empirical and theoretical interpretation methods for the CPT are very good for estimating geotechnical parameters in well behaved 'normal' soils. These papers illustrate the problems when interpreting CPT results in 'unusual' materials. Hence, it is important to be able to identify when the soil or material is unusual and then to modify the interpretation accordingly. Research is underway at several places to address this problem for CPT interpretation. One avenue is to add more independent measurements to the CPT in an effort to identify these unusual materials. One such approach uses the seismic CPT. The shear wave velocity can be measured during a CPT sounding to provide additional independent data. The shear wave velocity is a small strain measurement, unlike the penetration resistance which is a large strain measurement. The shear wave velocity is controlled by many of the same soil variables as the penetration resistance, such as, void ratio (density), effective confinement stresses, stress history, soil fabric, cementation and aging. However, the shear wave velocity is almost independent of soil compressibility or crushability. Hence, there exists the potential to identify crushable sands using the seismic CPT. Research has also shown that there exists the potential to identify cemented or aged soils using the seismic CPT since the shear wave velocity is influenced to a different degree than penetration resistance by these factors. Clearly there is a need for more research into the effects of unusual soil properties on in-situ test results and how we can identify the existence of these unusual soils from the test results. Although the term 'unusual' soils has been used to describe these materials, we may find that these materials are more common than we had previously expected.

SUMMARY
Papers to this part of Session IV cover a range of topics from equipment development, software, geo-environmental applications and applications in unusual or non-textbook materials. It is clear from these papers that the CPT is one of the most important in-situ tests. Equipment developments for the CPT are well advanced with reliable, robust penetrometers available. New sensors are being added to the penetrometer, especially in the area for geo-environmental purposes where sensor technology is rapidly developing. Software for data collection, processing and interpretation is also developing rapidly. Software is now available to process and interpret CPT data in real time during the sounding. Empirical correlations exist for a wide range of soil parameters in ideal soils. Several Papers to this session illustrate the problems associated with interpretation of penetration test data in unusual or non-textbook materials. However, it may be that many natural soils will fit into this category of 'unusual' soils and that it will be important to be able to identify these materials from the in-situ test results.

REFERENCES

Butcher, A.P , K. McElmeel and J. J. M. Powell, 1995, Dynamic Probing and its use in clay soils, Advances in Site Investigation Practice, London.

Campanella, R.G. and P.K. Robertson, 1991, Development and Use of a Research DMT, Canadian Geotechnical Journal, Vol. 28, No. 1, pp 113-126.

Jacobs, P.A. and A. P. Butcher , 1995, The Development of the seismic cone penetration test and its use in geotechnical engineering, Advances in Site Investigation Practice, London.

Kaggwa, W. S. ., M. B. Jaksa and R. K. Jha, 1995, Development of Automated dilatometer and cone penetration tests at the University of Adelaide, Advances in Site Investigation Practice, London.

Lunne, T., J. Powell and P. K. Robertson, 1995, Use of Piezocone tests in non-textbook materials, Advances in Site Investigation Practice, London.

O'Neil, D.A., G. Baldi and A. Della Torre, 1995, The Multifunctional Envirocone Test System, Advances in Site Investigation Practice, London.

Robertson, P. K. and Campanella R. G., 1988, Guidelines for Use and Interpretation of the CPT, Soil Mechanics Series, University of British Columbia, 1st edition, 1988.

Robertson, P. K., R.G. Campanella, D. Gillespie, and T. Rice, 1986, Seismic CPT to Measure In-situ Shear Wave Velocity, Journal of Geotechnical Engineering, ASCE, Vol. 112, No.8, pp 791-803

Robertson, P.K., 1990, Soil Classification using the CPT, Canadian Geotechnical Journal, Vol. 27, No.1, pp 151-158.

Robertson, P. K., T. Lunne and J. Powell, 1995, Applications of penetration tests for geo-environmental purposes, Advances in Site Investigation Practice, London.

Takesue, K., H. Sasao and Y. Makihara, 1995, Cone Penetration testing in volcanic soil deposits, Advances in Site Investigation Practice, London.

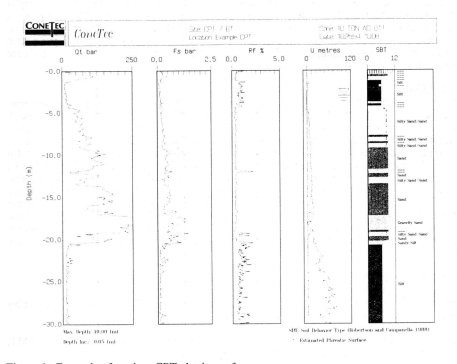

Figure 1. Example of modern CPT plotting software

Discussion on Session 4(a): penetrometers

Discussion reviewer C. H. ADAM
Fugro Scotland Ltd, Glasgow, Scotland

GENERAL

The discussion for this session centred on three main themes as emphasised by Professor Robertson in his moderator's report. These were:

- Equipment and Software Developments
- Environmental Applications
- Testing in Non-Textbook Materials

Each of the authors were invited to respond to questions raised in the Moderator's report. Following on from this the floor was opened to general discussion

EQUIPMENT AND SOFTWARE DEVELOPMENTS

Dr. W.S. Kaggwa, University of Adelaide (Paper entitled "Development of Automated Dilatometer and Comparisons with Cone Penetration Tests)

The moderator, Prof. Robertson, has raised a number of interesting issues in his presentation on dilatometer development, calibration and testing.

I agree with the comments regarding problems in manufacture and instrumentation to automate the dilatometer. The hysteric curve obtained during calibration is similar to the one that he presented.

I see the dilatometer, because of its small size, to be a useful device for measuring the horizontal behaviour in highly variable non-textbook materials. We have been successful in using the device in calcareous sediments when used in conjunction with the CPT, where we were able to determine the parameters in very soft sediments.

A key requirement of using the automated dilatometer is the incorporation of two calibration/checking stages; one before going out to the site and the second stage soon after returning from the field.

Dr. A. Butcher, Building Research Establishment (Paper entitled "The Development of the Seismic Cone Penetration Test and its use in Geotechnical Engineer)

The moderators report hints that the single geophone set seismic cone can give equally good results as the dual geophone set mentioned in Butcher and Powell (1995) and Jacobs and Butcher (1995). The advantages of the dual geophone seismic cone are as follows:

Advances in site investigation practice. Thomas Telford, London, 1996

(a) The geophones are a fixed distance apart (0.5 or 1.0m for the BRE version) so avoiding errors due to depth measurement of the single geophone set seismic cone.

(b) The seismic cone receives the signal to each geophone set from one source activation so avoiding the variable trigger delays that are influenced by the source input energy, i.e. the trigger delays depend on how hard the anvil of the source is struck by the operator.

(c) The travel time of the shear wave between the two geophone sets can be read off the signal record directly and checked by considering different parts of each signal, i.e. first arrival, first maximum positive response, first zero response, first maximum negative response, etc.

In general the dual geophone set seismic cone reduced the operators influence and thereby gives more consistent and reliable results.

Mr J.F. Nauroy, Institut Francais due Petrole
I would like to emphasise the use of the seismic cone to measure not only the shear wave velocity but also the compressional wave velocity. This is a very useful tool for calibrating seismic survey sections in offshore conditions.

Mr P.A. Jacobs, Fugro Ltd
Professor Robertson made reference to the possibility of a dual geophone array seismic cone with closer separation than one metre, providing more detailed information on measured in situ shear modulus. The seismic cone system used to collect the data presented in our Paper reviewed in this session (Jacobs and Butcher) included a dual array seismic piezocone with optional spacing of one or half a metre. The change over between these spacings can be done in the field.

Professor Robertson
Which spacing do you recommend?

Mr P.A. Jacobs
We mainly use the seismic dual array cone in one metre spacing, but when more detail of the ground is required in layered soil, we can use the half metre spacing to investigate.

Professor Robertson
I had asked the controversial question as to whether or not we needed to go to dual elements and I had shown some data from about 10 years ago that said you didn't need to, but certainly industry is moving in that direction. The single element cone works well provided you have an excellent trigger and source, you put the source close to the rods so that you get near vertical waves and you are in reasonably uniform ground conditions. If you cannot get access to a really good trigger and you are in more complex ground conditions the dual array system is the way to go and certainly there is a trend in industry to move in that direction

Dr. A. Butcher (Paper entitled "Dynamic Probing and its Use in Clay Soils")

Its strange really that dynamic probing should be featured in a conference on the advances in site investigation practice, since it is probably the oldest and simplest soil test and you would think we would have sorted it out by now.

The first thing I would like to address is standardisation. The German Standard DIN 4094 set out quite clearly in 1974 and which BS1377:1990 successfully omitted potentially the most useful combinations of drop weight and cone size. Tables 1 and 2 of dynamic probing test specifications included in Butcher et al (1995) show that BS1377:1990 only include the heaviest configurations of DPH and DPSH, DIN 4094 includes DPL, DPM, and DPH, and research at the Building Research Establishment has identified a configuration DMP15 as probably the most versatile configuration. DPM15 has a specific work per blow midway between DPL and DPH and uses a cone of base diameter significantly larger than the drive rod diameter which will help to reduce rod friction build-up.

The data contained in Figures 4 and 5 (from London clay and a Glacial till) from Butcher et al (1995) show significant depths of the profiles from the DPH and DPSH to be below the minimum recommended N_{10} (≥ 3, from Table 1). The DPM15 profiles, however have N_{10} values of greater than 3 and still penetrate to the same depths as the DPH.

The moderators report mentions the problem of rod friction which needs to be looked at carefully since the corrections using rod torque measurements vary with both configuration of equipment and soil type as shown in Table 4 of Butcher et al (1995). The effect of rod friction on penetration and profile is illustrated in Figure 1 below which shows the uncorrected data from Canons Park. The DPM15 profiles with the use of bentonite to reduce the rod friction clearly have greater penetration than those without the bentonite. The fully corrected data is included in Figure 3 of Butcher et al (1995) which clearly illustrates the ability of dynamic probing to identify different sections of the London clay deposit.

Professor S. Marchetti, University of L'Aquila

I remember a slide shown by Professor Robertson about ten years ago which showed a high concentration of stress on the friction sleeve of a cone penetrometer and the formation of a strong annulus of soil which was arching. I would like to know if this problem still exists. I believe that a flat plate should largely eliminate arching and improve the measurement of horizontal stress. I would ask Professor Robertson what is the recent thinking of arching around circular probes.

Professor Robertson

Some years ago I commented on the rapid stress relief as a cone penetrates the ground. Immediately behind a circular penetrometer tip there is large stress relief and the horizontal stress drops enormously. I think Professor Marchetti is raising the question "Is that still true and does it impact on the CPT and later on cone pressuremeter since the pressuremeter element is now in that zone. The implicit question is would a flat plate have less stress relief.

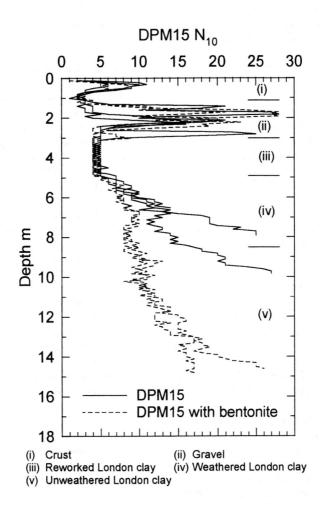

Figure 1: DPM15 profiles from Canons Park

Professor G. Houlsby, Oxford University
I am sure that there is a significant relaxation of stress on the shaft of a cone penetrometer immediately behind the cone shoulder. The problem is that the magnitude is highly non-repeatable. This arises because the reduction in stress is the product of a very small strain and a very high stiffness. I think that this means that any attempt to use measurements above the shoulder of the cone to deduce horizontal stress is doomed to failure.

This does not cause a problem for interpretation of the cone pressuremeter since for that device we concentrate on measurements at relatively large strain by which stage the above effects should have been largely eliminated. It is unlikely, however, that one could obtain any very useful data from measurements at small strains with the cone pressuremeter.

Professor Robertson
I agree absolutely with Professor Houlsby's comments regarding measuring lateral stresses close to the friction sleeves. The least accurate measurement from a CPT is the sleeve friction for many of these same reasons of its location. There has been discussion for many years on whether the friction sleeve should be moved back some distance to get it into a more stable area but standardisation has encouraged us to keep it where it is to maintain our history of data.

GEOENVIRONMENTAL APPLICATIONS
Ms D. O'Neill, ISMES SpA, (Paper entitled "The Multi functional Envirocone Test System")
Professor Robertson has raised several important points regarding METS, two of which I would like to take this opportunity to address. The first regards the time required to obtain ground water specimens especially from low permeability soils and the second point deals with the capability of the system with respect to potential cross contamination.

Sampling Time:
METS facilities environmental characterisation in its capacity as a sophisticated sampling tool permitting the study of groundwater in a much more controlled manner than conventional sampling techniques. For example, bore construction, choice of screened zone, pump types, location of pump with respect to the filtered zone and purging techniques will significantly influence the quality of the collected sample.

Penetrometers, on the other hand, such as that fitted with fibre optic laser induced fluorescence (LIF) sensors as presented by Fugro provide an immediate profile and, thus, probably best represent the environmental analog to geotechnical parameters such as tip resistance. Although certainly serving as useful investigative tools, these instruments, however, do not provide samples and tend to be limited with respect to component specificity. The measurements can be susceptible to chemical interference from wide site to site variations in chemical species and concentrations and frequently must be validated for different sample matrices.

Thus, although it is well recognised that significant time is required to physically collect ground water specimens, substantial advantages are associated with site specific flexibility and analytic accuracy regarding contaminant compounds. To help assure high quality, representative ground water specimens, METS tags the sample flow on line in real time as it passes through a cell currently equipped with five indicator parameter sensors. Recent software improvements permit the inspection of this data in graphic form in real time as seen in Figure 2.

Efforts are being directed at maximising sample collection velocity so as to provide component specific data from a greater number of sampling points in a given working day. The constraint for gas sampling velocity tends mostly to be a function of the components of interest (for example, aromatic hydrocarbon contaminated specimens should be collected at rates not exceeding about 100 ml/min although for most soil conditions it is not difficult to pump at rates of more than 3000 ml/min). METS ground water sampling rates, on the other hand, are much lower because maximum pumping rates (about 30 ml/min) are typically further reduced by the effective permeability of the soil. Total ground water sampling time is the sum of both sample arrival time (a function of pumping rate, line dead volume and 'mixed' volume prior to equilibration of pH and conductivity, etc. as illustrated in Figure 6 in the paper) and the time required to pump the volume of representative sample desired. In the present, redesigned METS configuration with a modified pump and sufficient line to penetrate to up to 50m, the 'mixed' volume is expected to be about 60 ml and the line dead volume is about 210 ml. With pumping rates of about 25 ml/min and 5 ml/min expected for sand and clay soils, respectively, a one litre sample - sufficient to effect a full series of hydrocarbon analyses or metals - will require about 50 min to be collected in sand and about 4 hours in clays. Seventy five percent of this time is attributable to the large volume of sample desired. Worst case (i.e. for clays) 'front' time (to satisfy dead and mixed volumes) is likely to be about an hour.

Conventional sampling options appear to be somewhat less attractive. Drilling a well is potentially problematic both from sample integrity points of view (effect of well construction, materials and development/purging) not to mention cost and time. Purging of wells alone may well require hours due to low recharge rates. Other penetrometer-based samplers are likely to encounter similar sampling times.

Research is underway on the use of soil phase micro extraction (SPME) as a method to measure volatile and semivolatile hydrocarbons and phenolics in ground water specimens on site as this method requires a specimen of less than 100 ml. Thus, the relatively low 'front' times are capitalised upon and total sampling time is then reduced to less than about 80 min in clays and to about 15 min in sands. The analysis is effected with a polidimethylsiloxane-coated microfibre housed in a syringe. The fibre is immersed for 15 min to 40 min in a 30ml ground water sample tapped from the sample line and maintained under continuous, gentle agitation with a PTFE-coated stirrer activated by a magnetic motor. The fibre is then introduced in a high temperature chamber of a field portable gas chromatograph, volatising absorbed components. The analysis provides a fingerprint for the sample from which specific components can be identified on the order of parts per billion as illustrated, for example, in Figure 3. While equilibration and chromatograph analysis proceeds, additional samples can be collected. In comparison with headspace analysis, not only are equilibration times reduced, but analytic accuracy is increased by mitigating problems associated with the temperature at which equilibration is achieved.

Further exploiting the relatively low 'front' times, it is proposed to equip the indicator parameter flow cell with ion selective and fibre optic sensors for such species as total hydrocarbons, trichloroethene and ammonia in order to enhance and tailor the tagging process to site specific contaminants. Once specific zones have been identified as warranting additional laboratory chemical analyses, the typically large samples can be obtained already

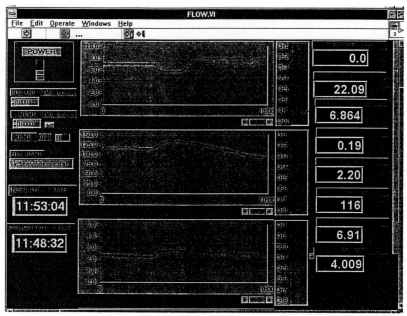

Fig. 2 - Example of real time computer screen display of flow cell indicator parameters.

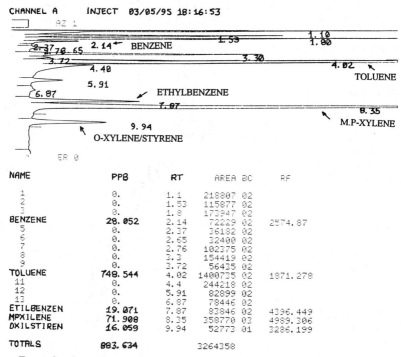

Fig. 3 - Example of gas chromatograph results for a contaminated ground water sample using solid phase microextraction.

with a precise idea of the solvents and other volatile components contained - particularly if analysed via SPME.

Cross Contamination:
METS, designed with decontamination in mind, is equipped with some valid features for maintaining sample and/or measurement integrity in light of possible cross contamination. All sensors are located at ground surface and can be both decontaminated and recalibrated (important for fibre optic and ion selective sensors) at any time. The flushing capability permits the flow of distilled water through the sampling lines, the pump and the filter zone while the envirocone is in the ground. Measurements by the sensor-equipped flow cell performed on the flushing fluid after having passed through the envirocone can be utilised in the assessment of the decontamination process. Should decontamination with distilled water be insufficient, the envirocone can be removed from the ground and flushed with a nitric acid solution. Analagously for gas sampling, the lines can be flushed with nitrogen. Bag samples can be obtained as the flushing gas exits the envirocone and these samples can then be analysed for various compounds. With this approach, one can evaluate quantitatively if compounds are being contributed to the specimen by a 'dirty' system and when the contaminants have been removed. In fact, between consecutive samples, several hundred millilitres of distilled water are monitored and recorded on the flow cell system. Given such measuring capabilities, METS is quite suitable to conduct research regarding cross-contamination phenomena.

Closure:
Professor Robertson emphasises the importance of involvement of the engineer in the definition of geo-environmental -related regulations. Such involvement would be greatly facilitated by active participation in the quantitative evaluation of contaminant distribution. Precise field measurements of contaminant components is critical to accurate site assessment and remediation.

Mr P.A. Jacobs - Fugro Limited
In his moderators report Professor Robertson made reference to the increased use of the CPT system for geo-environmental investigation purposes. I would like to bring to the attention of this conference the recent developments of the hydrocarbon contamination detection cone, the Laser Induced Fluorescence or LIF cone that my company have been involved in with UNISYS. The LIF cone was referred to in a paper presented in this session by Professor Robertson and his co-workers. The LIF cone is now routinely available in the field and we have worked with it in the UK, Europe and North America since spring 1994.

What is the basis of the system? Figure 4 shows the process of Fluorescence where a molecule is bombarded by laser light (or photons). When photons are absorbed the molecule goes into an unstable excited state and emits a photon to become stable again. This emission of photons from the molecule is known as Fluorescence.

The LIF cone system can be used with standard CPT equipment on any CPT vehicle. The LIF cone system comprises an additional two boxes of equipment, one for the portable laser and one for the logging and control computer. The two optical fibre cables are used for the excitation laser and the other for the Fluorescence detection system. These are strung with the normal electrical CPT umbilical cable inside standard CPT push rods. Figure 5 shows diagramatically the LIF system.

Fluorescence Process

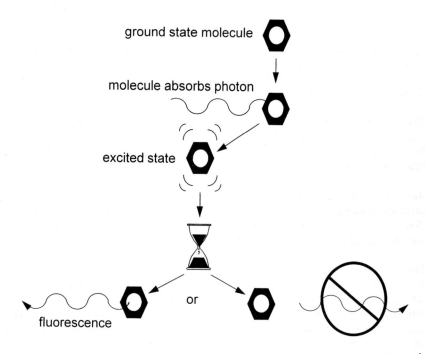

ground state molecule

molecule absorbs photon

excited state

fluorescence

or

Figure 4

The two optical fibres terminate at a sapphire window mounted in a crossover sub fitted between a standard cone and the first CPT push rod.

The Fluorescence of the ground is continuously monitored and logged at one second intervals during the normal CPT push which is at 2cm per second. This means that the normal production rate of about 150 metres of probing a day can be maintained.. The data is presented as Fluorescence Versus Depth or FVD as shown in Figure 5. When hydrocarbon contamination is detected the nature of the hydrocarbon can be characterised by collecting detailed information on the emission wavelength, lifetime and fluorescence intensity called a wavelength time matrix or WTM test. The WTM takes 1 to 2 minutes to perform and can identify the nature of the hydrocarbon i.e. if it is diesel fuel, petrol fuel, jet fuel or coal tar. This is possible because of the chemical phenomenon that molecules signify their concentration and identify via fluorescence. Also the concentration information is encoded in the amplitude or fluorescence intensity and the molecule type is encoded in the wavelength spectrum.

Typical results from WTM's carried out in diesel, petrol and jet fuel are shown on Figure 6 and it can be seen that they each show a unique shape. Our experience has shown during work in the UK, North America and Germany that a WTM in diesel and other hydrocarbon projects is the same at all these sites.

The LIF cone is very sensitive and this is illustrated on Figure 7, where measurements of fluorescence made at a site are compared with results of laboratory tests on samples taken from an adjacent sampling hole. This shows good correlation with LIF intensity and measured total semi VOCs.

Although the excitation wavelength of the laser is tuneable between 266 and 290nm using a particular wavelength will not cause non detection of hydrocarbon contamination. This is illustrated in Figure 8 where the two extremes of the LIF system were used on a sample of unleaded petrol and the sample fluoresced sufficiently for the system to detect it.

Professor Robertson
Experience in North America has shown the economic balance between continuous and discrete sensing. Continuous sensing such as the laser induced fluorescence cone as a screening device to identify possible areas of contamination and then the installation of discrete samplers to take small samples for careful analysis and also possibly long term monitoring such as say the BAT sampler are becoming the norm.

The other experience from North American practice, and it is something I raised in my Moderator's report, is the influence of regulations. In the USA practice is very much governed by the Environmental Protection Agency (EPA) and they have traditionally called the shots and told engineers what they have to do. My encouragement to practice in the UK is for engineers to take more of a proactive role in establishing the regulations and making sure that you have an input into how the regulations are drafted and to make sure that you have some flexibility to control what you want to do.

In North America one of the disadvantages is that there is no flexibility - you have to do what you are told and so there has been a tendency not to innovate because the EPA discouraged

LASER INDUCED FLUORESCENCE CONE

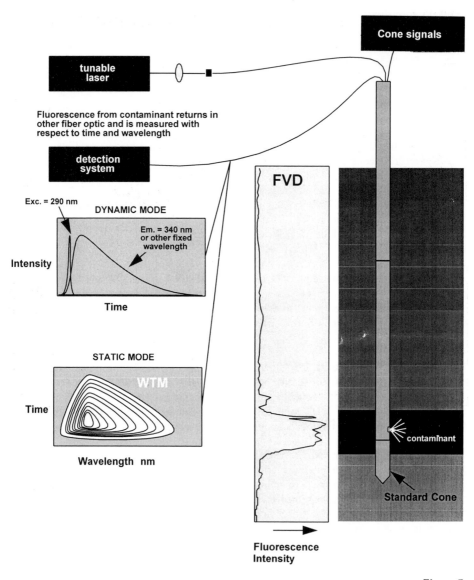

Figure 5

WTMs - TYPICAL HYDROCARBON PRODUCTS

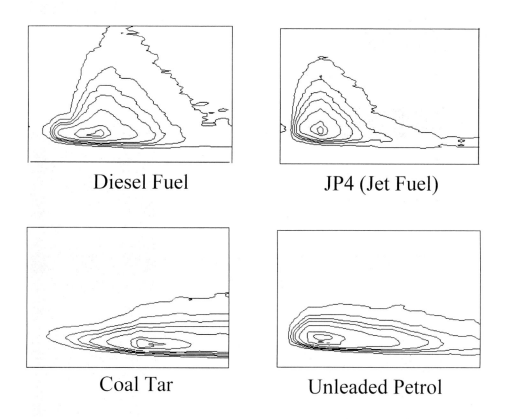

Diesel Fuel

JP4 (Jet Fuel)

Coal Tar

Unleaded Petrol

Figure 6

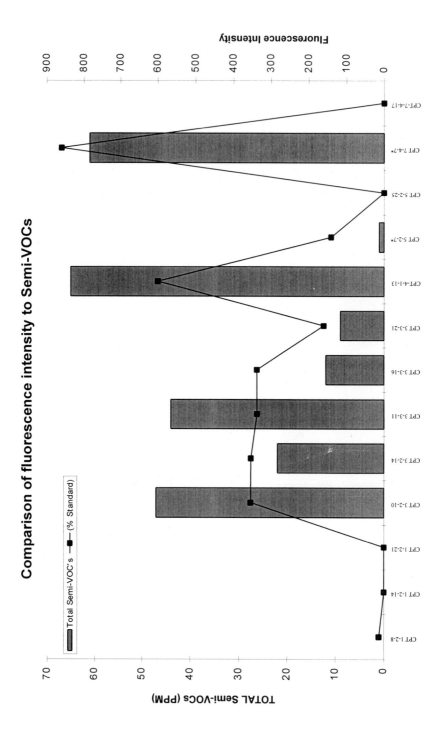

Comparison of fluorescence intensity to Semi-VOCs

Figure 7

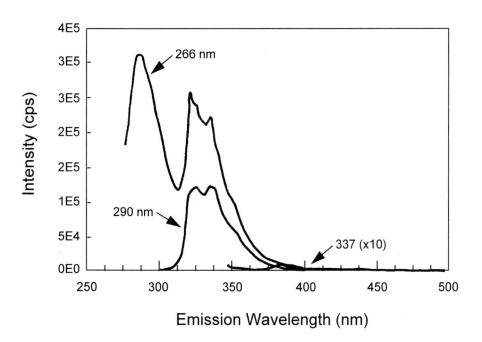

Fluorescence of unleaded petrol excited at different wavelengths

Figure 8

innovation since it was against their rules. That is changing fortunately but this experience has held up practice in North America for a number of years and my recommendation is to become more proactive. Regulation is going to come in the UK so get involved.

Mr R. Noakes - Norfolk County Council, Engineering Services Laboratory
A very simple question really, you talk about the costs of these instruments and the methods of analysis, e.g. the envirocone is a very complex instrument which has cost millions of pounds to develop and similarly the LIF cone. Is it possible to give approximate costs to have surveys done and how long will we have to wait for these instruments to become available for more general use.

Mr P.A. Jacobs
Certainly we have found that the pricing of the LIF cone system is about the same or slightly cheaper than drilling in contaminated land.

Professor Robertson
An interesting question Paul is if you break the LIF cone off and lose it what does it cost and is this chargeable to the Client.

Mr P.A. Jacobs
The actual cost of losing the LIF part of the cone string is less than that of a normal friction cone or piezocone and the cost is borne, as usual by the operator of the system.

Professor Robertson
Experience in North America has certainly shown that competition encourages improved price and competition comes very rapidly. So as long as there is work which encourages people to start companies you get competition rapidly. So it is a Catch 22 situation in terms of if there is enough work the price will come down.

Ms O'Neill
The rental cost of environmental site investigation with a system such as METS is approximately equal to that for a drill rig (about 20% greater if accompanied by a manned mobile laboratory vehicle) without taking into account the cost of disposal of contaminated materials. In the USA, for example, all drilling fluids generated must be disposed of properly which could result in considerable expense if they must be taken off-site. The question of cost, however, must also take into account the final product. Effecting an environmental investigation with conventional equipment without the help of such instruments as the LIF cone or the METS envirocone will undoubtedly result in the collection of many samples of ground water, soil gas and soil at significant expense, probably requiring more time with less points and elevations sampled. Much information regarding contamination will have been obtained but there is a very high probability that, for example, many samples will not be very representative due to purging and/or pump location errors, that some boreholes may have been located at less than optimum points and/or screened at inappropriate depths, etc. It may be necessary to return to the site to ascertain with greater precision (at the part per billion level) the actual level of contamination. Analytic field screening, recommended even for conventional borehole investigation, would add to the in situ cost, but help with laboratory costs. Upon completion of the program, evaluation of remediation schemes is likely to require geotechnical characteristics and yet another site investigation program must be effected. Taking into account all these costs, certainly collecting higher quality samples, quantitatively

tagged and even chemically analysed on site, and obtaining quantitative penetration characteristics could ultimately save millions of Lire.

Professor Robertson
Deidre raised a very good point about the costs of drilling. In many of the states in North America you now have to pay for the disposal of any cuttings from drilling from potentially contaminated sites. In fact, disposal is not the right word, you have to store it permanently and the owner of the site is permanently responsible for these cuttings. The cost of this storage is approximately £700 per day for cuttings disposal so Deidre's point about costs is quite relevant. In those states the idea of pushing a small penetrometer into the ground that produces no cuttings and then grouting that hole up while you retract the probe is very attractive and so they will spend over £3,000 a day to do these sorts of investigations.

TESTING IN NON-TEXTBOOK MATERIALS
Mr D.A. Baker, Balfour Beatty Projects and Engineering Limited
My contribution relates to the Jebel Ali G gas fired power station and desalination plant in Dubai for which both the design and construction of the civil works were carried out by Balfour Beatty Group companies. The subject is penetration testing in calcareous soils.

At tender stage the client supplied a shell and auger borehole investigation which showed calcareous sands which were locally shell-rich. SPT N values showed a wide scatter with a general increase with depth and showed the deposit to be medium dense or dense calcareous sand. After winning the contract, supplementary shell and auger boreholes were carried out with careful attention to avoiding disturbance during boring by all the usual means. SPT N values in these holes were generally very low at depth below the water table. Static cone penetration tests: (a) confirmed the results were too low, (b) showed very dense (cemented?) strata at shallower depths although cementation had not been identified in the borehole logging, and (c) highlighted thin zones of very low cone resistance, q_c, with high friction ratios of 3 to 4%, normally indicative of clay. To resolve the discrepancies in the borehole data, small diameter boreholes advanced by rotary means with drilling mud were put down. These produced sensible SPT 'N' values which agreed with the CPT data but no clay was encountered. The results, however were not compatible with the original shell and auger boreholes. We believe that investigation may have been executed without supervision. A possible explanation for the discrepancy is that perhaps the drillers did not faithfully report what they may have thought were unacceptably low SPT N values attributable to disturbance.

Lessons from the above are (a) the importance of supervision and (b) that small diameter rotary drillholes with mud support produce more reliable SPT results than shell and auger boreholes in calcareous sands beneath the water table.

Design was based on the CPT results. The site was divided into five zones of similar CPT profile and cut or fill depth. Settlement was predicted using Burland and Burbidge's method and CPT profiles were input faithfully following the actual profiles closely. q_c/N values were obtained by correlation with N values from the mud-supported boreholes. A comprehensive spreadsheet was developed for applying Burland and Burbidge's method including adjustments for cut or fill subsequent to the investigation. The spreadsheet could also determine the allowable bearing pressure for a given permissible settlement.

Ground beneath the power block was improved by vibroflotation. Control of the process was based on the achievement of a specified minimum q_c versus depth profile. This proved impossible to achieve within the thin bands of high friction ratio mentioned earlier. There was limited improvement in q_c in these bands although the friction ratio was generally reduced. Fortunately there was a giant trial pit through the full depth of the sand deposits in the form of an open excavation for the cooling water pumping station. Inspection showed that the zones where the CPT recorded apparent clay in the form of low q_c and high friction ratios were layers of shells. The results are understandable as crushing of the shells by the cone will have proportionally more effect on reducing the q_c value than on reducing the sleeve friction.

Settlement monitoring has confirmed that use of CPT results with Burland and Burbidge's method works well for predicting settlement in calcareous sands which include both shell layers and lightly cemented sand.

Professor Robertson

This raised some interesting points related to non-textbook soils i.e. the ability to have been able to identify that more crushable compressible sand layer rather than identifying it as a clay. What is also interesting is that even if you didn't identify it, the cone resistance was low because it was compressible and when you use that empirical technique it still predicted roughly the right settlement which was an interesting observation.

A new field test for the 'degree of compaction'

DW Cox, PhD., MSc., BSc., FGS., MICE.
University of Westminster, London, UK

SUMMARY

A new method of field density testing is proposed. Just sufficient soil is excavated from a hole to simultaneously fill a Proctor or CBR mould, using standard compaction. The soil is then recompacted back into the original hole, occupying the same volume as in the mould. The deficit (or surplus) volume remaining after backfilling is measured to indicate immediately how dense or loose the original soil was in comparison with the standard compaction density. See Fig 1.

INTRODUCTION

The field (in-situ) density of soils is routinely determined for both fills and natural soils, as a guide to settlement and shear strength characteristics. Typical density test applications are dam and road embankments and 'engineered' fills below structures. Density tests are also used to identify loose natural deposits such as loose wind-blown sands and silts (loess); dry slightly cemented sands; and clayey sands and gravels usually loosely deposited in dry climates by flash flooding.

All loose soils and particularly loose fills are likely to settle when vibrated (called compaction) or when wetted / inundated (called collapse), causing differential settlement at the surface which is usually sufficient to disrupt pavement or foundations. Where a loose soil attains a high degree of saturation before collapsing a more serious condition (liquefaction) occurs, where the pore water squeezed out by the collapsing voids cannot escape sufficiently quickly, and a rise in pore pressure occurs. The rising pore pressure reduces effective stress and can cause a temporary but severe loss in shear strength resulting in embankment or foundation failure. The collapse may be initiated at one small location and then spread through the whole deposit, as in landslips or embankment failures, or be generated everywhere simultaneously, as by an earthquake.

The changes in moisture content and density of soils undergoing compaction, collapse and liquefaction are shown on Fig.2. As can be seen compaction is the removal of air at constant moisture content and liquefaction is the removal of water at constant (usually near zero) air content following wetting. Collapse is a simultaneous combination of wetting and removal of air.

The settlement of loose fills or loose natural soils due to wetting or vibration and the failure of loose embankments or foundations due to liquefaction is common and widely documented. As simple examples the most common source of insurance claims on the National House Building Councils (NHBC) national guarantee scheme for new houses was settlement of fill under raised ground floor slabs, Ref 1. The Aberfan disaster in 1966 was due to a loosely tipped mudstone

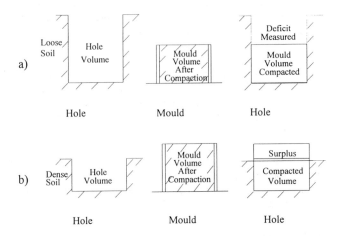

Fig. 1 Principle of Degree of Compaction Tests

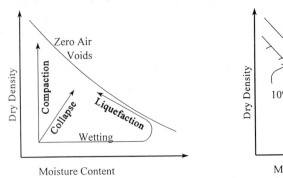

Fig. 2 Changes in Density after Placing

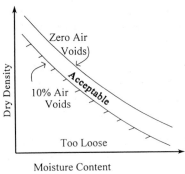

Fig. 3 Air Voids Specification for Compaction

which became saturated, slipped and temporarily liquefied, Ref 2. In both cases fill compaction to a high density is now specified to avoid such problems. Since failures of this type have been known and understood for many years and yet continue to occur, any improvement in density testing techniques should be considered worthwhile.

FIELD DENSITY TESTS

Field density is usually established by 'replacement' tests where a sample of soil is excavated, weighed and replaced by a measured volume of sand or water (eg the water balloon and sand replacement tests). The ratio of excavated weight to replaced volume gives the bulk density, and in combination with the moisture content (m/c) and mineral specific gravity (SG), gives the dry density (DD) and percentage air voids (%AV).

Another commonly used technique is to measure the amount of radiation transmitted or reflected from a radioactive point source placed on or in the soil (the nuclear density test). In fine grained cohesive soils an undisturbed core can be extracted, weighed and the volume measured directly. The tests are described in detail in Ref.3 (BS 1377)

All the replacement tests have practical performance difficulties and generally there is both insufficient volume of sample and insufficient numbers of tests for a representative assessment, with considerable variation between results, often up to 10 to 20% see Ref 4. The replacement and core tests require drying and weighing the sample, which usually means returning the sample to a laboratory, before a decision can be given on quality control acceptance for earthworks. The nuclear density test reads directly but has other operational restrictions such as calibration and a licensed operator etc.

This paper proposes an alternative test which determines directly, without weighing, the field density of the soil in relation to the density achieved after standard compaction in a mould (a variation of Relative Compaction).

Two versions of the test are suggested, a simple qualitative version requiring only a compaction mould and standard hammer, and a second more precise quantitative version additionally requiring the 'balloon density apparatus' and giving an exact value. For both cases the 'Degree of Compaction' can be determined immediately, which is a significant advantage over current methods.

SPECIFICATION AND INTERPRETATION OF FIELD DENSITY TESTS

The field value of density has to be compared with some reference value to determine whether it is sufficient to prevent settlement or liquefaction. The specification of the reference value depends on the site and soil characteristics.

THE AIR VOIDS SPECIFICATION

Using the Air Voids specification the amount of air left in the soil is calculated (from the measured field density, the m/c and the SG.) A typical specification for road embankments is a maximum of 10% air voids (reduced to 5% near the pavement), see Fig 3. This specification was used fairly successfully as a national standard for most UK road embankments until about 1970. It was found the cost of density testing was approaching or exceeding the cost of compaction and a method specification has been substituted, Ref 5 & 6. The wide variation in density results was a frequent problem in interpretation of the air voids specification. When compacting some dry soils (which are not too common in the UK) it was difficult to achieve the

10% air voids (or 5%) limit.

THE RELATIVE COMPACTION SPECIFICATION
After measuring the field density, the maximum dry density at optimum moisture content is determined by a series of ' Proctor' (or heavier standard) laboratory compaction tests at different moisture contents, on the same soil. The field density is then expressed as a percentage of the maximum dry density from teh series of tests.

Relative Compaction = Field dry density / Max.DD @ optimum m/c

Typical specifications are a minimum of 90% relative compaction for general filling and 100% or more for 'structural' fill. As can be seen from the diagram (Fig 4) there is an implied limit on moisture content as well as dry density, which prevents the use of wetter soils with low air voids, which might otherwise be acceptable. For this reason the relative compaction specification is more commonly used in drier climates eg the mid and southern states of America, as compared to those further north with wetter climates, where the specification would exclude much of the available fill unnecessarily. As with air voids specifications the wide local variation in density results makes interpretation of the specification a frequent problem.

Another problem with the specification is that either a maximum dry density is assumed for the whole deposit of the material, or a series of five or more laboratory compaction tests must be carried out for each in-situ field density determination. Even then the laboratory compaction tests cannot be carried out on exactly the same sample of soil as the density test and either method leads to wide discrepancies. This is discussed in Ref.7 where numerous field density and compaction tests by ten companies were compared at the same site. The Relative Compaction typically varied by over 20% (80-100%) against a 90% requirement.

OTHER RELATIVE COMPACTION SPECIFICATIONS - 'DEGREE OF COMPACTION'
To avoid the need to assume a maximum density or to determine it by a series of laboratory compaction tests, an alternative specification to Relative Compaction is to determine the field in-situ density and then carry out a single laboratory compaction test on the excavated sample, at the same moisture content, and compare the density results.

To avoid confusion of the two possible definitions for 'Relative Compaction' it is proposed to refer to this second method as the 'Degree of Compaction' for the purposes of this paper.-

'Degree of Compaction'   =   Field Dry Density DD
                               DD after standard compaction at field m/c

Since the same soil at the same moisture content is used for both field density and laboratory compaction tests, the variation in results is only that due to compaction. Site delay is also reduced since bulk densities can be compared without necessarily waiting for moisture determinations. However as shown by Fig 5, the field density is being compared with the laboratory density at the field moisture content, and the laboratory density necessarily will be slightly less than the maximum at optimum moisture content. However the test is still an indication of whether the field soil is dense or loose with respect to the density achieved by standard compaction in the laboratory, and this is often a sufficient specification when considered together with other classification test results.

490

The proposed new test measures the 'Degree of Compaction'. The concept of 'Degree of Compaction' is quite widely used in practice but has no formal name, to the writers knowledge. The term Relative Compaction is fairly widely used in practice but seldom discussed in the literature. By comparison the term Relative Density (see below) is widely referred to in textbooks but infrequently used in specification.

The Relative Density Test

$$\text{Relative Density} = \frac{\text{Field Density - Minimum Density}}{\text{Maximum Density - Minimum Density}}$$

The minimum density test is only suitable for free flowing sands and hence 'relative density' is not applicable to clayey soils. Its main application is as an approximate density calibration for the Standard Penetration Test (SPT), enabling penetration tests on different sands to be compared.

INDIRECT FIELD TESTS FOR DENSITY

Tests such as the SPT can be used to identify loose low density granular soils in boreholes, and give a rough indication of relative density. However particle size, water level, cementation etc. also affect the SPT results and it is only an approximate indication.

Liquefaction is often considered to be a risk for granular soils with an SPT value below about 15. However gravels can have much higher SPT values and still collapse, Ref 8. In dry climates the writer has performed SPT tests before and after adding water, to try to detect loose slightly cemented granular deposits which exhibit collapse settlement on wetting. These can give high SPT values when dry, but very low ones once the dry clay cementation is destroyed by wetting. The pressuremeter can also be used in a similar manner, before and after wetting, Ref 9.

The writer has carried out plate bearing tests at ground level, at depth in test pits; and in boreholes; both before and after wetting, and with wetting while loaded. Loose fill soils or dry soils undergo significant additional collapse settlement when first wetted. Typical test results are shown diagrammatically in Fig 6. Loose undisturbed soils which have previously been regularly inundated show less settlement on wetting, and lay somewhere between the dense behaviour and the loose fill or 'dry' soil behaviour, see Ref 10. Soaked plate bearing tests will therefore predict foundation performance on loose fills. However they are difficult and expensive to perform. Also the results are highly variable and difficult to interpret, as when scaling up the results of plate tests to predict the effects of wetting on foundations, both soil self weight and foundation load must be taken into account.

Other techniques in regular use to estimate density or compaction are to measure sound wave (shear or compression wave) velocity, or to measure the resistance to penetration (using the Proctor needle, Dutch Cone or California Bearing Ratio (CBR) tests.) Rebound devices such as the Clegg hammer and Schmidt hammer record a higher rebound for denser soils, and a similar principle is used by a meter attached to vibrating rollers to detect the improvement in density. Direct measurement of ground level change is often used to estimate the improvement in density achieved by falling weight dynamic compaction and can also be used with conventional roller compaction. The extent of these indirect techniques tends to confirm the need for an improvement in the efficiency of direct measurement.

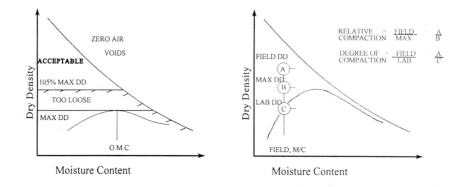

Fig. 4 Relative compaction specification

Fig. 5 Comparison of degree of compaction and relative compaction

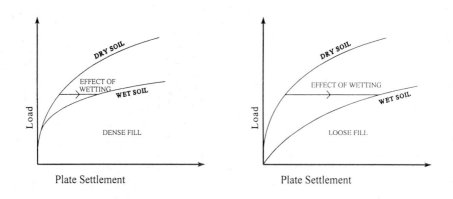

Fig. 6 Typical results of plate tests on dense and loose soils when wetted

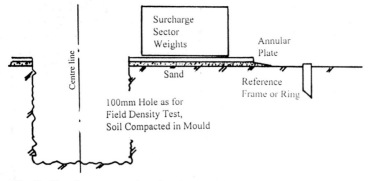

Fig. 7 Sketch across section showing arrangement.

COMPACTION INTO TEST PITS

The origin of the proposed new test technique was that in assessing the suitability of fill for road or dam embankments or foundation construction it has been the writers practice to carry out miniature compaction trials in test pits, Ref 11. The procedure is to excavate a test pit and then recompact the soil back into the pit in layers with a narrow trench roller. The layered compacted soil is tested at various stages for trafficability, shear strength, CBR etc. On completion of backfilling there is either a surplus of spoil (in dense soils) or there is insufficient spoil to refill the pit (in loose soils), because of the change in density. The deficit required to completely fill the pit (or the surplus to dispose of) is measured by volume or weight. Since the soil within the pit has been compacted to the same density as that to be achieved in an embankment, the remaining volume deficit or surplus is a good indication of the percentage bulking or shrinkage to be expected in the subsequent earthworks.

PROCEDURE FOR PROPOSED NEW FIELD 'DEGREE OF COMPACTION' TEST

The proposed new 'Degree of Compaction' test is a miniaturised test pit procedure. A hole is excavated as for the field density test, simultaneously compacting the excavated soil into a standard compaction test mould. Just sufficient soil is excavated to fill the mould after standard compaction. The same soil is then compacted back into the field density test hole. The deficit or surplus represents the difference between the in situ field density and the standard compaction density. In a second more precise version of the test the volume of the hole is recorded by the 'balloon density apparatus' before backfilling.

PROPOSED STANDARD TEST PROCEDURE

The equipment list is given below. Fig 7 shows the arrangement.

1) Install the reference frame (optional) and level the soil surface over an area about 0.3 x 0.3m. If necessary make up the surface with a thin (2 - 3mm) layer of sand.
2) Place the thin annular steel plate on the soil and steady with the surcharge weights.
3) Excavate just sufficient soil, simultaneously filling and compacting the soil into the mould, using standard compaction
4) Measure the volume of the hole using a balloon apparatus (for precise test).
5) Recompact the soil from the mould into the hole using standard compaction. The last layer may require the collar to hold the excess loose soil in place prior to compaction. The surcharge prevents any heave of soil outside the hole perimeter.
6) Remove the annular plate and surcharge.
7) Re-level the soil surface relative to the reference frame.
8) Make up the deficit with free flowing sand poured from a measuring cylinder, and record the required volume. The cylinder should be tilted end on end in the manner of the minimum density test, and the sand poured from a shallow height of a few cm to achieve a similar density as in the cylinder. (The balloon apparatus can also be used).
9) If there is a surplus, usually a small one, it can be measured by weight or compacted volume.
10) Sample the soil for moisture content (optional).
11) Express the volume of deficit or surplus as a ratio (x)

$$x = \frac{\text{Volume of deficit}}{\text{Volume of mould}}$$

$$\text{Degree of Compaction} = \frac{\text{Field Density of soil}}{\text{Compacted density of soil}} = 1/(1+x)$$

<u>Example</u> Volume of mould   =   1000cc
              Deficit remaining   =   200cc
              Ratio   x   =   0.2

$$\text{Degree of Compaction} = \frac{1}{1 + 0.2} = 83\%$$

Equipment

Reference Frame 330mm dia. steel circular ring (or bars) pinned or weighted at ground level to provide an edge for screeding beam

Screeding Beam + 500mm long x 50 x 15mm (any material)

Annular Plate, steel, 300mm dia, 1-3 mm thick with 110mm dia. central hole.

Surcharge weights   -   37Kg approx. (This quantity may be revised)

Compaction Mould   -   BS 'Proctor' (or other sizes or standards)

Compaction Hammer -   BS standard drop hammer (or other standards)

Measuring Cylinder   -   100cc glass or translucent.

Sand   -   uniform fine to medium, free flowing

Collar -   Thin walled plastic or steel tube, 100mm dia x 100mm long

Balloon Density Apparatus   -   (precise test)

Excavation of Hole

To excavate exactly sufficient material requires some care. The first layers in the mould can be used as a guide to the amount subsequently required. The final layer can be compacted in stages so that half the blows are used to compact the first half of the layer with progressive surface levelling and reductions until the last blows produce a full mould. A useful aid is to calibrate the inside wall of the mould and collar, or manufacture a blank cylinder calibrated on the side, which can be inserted into the mould to indicate the volume remaining.

Sample size

The simple version of the test can be performed with any size of hole and mould, provided the compaction is increased pro-rata. The precise version using the balloon replacement apparatus is limited by the current balloon size, which is in turn limited by the need to hold down the apparatus while pressurising the balloon. If larger precise tests are required a larger sand replacement or water replacement apparatus must be substituted.

TRIALS

A series of trials were conducted by undergraduate and graduate experimenters, mainly in the laboratory. A bed of well graded medium sand was compacted into a large (28 litre) steel container in 3 *No* x 100mm thick layers to a chosen density, which was measured by weighing the whole container and contents, as the layers were added and compacted to the required volume (by vibrating plate compaction).

The proposed 'Degree of Compaction' test procedure, as described previously, of excavating just sufficient material to compact into a standard 'Proctor' (1 litre) mould, was then performed. The volume of the hole was measured by the balloon density apparatus. (The bulk density by balloon test is compared with the bulk density by weighing in Fig 8). The soil from the mould was then re-compacted back into the hole and the deficit measured.

Measuring both the hole size and recompacting the soil into the hole gave both a precise 'Degree of Compaction' and, using the recompaction method, also an approximate indication. These

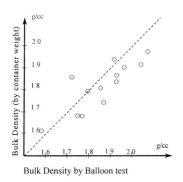

Bulk Density by Balloon test

Fig. 8   Comparison of bulk density by
balloon test and by overall
container weight

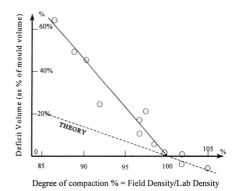

Degree of compaction % = Field Density/Lab Density

Fig. 9   Test results using recompaction
(simple method)

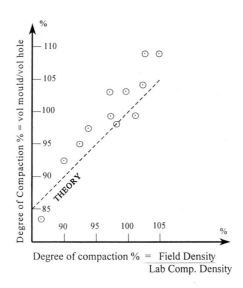

Degree of compaction % = Field Density / Lab Comp. Density

Fig. 10   Test results using balloon density
apparatus (precise method)

495

could then both be compared with the known value.

CALCULATION
The relationship between the measured deficit (or surplus) and the 'Degree of Compaction' is as follows

$$\text{Degree of Compaction} \quad = \quad \frac{\text{Field Dry Density}}{\text{Compacted Dry Density}} \quad = \quad \frac{Wf / Vf}{Ws / Vs}$$

where Wf = weight of soil extracted (dry)
      Vf = volume of hole left
      Ws = weight of soil in mould (dry)
      Vs = volume of standard mould
      D = volume of Deficit

Since the same soil as excavated is compacted into the mould
$$Wf = Ws \quad \text{and} \quad Vf = Vs + D$$

$$\text{Degree of Compaction} \quad = \quad \frac{Wf/Vf}{Ws/Vs} = \quad \frac{Vs}{Vf} \quad = \quad \frac{Vs}{Vs + D} \quad = \quad \frac{1}{1 + D/Vs}$$

Since Vs is a known constant and D is measured, the Degree of Compaction can be calculated for values of D/Vs, the ratio of the deficit to the mould volume. This is shown on Fig 9.

PROCEDURE FOR DEGREE OF COMPACTION WITH BALLOON APPARATUS
When the Balloon Replacement apparatus is used the size of hole is measured directly.

$$\text{Degree of Compaction} \quad = \quad \frac{\text{Field Dry Density}}{\text{Compacted Dry Density}} \quad = \quad \frac{Wf / Vf}{Ws / Vs} \quad = \quad \frac{Vs}{Vf}$$

Since the same soil is excavated and compacted into the mould Wf = Ws , Vf is measured by balloon test, and Vs is a known constant mould size, hence the Degree of Compaction (Vs/Vf) can be determined immediately and is shown on Fig.10

Procedure for Degree of Compaction and Field Density
Where the balloon replacement apparatus is used, a sample of the soil excavated can be retained and subsequently weighed wet and dry. The weight will then give the field density and standard compaction density, in addition to the 'Degree of Compaction' already determined immediately at the site. If the maximum dry density is determined by a series of compaction tests then the more conventional Relative Compaction can also be determined.

DISCUSSION OF RESULTS
The 'precise' method using the balloon density apparatus gave accurate values of the 'Degree of Compaction' (the ratio of the field density to the standard compaction density). The method gave immediate field results without the need for weighing or laboratory facilities and could be used for quality control of compaction, or for more quantitative records of field density if soil samples were retained for subsequent weighing.

The simpler recompaction method, requiring only the mould and hammer, gave reasonably accurate values of the 'Degree of Compaction' at 100% degree of compaction or higher densities. At field densities below the standard compaction value the disturbance of recompaction enlarged the hole, and the deficit, significantly exaggerating the lack of compaction. This test can be used to identify poorly compacted and loose soils, but would only be sufficient for a quantitative determination of the Degree of Compaction of dense soils. Both test methods measure the difference in density directly rather than comparing two separate values, which should reduce the variability of the results.

OTHER TRIALS
A number of other laboratory trials have been carried out on sand at different moisture contents in the laboratory. The results were similar to those discussed previously. A field trial on loosely compacted clay (London Clay) placed several years previously for a playing field gave reasonable results but with a much wider variation. A practical problem was that when using the sand replacement test to compare densities, the sand ran into large voids and cavities remaining open between the clay lumps, giving abnormally low field density results.

Initial trials have been with a 100mm diameter 1 litre hole as -for the conventional replacement tests. In trials to determine the effect of hole shape the hole depth has been approximately doubled, and also halved. The disturbance effect increases with hole depth and is mainly due to pushing out the sides rather than compacting the base of the hole. Research continues to determine an efficient depth to breadth ratio. Trials continue on different materials, on the necessary surcharge, on techniques for test pits, and possible recompaction into boreholes. A portable powered standard compaction hammer is being developed to enable convenient compaction of larger volumes.

Site comparisons of the test procedure by others would be welcome.

LIMITATIONS ON SPECIFICATION
Where a soil is well dry of optimum moisture content, the 'Degree of Compaction' will be with respect to a lower standard compacted density than the maximum, and this may have to be reflected in a higher specified 'degree of compaction' to give the same relative compaction with respect to the maximum dry density. Initial determination of the optimum moisture content and maximum dry density should enable this to be specified.

Where a low density granular soil is well wet of optimum and saturated it will register a 100% degree of compaction. Such soils cannot be further compacted but may be susceptible to liquefaction, which should be determined by a separate test procedure.

CONCLUSIONS
The 'Degree of Compaction' test using recompaction back into the same hole works well as a simple quick test to identify loose dry soil, or whether fill has been compacted sufficiently. The technique exaggerates the effect of low density, which is helpful in quality control and identification. The possible advantages over existing techniques are its simplicity, economy, flexibility and immediate evaluation.

The 'Degree of Compaction' test using the balloon density apparatus enables an immediate and precise determination of the 'Degree of Compaction', accurate for any field density.

As with any density quality control, the required value of density and Degree of Compaction must still be properly specified in relation to the proposed works.

ACKNOWLEDGEMENTS
Messrs Dar Al Riyadh provided the original inspiration for the work, with sites in Saudi Arabia requiring evaluation of the effects of water on loose partially cemented deposits. Messrs Frank Graham & Ptns and AE Genet Ltd helped develop the test pit procedures. Westminster University Students Mr Miles did the first preliminary tests, Miss Samar Z. Rahman and Mr Oriogun developed the first laboratory experimental procedures, Mr Rahimi did the first laboratory precise tests and Mr Fattal the first field tests. Messrs K. Johnson, S. Douglas, T. Dyer (workshop), J. Davis assisted with laboratory testing and facilities. Their assistance is much appreciated.

REFERENCES
1   Suspended Floor Construction, Practice Note No 6, 1974, National House Building Council, London & Amersham
2   Report of the Tribunal into the Disaster at Aberfan on October 21 1966, HMSO, London
3   BS 1377 1975 Methods of testing soil for civil engineering purposes, British Standards Institution, London
4   Privett 1990 Thick layers in Chalk earthworks, Proc Int Chalk Symp, ICE, T Telford London p429-36
5   Dept Transport, Specification for Road and Bridge Works, 1st Edition, 1951, HMSO London
6   ibid., 4th edition.
7   Noorany 1990 Variability of Degree of Compaction, J Geo Eng ASCE V116 No7 Jul 90 pp. ll32-36
8   Rollins Smith Beckwith 1994 Identification of Collapsible Gravels, J Geo Eng ASCE V120 No3 Mar 94 p528-43
9   Smith Deal et al 1991 Moisture Induced Collapse 4th Int Conf Ground Movements and Structures, Cardiff, Pentech Press p535
10   Mojabi 1984 Effect of pore water pressure on ultimate bearing pressure, MPhil thesis, University of Westminster/PCL
11   Cox Dawson Hall 1986 Small Scale Trial Compactions, Site Investigation Practice, SP2 Eng Group, Geol Soc, London p177-193

A new technique for in situ stress measurement by overcoring

A P WHITTLESTONE[1] AND C LJUNGGREN[2]
[1]   Soil Mechanics Limited, Wokingham, UK
[2]   Vattenfall Hydropower AB, Luleå, Sweden

INTRODUCTION

The *in situ* stress at the location of a proposed underground excavation in rock is a primary piece of geotechnical information required for both site assessment and design. Measurement of the *in situ* stress tensor can be carried out by measurement of hydraulic pressures to induce fracture (hydraulic fracturing) or measurement of induced strain changes during stress relief (overcoring). The overcoring technique is widely regarded as the preferable method as most instruments measure the complete three dimensional *in situ* stress tensor in a single measurement from a single borehole; calculation of the stress magnitudes from the measurement results also requires no qualifying assumptions to be made on the principal stress orientations.

The applicability of each of the well known methods of overcoring to general site investigations is, however, restricted. In general the instruments produced for overcoring were developed in the hard rock mining industry for measurement of *in situ* stress from existing underground openings. As such the instruments are designed for use in small diameter, short, dry drillholes and frequently utilise a set of strain gauge rosettes bonded to the borehole wall. The methods are therefore unsuitable for ground investigation in the familiar context; a greenfield site with vertical, fluid filled boreholes drilled from surface. As a result *in situ* stress is infrequently measured during initial site investigation for underground structures.

The Borre Probe, developed in Sweden, provides a solution to this problem. The Borre Probe was developed specifically for the performance of *in situ* stress measurements in fluid filled boreholes drilled from surface or underground. The equipment was used extensively during investigations for the Swedish nuclear waste repository at Forsmark (Carlsson and Christiansson, 1986) and since has been used on a variety of civil and mining projects worldwide.

This paper describes the Borre Probe and its operation in an attempt to highlight the potential benefit of the technique to conventional site investigation. Case histories are presented to demonstrate the results obtained with the equipment in a variety of situations demonstrating the applicability of the equipment.

DESCRIPTION OF BORRE PROBE
The Borre Probe was developed in Sweden to perform *in situ* stress measurements in near vertical, fluid filled boreholes drilled from surface. Development of the equipment was started in the late sixties by Hiltscher and Vattenfall, the Swedish State Power Board (Hiltscher *et al*, 1979). More recent development and commercial operation of the equipment has been carried out by Vattenfall Hydropower AB.

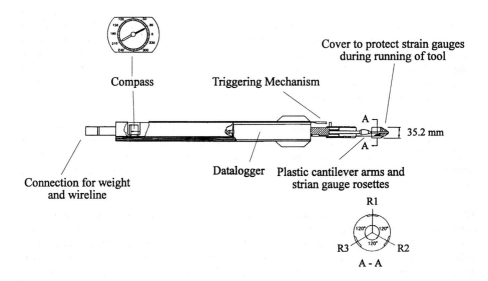

Fig. 1   Borre Probe in standard installation tool

The Borre Probe (Fig. 1) measures the *in situ* stress state from within a borehole using electrical resistance strain gauges bonded directly to the walls of a pilot hole drilled ahead of the main borehole. Nine strain gauges are mounted in three rosettes located on plastic cantilever arms at the down-hole end of the Probe. The arms are located 120° apart at a known orientation to the main body of the instrument. Besides the strain gauges the Probe also contains a thermistor and one dummy strain gauge to permit assessment of the environmental effects on the gauges and readings obtained during the measurement.

Epoxy or resin adhesives have been specially developed, capable of successfully bonding the strain gauges to the pilot hole walls in the presence of water. The adhesive composition is vital to the success of the measurement and depends upon the rock temperature and depth of the measurement location from the borehole collar. The time available for installation of the Probe in the pilot hole is limited by the adhesive pot life and therefore the adhesive composition used is carefully engineered to provide adequate installation time as well as effect a good bond to the rock.

A significant operational feature of the Probe is a data logger wholly contained within the body of the instrument. Generally overcoring measurements are not data logged and also rely on electrical connection to surface for manual reading of strains (resistances). The inclusion of a data logger within the instrument thus simplifies operational aspects of the measurement besides providing more detailed strain response data.

The data logger continuously and automatically records the strain changes and temperature variations occurring during overcoring and operates down the hole without connection to surface. The logger records the eleven channels of data at preset intervals from a preset start time. The logger is currently capable of storing 8 hours of data recorded at 60 second intervals and is powered by a battery also located within the Probe. Prior to the installation of the Probe, the data logger is connected to a portable computer and programmed with the measurement start time and recording interval. No further connection to ground surface is required after this programming. After overcoring the Probe is recovered with the overcore sample. The Probe is again connected to the portable computer, before removal from the sample and disconnection of the strain gauges, and the data recovered to a portable PC using standard communications software.

The logger is under going some modification at present to increase the storage capacity and reading rate to allow more frequent readings to be taken. An increased number of down-hole readings will allow a more detailed consideration of the strain response throughout the overcoring process permitting a better understanding of the stress measurement result (Blackwood, 1978).

The Probe has a maximum diameter of approximately 54 mm and can therefore be used in boreholes with a minimum diameter of approximately 76 mm (to provide clearance for overcore drilling). As with other overcore methods, the Probe is installed into a clean pilot hole using a special installation tool. Whilst the Probe was developed for use in near vertical down holes it can also be used in horizontal or up holes. The Probe and tool is advanced into the borehole either using a wireline system for down-holes or on rods for inclined or up-holes. A weight is attached to the installation tool when using a wireline installation method to assist tool descent.

The installation tool contains a mechanical latch that is triggered when the tool lands on the base of the main borehole (Fig.2). Triggering the latch releases the Probe from the tool and forces the cantilever arms and strain gauges against the pilot hole wall. The tool maintains the required pressure on the strain gauges during adhesive hardening to ensure a good bond between the gauges and the pilot hole wall. The installation tool also carries a magnetic compass, connected to the latch, that is mechanically fixed in its orientation when the latch is triggered and records the orientation of the Probe as installed.

The instrument has been pressure tested at 110 bar (equivalent to a 1100 m head of water) over a 10 hour period and successfully used, submerged, on numerous occasions. The maximum depth of a measurement has exceeded 700 m below ground level (BGL) under an equivalent hydrostatic pressure. To date over 600 successful measurements have been made with the Probe, of which one third have used the down-hole data logger. The components of the Probe are described in greater detail in Christiansson et al (1989) and Hallbjörn et al (1990). The process of installation, overcoring, Probe recovery and biaxial testing is described fully in Christiansson et al (1993).

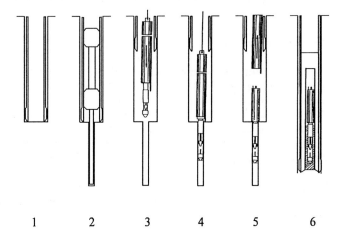

1   2   3   4   5   6

1   Advance main borehole to measurement depth

2   Drill 36 mm pilot hole and recover core for appraisal

3   Lower Probe down hole

4   Gauges bonded to pilot hole wall

5   Raise installation tool. Probe bonded in place

6   Overcore Probe and recover to surface in core barrel

Fig.2   Overcoring stress measurement procedure

CALCULATION OF *IN SITU* STRESS TENSOR

The calculation of the *in situ* stress tensor from the strain data and elastic properties follows the closed form solution described by Leeman (1968). The solution has been modified for the specific configuration of the Borre Probe which differs from the CSIR cell described in the Leeman paper.

The assumption is made that the overcore sample is homogenous and isotropically elastic. If the rock is neither isotropic or elastic in its behaviour then it follows that errors could occur in both the calculated magnitude and orientation of the *in situ* stress field (Amadei and Goodman, 1982). No assumption on the orientation of principal stresses is required to determine stress magnitudes unlike assessment of hydrofracture or borehole breakout information.

As up to nine strain readings are recorded by the Borre Probe for each measurement, statistical methods (least-squares analysis) are used to iteratively determine the stress tensor.

The methods described by Vreede (1981) form part of the computer program used for analysis and calculation.

STRESS MEASUREMENTS FOR DEEP NUCLEAR WASTE REPOSITORY, SELLAFIELD, UK

Investigations into the suitability of constructing a repository for disposal of low and intermediate level solid radioactive wastes by United Kingdom Nirex Limited (Nirex) included active consideration of sites at Sellafield, Cumbria and Dounreay, Caithness. These investigations included the drilling of boreholes to the depth of up to 2,000 m for geological, geotechnical and hydrogeological purposes (UK Nirex, 1993). Following preliminary boreholes at both sites Nirex have announced that it has chosen Sellafield as the site at which it would concentrate further investigations.

The conceptual design for a repository at a depth of about 650 m below Ordnance Datum at Sellafield involves a series of vaults serviced by access central and perimeter tunnels. Access to the disposal area would be by two inclined drifts from the Sellafield site.

The generalised geological succession is principally a Permo-Triassic sequence comprising abrasive fine to medium grained, moderately strong to strong, Calder and St Bees sandstones, shales and poorly sorted breccias overlying Carboniferous Limestone and a basement of the Lower Palaeozoic Borrowdale Volcanic Group. The Borrowdale Volcanic Group comprises welded tuffs, with interbeds of volcanic breccia and some intrusive andesite sills.

Hydrofracture stress measurements have been made at the Sellafield site in the Borrowdale Volcanic group (Heath, 1992) below 400 m below ground level (BGL). A programme of overcoring measurements using the Borre Probe was carried out in a vertical borehole drilled from surface (Borehole S10B), at depths between 50 m and 250 m in the Calder and St Bees sandstone strata to define the three-dimensional stress state in the shallow strata. The borehole was also required to provide high quality nominal 100 mm diameter core for lithological and geotechnical logging, and to form a well for a pumping test on the completion of drilling. The drilling was carried out using a Boyles BBS 56HD rotary drilling rig using a Geobor S wireline coring system producing a nominal 146 mm diameter borehole. Coring was carried out with a wireline double tube corebarrel similar to that used in the other cored boreholes drilled at Sellafield (Ball et al, 1993). These measurements were the first use of the Borre Probe in the United Kingdom.

Adaption Of Equipment

As described above the Borre Probe was developed for use in slim holes (typically 76 mm) drilled in very strong, crystalline rocks using conventional drilling techniques. The nominal 12 mm annular overcore width produced by the typical system was considered inadequate for the weaker sedimentary rocks at Sellafield. A 146 mm diameter borehole was selected to allow a 100 mm diameter overcore to be carried out with a nominal 33 mm annular width. Consequently the installation tool for the Borre Probe was modified slightly to allow operation in a larger diameter borehole. New tools were specially manufactured that allowed the Probe to remain unaltered. Adaption included the attachment of centralising fins to the installation tool body to centre the equipment in the wireline casing and borehole during lowering and guide the Probe into the pilot hole.

With the new tools the procedure for one measurement (Fig. 2), from commencement of drilling the pilot hole to recovery of the strain data from the logger after overcoring, took approximately 12 hours at the maximum measurement depth of 250 m BGL.

Measurement Details

Fifteen measurements were attempted in the borehole at levels of approximately 100, 150, 200 and 250 m BGL in groups of three to six measurements. This configuration of measurements provided two groups of stress measurements in the Calder Sandstone and two in the St Bees Sandstone.

Of the fifteen tests attempted, ten were successful allowing determination of the stress tensor. Of the other tests; four were unsuccessful due to problems in forming a pilot hole of appropriate tolerance in weak rock; on one occasion two strain gauges malfunctioned after installation. Both gauges were oriented in directions non-parallel to the borehole axis rendering the measurement indeterminable.

The elastic properties determined from biaxial testing of the recovered samples were highly variable and atypical for sandstone strata.

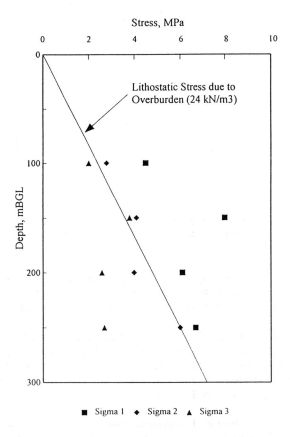

Fig. 3   Mean Principal Stress Magnitudes

Assessment Of Results

The direction and magnitude of the principal stresses was determined for each of the ten successful stress measurements. The two or three calculable measurements at each test level were combined to calculate a mean principal stress tensor at each level by transforming each measurement to a common coordinate system. The principal stress orientations were generally consistent for each level but where the strata was steeply cross bedded there was notably more scatter in the calculated results.

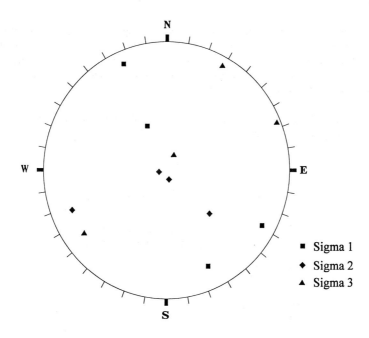

Fig 4.   Stereographic lower hemisphere projection showing
orientation of mean principal stresses

Figs. 3 and 4 show the mean principal stress magnitudes and orientations at each of the test levels in the borehole. The results indicate that generally the maximum principal stress direction is horizontal in a plane broadly parallel to the bedding direction of the strata.

The state of stress in the vertical $\sigma 2$ - $\sigma 3$ plane within the Calder sandstone is indicated to be close to isotropic and therefore the $\sigma 2$ and $\sigma 3$ directions are effectively interchangeable (Fig. 4). This is not the case in the St Bees sandstone strata where a more distinct difference in stress magnitudes can be seen. The results indicate the stress regime to change from thrust faulting conditions to strike-slip conditions at depth.

The indicated *in situ* stress state in the sandstone strata is reasonable when considering the geological setting and structures in the vicinity of Sellafield. The results indicate that the major principal, major horizontal stress direction seems to be controlled by the direction of one of the major fault sets, broadly parallel to the coastline. The results also seem to suggest that the anisotropy and porosity observed in the samples during the biaxial testing might have only a limited influence.

These results are undoubtedly of value to the repository design team providing:

- An important design parameter for the location, orientation and preliminary design of access tunnels to the proposed repository at the feasibility stage of the project.

- Confirmation of the assumptions made for assessment of hydrofracture measurements that one principal stress is orientated vertically.

STRESS MEASUREMENTS FOR ROAD TUNNELS IN STOCKHOLM, SWEDEN

In contrast to the deep and rather complicated overcoring stress measurements carried out at Sellafield, the Borre Probe can also be used for shallow overcoring measurements from surface or underground openings. A series of measurements from surface have recently been carried out in Stockholm City as part of the ground investigation for road tunnels that will form parts of a ring road around the down town area. The tunnel excavation and road construction work is scheduled to start in the beginning of 1995.

At the time of writing, overcoring stress measurements with the Borre Probe have been carried out in seven vertical, or near vertical, boreholes drilled from surface. All measurements were carried out at depths of between 10 m and 47 m BGL. The bedrock in the area consists mainly of a competent fine to medium grained grey to dark grey granite. A total of 34 successful measurements have been carried out in the seven boreholes to date.

The boreholes in which stress measurement was carried out were located at the positions of complicated tunnel geometry, including access ramp locations were narrow rock pillars between the main and access tunnels were planned. The overcoring stress measurements were to provide input data to two and three dimensional numerical modelling of the tunnel layouts at these specific locations to assess the magnitude and effect of stress concentration that will occur after tunnel excavation. The numerical modelling will ultimately permit assessment of excavation performance, deformation and allow selection and design of the final rock support for the tunnels.

Fig. 5 shows the magnitudes of the horizontal and vertical stresses determined from the measurements. The stresses shown are mean values for each level at which measurements were carried out and thus summarise all 34 measurements made. Each mean is generally based on three measurements.

The results are somewhat scattered, especially the magnitude of the vertical stress which is indicated to decrease with depth over the measurement interval. This phenomena is a result of high stress levels near ground surface. The measurements have allowed a reasonable understanding of the stress trends in the upper 50 m of bedrock across the site. The magnitude of the minimum horizontal stress is approximately 4 MPa within the measurement interval while the maximum horizontal stress has an indicated average value of 6.5 MPa and increases at approximately 0.05 MPa/m.

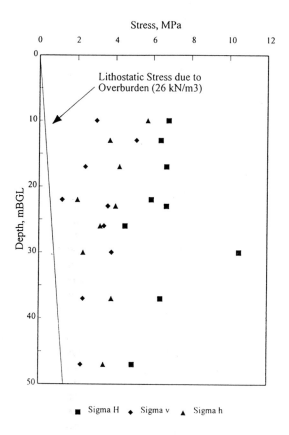

Fig 5. Mean vertical and horizontal stresses obtained from overcoring measurements in
seven boreholes, Stockholm during 1993 and 1994

The magnitudes of the horizontal stresses are consistent with other measurements made in
Sweden at shallow depth. Excessive horizontal stresses are frequently measured near ground
surface. Measurements at greater depth (below 100 m BGL) extrapolated to surface also
often indicate and excess horizontal stress of a few Megapascals if a linear relationship
between stress and depth is assumed.

The orientation of the maximum horizontal stress from all measurements is consistently north-
northwest with a mean orientation of approximately N 35° W.

MEASUREMENT ACCURACY
The accuracy and reliability of *in situ* stress measurements in the Earth's crust cannot be
assessed. It is not unusual for the results of stress measurements to be scattered and the level
of scatter indicated by the case study results is not atypical. As with many geotechnical *in*

situ measurements, engineering judgement is required to assess and interpret the results, and determine the *in situ* stress state for input into design calculations or numerical modelling.

Comparative measurements using different techniques have been made on a number of occasions to assess the results provided by the Borre Probe. Pitman et al (1992) and Stephansson and Ångman (1984) have both reported that an acceptable agreement in both magnitudes and orientations of the assessed stress field (considering horizontal stresses) is achieved between the Borre Probe and hydraulic fracturing stress measurement at a range of depths below surface.

IN SITU STRESS MEASUREMENT DURING SITE INVESTIGATION

Measurement of the *in situ* stress at the depth of proposed underground caverns or tunnels is infrequently carried out as part of conventional site investigation. Typically measurements are performed from trial excavations during construction. An advance knowledge of the *in situ* stress state at the location of a proposed underground opening would allow better preliminary design and possibly reduce construction/design problems.

The Borre Probe permits overcoring stress measurement to be performed in investigation boreholes drilled from surface. Overcoring *in situ* stress measurement could thus form part of the comprehensive site investigations carried out for any underground project in rock. A set of closely spaced stress measurements could be carried out in investigation boreholes drilled at proposed shaft positions, concentrated at the levels for proposed tunnels and caverns. Measurements could also be made in other boreholes in the vicinity and in the same tectonic unit to provide confirmation of the stress field. Additional measurements in other investigation holes in nearby tectonic units would also be of value for design and safety assessments. If underground construction commences, further stress measurements could then be carried out from the underground openings using the range of stress measurement instruments including the CSIRO HI stress cell as well as the Borre Probe.

ACKNOWLEDGEMENTS

Acknowledgment is given to the Managing Directors of the three constituent companies of the KSW Deep Exploration Group Joint Venture which comprise Kenting Drilling Services Limited, Soil Mechanics Limited and Bohrgesellschaft Rhein-Ruhr MbH (formerly Gewerkschaft Walter AS) and to the Directors of Vattenfall Hydropower AB for their support in publishing this paper. The authors are also grateful to the Sellafield project client; UK Nirex Limited for permission to publish information relating to the deep waste repository project and the Swedish National Roads Administration for their permission to publish the data from the Stockholm project.

REFERENCES
AMADEI B and GOODMAN R E : 1982 : The influence of rock anisotropy on stress measurements by overcoring techniques. Rock Mechanics, 15 (4), 167-80.

BALL T G, BESWICK A J and SCARROW J A : 1993 : Geotechnical investigations for a deep radioactive waste repository: drilling. Proceedings of the 29th Annual Conference of the Engineering Group of the Geological Society.

BLACKWOOD R L : 1978 : Diagnostic stress-relief curves in stress measurement by overcoring. Int J Rock Mech Min Sci & Geomech Abstr, Vol 15, 205-9.

CARLSSON A and CHRISTIANSSON R : 1986 : Rock stresses and geological structures in the Forsmark area. Proc Int Symp Rock Stress and Rock Stress Measurement, Stockholm, 457-65.

CHRISTIANSSON R, INGEVALD K and STRINDELL L : 1989 : New equipment for *in situ* stress measurement by means of overcoring. Proceedings of International Congress on Progress and Innovation In Tunnelling, Toronto, 49-53.

CHRISTIANSSON R, SCARROW J A, WIKMAN A and WHITTLESTONE A P : 1993 : Geotechnical investigations for a deep radioactive waste repository: *in situ* stress measurement. Proceedings of the 29th Annual Conference of the Engineering Group of the Geological Society.

HALLBJÖRN L, INGEVALD K, MARTNA J and STRINDELL L : 1990 : New automatic probe for measuring triaxial stresses in deep bore holes. Tunnelling Underground Space Technology, Vol 5, No 1/2, 141-5.

HEATH R J : 1992 : Deep discussions. A report on the British Geotechnical Society AGM meeting held on 24 June (1992). Ground Engineering, Vol 25, No 8 (October), 34-5.

HILTSCHER R, MARTNA J and STRINDELL L : 1979 : The measurement of triaxial rock stresses in deep boreholes. Proceedings of 4th International Congress On Rock Mechanics, Montreux, 227-34.

LEEMAN E R : 1968 : The determination of the complete state of stress in rock using a single borehole - laboratory and underground measurements. Int J Rock Mech & Min Sci, 5, 31-56.

PITMAN W, WIKMAN A and KLASSON H : 1992 : Rock stress measurements for the Pergau Hydroelectric Project, Kelantan, Malaysia. Swedpower Report Series, Stockholm, Sweden.

STEPHANSSON O and ÅNGMAN P : 1984 : Hydraulic fracturing stress measurements at Forsmark and Stidsvig, Sweden. Research Report TULEA 1984:30, Luleå University of Technology, Luleå, Sweden, 61 pp.

UNITED KINGDOM NIREX LIMITED : 1993 : Nirex deep waste repository project. Scientific update 1993.

VREEDE F A : 1981 : Critical study of the method of calculating virgin rock stresses from measurement results of the CSIR triaxial strain cell. CSIR Report ME 1679, Pretoria.

A pore water pressure probe for the in situ measurement of a wide range of soil suctions

A.M.RIDLEY AND J.B.BURLAND
Soil Mechanics Section, Department of Civil Engineering, Imperial College of Science, Technology and Medicine, London. England.

INTRODUCTION

Historically soil mechanics theory has concerned itself mainly with the behaviour of saturated soil. The positive pore water pressure which many of these soils exhibit whilst in the ground is largely due to the presence of the natural water table. Soil that lies above the water table is not subjected to a positive hydrostatic profile. In this instance the water, which is still subjected to gravitational forces, is in a state of tension.

Evaporation from the surface may eventually lower the pore water pressure enough for the water in the soil near the surface to partially drain out of the voids. The voids will subsequently be occupied by a combination of air and water. This process creates a zone of soil, part of which is saturated and part is unsaturated. The pore water pressures in this zone will be predominantly negative (figure 1).

A compacted soil consists of a number of highly compact 'clods' of soil (each 'clod' having a high degree of saturation). These 'clods' are compressed together to form a matrix of 'clods' and voids. Where a compacted soil lies above the water table it will be inherently unsaturated and the initial pore water pressure will be negative.

Many geotechnical problems are associated with either natural or man-made unsaturated soils (eg. the behaviour of earth dams and embankments, the movement of retaining structures, the stability of slopes and near vertical excavations, the movement of shallow foundations and the disposal of waste products). Over the last thirty years the theories for the behaviour of unsaturated soil have been developed. The key parameter in these theories is the soil moisture suction (ie. the negative pore water pressure). However, the application of these theories has been difficult because of the inability to measure the negative pore water pressure *in situ*.

SOIL SUCTION

Under normal conditions the pore air pressure in an unsaturated soil will be equal to the ambient atmospheric pressure and the pore water pressure will be negative for the reasons discussed above. It has been shown (Hilf, 1956) that if the pore air pressure is raised to a level above the ambient atmospheric air pressure, the pore water pressure will rise by an amount equal to the difference between the new air pressure and the atmospheric pressure. Therefore, the difference between the soil pore air pressure and the soil pore water pressure is independent of their absolute values. It has become known as the soil suction.

Advances in site investigation practice. Thomas Telford, London, 1996

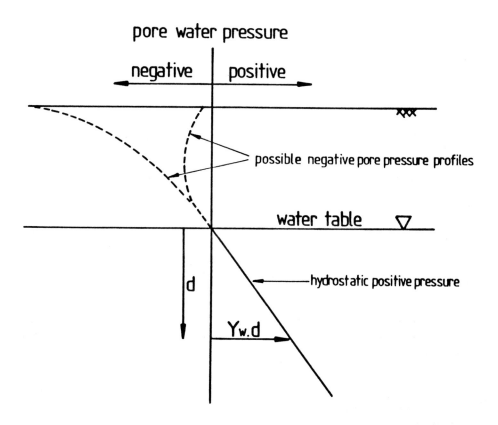

Figure 1 - Possible pore water pressure profiles

If the pore air pressure is raised to a level high enough to induce a positive pore water pressure in the soil the latter can be measured using conventional pore pressure measuring instruments. This technique is known as axis-translation. Its application *in situ* has not been pursued because of the difficulty of maintaining a stable and uniform high air pressure down a borehole and in the surrounding ground. In addition, although the use of axis-translation is common in laboratory testing, the effect of artificially changing the natural negative pore pressure into a positive pressure on the behaviour of an unsaturated soil, has not yet been satisfactorily assessed.

When the pore air pressure is at atmospheric level, the soil suction will be equal in magnitude to the negative pore water pressure in the soil. Furthermore, measuring the absolute negative pore water pressure in the ground (and in the laboratory) leaves the soil in its natural condition.

SUCTION MEASUREMENT DEVICES
Within soil science a few techniques have evolved which successfully measure soil suction (Fredlund & Rehardjo, 1988). However only the tensiometer is capable of making a direct measurement under normal atmospheric conditions (ie. measuring a negative pore water pressure).

The tensiometer (figure 2) is a device for directly measuring soil suction. It is principally used in the field but has recently found some application in the laboratory. It works by allowing water to be extracted from a reservoir in the tensiometer, into the soil, until the stress holding the water in the tensiometer is equal to the stress that is holding the water in the soil (ie: the soil suction). When equilibrium is reached no further flow of water will occur between the soil and the tensiometer. The suction will then manifest itself in the reservoir as a tensile stress in the water and can be measured using any stress measuring instrument (eg. manometer, vacuum gauge or electronic pressure transducer).

However, previous observation has noted that the range of operation of the normal tensiometer is limited to between zero and about -1 atmosphere (Stannard, 1992). At about this pressure air bubbles form in the reservoir and the tensiometer fails to record the pore water pressure accurately. The explanation of the formation of the air bubbles is linked to the phenomenon of cavitation.

CAVITATION
Water under tension is in a metastable condition and this metastability can be destroyed if nucleation occurs. Nucleation is what is known as cavitation - the formation of vapour cavities within the liquid itself or at its boundary with another material.

The theory of the tensile strength of pure liquids predicts that a *vapour* cavity will only form when the liquid is placed under extremely high tension (eg. about -50MPa), or when the liquid is superheated. Neither of these conditions exist within the tensiometer when an air cavity forms within the reservoir. Therefore the failure of the tensiometer to measure pore water pressures lower than -1 atmosphere is not due to the formation of a vapour cavity within the liquid.

The imperfections that exist in the surface of objects, even after the finest machining or polishing, provide the ideal trap for tiny amounts of air that can remain after thorough de-

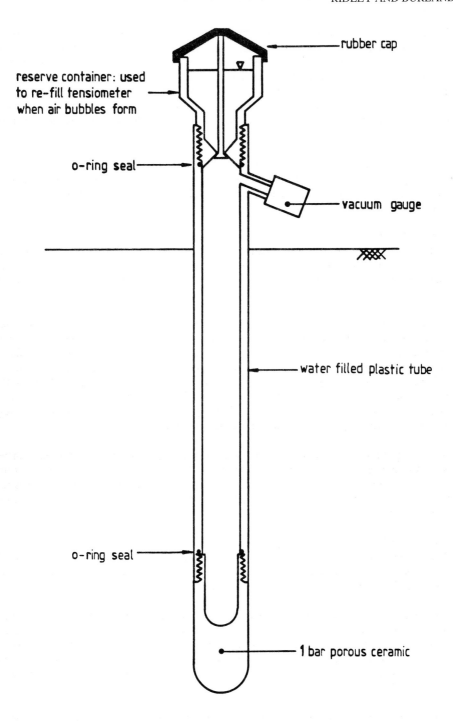

rubber cap

reserve container: used
to re-fill tensiometer
when air bubbles form

o-ring seal

vacuum gauge

water filled plastic tube

o-ring seal

1 bar porous ceramic

Figure 2 - Vacuum gauge tensiometer

airing using vacuum equipment (figure 3a). When the water in the reservoir of the tensiometer is placed in a state of tension the air trapped in a crevice can easily be sucked out of the crevice to form a bubble in the reservoir (figure 3b). Since the reservoir water has been preconditioned by the application of a vacuum (ie. -1 atmosphere) the formation of this bubble will occur shortly after the tension in the water is reduced below the evacuation pressure. In addition, once a bubble has formed in the reservoir water, the pressure recorded by the tensiometer will be -1 atmosphere.

AVOIDING CAVITATION

Harvey *et al.* (1944) found that if the water inside a vessel had been previously compressed to 100MPa it could withstand tension. By pressurising the water, the air trapped in a crevice is dissolved further into solution. When the high pressure is subsequently reduced back to atmospheric level the dissolved air finds it easier to come out of solution where the water meets the free atmosphere. Therefore, the air-water interface is pushed inside the crevice (figure 3c). Once inside the crevice the air-water interface will, after the application of a positive pressure, be convex on the air side, making it able to resist a tensile stress well in excess of 1 atmosphere. Using the technique of pre-pressurising, Richards and Trevena (1976) were able to apply a tensile stress of about 35 Bar, to water inside a closed system, without air bubbles forming.

A NEW TENSIOMETER

A tensiometer uses a saturated porous ceramic to transmit the soil suction from the soil to the water reservoir. An inherent characteristic of these ceramic materials is the air-entry pressure. The air-entry pressure is the pressure difference between two surfaces of the saturated ceramic that will cause air to pass into or through it. It is a material property that is determined by the pore size and the preparation technique.

The standard tensiometer uses a ceramic material with an air-entry pressure of about 1 Bar. Hence, when the pressure in the reservoir water is reduced to about -1 atmosphere and the pore air pressure in the soil is atmospheric (or zero), air will be drawn through the porous ceramic and an air bubble will form in the reservoir. Therefore, it would be insufficient to simply apply a high pressure to the water in the reservoir (á la Richards & Trevena) and expect the standard tensiometer to successfully measure pore water pressures less than -1 atmosphere. However porous ceramics can be manufactured with air-entry pressures of up to 15 Bar.

Ridley and Burland (1993) described a new tensiometer that is capable of measuring pore water pressures as low as about -15 Bar. The tensiometer was made small to reduce the compressibility of the system and ease the de-airing procedure. The 15 Bar ceramic inhibits the passage of air into the reservoir from outside the tensiometer. However, air can pass through the ceramic when the pressure difference across it exceeds the air entry pressure. When this happens the tension in the water breaks down and the tensiometer indicates a pressure of -1 atmosphere. A tension breakdown can also occur when the pressure difference across the ceramic is less than the air entry pressure, if the water in the reservoir has been insufficiently pre-pressurised.

The design of the tensiometer has recently been modified (figure 4). The overall size of the instrument has been increased slightly and the commercially available pressure transducer has been replaced by an integral strain gauged diaphragm. The volume of the reservoir has been

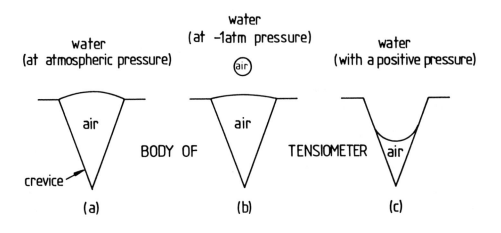

Figure 3 - A mechanism for cavitation

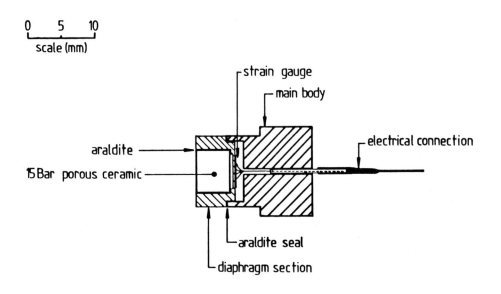

Figure 4 - The new tensiometer

reduced further and the thickness and diameter of the porous ceramic have been increased. As described in the following sections this device has been used to measure the negative pore water pressure at the bottom of a borehole sunk in an unsaturated compacted clay.

BOREHOLE FORMATION

The initial borehole is sunk using a 2.5 inch diameter helical hand auger (figure 5a). When the required depth is reached the borehole is cleaned out using an industrial vacuum cleaner and lined using durapipe™ with a 2 inch diameter internal bore (figure 5b). To make representative measurements of suction using the new tensiometer it is necessary to ensure a good all-over contact between the porous ceramic and the soil. To do this a smooth, relatively undisturbed, horizontal bottom is required for the borehole. Standard hand augering tools cause considerable disturbance to the soil at the bottom of the augered borehole, particularly in the centre of the hole. Therefore a tool has been designed (figure 6) that can be placed at the base of the initial borehole and used to carefully excavate up to 3 inches below the base.

The tool consists of a 1.375 inch diameter milling machine slot cutter that is fitted to a standard hand auger extension rod (gas pipe) using a connecting bar. The cutter is free to slide inside a metal barrel, with an external diameter of 2 inches, that is lowered inside the durapipe™ lining. During the lowering operation the barrel is prevented from falling (figure 5c) by a using a pin locking mechanism on the connecting bar. When the barrel arrives at the bottom of the borehole, tabs fitted to the inside of the durapipe™ locate in slots cut into the outside of the barrel (figure 5d). This prevents the barrel from rotating in the borehole whilst the secondary hole is being drilled. The cutter is then rotated and advanced downwards (figure 5d) until the pins in the connecting bar are flush with a stop fitted to the inside of the barrel. In a compacted clay, trimmings from the secondary hole are deposited inside the barrel (see figure 6). Thus, the cutter gives a clean secondary hole, with a smooth horizontal bottom surface, at the base of the initial borehole (figure 5e).

After completing the borehole excavation the barrel is extracted from the borehole and cleaned. Any trimmings taken from the barrel can be used to estimate the *in situ* moisture content of the clay at the depth of the suction measurement.

SUCTION MEASUREMENT

The cutter and connecting bar used to drill the secondary hole are replaced by another connecting bar fitted to a spring loaded mounting for the tensiometer. The spring holds the tensiometer proud of the mounting until the latter is resting on another flat surface (ie. the bottom of the secondary borehole). The mounting and the connecting bar are fitted inside the barrel in a similar manner to the cutter. The barrel is then lowered inside the durapipe™ and located once again by the tabs. Finally the tensiometer mounting is lowered to the bottom of the secondary hole (figure 5f) and the spring depresses until the mounting and the tensiometer are flush with the soil.

As part of a long term research programme on the measurement of soil suction, a 1.5m diameter reinforced PVC bin was part filled with London clay compacted at a moisture content of 26% (ie. 4% above the optimum water content) and the surface was covered with polythene. A long time after compacting the soil (ie. 3 years) a borehole was drilled in the soil (using the method described above) to a depth of 50cm. Soil taken from the borehole confirmed the compaction moisture content.

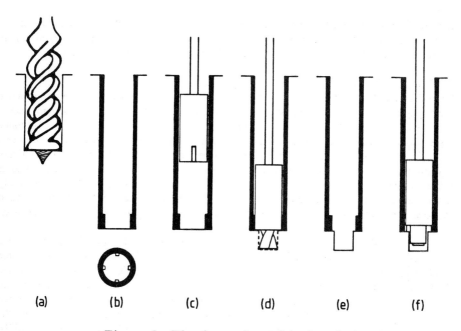

(a) (b) (c) (d) (e) (f)

Figure 5 - The formation of the borehole

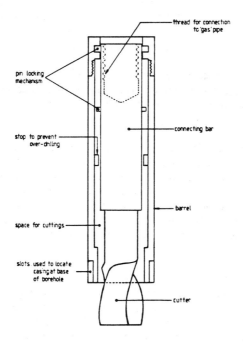

Figure 6 - Tool used to cut the secondary borehole

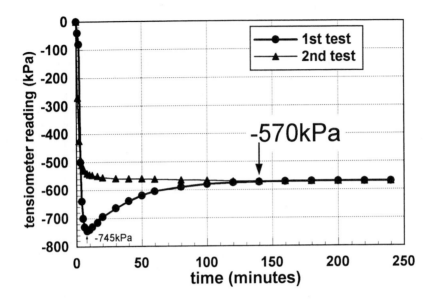

Figure 7 - *In situ* suction from new tensiometer

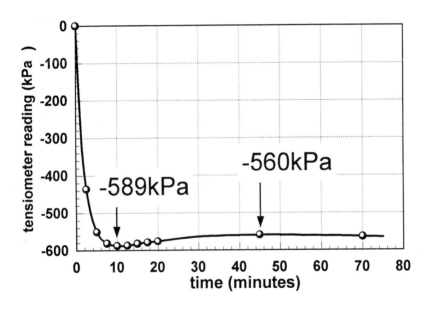

Figure 8 - Laboratory suction from new tensiometer

Suction measurements were made using the new tensiometer and the results were recorded directly onto a chart recorder. The results from two tests are shown in figure 7. In the first test, the excess water left on the surface of the porous ceramic (after saturating the tensiometer) was completely absorbed prior to inserting the tensiometer into the borehole. The tensiometer pressure decreased to -745kPa after about 8 minutes and subsequently increased to -570kPa after 140 minutes. It then remained constant for a further 100 minutes, after which the tensiometer was removed from the borehole. In the second test, the excess water on the surface of the ceramic was only partially absorbed prior to inserting the tensiometer into the borehole. The tensiometer pressure decreased to -500kPa in 4 minutes and -570kPa after 140 minutes. Again it remained constant at -570kPa and the tensiometer was removed from the borehole after 4 hours. Clearly the response of the tensiometer was affected by the amount of water left on the surface of the ceramic before making the measurement. However, the final pore water pressure, recorded by the tensiometer, was unaffected by the presence of excess water.

To confirm the validity of the *in situ* measurement a sample of the same London clay was mixed to 26% moisture content and compacted under Proctor conditions according to BS1377: Part 4:1990. The pore water pressure of this sample was measured using the same tensiometer. After ten minutes a minimum pressure of -589kPa was recorded (figure 8). This gradually increased and reached an equilibrium value of -560kPa after 45 minutes. The suction of this sample was also measured using the in-contact filter paper method (Chandler et.al. 1992). Two measurements were made; the first corresponded to a suction of 520kPa and the second corresponded to a suction of 470kPa, giving an average suction of 490kPa.

The *in situ* pore water pressure and that of the laboratory compacted sample, measured using the new tensiometer, are in excellent agreement. They are about 15% higher than the suction inferred from the filter paper tests, but that is within the range of accuracy of this test. Ridley and Burland (1994) have drawn attention to some problems of measuring suction in compacted material. Research is required into the discrepancies observed when using a variety of suction measuring techniques, but the new tensiometer does provide the first opportunity to directly measure *in situ* pore water pressures in the range 0 to -1500kPa.

CONCLUSIONS

The tensiometer device for measuring absolute pore water pressure in the range 0 to -1500kPa (Ridley and Burland, 1993) has been adapted for use *in situ*. A technique has been developed for excavating a borehole with a flat, horizontal bottom in compacted London clay.

The tensiometer has a rapid response and representative measurements of the *in situ* negative pore water pressure can be obtained in a few minutes. Equilibrium was reached, in this instance, in less than three hours.

The *in situ* pore water pressure, measured three years after compacting the clay to a condition slightly wet of the optimum moisture content, was -570kPa. This was validated by measuring the suction of a London clay sample compacted, in the laboratory, at the same moisture content.

ACKNOWLEDGEMENTS

The research presented here was jointly funded by the Engineering and Physical Sciences Research Council and the Transport Research Laboratory. The continued interest in the subject of soil suction measurement from Dr Ken Brady (TRL) is appreciated. The skill of Mr Steve Ackerley and Mr Graham Keefe (Imperial College) in making the instruments described herein was invaluable.

REFERENCES

BS 1377 (1990). Methods of test for Soils for Civil Engineering purposes. Part 4. Compaction - related tests. BSI.

Chandler, Crilly and Montgomery-Smith (1992). A low cost method of assessing clay desiccation for low-rise buildings. Proc. Institution of Civil Engineers, May. pp 82-89

Fredlund D.G. and Rehardjo H. (1988). State-of-development in the measurement of soil suction. Proc. International Conference on the Engineering Problems of Regional Soils, Bejing, China. pp 582-588.

Harvey E.N., Barnes D.K., McElroy A.H., Whiteley A.H., Pease D.C. and Cooper K.W. (1944). Bubble formation in animals, 1. Physical factors. Journal of Cellular and Comparative Physiology 24, No.1, August.

Hilf J.W. (1956). An investigation of pore water pressure in compacted cohesive soils. US Bureau of Reclamation, Technical Memo. 654, Denver.

Richards B.E. and Trevena D.H. (1976). The measurement of positive and negative pressures in a liquid contained in a Berthelot tube. Journal of Physics D: Applied Physics 9, L123-L126.

Ridley A.M. and Burland J.B. (1993). A new instrument for the measurement of soil moisture suction. Geotechnique 43, No.2, pp 321-324.

Ridley A.M. and Burland J.B. (1994). A new instrument for the measurement of soil moisture suction. Author's reply to discussion. Geotechnique 44, No.3, pp 551-556.

Stannard D.I. (1992). Tensiometers - Theory, Construction and Use. Geotechnical Testing Journal. Vol.15, No.1, pp 48-58.

Groundwater monitoring — the Sellafield approach

CHRIS ELDRED - Technical Director
Sir Alexander Gibb & Partners Ltd, Reading, England

JOHN SCARROW - Managing Director
Soil Mechanics Ltd, Doncaster, England

MIKE AMBLER - Principal
Sir Alexander Gibb & Partners Ltd, Sellafield, England

ADRIAN SMITH - Senior Engineer
Soil Mechanics Ltd, Doncaster, England

INTRODUCTION

Groundwater monitoring is a fundamental aspect of all civil, environmental and development projects. Traditionally groundwater data has been obtained from a variety of sources which range from the relatively crude to the highly sophisticated. These include drillers records and standpipe, pneumatic, electric and hydraulic piezometers.

The choice of instrumentation to be adopted is generally decided by the designer of the investigation recognising that different approaches have a variety of cost-benefit consequences.

Drillers records are notoriously inaccurate, being indicative only of the water levels noted during the course of hole construction. Standpipe piezometers are the most widely used as they are relatively inexpensive, both to install and monitor, with readily available measuring equipment. This type of piezometer relies upon the movement of water and are ideally suited to long term monitoring in relatively permeable formations.

Hydraulic piezometers have been used for many years for instrumentation of earthfill embankments and dams. They are able to accurately measure pore pressures and can be used to undertake permeability measurements in low permeability materials. Electric piezometers are capable of measuring pressure to a high level of accuracy, and with modern dataloggers can provide a continuous monitoring record. Pneumatic piezometers have a range of applications but have never been used as extensively as the other systems.

One of the most significant shortcomings of conventional piezometers is that it is normal practice to only install one per borehole. Often attempts are made to install two or more piezometers at different levels within a single borehole. Given careful installation procedures installation of dual systems can be made relatively successful although for more than two piezometers the success rate has generally been low.

Where boreholes are relatively shallow with only one or two target monitoring horizons such limitations have never been regarded as major constraints. However, where boreholes are deep and therefore expensive, the numbers of boreholes available for instrumentation are often more limited. Coupled to this deep boreholes are likely to intercept multiple horizons which must be monitored separately in order to understand and model the overall hydrogeological regime. In such a situation the approach to groundwater monitoring has to be very different.

SELLAFIELD INVESTIGATIONS

Overview

The Sellafield Geological Investigations being carried out by UK Nirex Ltd (NIREX) with Sir Alexander Gibb & Partners Ltd (GIBB) as management contractor. The investigations, which have been ongoing for some five years are being carried out to help determine whether the Sellafield site in Cumbria, England is suitable for the disposal of intermediate and certain low level radioactive wastes. A major part of the investigation has been the drilling and testing of a number of 156 mm diameter cored boreholes extending to depths of up to 2000 m.

To date a total of 24 boreholes have been drilled by KSW Deep Exploration Group, (a Joint Venture comprising Kenting Drilling Services Ltd, Soil Mechanics Ltd and BRR), from some 19 sites and two old boreholes, initially drilled in 1908 and 1962 have been refurbished. The generalised location of the boreholes are shown in Figure 1.

Within each fully cored borehole a detailed programme of geophysical logging and hydrogeological testing has been carried out; which in some instances has been ongoing for over three years following completion of the drilling of the borehole.

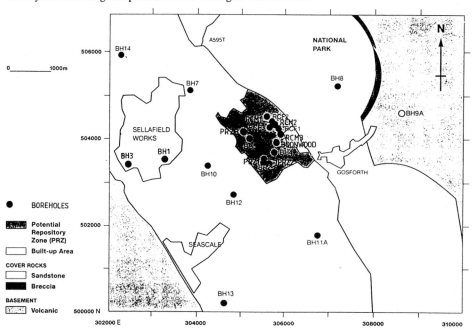

Figure 1 - Location of Boreholes

Geology of Site

Nirex is investigating the rocks of the Borrowdale Volcanic Group (BVG) as a potential host for a waste repository. The BVG underlies the whole of the Sellafield area outcropping to the east of the site and dipping steadily to the west where they are encountered at 1650 m depth at the coast. The BVG is overlain by sandstones and mudstones of the Sherwood Sandstone Group at the base of which is a breccia known as the Brockram. To the west and north of the site the St Bees Evaporite and Carboniferous Limestone are also present. Thin glacial superficial material overlies virtually the whole area. A simplified geological map of the site is presented in Figure 1.

Hydrogeology of Site

The definition of the geological units and structure of the rocks forms a framework which has an important influence on the hydrogeology of the site. The hydrogeology defines the pattern of groundwater flow beneath the site, which is assessed to be one of the mechanisms in the transport of radionuclides from a repository back to the surface.

Much of the testing within the boreholes has been concerned with measuring the distribution of hydraulic properties of the rocks, in particular the hydraulic conductivities, and the groundwater pressures.

The groundwater at depth within the BVG is saline, but is not related to present day seawater. Within the Potential Repository Zone (PRZ), the water has a salinity similar to that of sea water, whilst further west the salinity levels are up to six times that of sea water. Saline water is more dense than fresh water and so, in order to assess the possible driving force for a vertical component of flow, the measured groundwater pressures must be converted to 'environmental' pressures which take account of the influence of varying density.

The hydraulic conductivity of the rocks has been measured in the boreholes, initially in 50 metre long contiguous sections. Within the BVG the conductivity values are typically very low with half the values measured over 50 m lengths in the boreholes being less than 1×10^{-10} ms^{-1}, including tests over faulted and fractured zones.

The available information indicates that the flow of water through the BVG is likely to be controlled largely by the presence of fractures in the rock. A similar pattern is noted for the lower part of the sandstone formations, whereas in the upper, more conductive part of the sandstone sequence, the field and laboratory values are very similar, indicating that the rock is behaving as a porous medium and the groundwater flow is not primarily through fractures.

Objectives of Monitoring System

From the foregoing it can be appreciated that an accurate measurement of 'environmental' pressures is fundamental to assessing the suitability of the site.

On completion of the drilling and borehole testing programme, it is therefore a key objective of Nirex's site characterisation programme, to install a series of groundwater monitoring systems which will provide.

1) baseline pressures within the rockmass, which can be converted to environmental pressure, given knowledge of the density of the fluid column.

2) the ability to identify and characterise natural changes in groundwater pressure.

3) the ability to assess the impact of interborehole testing and identify hydraulic interference paths within the rockmass.

4) a network of monitoring boreholes near to the proposed location of the proposed underground laboratory, the Rock Characterisation Facility (RCF), in order to determine baseline groundwater conditions prior to construction of the RCF.

5) subsequently to monitor the impact of shaft sinking on the hydrogeology.

As can be appreciated from the above, the monitoring systems are required to provide data to meet a variety of objectives and must operate in the longterm in instances of hostile saline conditions.

In order to meet the above objectives alternative commercially available and bespoke monitoring systems were evaluated. Based upon technical and cost-benefit assessments the Westbay MP System offered by Soil Mechanics Ltd, a partner of the KSW Joint Venture was selected as the most appropriate system, although it was acknowledged that in terms of the specific Sellafield objectives installations would be required to be installed to depths outside previous experience.

EQUIPMENT SPECIFICATION AND DESIGN

Outline Design and Operation

The Westbay MP System is a multi-level groundwater monitoring system utilising a closed access casing comprised of modular casing sections, packers and valved port couplings. The valved port couplings are used to provide a hydraulic connection through the casing for access by proprietary pressure monitoring and groundwater sampling tools. The modular nature of the system permits as many monitoring zones as desired to be established in a borehole, commensurate with strength constraints, with complete flexibility in the configuration of the system during the design and installation stages. Figure 2 illustrates the principles of the system which are described below.

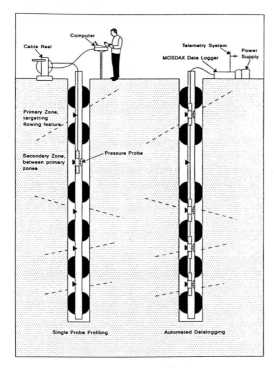

Theoretically the number of zones is only dictated by the minimum length of the individual system components. However, based upon depth and strength criteria and cost-benefit considerations, the maximum number of zones achieved at Sellafield is 30 for a 1200m deep borehole. The casing components can be manufactured from unchlorinated polyvinyl chloride (UPVC), chlorinated polyvinyl chloride (CPVC) or in special circumstances of great depth and/or aggressive groundwater conditions, stainless steel. There are two sizes of casing systems, MP55 which is nominally 80 mm OD, 55 mm ID, and MP38, which is nominally 48 mm OD and 38 mm ID. For the Sellafield environment, consideration was given to both the plastic and steel system.

Based upon cost-benefit and technical considerations, a decision was made to proceed with principally the MP55 plastic systems which to date have been installed successfully to

Figure 2 - The Westbay MP System

depths of 1200 m in a vertical borehole, in deviated boreholes (up to 36° from vertical) and to lesser depths in inclined boreholes (30° from vertical).

Installation

Prior to installation, zones of potential interest are targeted on the basis of geology and hydrogeological and geophysical testing carried out in the boreholes during and following completion of all drilling works. These zones are then reviewed against the objectives of the testing strategy for that borehole to derive a proposed monitoring system configuration. Calliper logs are run in the hole prior to completing the exact design of the system to assist in the selection of the best available packer seat locations commensurate with the dimensions of the potential zone of interest. An installation log is then prepared showing the locations of all the individual casing components. If only groundwater pressure measurements and small volume sampling are required, a measurement port coupling is installed in each monitoring zone. If large volume sampling, fluid withdrawal or fluid injection is anticipated, a pumping port and a measurement port coupling are installed in selected zones. Duplicate measurement port couplings are also installed in zones of particular interest, designated as primary zones, to allow relatively low cost redundancy in the system.

Measurement port couplings are also installed beneath every packer to correctly locate the packer inflation tools and to act as a vent to relieve squeeze pressures generated by packer inflation. Thus monitoring zones, known as secondary zones, are generated over the sections of borehole between adjacent packers. Casing lengths which do not have either a monitoring or pumping port coupling are connected to their neighbour by a standard blank (regular) coupling. All couplings contain two 'O' rings which together with an end plug make positive hydraulic seals and flexible shear wires assure a tensile connection with adjacent casing components.

A unique and cost saving aspect of the system is that it can be installed by flotation without the use of a rig to depths of approximately 700 m, below this depth casing strings are installed generally using a small crane. Loads of up to 900 kgs being measured whilst handling installations up to 1200 m depth.

Packer Inflation
Once the casing string has been positioned in the borehole, the packers are inflated from the bottom upwards using a packer inflation tool. The tool is lowered down the inside of the MP casing and located in the correct position by the location arm which seats in the correct orientation in the integral helical landing ring within the measurement port coupling located directly beneath each packer. Water is pumped under pressure from a surface supply vessel to the tool, the tool activates opening the measurement port valve and sealing the tool injection valve to the internal packer valve, water is then pumped into the packer. The volume of water pumped is measured at the surface together with the inflation pressure, measured by a pressure transducer mounted within the tool. Inflation is stopped when either the inflation volume or pressure criteria is reached.

The absence of permanent surface inflation lines leading to each packer, provides greater flexibility to the number of packers that can be accommodating within an installed system.

Pressure Measurements
Groundwater pressure and temperature measurements can be made at each location in the borehole where a measurement port coupling has been installed. The environmental pressure is measured using a Mosdax pressure probe, incorporating a location arm, a backing shoe, a face seal 'O' ring and an appropriate range fluid pressure and temperature transducer. The probe is operated on a wireline single core cable connected to a Mosdax Personnel Computer Interface (MPCI) and either a Mosdax hand held controller or a portable computer, running a proprietary Mosdax monitoring program MLog, at the top of the borehole.

To perform pressure measurements, the probe is lowered to a point below the first measurement port to be accessed (usually the deepest). The location arm is released from within the probe body. The probe is raised to just above the measurement port coupling and then lowered until the location arm rests into the helical landing ring in the coupling. The backing shoe is activated, pushing the probe to the wall of the measurement port coupling so that the face seal 'O' ring on the probe seals around the measurement port valve at the same time as the plunger within the 'O' ring pushes the valve open. The transducer is now hydraulically connected to the fluid outside the coupling and isolated from the fluid inside the casing. The reading displayed on surface will be the environmental pressure in the formation outside the measurement port. After the reading has been recorded, the probe backing shoe is deactivated (retracted) and the valve in the coupling reseals and the probe moved up to the next port.

The two principle methods of monitoring environmental pressure with the MP System are single probe profiling and automatic multi probe monitoring. Single probe profiling usually involves a team of two technicians travelling to each borehole where they manually locate the probe at each measurement port and carry out fluid pressure and temperature measurements one at a time. The Westbay MP systems at Sellafield which do not currently have an automatic monitoring facility are generally profiled once per week.

The policy at Sellafield is to install automated multi probe monitoring systems in the majority of the Westbay MP systems to allow continuous observation and recording of environmental pressure and temperature in selected zones. To configure a Westbay system for automatic monitoring, a number of Mosdax probes are linked together in series on lengths of cable to form a single cable and probe string, which is then lowered into the borehole casing, (see Figure 2) connected at surface to the Mosdax datalogger. Recording of pressure measurements may be carried out on a simple time basis, or the datalogger may be programmed to continually scan each probe and record pressures if a specific threshold value is exceeded or the rate of change of pressure exceeds a prescribed threshold.

The datalogger, which is powered by a 12 volt battery, works unattended, requiring periodical downloading of stored data either by visiting the site with a portable computer, or via a land line or cellnet telephone link via a modem.

Accuracy and Precision of Pressure Probes
The transducers installed in the Mosdax pressure probes incorporate silicon strain gauge transducers with an accuracy of $\pm$ 0.1% Full Scale CNLRH (Combined Non-Linearity, Repeatability and Hysteresis). The resolution of the gauges is determined by the Analogue/Digital conversion that takes place in the Mosdax module. The resulting resolution is typically 1/30,000 (i.e. Full Scale/30,000).

In order to, standardise as much as possible and allow flexibility in any future restringing of probes all the Sellafield transducers have a 0 to 2000 psi range. These transducers were chosen because they could accommodate the depth range of a majority of the Sellafield boreholes and there was considered to be little technical advantage in reducing the resolution below 0.5 kPa. Key advantages of the Westbay system for the Sellafield works is that each probe may be calibrated by removal from the borehole at appropriate intervals and can be function checked against the internal fluid column of fresh water at any time by deactivation from the measurement port coupling, both facilities enabling the stringent quality standards of the project to be met.

SELLAFIELD MONITORING NETWORK

The Sellafield monitoring network currently comprises 22 instrumented boreholes within which there are some 270 zones capable of providing pressure measurements. These zones can be divided into primary zones which are designed to monitor specific features and the intervening secondary zones within which there are no specific hydrogeological targets.

Of these zones some 160 are being monitored utilizing dedicated probes providing data at 2 minute intervals in the RCF area and 30 minute intervals in the regional boreholes.

At Sellafield it is desirable to centralise the gathering of data. In order to meet this objective either mains electricity supplies or solar panels with batteries will be provided to power the system at each borehole site. In addition, in order to download the data remotely, modems connected to BT land lines or cellnet will be installed. This data management system, which is currently being completed, will allow the majority of the monitoring systems to be accessed and downloaded from a central single computer remote from the site.

The modular design of the monitoring system enables the system to be tailored to meet specific objectives. For the long term pressure monitoring, which will continue for at least ten years, in general, only the primary zones will be monitored. However, in order to meet specific requirements, such as monitoring the effects of pump tests, the monitoring strings may be reconfigured, by the addition of probes, to monitor both primary and secondary zones. Such reconfiguration was carried out for a crosshole test where 29 zones were instrumented in Borehole No. 4 during a series of pump tests carried out in Borehole No. 2 some 125 m away. Subsequently this string was reconfigured to monitor 9 zones only and the remaining probes installed in three other boreholes.

DATA ASSESSMENT

When the data from the first borehole to be instrumented, Borehole No. 14, was analysed it was evident from the presence of cyclic fluctuations in the data that the pressure was being influenced by a variety of natural phenomena. A typical set of records from Borehole No. 14 are shown in Figure 3.

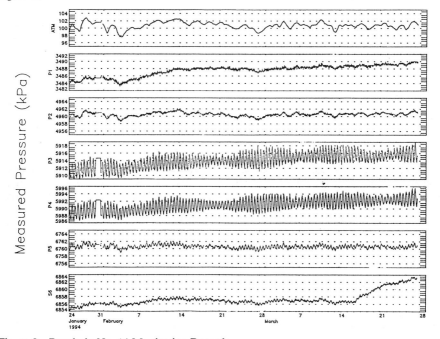

Figure 3 - Borehole No. 14 Monitoring Records

In order to clearly establish fluctuations in groundwater pressure it is necessary to understand these other phenomena.

Many physical phenomena may cause changes in the measured pressure in deep boreholes. Some phenomena relate to processes that create pressure disturbances that may be used to derive hydrogeological properties; others are more clearly classified as instrument noise. The main phenomena which are considered to be acting are outlined below.

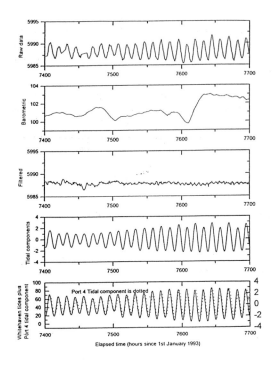

Aquifer Response to Barometric (Atmospheric) Pressure

It has been frequently observed that water levels in boreholes respond to barometric pressure.

It is assumed that this is due to the variation in load on rocks overlying such aquifers and the strains induced by these loads. Authors have calculated atmosphere efficiencies of aquifers and have related these to the bulk elastic properties of the aquifer (Narasimhan et al 1984).

Ocean Loading

In coastal areas water levels in aquifers may fluctuate in response to the ocean/sea tides. This can either be due to hydraulic connection of the aquifer to the tidal regime or to strain induced by the varying loads placed on overlying rocks.

Figure 4 - Filtering of Data to Extract Tidal Components

Earth Tides

Periodic aquifer dilations and associated water level fluctuations are known to result from lunar and solar tidal forces. Various authors have used spectral analysis methods to estimate aquifer properties from the responses observed in boreholes to these tides. Models for both homogeneous and fractured aquifers have been considered.

From a broad assessment of the Sellafield data set the maximum magnitude of the impact of these phenomenon are as follows:

Instrumental Noise	± 0.5 kPa
Barometric Effects	± 1.5 kPa
Tidal Effects	± 3.0 kPa

The scale of these phenomenon is relatively small when one is considering a large scale hydrogeological system over an extended period of time. However, in the Sellafield environment an understanding of these phenomena is important as such an understanding will:

*   Allow removal of these effects to ensure identification of short small scale transient responses to pumping from adjacent boreholes.

*   Analysis of such phenomena may provide important information on the bulk hydrogeological properties of the rock mass.

Data Processing

In order to remove these phenomena the data is processed on a Sun workstation by removing the barometric effects, passing the data into the frequency domain, applying a Fast Fourier Transform (FFT) followed by an appropriate filter (low pass, band pass or notch) and returning the data to the time domain for hydrogeological analysis.

In order to facilitate the removal of barometric effects from pressure data measured in an interval, it is necessary to calculate the Atmospheric Efficiency Factor (AEF) of the interval in question. The AEF of a zone is the change in interval pressure caused by a 1 kPa change in barometric pressure. In order to provide the best estimate of AEF, barometric pressure records are provided at each borehole site as an integral part of the Mosdax system.

In applying the AEF, the data set is normalised to a mean atmospheric pressure of 100 kPa.

The AEFs for all of the zones in Borehole No. 4 lie between 0.3 and 0.65. In the process of normalising the data set to 100 kPa, with an average variation in atmospheric pressure of ±2.5 kPa, the maximum correction to the data has been between ±0.6 and ±1.3 kPa.

The procedure used to filter the Borehole data is summarised below. In order to pass data through a FFT it is necessary to have data points at regular time intervals. In practice this has not always been possible and it is necessary for data sampled with an uneven time interval, with null points or data gaps, to be interpolated onto a regular time grid to make it suitable for passing through a FFT.

The adopted procedure is to:

1)   Identify null points and eliminate them from data array.

2)   Calculate AEF and remove atmospheric pressure effects, if required.

3)   Calculate average of first and last ten points in array, generate straight line and detrend data.

4)   Pass data to FFT to transform it to the frequency domain.

5)   Generate filter with required frequency cutoffs.

6)   Apply filter.

7)   Apply reverse FFT to go back into the time domain.

8)   Re-trend the data by adding line from (3) above, if required.

For frequency domain filtering to be effective, the response being examined should have a characteristic frequency significantly different to the frequency of the background events being filtered out. Four general types of filters can be used in the frequency domain, low pass, high pass, band pass and notch filters.

It is possible, by the use of notch filters, to extract data of specific frequencies. Notch filters were used to extract the 12 and 24 hour tidal components from the Borehole No. 14A data. These tidal components have been displayed in Figure 4 alongside the barometric signal and sea tide data from Whitehaven which is situated just North of the Sellafield site. It can be noted that the tidal component of the data is in phase with the sea tide.

The long term purpose of the monitoring network is to determine the natural environmental pressure gradients and natural changes to those pressure gradients. However, over the next year the monitoring network within the area of the PRZ will play a key role in the site characterisation process providing data on the response of the hydrogeological system to imposed changes. The data derived from such testing will be used to support the hydrogeological models of the site.

Three phases of active testing have been scheduled, these are:

*   Crosshole interference testing between Borehole No. 2 and 4
*   Groundwater abstraction test in Borehole No. RCF3
*   Construction of the RCF

To date only the first phase has been completed and the data is currently being analysed, the second phase is being planned for implementation in early 1995. The third phase has yet to be planned in detail and will not be described here.

Crosshole Testing Borehole Nos. 2 and 4

Borehole No. 2 and 4 are two deep boreholes some 125 m apart and 1600 m and 1250 m deep respectively. In both boreholes the sandstone sequence has been permanently cased off during drilling leaving only the BVG exposed. Both boreholes have been well characterised by an extensive programme of testing which includes geological assessment of stratigraphy and fracture orientation and hydrogeological studies to identify the location and properties of hydrogeologically transmissive fractures.

In addition, a high resolution seismic tomography survey has been carried out between the two boreholes in order to provide details of the geological structure between the two boreholes.

At an early stage it was agreed that Borehole No. 2 should be the source hole and Borehole No. 4 the monitoring hole. To this end a 30 zone Westbay system was installed in Borehole No. 4 within which the monitoring zones were targeted on known hydrogeologically conductive features, and zones adjacent to structural features that intercepted both boreholes.

In Borehole No. 2 a conventional oilfield surface inflate packer system was deployed on steel testing pipe which comprised twelve packers which isolated the six known hydrogeologically conductive features. Within each of these six target zones a hydraulically activated sliding sleeve was installed which once opened allowed the extraction of groundwater from the target zone.

Within each of the target zones and the intervening zones were installed pressure gauges which provide a real time surface readout of pressure and temperature.

Each test comprised a period of pumping from the target zone followed by the shutting in of the system to allow full recovery. Due to the low permeability of the formation constant rate pumping tests were technically difficult with a lower rate cutoff of 0.015 litres/minutes. The majority of the tests were therefore carried out as constant head extraction tests with imposed drawdowns of up to 200 m.

The configuration of the testing allowed the drawdown to be applied and held for approximately 10 days whilst the responses in the adjacent 11 zones, within Borehole No. 2 and 29 of the zones in Borehole No. 4 were monitored. In this manner the rate and magnitude of the propagated signal could be measured. A diagrammatical layout of a section of a typical test is shown in Figure 5. The left hand column represents Borehole No. 2 and the imposed drawdowns, the right hand side the monitoring system with typical drawdowns in selected zones. Between the two boreholes has been superimposed a part of the seismic tomogram.

The layout above has been presented for illustrative purposes only, the acquired data set is currently being analysed.

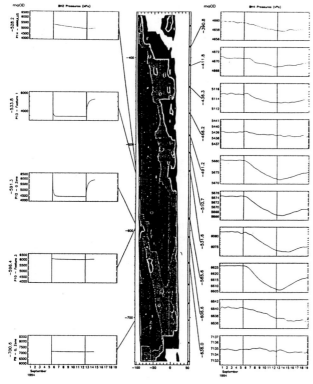

RCF3 Pump Test

Following on from the cross-hole testing programme, it is currently proposed to carry out a large scale pumping test within the RCF area. The principles developed for the cross-hole testing programme will be retained but applied on a larger scale. The 12 packer string previously installed in Borehole No. 2 has been withdrawn and reinstalled in Borehole RCF3 which has been drilled down the centreline of the proposed RCF shaft.

Within all of the boreholes adjacent to Borehole RCF3, monitoring systems have been installed with Mosdax probes installed in the primary zones. Within this rock mass there will be in excess of eighty individually monitored zones which will provide data at two minute intervals throughout the test.

A diagrammatical representation of the pumped well Borehole No. RCF3 and the adjacent monitoring holes is presented in Figure 6.

Figure 5 - Cross Hole Hydrogeological Testing

CONCLUSIONS

The NIREX Sellafield Geological Investigations are one of the largest and most comprehensive site characterisation programmes currently being carried out in the world. The adoption of the described monitoring system has provided a very flexible and cost beneficial approach to groundwater monitoring and testing. The quality of the data that is being acquired is of the highest order and has enabled the identification of pressure changes of down to 1 to 2 kPa in response to natural and imposed drawdowns in adjacent boreholes. The monitoring network is currently being integrated with other time series data derived from weather stations, river gauging and NRA boreholes. When complete the system will be accessible from a single computer workstation if required located off site that will be capable of integrating and processing the data, not only as simple time series graphs, but also as 2D and 3D visualisation of pressure changes.

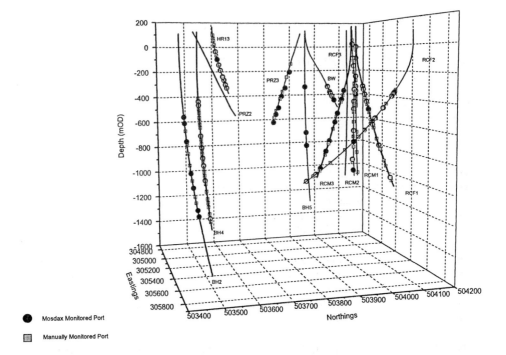

Figure 6 - Layout of Monitoring Network in RCF Area

REFERENCES

CHIEN Y M, BRYCE R W, STRAIT S R, YEATMAN R A (1986). Elimination of frequency noise form groundwater measurements. In Disposal of High Level Nuclear Waste Burkholder (ed).

CLARK W E (1967). Computing the Barometric Efficiency of a Well, Proc. ASCE, J. Hyd, Div., pp 92-98.

MARINE I W (1975). Water level fluctuations due to earth tides in a well pumping form slightly fractures crystalline rock. Water Resources Research Vol. 11, No. 1 February 1975.

MELCHIOR P (1978). The tides of planet earth.

NARASIMHAN T N, KANEHIRO B Y, WITHERSPOON P A (1984). Interpretation of earth tide response of three deep confined aquifers.

NIREX Report 524

ACKNOWLEDGEMENTS
The authors would like to thank UK Nirex Ltd for providing support in preparing this paper and funding the work described herein, the Directors of Sir Alexander Gibb & Partners Ltd and Soil Mechanics Ltd and colleagues from Entec Hydrotechnica Ltd, GeoScience Ltd and Golder Associates.

Recent developments in the cone pressuremeter

R. W. WHITTLE
Cambridge Insitu, Cambridge, England

INTRODUCTION
As the name implies, the Cone Pressuremeter (CPM) described in this paper consists of a pressuremeter module which is tipped with a 15CM^2 cone. The instrument is pushed into soil using a cone penetrometer rig, and adds to the cone profile data for insitu stiffness and strength at chosen locations in a probing.

Pushing a pressuremeter is the fastest insertion method and the CPM was developed in 1985 for offshore site investigation. The first commercial use of the instrument took place in 1989, not offshore but on land. Since then interest in the instrument has increased. It has the potential to gather data faster and ultimately more economically than other pressuremeters, and can be placed in a greater variety of materials.

The analysis of the test curve that the instrument produces is not yet satisfactorily resolved, but indications are that it is as good a tool as any other pressuremeter for determining soil stiffness. This paper concentrates on the measurement of stiffness from expansion tests in clay. The mechanics of the instrument which influence the measurement are assessed, and the calibration procedures in general use are critically examined.

Problems with analysis are avoided by presenting the fieldwork results as visual comparisons between the measured test curves for CPM tests and self boring pressuremeter (SBP) tests at the same location. The purpose is not to quote values for strength and stiffness, but to demonstrate the degree to which the two devices plot the same curve.

COMPARING THE CPM TO OTHER PRESSUREMETER TYPES
The CPM is one of a number of pressuremeter types in use in the United Kingdom. There are three types of pressuremeter test, categorised by the manner in which the pocket for the instrument is formed, i.e. by pre-boring, by self-boring or by pushing. The insertion method determines the state of stress in the soil before pressure is applied to the internals of the probe, and hence conditions the measured boundary which provides an identifiable point for subsequent analysis.

Pre-boring.
In this test the instrument is placed in a hole that has been made by rotary drilling, coring or augering. The cavity sees complete stress relief before the instrument is placed. Some preliminary expansion of the borehole is required to restore the stress in the surrounding soil to the insitu state, and the strain origin for the test is interpolated from this stress. The measured boundary is the material yield stress, and cavity expansion theory is used to interpret the test.

Advances in site investigation practice. Thomas Telford, London, 1996

synonymous with the external diameter of the probe. In principle, the insitu lateral stress and the material yield stress are measured boundaries, and cavity expansion theory is used to interpret the test.

Pushing

Pressuremeters of this type are pushed into the soil. The hoop stress in the material adjacent to the probe is raised by this procedure, and the disturbance is such that there is no possibility of restoring the soil to the state of stress before the arrival of the instrument. Analysis of the loading curve depends on erasing this initial stress state by establishing the radial stress as the principle stress for a significant volume of material around the probe.

The CPM is a special case of a push in pressuremeter in that the passage of the cone has grossly disturbed the material in a manner that approximates a cylindrical cavity expansion from zero to a very large strain. Potentially the measured boundary for this test is a limiting pressure for the soil.

The expansion phase of the test is used merely to establish this limit pressure. Cavity contraction theory is applied to the unloading part of the measured curve in order to obtain soil strength parameters, with some success for undrained tests in clay (Houlsby & Withers 1988) but not for drained tests in sand (Withers et al 1989). A semi-empirical strategy has been suggested to obtain strength parameters for sands (Schnaid & Houlsby 1990) using the measured limit pressure and cone tip resistance.

DETERMINING SHEAR MODULUS

In principle any pressuremeter test can provide reliable and repeatable data for the elastic properties of the soil, specifically the shear stiffness of the material being tested. This is achieved by taking small rebound cycles. This simple but powerful technique seems to be independent of the initial disturbance to the soil provided that the radial stress is the major stress at the point where the cycle is initiated (Withers et al 1989). The method demands good resolution of displacement and pressure changes, which in practice restricts it to pressuremeters which measure the radius or diameter of the membrane rather than the volume.

The CPM was developed primarily to measure soil stiffness, and the data quoted here is taken from two sites where comparisons can be made with good quality SBP tests. This restriction means that the quantity of data is small, although there have been many commercial CPM tests taken. There are similarities between the mechanics of the SBP and CPM which makes the comparisons more than usually pertinent.

DESIGN OF THE CONE PRESSUREMETER.

The model of cone pressuremeter described here was built by Cambridge Insitu as part of a joint venture with Fugro BV in 1985. It is described in Withers et al (1986), and is shown in figure 1. Much of the detail of the prototype applies to the current instrument, but there are some important modifications. The basic details are these:

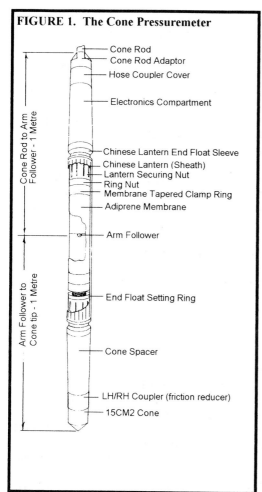

FIGURE 1. The Cone Pressuremeter

- Cone Rod
- Cone Rod Adaptor
- Hose Coupler Cover
- Electronics Compartment
- Cone Rod to Arm Follower - 1 Metre
- Chinese Lantern End Float Sleeve
- Chinese Lantern (Sheath)
- Lantern Securing Nut
- Ring Nut
- Membrane Tapered Clamp Ring
- Adiprene Membrane
- Arm Follower
- Arm Follower to Cone tip - 1 Metre
- End Float Setting Ring
- Cone Spacer
- LH/RH Coupler (friction reducer)
- 15CM2 Cone

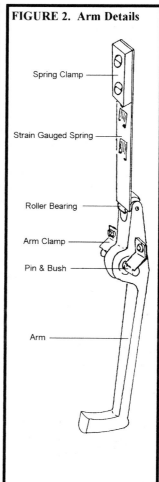

FIGURE 2. Arm Details

- Spring Clamp
- Strain Gauged Spring
- Roller Bearing
- Arm Clamp
- Pin & Bush
- Arm

• The nominal diameter of the instrument is 43.7mm, and the length to diameter ratio of the expanding part of the probe is 10. From the cone tip to the centre of the expanding part is 1 metre.

• The total pressure inside the instrument is measured by a strain gauged transducer.

• There are three displacement sensors positioned 120° apart and located centrally behind the membrane. The displacement sensors are similar to those in the current Cambridge Self Boring Pressuremeter. A sensor consists of a light pivoted lever whose long end is forced to follow the movements of the inside wall of the membrane by a strain gauged leaf spring acting on the short end.

Changes in the latest instrument are described below:

a) In the prototype instrument the leaf spring bore directly on the lever, so that the resolution of small displacements after reversing the direction of straining was affected by a frictional component. In a rebound loop the strain direction changes twice, so that the contribution of friction is a significant source of error. In the current instrument the spring is decoupled from the lever by a small ball race, so reducing the friction to an insignificant level (Fahey & Jewell 1990). Figure 2 shows an arm in detail.

b) Pushing a pressuremeter results in high stresses bearing on the surface of the instrument during insertion. The metal sheath provides a considerable amount of protection to the vulnerable elastic membrane but nevertheless granular materials can be forced through the necessary gaps in the sheath possibly damaging the membrane. This was a frequent occurrence on the prototype instrument in hostile materials. To alleviate this, the part of the assembly which connects the cone to the pressuremeter has been made slightly larger than the diameter over the membrane. This friction reducer has eased greatly the difficulties experienced when pushing the instrument into dilatant material. The increase is small (2% of the instrument diameter) but it means that the material deformed by the cone has been unloaded before the pressuremeter starts to inflate. It happens that the geometry of the cone is such that some unloading occurs anyway, so that the friction reducer does not fundamentally modify the stress concentrations in the material due to the passage of the cone or the assumptions governing the expansion phase of the pressuremeter test.

c) The resolution of the radial displacement measuring system has been increased, so that a change of 1 micron can be resolved in a total movement of more than 11 millimetres.

d) Readings of pressure and displacement are now logged at five second intervals instead of ten seconds. This increases the quantity of information for significant parts of the curve such as unload/reload loops.

e) Aspects of the assembly and construction have been revised for ease of maintenance but this does not of itself improve the data although it has made the system more reliable.

The performance of the displacement measuring system

Table 1 gives the results of a recent calibration on one of the displacement sensors (normally referred to as ' arms') of a current CPM. The performance is being examined over the range '0 to 10' millimetres, '0 to 5' millimetres and '0 to 1' millimetre. The method employed was to fix a micrometer head with a 13 millimetre range above the arm, and to take 10 readings over each displacement range for an increasing and reducing deflection. One rotation of the micrometer barrel is a change of 0.5 millimetres.

Linearity is here defined as the difference in output between intervals expressed as a percentage of the average output per millimetre obtained by linear regression.

Hysteresis is defined as the difference in measured output for the same micrometer position, there being a reading taken for increasing and reducing deflections. The hysteresis figure is expressed as a percentage of the measured output range.

The action of rotating the micrometer stem on the end of the arm introduces a component of friction into each reading that does not occur when the arm follows a membrane. To minimise this the micrometer head coupled to the end of the arm via a low friction link, consisting here of a long double-ended needle.

Table 1 - Calibration data for a cone pressuremeter displacement follower

mm	0-10 (mV)	linearity %	hyster. %	mm	0-5 (mV)	linearity %	hyster. %	mm	0-1 (mV)	linearity %	hyster. %
0	-1162.0	100.2	0.17	0.0	-1164.1	100.8	-0.02	0.0	-1162.8	94.0	-0.39
1	-930.8	99.7	0.28	0.5	-1047.6	99.2	0.20	0.1	-1140.9	103.0	-0.26
2	-700.7	100.2	0.31	1.0	-933.0	100.2	0.15	0.2	-1116.9	108.2	-0.17
3	-469.5	99.6	0.35	1.5	-817.2	100.6	0.26	0.3	-1091.7	103.5	0.09
4	-239.6	99.8	0.34	2.0	-701.0	99.8	0.35	0.4	-1067.6	91.9	0.22
5	-9.3	100.3	0.35	2.5	-585.7	100.7	0.35	0.5	-1046.2	92.7	0.30
6	222.1	99.5	0.39	3.0	-469.4	99.2	0.36	0.6	-1024.6	103.5	0.17
7	451.7	100.6	0.33	3.5	-354.8	100.4	0.25	0.7	-1000.5	106.0	0.34
8	684.0	100.5	0.39	4.0	-238.8	99.7	0.28	0.8	-975.8	102.2	0.09
9	915.9	100.3	0.23	4.5	-123.6	100.1	0.22	0.9	-952.0	92.3	0.09
10	1147.4	-102.6		5.0	-8.0	-102.3		1.0	-930.5	-93.2	
9	910.7	-102.1		4.5	-126.2	-100.2		0.9	-952.2	-102.2	
8	675.0	-100.0		4.0	-242.0	-100.1		0.8	-976.0	-108.6	
7	444.1	-100.0		3.5	-357.7	-100.3		0.7	-1001.3	-101.8	
6	213.2	-99.9		3.0	-473.6	-100.5		0.6	-1025.0	-94.0	
5	-17.3	-99.7		2.5	-589.7	-99.9		0.5	-1046.9	-91.0	
4	-247.5	-99.7		2.0	-705.1	-99.6		0.4	-1068.1	-102.2	
3	-477.5	-99.8		1.5	-820.2	-99.1		0.3	-1091.9	-105.6	
2	-707.9	-99.3		1.0	-934.7	-99.7		0.2	-1116.5	-102.2	
1	-937.2	-99.1		0.5	-1049.9	-98.7		0.1	-1140.3	-92.7	
0	-1165.9			0.0	-1163.9			0.0	-1161.9		
Zero	-1165.8 mV			Zero	-1164.7 mV			Zero	-1162.8 mV		
Slope	230.8 mV/mm			Slope	231.1 mV/mm			Slope	232.9 mV/mm		

The readings taken are to some extent operator sensitive. For hysteresis readings to be representative the micrometer must be turned consistently in one direction, so that arriving precisely on a mark takes practice and will influence the repeatability of a calibration. However despite this reservation a number of conclusions can be drawn from this table:

• The *slope* figure quoted in the last line of the table is the result of the linear regression calculation. The change in magnitude is small, despite the fact that the lever arm moves on an arc.

• The columns for linearity reveal that the arm is performing better than the standard. It appears that the linearity when looking at the '0 to 1' range fluctuates badly. In fact what is shown is the imperfections of the thread of the micrometer barrel. To gather this data took 4 revolutions of the micrometer barrel, and four cycles of trough to peak can be seen in the linearity figures. The steps in the '0 to 10' and '0 to 5' ranges are comprised of complete turns of the micrometer barrel so that the thread error is not apparent. A check with the manufacturer of the micrometer confirmed an uncertainty figure of plus or minus 2 microns in any reading.

• The hysteresis percentage remains fixed for all ranges, which implies that as an absolute error it is dependant on the magnitude of the deflection that has occurred since the last reversal of the direction of measurement. This means that when considering the unloading part of the first unload/reload cycle in a test, the path followed will be different from the true path due to hysteresis whose magnitude depends on the total expansion of the cavity until the point when the loop was initiated. However, when considering the reload path in the same loop, the error due to hysteresis will be related only to the strain amplitude of the loop.

CALIBRATING THE MEMBRANE

The measurements that the instrument makes represent pressure behind and the movements of the inside wall of the membrane. To determine the applied radial stress and resulting displacements of the borehole, an assessment has to be made of the contribution of the membrane.

The measured pressure needs to be corrected for membrane stiffness; the measured displacements require correction for changes in the thickness of the membrane due to expansion and pressure related instrument compliance.

Membrane correction for pressure

To correct the pressure readings for the stiffness of the membrane, the accepted procedure is to inflate the assembled instrument in air and so obtain a plot of pressure versus displacement. The slope and intercept on the pressure axis of the measured curve are assumed to describe the membrane modulus and initial stiffness. In figure 3 the two correction factors are expressed as a zero offset in kPa to start the membrane moving, and a linear correction in kPa/mm to account for the increasing resistance with displacement. Although the slope is treated as being linear, it is noticeably curved, and this has persuaded some users to fit a hyperbolic function to the correction plot.

The expansion in air is not a good analogue for what occurs in the ground. In air the membrane sees negligible resistance to expansion, and hence local variations in the properties of the membrane can dominate the correction test. In an extreme case, it is possible to see large expansion of the membrane at one point, but little or no movement where the arms are located. It is even possible for uneven expansion to make the membrane over the arms contract. Yet in the ground this same membrane will give a test curve that shows none of this extreme behaviour because soil has stiffness. An attempt by the membrane to inflate more at one point than another would represent a difference in strain that in practice is made impossible by the magnitude of the stress that would result. If the ground has isotropic properties the membrane must inflate in a uniform manner. Unfortunately it is not easy to model this behaviour out of the ground.

It appears that calibration techniques that seem to work for similar pressuremeters such as the SBP give problems when applied to the CPM. The hoop stress is higher in the CPM membrane, the length to diameter ratio is larger (10 as opposed to the 6 of the SBP) and the required expansion is greater, all of which amplify the difficulties of calibrating a membrane which cannot have perfectly uniform properties. Unless the weakest part of the membrane happens to lie over the arms, then the calibration curve in air will result in an over large correction factor being derived for membrane stiffness. The form of the CPM test makes this an undesirable characteristic, because close to the limit pressure large displacements occur for little change in pressure. Hence an over-correction for membrane stiffness will make it appear that the limit pressure is falling with increasing displacement.

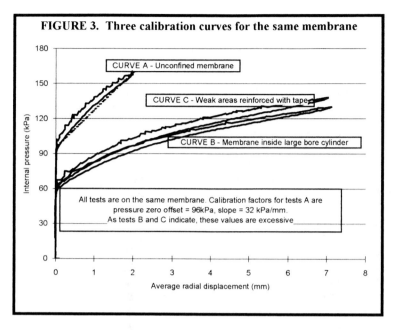

FIGURE 3. Three calibration curves for the same membrane

CURVE A - Unconfined membrane

CURVE C - Weak areas reinforced with tape

CURVE B - Membrane inside large bore cylinder

All tests are on the same membrane. Calibration factors for tests A are pressure zero offset = 96kPa, slope = 32 kPa/mm. As tests B and C indicate, these values are excessive

Internal pressure (kPa)

Average radial displacement (mm)

Figure 3 is a composite of three calibration tests on a membrane known to have markedly non-uniform properties. Test A is an inflation in air. It appears to have been terminated at an early stage, but part of the membrane away from the measuring point had inflated to about 20mm radial expansion. Test B is an inflation inside a large bore cylinder. The membrane still inflates in air , but the weakest point eventually contacts the inside wall of the cylinder and thereafter ceases to influence the calibration. Test C is an inflation in air, but the weak points on the membrane have been identified and reinforced with a ring of packaging tape. Tests B and C give similar results, with a zero stiffness correction of about 65kPa but with a curved relationship between pressure and displacement. Test A is very different from both B and C.

A succession of tests were carried out with a large bore cylinder fitted over the membrane and sealed to the instrument body. An external pressure was then applied to the membrane, and was maintained at a constant level by an automatic control system. A number of membrane correction tests were run with different external loads applied. As expected, all these tests gave curves very similar to test B but with a pressure offset corresponding to the external pressure.

Figure 4 shows the results of a simple experiment to generate some reaction to the expansion of the membrane. Instead of maintaining an external pressure at a fixed level, it was allowed to rise as the external volume became smaller. The external pressure was read off a test gauge at zero displacement, at maximum displacement, and at the return to zero displacement once more. There was a small loss of external pressure due to imperfect sealing, but the experiment gave an approximately linear inflation. The zero figure is consistent with tests B and C, but the derived slope is more like the latter part of the curves for those two tests.

None of these experiments are recommended as being definitive practice for future calibration. However there is evidence here that an inflation in air gives unreasonably high correction parameters and a better procedure is required.

Recently a softer membrane made of natural rubber has been made available for work in soft to firm deposits. This membrane has been used with success at the Bothkennar test site. The difficulties outlined above still apply in principle, but the magnitude of the uncertainty is much reduced. The strength of the standard polyurethane membrane is necessary for tests where the applied pressure is likely to exceed 1MPa.

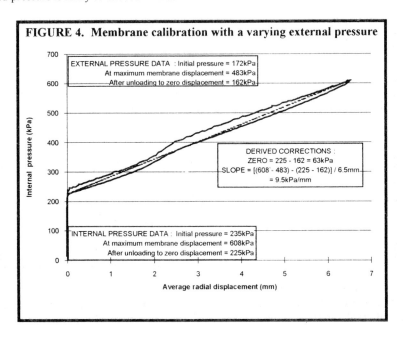

FIGURE 4. Membrane calibration with a varying external pressure

Membrane thinning with expansion

The thickness of the membrane changes with displacement. This is allowed for by assuming that the cross-section area of the membrane remains constant. The cross-section area is known, as is the inner radius of the membrane at all times because this is what the arms measure. Hence the thickness of the membrane can be calculated and an appropriate correction made, a procedure that is automatically implemented in the data logging software.

Instrument compliance

If the instrument is placed inside a close fitting metal cylinder and pressurised to the limit of its range, then once the membrane contacts the sides of the cylinder no further movement should be seen from the displacement followers, apart from a small amount representing the elastic properties of the cylinder. Any movement greater than this is a pressure dependant error, sometimes attributed to membrane compression but more likely to be instrument compliance.

Despite the reports of other users (Schnaid 1990) the error is very small. However the procedure outlined in the preceding paragraph will show an apparently large compliance if the outer metal sheath (known as the 'Chinese Lantern') is not first removed. The sheath is made up of eighteen longitudinal strips of length 705 mm, width 7.5mm and 0.4mm thick. These are joined at the ends, so that the sheath looks like a cylinder but in reality is a succession of flat plates. Pressurising the fully assembled instrument inside a tube shows the strips curving to the shape of the container, not instrument compliance. This is illustrated in figure 5.

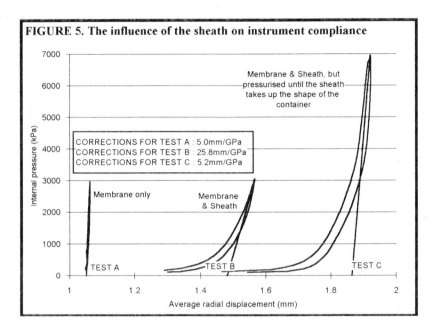

FIGURE 5. The influence of the sheath on instrument compliance

Here three compliance tests are shown, all on the same instrument and membrane. There has been some adjustment to the data to make the three tests lie within a 1mm wide displacement range, and the elastic slope of the cylinders used has been subtracted from the readings.

The normal operating pressure range of the CPM is from 0 to 3MPa. It happens that it has been designed to go to 10MPa, which is fortunate because it was necessary to go to 7MPa to demonstrate the effect of the metal sheath curving. As can be seen, once the sheath has taken up the shape of the tube (at about 4.5MPa) then the compliance values obtained correspond to the test in which the sheath was removed. The values are quoted in units of mm/GPa, and in this form a subtraction for compliance is implemented automatically by the data logging system.

The behaviour apparent in the cylinder will not occur in the ground. The soil will be yielding before pressures are applied that would force the metal sheath to change shape.

If the compliance values quoted here were converted to shear modulus parameters, 5mm/GPa represents a modulus of about 2.5GPa. 25mm/GPa, on the other hand, represents a modulus of about 500MPa, not very much greater than might be measured in dense sand deposits. The true compliance correction is almost negligible, the erroneous value if applied might have a significant and unsafe influence on the results.

In the SBP where a similar sheath is used the strips are pre-formed to the circumference of the instrument at rest. It is not easy to do this for the CPM, where the strips are both narrower

In the SBP where a similar sheath is used the strips are pre-formed to the circumference of the instrument at rest. It is not easy to do this for the CPM, where the strips are both narrower and thicker.

THE RESULTS OF FIELDWORK

There is relatively little field work carried out to date with the CPM that allows definitive comparison to be made with other pressuremeter tests. Where such evidence is available, it is even harder to obtain the measured test curve - often comparisons must be made with derived parameters, which is of little help when a conflict arises. Hence the tests presented here are valuable. They are from two sites, where CPM and SBP tests were carried out in adjacent boreholes. Site A is near Paddington Station in London, and the tests were carried out in 1991 as part of the CrossRail site investigation. The CPM tests were carried out by Fugro UK with the original 15CM² prototype instrument fitted with a dummy cone. The SBP tests were carried out by Cambridge Insitu on the same day.

Site B is the Madingley test site where the prototype CPM was first used (Houlsby & Withers 1986). The CPM tests were carried out by the Building Research Establishment (BRE) in January 1994. The purpose was to gain some experience in the handling and use of the instrument, which in this case was an improved version of the prototype tool incorporating the modifications previously outlined. A live cone with a friction sleeve was used. The SBP tests were again carried out by Cambridge Insitu, but in 1991 and 1992.

Figure 6 is typical . It shows a CPM test and a SBP test plotted on the same axes. For each instrument the initial radius has been used as the origin for the expansion of the borehole and logarithmic strain is plotted against corrected total pressure.

Between the two tests there is a general similarity in the slopes of the unload/reload loops and in the form of the final unloading. The comparison also reveals that, somewhat unexpectedly, the part of the CPM loading path between 0.02 and 0.10 strain will in this example give a plausible value for the undrained shear strength. The possibility that the volume of soil disturbed by the passage of the cone is comparatively small has been remarked by other users of the CPM (Campanella et al 1990). In all tests where a comparison can be made with the SBP this correspondence has been noted. In general, it appears that the radial stress is established as the principal stress at an early stage in the expansion, and a plateau pressure is reached at about 0.14 strain.

Some defects of the CPM tests are also apparent in this plot. The contraction was probably conducted at too fast a rate, so that the membrane lost contact with the borehole wall at about 250kPa on the final unload. In addition it is clear that the taking of an unload/reload loop becomes increasingly difficult as the radial stress approaches the plateau pressure.

For all pressuremeter undrained expansion tests attempting to reverse the direction of straining involves a compromise. Ideally before changing direction the pressure should be held at a constant level until the cavity stops expanding. However too long a pause will encourage consolidation. In the CPM test shown here, it is unlikely that holding the pressure constant before the second loop would of itself have stopped the expansion. The loop is carried out by reducing the internal pressure, and in this example the reduction is made large and is implemented rapidly in the hope that eventually the unwanted deformation, whatever its cause,

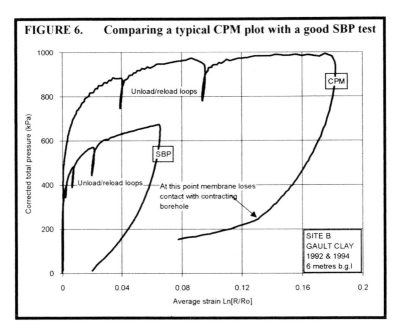

FIGURE 6. Comparing a typical CPM plot with a good SBP test

To demonstrate the extent to which the two tests are giving similar parameters for stiffness, seven reloading paths from a test at 3 metres in the Gault clay are plotted together in figure 7. Four of the loops were taken in the SBP test, and three in the CPM test. For each path the lowest displacement and pressure recorded set the origin. The extent to which all reloading paths plot the same curve is clear and persuasive. The plot can be used to determine a secant modulus by taking a chord from the origin to intersect the curves at any desired strain level.

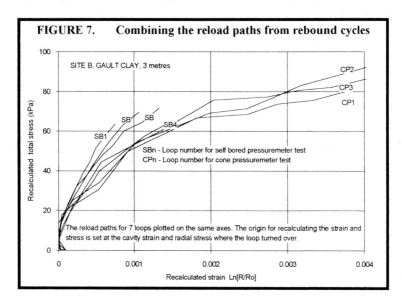

FIGURE 7. Combining the reload paths from rebound cycles

This technique has been suggested before (Reid et al 1982, Whittle et al 1993) and is used here to emphasise the repeatability of unload/reload loops both within a test and between instruments that are very different in terms of the disturbance they cause to the ground.

Not all CPM tests are so convincing. Figure 8 is an example of a test at 6 metres depth in London clay. The contraction curve shows the membrane almost snapping back at about400kPa, so that there is little data left to form the basis of an analysis. Furthermore, as figure 9 indicates, the test is in difficulties at a much earlier stage, so that an attempt to take an unload/reload loop at 15% strain fails. It appears that after a modest pressure reduction, less than was used to take the first unload/reload loop, the radial stress is no longer the major stress giving the loop an unmistakable kink at about 705kPa. However the first unload/reload loop in the CPM test appears to be reasonable and compares very well with the SBP test plotted below it.

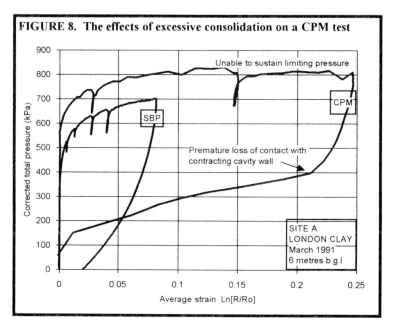

FIGURE 8. The effects of excessive consolidation on a CPM test

In this borehole only CPM tests down to 4 metres gave a full contraction curve. All other tests had a diminished contraction curve and appeared to be in distress after an expansion of about 10%. It is thought that this is partly due to the CPM tests being too closely spaced at 1 metre intervals. One metre is about the spacing of the tip of the cone to the centre of the expanding membrane. Hence whilst a CPM test was being carried out, some of the material which formed part of the next test zone was consolidating. This effect has been seen at other sites.

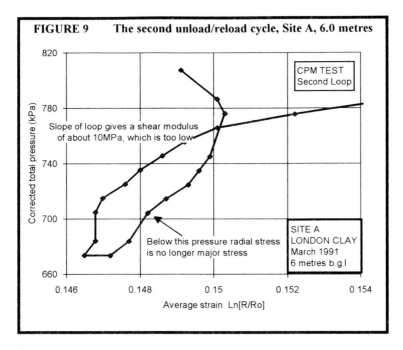

FIGURE 9   The second unload/reload cycle, Site A, 6.0 metres

Corrected total pressure (kPa)

820

780

Slope of loop gives a shear modulus
of about 10MPa, which is too low

740

700

660

CPM TEST
Second Loop

Below this pressure radial stress
is no longer major stress

SITE A
LONDON CLAY
March 1991
6 metres b.g.l

0.146   0.148   0.15   0.152   0.154

Average strain Ln[R/Ro]

CONCLUSIONS

• The minimum spacing between tests should be no less than 1.5 metres.

• Drainage will spoil the contraction curve, so it is important to commence tests as soon after the passage of the cone as possible. Excessive consolidation can result in a situation where it is not possible to take unload/reload loops.

• As a precaution, stiffness data from unload/reload loops should be gathered as early in the test as is feasible.

• There are similarities between the form of the early part of the CPM test curve and the SBP test curve, which implies that shear strength parameters could be derived from the CPM loading path.

• In a good CPM test, shear modulus parameters from unload/reload loops in CPM tests agree closely with those from SBP tests if the reloading path is used as a basis for comparison.

• The standard methods for calibrating the membrane for stiffness lead to over correction.

• The sheath should be removed before attempting to calibrate for instrument compliance. This reveals that instrument compliance is a much smaller problem than has been imagined.

• The conclusions concerning the calibration of the membrane also apply to other pressuremeters such as the SBP and high pressure dilatometer (HPD). The methods and physical arrangements used to calibrate these instruments must reflect the conditions that will be experienced in the ground.

ACKNOWLEDGEMENTS

None of the CPM tests presented here were carried out by Cambridge Insitu, so grateful thanks are due to Fugro-McClelland Ltd, Building Research Establishment and the CrossRail Project for making available cone pressuremeter test data.

REFERENCES

CAMPANELLA, R.G., HOWIE, J.A., SULLY, J.P., HERS, I. & ROBERTSON, P.K., *Evaluation of cone pressuremeter tests in soft cohesive soils* 1990.Proc.3rd Int.Symposium on Pressuremeters, pp. 125-136.

FAHEY, M. & JEWELL, R.J.,*Effect of pressuremeter compliance on measurement of shear modulus* 1990. Proc.3rd Int.Symposium on Pressuremeters, pp. 115-124.

HOULSBY, G.T. & WITHERS, N.J., *Analysis of cone pressuremeter test in clay* 1988. Geotechnique 38, No.4, pp. 575-587.

REID, W.M., FYFFE, S., ST. JOHN, H.D. & RIGDEN, W.J., *The push-in pressuremeter* 1982. Proc. Symposium on the Pressuremeter and its Marine Applications, pp 247-262.

SCHNAID, F., *A study of the cone-pressuremeter test in sand* 1990. DPhil thesis, Oxford University.

SCHNAID, F. & HOULSBY, G.T., *Calibration chamber tests of the cone-pressuremeter in sand* 1990. Proc.3rd Int.Symposium on Pressuremeters, pp. 263-272.

WHITTLE, R.W., DALTON, J.C.P. & HAWKINS, P.G., *Shear modulus and strain excursion in the pressuremeter test* 1992.Proc. Wroth Memorial Symposium, pp. 767-782.

WITHERS, N.J., HOWIE,J., HUGHES, J.M.O. & ROBERTSON, P.K.,*The Performance and Analysis of Cone Pressuremeter Tests in Sands* 1989. Geotechnique 39, No.3, pp. 433-454.

WITHERS, N.J., SCHAAP, L.H.J. & DALTON, J.C.P., *The Development of a Full Displacement Pressuremeter* 1986. Proc. 2nd Int.Symposium on the Pressuremeter and its Marine Applications, pp. 38-56.

Ménard pressuremeter test to foundation design — an integrated concept

Prof J NUYENS
Université Libre de Bruxelles, Belgium
F BARNOUD
Geotec, Dijon, France
M GAMBIN
Apageo-Segelm, Magny-les-Hameaux, France

SYNOPSIS

From 1958, Louis Ménard strove to formalise his new prebored pressuremeter test and concomitant foundation design method. This combined test-and-design approach, officially sanctioned by the French Ministry of Planning, Housing and Transport and more recently by the AFNOR Standards Association, has culminated in an integrated design approach through refinements developed by Ménard's disciples: SPAD test data logger, PRESSIO data reduction program deriving the usual Ménard pressuremeter parameters, and ISPRESS foundation design program.

Regarding pressuremeter test operation, the recent AFNOR Standard NF P94-110 (1991) is very similar to ASTM D4719-87, both following the early Ménard test specification. Regarding data reduction, the PRESSIO software is now in standard use with French state-run and approved bodies. Foundation design codes based on the Ménard approach are embodied in the standard specification for government contracts and have been officially adopted in the French building industry's DTU's, to be incorporated in a series of AFNOR Standards.

The data logger and software described below now make the test, data interpretation and foundation design procedures quick and simple.

GENERAL

The Ménard pressuremeter test has been used in France and French-speaking countries since the late nineteen-fifties. Its inventor tried to build up a design philosophy around his instrument, as follows.

Testing was to follow a standard procedure for drilling the borehole, keeping soil disturbance to a minimum, under stress-controlled conditions. An E modulus was obtained from a G modulus at microplastic deformations. The limit pressure was taken as the pressure at which probe volume is twice the original cavity volume (for comparison with footing bearing capacity, taken as the stress at which footing settlement is one-tenth of footing width or diameter).

Foundation design is based on assumed elastic-plastic soil behaviour, and correlation between soil behaviour (i) around the pressuremeter in the borehole and (ii) under a shallow footing or around a pile tip.

Ménard's first rules appeared in 1962-1965 for calculation of pile and footing settlement, pile deformation under lateral loading, and footing and pile bearing capacity [1]. The foundations of hundreds of thousands of high-rise and other buildings, bridges, harbour works, etc. were successfully designed to these rules with significant cost savings as compared to more academic methods based on laboratory data. In 1971, the Ménard pressuremeter test was the subject of French Ministry of Planning Standard MS-IS-2, later issued as AFNOR Standard P94-110.

Subsequently, two engineering departments of the French Ministry of Planning, the *Laboratoire Central des Ponts et Chaussées* and *Service des Etudes Techniques des Routes et Autoroutes*, adopted Ménard's rules in their 1972 foundation design manual FOND 72. More recently, a new specification referring to hundreds of pile load tests with strain gauges at various depths, and long-term records of experimental footing performance [2] has been issued as Section 62-V of the standard specification for public works contracts [3], which has been reviewed in English [4].

The French federation of building industry contractors also adopted the Ménard rules in their unified technical documents (DTU) series, to be incorporated into AFNOR Standards P94-261-1, P94-262-1, P94-271-1 & 2, P94-273-1. At the present time, AFNOR P11-212 (1992) makes the link between the DTU's and the standards under preparation.

For those familiar with these Standards, the next step had to be automation and computerisation.

THE INTEGRATED CONCEPT IN ITS FINAL STAGE
Computers have brought great improvements to every stage of the process. During the test, pressure and volume are continually recorded, including pressure lag between the measuring cell and guard cells (to compensate for hydrostatic pressure and expansion resistance of central cell membrane). A computer reduces the data and derives conventional Ménard pressuremeter parameters (E modulus and limit pressure) with allowance for volume and pressure losses (system compliance) during the test. Lastly, Ménard's original rules can easily be turned into software instructions [5].

DATA LOGGER
The SPAD data logger (Fig.1) records pressure and volume at the level of the Ménard pressuremeter monitoring box. It does not interfere with the running of the test, unlike the computer-aided pressuremeter which was proposed in the early 80's but enjoyed little success.

SPAD can record pressures up to 10 MPa to within 10 kPa, which is much better than what can be achieved with a Bourdon gauge, even for the smallest range 0-2 MPa. Accuracy on volume readings is 1 cm³, better than the operator can read by eye. It incorporates a clock and timer, for automatic recording of data at 15, 30 and 60 seconds at each pressure level. It is powered from a 12 or 24 volt car battery, weighs about 10kg, and can operate between -20°C and +70°C. It can be connected to any Ménard pressuremeter box having two transducers built into it.

SPAD improves the quality of readings, with greater pressure, volume and time accuracy. It is not surprising that plotting SPAD-acquired data yields much smoother pressuremeter curves.

Once all the physical data for the borehole, probe type and test site depth has been entered, the program advises the operator of the expected pressure lag between central and guard cell circuits at the elevation of the monitoring box and simultaneously displays the actual pressure lag. The operator can then make the two values coincide.

He is then ready to apply the first pressure increment. Once the set pressure is reached, he presses the timer button to start recordings at this pressure level. He can compare the values displayed on the screen with readings from the pressure gauges and volumeter sight tube. He can call up theoretical and actual pressure lags for display at any time. SPAD also stores the calibration tests needed for measuring system compliance (pressure loss and volume loss calibrations).

On completion of a test, the built-in thermal printer lists the main pressure and volume readings and plots the raw-data pressuremeter curve. This is very helpful for the operator preparing the next test and for the supervising geotechnical engineer.

A solid-state memory card can be inserted for transferring readings to a PC with a special card reader. The card is produced by the data logger manufacturer and is compatible with a variety of models (e.g. Explofor drilling data logger). It can store results from up to 700 pressuremeter tests.

PRESSIO SOFTWARE
Pressuremeter test data is processed by the PRESSIO software after being entered into the computer from the memory card (or via the keyboard for pre-computerised tests). The program reduces the data using calibration tests for system compliance and derives the typical Ménard pressuremeter data: Ménard E modulus, Ménard limit pressure, and Ménard creep pressure.

The process is fully automatic when working to French AFNOR Standard NF P94-110. With ASTM D4719-87, the pressuremeter curve must be displayed on screen to determine the so-called pseudo-elastic range.

Each pressuremeter test can be printed either as tabulated data or pressuremeter and creep curves (Fig.2) in black and white, or colour with HPGL printers and plotters.

Twenty-five tests per hour can be processed in automatic mode (French Standard). Test data can be stored in special files to produce borehole logs, which can be supplemented with data from other sources, such as Explofor drilling data (Fig.3), via the keyboard.

Efficient and user-friendly, the PRESSIO software yields documents of high quality. Company logos, letterheads and graphic formats can be entered, horizontal axes can be linear or log scale, the vertical scale can be varied, column size and numbers can be stipulated, and graphic symbols can be added to soil descriptions if required.

ISPRESS SOFTWARE
ISPRESS shareware software is a preliminary design code using digital data from the SPAD data logger processed by the PRESSIO software. The data input format conforms to the earlier ISSMAFE integrated system [6]. The general parameters (number, name, x, y and z coordinates, water level, date of boring) and geotechnical parameters (limit pressure and Ménard modulus for each depth) are entered in the data input file (*.ini). This can also be done separately, starting from an ASCII format, if necessary. After this, ISSMAFE general

procedures allow the user to create a layer profile file (*.trt), needed for the calculation. These procedures are also available with ISPRESS. Both programs work under MS-DOS but may also be called from Windows.

Working directories must be created and foundation parameters are entered in a foundation file (*.fon). The user is supposed to enter the characteristics of either a shallow or deep foundation, but since the program produces values for any depth, two sets of parameters are simultaneously calculated, one where the foundation is considered as shallow, one where it is considered deep. It is important to use only the values relevant to the shallow or deep assumption for a given foundation. For example, a shallow foundation will be defined by the main excavation line around the footing, the depth of the competent layer, footing shape, width and length, load eccentricity and angle, and distance from any deeper trenches and pits. A deep foundation will be defined by the depth of the competent layer, shape, width, length and (if enlarged) height of the bulb, width and length of the shaft, type, depth, unit weight of the pile, and range and increment of the possible pressure at the pile base for settlement calculation. However, all soil parameters for both assumptions must be defined to sufficient depth, using default values if necessary.

The calculation may proceed according to Fascicule 62-V of the French Ministry of Planning design rules [3]. When the user selects the *bearing capacity* option, the program asks for the number of the test location and displays values of *limit pressure pl**, *equivalent limit pressure ple**, *allowable pressure (qa)*, *equivalent depth (he)* and *settlement (w)* of a shallow foundation, or the *limit load (Qpl)* at the base, *limit load (Qsl)* along the shaft, *yield load (Qc)* and *equivalent depth (he)* of a single pile. All these values are referred to the foundation depth and are presented as curves (Figs.5, 6) with numerical values along the bottom applying to the position of the horizontal cursor line.

It should be emphasised that results are presented continuously and give values for any level of the base of the foundation, independently of foundation type (deep or shallow). Only part of the graph is of course meaningful for each foundation file. For example, settlement (*wsup*) is only relevant for footings. Settlement of a single pile is calculated with the second option. All values at depth may be disregarded for a 'shallow' foundation.

Some parameters such as *ple** cannot be calculated close to the bottom of the borehole, leading to lack of data at greater depth. This problem arises from the design method and would be the same with hand calculation. It is necessary to add *assumed values* at depths below borehole final depth at the time of the initial file preparation.

The interactive graphic screen allows the user to move the cursor line until the optimum level is found, the foundation parameters must then be amended in the *.fon file and a new calculation and cursor search performed.

When the settlement option is selected, the program displays a loading curve with pile head settlement plotted against pile head load. This calculation follows the models proposed by Frank & Zhao [7, 8]. All the numerical results are stored in temporary ASCII files which can be used for reports.

Regarding other more elaborate design problems such as laterally-loaded piles, diaphragm walls, etc., the Ministry of Planning's Central Laboratory has already developed specific software available to the public [9].

WORKED EXAMPLE

Figure 4 refers to a pressuremeter borehole (F12) at a bridge abutment site on a motorway in the French Vosges area. It was decided to use bored piles down to marl. Pile diameter had to be 0.8m, 0.9m, 1.0m or 1.2m with a maximum concrete compressive stress of 5 MPa. This represents the maximum loads shown in Table 1.

Table 1

Boring F12, Max Concrete Pressure 5 MPa			
Diameter (m)	Load (kN)	Pile Depth (m)	Settlement (mm)
0.8	2513	26.88	2.85
0.9	3181	27.60	3.29
1.0	3927	28.05	3.59
1.2	5655	28.64	4.43

Examples of the corresponding calculations by ISPRESS are shown in Figures 5 and 6. The three main columns in Figure 5 show:
First column
- solid line: limit pressure in MPa at each test depth
- dotted line: equivalent limit pressure in MPa for footing option (size in working directory)
Second column
- solid line: allowable bearing pressure in MPa for footing
- dotted line: yield load in kN for pile option (size in working directory)
Third column
- solid line: equivalent depth in metres for pile option
- dotted line: settlement in millimetres for footing option.

Note that the horizontal cursor line across the screen is located at the chosen depth for the pile option (see Table 1).

In Figure 6, note that Q_c (yield load) vs depth shows a similar trend as in Figure 5 but the load scale (this time in MN) is much smaller.

The horizontal cursor lines in Figures 5 and 6 coincide with the load values from Table 1. Pile depth can be selected from this information. A typical settlement vs load curve is shown in Figure 7.

CONCLUSION

The goals of Louis Ménard, expressed in the mid-sixties, have now been achieved:
i) The pressuremeter test data can now be conveniently logged.
ii) Test data can be reduced and Ménard parameters derived to existing Standards, using the PRESSIO software.
iii) Foundation design proceeds from test parameters used with ISPRESS software.

REFERENCES

[1]   GAMBIN, M, *The History of Pressuremeter Practice in France*. Pressuremeters, BGS Symposium (ISP3), Thomas Telford, London 1990
[2]   FRANK, R, *Some Recent Developments on the Behaviour of Shallow Foundations*. Proc. X ECSMFE, Vol. 4, Florence 1991
[3]   Ministère de l'Equipement, du Logement et des Transports, *Règles Techniques de Conception et de Calcul des Fondations des Ouvrages de Génie Civil, Cahier des*

Clauses Techniques Générales applicables aux Marchés Publics de Travaux, Fascicule 62, Titre V, Textes Officiels No.93-3 T.O., (182 pages, 1993

[4]   FRANK, R, *The New Code of Practice for Foundation Design of the French Highways Authority,* Proc. Novel Foundation Techniques, Prog. for Industry, Cambridge 1993

[5]   ISSMFE European Regional Committee, *The Application of Pressuremeter Test Results to Foundation Design in Europe,* A A Balkema, Rotterdam 1991

[6]   NUYENS, J, *Integrated System in Soil Mechanics and Foundation Engineering,* ISSMAFE, A A Balkema, Rotterdam, 1989

[7]   FRANK, R, & ZHAO, S R, *Estimation par les Paramètres Pressiométriques de l'Enfoncement sous Charge Axiale de Pieux Forés dans des Sols Fins,* Bull Lab. des Ponts et Chaussées, 119, 17-24, Paris 1982

[8]   FRANK, R, *Etudes Théoriques des Fondations Profondes et d'Essais en Place par Autoforage dans les LPC et Résultats Pratiques, 1972-1983,* LPC Research Report 128, Paris 1984

[9]   M.E.L.T, *Catalogue des Logiciels des LPC,* Paris 1994

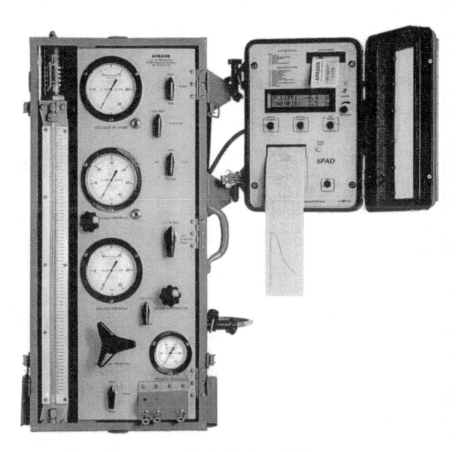

Figure 1 - Menard Pressuremeter Monitoring Box with SPAD Data Logger. Note memory card protruding from its slot and tabulated and plotted data on heat-sensitive paper

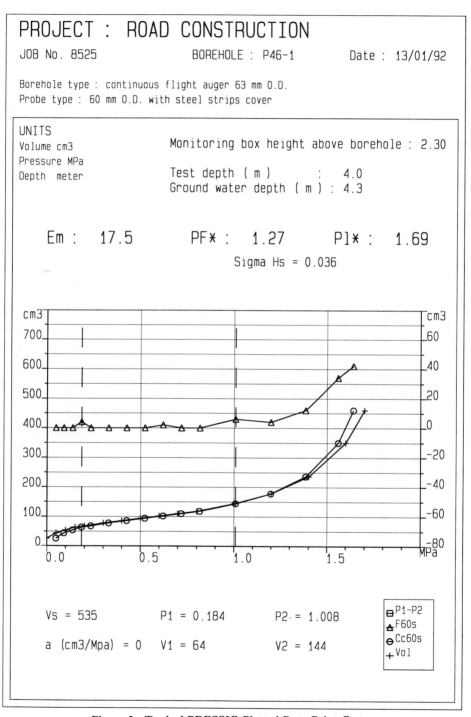

Figure 2 - Typical PRESSIO Plotted Data Print-Out

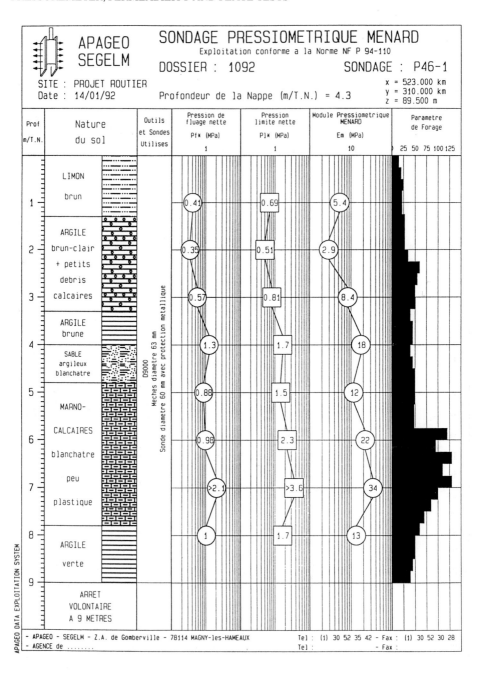

Figure 3 - Typical PRESSIO Log

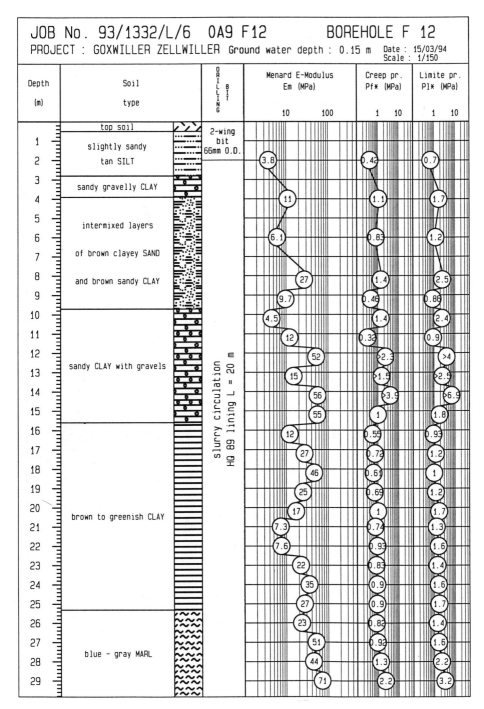

Figure 4 - Worked Example. Boring F12. Note log scale for Ménard E modulus, yield pressure and limit pressure

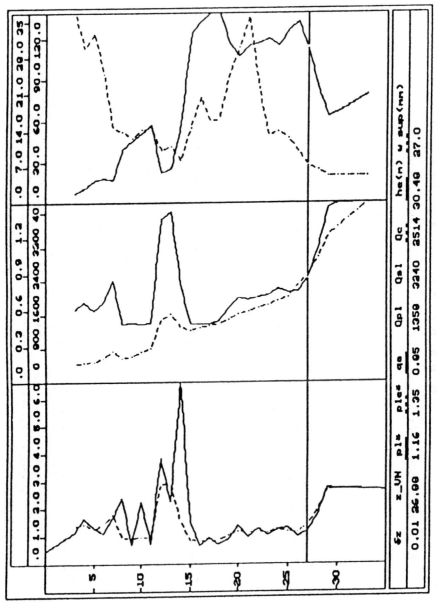

Figure 5 - Worked Example. PC Screen Display for 0.8m dia. Pile Option

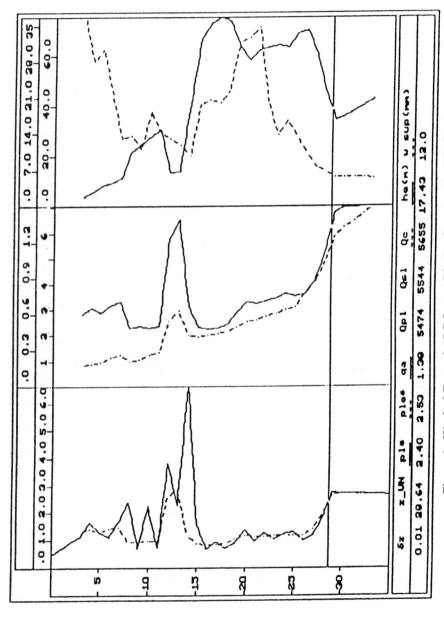

Figure 6 – Worked Example. PC Screen Display for 1.2m dia. Pile Option

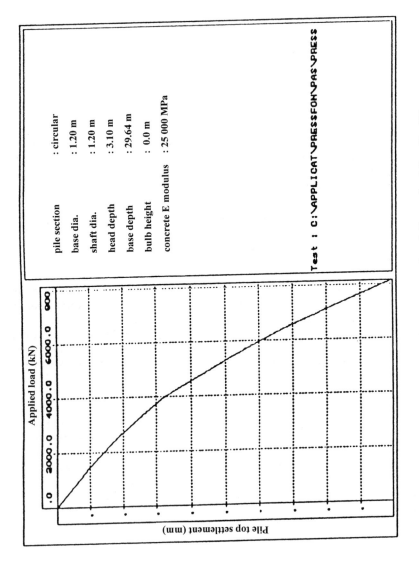

Figure 7 – Worked Example. Settlement vs Load Curve for 1.2m dia. Pile

A practical guide to the derivation of undrained sheer strength from pressuremeter tests

B. G. CLARKE and J. A. SADEEQ
University of Newcastle upon Tyne, UK

INTRODUCTION

The expansion curve obtained from an ideal pressuremeter test in clay can be interpreted to give in situ stress, stiffness and strength. In practice, the values obtained are a function of the type of pressuremeter (and hence installation technique), the quality of the test procedure and the model chosen to interpret the data.

The in situ stress can be determined directly from a self-boring pressuremeter (SBP) test curve using the lift off method (Wroth, 1982) but only by empirical or secondary correlations from prebored (PBP) (Marsland and Randolph, 1977) and full displacement or push in (PIP) pressuremeter tests (Houlsby and Nutt, 1993).

Shear modulus, usually taken directly from an unload-reload cycle using the theory of elasticity, represents an average stiffness of the ground provided the cycle is carried out on ground undisturbed by installation of the pressuremeter. The initial ground response represents the behaviour of ground affected by installation and is often neglected unless, as in the Ménard method, it is used to produce design parameters. Bellotti et al (1989) and Jardine (1991) have shown that the average modulus from an unload-reload cycle can be converted to an elemental modulus allowing direct comparisons with results from other tests, for example triaxial tests.

There are three methods used to obtain shear strength of clays. Amar and Jézéquel (1972) and others proposed that shear strength is a function of the modified net limit pressure and soil type where the modified limit pressure is the difference between the overburden pressure and the pressure required to expand a prebored pressuremeter cavity by 41% or a self-boring pressuremeter by 20%. Palmer (1972) and others developed a method to produce a stress-strain curve that is based on the assumption that undrained conditions apply. Others (eg Gibson and Anderson, 1961, Denby and Clough, 1980, Prévost, 1979, Jefferies, 1988, Houlsby and Withers, 1988 and Ferreira and Robertson, 1992) have developed expressions for the pressuremeter test curve using a number of models including perfectly plastic and hyperbolic soil models.

Results from pressuremeter and other tests are compared to justify the test, gain confidence in the results or develop profiles with depth. Most studies of pressuremeter tests have focused on shear modulus and in situ stress and have concluded that good SBP quality tests can be interpreted to give in situ horizontal stress and stiffness which represents the true ground response.

Profiles of strength often vary with depth and differ between test types, leading to general conclusions such as;- the soil is variable; undrained shear strengths from pressuremeter tests in stiff clay are the same as those from triaxial tests, whereas for soft clay pressuremeter test results are greater; and, the pressuremeter results do not represent the soil response.

Clarke (1994) has shown that the scatter in pressuremeter shear strength profiles may be due to natural variations, the length and shape of the expanding section, the installation technique and the strain range over which the strength is measured. The stress path followed in a pressuremeter test differs from that followed in other tests. Results from all types of tests are affected by the method of installation or sampling used prior to testing. For these reasons the results from different tests will not necessarily be the same.

In this paper the techniques used to derive undrained shear strengths from pressuremeter tests in clays are reviewed, together with corrections to allow for the length and shape of the expanding section. A practical method is proposed whereby consistent values of strength should be obtained thus allowing a true assessment of the natural variations in strength.

EFFECTS OF INSTALLATION AND TEST PROCEDURE
The interpretation of a pressuremeter test in clay to obtain a value of strength is based on the assumption that the test is carried out in a homogeneous deposit which is unaffected by installation; that is the pressuremeter is "wished" into place thus the test could be described as an ideal test. This does not happen; preboring allows stress relief and softening, self-boring induces shear, pushing in causes a cavity expansion. The installation process will change the properties of the clay adjacent to the probe.

Very few clays are homogeneous. The deposition and ageing process produces a fabric, including discontinuities, which will affect the response of the clay (Burland, 1990). Further, during a test if the circumferential stress at the cavity wall reduces to the tensile strength of the clay, cracks will propagate from the cavity wall. This can occur during tests at shallow depths and in stiff clays.

Many soils are layered and since these layers having different stiffnesses the expansion of the cavity will be non uniform. Volume displacement type probes will give an average response, whereas radial displacement type probes will give the response of the clay at the level of the displacement transducers. This will be further complicated if that layer is soft since the response will include a component of shear between the harder layers above and below.

The time between creating a test pocket and start of testing can affect the results obtained if any pore pressures are developed during installation. Negative pore pressures due to stress relief caused by preboring lead to an increasingly softer zone around the pocket. This will always occur but keeping the time to a minimum will reduce the effect. Mair and Wood (1987) recommend that a PBP probe should be installed as soon as possible after the pocket is created. Clarke and Smith (1992) suggest 15 min. is sufficient time for most tests within a twenty metre deep borehole.

Pore pressures generated during installation of a PIP will either be negative or positive depending on the over consolidation ratio of the clay. It is normal practice to start tests immediately the test depth is reached to prevent excess pore pressures dissipating. Thus, properties of the ground around the probe will be the same for all tests in that clay deposit; that is, the amount of disturbance is consistent.

Windle and Wroth (1977) showed that pore pressures are generated during the installation of a SBP but, for practical purposes, these dissipate within 30 minutes. It has been common practice to wait 30 minutes after installing the probe before testing to allow any excess pore pressures time to dissipate.

The test procedure will affect the strength since results are a function of strain rate which in turn influences the effects of creep and drainage. Drainage will occur during a test, though Baguelin et al (1986) and Clarke (1990) have shown that this will have little effect on the undrained shear strength since the effects of consolidation and increasing strength cancel each other out. Consolidation results in further expansion, but consolidation causes an increase in strength, therefore the applied pressure must be increased to cause further expansion.

Further consolidation occurs during an unload-reload cycle resulting in an increase in ground stiffness. This can be avoided by carrying out a cycle prior to yield or toward the end of loading. It is normal to carry out a cycle at 1% cavity expansion for SBP tests since yield occurs at between 0.5% and 1.5% in most clays. Additional cycles or cycles at greater strain levels will affect the strength obtained. Cycles in PBP and PIP tests can be at any strain level since strengths are obtained from empirical correlations or the unloading curve.

Table 1   Summary of commonly used test procedures to measure ground response

Test Procedure	Type	Probe	Ground Conditions	Stress Rate	Strain Rate
Menard	stress	MPM other PBPs with max strain of 55%	all	$p_{lm}/10$	
ISRM	stress	PBP	rock	not specified	
ASTM	stress	PBP	all	25 - 200 kN/m^2	
ASTM	strain	PBP	all		0.05 to 0.1 x V_o
GOST	stress	PBP	soils	25 kN/m^2	
Stress	stress	all	all		varies
Strain	strain	PBP, SBP	all		1%/min
		PIP	soils		5%/min

Stress controlled tests, in which the strain rate varies, give different strengths from strain controlled tests and varying the stress or strain rate will vary the ground response (Anderson and Pyrah, 1986). Table 1 gives a summary of commonly used loading rates which shows that the tests vary according to the probe used and the specification followed. Strain controlled tests are either at 1%/min for SBP and PBP tests or 5%/min for PIP tests. The rate used in stress controlled tests is a function of the maximum pressure required. The strain rate will vary in a stress controlled test and, in one borehole, the strain rate will vary with depth of test because the maximum pressure required will most probably increase as the test depth increases. It is for this reason that tests designed to obtain ground properties should be strain controlled and the rate of strain should be fixed possibly to 1%/min though a maximum rate of increase of stress should be set to prevent instability prior to yield and ensure there is sufficient data to define the initial portion of the curve.

THE INFLUENCE OF THE METHOD OF INTERPRETATION

The installation of any form of pressuremeter into a test pocket in clay inevitably disturbs the clay surrounding the pocket, the amount of disturbance being a function of the installation technique. Clarke (1994) suggested if the latter part of a PBP or SBP test curve is used and it is also the initial loading of undisturbed ground, the result obtained using cavity expansion theory represents a post peak strength. In general, however, the strength from a PBP test is obtained from empirical correlations with a modified limit pressure, p_{lm}. There is no unique correlation as shown in Table 2 and Figure 1. Thus, at best, the strength obtained from empirical correlations can only be an estimate since for a given value of modified limit pressure the undrained shear strength can be between 0.69 and 1.32 of the average value of all the correlations. The equations in Table 2 could be used for SBP tests since Baguelin et al (1978) suggested that the applied pressure at 20% cavity strain in an SBP test is the same as the modified limit pressure from a PBP test.

Table 2   Empirical relations between undrained shear strength and net limit pressure

s_u	Clay Type	Reference
$(p_{lm} - \sigma_h)/k$	k = 2 to 5	Ménard (1957)
$(p_{lm} - \sigma_h)/10 + 25$		Amar and Jézéquel (1972)
$(p_{lm} - \sigma_h)/5.5$	soft to firm clays	Amar and Jézéquel (1975)
$(p_{lm} - \sigma_h)/8$	firm to stiff clays	
$(p_{lm} - \sigma_h)/15$	stiff to very stiff clays	
$(p_{lm} - \sigma_h)/6.8$	stiff clay	Marsland and Randolph (1977)
$(p_{lm} - \sigma_h)/5.1$	all clays	Lukas and LeClerc de Bussy (1976)
$(p_{lm} - \sigma_h)/10$	stiff clays	Martin and Drahos (1986)
$p_{lm}/10 + 25$	soft and stiff clay	Johnson (1986)

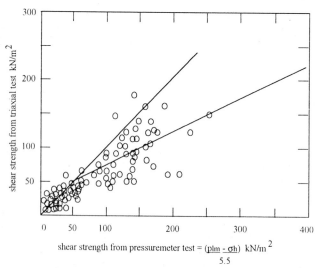

Figure 1   A correlation between modified limit pressure and undrained shear strength for clays with classification ranging between CL and CH (after Amar et al, 1975)

Push in pressuremeters cause an expansion during installation which, theoretically, implies that the pressure on the probe is equal to the limit pressure. In practice, there is some unloading due to the geometry of the probe but during a test the limit pressure is rapidly reached such that strength is taken from the unloading curve (Houlsby and Withers, 1988). Alternatively, the maximum pressure reached could be related to results from other tests to produce an empirical correlation similar to those given in Table 2. This device is relatively new in ground investigation practice and as yet there are insufficient tests to develop confidence in the results but, as with all tests, this will come with practice.

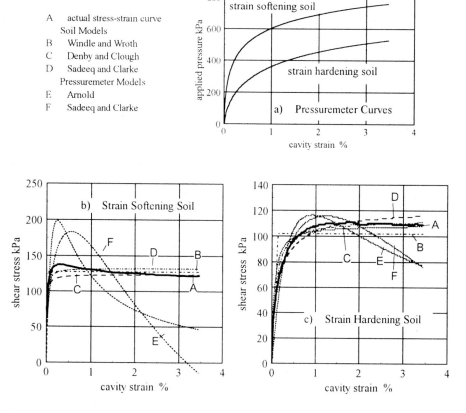

A	actual stress-strain curve
	Soil Models
B	Windle and Wroth
C	Denby and Clough
D	Sadeeq and Clarke
	Pressuremeter Models
E	Arnold
F	Sadeeq and Clarke

Figure 2   A comparison between the results of different mathematical models used to fit curves to theoretically derived pressuremeter curves; (a) pressuremeter curve, (b) strain softening soil and (c) strain hardening soil

A curve produced during a SBP test in clay is often likened to an ideal expansion of a cylindrical cavity and therefore is suited to rigorous analysis. Palmer (1972) and others have shown that a test can be interpreted directly to give a stress-strain curve assuming the clay behaviour is undrained but Baguelin et al (1978) suggested that the shear stress strain curve, in particular the peak stress, is a function of the amount of disturbance and therefore this method has to be treated with caution. This method is also very sensitive to fluctuations in data as shown in Figure 3. Note that a small amount of installation disturbance only affects the clay adjacent to the probe thus the shape of a test curve would still look "correct". It is for these reasons that it is more usual to use some form of curve fitting or computer aided modelling (CAM).

There are two approaches to curve fitting; the first is to fit a curve to the test data, assume an undrained expansion and analyse the test using Palmer's method. The second is to assume a stress-strain model, integrate it to produce a curve which is then adjusted to fit the recorded data. Table 3 is a summary of the mathematical models proposed for the analysis of ideal pressuremeter tests. They take several forms; loading and unloading linear elastic perfectly plastic, loading and unloading hyperbolic, and parabolic stress strain curves and hyperbolic and parabolic pressuremeter curves. The constants used in the models are either in situ stress, initial tangent or secant shear modulus and strength or are related to those properties.

Figure 2 shows the results of applying the equations for loading given in Table 3 to model tests on normally and over consolidated reconstituted London Clay. The curves were generated using Microsoft Excel Solver, an optimisation routine based on numeric methods. The variables given in Table 3 are adjusted to obtain the best fit to the pressuremeter test data and those variables are used to derived the stress strain curves. There are either two or three variables since the horizontal stress is zero. A number of combinations of those variables will satisfy the best fit criteria therefore constraints based on a realistic initial assessment of the soil properties are used to limit that number. Details of the method used and the constraints imposed are given by Clarke (1995).

Table 4   Results of computer aided modelling on pressuremeter test curves generated from triaxial tests and measured in the field

Method			harden		soften		field		
			G MPa	S_u kPa	G MPa	S_u kPa	σ_h kPa	G MPa	S_u kPa
	post rupture strength			110		122			
B	Windle and Wroth	all	40.0	102	27.7	128	1118	26.9	120
		>1%	31.0	113	27.7	128	1200	16.2	116
C	Jefferies						1085	66.0	105
D	B + C	all					1056	55.0	117
		> 1%					1183	61.5	85
E	Denby and Clough	all	77.3	113	60.6	133	1200	32.4	103
		> 1%	74.9	114	60.6	133	1200	69.9	92
F	Ferriera and Robertson						1200	14.8	84
G	E + F	all					1075	94.2	115
		> 1%					1162	11.7	94

The quality of the fit is such that it is difficult to distinguish between the pressuremeter curves but the interpreted stress-strain curves do differ. The difference in shear moduli arises because they are either initial loading secant stiffness, initial unloading secant stiffness or initial loading and unloading tangent stiffness. They are a consequence of the model used and may not represent the stiffness of the soil even though they could be realistic estimates. Note that the fit is improved if the latter part of the pressuremeter curve is used to such an extent that the fit gives very similar values of shear strength, comparable to the actual strength as shown in Table 4. This conforms with the authors' views but the reason is different. They claimed that the lack of fit was due to installation disturbance or strain softening whereas in practice this is only part of the reason since a single model cannot be used to model the whole curve because there is an inconsistency between post and pre yield behaviour.

The results show that all the stress strain models tend to over predict the post rupture strength of an over consolidated soil since they do not include an element of strain softening. All models produce a similar post rupture or average strength if the initial portion of the pressuremeter curve is neglected. The hyperbolic and parabolic pressuremeter curves do not yield sensible stress-strain curves in this ideal case.

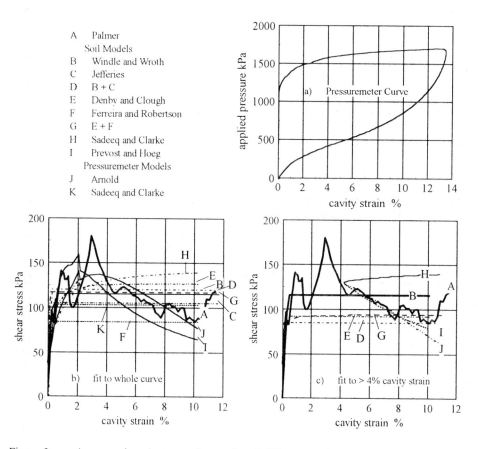

Figure 3   A comparison between the results of different mathematical models used to fit curves to a quality pressuremeter test curve (unload-reload cycles have been removed); (a) pressuremeter curve, (b) fit to whole curve and (c) fit to curve > 4% cavity strain

Figure 3 shows the results of applying all the equations in Table 3 to a quality test in stiff clay which includes a complete unloading curve. The peak strengths shown on the integrated curve (A) arise because of unload-reload cycles. One extra variable is required to fit the data that is the horizontal stress. The properties obtained from this study have not been interpreted with respect to the geological history since they are only being used to demonstrate the techniques. All the models predict an average strength since they do not include an element of strain softening. Figures 2 and 3 show clearly that all strain hardening models give a maximum strength which is the average of the integrated curve (A) but, if only the latter part of the loading and the unloading curves are used, the models, except for the loading perfectly plastic and parabolic models, give a strength equal to the minimum strength at the maximum cavity strain. The loading perfectly plastic model gives an average over the strain range used in the calculations. The parabolic pressuremeter model gives a better fit to the integrated stress strain curve though if the latter part of the curve is used both pressuremeter models give a reasonable fit.

It can be concluded that, the simple, perfectly plastic model gives a realistic estimate of undrained shear strength provided the model is only applied to the latter part of the test curve. Unloading models tend to underestimate the strength. Fitting a parabolic curve to the data and interpreting it using Palmer's method produces the closest fit to a direct interpretation of the pressuremeter test curve suggesting that there is potential development of this method. The conclusions are based on a modelling process which is constrained but there is no independent validation of the parameters obtained, that is they may not be representative of the ground properties.

EFFECTS OF STRAIN LEVEL AND RANGE

It is possible to derive a realistic shear stress-strain curve from a good quality test but the peak strength may be a function of the installation disturbance and not the property of the clay. A design engineer normally requires a single value of strength, usually the post rupture value (Burland, 1990) and the discussion above shows that this would be derived from the latter part of the test curve using a simple rigid model. This has the advantage that errors inherent in the instrumentation are averaged out.

Figure 3 and Table 4 confirm that the average undrained shear strength derived from a pressuremeter test curve will depend on the range of strain over which it is derived. This is a consequence of installation disturbance, strain hardening or softening, possible cracking and non-cylindrical expansion. Clarke (1994a) found that consistent results are obtained using the simple rigid model from tests in London Clay if the lower bound to the cavity strain is taken as 6% and suggested that the maximum cavity strain should be at least 10% for SBPs and 40% for PBPs and PIPs and for radial displacement type probes in which displacement is measured on more than one plane. The minimum cavity strain for estimating strength should be 6% for SBPs and 30% for other probes.

EFFECTS OF LENGTH AND SHAPE OF EXPANDING SECTION

It is generally assumed, if the length to diameter ratio of the expanding section is at least six, the expansion can be modelled as the expansion of an infinitely long cylinder. Borsetto et al (1983) investigated length to diameter ratios of 2, 4 and 6 and showed the interpreted value of strength reduced as the ratio increased. Houlsby and Carter (1993) undertook a numerical study to show that the interpreted shear strength could be up to 43% greater than the actual because of non-cylindrical expansion if the expansion of a probe with length to diameter ratio of six was measured at the centre. Yeung and Carter (1990) suggested that, for practical purposes, the correct value of undrained shear strength is three quarters of the interpreted value because of restraints imposed at the ends of the test section.

Clarke (1994a) used the work of Houlsby and Carter to modify the pressuremeter curve to take into account non-cylindrical expansion. This was based on the linear elastic perfectly plastic model such that the pressuremeter curve is given by:

$$p = \sigma_h + s_u \left[1 + \ln\left[(G\,\Delta V)/(s_u\,V) + 0.025\,(G\,\Delta V)/(s_u\,V)^2 \right] \right] \qquad (1)$$

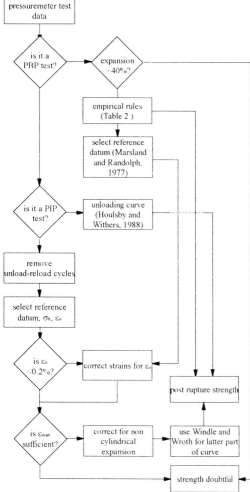

The second term in the bracket represents the effect of non-cylindrical expansion. An estimate is made of G from an unload-reload cycle measuring the secant over a stress range equal to s_u which is taken from the latter part of the curve using a perfectly plastic model. The correction is insensitive to the ratio (G/s_u). A plot of p-$\ln[(G\,\Delta V)/(s_u\,V) + 0.025\,(G\,\Delta V)/(s_u\,V)^2]$ is very nearly linear for a perfectly plastic soil.

RECOMMENDED PRACTICE
Site Operations
There is no international standard for pressuremeter testing but there are a number of national standards and recommended procedures including LCPC, ASTM, GOST, and Clarke and Smith (1992). They cover the probe, calibrations, installation and test procedure and if adhered to will produce consistent test curves.

The engineer can specify the probe, number, type and frequency of calibrations and test procedure. Further guidelines are given by Clarke (1994b). The operator controls the installation procedure and, since results depend upon this, the operator must have experience of creating test pockets and using pressuremeters. Every effort must be made to ensure that there is minimum or repeatable disturbance to the clay during installation.

Figure 4   A procedure for interpreting pressuremeter test results to obtain undrained sear strength

Method of Interpretation

latter part of the loading curve or the unloading curve. Peak strengths can only be obtained from good quality SBP tests, though they must be treated with caution because of their sensitivity to any disturbance.

A strength can be obtained from the latter part of a PBP curve using a simple linear elastic perfectly plastic model. It is necessary to choose the reference datum to apply this. This can be selected either using Marsland and Randolph's (1977) iterative procedure linking the reference datum to the strength or inspection of the curve to select a suitable point representing the in situ stress or assuming an in situ stress based on other information from the site. The derived strength is not too sensitive to the chosen reference datum provided the expansion is sufficient to test soil undisturbed by installation. Typically, if the pocket diameter is less than 10% greater than the probe diameter, then 40% expansion should be sufficient. Alternatively, the strength could be taken from Figure 1 using the modified limit pressure and estimated horizontal stress.

A PIP test should include a complete unloading curve after it has been expanded by 50%. The shear strength can be derived from the linear portion of the unloading curve plotted to a logarithmic scale as described by Houlsby and Withers (1988).

The maximum cavity strain in a quality SBP test should be at least 10% with no unload-reload cycles between 2 and 10%. If cycles are included the maximum cavity strain should be increased. The post rupture strength should be taken from data at strains greater than 6% corrected for non cylindrical expansion using Equation 1.

CONCLUSIONS

To obtain consistent results from pressuremeter tests which represent the ground conditions it is necessary to follow a standard procedure using operators who can demonstrate experience with pressuremeters. A number of specifications exist which fall into three groups; the Menard stress controlled method, SBP and PBP strain controlled tests and PIP strain controlled tests.

There are two commonly used methods for the interpretation of pressuremeter data and these are the empirical method based on a modified limit pressure (PBP and SBP) and a theoretical method based on cavity expansion theory. A number of models for either the soil or the pressuremeter curve exist but the simple perfectly plastic model for the loading cure gives a strength which represents the average of the integrated curve and is the post rupture strength of the soil if it is taken from the latter part of the loading curve.

REFERENCES

AMAR, S and JÉZÉQUEL, F.J. (1972) Essais en place et en laboratoire sur sols cohérents comparaison des résultats, Bulletin de Liaison de LCPC, Paris, No 58, pp 97 - 108
AMAR, S, BAGUELIN, F, JÉZÉQUEL, J F and LE MÉHAUTÉ, A (1975) In situ shear resistance of clays, Proc ASCE Spec Conf on In Situ Measurement of Soil Properties, Raleigh, Vol 1, pp 22 - 45
ANDERSON, W. F. and PYRAH, I C (1989) Consolidation and creep effects in the PMT in clay., Proc 12th Int Conf SMFE, Rio de Janeiro, Br, Vol 1, pp 153-156

ARNOLD, M (1981) Empirical evaluation of pressuremeter test data, Can Geotech J, Vol 18, No 3, pp 455-459

ASTM Standard D4719-87 (1988) Standard Test Method for Pressuremeter Testing in Soils

BAGUELIN, F, JÉZÉQUEL, J F and SHIELDS, D H (1978) The Pressuremeter and Foundation Engineering, Trans Tech Pbl

BAGUELIN, F., FRANK, R A. and NAHRA, R. (1986b) A Theoretical Study of Pore Pressure Generation and Dissipation around the Pressuremeter, Proc 2nd Int Sym Pressuremeter Marine Appl, Texam, USA, ASTM No STP 950, pp 169 - 186

BELLOTTI, R., GHIONNA, V., JAMIOLKOWSKI, M., ROBERTSON, P. K. and PETERSON, R. W. (1989) Interpretation of moduli from self-boring pressuremeter tests in sand., Geotechnique, Vol 39, No 2, pp 269 - 292

BORSETTO, M, IMPERATO, L, NOVA, R and PEANO A (1983) Effects of pressuremeters of finite length in soft clay, Proc Int Sym on Soil and Rock Investigations by In-situ Testing, Paris, Vol 2, pp 211-215

BURLAND, J B (1990) Thirtieth Rankine Lecture: On the compressibility and shear strength of natural clays, Geotechnique, Vol 40, No 3, pp 327 - 378

CLARKE B G (1990) Consolidation characteristics of clays from self-boring pressuremeter tests, Proc 24th Ann Conf of the Engng Group of the Geological Soc: Field Testing in Engineering Geology, Sunderland, pp 19-35

CLARKE, B G (1994a) Peak and post rupture strengths from pressuremeter tests, Proc 13th Int Conf SMFE, Delhi, India, pp 125-128

CLARKE, B G (1994b) *Pressuremeters in Geotechnical Design*, Blackie, Glasgow

CLARKE, B G (1995) The practical application of a numeric technique for the interpretation of self boring pressuremeter tests, to be published

CLARKE, B G and SMITH, A (1992) A model specification for radial displacement measuring pressuremeters, Ground Engng, Vol 25, No 2, pp 28 - 38

DENBY, G M and CLOUGH, G W (1980) Self-boring pressuremeter tests in clay, J Geotech Engng Div, ASCE, Vol 106, No GT12, pp 1369-1387

FERREIRA, R S and ROBERTSON, P K (1992) Interpretation of undrained self-boring pressuremeter test results incorporating unloading, Can Geotech J, Vol 29, pp 918 - 928

GIBSON, R E and ANDERSON, W F (1961) In situ measurements of soil properties with the pressuremeter, Civ Engng Public Wks Rev, Vol 56, pp 615 - 618

GOST 20276 - 85 (translated by FOQUE, J B and SOUSA COUTINHO, A G F) (1985) Soils methods for determining deformation characteristics

HOULSBY, G. T. and WITHERS, N. J. (1988) Analysis of the cone pressuremeter test in clay., Geotechnique, Vol 38, No 4, pp 575-587

HOULSBY, G T and CARTER, J P (1993) The effects of pressuremeter geometry on the results of tests in clay, Geotechnique, Vol 43, No 4, pp 567-576

HOULSBY, G T and NUTT, N R F (1993) Development of the cone pressuremeter, Predictive Soil Mechanics, Proc Wroth Memorial Symposium, Oxford, pp 254 - 271

ISRM (1987) Suggested methods for deformability determination using a flexible dilatometer, pp 125-134

JARDINE, R J (1991) Discussion on "Strain dependent moduli and pressuremeter tests", Geotechnique, Vol 41, No 4, pp 621 - 626

JEFFERIES, M. G. (1988) Determination of horizontal geostatic stress in clay with self-bored pressuremeter, Can Geotech J, Vol 25, No 3, pp 559 - 573

JOHNSON, L D (1986) Correlation of soil parameters from in situ and laboratory tests for Building, Proc In situ'86 Use of In Situ Tests in Geot Engng, Blacksburg VA, pp 635-648

LUKAS, G.L. and LeCLERC De BUSSY, B. (1976) Pressuremeter and Laboratory Test Correlations for Clays, J Geotech Engng Div, ASCE, Vol 102, No GT9, pp 954 - 963

MAIR, R J and WOOD, D M (1987) Pressuremeter Testing - Methods and Interpretation, Butterworth

MARSLAND, A and RANDOLPH, M F (1977) Comparisons on the results from pressuremeter tests and large in-situ plate tests in London clay., Geotechnique, Vol 27, No 2, pp 217-243

MARTIN, R E and DRAHOS, E G (1986) Pressuremeter correlations for preconsolidated clay, Proc In situ'86 Use of In Situ Tests in Geot Engng, Blacksburg VA, pp 206-220

MÉNARD, L (1957) Measures in situ des propriétiés physiques des sols, Annales des Ponts et Chaussees, Paris France, No 14, pp 357 - 377

PALMER A C (1972) Undrained plane-strain expansion of a cylindrical cavity in clay: a simple interpretation of the pressuremeter test, Geotechnique, Vol 22, No 3, pp 451-457

PRÉVOST, J H (1979) Undrained shear tests on clays, J Geotech Engng Div, ASCE, Vol 105, No NGT1, pp 49-64

PRÉVOST, J H and HOEG, K (1975) Analysis of pressuremeter in strain softening soil., J Geotech Engng Div, ASCE, Vol 101, No GT8, pp 717-732

SADEEQ, J A and CLARKE, B G (1995) Non linear stress strain behaviour of clay from self-boring pressuremeter tests, Proc 4th Int Sym on Pressuremeters, Quebec, to be published

WINDLE D and WROTH, C P (1977) Use of self-boring pressuremeter to determine the undrained properties of clays, Ground Engng, Vol 10, No 6, pp 37-46

WROTH, C P (1982) British experience with the self-boring pressuremeter, Proc Int Sym on the Pressuremeter and its Marine Appl, Paris, pp 143 - 164

YEUNG, S K and CARTER, J P (1990) Interpretation of the pressuremeter test in clay allowing for membrane end effects and material non-homogeneity, Proc 3rd Int Sym on Pressuremeters, Oxford, pp 199 - 208

Reliable parameters from imperfect SBP tests in clay

D A SHUTTLE & M G JEFFERIES
Golder Associates (UK) Ltd, Nottingham, UK

INTRODUCTION

The self-bored pressuremeter (SBP) is unique amongst soil testing technologies in that it tests almost undisturbed soil at its natural state. Given the propensity of soil to be affected by both disturbance and initial state, this should have been an overpowering reason to adopt the SBP as the basic reference test for geotechnical engineering. Particularly as the SBP test is amenable to theoretical analysis rather than reliance on correlations. But, such adoption of the SBP has not happened in engineering practice, at least within the English speaking world: most geotechnical engineers regard the SBP as an unreliable test for design parameters.

In discussing the use of SBP data with practising engineers, three broad themes emerge as to why SBP tests in clay are treated cautiously:

- the SBP is thought to overestimate undrained strength, perhaps by as much as a factor of two
- perfect self-boring is a myth and real SBP results are affected by soil disturbance
- interpretation is a specialized skill with only a few people having the requisite knowledge

While acknowledging that some ground disturbance will occur during self-boring, we assert that the principal problem in obtaining reliable parameters from the SBP is the use of conventional methods to interpret the data. We illustrate this point by two case histories where experts interpreted SBP data and yet the recovered parameter sets were not self consistent with the data on which they were based. The concerns of practitioners are real, but they are not caused by intrinsic defects in the SBP itself.

Difficulties with conventional SBP interpretations can be avoided by analyzing the data using the technique of iterative forward modelling (IFM). IFM involves simple image matching carried out on a personal computer, generally using special software although the method can be implemented using a spreadsheet. The technique can readily incorporate advanced aspects of soil constitutive behaviour transparently to the end user. IFM is easily used by any geotechnical engineer and the parameters estimated are essentially independent of their experience provided they are diligent in seeking an image match, so avoiding the concern that only a few people have the requisite experience to properly interpret SBP data.

It is essential, however, to base IFM on SBP data that includes the full contraction phase as the instrument is depressurized from maximum expansion. Contraction data allows reasonably unique parameter estimates which are insensitive to imperfect self-boring. Finally, we compare the undrained strengths estimated by IFM on SBP data at Bothkennar with research triaxial data and strengths back calculated from a footing failure test. The parameters estimated from the SBP by IFM are reliable by both these measures.

DIFFICULTY WITH CONVENTIONAL ANALYSIS

By far the most common approach to SBP interpretation is to determine the three parameters of interest from three separate aspects of the measured data. The horizontal geostatic stress (σ_{h0}) is commonly estimated by one of several techniques (see eg. Lacasse & Lunne, 1982) which are weighted to the initial part of the SBP expansion. The shear modulus (G) is usually taken from the slope of pressure-displacement data; either for the initial loading or from an unload-reload loop. And the undrained strength (s_u) is estimated from the slope of a semi-logarithmic plot of the pressure-strain data (which assumes a Tresca model). Underlying these parameter estimates is the concept that a SBP test can be treated as the expansion of a cylindrical cavity. The clay is commonly idealized as an elastic-perfectly plastic material with a Tresca yield criterion (subsequently referred to simply as a Tresca material) using the theory set out by Gibson & Anderson (1961). This is rational as the Tresca model that has been used in developing most design formulae or upper bound (limit equilibrium) analysis.

Two case histories of expert SBP testing and analysis serve to illustrate the problems with the parameter isolation approach. These case histories are the testing carried out at the SERC Bothkennar research site and the testing carried out at Paddington for the CrossRail project.

The Bothkennar case is interesting because this site has many different tests of undrained strength, lending itself to the assessment of strength bias presented later. Paddington data is included because there are two adjacent SBP borings allowing an assessment of repeatability under field conditions. Expert interpretation was tabled for each site by separate UK testing companies, each with extensive experience in this field.

Six tests have been selected from each site spanning the investigated depth in more or less equal steps, plotted on Figures 1 and 2 for Bothkennar and Paddington respectively. Only six examples are shown because of the space limitations here, but these are sufficient to illustrate the issue and show that it is quite general.

On Figures 1 and 2, the measured data has been plotted for each test and contrasted with the theoretical curve calculated using the reported best estimate parameters for that test based on expansion/contraction of a cylindrical cavity in a Tresca material (ie. consistent with the parameter interpretation). The Paddington data from tests 4 and 15 have been corrected for disturbance effects by matching the lift-off pressure on data and fit; consistent with the method used by the expert interpreters.

As can readily be appreciated from Figures 1 and 2, the interpreted parameter sets poorly represent the measured pressure-displacement relationships of the SBP, even though based on the same data. There can be no question about the experience of the interpreters, and indeed the Bothkennar data has been considered by many of the leading UK researchers in clay behaviour (see Symposium in Print, Geotechnique July 1992). Clearly, something fundamental is missing as the interpretation procedures do not recover the pressure-displacement data upon which they have been derived. And, correspondingly, there is no reason to accept conventionally determined parameter sets as reasonable.

Moreover, no pattern of misfit is observed on Figures 1 and 2. All three parameters are

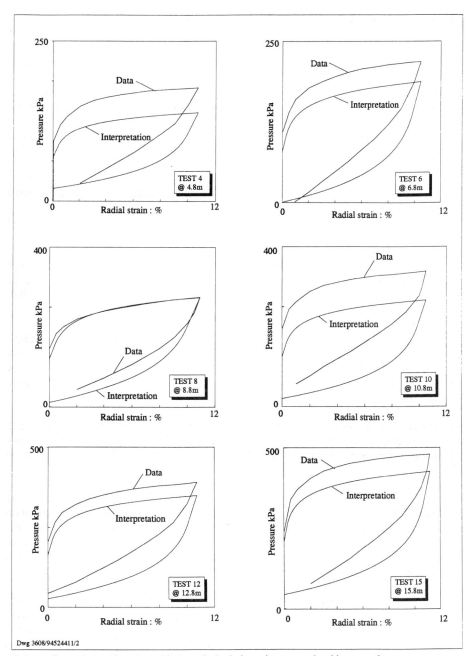

Dwg 3608/94524411/2

Figure 1: Comparison of SBP data with theoretical solution using conventional interpretation, Bothkennar Clay

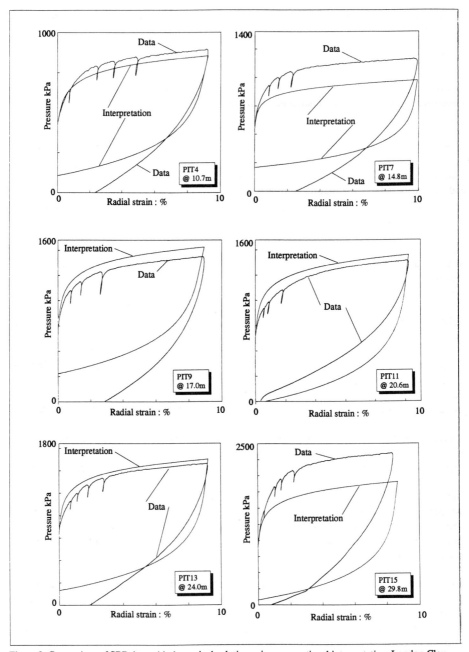

Figure 2: Comparison of SBP data with theoretical solution using conventional interpretation, London Clay

apparently poorly estimated much of the time, and the estimates can be both high and low on the same site. There is no systematic bias.

ITERATIVE FORWARD MODELLING OF SBP DATA

Civil engineers are familiar with boundary value problems in which the properties and constitutive law for the domain are specified, together with the boundary tractions and kinematic constraints. The challenge is to find the distribution of field variables such as stress, strain, and pore water pressure consistent with the applied loading.

A very different class of problem arises in engineering practice, referred to as inverse boundary value problems (IBVP). In a typical IBVP both stress (load) and displacement are concurrently measured for part of the domain, from which we are to infer both the system and its properties. Estimating soil properties from SBP data is an IBVP. There are significant mathematical limitations on extracting information from an inverse problem and it is these limitations that have been neglected in traditional methods of pressuremeter interpretation, and which have caused the difficulties illustrated on Figures 1 and 2.

Broadly, the real SBP data does not fully correspond to the theoretical assumptions. Isolating parameters one at a time does not allow a consistent pattern of idealization, and by ignoring how the data deviates from the theoretical idealization throughout the test estimating errors are amplified. The key to getting reliable parameters from the SBP is to avoid parameter isolation during interpretation.

A well known approach to solving inverse problems is iterative forward modelling (IFM). As implied by the name, IFM involves estimating the true parameter set, solving the corresponding forward boundary value problem, and then comparing the computed response with that actually measured. Subsequently, the estimated parameter set is refined and the process iterated until an acceptable image match is achieved between theory and data.

Analysis of the SBP is eminently suited to the IFM approach, which allows the determination of consistent and reliable results from the SBP. The availability of cheap personal computers with good graphics make IFM of the SBP feasible for general geotechnical engineering.

Throughout this paper we use average radial displacement on the central axis of the SBP for two reasons. First, the SBP is not a fixed body but floats in the soil mass - it is an error to treat each strain arm as an inertially isolated measurement (see Whittle, 1993). Second, the theoretical solutions assume axisymmetry, so that only the symmetric component of the data has any meaning in the context of these theories.

It is unarguable that strain rate affects the undrained strength of some clays, and several workers have considered this factor in the context of the SBP (Prevost, 1976; Anderson et al, 1987; Prapaharan et al., 1989). We have neglected strain rate for the SBP analyses shown in this paper, primarily relying on Prapaharan et al. (1989) that strain rate affects estimated strength by less than about $\pm 10\%$ of Su.

The concern about strain rate naturally lead to the wider question of constitutive model. Because IFM does not invert the boundary value problem, the method does not have to use a closed form solution; indeed a full finite element analysis using an advanced constitutive model can be used in principle. However, for IFM to be a usable proposition the analyst needs reasonably quick model updates. Updates in one second or less are more than

575

adequate, but the method becomes impaired in practice once the delay increases to about five seconds. For present personal computers (50 MHz Intel 486 equivalent) this necessitates closed form solutions.

Imperfect self-boring must now be considered. Self-boring disturbance was recognized at the outset by both Wroth & Hughes (1973) and Denby (1978) who reported that self-boring produced the equivalent outward movement of the soil comparable to ~0.5% radial strain. Mair & Wood (1987) suggest that disturbance during self-boring is inevitable, even with skilled operators. Presently, there is no means of monitoring the soil response during self-boring (using a non-intrusive technique such as resistivity) so that how well a particular insertion procedure works is largely speculative. We can minimize the influence of imperfect self-boring on estimated parameters by constraining the interpretation with the contraction phase of the test as the SBP is depressurized from maximum expansion. Theoretically, the contraction phase of the SBP test is insensitive to initial disturbance (at least for the Tresca idealization of clay).

Three groups have provided the required theoretical understanding to use this contraction stage information in a closed-form framework. In two concurrent studies Houlsby & Withers (1988) and Jefferies (1988) explored the value of unloading information in the context of a Tresca soil model, considering the full-displacement and self-bored pressuremeters respectively. In effect, both extend the well known Gibson & Anderson (1961) solution to include the contraction data and, with the exception of one second order term, the two solutions are identical if the required substitution of initial cavity radius is made.

Ferreira (1992) and Robertson & Ferreira (1992) suggest that the Tresca model is too simple for real soils and that a better approach is to approximate the stress strain behaviour with a hyperbolic curve; they give a closed form solution for the SBP using the hyperbolic model. For this study, the hyperbolic solution was tried, but surprisingly the Tresca model fitted some data better, at Bothkennar in particular. We have therefore used the Jefferies (1988) solution with the assumption of no strength change in the transition from expansion to contraction.

Existing closed form solutions for the SBP assume that the test is equivalent to expanding a cylindrical cavity, thereby idealizing the SBP as infinitely long. Real pressuremeters are quite short, most commercial devices having length to diameter (L/D) ratios of about 6. It has been appreciated for some time that this finite geometry affects the parameters estimated from SBP data (Laier, 1973; Baguelin et al, 1978; Borsetto et al, 1983; Yeung & Carter, 1990; Houlsby & Carter, 1993; Shuttle & Jefferies, 1994). However, the recent paper by Houlsby & Carter (1993) indicated that this L/D effect is far more important than previously appreciated; correction of cylindrical cavity theory for finite geometry is mandatory to estimate reliable parameters. We have followed the suggestion of Houlsby & Carter (1993) and scale the undrained strength to reflect the L/D factor, although we use correction factors developed by Shuttle & Jefferies (1994).

A key step in IFM is comparing the image match of theory to data. While it is possible to conceive of various measures to describe the fit between the two curves, a weighting scheme is required because we know that not all parts of the curve have the same value. For example, one is not surprised with a lack of fit near the origin because of unquantified disturbance. On the other hand there should be a good match to the pressure at ⅔ of maximum strain as required by the L/D correction (Shuttle & Jefferies, 1994). Further, the fit should be constrained to closely match contraction data because this phase of the test is insensitive to disturbance (Jefferies & Shuttle, 1995). We have not encountered a good scheme to capture these factors and quantify the fit. Rather, we use several independent interpreters and accept the range of solutions as being a measure of uncertainty about ground truth. Hence, results given in this paper are in the form of bars showing uncertainty but without statistical confidence estimates.

Examples of how well our approach matches measured data are shown on Figures 3 and 4, these figures showing the same tests as Figures 1 and 2 but now iterating all three parameters concurrently to achieve a fit. Clearly, the IFM approach is a substantial improvement on the conventional method as can be seen by inspection.

IFM WITH IMPERFECT DATA (REPEATABILITY)
An opportunity to ascertain the repeatability of the IFM approach is provided by some SBP data obtained as part of the site investigation for the CrossRail project. During the testing at Eastbourne Terrace (Paddington Station) the borehole deviated from vertical sufficiently that the five tests below 30m became regarded as disturbed and hence unreliable. Indeed, the factual report does not give any assessment of these tests. A second boring was put down a few metres from the original and a further ten SBP tests were carried out below a depth of 20m. These two borings provide a data set which allows us to establish the repeatability of IFM of SBP data as a total approach to estimating ground truth.

There is some variation of properties within the London Clay so that direct comparison of two tests requires they be at about the same depth. The pair of tests P1T18 @ 34.2m (original borehole and "disturbed") and E201T9 @ 33.5m (second borehole and "undisturbed") provide a suitable comparison.

A fit of theory to the disturbed data is shown on Figure 5a. Note the way the model is constrained to match the peak expansion pressure and the contraction phase. By comparison with the theory, the test data suggests the soil has been overcut (a skewed SBP ?) so that some stress relief occurred. It is possible to correct the data for estimated disturbance, thus improving the image match. However, our experience is that there is not a great change in the estimated ground properties during this disturbance correction, as the slope of the expansion pressure is near constant at typical peak strains and the unloading curve is insensitive to disturbance. Disturbance correction appears more cosmetic in improving the image match than causing a real change in ground truth estimates (Jefferies & Shuttle, 1995). We have neglected data correction for estimated disturbance in this paper except for Paddington tests 4 and 15, where it was included for consistency with the previous interpretations.

The corresponding replicate test is shown on Figure 5b. As can be seen, the parameter values are very similar to those estimated from the disturbed test.

A complete comparison of the disturbed and replicate data is shown on Figure 6 which plots profiles of horizontal geostatic stress, undrained strength and rigidity index versus depth. The stratigraphic sub-division of the London clay is based on an adjacent piezocone (CPTu). These profiles were developed from the SBP data by three independent interpreters, none of whom plotted trends with depth during the IFM process, so as to get an unbiased estimate of uncertainty in ground truth. The uncertainty in the horizontal geostatic stress, undrained strength and stiffness are about $\pm 2\frac{1}{2}\%$, $\pm 5\%$ and $\pm 12\%$ of the average value respectively.

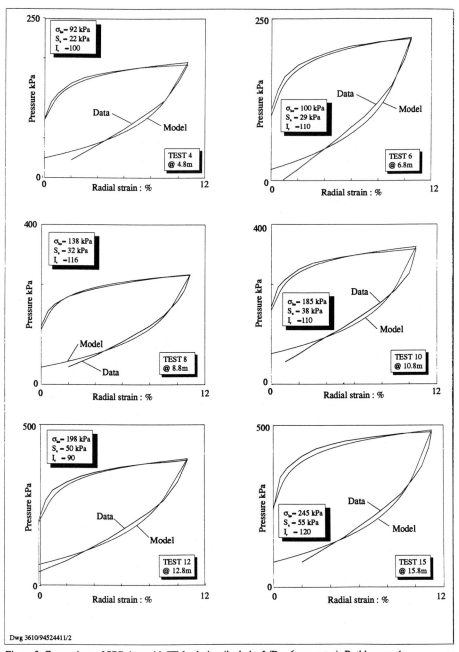

Dwg 3610/94524411/2

Figure 3: Comparison of SBP data with IFM solution (includes L/D = 6 geometry), Bothkennar clay

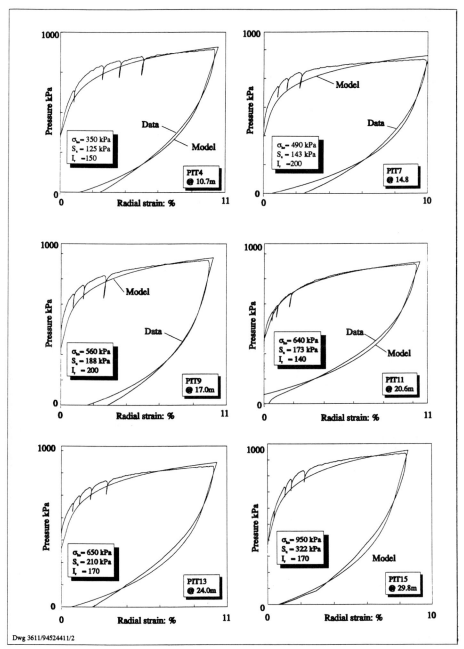

Dwg 3611/94524411/2

Figure 4: Comparison of IFM solution (includes L/D – 6 geometry) with SBP data,
London Clay

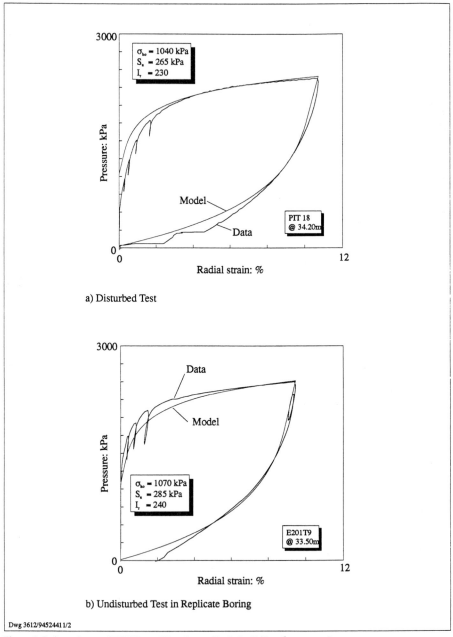

a) Disturbed Test

b) Undisturbed Test in Replicate Boring

Dwg 3612/94524411/2

Figure 5: Independence of IFM From Disturbance (includes L/D = 6 geometry)

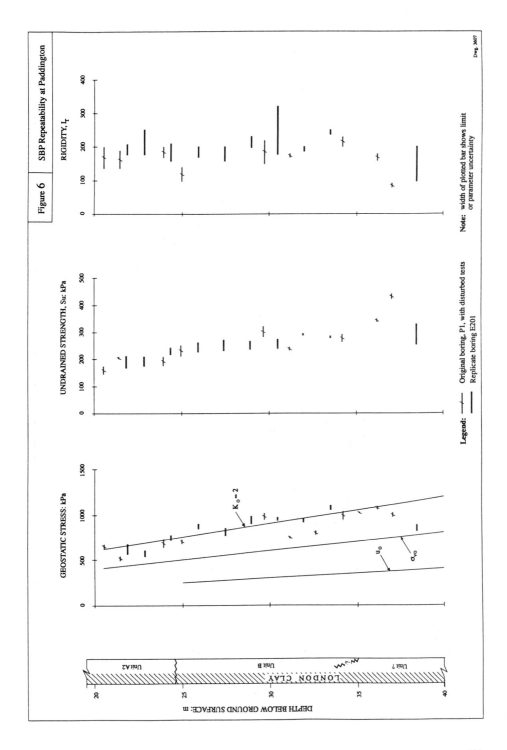

STRENGTH BIAS

The question of apparent overestimation of strength was the issue that has apparently dominated doubts about the reliability of the SBP. The Bothkennar site offers the opportunity to explore this question of strength bias because of the considerable research quality strength testing carried out here.

Some examples of IFM fits to Bothkennar data were shown on Figure 3. The profiles of horizontal geostatic stress and undrained strength developed from analyzing all 18 SBP tests in borehole PR1 are shown on Figure 7, and contrasted with the stratigraphic sub-division reported by Paul et al (1992). As in the case of Paddington, three independent interpreters were used to get an appreciation of the parameter uncertainty.

The SBP tests in soil Unit L3 are reasonably well fitted by the Tresca model and the uncertainty in the undrained strength is about ± 5% of the average value. Below a depth of 10m, in the L4 soils, fits are more difficult to achieve and the adequacy of a simple Tresca model looks questionable. Correspondingly, the uncertainty in the undrained strength increases to ± 10%. The horizontal geostatic stress remains within a narrow band, however, indicating considerable confidence in this parameter.

Also shown on Figure 7 are the undrained strength profile determined by triaxial testing using five different protocols, the profiles being those summarized by Hight el al (1992) here simplified to straight lines to emphasize bias in comparison to clay variability. Overall, the uncertainty in triaxial strength is comparable to, or slightly larger, than the uncertainty in SBP strength. There is a slight skew between the two tests , however, with the SBP strength being biased lower than the triaxial at a depth of about 6m and biased slightly high at a depth of more than 17m.

To truly establish ground truth we need full scale load tests (both large footings and embankments), but these have not yet been carried out at Bothkennar. However, Gildea (1993) has reported some tests of instrumented footings loaded to failure. Back analysis of these footing tests showed an operating strength $Su = 20$ kPa at a depth of 2.8m. This strength is the same as the centre of the SBP uncertainty band at 2.8m depth and also corresponds to the two higher strength triaxial protocols.

CONCLUSION

Conventional methods of analyzing SBP data in clay, even when applied by expert analysts, do not lead to parameter sets for ground truth that are consistent with the measured SBP response from which they were derived. This inconsistency is readily overcome if SBP data is recognized as an inverse problem and analyzed using IFM. In addition, to establish ground truth we need to;

- account for the geometric limitation of common SBPs with $L/D \approx 6$
- use contraction phase data to make interpretations insensitive to imperfect self-boring.

Using geometry corrected IFM, some SBP data previously disregarded as obviously disturbed produces estimates of ground truth that are similar to high quality undisturbed tests in an adjacent replicate borehole.

Comparing the strength profile obtained from SBP with IFM at the Bothkennar research site shows similar uncertainty to the research triaxial results. There appears little bias in

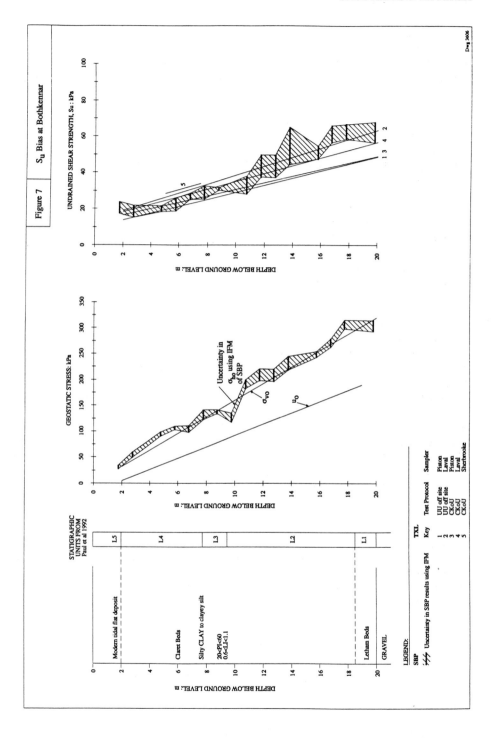

Figure 7 | S_u Bias at Bothkennar

estimates of undrained strength between the SBP and the best triaxial data.

In summary, provided the SBP is analyzed by the IFM approach, and provided this interpretation is constrained by the contraction phase, then the estimate of ground truth obtained is as good as the very best laboratory based research quality studies. The SBP has the advantage of being readily available in commercial practice and the results can be analyzed on-site in near real time.

Looking to the future, we note that the uncertainties in estimated ground truth in the present study are largely attributable to the over-idealization of real clay behaviour inherent in the Tresca model, and are unrelated to the SBP itself. IFM is not restricted to closed-form solutions. Increased confidence in real site conditions can be gained by developing optimized finite element procedures for use within IFM incorporating realistic constitutive models.

REFERENCES
Anderson, W.F., Pyrah, I.C., Haji Ali, F. (1987); Rate effects in pressuremeter tests in clays. Proc. ASCE J of Geotechnical Engineering, Vol. 113, GT11, 1344-1358.
Baguelin, F., Jezequel, J.F. & Shields, D.H. (1978); The pressuremeter and foundation engineering. Claustal: Trans Tech.
Borsetto, M., Imperato, L., Nova, R. & Peano, A. (1983); Effects of pressuremeters of finite length in soft clay. Proc Int Symp In Situ Testing, Paris, 2, 211-215.
Denby, G.M. (1978); Self-boring pressuremeter study of San Francisco Bay mud. PhD thesis, Stanford University.
Ferreira, R. (1992); Interpretation of pressuremeter tests using a curve fitting technique. PhD thesis, University of Alberta.
Gibson, R.E. & Anderson, W.F. (1961); In situ measurement of soil properties with the pressuremeter. Civil Eng and Public Works Review, 56(658), 615-618.
Gildea, P. (1993); Instrumented footing load tests on soft sensitive Bothkennar Clay. Ground Engineering, July/August, 30-38.
Hight, D.W., Bond, A.J., & Legge, J.D. (1992); Characterization of the Bothkennar clay: an overview. Geotechnique 42, 303-347.
Houlsby, G.T. & Carter, J.P. (1993); The effects of pressuremeter geometry on the results of tests in clay. Geotechnique 43, 567-576.
Houlsby, G.T. & Withers, N.J. (1988); Analysis of cone pressuremeter test in clay. Geotechnique 38, 575-587.
Jefferies, M.G. (1988); Determination of horizontal stress in clay with self-bored pressuremeter. Can Geot J 25, 559-573.
Jefferies, M.G. & Shuttle, D.A. (1995); Disturbance effects in the interpretation of reliable parameters from SBP tests in clay, Proc. 4th International Symposium on Pressuremeters, Sherbrooke, Quebec, Canada.
Lacasse, S. & Lunne, T. (1982); In situ horizontal stress from pressuremeter tests. Symposium on the pressuremeter and its marine applications, Editions Techniq, Paris, France, 187-208.
Laier J.E. (1973); Effects of pressuremeter probe/diameter ratio and borehole disturbance on pressuremeter test results in dry sand. PhD thesis, University of Florida.
Mair, R.J. & Wood, D.M. (1987) Pressuremeter testing: methods and interpretation. London: Butterworths.
Paul, M.A., Peacock, J.D., & Wood, B. F. (1992); The engineering geology of the Carse clay of the National Soft Clay Research Site, Bothkennar. Geotechnique 42, 183-198
Prapaharan, S. , Chameau, J.L., & Holtz, R.D. (1989); Effect of strain rate on undrained

strength derived from pressuremeter tests. Geotechnique 34, 615-624.

Prevost, J.H. (1976); Undrained stress-strain-time behaviour of clays. J Geot Eng Div ASCE 102, 1245-1259.

Robertson, P.K. & Ferreira, R.S. (1992); Seismic and pressuremeter testing to determine soil modulus. Predictive Soil Mechanics (Wroth Symposium), pbl Thomas Telford, 562-580.

Shuttle, D.A. & Jefferies, M.G. (1994); A practical geometry correction in the interpretation of pressuremeter tests in clay, Geotechnique in press.

Whittle, R. (1993); Separate arm analysis is unsafe. Ground Engineering, September, 19-20.

Wroth, C.P. & Hughes J.M.O (1973); An instrument for the in situ measurement of the properties of soft clays. Proceedings of the 8th International Conference on Soil Mechanics and Foundation Engineering, Moscow, vol. 1.2, 487-494.

Yeung, S.K. & Carter, J.P. (1990); Interpretation of the pressuremeter test in clay allowing for membrane end effects and material non-homogeneity. Pressuremeters, 199-208, London, Thomas Telford Limited.

Pressuremeter tests in unsaturated soils

F. SCHNAID,[1] G. C. SILLS[2] and N. C. CONSOLI[1]
[1]Federal University of Rio Grande do Sul, Porto Alegre, Brazil
[2]University of Oxford, England

ABSTRACT
Geotechnical design for unsaturated soils requires an understanding of their fundamental behaviour, and, in particular, the role of suction. However, it is complicated to carry out laboratory tests that take due account of suction, and it is therefore difficult to characterise these soils from laboratory tests alone. This paper discusses a way in which pressuremeter measurements can be made and analysed to provide soil parameters appropriate to the suction existing in the site.

INTRODUCTION
Traditionally, in both laboratory and in situ tests, strength parameters for soils are classified either as undrained or as fully drained. The time taken to move from one condition to the other in response to a change in stress depends on the soil itself (its permeability and compressibility) and on the boundary conditions (which determine the drainage path length). For saturated soils, the distinction between undrained and drained conditions is unambiguous, and may be defined in terms of volume change. The undrained condition corresponds to the condition of no volume change, before any water has flowed, and the drained condition exists after the volume change is complete and all excess pore water pressures have dissipated. However, an unsaturated soil contains pore water and pore air, both in continuous voids, and any change in stress will, in general, cause immediate volume change due to compression and/or flow of the air, followed by further changes whose time scale is determined by comparatively short water drainage path lengths between air voids rather than the typically much larger drainage path lengths existing in a saturated soil. It is, therefore, essential to specify carefully the appropriate conditions for testing unsaturated soils both in the laboratory and in the field and to recognise the complexity of behaviour of these soils in interpreting the results.

Experience with unsaturated soils has shown that many of them appear to be cemented, and such soils therefore demonstrate a higher shear strength due to their bound structure. A further increase in strength occurs due to the pore water suction caused by surface tension effects with the pore air, the pressure of which is maintained at atmospheric, due to continuity of air voids. For foundation design, it is necessary to distinguish between these two effects of cementation and suction, since the magnitude of the pore water suction depends on the available water supply as well as on the structure of the soil, and the strength benefit that it provides may therefore be transient.

Recently, considerable progress has been made in understanding the behaviour of unsaturated soils. In particular, work by Fredlund and Morgenstern (1977), Alonso et al (1990) and Wheeler and Sivakumar (1993) has provided a Critical State framework, in which it is shown that two independent stress parameters are necessary for a complete description. These parameters may be taken as the net stress, σ-u_a and the suction u_a-u_w, where σ is the total stress, u_a is pore air pressure and u_w is the pore water pressure. These two variables replace the single effective stress variable σ-u_w that is assumed to control the behaviour of a saturated soil. For a saturated soil, the shear strength is described by a frictional model, incorporating the normal effective stress, a friction angle ϕ' and a cohesion intercept c'. For an unsaturated soil, dependence of the shear strength on effective stress is replaced by dependence on the net stress and suction. The familiar Mohr-Coulomb envelope for saturated soils, plotted on the axes of shear or deviator stress and normal effective stress, is replaced by a surface in three dimensions, with axes of shear or deviator stress, normal net stress and suction. Each constant suction plane is parallel to the familiar saturated (or zero suction) plane, and each has its own cohesion intercept and friction angle. Therefore, in order to predict the strength of unsaturated soil it is necessary to measure the cohesion and friction angle as a function of suction, as well as assuming appropriate values of stress and suction in situ. As yet, few laboratory tests have been undertaken under rigorous conditions of controlled suction, and some controversy exists about the conclusions that may be drawn. However, perhaps surprisingly, it appears that the friction angle ϕ' can be taken to be independent of suction, at least for limited ranges of stresses and suction values (Fredlund et al 1978, Escario and Saez 1986, Alonso et al 1990, and Wheeler and Sivakumar 1993). The friction angle can therefore be measured in a drained laboratory test on an unsaturated sample that has been flooded to saturation, thereby reducing the suction to zero. It is still necessary to measure the in situ suction value, and the corresponding cohesion, if the field strength of the unsaturated soil is to be calculated from these results and the in situ net stress.

In principle, laboratory tests could be carried out on samples under carefully controlled conditions of suction, in order to provide a correlation between the cohesion intercept and the suction value, with or without the assumption that the friction angle is independent of suction, and it is recommended that more tests of this type should be carried out in future. The only problem then remaining would be that of measuring the in situ suction. In the past, however, the practice has generally been to carry out standard "undrained" triaxial tests, or standard drained tests on flooded samples, to produce the c' and ϕ' parameters used in design. The results of the undrained test are particularly misleading, since the prevention of water and air flow from the sample will cause changes in both pore water and pore air pressures, with corresponding suction changes. None of these changes is measured, so that failure, when it occurs, will be at some unknown suction. In two different tests, it is likely that the final states will correspond to different suction values, and any interpretation therefore spans different constant suction planes, so that the use of such results is questionable. The drained test on a flooded sample, on the other hand, is more useful, since the results apply to a single constant suction plane, of zero suction, so that it is possible to calculate the friction angle ϕ' from two or more such tests. This may then be assumed to be a suitable value for all suction conditions. The cohesion intercept is valid only for this zero suction condition, so is not the appropriate value for field use.

Even if rigorous laboratory testing with suction measurement is carried out, there are serious difficulties associated with suction measurement in situ, and a better approach is to try to make a more direct in situ strength measurement. This paper explores the use of the cone penetrometer and the pressuremeter, the former as a tool for soil profiling and the latter to provide a shear strength, based on the assumption of a suitable soil model and drained, flooded direct shear test results. These techniques are illustrated by field results.

TEST SITE

An unsaturated field site at Cachoeirinha, in the south of Brazil has been tested in a comprehensive field programme carried out in the Federal University of Rio Grande do Sul. The programme includes the use of cone tests, SPT, Menard pressuremeter tests and plate loading tests, as well as various laboratory shear tests.

The site investigation revealed features that are common to almost every site which has been subjected to severe weathered conditions. The degree of weathering varies gradually with depth and distinctive layers, or horizons, can be identified. Various degrees of inhomogeneity of the soil mass can be expected, being often described by a linear variation of the soil properties with depth. The water table in this deposit is at approximately 4.0m depth. In the experimental site it was possible to identify three subsoil layers typically described as:

Horizon A - a superficial 0.75m layer composed of silty-clayey sand, containing roots and other organic materials.

Horizon B - an approximately 2.5m layer of sand-silty red clay (identified as a lateritic soil).

Horizon C - a 4.0m layer of red silty clay with grey and yellow intrusions (identified as a saprolitic soil).

Variations of water content with depth are presented in Figure 1. Measurements

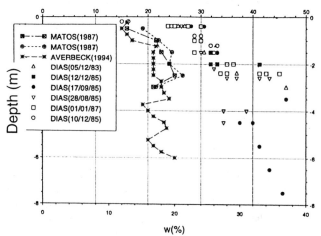

Fig. 1   Variation of water content with depth at Cachoeirinha

suggest that the water content does not vary significantly with depth, but it may well change during the seasons of the year and with local variability. A significant scatter on measured values of water content is observed at a depth of approximately 2.5m (Dias, 1987) in which a 12% variation is shown at a single depth for samples taken at about the same time. These variations may reflect the macro and micro structural complexity of the soil but they are not directly associated to the position of the water table, nor, therefore, to capillarity effects.

SOIL PROFILING

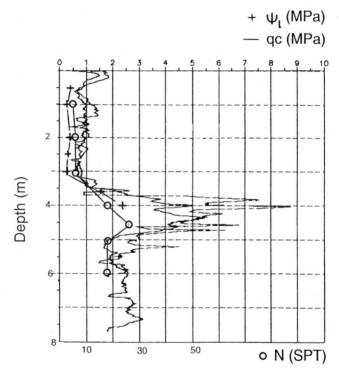

Fig. 2   Soil profiling

A typical soil profile from the experimental field is presented in Figure 2, in which two independent measurements of tip cone resistance q_c, pressuremeter limit pressure ψ_l (both referred to the upper axis) and SPT blow count numbers N (on the lower axis) are plotted against depth. It can be seen that all three types of measurement provide consistent results, but that the quasi-continuous cone penetration provides the most detailed information on the soil profile and demonstrates much more clearly the identification of different layers. The potential of the cone penetrometer to identify spatial variability in saturated soils is already well established, and it appears that a similar potential exists also for the inherent vertical and spatial heterogeneities of the soil properties observed in structured unsaturated deposits.

However, even in saturated soils, the interpretation of a penetrometer profile in terms of strength parameters relies largely on empirical correlation through a cone factor N_k, and it can be anticipated that, given the greater complexity of unsaturated soil behaviour, and the difficulties of expressing the constitutive relationships, the penetrometer is unlikely to be a satisfactory instrument for assessing soil parameters in such soils. From this point of view, the pressuremeter and plate loading tests appear to be the best possible alternatives, leading to strength and stiffness estimates respectively.

SOIL PROPERTIES

At present there is an almost complete lack of experience of the assessment of soil properties in unsaturated soils from in-situ tests. A small number of comparisons of laboratory and field data has been previously reported (e.g. Brand, 1985; Rocha Filho and Queiroz, 1990; Schnaid and Rocha Filho, 1994), with the SPT being the most frequently used tool for site investigation and foundation design. This paper describes new methods of assessing shear strength and stiffness from measurements taken from both pressuremeter and plate loading tests.

Shear strength

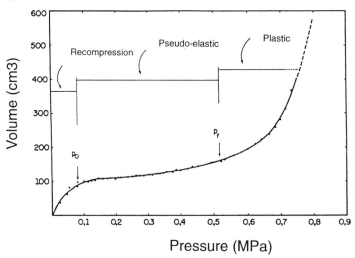

Fig. 3   Typical Menard pressuremeter pressure-expansion curve

Figure 3 shows a typical set of Menard pressuremeter results from the site at Cachoeirinha. Various points can be identified, such as the estimated in situ horizontal stress, marked as p_0, at the first curvature of the S-shaped curve. This point is not always well defined, and sometimes has to be judged from an interpretation of the data. The onset of plastic behaviour is marked by p_f, and the pressuremeter limit pressure ψ_1 corresponds to the end of this plastic phase. Again, this is not always easy to identify directly, and may have to be obtained by extrapolation methods. Between the pressures p_0 and p_f, the behaviour is described as pseudo-elastic. The pressuremeter modulus G is taken from the pseudo-elastic region, and relates the pressure increment to the volumetric strain. Interpretation of results such as these must be based on a suitable model of soil behaviour, two of which are considered here.

The first approach uses the analytical cylindrical cavity expansion solution for an ideal cohesive frictional material presented by Carter et al (1986). This is an elasto-plastic model, developed for small strain deformations, in which it is suggested that cavity expansion limiting values can be explicitly determined. In this case, the formulation can be used to predict the cohesion intercept c for specific values of measured pressuremeter limit pressure ψ_l, measured pressuremeter modulus G, estimated in-situ stress p_o (taken also from pressuremeter data), Poisson's ratio, dilatancy (taken as zero for a first estimate) and friction angle (obtained from direct shear tests on flooded samples). Unfortunately, this analytical formulation is very

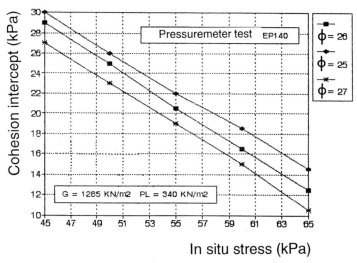

Fig. 4   Values of cohesion intercept expressed as a function of in situ stress and friction angle

sensitive to the magnitude of the in-situ stress, as shown in Figure 4, in which the cohesion intercept is expressed as a function of in-situ stress and friction angle. This imposes a limitation on predictions based on the Carter analysis of test results from the Menard pressuremeter.

The second methodology is based on a non-linear elastic hyperbolic model (Duncan and Chang, 1970). Cavity expansion was assumed to occur under conditions of axial-symmetry and a numerical finite element approach performed to obtain the complete pressure-expansion relationship. The hyperbolic parameters and friction angle were taken from laboratory stress-strain data from flooded samples. These data were used in the numerical model to predict the pressuremeter output, and the values were then optimised to give the best overall fit to the actual experimental curve. Several tests have been interpreted in this study, a typical example being given in Figure 5.

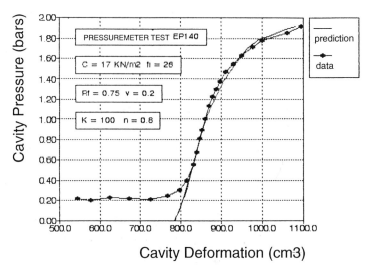

Cavity Deformation (cm3)

Fig. 5 Comparison between measured and simulated pressure-expansion relationships of a typical pressuremeter test

By making use of both solutions it was possible to predict the variation of the cohesion intercept with depth for various values of friction angles (chosen within the range of 25° to 27°, as obtained from direct shear tests). The results are presented in Figure 6, and although some scatter is observed, the predictions from the FE analysis (non-linear elastic model) and the analytical predictions (elasto-plastic model) for different ϕ' values fall within the same order of magnitude. The difference between Figures 6a and 6b for the analytical solution is related to the procedure used for deriving the in situ stress p_0. In Figure 6a p_0 is obtained by fitting the whole curve, as required by the French Standards analysis of the Menard Pressuremeter, whereas in Figure 6b p_0 is taken as a single point representing the beginning of the pseudo-elastic segment of the pressure-expansion curve. It can be seen that the main differences occur at a depth of 1.9m, and that, by chance, the values at the other depths are hardly affected. It is interesting to note that cohesion values obtained from the French Standard approach closely agreed with the FE predictions. There is a general argument for suggesting that the prime requirement for a successful prediction is to fit the complete curve and not rely on a single value assessed from a graphical method for deriving the in situ stress. On this basis, the analysis based on the French method is to be preferred.

Results from drained direct shear tests carried out on flooded samples are also presented in Figure 6. Since suction values should be close to zero the measured cohesion intercept reflects the effects of bonding and structure, and therefore provides a lower bound to the values to be used in predictions for field use where suctions exist.

The theoretical models used in this analysis cannot be applied generally to unsaturated soil conditions, as they do not incorporate the effect of suction. However, when the pressuremeter is expanded in unsaturated soils, it is assumed that

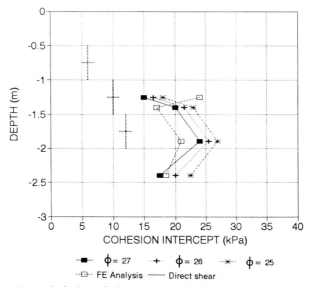

(a) Analytical predictions based on in situ stress values
obtained from French Standards approach

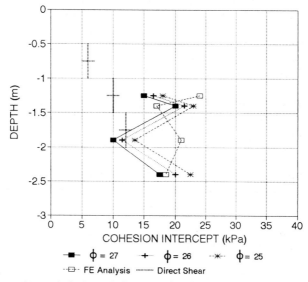

(b) Analytical predictions based on in situ stress values
obtained from a single point corresponding to the beginning
of the pseudo-elastic segment

Fig. 6   Variation of choesion intercept expressed as a function of in situ
stress and friction angle

drained conditions exist. This may be justified by considering that the drainage path
lengths are comparatively short, so that drainage of pore water will occur quickly into

the air-filled voids, which themselves remain at atmospheric pressure due to the very high air permeability associated with the interconnected air voids. However, although this situation seems physically plausible, it needs to be confirmed for individual soils by pressuremeter tests at different rates, to identify an upper bound speed below which the results are not rate dependent. Once drained conditions are verified, the net stress is numerically equal to the total stress, and the suction remains constant, so that it is not necessary for the model to incorporate potential changes in suction.

Preliminary results demonstrated the potential application of the proposed approach encouraging its use to give reliable information on the shear strength of structured unsaturated soils.

Measurements of stiffness
Empirical correlations between reference values of penetration resistance and soil compressibility are abundant in the literature. It is not within the scope of this work to review this approach. It should however be questioned if there is any satisfactory reasons why end bearing measurements should be related to stiffness at all, since there is no physical appreciation of why this properties can expected to be correlated (e.g Wroth, 1988).

A significant improvement can be achieved by the use of the pressuremeter which enables a direct measurement of soil compressibility. Comparisons between pressuremeter data and standard plate loading data should be encouraged to envisage its application to foundation design, since current techniques of estimating settlements in granular drained materials are simply based on the extrapolation of plate loading tests.

To compare values of compressibility obtained from pressuremeter tests and plate loading tests, a reference study was carried out at the experimental site at Cachoeirinha. Figure 7 illustrates the variation of the values of Young's Modulus with depth. The pressuremeter modulus is seen to be very similar to the large strain Young's modulus obtained from plate loading tests. It is then possible to suggest that Menard pressuremeter stiffness measurement can be used to define empirical upper limit settlements for this unsaturated site. In the same way that the SPT blow count N is used to define an upper limit on settlement in saturated sands (Burland and Burbidge 1985), so, in this case, the measured moduli could be linked to a maximum possible settlement. For the particular geotechnical design problem, a lower settlement would be assumed to take account of the relevant stress and strain level.

CONCLUSIONS
It is important to distinguish between the study of a soil to identify its general type and a study to measure the parameters describing its behaviour. In site investigations in saturated ground, the cone penetrometer is recognised as a powerful tool, and it appears that it has an equally important role in profiling unsaturated soils. However, other methods must be used to characterise the soil behaviour quantitatively.

Many of the general comments that apply to measuring the design parameters of saturated soils, such as the problems caused by sampling disturbance and stress relief,

Deformation Moduli (MPa)

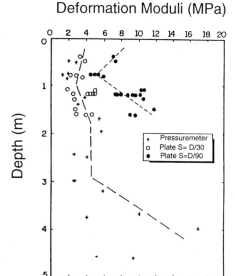

Fig. 7   Deformation moduli obtained from pressuremeter tests and plate loading tests

and the difficulty of ensuring that results from a small sample are representative of the field condition, apply equally to measurements on unsaturated soil. There are, however, a number of further problems with unsaturated soils that severely limit the applicability of laboratory testing. The most fundamental of these is related to the greater complexity of unsaturated soils, and the fact that the basic parameters of net stress ($\sigma - u_a$) and suction ($u_a - u_w$) are needed to characterise its behaviour. It is strongly recommended that any discussion of the behaviour of unsaturated soils should be couched rigorously in terms of these parameters, so that their significance is not accidentally overlooked. It has already been noted that measurements made at unknown suction levels or spanning different suction values cannot be sensibly interpreted. Thus, in order to correlate measurements with the corresponding net stress and suction values, the minimum requirement of the test apparatus is that it should allow the control and/or measurement of two of the three parameters pore air pressure, pore water pressure and suction. The simplest method of suction measurement is by measuring pore water pressure in a test in which air is free to drain, and is therefore always at atmospheric pressure. Under these conditions, the net stress is equal in magnitude to the total stress. To provide controlled suction conditions, it is necessary to control pore water and pore air pressures independently.

There is therefore considerable advantage in using in situ techniques to measure unsaturated soil parameters. There is less danger of the gross disturbance that is difficult to avoid in sampling, and the calculated soil parameters automatically correspond to the appropriate suction level. However, the interpretation of field measurements requires the assumption of a soil model, which is applied to a problem with complicated boundary conditions. It has already been noted that a realistic model for unsaturated soil is very complex, and in practice considerable simplification is necessary for a practical solution. In this paper, two simplified models have been examined, the first based on an elasto-plastic model and the second on a non-linear elastic model. It is recognised that in general an elasto-plastic model is to be preferred to a wholly elastic one for the case of cavity expansion. However, there is considerable advantage in a formulation which allows the whole of the data curve to be incorporated in the analysis, and this advantage of the non-linear elastic approach appears to override its limitation. It should also be remembered that neither model is adequate to predict the behaviour of unsaturated soil except under conditions of constant suction.

In conclusion, then, a recommendation is made for measurement of parameters for design in unsaturated soils. Simple, standard drained shear tests should be carried out on samples recovered as carefully as possible, and then flooded to achieve conditions of zero suction. The friction angle obtained from such tests may be assumed to be independent of suction, and therefore suitable for design use in the field. The pressuremeter is then the best in situ instrument, and the results can be interpreted to provide an assessment of the appropriate cohesion intercept for the field conditions. The Menard pressuremeter has the additional benefit that it provides a lower bound to the soil stiffness.

REFERENCES

Alonso, E.E., Gens, A. and Josa, A. (1990) - A constitutive model for partially saturated soils. Geotechnique, Vol 40, No 3, 405-430.

Averbeck, J.H.C. (1994) - Study of the behaviour of small diameter excavated piles in partially saturated soils. M.Sc. Thesis, Federal University of Rio Grande do Sul, Brazil. (In portuguese)

Brand, E.W. (1985) - Geotechnical engineering in tropical Soil. International Conference in Tropical Lateritic and Saprolitic Soils, Brazil, Vol 1, 23-91.

Burland, J.B. and Burbidge, M.C. (1985) - Settlements of foundations on sand and gravel. Proc. Centenary Celebration of Glasgow and West of Scotland, Association of Institution of Civil Engineers.

Carter, J.P, Broker, J.R. and Yeung, S.K. (1986) - Cavity expansion in cohesive frictional soils. Geotechnique, Vol 36, No 3, 349-358.

Dias, R.D. (1987) - The applicability of geotechnology and pedology to the foundation design of electric transmission lines. Ph.D. Thesis, Federal University of Rio de Janeiro, COPPE, Brazil. (In portuguese)

Duncan, J. and Chang, C.H. (1970) - Nonlinear analysis of stress and strain in soils. Proc. ASCE, Vol 96, No 5.

Escario, V. and Saez, J. (1986) - The strength of partially saturated soils. Geotechnique, Vol 36, No 3, 453-456.

Fredlund, D.G. and Morgenstern, N.R. (1977) - Stress state variables for unsaturated soils. Proc. ASCE, Vol 103, No GT5, 447-466.

Fredlund, D.G., Morgenstern, N.R. and Widger, R.A. (1978) - The shear strength of unsaturated soils. Canadian Geotechnical Journal, Vol 15, 313-321.

Matos, L.F.S. (1989) - Ultimate load of small diameter piles in traction in partially saturated soils. M.Sc. Thesis, Federal University of Rio Grande do Sul, Brazil. (In portuguese)

Rocha Filho, P. and Queiroz, C. (1990) - General Report: Building foundation in tropical lateritic and saprolitic soils. International Conference on Geomechanics in tropical Lateritic and Saprolitic Soils. Singapore, Vol 1, 587-601.

Schnaid, F. and Rocha Filho, P. (1994) - Experience in applying pressuremeter test results to structured partially saturated soils. XI Brazilian Conference on Soils Mechanics and Foundation Engineering. Foz do Iguacu, Brazil, In Press. (In Portuguese)

Wheeler, S.J. and Sivakumar, V. (1990) - Development and application of a critical state model for unsaturated soil. Predictive Soil Mechanics, Proc. of the Wroth Memorial Symposium, Thomas Telford, London, 709-728.

Wroth, C.P. (1988) - Penetration testing - A more rigorous approach to interpretation. ISOPT-1, Orlando, USA, 303-314.

The determination of deformation and shear strength characteristics of Trias and Carboniferous strata from in situ and laboratory testing for the Second Severn Crossing

J. D. MADDISON, S. CHAMBERS, A. THOMAS and D. B. JONES
Halcrow – SEEE, Cardiff, UK

SYNOPSIS

The ground investigation for the 5.2km long Second Severn Crossing was one of the largest single contracts carried out in the UK to date. The main aims of the investigation included the determination of the engineering properties of the strata with particular emphasis on the deformation and shear strength characteristics. The investigatory work was both ambitious and innovative employing extensive in situ pressuremeter (HPD) testing, geophysical testing, rotary rock coring at 100 and 300mm diameters and large diameter plate bearing tests in boreholes both onshore and offshore. Large scale (700mm square) in situ direct shear tests were also performed. These works were complemented by comprehensive rock laboratory testing on samples of 100 and 300mm diameters. These ground investigation works are described below and the advances made in field and laboratory operations are highlighted. The test results obtained, their interpretation and the derivation of engineering parameters for design are also discussed.

INTRODUCTION

The Second Severn Crossing provides a second motorway link between South Wales and England across the River Severn estuary. At the site the estuary has a tidal range of some 14m, the second largest range in the World. The concession for the crossing was let to Severn River Crossing plc, a consortium led by John Laing Construction and GTM Entrepose. The designer was Sir William Halcrow & Partners Ltd in joint venture with Societe d'Etudes et d'Equipements d'Entreprises. The new 5.2km long crossing is sited some 5km down stream of the existing Severn Bridge which opened in 1965, ref Figure 1. The alignment crosses rock outcrops which are exposed at low tide which lie east and west of the Shoots navigation channel and mudflats extending some 1km out from the Welsh shore. About 250m off the Avon shore the alignment crosses the 120 year old British Rail tunnel which carries the London - South Wales main line. The crossing comprises approach viaducts of 99m span in general with a 912m long cable stayed main bridge spanning some 456m across The Shoots navigation channel. Caisson foundations founded on Trias Sandstone strata were adopted for the main bridge and most of the viaduct piers. Pile foundations were employed for fourteen viaduct piers including those of the Gwent mudflats and those adjacent to the British Rail tunnel. Details of the crossing structures and foundations are presented in Table 1.

Advances in site investigation practice. Thomas Telford, London, 1996

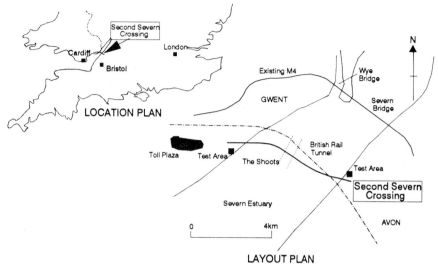

Figure 1: Site location & layout plans

Structure	General Dimensions	Foundation Details
Main Cable Stayed Bridge	overall length 912m central span 456m	Main caissons 53m long by 13m wide Back span caissons 34m long by 13m wide
Gwent Viaduct	2.2km length approx 22 spans of 99m in general	Viaduct piers (offshore): caissons 34m long by 10m wide Viaduct piers (inshore - Gwent mudflats): piles 2m dia. 20m max. length (8No per pier) Abutment: piles 1m dia. 15m max. length
Avon Viaduct	2.3km length approx 23 spans of 99m in general	Viaduct piers (offshore): caissons 34m long by 10m wide & 28m long by 6m wide with 2m 'feet' Viaduct piers (onshore, inshore & adjacent to BR tunnel): piles 2m dia. 30m max. length (7/8 No. per pier) Abutment: piles 1m dia. 15m max. length

Table 1: Details of crossing structures and foundations

GEOLOGY

The geological sequence at the site consists of Trias strata which dip south to south east at about 1 to 2 degrees and crop over the full length of the crossing. These strata rest unconformably over Carboniferous strata dipping at between 10 and 20 degrees south east. A geological section along the crossing is presented as Figure 2. The Trias strata predominantly comprised light yellowish brown, slightly to moderately weathered, moderately weak to strong sandstones and reddish brown, occasionally greenish grey, very weak to moderately strong mudrocks (siltstones and claystones). The mudrock strata were classified generally as being of weathering grades IVb to II defined in accordance with Chandler and Davies (1973). The Carboniferous strata comprised strong Pennant sandstones with subsidiary mudstones containing coal seams and seatearths of the Upper Coal Measures. The strata are cut by three generally north-south trending faults. The two faults west of The Shoots result in the downthrow of the strata to the west by some 26m. The Bull Fault which crosses the English Stones east of The Shoots downthrows the strata to the east

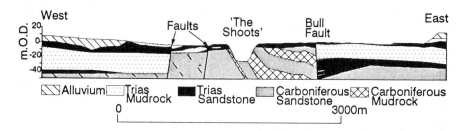

Figure 2: Geological section

in excess of 40m. The superficial deposits overlying bedrock on the Gwent bank and foreshore were up to 13m thick and predominantly comprised soft and very soft estuarine silts and clays in general overlying sand and gravel deposits of some 3m thickness. Similar deposits up to 6m depth were encountered on the Avon bank.

GROUND INVESTIGATION WORKS

A preliminary ground investigation comprising some 54 rotary drilled boreholes concentrated in the region of the main bridge and backspan piers was carried out by Soil Mechanics Limited in late 1988. This was part of a feasibility study performed on behalf of the Department of Transport. The detailed ground investigation for the actual crossing was to lie some 80m south west of the previous investigation and was performed by Wimpey Geotech Limited between September 1990 and June 1991. The ground investigation works undertaken are summarised in Table 2. There were fifty one major foundations to consider. Some 167 boreholes were sunk for the investigation. It was designed to obtain high quality representative data for each of the four basic strata categories at the site with complete coverage for all foundations across the estuary while optimising the volume of data obtained to facilitate design and construction within a six year period. In general one wireline rotary cored borehole was planned at each viaduct pier foundation with the option to sink additional boreholes as the investigation might require. In areas where the ground conditions were relatively complex such as at and adjacent to faults, up to 7 boreholes were planned. At each of the main bridge piers up to 11 boreholes were drilled and a maximum of 6 boreholes were sunk at the viaduct piers either side of the British Rail tunnel. Sites for large scale in situ testing were also identified on the Avon and Gwent shores.

Investigations	Scope of Works
Field surveys	Geological mapping & discontinuity surveys.
Exploratory holes	161 No. 100mm dia. rotary core boreholes to 50m max depth, 6 No. 300mm dia. rotary core boreholes to 20m max depth , 3 No. trial excavations to 3m max depth.
In situ testing	High pressure dilatometer tests (239 No.), impression packer tests (75 No.), packer permeability tests (31 No.), 500mm dia. borehole plate bearing tests (5No.), 1m dia. trial pit plate bearing tests (5 No.), 700mm square direct shear tests (5 No.).
Geophysics	Borehole geologging (47 No.), cross hole 'shooting' (2 No.), marine seismic reflection (12km) and refraction (6.5km), land seismic refraction (4.5km), echo sounding (12km) and side scan sonar (12km).
Laboratory testing	Classification suite (400 No.), direct shear strength (100mm dia.- 248No, 300mm dia.- 13 No.), unconfined compressive strength (100 & 54mm dia.- 438 No.), Young's modulus (214 No.), Poisson's ratio (166No.).

Table 2: Summary of ground investigation works

Correlation tests showed the Trias Sandstone and Mudrock strata at these sites to be similar to those encountered across the estuary.

Predominantly the borehole cores were of 100mm nominal diameter and were sunk using a wireline double tube core barrel with a continuous core liner and estuary water flush. This core diameter was chosen to obtain the best practical quality core in the anticipated strata. A limited number of boreholes drilled onshore and on the Gwent mudflats were sunk using a conventional double tube core barrel with plastic core liner and drill rods. Six 300mm diameter rotary drilled boreholes sunk at the main bridge foundations and in two onshore test areas located on the Gwent and Avon banks as shown on Figure 1. It was considered that this core size was the maximum diameter which could be practically handled and was likely to recover better quality core compared to the other boreholes, particularly in the weaker strata such as the Trias Mudrocks. These boreholes also provided large diameter samples for laboratory direct shear testing which it was considered were likely to give results which were more representative of the in situ strata than tests on conventional 60 and 100mm diameter samples. Following reaming out to 600mm diameter these boreholes facilitated the execution of in situ plate loading tests at depth within the founding strata. Extensive literature searches and enquiries with major ground investigation contractors showed that there was little experience with large diameter rotary core drilling in the UK and Europe. The 300mm diameter double tube core barrel, associated equipment and reaming bits used on the project were designed and manufactured by Core Drill (UK) Ltd. A core liner was not used.

The deformation characteristics of the strata were key engineering parameters to be determined by the ground investigation works. Extensive in situ testing was performed using the Cambridge Insitu high pressure dilatometer to determine the shear modulus and the elasticity moduli of the strata principally in the horizontal plane. Large diameter plate loading tests were also undertaken to directly measure the strata elasticity moduli in the vertical plane. Four 500mm diameter plate loading tests were specified offshore in boreholes at the main bridge foundations. An extensive literature search revealed no published information on plate bearing tests performed in similar conditions. A further two 500mm diameter borehole plate bearing tests and five 1000mm diameter trial pit tests were performed on Trias Sandstone in the Gwent Test Area. A single borehole plate bearing test was performed on Trias Mudrock in the Avon Test Area. These tests complemented laboratory measurements of elasticity moduli on cores of 54 to 100mm diameter.

Extensive geophysical testing was also performed at the site including cross hole and down hole seismic measurements in dedicated borehole arrays and down hole geologging using sidewall sonic and density sondes. This testing provided dynamic deformation properties of the strata and was used to assist in the correlation and extrapolation of data between boreholes.

The rock mass shear strength was a further key engineering parameter to be determined from the ground investigation because the crossing and its foundations were designed to withstand possible ship collisions which could impose substantial lateral loading. Field surveys revealed the Trias Sandstone, which forms the founding stratum for the majority of the caissons, to be thinly to medium bedded but with relatively few persistent vertical or sub-vertical discontinuities. The in situ strata could not therefore, be classified as being a closely jointed rock mass in accordance with Hoek's empirical failure criteria (1983). Shearing of the in situ rock mass would accordingly most likely involve a combination of shearing along existing discontinuities and through intact rock material. It was considered therefore, that the behaviour of the in situ rock mass would generally be controlled by the discontinuity shear strength.

Discontinuity shear strengths were determined using laboratory direct shear tests on core samples of both 100 and 300mm diameter and were also measured in five in situ (700mm square) direct shear tests on Trias Sandstone strata in the Gwent Test Area. The investigation philosophy was to create a large data set based on the smaller diameter test samples and to enhance these data with a smaller number of tests on larger samples with discontinuity morphology closer modelling the full scale engineering geological situation. The intact rock material strength was measured in the laboratory using triaxial testing on samples of 54 and 100mm diameter.

In addition to measuring the deformation and shear strength characteristics of the strata in the laboratory, suites of classification tests were performed. The median engineering properties determined from these tests are presented in Table 3.

Properties (median values)	Trias Sandstone	Trias Mudrock	Carboniferous Sandstone	Carboniferous Mudrock
Moisture content (%)	7.7	7.5	2.3	5.9
Dry density (Mg/m³)	2.23	2.26	2.52	2.37
Porosity (%)	17.3	17.4	7.1	12.7
Specific gravity	2.74	2.71	2.78	2.64
Saturated density (%)	81.8	79.7	-	-
Point load $Is_{(50)}$ axial (MPa)	0.92	0.53	4.01	0.88
Point load $Is_{(50)}$ dia'l (MPa)	0.81	0.38	2.99	0.67
UCS (MPa)	31.1	16.6	75.6	18.3
Poisson's ratio	0.39	0.36	0.22	0.38
Tensile strength (MPa)	1.93	1.42	4.47	1.04

Table 3: Summary of laboratory tests

ENGINEERING GEOLOGY

The philosophy of the ground investigation was to classify the rock strata based on geological and discontinuity mapping prepared by experienced engineering geologists, rock quality parameters determined from rock cores and engineering tests performed in situ and in the laboratory. A main aim was to establish the inter-relationship between the engineering properties of the rock material and the geological environment. Numerous authors, notably Pinto de Cunha (1990) and Hobbs (1975) have studied the scale effects between rock material properties measured in the laboratory and in situ performance of the rock mass. Hobbs (1975) proposed a correlation using a mass factor 'j' related inter alia to the average fracture spacing (AFS) and rock quality designation (RQD) of the strata. These parameters were determined for all core recovered at the site and were used to derive a mass factor 'j' for the strata encountered at each borehole location. The average rock quality parameters and typical values of mass factor 'j' determined for the 100 and 300mm diameter cores from similar locations are presented in Table 4. In general the rock quality parameters for both core diameters were comparable. A substantial improvement in total core recovery in the relatively weaker Trias Mudrock strata was however achieved with 300mm diameter coring compared with 100mm diameter sampling. It is possible that further improvement in core recovery of 300mm diameter core may be achieved by use of a core liner.

The handling of rotary cores is particularly important to maintain their integrity and to ensure that they are not subjected to undue disturbance which might make the data retrieved unrepresentative of the in situ strata. In common with industry practice for standard diameter cores, the 300mm diameter core was extruded horizontally. As a core liner was not used it was necessary to extrude the core into a 300mm diameter semi-circular steel trough. The core was photographed and recorded on board the jack-up platform, offshore and at the drilling sites onshore. Representative samples were taken for laboratory testing and the remaining core was preserved. To enable the core boxes to be man-handled it was necessary to restrict the weight of the core and box to 60kg maximum. Where core lengths were longer than about 0.35m they were cut, using a circular concrete saw, to a suitable length for transportation. Owing to the size and weight of the 300mm diameter core, difficulties were experienced with its extrusion and handling on site, particularly offshore. It is considered that significant improvements in the handling of 300mm diameter core may be achieved through the use of a core liner to which end caps could be securely fitted. A suitable core liner could allow transfer and transportation of the core in the vertical position, this being more suitable to the employment of mechanical equipment for lifting. The core could then be extruded and examined in the controlled environment of a laboratory.

Through the application of the appropriate mass factor 'j' to the material deformation properties determined in the laboratory it was possible to directly compare those results with the mass deformation properties measured in situ. This also enabled all available data to be used to derive the design parameters which was particularly important for economic as well as technical reasons.

Property	Trias Sandstone		Trias Mudrock		Carboniferous Sandstone		Carboniferous Mudrock	
Core diameter (mm)	100	300	100	300	100	300	100	300
Total core recovery (%)	85	82	80	100	99	98	99	-
Solid core recovery (%)	63	61	53	46	74	48*	46	-
Average fracture spacing (mm)	200	325	158	101	291	221	81	-
Rock quality designation (%)	46	46	40	40	64	46*	27	-
Typical mass factor 'j'	0.2	0.2	0.2	0.2	0.5	0.5	0.2	-

* borehole encountered vertical fissure

Table 4: Comparison of rock quality parameters for 100 & 300mm diameter cores

Based on the engineering geological assessment of the site and the imposed foundation loads it was possible to group the 49 foundation locations into 10 regions to facilitate efficient foundation analysis and design. The design parameters for each group of foundations were therefore derived taking direct and quantitative account of the local engineering geology.

DEFORMATION PROPERTIES

Deformation properties were derived from plate bearing tests in boreholes and trial pits onshore, pressuremeter tests offshore and onshore and from laboratory testing of core samples. These methods yielded 'static' parameters. Cross hole and down hole geophysics was also carried out as correlation between boreholes. This also provided data on 'dynamic' properties of the strata. Summaries of the static and dynamic test data are presented in Tables 5 and 6 respectively.

The borehole plate bearing tests essentially followed the methodology set out in BS5930 (1981) and the ISRM Suggested Methods (1981). The tests were performed using a 500mm diameter plate in boreholes of 600mm diameter. The boreholes were initially sunk at 300mm diameter by rotary coring and were subsequently reamed out to 600mm diameter using a purpose designed rock roller bit. The test surface was then 'smoothed' using a flat faced bit which achieved a test surface with smooth curved undulations of generally less than 5mm. The optimum cementitious grout mix for bedding the test plate on the prepared surface was determined from a series of trial mixes. The test results are presented in Table 5.

Method of derivation	Deformation Properties (MPa)	Trias Sandstone	Trias Mudrock	Carboniferous Sandstone	Carboniferous Mudrock
In situ plate bearing tests	E'_{vi}(stress 0-3MPa)	193 to 540	96	-	-
	E'_v (stress >3MPa)	416 to 1514	328	-	-
Mass factor 'j' & UCS test E	E'_{vi}(stress 0-3MPa)	150	70	4000	600
	E'_v (stress >3MPa)	1000	160	6000	1200
In situ HPD tests	G_i	80	25	300	50
	G_{ur}	500	125	900	150
HPD G & lab Poisson's ratio	E'_{hi}(stress 0-3MPa)	220	70	730	140
	E'_h(stress >3MPa)	1390	340	2200	410

Table 5: Typical design static deformation parameters

A typical load-displacement plot for the plate bearing tests is shown as Figure 3. Initial Young's

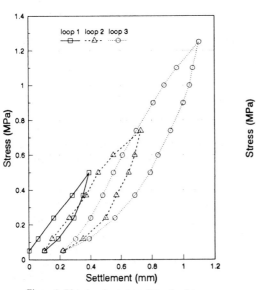

Figure 3: Plate bearing test on Trias Sandstone

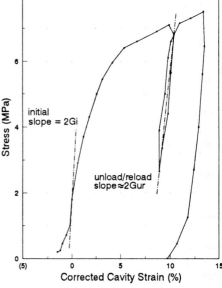

Figure 4: HPD test on Trias Mudrock

modulus (E'_{vi}) values were obtained from the initial parts of the curves, within the stress range 0 and 3MPa which encompasses the maximum design loadings below the caisson foundations. Unload-reload Young's modulus values were obtained from the gradients of the load cycling curves. The E'_{vi}/E'_v ratio was in the range 0.36 to 0.46 in the Trias Sandstone. This is considered to be due to irreversible closure of micro-fractures and open discontinuities within the rock mass on application of load. The single test on Trias Mudrock gave an E'_{vi}/E'_v ratio of 0.29.

Pressuremeter tests were performed in pre-drilled pockets of 76mm nominal diameter using the Cambridge Insitu high pressure dilatometer (HPD). Pressuremeter test data are summarised in

Table 5. A typical load-displacement plot for the HPD tests on Trias Mudrock is presented as Figure 4. Where the test data indicated minimal installation disturbance, values of initial shear modulus (G_i) were derived. Unload-reload shear moduli (G_{ur}) were obtained using one or more loading loops in each of the tests. The majority of the unload-reload loops were carried out over a 0.05 to 0.2 per cent change in cavity strain corresponding to the typical engineering strains anticipated beneath the shallow caisson foundations. Mair and Wood (1987) however, concluded that if installation disturbance on the rock strata is minimal then the initial shear modulus (G_i) is more likely (than G_{ur}) to be applicable to most foundation problems in jointed or fissured weak rocks. Meigh (1976) in his study of the Trias strata also states that the initial shear modulus is the parameter which most closely models the response of the (Trias) rock mass to foundation loading. The initial shear modulus was used to determine the initial Young's modulus for use in design. The values of initial shear modulus determined for the Trias Sandstone plotted against depth are shown on Figure 5.

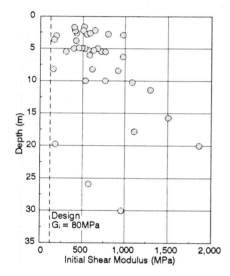

Figure 5: Trias Sandstone initial shear modulus

From the values of initial shear modulus determined in the HPD tests and drained Poisson's ratio measured in laboratory tests values of horizontal initial Young's modulus (E'_{hi}) were calculated using the elastic theory relationship $E' = 2G(1+v')$ (Jaeger and Cook (1976)). This equation assumes an isotropic material and its use was considered justified in the context of overall material variations (Wilson and Corke (1990)). Good agreement between the Young's moduli values determined using the different methods of derivation indicated in Table 5, confirmed the validity of this approach. The results indicate an anisotropy ratio (E'_{vi}/E'_{hi}) of between 0.9 and 2.5 for Trias Sandstone and 1.4 for Trias Mudrock strata.

In addition to the in situ testing, deformation parameters were also derived in the laboratory. In excess of 400 uniaxial compressive strength tests were carried out on samples of rock core, half of which were instrumented with strain gauges in order to measure stress-strain response. The values of Young's modulus so obtained were only representative of the intact rock material owing to the small sample size. In order to apply these data to the field situation as measured in the plate bearing and pressuremeter tests mass factor 'j' (ref Engineering Geology section) was applied to the results. Typical *mass* Young's moduli obtained in this way from laboratory testing are shown in Table 5.

Dynamic deformation moduli derived from the geophysics testing are summarised in Table 6. There is general agreement between the dynamic deformation properties obtained from the various methods of test. The dynamic deformation moduli are however, at least one order of magnitude greater than the static moduli referred to above.

Geophysical Technique	Deformation Properties	Trias Sandstone	Trias Mudrock	Carboniferous Sandstone	Carboniferous Mudrock
Cross hole seismic	E_d (GPa)	6.49	-	29.3	7.93
	G_d (GPa)	2.30	-	10.87	2.82
	v_d	0.40	-	0.34	0.40
Down hole seismic	E_d (GPa)	4.09	-	11.77	7.63
	G_d (GPa)	1.45	-	5.72	2.76
	v_d	0.42	-	0.43	0.38
Geologging	E_d (GPa)	15.75	11.93	27.89	20.01
	G_d (GPa)	6.89	4.8	10.64	7.71
	v_d	0.32	0.32	0.31	0.31

Table 6: Typical dynamic deformation parameters

SHEAR STRENGTH CHARACTERISTICS

During the logging of the rock cores, samples containing discontinuities representative of the forms encountered in the strata were identified and preserved for laboratory direct shear testing. The two halves of the test specimens were secured together and wrapped in plastic film and aluminium foil and coated in microcrystalline wax in accordance with ISRM 'Suggested Methods' (1981) and BS5930 (1981) to preserve the condition of rock and the discontinuity. The samples were then transported and stored in accordance with the recommendations in BS5930 (1981).

The ISRM suggested method of test was used to determine the direct shear strength of the rock core discontinuities. The 100mm diameter specimens were tested in a Robertson Research portable rock shear box. Extensive modifications to the basic equipment were made to achieve accurate control and monitoring of the test. The improvements included, use of a triaxial machine pressure maintainer to apply the normal and shear loads and to regulate the rate of shear, a low voltage digital transducer (LVDT) to measure horizontal displacements and two LVDTs diagonally opposed to measure vertical movements, all of which were linked to a data logger. The 300mm diameter specimens were tested in a 300mm square Wykeham Farrence 'soils' shear box.

To complement the laboratory tests on the Trias sandstone founding strata and to provide data on substantially larger discontinuity lengths closer to the full scale situation, five in situ direct shear tests were performed in the Gwent Test Area. The discontinuities for test were selected from trial excavations sunk adjacent to the proposed test locations. The excavations were then extended to prepare the 700mm square, 350mm high test block. Initial excavations were made using a hydraulic pecker. The block was then cut out using a circular concrete saw and trimmed using hand held tools to minimise disturbance to the test block and discontinuity. The ISRM suggested method was adopted for the tests. Each test comprised four shearing stages at normal stresses of between 0.25 and 1.25MPa. Following testing, the discontinuity faces of the laboratory and in situ test specimens were profiled and accurately recorded to determine the contact area between the specimen halves. Correction of the test results for contact stress and for dilation in accordance with Hencher and Richards (1989) was found to significantly reduce data scatter. The corrected results for Trias Sandstone are presented on Figure 6 as a graph of shear stress v normal stress. The results of the large scale in situ tests which may more closely represent the en masse behaviour (ref Bandis (1990)) were found to lie in a narrow band within the overall data spread for the smaller scale laboratory tests. A greater weighting was therefore

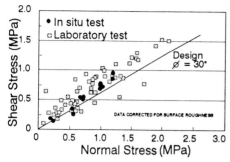

applied to in situ test results in determining a moderately conservative shear strength (as defined in CIRIA Report 104) for use in design. The typical design peak and residual discontinuity shear strengths determined for the four strata categories are summarised in Table 7. Also presented in the table are the typical equivalent intact shear strengths determined from p' v q' plots from triaxial tests and undrained shear strengths determined from HPD tests.

Figure 6: Trias Sandstone shear strength test data

Shear Strength	Trias Sandstone	Trias Mudrock	Carboniferous Sandstone	Carboniferous Mudrock
Discontinuity peak shear strength	$c = 0, \phi = 30°$	$c = 0, \phi = 26°$	$c = 0, \phi = 30°$	$c = 0, \phi = 27°$
Discontinuity residual shear strength	$c_r = 0, \phi_r = 28°$	$c_r = 0, \phi_r = 24°$	$c_r = 0, \phi_r = 28°$	$c_r = 0, \phi_r = 21°$
Equivalent intact shear strength (p'-q')	$c = 0, \phi = 57°$	$c = 0, \phi = 51°$	$c = 0, \phi = 75°$	$c = 0, \phi = 60°$
Undrained shear strength (c_u)	2.5 MPa	1.0 MPa	7.5MPa	1.0MPa

Table 7: Summary of typical design shear strengths

CONCLUSIONS

The geotechnical design parameters for the new 5.2km long crossing of the Severn estuary were derived from geological and discontinuity mapping, rock quality parameters determined on recovered core of 100 and 300mm diameters, conventional and specialist geotechnical investigations. These investigations included extensive in situ HPD pressuremeter testing and large scale plate bearing and direct shear tests and wide field geophysical testing complemented with laboratory testing. The combined qualities of experienced engineering geologists and geotechnical engineers were important to achieve the most pragmatic approach to bridge foundation design. The accurate recording of field data should, as has been proved at this site minimise the risk of unexpected conditions being encountered during construction. This can be particularly important in design and build schemes where the provision for additional payments to overcome unforeseen ground conditions may be removed.

ACKNOWLEDGEMENTS

The concession company for this project was Severn River Crossing plc, a consortium lead by John Laing Construction in joint venture with GTM Entrepose. The designer was a joint venture between Sir William Halcrow & Partners Ltd and Societe d'Etudes et d'Equipements d'Entreprises and the ground investigation was performed by Wimpey Geotech Ltd. The authors express their thanks to their colleagues in these organisations.

NOTATION INDEX

E'_{vi} drained initial vertical Young's modulus
E'_v drained vertical Young's modulus
E'_{hi} drained initial horizontal Young's modulus
E'_h drained horizontal Young's modulus
E_d dynamic Young's modulus
G_i initial shear modulus

G_{ur} shear modulus (unload/reload loop)
G_d dynamic shear modulus
v' drained Poisson's ratio
v_d dynamic Poisson's ratio
j mass factor
c peak cohesion intercept
ϕ peak angle of shearing resistance
c_r residual cohesion intercept
ϕ_r residual angle of shearing resistance
c_u undrained shear strength
p' $\frac{1}{2}(\sigma_1'+\sigma_3')$
q' $\frac{1}{2}(\sigma_1'-\sigma_3')$

REFERENCES

Bandis S C. (1990) Scale effects in the strength and deformability of rocks and rock joints. Proceedings First International Workshop on Scale Effects in Rock Masses. Loen, Norway.

British Standards Institution. (1981) BS5930 Code of Practice for Site Investigations

Chandler R J. and Davies A G. (1973) Further Work on the Engineering Properties of Keuper Marl. CIRIA Report 47.

Hencher S R. and Richards L R. (1989) Laboratory direct shear testing of rock discontinuities. Ground Engineering. March

Hobbs NB. (1975) Factors affecting the prediction of settlement of structures on rock: with particular reference to the Chalk and Trias. British Geotechnical Society Conference on Settlement of Structures.

Hoek E. (1983) Strength of jointed rock masses. Geotechnique 33 No. 3 pp187-223.

International Society for Rock Mechanics Testing (1981). Rock Characterisation Testing & Monitoring : ISRM Suggested Methods. pub Pergamon Press.

Jaeger J C. and Cook N G W. (1976) Fundamentals of Rock Mechanics 2Ed. pub Chapman and Hall Ltd.

Mair R J. and Wood D M. (1987) Pressuremeter Testing, Methods and Interpretation. CIRIA

Meigh A C. (1976) The Triassic rocks with particular reference to predicted and observed performance of some major foundations. Geotechnique 26 No. 3 pp391-452

Padfield C J. and Mair R J. (1984) Design of retaining walls embedded in stiff clay. CIRIA Report 104.

Pinto de Cunha A. (1990) Scale effects in rock masses. Proceedings of First International Workshop on Scale Effects in Rock Masses. Loen, Norway.

Wilson W. and Corke D J. (1990) a comparison of modulus values of sandstone derived from high pressure dilatometer, plate loading, geophysical and laboratory testing. Pressuremeters, pub Thomas Telford Ltd, London

M5 Avonmouth bridge strengthening — in situ testing of Mercia Mudstone

N V PARRY, A J JONES
Acer Consultants Limited, Bristol, England
M R DYER
Mark Dyer Associates, Bristol, England*

SYNOPSIS

This paper describes the results of in situ tests carried out as part of a ground investigation for the Avonmouth Bridge Strengthening works near Bristol. The in situ tests comprised plate bearing tests and pressuremeter tests carried out in Mercia Mudstone. Analysis and correlation of the test data has provided shear strength, in situ horizontal stress, shear and elastic modulus profiles with depth for the Mercia Mudstone. The results are correlated with laboratory test data and Standard Penetration Tests. Practical factors relating to performing tests in the Mercia Mudstone are also discussed.

INTRODUCTION

A ground investigation was carried out for the Avonmouth Bridge Strengthening Works between Junctions 18 and 19 of the M5. The aim of the investigation was to provide additional geotechnical data for the design of the proposed motorway improvement options. The project involves an assessment of the Avonmouth Bridge to Departmental Standard BD37/88 considering:-

- increased net bearing pressures for spread foundations which support the majority of bridge piers.
- reduction of bending moments at piers due to foundation flexibility permitting some rotation of the bases.
- increased pile loads at three pier locations.

To provide engineering recommendations on the proposed increase in foundation loads and the stiffness of the founding bedrock the following in situ tests were carried out:-

a) Selective pressuremeter tests in additional rotary drilled and self-bored holes at depths within the zone of influence for pier foundations. Due to the variations in lithology expected in Mercia Mudstone, two types of pressuremeter equipment were mobilised to the site. These were:-

   i) High Pressure Dilatometer (HPD), and
   ii) Weak Rock Self Boring Pressuremeter (SBP)

*Formerly of Acer Consultants Limited

The pressuremeters were used to measure horizontal shear moduli (G) for the Mercia Mudstone.

b)   In addition plate bearing tests (PBT) were carried out in deep trial pits to measure the vertical elastic modulus (E) of the Mercia Mudstone. To investigate scale effects, two plate sizes of 300mm and 600mm diameter were used.

A comparison of the PBT and pressuremeter tests in the Mercia Mudstone are made together with conventional data used to obtain stiffness moduli.

AVONMOUTH BRIDGE STRENGTHENING

Avonmouth Bridge is a high level 1350m long twin steel box girder bridge located to the west of Bristol, as shown in Figure 1. Twenty spans of between 30m and 173m are supported by 2 abutments and 19 piers, 3 of which are piled. An additional lane is to be added asymmetrically to each side of the bridge deck as part of the M5 motorway widening scheme between Junctions 18 and 19. The proposed works will result in an increase in foundation pressures and stresses in the piers.

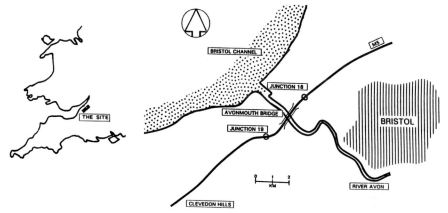

Fig. 1 - Location of Avonmouth Bridge.

GEOLOGY AND GROUND CONDITIONS

The study area is underlain by the general stratigraphy presented in the table below:-

Stratum	Age
Estuarine Alluvium Head Deposits	Recent and Pleistocene
Mercia Mudstone Dolomitic Conglomerate	Triassic

General Stratigraphy

The Dolomitic Conglomerate, is a dolomite-cemented sandy limestone gravel which passes laterally into the sandstones, siltstones and mudstones of the Mercia Mudstone Group.

Mercia Mudstone represents the finer grained sediments of the Triassic Period. It is generally a very weak to moderately strong mudstone which has a varying proportion of silt and sand sized particles within it. Thin layers of red and yellow sandstone are known to occur within the strata. The mudstones generally weather to stiff to very stiff clays, which may become substantially softened in excavations due to water issuing from fissures and sandstone layers.

Previous boreholes (dated 1961-1967) at the site generally indicated three to six metres of alluvium overlying the Mercia Mudstone and Dolomitic Conglomerate.

The scope of the ground investigation performed in 1994 is shown in Figure 2.

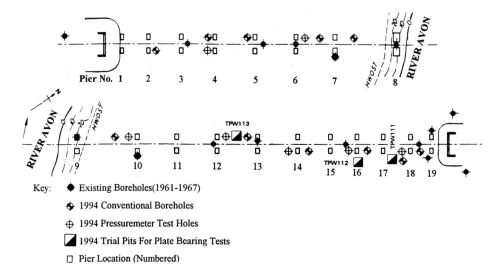

Fig. 2 - Avonmouth Bridge Ground Investigation

CABLE PERCUSSIVE/ROTARY BOREHOLE DATA
200mm diameter cable percussive boring and 112mm diameter rotary coring were completed prior to commencing pressuremeter testing. This allowed the ground profile to be identified before carrying out PBT and SBP work.

In addition, in situ Standard Penetration Tests (SPTs) were taken and laboratory Unconfined Compressive Strength and Point Load Tests performed on selected samples of the 112mm diameter core recovered.

The stratigraphy identified is shown in Figure 3.

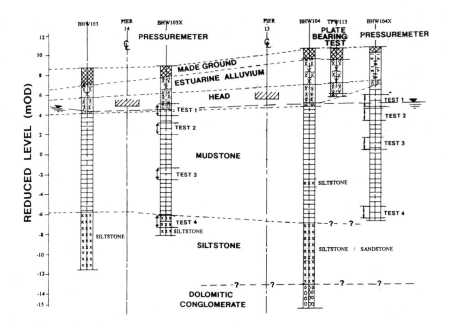

Fig 3. - Ground Profile beneath Avonmouth Bridge.

PLATE BEARING TEST DATA

Three sets of plate bearing tests were performed at Avonmouth Bridge using 300mm and 600mm diameter plates. The tests in TPW111 and TPW112 were performed on material considered to be Head Deposits. The tests in TPW113 were on Zone III Mudstone. Reaction was supplied by ground anchors grouted into the mudstone to a depth of up to 8 m. The 8.5m x 4m x 5m deep pits were supported by trench boxes and where necessary dewatered using sump pumping. The 600mm diameter plate was 75mm thick and the 300mm diameter plate 35mm thick to provide sufficient rigidity under the anticipated loads. The plates were bedded on a thin layer of Plaster of Paris which was tested to give a compressive strength of 1.5 MN/m^2.

Water issuing from fissures within the ground at the test levels occurred in Trial Pits TPW111 and TPW112 where a 2m head of water was present. This resulted in softening of the ground. Testing in TPW113 was performed above the water table and no softening due to ingress of water occurred.

The PBT's in TPW113 were conducted on a red brown moderately weathered mudstone, weak to moderately weak with very closely spaced fissures and a blocky structure.

Each increment of load in the PBT was maintained until the rate of settlement reduced to less than 0.05mm in half an hour. Once the settlement had stabilised at this rate, the primary consolidation was considered to have been completed. The following loading sequence was used:

First Cycle   -   Incremental Loading to Designated Load, 4 increments,
Second Cycle   -   Incremental Loading to 1.5 times Designated Load, 3 increments,
Third Cycle   -   Incremental Loading to 2 times Designated Load, 3 increments
                   (maintained for 24 hours),
Fourth Cycle   -   Incremental Loading to stipulated Maximum Reaction, 5/6
                   increments.

For TPW113 the designated Loads were 0.05 MN and 0.21 MN and the Maximum Reactions were 0.16 MN and 0.63 MN for the 300mm and 600mm plates respectively. The PBT results for the plates are presented in Figures 4 and 5 below.

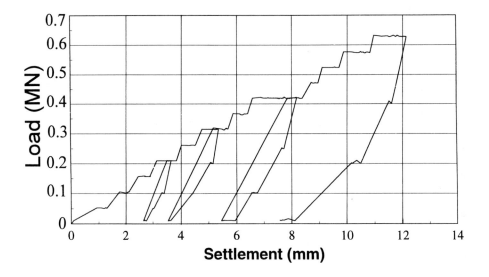

Fig. 4 - Plate Bearing Test Results (600mm diameter plate)

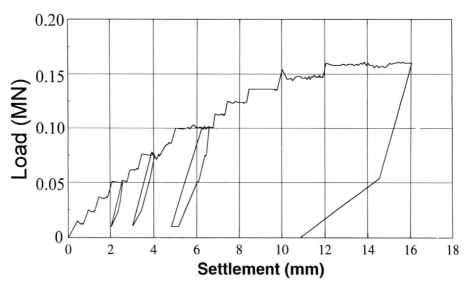

Fig.5 - Plate Bearing Test Results (300mm diameter plate)

The elastic moduli have been determined by assuming a drained Poisson's ration (μ) of 0.2 which is typical of weak rock (Mair and Wood 1987).

The elastic moduli have been determined assuming a homogenous elastic isotropic soil penetrated by a rigid circular plate on a semi infinite plane surface, using the following equation:-

$$S = \frac{\pi q}{4} \cdot \frac{B \, (1 - \mu'^2)}{E'}$$

Where E' is the drained elastic modulus
q is the pressure applied to the plate
S is the average settlement of the plate
B is the diameter of the plate
μ' is the drained Poisson's ratio

A summary of the elastic moduli from the tests in TPW113 is given in the table below:-

600mm diameter Plate		300mm diameter Plate	
Stress Range (MN/m^2)	E' (MN/m^2)	Stress Range (MN/m^2)	E' (MN/m^2)
Loading 0-0.743	$E'i_1 = 91.8$	Loading 0-0.707	$E'i_1 = 62.0$
0.743-1.114	$E'i_2 = 110.5$	0.707-1.061	$E'i_2 = 62.0$
1.114-1.485	$E'i_3 = 103.1$	1.061-1.415	$E'i_3 = 47.6$
1.485-2.228	$E'i_4 = 89.1$	1.415-2.087	$E'i_4 = 29.5$
Reloading 0.035-0.743	$E'ur_1 = 367.8$	Reloading 0.141-0.707	$E'ur_1 = 237.0$
0.035-1.132	$E'ur_2 = 300.6$	0.141-1.061	$E'ur_2 = 231.1$
0.035-1.485	$E'ur_3 = 271.1$	0.141-1.415	$E'ur_3 = 201.4$
Creep: at 1.485 MN/m^2 for 12 hrs = 0.35mm		Creep: at 1.415 MN/m^2 for 12 hrs = 0.23mm	

Summary of Moduli and Creep from PBTs

The results indicate that the overall response of the supporting ground beneath the 300mm plate was less stiff than that beneath the 600mm plate. This can be expected since the theoretical bulb of pressure from the 300mm plate will be mainly within the more weathered rock.

The maximum loads applied to the plates did not result in shear failure of the rock. The undrained shear strengths were estimated in accordance with the method proposed by Chin (1970). The computed undrained shear strengths were 0.43 and 0.60 MN/m^2 for the 300mm and 600mm plate respectively. These values reflect the scale effects noted in the resultant drained elastic modulus (E') values.

PRESSUREMETER TEST DATA
Both High Pressure Dilatometer (HPD) and Weak Rock Self Boring Pressuremeter (SBP) equipment were utilised to provide results for the varying lithologies of Mercia Mudstone. Four tests were performed in each borehole within the anticipated pressure bulb of the spread foundation. 100mm diameter rotary open-hole drilling was used to progress the holes between tests, HPD tests were taken within pockets from 76mm diameter rotary cored holes and SBP holes were 75mm diameter. Calibration of the equipment was performed before and after the completion of each borehole.

21 No. Pressuremeter Tests were carried out in seven boreholes within the Mercia Mudstone, three of these tests were within, or partly within, the Head Deposits.

The undrained shear strength (Su) from the pressuremeter tests are plotted against depth below rock head in Figure 6. The results indicate that there is a gradual increase in the mudstone Su values as the degree of weathering reduces. The computed Su values assume that no drainage occurs during failure and they may therefore be an overestimate of the undrained shear strength.

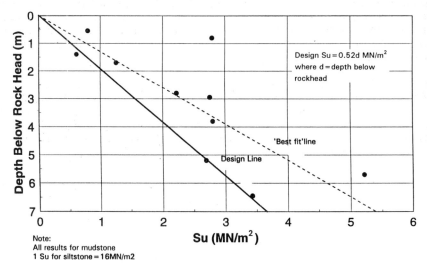

Fig. 6 - Shear Strength (Su) against Depth Below Rock Head.

The assessed in situ total horizontal stress (σ_{ho}) for each test within mudstones or Head Deposits are plotted against depth below rock head in Figure 7. The best fit line indicates that there is a gradual increase in σ_{ho} with depth.

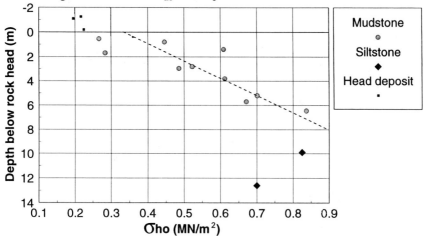

Fig. 7 - In Situ Horizontal Stress (σ_{ho}) against Depth Below Rock Head

Shear moduli were computed for initial loading (Gi) and for unload-reload loops (Gur). The ratio of Gur/Gi is a measure of the degree of fissuring or fracturing and of initial disturbance during installation of the pressuremeter in weak rock. Mair and Wood (1987) report that the ratio of Gur/Gi is often as high as 3 for moderately intact rocks, and values of up to 10 may be obtained for more heavily fractured materials. The ratio of Gur/Gi for mudstones at this site generally ranges between 8 and 3, decreasing with depth, indicating that the rock ranges from heavily fissured to near intact. The Gur values for mudstone also tend to increase with increasing applied stress which indicates that fissures were being closed as the cavity expanded. The Gur values are dependent on both the applied stress level and the cavity strain range.

The Gi and Gur_1 (for the first loop) are plotted against depth below rock head in Figure 8 and 9 respectively for the Mercia Mudstone. From the best fit line of the mudstone, the relationship between Gi and depth is as shown in Figure 8. Stiffer modulus values are required for a conservative analysis of the redistribution of bending moments in the bridge superstructure. Thus the Gur values are recommended for analysis of the pier bases founded on the mudstone.

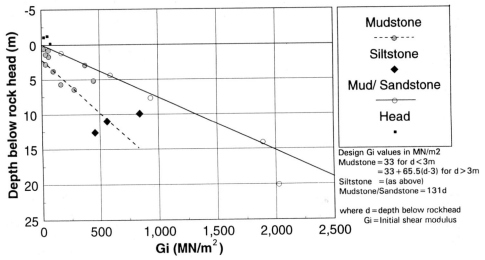

Fig.8 - Gi against depth below rock head

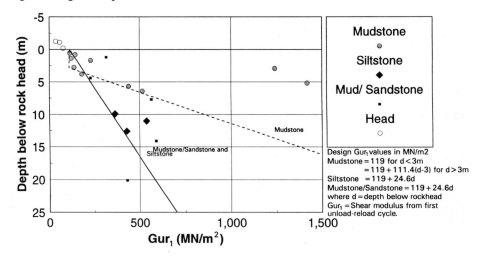

Fig. 9 - Gur against depth below rock head

REVIEW OF RESULTS
Correlation of PBT and Pressuremeter Results
The PBT and pressuremeter results are correlated using the in situ PBT at 5.9 mOD in TPW113 and the pressuremeter results at 5.2m mOD in BHW104X which are adjacent to each other.

The shear modulus results obtained from the pressuremeter tests are presented below together with computed E' values.

Loading Stage (MN/m^2)	G (MN/m^2)	E'h (MN/m^2)
Loading to 0.85	Gi = 33	E'hi = 79
Unloading from 0.98	Gur$_1$ = 121	E'hur$_1$ = 290
Unloading from 1.20	Gur$_2$ = 155	E'hur$_2$ = 372
Unloading from 1.5	Gur$_3$ = 1042	E'hur$_3$ = 2500

The above E' values have been calculated using:-

$$E'h = 2G(1 + \mu')$$

The value of the drained Poisson's ratio (μ') is again taken as 0.2.

The 600mm diameter PBT produced the stiffer response as shown in Figures 4 and 5.

The initial modulus E'hi and unload reload modulus E'hur$_1$ (for the first loop) from the pressurement test are correlated with the corresponding E'i and E'ur$_1$ values from the 600mm diameter PBT. This correlation indicates that the vertical moduli are 15% to 30% higher than the equivalent horizontal moduli. On the basis of this correlation, the horizontal moduli for all the pressuremeter tests are converted to vertical moduli assuming the rock is homogeneous and isotropic. These vertical moduli, derived from Gi and Gur$_1$ values, are shown in Figure 10.

Correlation with UCS and SPT Results
E' values are derived from the unconfined compressive strength (UCS) using the relationship:-

$$E' = j. Mr. UCS \text{ (Reference BS8004)}$$

where j is the mass factor relating to the discontinuity spacing (taken as 0.2)
Mr is the ratio between the elastic modulus and the UCS (taken as 150)

E' values have also been derived from Standard Penetration Test (SPT) N values using the relationship which was proposed by Stroud (1988) of E' = 1.0 N as shown below in Figure 10. The correlation between the results is in good agreement based on Gi values. As expected, the E' (unloading - reloading) values derived from Gur values are generally higher than the other E' (loading) values. Figure 10 further identifies the upper 3m of rock to be less stiff than the underlying rock.

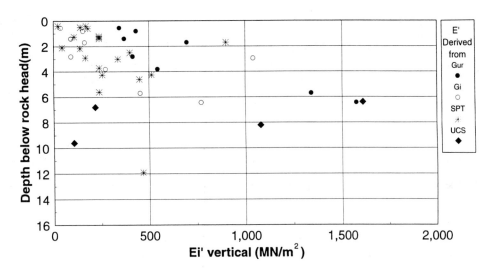

Fig. 10 - Correlation of E' Values

In addition the correlation of shear strengths is presented below in Figure 11. Shear strengths have been computed from SPT (N) values using the following relationship which was developed using Su = 5 N after Stroud (1988).

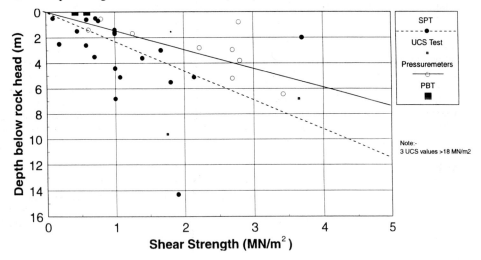

Fig. 11 - Correlation of Shear Strength.

The correlation between the best fit lines of shear strengths computed from the SPT (N) values and pressuremeter tests is generally in agreement.

Determination of k values

Following the estimation of modulus values, upper and lower bound coefficients of subgrade reaction (k) have been determined for each pier foundation. The k values were computed from the analysis of a flexible foundation within an elastic layered half space using Mindlin's equations. A factor of 0.8 was applied to the maximum central displacement to determine the approximate settlement of a rigid footing. k values were then calculated by dividing the applied loading intensity by the deformation.

The upperbound k values range from 0.06 to 0.284 MN/mm which represent the relatively stiff unload-reload response of the footing under variable loading conditions. The lowerbound k values range from 0.0015 to 0.159 MN/mm which represent the settlement response under an increase in load above existing pressures. For conservative design, upperbound values were used to determine reduction in bending moments in the piers and lowerbound for settlement of the foundations.

CONCLUSIONS

Comparison of Plate Bearing and Pressuremeter Tests indicated a ratio of typically 1.16 to 1.27 for vertical to horizontal moduli in the range of net foundation pressures.

Comparison of Pressuremeter Tests and SPTs showed Stroud's correlation to be in good agreement on initial loading. However, moduli values from reload tests were significantly higher and these values were used to calculate the upperbound k values.

The upper bound range in k values corresponded to a reduction in bending moments for the bridge piers of between 5 and 20 percent. Likewise the lower bound range in k values corresponded to predicted settlements of between less than 1mm and 25mm.

On a practical level, the plate bearing tests illustrated the difficulty of carrying out the tests below the groundwater table. The main difficulty arose from a significant seepage of ground water through fissures in the mudstone. Nevertheless the tests were valuable in correlating the pressuremeter test results for vertical foundation loads at this particular site. In the event, the initial and first unload-reload horizontal modulus values for the pressuremeter tests tended to be up to 30% lower than the corresponding vertical modulus values from the PBTs. In comparison, the variation in moduli values were significantly greater than the differences in values between plate bearing and pressuremeter tests.

ACKNOWLEDGEMENTS

The authors wish to thank the Highways Agency for permission to publish this paper. The main ground investigation and plate bearing tests were undertaken by C J Associates with pressuremeter tests performed by Exploration Associates. Thanks are extended to Peter Allan of Exploration Associates for his advice in interpreting the pressuremeter results. Thanks also to Justin Wilmott who produced the figures.

REFERENCES

Chin F.K. (1970), Estimation of the ultimate load of piles not carried out to failure. Proceedings 2nd S.E. Asia Conference on Soil Engineering, pp 81-90.

Broch E. and Franklin J.A. (1972); The point load strength test, International Journal of Rock Mechanics and Mining Science,9.

BS8004: 1986 Foundations, British Standards Institution.

Mair R.J. and Wood D.M. (1987); Pressuremeter Testing - Methods and interpretation, CIRIA Ground Engineering Report: In Situ Testing, Butterworths, 1987.

Meigh A. C. (1976); The Triassic rocks with particular reference to predicted and observed performance of some major foundations. 1976 Rankine Lecture, Geotechnique 26, No. 3 391-452.

Stroud M. A. (1988); The Standard Penetration Test - its Application and Interpretation. Penetration Testing in the UK - Proceedings of the Geotechnology Conference organised by the Institution of Civil Engineers. Birmingham, July 1988.

Moderator's report on Session 4(b): pressuremeter, permeability and plate tests

B G CLARKE
Department of Civil Engineering, University of Newcastle, Newcastle upon Tyne, UK

INTRODUCTION

In situ tests can be intrusive or non intrusive depending on whether the properties of the ground are changed during installation and/or testing. Non intrusive tests are predominantly surface geophysical tests though borehole geophysical tests are often assumed to be non intrusive since the majority of the ground tested is unaffected by the installation of the borehole and none is affected by the test.

Intrusive tests can be subdivided into profiling tests and others. Profiling tests with penetrometers are generally used to obtain semi continuous or continuous profiles of ground properties and/or strata. The ground is tested as the profiling tool is installed. Usually measurements are taken at set intervals independent of the strata encountered though tests can be specified at selected horizons (e.g. Standard Penetration Test).

Other intrusive tests can be subdivided into those that measure a single property and those that give a ground response curve. The former includes devices to measure density, total horizontal stress, pore pressure and permeability; the latter includes devices used to obtain vertical and horizontal stiffness and strength. In both cases, tests are usually specified at particular horizons. This session is devoted to intrusive in situ tests excluding penetrometers. This paper covers the background to this form of intrusive testing and highlights features of recent developments including those issues raised in the papers reported in the session. Seven of the eleven papers in the session refer to pressuremeter testing and cover equipment design, interpretation and application.

INTRUSIVE TESTS EXCLUDING PENETROMETERS

In situ tests are commonly known by the description of the equipment used rather than the test procedure or parameter obtained. Table 1 is a summary of intrusive tests, their usefulness and applicability. The table was introduced by Robertson (1986) and has been updated and added to by others as improvements in equipment and interpretation were introduced. Unlike the majority of laboratory tests there are few British Standard procedures for in situ tests, though standard procedures are followed. For example a commonly used procedure for prebored pressuremeter tests has been specified by LCPC.

In situ tests have been the subject of conferences, speciality sessions, and seminars both at national and international level. Each of the recent International Society of Soil Mechanics and Foundation Engineering (ISSMFE) Conferences, both at regional and international level, included themes and special reports on in situ testing; there have been three conferences and a fourth planned (Sherbrooke, May 1995) on pressuremeter testing; four books have been

Table 1   The applicability and usefulness of in situ tests (after Robertson, 1986 and Wroth, 1984)

Group	Device	Parameters													Ground Type						
		soil	profile	u	ϕ'	s_u	Dr	m_v	c_v	k	G	σ_h	OCR	σ-ϵ	hard rock	soft rock	gravel	sand	silt	clay	peat
penetrometers	dynamic	C	A	-	C	C	B	-	-	-	C	-	C	C	-	C	B	A	B	B	B
	mechanical	B	A	-	C	C	B	C	-	-	B	C	C	-	-	C	-	A	A	A	A
	static (CPT)	B	A	-	C	C	B	C	-	-	B	C	C	-	-	C	-	A	A	A	A
	piezocone (CPTU)	A	A	A	B	B	A	B	A	B	B	C	A	B	-	C	-	A	A	A	A
	seismic (SCPTU)	A	A	A	B	B	A	B	A	B	A	B	A	B	-	C	-	A	A	A	A
	flat dilatometer (DMT)	B	A	C	B	B	C	B	-	-	B	B	B	B	-	C	-	A	A	A	A
	acoustic probe	C	B	-	C	C	B	B	-	-	C	B	C	-	-	C	-	A	A	A	A
	SPT	B	B	-	C	C	B	-	-	-	-	-	-	-	-	C	B	A	A	A	A
	resistivity probe	B	B	-	B	C	A	C	-	-	C	-	-	-	-	C	-	A	A	A	A
pressuremeter	PBP	B	B	-	C	B	C	B	C	-	B	C	C	C	A	A	B	B	B	A	B
	SBP	B	B	A	A	A	A	A	A	B	A	A	A	A	A	A	-	B	A	A	A
	PIP	A	B	B	C	B	C	C	A	B	A	C	C	C	-	-	-	B	A	A	B
	cone PIP	C	B	B	C	B	C	C	A	B	A	C	C	C	-	-	-	A	A	A	A
other	vane	B	C	-	-	A	-	-	-	-	-	-	-	-	-	-	-	-	B	A	B
	screw plate	C	C	-	C	B	B	B	C	C	A	C	B	B	-	-	-	A	A	A	A
	plate	C	-	-	C	B	B	B	C	C	A	C	B	B	B	A	B	B	A	A	A
	borehole permeability	C	-	A	-	-	-	-	B	A	-	-	-	-	A	A	A	A	A	A	B
	hydraulic fracture	-	-	B	-	-	-	-	C	C	-	B	-	-	B	B	C	C	B	B	C
	crosshole/downhole /surface seismic	C	C	-	-	-	-	-	-	-	A	-	-	B	A	A	A	A	A	B	A

applicability   A - high; B - moderate; C - low; - not

624

published on pressuremeters (Baguelin, et al, 1978, Mair and Wood, 1987, Briaud, 1992, Clarke, 1995). The published work reflects the development of the subject of in situ testing but many of the developments are not applied in practice. For example, there are numerous published theories of cavity expansion but very few are used in practice for the interpretation of pressuremeter tests.

In situ tests are carried out to:-

1) determine a specific parameter (e.g. density);
2) determine a ground response curve;
3) act as a control for engineered fill;
4) assess improvements in ground properties, and
5) calibrate other instruments.

The majority of tests are carried out to determine properties for design (items 1 and 2) or as a control.

There are four methods of interpreting tests:-

1) Tests can give a parameter directly, e.g. in situ density by sand replacement.
2) Tests can be interpreted using empirical rules to give test specific parameters, e.g. the interpretation of prebored pressuremeter tests using the rules developed by Ménard (1957).
3) Tests can be interpreted using semi empirical rules based on observation and theory.
4) Tests can be interpreted theoretically.

Atkinson and Sallfors (1991), in their general report to the 10th European Conference classified methods of interpretation of tests as:

1) primary correlation   sound theoretical basis with few assumptions
2) secondary correlation   theoretical basis with major assumptions and few approximations
3) empirical correlations   no theoretical basis.

Baguelin (1989), in his report to the 11th ISSMFE Conference, considered that primary correlations could only be justified if the application of the derived parameters in design could be validated against an extensive database of the behaviour of full scale structures. He concluded that only secondary and empirical correlations could be justified. These are possibly the most common methods of interpreting in situ tests.

Methods of design are calibrated against observations of full scale structures but the methods are often based on results of laboratory tests. For example, bearing capacity of clays is found from results of quick undrained triaxial tests. Thus, if in situ test results are to be used in those design methods the results of the in situ tests have to be similar to those of the laboratory tests. This is unlikely since, for example, the stress path and rate of testing are different. Unless a design method is specifically developed for an in situ test, as is the Ménard method for design based on pressuremeter tests, secondary or empirical correlations will be the preferred method and those correlations will be based on results of laboratory tests.

Table 2   The applicability of intrusive in situ tests (excluding penetrometers) to soil modelling (after Jamiolkowski et al, 1985)

Soil Behaviour-Parameter	Equipment and/or Procedures		Comments
In situ σ_h	2.1 SBP	2.1a	"Proven" to be successful in soft clays and stiff clays; less experience in sands
		2.1b	Greatest potential among in situ methods
	2.2 Spade total stress cells	2.2a	Limited positive experience only in soft to stiff clays; successfully use in other soils unlikely
		2.2b	In stiff clay overestimates σ_h; requires correction for bedding error
		2.2c	Vertical installation essential
	2.3 Hydraulic fracturing	2.3a	Applicable only to cohesive soils having $K_o < 1$
		2.3b	Interpretation uncertain.
	2.4 Overcoring	2.4a	Applicable to rocks
		2.4b	Interpretation dependent on results of laboratory tests and assumption that rock is linear elastic
In situ vertical effective yield stress	3.1 Plate tests	3.1a	Limited experience
		3.1b	Possible applications limited to relatively homogeneous cohesionless deposits at shallow depths in which tests are performed under fully drained conditions.
		3.1c	For screw plate tests the influence of plate shape and disturbance due to its installation on the load/settlement relationship not well understood
Deformability characteristics	4.1 Plate tests	4.1a	Application limited to shallow depth
		4.1b	Proven in cohesionless deposits to determine average drained Young stiffness E' within the depth of influence of the plate
		4.1c	In cohesive soils, despite uncertainty about drainage conditions, it is assumed to yield average undrained Young stiffness E_u
		4.1d	Since E is obtained from load displacement measurements, an a priori assumption regarding soil constitutive model is necessary
		4.1e	Very difficult to relate the E obtained from plate tests to the behaviour of a soil macro element, hence to strain or stress levels
	4.2 SBP tests	4.2a	Great potential for direct measurement of shear modulus G_h in horizontal direction
		4.2b	G_h describing the elastic soil behaviour can be assessed from small unloading-reloading cycles whose role is to minimise the soil disturbance due to probe insertion
		4.2c	Potential to measure non linear stiffness profiles in clays, sands and weak rocks
Flow and consolidation	5.1 Borehole	5.1	Outflow tests at constant head preferred; interpretation above water level extremely complex
	5.2 Large scale pumping	5.2	Very reliable but also very expensive test; accurate well installation and drawdown measurements with piezometers are required
	5.3 Piezometers	5.3	Constant head tests with Δu small to avoid fracturing are preferred; parameters from outflow tests relevant to OC conditions, inflow tests appropriate for NC conditions
	5.4 Self-boring permeameter	5.4a	Careful installation required
		5.4b	Potential to measure parameters from inflow and outflow tests
	5.5 Holding test (SBP)	5.5	Careful installation required; difficult interpretation due to non monotonic changes of effective stress

Numerical methods based on more realistic ground models permit the response of the ground due to stress changes to be predicted more accurately. Further, they allow parametric studies to be undertaken to calibrate and validate the more simple design methods. An example of this approach is the recent development of design rules for retaining walls. Numerical methods also permit the development of more universal methods of interpretation of tests. Jamiolkowski et al (1985), in their report to the 11th ISSMFE Conference, gave a summary of the capabilities of in situ testing for soil modelling. Table 2 is a development of that summary but only considering intrusive tests and excluding penetrometers.

There are many advantages and disadvantages to in situ tests as there are with all forms of testing. The four main advantages are that:-

1) the volume of ground tested is large enough so that the results reflect the influence of fabric,
2) the results are obtained at the time of the site operations,
3) tests can be carried out in ground from which it is difficult to obtain undisturbed samples, and
4) tests are carried out at the in situ stress.

The main disadvantages are:-

1) poorly defined boundary conditions,
2) unknown drainage conditions,
3) disturbance of the ground during installation of the test equipment, and
4) limited and often fixed stress paths.

In conclusion, non penetrating intrusive tests have been developed to obtain a single parameter directly or a ground response curve. In the latter case tests are interpreted using theories supplemented by observations or empirical correlations. Site operations, equipment and methods of interpretation are, in practice, usually standard though national or international standards may not exist. Improvements in equipment, site operations, interpretation and application are continually taking place. This is driven by the need to obtain rapidly, quality information on the properties of the ground relevant to design.

GROUND PROPERTIES
In Situ Stress
In situ stresses can be either measured passively using instruments or actively using in situ tests. Instruments, however, are more usually used to observe changes in stress with time as compared to in situ tests which are used to obtain design information. Instruments developed to measure in situ stress rather than changes in stress are classed as in situ tests here.

Vertical Stress   Total vertical stress is simply obtained from measurements of in situ density based on the weight and dimensions of undisturbed samples usually of cohesive soils and rocks, and empirical correlations with results of penetration tests, usually in cohesionless soils. Densities of natural ground range between 1.6 Mg/m^3 and 2.4 Mg/m^3, therefore the accuracy of the measurements is often not too important.

Measurements of in situ density, not to determine total vertical stress, but to use as a control on compacted fill, are routinely carried out as a requirement of a specification. Tests include sand replacement, core cutter and nuclear density probe tests which are described in detail in BS1377: 1990. Density is measured to determine indirectly the potential for the fill to settle or become unstable. It is usual to specify in the UK a minimum relative compaction which is the ratio of the actual field dry density to the maximum achievable in the laboratory. It is not unusual to obtain relative compaction values greater than 100% since the compactive effort in the field is greater than that in the laboratory. Any material that has a dry density less than 95% of the maximum is rejected. The financial consequences of this can be great given the time between testing and obtaining the results especially when fill is continually being placed.

Thus, the test procedure has to be correct and the results obtained quickly. Fill is variable and the target maximum dry density varies throughout a contract so relative compaction should be based on the maximum dry density of the fill being tested rather than that of fill placed at some earlier time. Further, density measurements are made in the surface layer, a layer that may be overstressed and loosened during compaction. Figure 1 shows typical results of a 2E ash fill. Density measurements were used as a control in accordance with the specification even though the engineer based the design on the stiffness and strength characteristics of the fill. The in situ CBR values exceeded 35% and the compressive strength was greater than 400 kN/m². This fill was acceptable to the design engineer but failed to meet the specified acceptance criteria.

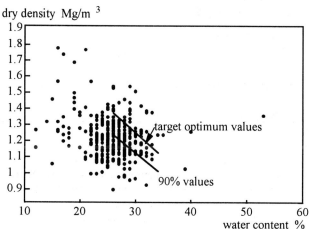

Figure 1   Variation in dry density with water content for a 2E ash fill

The degree of compaction proposed by Cox (1995) gives a relatively quick method of determining whether a fill has been compacted since the sample does not have to be dried. It does not indicate whether the dry density is acceptable, therefore there is a need for a further test such as the MCV which is designed to determine quickly the acceptability of a fill. The use of both tests, if end product control is specified, could ensure that a fill is acceptable and has been placed correctly without the need to measure water content.

Horizontal Stress   Horizontal stress can be measured

a)   passively using instruments such as a total stress cell,
b)   actively using intrusive in situ tests such as the DMT and SBP,

c)   indirectly using hydraulic fracture, and

d)   indirectly using stress relief.

In all cases the process of installing the instrument or equipment will change the in situ stresses as shown in Figure 2, therefore any method must be subjective. A summary of the methods available is given in Table 3.

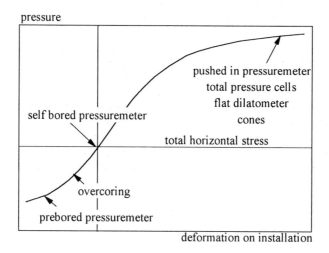

Figure 2   Deformation during installation of in situ devices to measure total horizontal stress

Equipment	Reference
model pile	Kenny (1967)
overcoring	Leeman (1968)
load cells	D'Appolonia et al (1969)
hydraulic fracture	Bjerrum and Anderson (1972)
self-boring load cell	Wroth and Hughes (1973)
push in cells	Massarsch (1975)
prebored pressuremeter	Baguelin et al (1978)
flat dilatometer	Marchetti (1980)
self-boring pressuremeter	Wroth (1982)
Ko stepped blade	Handy et al (1982)
lateral sensing cone penetrometer	Huntsman (1985)
sleeve friction cone penetrometer	Masood and Mitchell (1992)

Table 3   Methods of measuring in situ total horizontal stress

Total stress cells (Massarsch, 1975) are pushed into the ground, therefore they displace the soil. With time the pressure on the cell reaches equilibrium. Tedd and Charles (1983) suggested that the measured stress in clay is equal to the total horizontal stress plus the undrained shear strength of clay. This correlation was developed from observations made with other forms of in situ tests in stiff clay.

629

Penetrometers and pressuremeters either displace, replace or unload the ground during installation. Secondary and empirical correlations have been developed from correlations with other tests and tests in chambers (e.g. Schnaid and Houlsby, 1990). One exception to this is the SBP test since theoretically it should be possible to install the probe without changing the magnitude of the total horizontal stress (Clarke and Wroth, 1984). Comparisons such as those shown in Figure 3 suggest that realistic values of horizontal stress can be obtained from quality SBP tests provided allowances are made for installation disturbance.

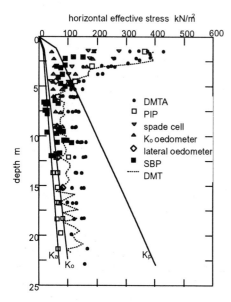

Figure 3   Measurements of total horizontal stress in soft clay (after Lutenegger and Benoit)

Hydraulic fracturing (Bjerrum and Anderson, 1972) only applies if the horizontal stress is the minor principal stress (that is Ko <1). The horizontal stress is the closure pressure of a crack created as the pore pressure is increased. This method can be applied to soils and rocks but there are a number of problems in interpretation and it is not used in routine investigations. The phenomenon must be understood as the consequences of hydraulic fracturing can affect other tests, e.g. in situ permeability tests.

Overcoring gives a measurement of in situ stress indirectly by relating the deformation created during overcoring to the change in stress assuming the rock is homogenous and isotropically elastic. The elastic properties are determined from laboratory tests on the samples retrieved from the core. Whittlestone and Ljunggren (1995) describe a piece of equipment designed to operate in 76 mm diameter boreholes and to depths of 700 m below ground level. This technique has been in use for many years but the recent major advances have been in improved equipment design permitting more accurate measurements and simplifying installation and recovery. Errors are introduced because of stress relief created during installation of the measuring device, the assumption that the stiffness (usual compression) measured in the laboratory is the same as the in situ stiffness (extension), and the assumption the rock is an homogeneous, isotropically elastic continuum.

Pore Pressure   Most positive pore pressure measurements are made using piezometers either installed in prebored holes or pushed into place. In both cases there is a delay between installation and equilibrium conditions which varies with installation disturbance and ground permeability. Some in situ testing probes, such as the piezocone and Cambridge SBP, include piezometers to enable pore pressures to be measured during installation and testing. Unless testing, cohesionless soils they would not normally be used to measure ambient pore pressures because of the time taken to reach equilibrium.

Negative or suction pressures are measured in a variety of ways (e.g. hydraulic piezometers, tensiometers) though the success depends on the pressures to be measured and the quality of installation. An understanding of the behaviour of partially saturated soils has become increasingly important in the UK because of problems of near surface deformations associated with clays and fills subject to changes in water content. The effect negative pore pressure has upon the behaviour of a clay is described by Wheeler and Sivakumar (1993) for example. It has been possible to determine suction pressures in the laboratory indirectly using the filter paper test (Chandler and Gutierrez, 1986) or fuse wire technique. Ridley and Burland (1993) describe a tensiometer to measure suction pressures directly in the laboratory which has now been adapted for field use (Ridley and Burland, 1995). The tensiometer is installed in a carefully drilled hole and, as with any device measuring in situ stress, there is a delay before steady conditions are achieved.

A major disadvantage of most pore pressure measuring devices is the limitation on the number per borehole. The West Bay system is designed to overcome that since it permits the installation of a number of probes in one borehole. It also permits samples of groundwater to be taken, permeants to be introduced and permeability tests to be carried out at different times and at different levels since each probe is self sealing and isolated from the other probes. The system has been used on a number of major investigations in the UK. Eldred et al (1995) describe the successful installation of a number of these systems down to depths of 1200 m. The quality and quantity of data produced using a data logger connected to transducers within the system, allowed an assessment of the effects of natural and drilling related phenomena to be made. While it may not be cost effective for smaller investigations, the potential of the West Bay system in larger investigations, especially when long term monitoring for legal or safety reasons is a requirement, could be increasingly important.

Stress History   Direct measurements of horizontal stress are difficult. Horizontal stresses are often estimated from K_o, either assumed or determined from secondary correlations with overconsolidation ratio. Overconsolidation is defined in terms of vertical stress but the effect of overconsolidation can arise because of other factors such as cementation and environmental changes as well as reduction in horizontal stress. A number of empirical correlations have been established between results of in situ tests and OCR. For example, parameters obtained from pressuremeter tests have been shown to relate to preconsolidation pressure (Mayne and Bachus, 1989); vane shear strength has been related to OCR (e.g. Aas et al, 1986, Chandler, 1987).

GROUND PROPERTIES - MECHANICAL PROPERTIES
Some in situ tests, such as pressuremeter tests, give a ground response curve which is then interpreted to produce either a single parameter (eg a CBR value from an in situ CBR test) or a number of independent parameters. An example of the latter is the selection of ground properties from pressuremeter test curves. In situ horizontal stress, shear modulus and strength are selected or derived independently from different parts of a pressuremeter ground response curve without checking that they are related to one another through known soil behaviour. The results obtained depend on a number of factors, most importantly the method of analysis, the section of the curve over which the results are taken and the assumptions made. Despite this many authors have used correlations with other data to demonstrate the quality of these independent parameters. Examples of this approach are widely published and in this proceedings Schnaid et al (1995) give an example using PBP tests in partially saturated soils and Parry et al (1995) and Maddison et al (1995) show how this technique was used to develop design parameters.

An alternative approach is to interpret the ground response curve in a specified manner to produce design parameters directly. The most comprehensive application of this approach is that developed for the design of vertically and horizontally loaded shallow and deep foundations and ground anchors from pressuremeter tests. Nuyens et al (1995) describe an automatic method to control and interpret a test and apply the rules developed from theoretical and experimental observations to produce directly, at the time of testing, a design.

Ground Response Curves

It should be theoretically possible to analyse any ground response curve to obtain a shear stress-strain curve. Palmer (1972) and Manessero (1989) describe methods for interpretation of pressuremeter tests in clay and sand respectively to produce shear stress-strain curves which are independent of any preconceived soil model. In practice, the shear stress-strain curve obtained is affected by the unknown installation disturbance and fluctuation in recorded data.

Two methods can be used to overcome these problems; a soil model can be assumed and integrated to produce a pressuremeter test curve; a model for the pressuremeter test curve can be assumed and differentiated to produce a shear stress-strain curve. Soil models include linear elastic perfectly plastic loading (Gibson and Anderson, 1961) and unloading (Jefferies, 1988), hyperbolic loading (Denby and Clough, 1979) and unloading (Robertson and Ferriera, 1986), parabolic loading (Sadeeq and Clarke, 1995) and trignometric functions (Prévost and Hoeg, 1976). Pressuremeter test curves include hyperbolic (Arnold, 1981) and parabolic (Sadeeq and Clarke, 1995) models. Computer aided modelling (CAM) by which a mathematical relationship, such as those referred to above, is made to fit field data, is being developed, especially with the development and availability of the more advanced spreadsheets on PCs. This technique is discussed in two papers in this Conference dealing with SBP tests in clays (Clarke and Sadeeq, 1995, and Shuttle and Jefferies, 1995).

The variables, a minimum of three representing in situ stress, and functions of stiffness and strength, are adjusted to obtain the best fit. A number of combinations of those variables would give a similar fit and therefore CAM may not be a valid approach. Shuttle and Jefferies (1995) suggest that the range of parameters that would produce a fit was small and within experimental accuracy; Clarke and Sadeeq (1995) imposed limits on the variables which ensure the parameters obtained conformed with theories of soil behaviour. The main concern with this approach is that the soil model chosen may not reflect the true behaviour of the ground. This is especially the case at small strains where the stiffness dominates the behaviour and the curve is a function of installation disturbance. Clarke and Sadeeq (1995) confirmed earlier findings that post peak shear strengths obtained from interpretation of SBP tests in clays are independent of the method of interpretation. Since these are often the required parameters for design, the simple approach in which the strength is selected from a part of a test curve assuming a perfectly plastic model for the soil is the best; that is it is unnecessary to use CAM.

Shuttle and Jefferies (1995) compare the actual pressuremeter test curves and those generated from parameters derived using the independent approach. They found that there are significant differences. They then go on to suggest that an integration of a simple perfectly plastic soil model produces a better fit to the measured curve. They quote values of horizontal stress, stiffness and strength taken from the curve fitting routine. A comparison between the parameters derived by Shuttle and Jefferies and those derived independently is given in Figure

4. The evidence from this Figure is that the shear strength is the same which is consistent with the findings of Clarke and Sadeeq (1995), that is the derived value is independent of the method used; shear moduli are not the same but the reason for this may be due to the strain range over which the moduli are measured. The shear moduli quoted by Clarke (1990) are average secant moduli from unload-reload cycles whereas the CAM method will produce an elemental secant modulus over a stress range equivalent to the peak undrained shear strength. The main difference, however, is in the derived value of horizontal stress. Shuttle and Jefferies (1995) quote greater values than the lift off method originally used. Any modelling technique must include an engineering assessment of the derived parameters to ensure the values obtained are realistic and conform with known ground behaviour. The fit to the first part of a test curve is dominated by the shear modulus (Jefferies, 1988) but that part of the curve is also affected by installation disturbance. Thus, the inconsistencies between the independently derived parameters and those derived using CAM may be due to the fact that the model chosen cannot adequately describe the start of the pressuremeter test curve.

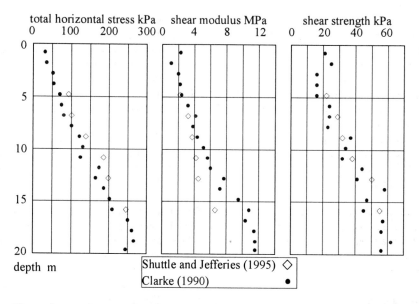

Figure 4   A comparison between an independent selection of parameters from SBP test curves and those derived from a model made to fit the data

The alternative method of using a soil model is to fit a curve to pressuremeter test data and then interpret that curve using the methods proposed by Palmer (1972) or Manessero (1989). Arnold (1981) suggested a hyperbolic pressuremeter test curve which can be adjusted to take into account installation disturbance. Sadeeq and Clarke (1995) found a parabolic model gave a fit to a SBP curve from a test in clay and the resulting shear stress-strain curve was reasonable, especially since the initial secant moduli were similar to those from unload-reload cycles. No account was taken of installation disturbance but the derived stress-strain curve appeared to be independent of that when a parabolic pressuremeter curve was assumed.

Deformation Properties

It is more usual practice to derive values of stiffness either from the initial loading portion of a pressuremeter test curve or an unload-reload cycle. The need to determine the stiffness of a soil has become increasingly important with improvements in analytical techniques to model the behaviour of the ground following loading or unloading. The developments were highlighted, in particular, in the European Conference which was devoted to displacements (Florence, 1991). Jamiolkowski et al (1985) concluded that the use of in situ tests to determine stiffness especially of sands and overconsolidated clays was important because of the difficulty of obtaining quality samples from those soils though Hight reports in this conference on recent developments in obtaining undisturbed samples which with improved laboratory techniques give realistic values of stiffness (e.g. Jardine et al, 1986).

Ménard (1957) produced an analysis of a pressuremeter test which included a linear elastic component. The installation disturbance, however, of a PBP is significant therefore Ménard developed the direct method of selecting design parameters from a pressuremeter test curve rather than interpreting a test to produce ground properties. The initial slope of a PBP test curve was subsequently referred to as the pressuremeter modulus, a form of drained modulus of elasticity which is used with design rules developed from theory and observation to predict deformation of shallow and deep foundations. This method was developed from studies of model foundations but the database now includes substantial records of performance behaviour recorded over the last thirty years. Predictions agree reasonably well with observations as shown in Figure 5. The method, however, is restricted to soils and foundations that appear within the database. It was developed in France and has been used elsewhere but using locally developed rules (e.g. Hansbo and Pramborg, 1990)

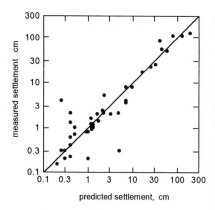

Figure 5   A comparison between predicted and measured settlement (after Clarke, 1995)

SBPs were developed in the seventies to enable tests to be carried out on minimally disturbed soil. There is still some disturbance, therefore the initial shape of the expansion curve reflects the stiffness of the disturbed ground. An unload-reload cycle is included in a test and this is used to determine the deformation properties since the slope of the cycle is approximately twice the shear modulus. This modulus is an average secant modulus which, if selected over the correct strain range (usually 0.1%) can be used to predict deformation (Clarke, 1995). Muir Wood (1990) showed that it is possible to determine an elemental stiffness from a cycle and Jardine (1991) showed that a profile of non linear elemental stiffness derived from an unload-reload cycle was similar to that obtained from triaxial tests on quality specimens of overconsolidated clays. This agreement has been consistently noted by a number of authors and is surprising because the direction of loading and the stress changes prior to testing are different.

Figure 6 shows the types of moduli and the derivation of the elemental stiffness. An unload-reload cycle can be included in any form of pressuremeter test and is now routinely specified for prebored, self-bored and pushed in pressuremeter tests and will be recommended in the revised version of BS5930:1981 Code of Practice for Site Investigations. The derived modulus is a shear modulus at the mean stress at which unloading starts and, providing the unloading is limited (Wroth, 1982), is an elastic modulus. More consistent results are obtained from the reloading curve since the unloading curve can include elements of creep (Robertson and Hughes, 1986, Clarke and Smith, 1992).

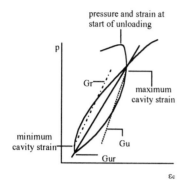
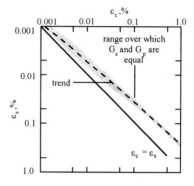

Figure 6   The definition of shear moduli from unload-reload cycles in pressuremeter tests and the derivation of elemental shear modulus from the average secant modulus form pressuremeter tests (after Clarke, 1995, and Jardine, 1991)

The mean effective stress during a test in clay does not change significantly once yield occurs, therefore the modulus represents an in situ modulus. In sands, the effective stress increases as the membrane expands, therefore the modulus increases. Bellotti et al (1989) showed that it is possible to determine the modulus at the in situ effective stress.

Thus, it is possible to obtain from an unload-reload cycle carried out as part of any pressuremeter test, either an average secant modulus or a non linear stiffness profile, and these can be corrected to the in situ stress. The former, if measured over 0.1% cavity strain range, produces an increasing stiffness with depth which is similar to that calculated from observations of ground behaviour (e.g. Clarke, 1993). The latter can be used in a numerical analysis to predict deformation (e.g. St John et al, 1993).

A pressuremeter loads the ground horizontally; a structure imposes a vertical load. A number of authors have demonstrated experimentally and numerically that, for practical purposes, ground can be considered isotropic (see Clarke, 1995, for further details). In practice, the vertical loading produces an increase in mean and deviatoric stresses, hence the settlement is formed of components of vertical and horizontal stiffness. Thus, the use of a pressuremeter stiffness is as valid as the use of stiffness obtained from a plate test, for example. Indeed the results are often comparable (e.g. Wroth et al, 1979). The stiffnesses derived from most intrusive tests are average stiffnesses of the ground local to the testing equipment. Schuldt and Foged (1994) suggest that the storage coefficient for an artesian aquifer, measured in an hydrogeological investigation, is a function of the bulk deformation properties of the soil mass. They found that the constrained modulus of a soil is approximately equal to the specific storage coefficient divided by the unit weight of water, where the specific storage coefficient is the amount of water expelled or taken into the soil mass of unit cross-section area per unit change in piezometric head. Eldred et al (1995) propose to carry out large scale pumping tests which could be analysed to give an indication of the constrained modulus for the rock mass.

Strength

The strength of ground can be measured in a variety of ways, each method giving a different value (Wroth, 1984). The actual value mobilised at failure depends upon the rate and direction of loading; the values obtained from tests may not equal the mobilised value. It is now generally accepted that the critical state value should be used in many cases for design (e.g. BS 8002:1994). Burland (1990) suggests that the post peak value is more useful than the peak value since that is the mobilised strength within a soil mass at failure.

In situ tests can be interpreted either directly to give a strength (e.g. vane) or indirectly to give a strength based on correlations with laboratory measurements of strength (e.g. strength from PBP tests) or to interpret a ground response curve to give the shear stress-strain curve (as explained above for pressuremeters). Generally it has been found that results of in situ tests in soft clays are greater than results of laboratory tests; results of in situ tests in stiff clays are equal to those from laboratory tests. Many design formulae have been developed using results of triaxial tests, thus, the strengths from in situ tests have to be adjusted if they are to be used in those design formulae. The only advantages of using in situ tests to obtain strengths are when testing ground that is difficult to sample or if samples are likely to undergo major changes during sampling, transporting, storage and preparation. Thus, in situ tests of soft clay, sands and weathered rock are routinely specified for strength determination. Lunne et al (1989), in reviewing in situ testing for the 12th ISSMFE Conference, suggested that interpretation of in situ tests will require some empirical content to take into account anisotropy, strain rate and disturbance and this should be established with consistent data sets (e.g. empirical corrections for vane tests such as those proposed by Chandler, 1987 or Aas et al, 1986).

Lunne et al (1989) concluded from a review of SBP and PIP tests in soft to stiff clays that it is more cost effective to use other in situ tests to obtain strength since more reliable results are given. More recent work suggests that reliable data can be obtained from pressuremeter tests. Clarke (1993) showed that the latter part of a SBP loading test curve should be used to derive the undrained shear strength of clays since it is unaffected by disturbance and represents the post peak strength. Shuttle and Jefferies (1995) suggest that if the strength from the loading curve is constrained by that from the unloading curve then profiles of undrained shear strength from SBP tests are as good a quality as those from laboratory tests. Clarke and Sadeeq (1995) confirm that the strength obtained from the latter part of an SBP loading curve is independent of the analysis and recommend that this should be the standard procedure. Houlsby and Withers (1988) show that the strength can be found from a PIP unloading curve and this will be independent of the amount of installation disturbance. Undrained strengths are often quoted from pressuremeter tests in weak rocks but as Parry et al (1995) point out these must be treated with caution since tests are actually partially or fully drained due to the stiffness of the rock (Johnston and Haberfield, 1990).

Drained tests in sands and partially saturated soils are often interpreted to give an angle of shearing resistance assuming cohesion is zero. This may be acceptable for recently deposited clean sands but in cemented sands and partially saturated soils the cohesion component will be significant. It is possible to develop an analysis that includes cohesion and angle of shearing resistance (e.g. Carter et al, 1986, Johnston and Haberfield, 1990) but a number of combinations of c' and ϕ' will produce a fit to the test data. Schnaid et al (1995) suggest that ϕ could be determined from fully saturated specimens of residual soil and then that ϕ' used in the interpretation of PBP tests to determine the in situ cohesion.

An alternative method is to correlate the limit pressure to strength. There are a number of correlations for clays and as Clarke and Sadeeq (1995) suggest these will give different results. Schnaid and Houlsby (1990) provide a correlation between limit pressure, tip resistance, horizontal stress and angle of shearing resistance for interpreting PIP tests in sands. This correlation has been developed from chamber tests and limited field experience therefore the database will have to be improved. It is likely that this will be the only method of obtaining ϕ' from PIP tests because of the installation disturbance.

EQUIPMENT AND SITE OPERATIONS

Unlike laboratory and penetrometer testing there are no UK standards for the majority of in situ tests. Thus, manufacturers have developed equipment and recommended calibration, installation and testing procedures which has led to a number of devices and procedures. For example, there are in the UK at least eight types of pressuremeter available for commercial work. It is generally assumed that calibrations for laboratory testing equipment can be undertaken at yearly intervals, perhaps, and be checked against national standards. Calibrations for field testing equipment, particularly pressuremeters, are required at very frequent intervals. No account is taken of the history of the equipment despite many contractors having evidence to show that the calibrations are, in practice, reasonably consistent. Calibrations are often specified without taking into account the effect they will have upon the interpreted results. For example, membrane stiffness for pressuremeter tests is unnecessary for tests in weak rocks, firm to stiff clays, medium dense to dense sands; membrane compression is unnecessary for all tests in soils unless non linear stiffness profiles are required in the stronger, denser soils.

Whittle (1995) gives details of the effects of procedure on the calibrations of membranes of a PIP, and recommends that the procedure should reflect the conditions in the ground during a test. Calibrations of different membranes and membranes after they have been used are different. There is a need to identify whether the accuracy of the calibration procedure is needed if the improvement in the results obtained is less than the natural variability of the properties of the ground and the experimental errors introduced because of installation. Calibrations are undoubtedly important but, in practice, if the cost has a significant effect on the total cost of the test and the results obtained are not significantly improved then their importance becomes greater than the test itself.

The improvements in instrumentation have led to the doubts expressed about calibrations. These improvements have also enabled measurements to be made downhole rather than at the surface. Examples of downhole instrumentation in in situ testing equipment include pressuremeters, permeameter (Harwood et al, 1995) and the West Bay system (Eldred et al, 1995). The use of transducers allows data logging to be automated and tests to be analysed at the time of testing. This has been common practice with penetrometers. Automatic data acquisition has been used with pressuremeters for a number of years but the interpretation, while undertaken using computers, has generally been office based because of the need to have an engineering input in selecting key data points and applying the correct method of interpretation. Standardisation of test procedures, calibration and methods of interpretation, together with a consistent framework for the selection of ground properties, could permit automatic interpretation of all pressuremeter tests. This is the approach described by Nuyens et al (1995) for the interpretation of PBP tests.

CONCLUSIONS
The main points arising from these papers and advances that have taken place in the last few years include the following.

1. There is a continuing development in instrumentation and data acquisition which permits more accurate recordings to be made of ground response curves. This, together with improvements in methods of analysis based on improved models and numerical techniques, allows more consistent results to be obtained from in situ tests.

2. Very often in situ tests are specified without advance knowledge of the ground conditions. This is acceptable for profiling tests but should not be recommended practice for intrusive tests that give single properties or a ground response curve. The interpretation of the test depends on the ground being tested. It is recommended that phased site investigations become routine. This may remove some of the criticism of in situ tests.

3. Improvements in data acquisition allowing on site analyses to produce ground response curves, non linear stiffness profiles and design parameters directly from pressuremeter tests at the time of testing.

4. There is a need to produce a standard for pressuremeter testing to cover calibrations, equipment, installation, test procedure and interpretation. This should apply to tests which are carried out in routine investigations. More sophisticated tests should be permitted as part of research activities and on larger investigations (cf. routine triaxial and stress path tests).

5. Methods of interpreting pressuremeter tests will either be based on empirical rules such as those developed for the PBP and PIP tests or on theories of cavity expansion. In the former case the database may be limited to particular soils and particular countries. Care will have to be taken when applying these rules elsewhere. In the latter case the use of CAM within a framework offers a theoretically sound method to interpret tests to produce reliable, consistent results for use in design.

REFERENCES
AAS, G, LACASSE, S, LUNNE, T and HOEG, K (1986) Use of in situ tests for foundation design on clay, ASCE Spec Conf In Situ'86, Use of In Situ Tests in Geotechnical Engineering, Blacksburg, Virginia, pp 1-30

ARNOLD, M (1981) Empirical evaluation of pressuremeter test data, Can Geotech J, Vol 18, No 3, pp 455-459

ATKINSON, J H and SALLFORS, G (1991) Experimental determination of stress strain time characteristics in laboratory and in situ tests, Proc 10th European Conf on Soil Mech and Found Engng, Florence, Italy, pp 915 - 956

BAGUELIN, F (1989) Discussion leader's report: Direct versus indirect use of in situ test results, Proc 12th Int Conf SMFE, Rio de Janeiro, Br, Vol 3, pp 2799 - 2803

BAGUELIN, F, JÉZÉQUEL, J F and SHIELDS, D H (1978) The Pressuremeter and Foundation Engineering, Trans Tech Pbl

BELLOTTI, R., GHIONNA, V., JAMIOLKOWSKI, M., ROBERTSON, P. K. and PETERSON, R. W. (1989) Interpretation of moduli from self-boring pressuremeter tests in sand., Geotechnique, Vol 39, No 2, pp 269 - 292

BJERRUM, L and ANDERSON, K H (1972) In situ measurement of lateral pressures in clay, Proc 5th Eur Conf on Soil Mech and Found Engng, Vol 1, pp 11-20

BRIAUD, J L (1992) The Pressuremeter, Balkeema, Rotterdam

BS1377:1990 Methods of Testing Soil for Civil Engineering Purposes, HMSO

BS5930:1981 Code of Practice for Site Investigation

BS8002:1994 Retaining Walls

BURLAND, J B (1990) Thirtieth Rankine Lecture: On the compressibility and shear strength of natural clays, Geotechnique, Vol 40, No 3, pp 327 - 378

CARTER, J P, BOOKER, J R and YEUNG, S K (1986) Cavity expansion in cohesive frictional soils, Geotechnique, Vol 36, No 3, pp 349-358

CHANDLER, R J (1987) The in situ measurement of undrained shear strength using the field vane, ASTM STP 1014, Int Sym Laboratory and Field Vane Strength Testing, Tampa, Florida, pp 13-44

CHANDLER, R J and GUTTIERREZ, C I (1986) The filter paper method of suction measurement, Geotechnique, Vol 36, No 2, pp 265-268

CLARKE B G (1990) Consolidation characteristics of clays from self-boring pressuremeter tests, Proc 24th Ann Conf of the Engng Group of the Geological Soc: Field Testing in Engineering Geology, Sunderland, pp 19-35

CLARKE, B G (1993) The interpretation of pressuremeter tests to produce design parameter, Predictive Soil Mechanics, Proc Wroth Memorial Symposium, Oxford, pp 75 - 88

CLARKE, B G (1995) Pressuremeters in Geotechnical Design, Blackie

CLARKE, B G and WROTH, C P (1984) Analysis of Dunton Green retaining wall based on results of pressuremeter tests, Geotechnique, Vol 34, No 4, pp 549-561

CLARKE, B G and SMITH, A (1992) Self-boring pressuremeter tests in weak rocks, Construction and Building Materials, Vol 6, No 2, pp 91 - 96

CLARKE, B G and SADEEQ, J A (1995) A practical guide to the derivation of undrained shear strength from pressuremeter tests, Proc Int Conf Advances in Site Investigation Practice, ICE, London, to be published

COX, D (1995) A new field test for the 'degree of compaction' Proc Int Conf Advances in Site Investigation Practice, ICE, London, to be published

DENBY, G M and CLOUGH, G W (1980) Self-boring pressuremeter tests in clay, J Geotech Engng Div, ASCE, Vol 106, No GT12, pp 1369-1387

ELDRED, C D, SCARROW, J A, AMBLER, M G and SMITH, A (1995) Groundwater monitoring - the Sellafield approach, Proc Int Conf Advances in Site Investigation Practice, ICE, London, to be published

GIBSON, R E and ANDERSON, W F (1961) In situ measurements of soil properties with the pressuremeter, Civ Engng Public Wks Rev, Vol 56, pp 615 - 618

HANDY, R L, REMMES, B, MOLDT, S, LUTENEGGER, A J and TROTT, G (1982) In situ stress determeinations by Iowa stepped blade, Jour Geotech Engng Div, ASCE, Vol 108, No GT 11, pp 1405-1422

HANSBO, S and PRAMBORG, B (1990) Experience of the Ménard pressuremeter in foundation design, Proc 3rd Int Sym on Pressuremeters, Oxford, pp 361 - 370

HARWOOD, A H, CLARKE, B G and ARARUNA, J T(1995) The design of a new self-boring permeameter, 11th Eur Conf Soil Mech and Found Engng, to be published

HOULSBY, G. T. and WITHERS, N. J. (1988) Analysis of the cone pressuremeter test in clay., Geotechnique, Vol 38, No 4, pp 575-587

HUNTSMAN, S T (1985) Determination of in situ lateral pressure of cohesionless soils by static cone penetrometer, PhD thesis, Univ California

JAMIOLKOWSKI, M, LADD, C C, GERMAINE, J T and LANCELLOTTA, R (1985) New developments in field and laboratory testing of soils, Proc 11th Int Conf SMFE, San Francisco, Vol 1, No , pp 57-154

JARDINE, R J (1991) Discussion on "Strain dependent moduli and pressuremeter tests", Geotechnique, Vol 41, No 4, pp 621 - 626

JARDINE, R J, POTTS, D M, FOURIE, A B and BURLAND, J B (1986) Studies of the influence of non linear stress strain characteristics in soil structure interaction, Geotechnique, Vol 36, No 3, pp 377 - 396

JEFFERIES, M. G. (1988) Determination of horizontal geostatic stress in clay with self-bored pressuremeter, Can Geotech J, Vol 25, No 3, pp 559 - 573

JOHNSTON, I W and HABERFIELD, C M (1990) Pressuremeter interpretation for weak rock, Proc 24th Ann Conf of the Engng Group of the Geological Soc: Field Testing in Engineering Geology, Sunderland, pp 85 - 90

LEEMAN, E R (1968) The determination of the complete state of stress in rock using a single borehole - laboratory and underground measurements, Int J Rock Mech and Min Sci, Vol 5, pp 31-56

LUNNE, T, LACASSE, S and RAD, N S (1989) SPT, CPT, pressuremeter testing and recent developments in in situ testing - Part 1: All tests except SPT, Proc Int Conf ISSMFE, Vol 4, pp 2339-2404

MADDISON, J D, CHAMBERS, S, THOMAS, A and JONES, D B (1995) The determination of deformation and shear strength characteristics of Trias and Carboniferous strata from in situ and laboratory testing for the Second Severn Crossing, Proc Int Conf Advances in Site Investigation Practice, ICE, London, to be published

MAIR, R J and WOOD, D M (1987) Pressuremeter Testing - Methods and Interpretation, Butterworth

MANASSERO, M. (1989) Stress-strain relationships from drained self-boring pressuremeter tests in sands., Geotechnique, Vol 39, No 2, pp 293-307

MARCHETTI, S (1980) In situ tests by flat dilatometer, Jour Geotech Engng Div, ASCE, Vol 106, No GT 3, pp 299-321

MASOOD, T and MITCHELL, J K (1992) Estimation of in situ lateral stress by cone penetration test, Jour Geotech Engng Div, ASCE, Vol

MASSARSCH, K R (1975) A new method for measurement of lateral earth pressure in cohesive soils, Canadian Geotech J, Vol 12, No 1, pp 142-146

MAYNE, P. W and BACHUS, R C (1989) Penetration pore pressures in clay by CPTU, DMT, and SBP., Proc 12th Int Conf SMFE, Rio de Janeiro, Br, Vol 1, pp 291-294

MÉNARD, L (1957) Measures in situ des propriétiés physiques des sols, Annales des Ponts et Chaussees, Paris France, No 14, pp 357 - 377

MUIR-WOOD, D (1990) Strain dependent moduli and pressuremeter tests, Geotechnique, Vol 40, No 26, pp 509 - 512

NUYENS, J, BARNOUD, F and GAMBIN, M (1995) Ménard pressuremeter test to foundation design - an integrated concept, Proc Int Conf Advances in Site Investigation Practice, ICE, London, to be published

PALMER A C (1972) Undrained plane-strain expansion of a cylindrical cavity in clay: a simple interpretation of the pressuremeter test, Geotechnique, Vol 22, No 3, pp 451-457

PARRY, N V, JONES, A J and DYER, M R (1995) M5 Avonmouth Bridge strengthening in situ testing of Mercia mudstone, Proc Int Conf Advances in Site Investigation Practice, ICE, London, to be published

PRÉVOST, J H and HOEG, K (1975) Analysis of pressuremeter in strain softening soil., J Geotech Engng Div, ASCE, Vol 101, No GT8, pp 717-732

RIDLEY, A and BURLAND, J B (1993) A new instrument for the measurement of soil moisture suction , Geotechnique, Vol 43, No 2, pp 321-324

RIDLEY, A and BURLAND, J B (1995) A pore pressure probe for in situ measurement of soil moisture suction, compaction' Proc Int Conf Advances in Site Investigation Practice, ICE, London, to be published

ROBERTSON, P K (1986) In situ testing and its application to foundation engineering, Can Geotech J, Vol 23, pp 573 - 594

ROBERTSON, P K and HUGHES, J M O (1986) Determination of properties of sand from self-boring pressuremeter tests, Proc 2nd Int Sym Pressuremeter Marine Appl, Texam, USA, ASTM No STP 950, pp 283 - 302

ROBERTSON, P K and FERREIRA, R S (1993) Seismic and pressuremeter testing to determine soil modulus, Predictive Soil Mechanics, Proc Wroth Memorial Symposium, Oxford, pp 434 - 448

SADEEQ, J A and CLARKE, B G (1995) Non-linear stress-strain behaviour of clay from self-boring pressuremeter tests, Proc 4th Int Conf on Pressuremeters, Sherbrooke, Canada, to be published

SCHNAID, F and HOULSBY, G T (1990) Calibration chamber tests of the cone pressuremeter in sand, Proc 3rd Int Sym on Pressuremeters, Oxford, pp 263 - 272

SCHNAID, F, SILLS, G C and CONSOLI, N C (1995) Pressuremeter tests in unsaturated soils, Proc Int Conf Advances in Site Investigation Practice, ICE, London, to be published

SCHULDT, J and FOGED, N (1994) Soil deformation properties evaluated from hydrogeological tests

SHUTTLE, D A and JEFFERIES, M G (1995) Reliable parameters from imperfect SBP tests in clay, Proc Int Conf Advances in Site Investigation Practice, ICE, London, to be published

ST JOHN, H D, POTTS, D M, JARDINE, R J and HIGGINS, K G (1993) Prediction and performance of ground response due to construction of a deep basement at 60 Victoria Embankment, Predictive Soil Mechanics, Proc Wroth Memorial Symp, Oxford, pp 581-608

TEDD, P and CHARLES, J.A. (1983) Evaluation of push-in pressure cells results in stiff clay, Proc Int Sym on Soil and Rock Investigations by In-situ Testing, Paris, Vol 2, pp 579 - 584

WHEELER, S J and SIVAKUMAR, V (1993) A state boundary for unsaturated soil, Proc Int Conf on Engineered Fill, Newcastle, pp 66-77

WHITTLE, RW (1995) Recent developments in the cone pressuremeter, Proc Int Conf Advances in Site Investigation Practice, ICE, London, to be published

WHITTLESTONE, A P and LJUNGGREN, C (1995) A new technique for in situ stress measurement by overcoring, Proc Int Conf Advances in Site Investigation Practice, ICE, London, to be published

WROTH, C P (1982) British experience with the self-boring pressuremeter, Proc Int Sym on the Pressuremeter and its Marine Appl, Paris, pp 143 - 164

WROTH, C P (1984) The interpretation of in situ soil tests. 24th Rankine Lecture, Geotechnique, Vol 34, No 6, pp 449 - 489

WROTH, C P and HUGHES, J M O (1973) An instrument for the in-situ measurement of the properties of soft clays., Proc 8th Int Conf SMFE, Moscow, Vol 1.2, No , pp 487-494

WROTH, C P. et al (1979) A review of the engineering properties of soil with particular reference to the shear modulus, University of Cambridge

Discussion on Session 4(b): pressuremeter, permeability and plate tests

Discussion reviewer JOHN J. M. POWELL
Building Research Establishment

The format of the discussion session was to take presentations prompted by the questions raised in the moderator's presentation and to include discussion from the floor.

<u>R Whittle</u> responded to the moderator's request to discuss some of the problems resulting from membrane compression in pressuremeter tests.

Figure 1 shows what happens when a cone pressuremeter (CPM) is inflated inside a thick wall metal cylinder. After contact with the cylinder wall the expansion should virtually stop. In practice, there is a significant non-linear response which is not very stiff. In this plot the steepest part is quoted as a slope in units of mm/GPa. Converting this to stiffness gives a modulus of only 500 MPa.

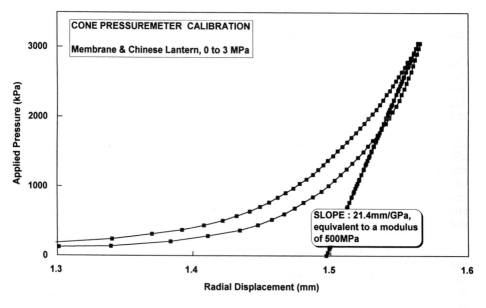

Fig. 1. Cone pressuremeter calibration in thick cylinder

Schnaid (1990) was concerned by the implications of this behaviour for the measurement of shear modulus. He developed 'correction' algorithms for his CPMs relating applied pressure to varying system stiffness. In Fig. 2 is plotted a visual representation of his correction curves. He didn't query the premise that instrument stiffness varies with pressure. Now, this is odd behaviour for something that is essentially a lump of steel, so what is going on?

Advances in site investigation practice. Thomas Telford, London, 1996

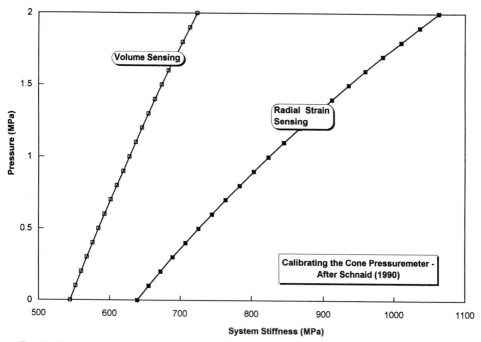

Fig. 2. Cone pressuremeter calibration (after Schnaid, 1990)

Figure 3 shows three calibration tests on the same CPM. Like many pressuremeters in current use, the membrane is in two parts. The inner part is an elastic sleeve which contains pressure. The outer part is a protective steel sheath, commonly known as a Chinese lantern. The left-hand plot is the calibration obtained when the sheath is removed. The response is almost ideal, being linear and very stiff, an implied modulus of over 3GPa.

The middle plot in Fig. 3 was seen in Fig. 1. The addition of the sheath has markedly affected the behaviour of the probe. The problem is that the sheath has a smaller radius of curvature than the calibration cylinder. After initial contact with the cylinder wall the sheath continues to deform as it takes up the shape of the calibration tube. Not until this is done will 'true' instrument stiffness be measured, as shown in the right-hand plot. Once past a critical pressure, about 45 bars in this example, the fully assembled instrument shows similar characteristics to the 'membrane only' test.

The CPM is a special case, because the loads necessary to form the sheath lie outside the normal operating range of the device. Similar behaviour at lower pressures is seen in other instruments. Fig. 4 shows a 'membrane only', and 'membrane plus sheath' calibration test on a normal self-boring pressuremeter (SBP). Incidentally, the width of this plot is only about 1 thou. The 'membrane only' response is very linear and stiff - equivalent to a system stiffness of 5 GPa. But once more, and for the same reason, the addition of the Chinese lantern spoils things.

All the non-linear behaviour is an artefact of the calibration procedure. In soils this will not happen. An unsung virtue of any self-boring or push-in device is that the probe is an excellent fit to the borehole. In this respect a pre-bored pressuremeter test is inferior, although the problem is only to establish a pressure limit below which data should be ignored. We have calculated that in good rock, with our HPD, reload loops will be spoilt if the applied pressure is allowed to fall below about 50 bars.

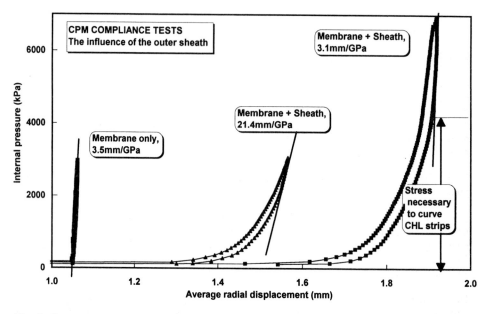

Fig. 3. Cone pressuremeter compliance tests

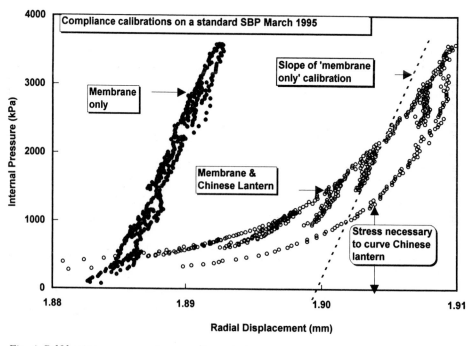

Fig. 4. Self-boring pressuremeter compliance tests

Nowhere in this presentation has true membrane compression been mentioned. We have never managed to measure any.

In conclusion, membrane compression calibrations

- have nothing to do with the membrane
- instead, describe instrument compliance
- are a fixed characteristic of any instrument
- the results are affected by unrealistic modelling of the true test conditions
- which result in excessive correction
- so don't bother carrying them out.

<u>P G Allen</u> made a presentation on small strain stiffness of clay using the self-boring pressuremeter.

It is recognised that the stress strain behaviour of clay is non-linear and the importance of this has been demonstrated by various workers (e.g. Jardine *et al*, 1991). For example, London Clay may be shown to be approximately five times stiffer at strains of 0.01% that at strains of 1%. This is important for the reliable and accurate prediction of ground movements, as most structures impose a strain of 0.1% or less on the ground.

Determination of small strain stiffness may be carried out by use of local strain measuring devices mounted on soil samples in triaxial cells. This normally enables strain to be determined down to 0.001%. This discussion aims to demonstrate that small strain stiffness may be determined from unload-reload loops of a self-bored pressuremeter test.

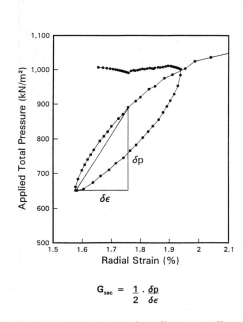

$$G_{sec} = \frac{1}{2} \cdot \frac{\delta p}{\delta \epsilon}$$

Fig. 5. Determination of small strain stiffness from a pressuremeter reload loop

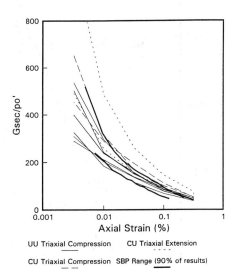

UU Triaxial Compression   CU Triaxial Extension

CU Triaxial Compression   SBP Range (90% of results)

Fig. 6. Comparison of normalised (by p'$_0$) SBP and triaxial shear moduli

Small strain stiffness may be determined from an unload-reload loop as shown in Fig. 5. Prior to carrying out a loop the stress is held constant to reduce creep and drainage effects. The strain magnitude of the loop is also specified as this may be used to ensure that the loop is performed over the full elastic range of the clay under test. Typical ranges used are between 0.3% and 0.4% radial strain. A secant shear modulus is calculated from the base of the loop to points on the reloading curve. This may be easily achieved by use of a computer spreadsheet programme. The validity of this approach may be confirmed by comparing the results from two loops in the same test. A good correlation has been found to exist between the calculated secant modulus for two loops in numerous tests. One conclusion which may be drawn is that one loop, controlled to extend to the elastic limit of unloading may be adequate. This may help reduce drainage effects on calculation of undrained shear strength.

In a triaxial test, all soil elements experience essentially the same strain. In a pressuremeter test the radial stress approaches the in situ horizontal stress with increasing radius from the pressuremeter. Therefore the strain experienced by the clay reduces with increasing radius from the pressuremeter. Muir Wood (1990) identified this with a simplified method suggested by Jardine (1992). This allows an equivalent axial stiffness to be determined.

A programme of 33 pressuremeter tests was carried out at a site in Kent for Kent County Council. Small strain stiffnesses of the London Clay were determined from this test programme. For a comparison of the results, the calculated shear modulus was normalised by the mean effective stress (p). Fig. 6 illustrates the envelope (of 90%) of results obtained. Superimposed on this plot are the results of small strain triaxial tests on pushed thin wall samples. The results demonstrate the good correlation obtained.

So, in conclusion, small strain stiffnesses of clay are an important consideration when predicting ground movements associated with new development. The ability of the self-boring pressuremeter to determine small strain stiffness, using Jardine's simplified approach, has been demonstrated. The results are found to be directly comparable with small strain stiffness determined in the laboratory.

I would like to thank Kent County Council and in particular Mr S Huntley for supplying the details of the triaxial tests and for permission to publish this data and data from the SBP tests at this site.

A presentation was then made by Y Koike on cyclic pressuremeter tests in soft rock.

The purpose of the investigation was to examine the data analysis method of strain dependent shear moduli using the pressuremeter test. A model test using a weak rock SBP was conducted in artificial soft rock. The undrained triaxial strength of this material is 4.4 MPa. Sinusoidal cyclic loading was applied and the detailed distribution of strain in the rock was examined using strain gauges.

Figure 7 shows an outline of a model test which is 80 cm in diameter and 100 cm in height. Strain gauges were set in the ground just around the cavity wall. Fig. 8 shows the pressuremeter curve. A total of about 250 cycles were applied with changing pressure amplitudes. Once the upper pressure reached about 3.8 MPa, corresponding to yield pressure, the accumulation of strain can be seen.

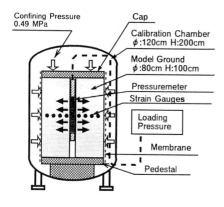

Fig. 7. Diagram of model tests

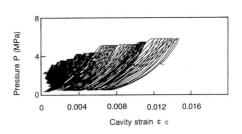

Fig. 8. Pressuremeter curve for sinusoidal cyclic loading

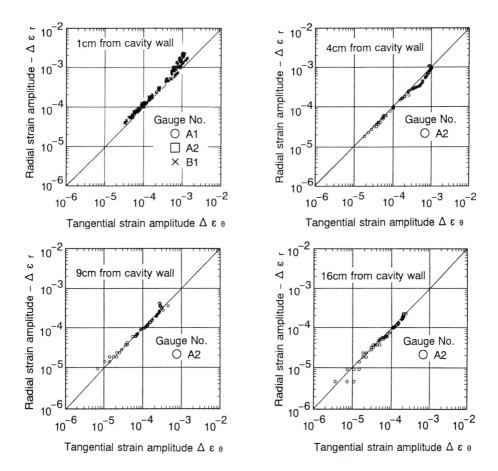

Fig. 9. Tangential and radial strain amplitude (frequency 0.1 Hz)

Strains in the tangential and radial directions relative to the cavity wall were measured at 1 cm, 4 cm, 9 cm and 16 cm from the cavity wall. Fig. 9 shows the relationship between strain amplitudes in the tangential and radial directions plotted to a logarithmic scale. The test starts from lower left and progresses towards upper right of the figure. During the whole test, at any position, the amplitudes of strain for the two directions show almost the same value. Therefore during cyclic pressuremeter tests a no-volume change condition exists. Therefore, the very simple assumption of undrained conditions to be used in data analysis is thought to be valid in the case of soft rock.

Figure 10 shows the relationship between the unload-reload modulus G_{ur} and shear strain. The non-linear property is quite clear and repeatable values of G_{ur} are observed down to a minimum shear strain of 10^{-4}. This curve is the relationship of the secant shear modulus and shear strain. Secant moduli were calculated using the relationship proposed by Muir Wood (1990). In this case a very simple hyperbolic function is applied. Results of cyclic triaxial tests under undrained conditions are not shown in Fig. 10 but the value of the shear modulus and its strain dependency were quite similar to that of the cyclic pressuremeter test. More details of the work can be found in Koike et al. (1995).

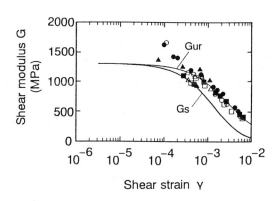

Fig. 10. Shear modulus G and shear strain γ from cyclic pressuremeter tests (G_{ur} - unload-reload modulus, G_S - secant shear modulus)

M Jefferies then responded to the fact that, in his introduction to the session, the Moderator had noted that the strengths inferred for the Bothkennar clay in the paper by Shuttle and Jefferies (Session 4b) were quite similar to those previously developed by conventional analysis of the SBP data as reported in the Bothkennar Symposium. However, the estimates of horizontal geostatic stress were markedly different between these two sets of work based on the same data, and the Moderator suggests that this is caused by the poor fit of the Tresca model to the initial part of the data (say, less than 2% cavity strain) where the data may have been affected by disturbance.

The idea that only the initial part of the SBP data carries horizontal stress information is a common confusion and quite wrong. Considering the equations for expansion of a cylindrical cavity in a Tresca material, the horizontal stress appears throughout the cavity expansion and subsequent contraction as a simple offset term on the cavity pressure at any strain (see the summary of governing equations in Shuttle and Jefferies 1995 for example). Thus the horizontal geostatic stress can be determined quite accurately by constraining the fit of theory to data at the dual asymptotes of maximum expansion and final contraction.

With the understanding that geostatic stress is not the problem, it is helpful to examine test data to determine where the errors in conventional SBP analysis arise. Fig. 11 shows the fit of the conventional Tresca theory to one of the London Clay tests with particular care taken to get an appropriate elastic unload and late-time expansion fit. No attention has been paid to the contraction data, but loading is well fitted. This figure corresponds to an excellent conventional analysis.

The first error in Fig. 11(a) is that cylindrical cavity theory has been used. Invoking finite geometry correction (here we use that proposed by Shuttle and Jefferies, 1995) leads to an equally good fit, Fig. 11(b), but with significantly different strength. However, inspection of both Figs 11(a) and (b) shows that neither parameter set is in any way a sensible estimate of ground truth. But don't blame the theory: if you only fit half the data, you have no right to expect reasonable results.

The way forward is to require that the theory fit all the data. This is done by realizing that the rigidity in a Tresca idealization of real clay is not an elastic parameter. Treating rigidity as a free parameter then leads to the fit shown as Fig. 11(c). This is a sensible representation of measured ground response. And, accordingly, the parameter set is a reasonable estimate of ground truth.

To illustrate that the solution space is not sensitive to the constitutive model, we can fit using the Robertson & Ferreira (1992) theory (but also using the finite geometry correction) which leads to Fig. 11(d). There is not a great deal to choose between Figs 11(c) and (d) in terms of theory to data match, and both solutions give similar estimates of ground truth.

Now if one compares the Fig. 11(c) solution with Fig. 11(a), it will be seen that the strength is actually quite similar. What has happened is that presently conventional procedures for analysing SBP have two opposing errors:

- an over-idealized geometry
- an erroneous representation of real soil rigidity in a Tresca model.

What you get in conventional analysis of any SBP test depends on the balance between these two errors, and this depends on the particular circumstances and the interpreter.

Reliable estimates of ground truth must obviously avoid arbitrary error. This is easily done using Iterative Forward Modelling, IFM, of the SBP data. As a final point, we note that IFM is really only practical using modern computers but this should not impede the wider use of the technique.

M Gambin spoke from the floor. Computer aided foundation design must not prevent the Engineer from using his or her judgement. For instance, a software design programme does not include information on the location of buried constructions or serious ground water conditions which must be kept in mind at the time of the final design choice. Everybody knows, and Mr Trenter recalled yesterday, that the above mentioned factors are, among others, the ones which are more likely to create overrun costs during foundation construction.

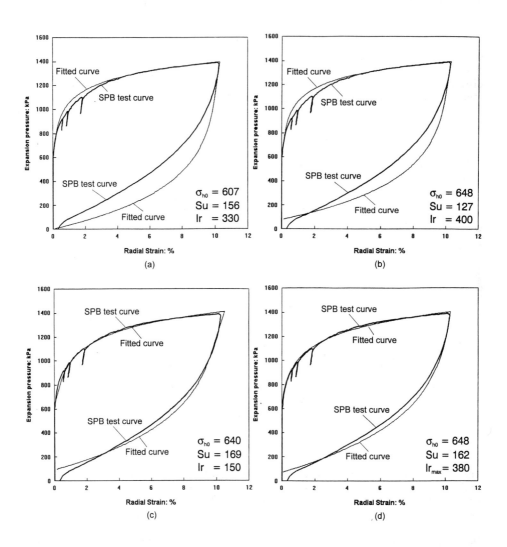

Fig. 11. Examples of curve fitting to field data: (a) using Tresca theory and elastic unload and late-time expansion fit; (b) invoking finite geometry; (c) allowing freedom of rigidity; (d) using Robertson and Ferreira (1992)

A presentation was then made by <u>R Whittle</u> on the approach adopted in the Shuttle and Jefferies paper to this conference.

The figures used in this example are from a test which is an example in their paper. The comparison between the conventional approach and that of Iterative Forward Modelling, will be interesting.

Conventional analysis begins by establishing the origin for the expansion. Fig. 12 shows the first part of the test. This is a good example of lift-off analysis. There is little sign of disturbance, and the lift-off stress of 735 kPa is assumed to be the in situ lateral stress. Shuttle & Jefferies derived 560 kPa for the same parameter.

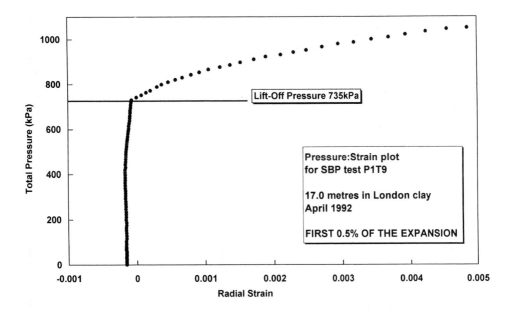

Fig. 12. Example of 'lift-off' in an SBP plot

The next step is to determine the undrained shear strength. Fig. 13 shows pressure plotted against current volumetric strain; note that strain is plotted on a log scale. It is a modified implementation of the Gibson & Anderson analysis. If the material deforms as a Tresca soil then the slope of this semi-log plot will give the undrained shear strength.

However, the Gibson & Anderson result is a complete solution for the pressuremeter test and this plot can be made to give more than the shear strength. The intercept on the pressure axis when volumetric strain = 1 gives a limit pressure. Knowing this, the solution can be re-arranged to produce the rigidity index and shear modulus.

This shear modulus will be lower than that derived from reload loops. It is a minimum value as it is pertinent to the yield strain. Most unload/reload loops are taken over smaller strain ranges than this and so give stiffer results. It follows that using reload shear modulus to obtain rigidity index gives misleading answers. This is one reason why Shuttle & Jefferies generate ill-fitting curves from the reported parameters.

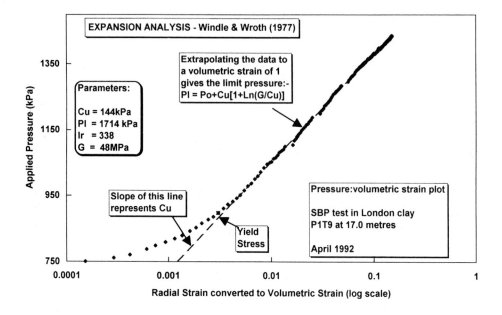

Fig. 13. Interpretation of an SBP plot

We can now test the conventional analysis to see if it recovers the measure pressuremeter curve.

In Fig. 14 the measured curve is shown with unload/reload loops removed. The solid line is the calculated curve using conventional analysis parameters. The dotted line is the published IFM result.

The conventional curve fits the expansion data like a hand in a glove. It also fits the elastic contraction and the first part of the plastic contraction. Thereafter it is in trouble. The Shuttle & Jefferies curve, in contrast, touches the loading phase in one place only, but gives a rough fit to the plastic contraction. What to choose?

The paper argues that data which reproduce the contraction curve are the best choice. However, pressuremeter contraction is not without problems.

Figure 15 shows the contraction part of the pressuremeter curve plotted in a manner similar to that used earlier for the expansion. The solution proposed by Mike Jefferies is used to plot the points. This is what should happen.

- When the pressuremeter is unloaded the ground at first responds elastically. This response ceases after a stress reduction of twice the undrained shear strength. The points are marked on the plot where elastic contraction ceases according to Shuttle & Jefferies, and according to me.

- Thereafter the ground is assumed to deform plastically at a constant strength, the gradient of the semi-log plot being twice the shear strength.

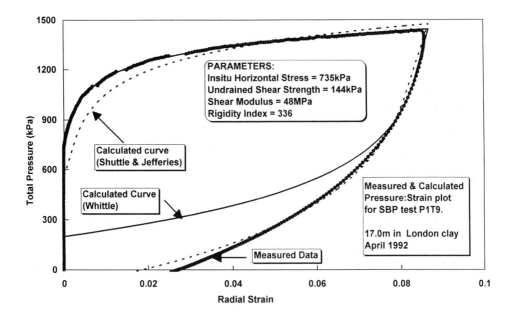

Fig. 14. Curve fitting an SBP test

There is no linear part to this contraction curve. Instead the response becomes increasingly stiff. This probably indicates severe drainage, so invalidating the assumption of constant volume deformation. Biasing the IFM procedure to the contraction curve masks this behaviour and can, as this example shows, lead to inappropriate parameters being selected.

With regard to reliable parameters from imperfect SBP tests <u>P Smart</u> suggested that (a) a least squares fit should be used to extract the parameters automatically rather than interactively; and (b) that when using curve fitting methods to establish the parameters then a sensitivity analysis should be done on the subsequent design calculations.

<u>P Jacobs</u> responded from the floor to the Moderator's comments on the lack of technical papers on the cone pressuremeter (CPM).

As an operator of the Cone Pressuremeter (CPM) we would have liked to have had a paper presented but unfortunately this was not possible. I would like to bring to the conference's attention the status of the CPM as an in situ test by way of a summary of the CPM system and its capabilities.

The system comprises a standard Fugro cone with a pressuremeter section 70 cm long behind this, being 44mm in diameter. The pressuremeter has a rubber membrane sheathed in a stainless steel lantern with internal pressure and displacement measurements (displacement measured at 120° around the circumference).

The CPM has been used on commercial contracts in the UK and Europe since 1991 and the use is increasing every year. The device is a full displacement push in pressuremeter capable of carrying out up to 10 inflations a day on site. The analysis of the test results in clay and sand have been thoroughly researched and the results published by Dr Guy Houlsby and his post graduates at Oxford University.

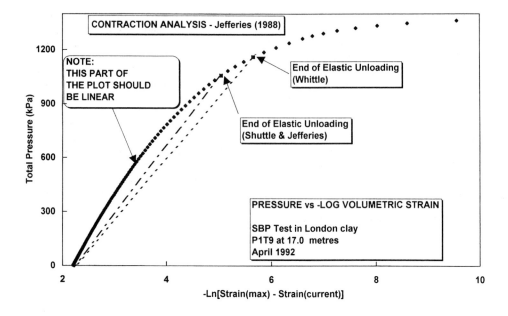

Fig. 15. Contraction analysis on an SBP test

The inclusion of a standard cone in the device allows precise determination of the soil strata being tested by the CPM. The analysis of undrained shear strength and shear modulus in clay is based on the unloading position of the test results by means of a closed form solution.

In sands the determination of D_r, ϕ and K_o is based on the results of a large programme of CPM Calibration Chamber tests. The use of unload reload loops also permits determination of shear modulus in both sand and clay soils.

The latest development of the CPM is a CPM which is inflated using hydraulic oil and only has a thick reinforced rubber membrane. This system no longer has the steel lantern, this has been done to permit rapid changeovers of the pressuremeter test section during offshore marine tests.

H Kolk continued the discussion on the CPM. About 400 Cone Pressuremeter Tests have been made with the improved Fugro CPM since its completion late 1994. These tests were done at six sites in the Netherlands and at one site in Bangladesh. The pressuremeter tests were performed in 25 Cone Penetration Test profiles to maximum penetrations of about 50 m. The sites consisted predominantly of sands. The Interpretation method described by Houlsby and Nutt (1992) was applied to these data to provide estimates of relative density, in-situ horizontal stress and shear modulus. These results compare favourably with estimates made using other soil investigation data and geological data.

Ménard Pressuremeter Test Data were available at three sites. Ménard parameters determined from the Cone Pressuremeter Tests compare favourably with those from the Ménard Pressuremeter. This suggests that the stress conditions around the new recessed Cone pressuremeter are similar to those around a Ménard pressuremeter. A summary of our experience with the new Fugro Pressuremeter will be presented by Zuidberg and Post (1995) at the May 1995 International Symposium on pressuremeters in Canada.

A further development in Cone Pressuremeter Testing is a three-year research programme which started in 1995. The objective is to improve interpretation procedures of Cone Pressuremeter data from soft clays relative to the currently used procedures developed by Houlsby and Withers (1988). This study will consist of numerical and analytical analyses, complemented by field testing. This study will be performed by Fugro Engineers, The Netherlands, and is financially supported by the Public Works and Water Management Authority of the Dutch Ministry of Transport.

J Powell continued on the theme of the cone pressuremeter. Despite being a promising concept very little field data from the CPM appear to be available to allow its potential to be fully assessed. However, despite its apparent limited use, considerable effort has been given to developing theoretical frameworks for its interpretation (these being required because of the full displacement method of installation). At the Building Research Establishment (BRE) a study of the CPM, in clay soils, is underway to try to resolve some of the questions that exist on the methods of interpretation of the test data as well as the influence of the testing method on the measured and derived parameters. The findings to date are reported in detail by Powell and Shields (1995) but may be summarised as follows.

The CPM has, in general, been found to be easy to use and to give repeatable results.

Equipment modifications resulted in a better control of the test, particularly in the unloading phase with a resultant beneficial effect on data interpretation especially in stiff clays.

Excellent agreement has been found between the shear modulus values derived from the unload-reload loops of the CPM and other pressuremeter devices in the stiff clays, as shown in Fig. 16. However, in the soft slightly cemented clays at Bothkennar it would appear that the effects of disturbance due to inserting the CPM had some effect on the soil structure, although repeatability was good.

Shear strength proved to be the most variable parameter studied. The Houlsby and Withers (1992) analysis based on the unloading curve was found to be affected by the rate of unloading in soft clays and by the equipment performance in stiff clays. However, the values calculated were generally of the same order as those from other test methods. Further study is needed in this area.

The Houlsby and Withers analysis does not appear to give realistic values of the in situ horizontal stress σ_{ho}. However, there appears to be a consistent ratio between those obtained from the CPM and the best estimated values. In two stiff clays this is less than one (this value differs greatly from that of Houlsby and Withers) and in the soft clay significantly greater than one (about 1.6). There is a need to try to develop these correlations further.

In both soft and stiff clays the limit pressures derived from the BRE CPM tests generally agreed well with values extrapolated from other pressuremeter devices. However, at one site there were significant differences between the BRE CPM limit pressures and those obtained by Houlsby and Withers (1992). The reason for this has not been found, although rate effects may be important.

In overall conclusion there is no doubt that the CPM has a great deal of potential as an SI tool.

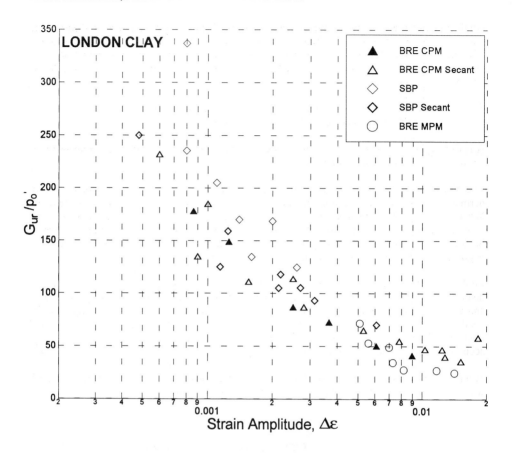

Fig. 16. Comparison of CPM unload-reload shear moduli (normalised by vertical effective stress, p_0) with those from other pressuremeter types (Powell and Shields, 1995)

R J Hutchinson responded to the Moderator's comment on the need for improved specifications for the more sophisticated in situ tests. In such areas of rapidly developing technology it is difficult, if not impossible, to provide an effective all encompassing specification as these are often written by parties unaware of current developments and capabilities. Without direct reference to the specialists such specifications are in danger of writing out potentially valuable results.

Instead, procurement methods should be reviewed to bring the specification in at an early stage. They should also encourage operation as an integrated team as it is only in this way that the level of communication necessary to ensure the optimum result can be achieved.

Unfortunately this could obviate the need for any specification but, at the same time, optimise the benefits to all parties, the ultimate client in particular.

J H Atkinson spoke from the floor. In the papers and in the discussion the terms soil model and mathematical model have often been used to mean different things.

On the one hand, a constitutive model is a piece of mathematics which describes shearing and volumetric effects together. The theory of elasticity, Cam clay, bricks on strings are examples of mathematical models. On the other hand, fitting a curve, say a parabola or hyperbola, to an observed relationship is also called modelling. These are different techniques and should not be confused.

This leads to the question of whether information about the basic soil model (i.e. its constitutive equations) can be determined from in situ tests.

A Harwood presented an outline of the design of a new self-boring instrument to measure the permeability of a wide range of soils in situ.

The objectives of the project were four-fold: to conduct a review of existing techniques of permeability measurement in both the laboratory and field and also the available technology. We then compiled a specification for the test, bearing in mind that it was intended for commercial use. We have designed and built two prototypes and written the logging software. So far we have bench tested the individual components and the assembly and are about to begin laboratory testing to be followed by field testing at some time.

Looking in more detail at the specification, the device should be self-boring to reduce disturbance. A major aim was to reduce system compliance using downhole instrumentation and valves and to ensure a hard flow cavity. The probe has to be robust and suitable for site conditions and be compatible with likely leachate permeants. The porous element should not clog and should be shut during installation and open during testing. The device must be accurate under various flow conditions for a range of permeabilities.

Figure 17 shows a schematic of the device. The instrument is approx. 1.5 m in length over all, with a diameter of 89 m. The length of the porous section is 350 mm.

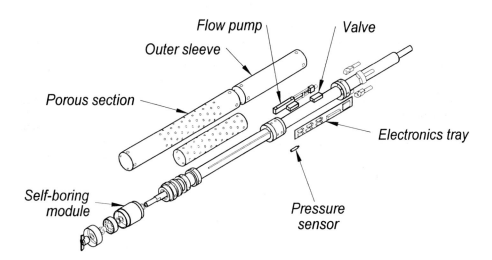

Fig. 17. Self-boring permeamater

The probe uses the same installation method as the Weak Rock Self-Boring Pressuremeter and has the ability to apply a variety of flow conditions to the relatively undisturbed pocket.

We have achieved downhole instrumentation and fluid control valves thus minimising effects of compliance of lines to the surface, and the instrument is being controlled and logged successfully by a PC.

References

Houlsby, G.T. and Withers, N.J. (1988). Analysis of cone pressuremeter test in clay. *Géotechnique*, 38, 575-587.

Jardine, R.J. (1992). Non-linear stiffness parameters from undrained pressuremeter tests. *Canadian Geotechnical Journal,* Vol. 29, 436-447.

Jardine, R.J., Potts, D.M., St John, H.D. and Hight, D.W. (1991). Some practical applications of a non-linear ground model. Proc. 10th ECSMFE, Florence, 223-228.

Koike, Y., Furuta, I., Fujitani, M. and Shimada, M. (1995). Strain distribution of artificial soft rock induced by cyclic pressuremeter testing. Proc. 4th. Int. Symp. on Pressuremeters (ISP 4) 'The Pressuremeter and its New Avenues'. Sherbrooke, Canada, Balkema. 281-288

Muir Wood, D. (1990). Strain dependant moduli and pressuremeter tests. *Géotechnique*, Vol. 30, 509-512.

Powell, J.J.M. and Shields, C.H. (1995). Field Studies of the Full Displacement Pressuremeter in clays. Proc. 4th. Int. Symp. on Pressuremeters (ISP 4) 'The Pressuremeter and its New Avenues'. Sherbrooke, Canada, Balkema. 239-248.

Robertson, P.K. and Ferreira, R.S. (1992). Seismic and pressuremeter testing to determine soil modulus. Predictive Soil Mechanics (Wroth Symposium). Thomas Telford, London. 562-580.

Shuttle, D.A. and Jefferies, M.G. (1995). A practical geometry correction in the interpretation of pressuremeter tests in clay. *Géotechnique*, Vol. 45, No. 3, 549-553.

Zuidberg, H. M. and Post, M.L. (1995). The Cone Pressuremeter: An Efficient way of Pressuremeter testing. Proc. 4th. Int. Symp. on Pressuremeters (ISP 4) 'The Pressuremeter and its New Avenues'. Sherbrooke, Canada, Balkema. 387-395

Better correlation between geophysical and geotechnical data from improved offshore site investigations

J. F. NAUROY, J. L. COLLIAT, A. PUECH, D. POULET, J. MEUNIER and F. LAPIERRE
IFP, ELF, GEODIA, TOTAL, IFREMER, ISM, France

ABSTRACT

The objective of the "GEOSIS" research project was to improve the correlation between geophysical and geotechnical data for offshore site surveys. In the first phase of the project, two improved geophysical site investigation techniques were developed and tested, namely very high resolution multichannel seismic surveying and vertical seismic profiling (VSP) in geotechnical boreholes. These techniques were used in an experimental site investigation campaign, offshore Monaco. The VSP profiles were used to calibrate the results of the surface geophysical survey. Relationships between seismic velocities and standard geotechnical parameters, such as cone resistance from in situ penetrometer tests and porosity or carbonate content, was attempted. The promising results obtained provide evidence that it is possible to derive and to quantitatively extrapolate consistent geotechnical data around a borehole from seismo-acoustic measurements. Until today, this was considered as theoretically possible only but not proven. The combination of geophysical and geotechnical data provided a major improvement in the offshore site investigation and, as a result, it is believed such investigation procedure will become typical of future offshore practice.

INTRODUCTION

The installation of offshore facilities requires precise assessment of the soil stratigraphy, as well as a reliable estimate of the geotechnical properties for foundation design. It is recognised that current offshore site investigation practice leads to relatively poor correlations between seismic and geotechnical data, involving large uncertainties or geotechnical engineering problems when the structure location is some distance away from the location of boreholes.

Current site investigation practice in the offshore oil industry, for the investigation of about 100m below seabed for foundation design purposes, involves two steps: (1) an initial geophysical survey, including bathymetry, side-scan sonar and very high resolution seismic, and (2) a dedicated geotechnical investigation including several boreholes at the intended location of the future structure. These two surveys are clearly distinct, generally performed at different periods by different contractors and involve different equipment and scopes of work (a surface area of about 5 sq. km is covered by the geophysical survey, while local boreholes are provided by the geotechnical investigation).

The geophysical and geotechnical surveys are generally difficult to integrate and cross-correlate, although it is obvious that relationships exist between geological processes (as highlighted by seismic reflectors, for example) and geotechnical properties of the soils used for the foundations assessment of a structure. It should also be clearly understood that the degree of correlation between geophysical and geotechnical data, or the distance over which one might be able to extrapolate the stratigraphy from a borehole using geophysical results, depends on the nature and quality of the data as well as on the geology of the site and the soil characteristics themselves.

A large number of empirical and theoretical relationships between the seismic and geotechnical properties of marine sediments exist in the literature. They generally involve three basic seismic parameters, i.e., acoustic impedance, wave velocity, and attenuation (Hamilton and Bachman, 1982, Haynes et al, 1992). Due to the poor quality of analogue geophysical data, the few existing attempts of actually studying such relationships have proven difficult. With the development of digital acquisition techniques for geophysical surveys, it is now considered as increasingly possible to extrapolate consistent geotechnical engineering information from seismo-acoustic measurements (Haynes et al, 1992).

The GEOSIS project, initiated by a group of French research organisations (IFP and IFREMER), oil companies (ELF and TOTAL), and offshore contractors for site investigations and foundation engineering (GEODIA and ISM) within CLAROM (Club for Research Activities on Offshore Structures), was aimed at improving the correlations between geophysical and geotechnical data by integrating offshore seismic and geotechnical investigations. This paper presents the results obtained on an experimental site offshore Monaco. Considering the high quality of results obtained, it is believed that the techniques developed can be used routinely in future offshore practice. It will enable the degree of correlation between geophysical and geotechnical data to be significantly improved.

IMPROVED SITE INVESTIGATION PRACTICE
Very High Resolution Multichannel Seismic Acquisition Equipment
When the GEOSIS project was initiated in 1990, it was obvious that the first step necessary to improve the correlations between geophysical and geotechnical data was to improve the seismic acquisition systems. Existing equipment generally resulted in a serious degradation of the output quality, which rendered the seismic data difficult (if not impossible) to interpret adequately. The advent of new fast A/D converters capable of operating at rates higher than 10 kHz and the availability of relatively inexpensive and powerful computers provided the possibility of development of a very high resolution multichannel digital seismic acquisition system.

The development of the "DELPH 24" system by ELICS and IFREMER has been an important step for the GEOSIS project. It allowed great improvement in both the quality of the seismic data and the reliability of its interpretation in terms of stratigraphy. The system is described in detail in the paper from Marsset et al (1994). Its main features are as follows: (i) improvement of the signal/noise ratio using the principle of summation (24 traces available), (ii) access to true depth information with migration algorithms, (iii) possibility of real time processing of the seismic data on site (data quality control in terms of data filtering and amplification...), as well as (iv) post processing of the digitally stored data using available dedicated software.

Furthermore, taking into account the specific requirements of the GEOSIS project (depth of investigation about 100m, spatial resolution better than 0.5m), and the likely range of P-wave velocities in offshore soils, a specific streamer was designed and developed to highlight a minimum frequency bandwidth of 3000 Hz, centred on 1500 Hz. The streamer contains two elements, 50-m long and with 24 channels each. The streamer and the seismic source must be towed close to the sea surface (0.50m of immersion approx.) in order to highlight a central frequency of 1500 Hz. Deeper or irregular immersion resulting from bad weather conditions would degrade the seismic data obtained.

Integration of Seismic and Geotechnical Data
Apart from processing of the multi-channel digital seismic data, the methods available for evaluating P-wave velocity (Vp) in soils include: (i) sonic logging in boreholes, (ii) vertical seismic profiling in boreholes, and (iii) velocity measurements on samples in the laboratory (Nauroy et al, 1994a). Both sonic logging and VSP profiling are routinely used in oil reservoir studies. However, because an uncased borehole is required for high quality velocity measurements, it is seldom used in geotechnical investigations. On the other hand, determination of the P-wave velocity on recovered samples can be considered as routine laboratory testing, although the results can be influenced by sample disturbance.

Vertical seismic profiling (VSP) provides different levels of information for the integration of seismic and geotechnical data. Firstly, P and S-wave velocities can be obtained by picking of the first arrivals. Secondly, further processing of the subsequent seismic events allows the primary and multiple reflections to be distinguished and the reflectors to be positioned along the well axis, both in terms of time and depth. Therefore calibration of the conventional seismic sections is possible without assuming wave velocity values. This important result was developed in previous papers (Nauroy et al, 1993 and 1994b).

In the GEOSIS project, great emphasis was placed on vertical seismic profiling (VSP) in geotechnical boreholes. The present paper concentrates on the work performed for studying the relationships between the geotechnical properties of the marine sediments and the P and S-wave velocities. P-wave velocity measurements were performed in geotechnical boreholes, using the same seismic source used for the very high resolution seismic survey with the "DELPH 24" system.

To the authors knowledge, the sole previous attempt at performing VSP profiling in geotechnical boreholes was made by Justice et al (1984), using a standard VSP probe. An alternative test procedure was also used in the GEOSIS project, namely in situ seismic cone testing. The incorporation of geophones in the cone penetrometer probe offered the possibility of combining in situ geotechnical testing and seismic profiling (measurement of P and S-wave velocities) within the same operation (Robertson et al, 1986). The seismic cone penetrometer has already been used in offshore site investigations for measuring S-wave velocity (Vs) and determining the shear modulus of soils (De Lange, 1991), but experience with P-wave velocity measurement is very limited.

The main advantages of the seismic cone over the standard VSP probe are : (i) it can be used in any type of soil normally encountered offshore (provided the soil can be penetrated by the cone tip, at least 1m); (ii) it does not require a predrilled and uncased borehole (the probe being pushed into the soil below the drilling bit, it also offers a perfect soil-probe coupling); and (iii) last but not least, it only requires minimum equipment in addition to the standard equipment currently used for offshore geotechnical site investigations.

EXPERIMENTAL SITE INVESTIGATION OFFSHORE MONACO
As part of the soil investigation for the construction of a piled breakwater structure in La Condamine harbour offshore Monaco, both experimental geophysical and geotechnical site investigations were carried out at a site with 50 m waterdepth. The geophysical site survey was performed using the "DELPH 24" very high resolution multichannel seismic acquisition system. The geotechnical investigation included several boreholes, using standard offshore equipment deployed from the dedicated geotechnical drilling vessel "M/V Whitethorn".

They included :
- 2 boreholes with P and S-wave vertical seismic cone profiling, down to 80m below mudline (boreholes P5 and P7),
- 1 continuous coring borehole in HQ piggy-back conditions (core diameter 63 mm), followed by sonic logging and P and S-wave vertical seismic profiling, using a standard VSP probe (borehole P'5).
The location of the boreholes with respect to the seismic lines is given in Fig. 1.

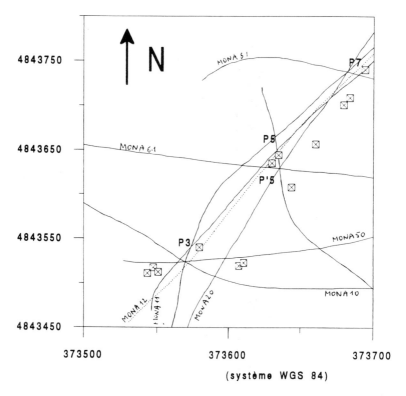

Fig. 1 Location of seismic lines and geotechnical boreholes offshore Monaco

Soil Conditions and Stratigraphy from the Seismic Survey

The soil conditions offshore Monaco are summarised in Fig. 2, showing the CPT cone resistance qc, wet density and carbonate content profiles from boreholes P5, P'5 and P7. The soil conditions at the site mainly consist of weakly cemented sandy to clayey silt. Four main units were identified from the seabed to 80m penetration:
- unit 1 (0 to 8m): soft mud and loose silt,
- unit 2 (8 to 36m): stiff carbonate silt, locally weakly cemented or with some thin strongly cemented beds,
- unit 3 (36 to 52m): stiff carbonate silt, weakly cemented,
- unit 4 (below 52m): heterogeneous formation with gravels and pebbles within a stiff carbonate silty matrix.

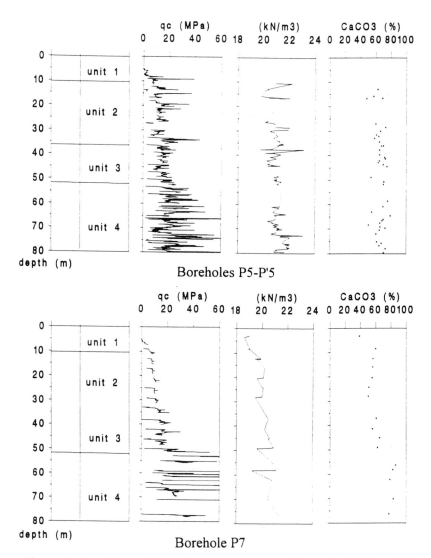

Fig. 2 - Soil conditions offshore Monaco at locations of boreholes P5-P'5 and P7

From Fig. 2, one can see that the soil conditions offshore Monaco are very heterogeneous and/or complex. The local strong cemented beds in unit 2 are clearly highlighted by the peaks in cone resistances (qc) in the CPT profile. At borehole P7 location, unit 2 (between 8 and 35 m approx.) is more homogeneous with more uniform qc values measured. In unit 4, the peak cone resistances highlight the thin cemented beds and the gravels/pebbles layers in the silt matrix.

The results of the very high resolution seismic survey are illustrated by Fig. 3, giving time sections corresponding to two lines passing through the location of the boreholes: line 12 (South-North) passing through all boreholes locations, and line 61 (West-East), passing through boreholes P5 and P'5. The recording conditions were as follows: frequency bandwidth 2.4 kHz, sampling rate 125 ms, fixed gain adjustable between 0 and 72 dB, resolution 16 bits. The seismic source was a 15 cubic inch water gun. The seismic sections are time migrated after standard processing. More details about acquisition and processing of the "DELPH 24" seismic data can be found in Marsset et al (1994) and Nauroy et al (1993, 1994a and 1994b).

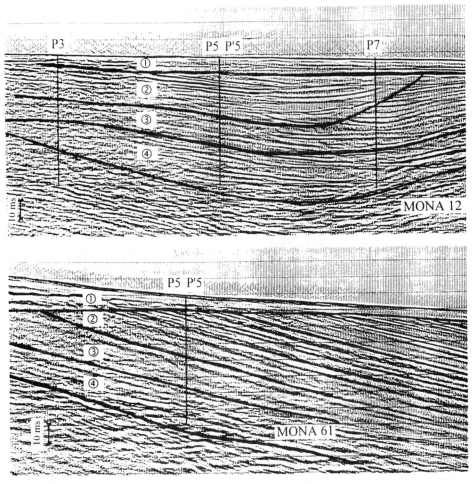

Fig. 3 - DELPH 24 seismic sections. lines 61 and 12

The quality of the seismic results has been significantly improved by the use of the multichannel acquisition system. The seismic sections of Fig. 3 clearly illustrate the complex stratigraphy encountered offshore Monaco, with a general dipping angle of 20° to the North-East. Numerous seismic reflectors are depicted by the very high resolution seismic results, some of them clearly corresponding to thin cemented beds, while others do not necessarily correspond to identified geotechnical interfaces. One may also notice that a powerful reflector is highlighted on line 61 some 10 ms two-way travel time (about 8m) below the end penetration point of boreholes P5 and P'5.

Vertical Seismic Profiling and Sonic Logging in Boreholes

Standard VSP profiling and sonic logging were performed in borehole P'5. Measurements were taken after completion of the coring operations while retrieving the drillstring and the probe to the surface. P-wave velocity measurements were performed on selected samples of the same borehole for purpose of comparison with the in situ measurements.

P and S-wave velocity measurements by seismic cone testing was performed in boreholes P5 and P7. The operational procedure is illustrated by Fig. 4. The standard seismic cone used was equipped with a triaxial set of geophones (1 for vertical motion, 2 for horizontal motions in two directions) in addition to the usual sensors for measurement of cone resistance, sleeve friction and pore pressure. Seismic measurements were made every 1m or 1.5 m after the penetration of the probe was stopped at the required depth. The P-wave source (15 cubic-inch water gun) was operated over the side of the vessel. The S-waves were produced by an hydraulic source, installed below the seabed frame, and generating horizontal shock waves. For both P and S-waves, several shots were recorded at each level, thus enabling further stacking.

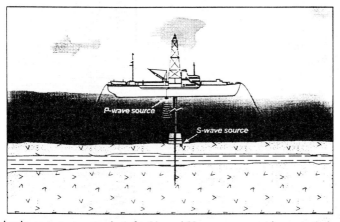

Fig. 4 - Seismic cone test procedure for Vp and Vs measurement in geotechnical boreholes

The P-wave velocities obtained in the same borehole P'5 by means of the VSP probe (automatic picking of the first arrivals), the sonic logging and the measurements on samples are combined in Fig. 5. Generally good agreement is found between all Vp values obtained and it can be concluded that the three different procedures can be used to obtain the P-wave velocity in marine sediments. The thin cemented layers are clearly highlighted by peak values of the Vp velocity from VSP and sonic logging. It can also be noticed that the measurements on samples in the laboratory give lower bound values of Vp velocity, probably due to decompression of the samples.

Fig. 6 gives an example of a velocity profile obtained by stacking velocity analysis on a CDP gather. At this stage, it is considered this only gives an initial estimate of P-wave velocity. However the results are encouraging and, with more refined processing to improve the accuracy, such an analysis of the multichannel digital seismic results can provide a better calibration of the very high resolution seismic data. In particular, a better time-depth correlation will allow a more accurate comparison of seismic reflectors and geotechnical formations.

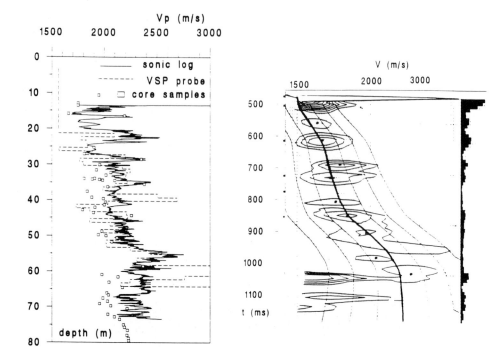

Fig. 5 - Measurements of P-wave velocity in borehole P'5

Fig. 6 - Stacking velocity analysis from a 24-channel seismic CDP gather

The high quality of P-wave signals allowed the processing procedure, routinely used in oil exploration well seismic surveying, to be applied. Fig. 7 shows the results of such processing applied to both the VSP probe data and the seismic cone data. The vertical axis represents the two-way travel time of standard seismic sections, while the location of the probe below the mudline is given by the horizontal axis. At the start of each trace, the small horizontal line represents the picking of the first arrival and provides a direct correlation between true depth and the reflector position expressed in time, without assuming any Vp velocity value. This chart is a useful tool for determining the geotechnical meaning of seismic events.

It should also be noticed that the VSP profiling results can disclose reflectors located below the bottom of the borehole. Such a reflector appears to be found about 10 ms two-way travel time (about 8 m) below the final depth of the CPT sounding in P5 and P'5 (as was seen on the multichannel seismic survey results, see Fig. 3).

CORRELATIONS BETWEEN SEISMIC AND GEOTECHNICAL DATA
One of the objectives of the GEOSIS project was to study the possible correlations between the P and S-wave velocities and the geotechnical engineering properties of the sediments (CPT cone resistance, porosity, or void ratio or wet density, and the carbonate content) from offshore Monaco.

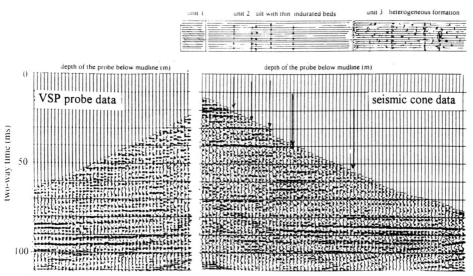

Fig. 7 - Results of VSP processing applied on the data from two boreholes 10 m apart

Correlation Between P and S-Wave Velocities and CPT Cone Resistance

A comparison between the CPT cone resistance qc profile and the P and S-waves velocity profiles, for boreholes P'5 and P7, is given in Fig. 8. The seismic velocities were obtained with the seismic cone and sonic logging (Vp profile in borehole P'5). It can be seen that the qc and Vp profiles in borehole P'5 show similar trends with peak Vp velocities generally corresponding to peak cone resistances. Correlation is also observed with the Vs profile, although the comparison is less easy due to a more discontinuous S-wave velocity profile obtained with the seismic cone. This qualitative correlation is not surprising since both cone resistance and velocity can be considered as parameters that integrate bulk properties of the sediment although they concern different scales of soil deformation.

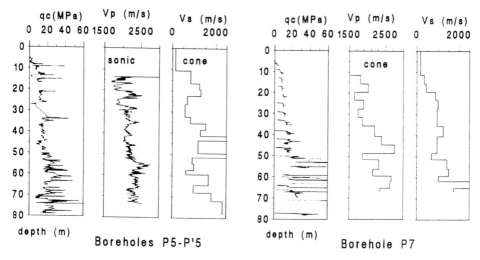

Fig. 8 - Comparison between cone resistance qc and velocity Vp and Vs profiles

<u>P-wave velocity vs. cone resistance Vp-qc correlation:</u> The Vp-qc comparison for the calcareous silts from offshore Monaco is given in Fig. 9. The following relationship is obtained:

$$Vp = 1500 + (25 \text{ to } 60) \cdot qc \qquad \text{(with qc in MPa and Vp in m/s)}.$$

Two other correlations are also given in Fig. 9:
- the relationship proposed from data of Campanella et al (1987) for a medium dense to dense, fine silty sand, i.e.,

$$Vp = 868 \cdot qc^{0.278}$$

- and the correlation obtained at an onshore site (site A in Calais), in medium dense fine sand, where the seismic cone equipment and procedure have been tested before performing the experimental   soil investigation offshore Monaco, i.e.:

$$Vp = 1500 + (5 \text{ to } 15) \cdot qc$$

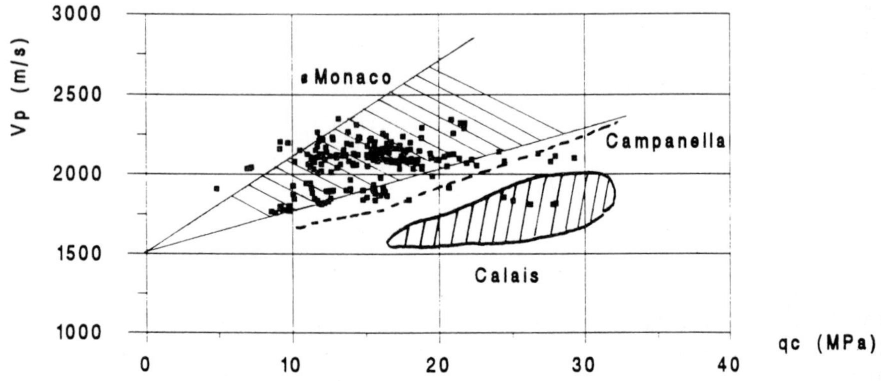

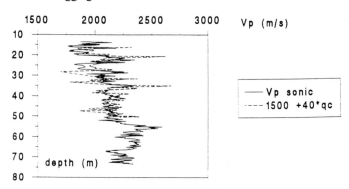

Fig. 9 - Correlation Vp -qc

Using the proposed Vp-qc relationship, the P-wave velocity for Monaco soils was back-calculated from the qc profile in borehole P5, for the purpose of comparison with the Vp profile obtained from sonic logging in borehole P'5 (about 10 m from P5). This comparison is given in Fig. 10. Good agreement is found, in particular in the first 50 m of penetration below mudline. Below this depth, the cone resistance profile was too erratic to allow a consistent comparison with the sonic logging.

Fig. 10 - Comparison between Vp from sonic logging and back-calculated Vp from qc profile

S-wave velocity vs. cone resistance Vs-qc correlation: The proposed relationship between the S-wave velocity and the cone resistance is given in Fig. 11. This figure is also completed by results obtained at other onshore sites (site A at Calais and site B with dense to very dense fine silty sand). The following relationships were obtained :

Offshore Monaco: Vs = (50 to 100) . qc
Onshore site A : Vs = (7 to 10) . qc
Onshore site B : Vs = (7 to 8) . qc (with qc in MPa and Vs in m/s).

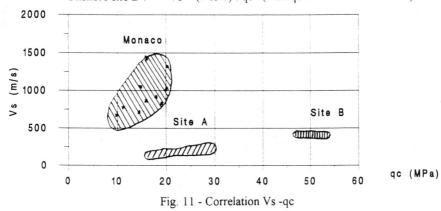

Fig. 11 - Correlation Vs -qc

It should be noted that the soils encountered offshore Monaco have complex properties, with carbonate silts showing a highly variable degree of cementation, both in the vertical and horizontal directions. Taking this into account, there are unavoidable difficulties in deriving clear relationship. As a result, the relatively good relationships obtained between the cone resistance qc and Vp and Vs velocities are very promising. Nevertheless, one should remember that the reliability of the relationships is greatly dependent on the quality of the data and on the type of equipment and procedure used to obtain P and S-wave velocities in the geotechnical boreholes.

For both results presented in figures 10 and 11, it is believed that the higher Vp/qc and Vs/qc ratios obtained for the offshore site in Monaco are due to the degree of cementation of the carbonate silts. Therefore, it is obvious that the relationships between the cone resistance qc and Vp and Vs are highly site dependent and should be carefully developed from local measurements.

Correlation Between P and S-Wave Velocities and Porosity
A tentative correlation between the P and S-wave velocities and the porosity of Monaco sediments is presented in Figs. 12 (Vp-n) and 13 (Vs-n).

The relationships extracted from data of Fumal (1985) and Davis et al (1989) have also been drawn for purpose of comparison. It should be recalled that the data of Davis et al concern superficial sediments only (about 1 m below mudline) from different offshore sites. A tendency for a (logical) decrease in Vp velocity with increasing porosity could be tentatively proposed from the results given in Fig. 12. However given the scatter in the results and the lack of data, a quantitative correlation would be meaningless. The database needs to be extended, with a larger range of porosity (i.e., a larger number of type of soils), in order to allow a clearer allow a clearer correlation between the porosity (or void ratio, or density) of the soil and the P and S-wave velocities to be proposed.

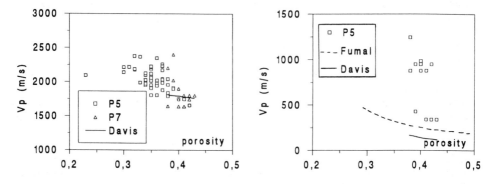

Fig. 12 - Correlation Vp-porosity   Fig. 13 - Correlation Vs-porosity

Correlation Between P and S-Wave Velocities and Carbonate Content
The carbonate contents of the silts encountered offshore Monaco are relatively high, generally ranging between 50 and 85%. On Figs. 14 and 15 the comparisons between P and S-wave velocities and the carbonate content determined from laboratory tests on samples are given. From these results, one could propose relationships giving increasing P and S-wave velocities with carbonate content. Such a correlation could logically be expected with such silts, since higher carbonate contents are related to higher degrees of cementation and lower water contents. Nevertheless more data are needed before drawing any final conclusion about the possible relationships between P and S-wave velocities and carbonate content of the soil.

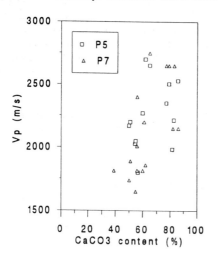   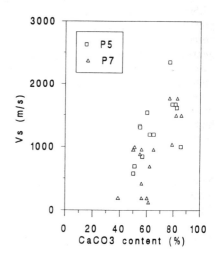

Fig. 14 - Correlation Vp - CaCO3   Fig. 15 - Correlation Vs - CaCO3

CONCLUSIONS
The objective of the GEOSIS research project was to improve the correlation between geophysical and geotechnical data for offshore site investigations. The problem of deriving geotechnical information from geophysical data has preoccupied geophysicists and

geotechnical engineers for many years. Until today, this was considered as theoretically possible but not proven, mainly because of the relatively low quality of seismic data obtained from geophysical site investigations. It has been recognised as a specific challenging issue for the design of the foundations of future offshore structures (Lacasse and Nadim, 1994).

The development of the "DELPH 24" very high resolution multichannel seismic acquisition equipment has been an important step forward, since it allowed a significant improvement in the quality of the seismic data and in the reliability of its interpretation. Vertical seismic profiling in geotechnical boreholes allows (1) the results of the shallow surface seismic survey to be calibrated and (2) seismic information to be obtained that can be directly correlated with geotechnical information obtained at the same depth in the same borehole. Furthermore the advent of a very high resolution digital multichannel seismic acquisition equipment for offshore site surveys allows the application of dedicated processing software, routinely used in oil exploration well seismic surveying, to be considered. Work is presently on-going on this subject with the continuation of the GEOSIS project. In particular, seismic processing should allow the potential application of VSP inversion for stratigraphic interpretation to be considered. It should also allow further correlations with other seismic parameters, such as attenuation and acoustic impedance, to be studied.

An experimental geophysical and geotechnical site investigation has been conducted offshore Monaco. Some relationships have been proposed between P and S-wave velocities and standard geotechnical parameters, such as CPT cone resistance, porosity and carbonate content. Taking into account the complex soils conditions encountered offshore Monaco, the high quality results obtained are very encouraging. However, one needs to extend the database to improve the relationships proposed.

The present paper plainly demonstrates that the equipment and procedure developed and tested within the GEOSIS project can be used routinely for offshore site investigations and can provide the necessary data to quantitatively extrapolate consistent geotechnical data around a borehole. However, it has been clearly shown that the relationships between geophysical and geotechnical parameters are highly site dependent and they should be developed on a case by case basis rather than by tentatively proposing general rules. Also, the distance over which one might be able to extrapolate the stratigraphy and associated engineering properties from a borehole, using the geophysical results, will depend on the geology of the site and the local soil properties. It should be emphasized that geotechnical boreholes will not be replaced by a very high resolution seismic survey and a set of relationships. Geotechnical data profiling and seismic measurements in geotechnical boreholes will always remain necessary for the purpose of calibration of the seismic survey.

Nevertheless the improved understanding of the soil stratigraphy from the improved seismic results should allow optimisation of the number of geotechnical boreholes needed for foundation assessment of a structure. In this respect, it is believed that the procedure developed within the GEOSIS project represents a major improvement and will become typical of future offshore site investigation practice.

ACKNOWLEDGEMENTS
The "GEOSIS" research project is sponsored by CLAROM (Club for Research Activities on Offshore Structures) in France. CLAROM, IFP, IFREMER, ELF, TOTAL, GEODIA and ISM are acknowledged for granting the permission to publish this paper.

REFERENCES

Campanella, R.G., Robertson, P. K., Gillespie, D., Laing, N. and Kurfurst, P. J., (1987), "Seismic cone penetration testing in the near offshore of the MacKenzie Delta", Canadian Geotechnical Journal, V24, n°1, pp 154-160.

Davis, A.M., Bennel, J.D., Huss, D. G. and Thomas, D., (1989) "Development of a seafloor geophysical sledge", Marine Geotechnology V8, pp. 99-109.

De Lange G. ,(1991), "Experience with the seismic cone penetrometer in offshore site investigation", In "Shear Waves in Marine Sediments", J.M. Hovem, M.D. Richardson and R. Stoll ed., Kluwen Academic Publ., pp. 275-282.

Fumal, T. E. and Tinsley, (1985), USGS Professional paper 1360 -pp.127-150.

Haynes, R., Davis, A.M., Reynolds, J.M. and Taylor, D.I. (1992) "The extraction of geotechnical information from high-resolution seismic reflection data", Proc. of Society of Underwater Technology Conf. on "Offshore Site Investigation and Foundation Behaviour", London, Kluwer Academic Publ., Vol. 28, pp. 215-228.

Hamilton, E. L. and Bachman, R.T. (1982), "Sound velocity and related properties of marine sediment", J. Acoustic Society of America, V. 72, pp. 1891-1904.

Justice, J.M., Hinds, R. and Sirbys, A.F., (1984), "The use of vertical seismic profiling in geotechnical site investigation", Proc. 16th Offshore Technology Conf., Houston, OTC Paper 4756, pp. 391-396.

Lacasse, S. and Nadim, F. ,(1994),"Reliability issues and future challenges in geotechnical engineering for offshore structures", Proc. BOSS'94 Behaviour of Offshore Structures Conf., Boston, Invited papers, pp 9-38.

Marsset, B., Blarez, E. and Girault, R. (1994), "Very high resolution multichannel recording for shallow seismic", Proc. 26th Offshore Technology Conference, Houston, OTC Paper 7371, Vol. 1, pp. 15-19.

Nauroy, J.F., Dubois, J.C., Puech, A., Poulet, D., Colliat, J.L. and Marsset, B. (1993), "Vertical seismic profiling in offshore geotechnical boreholes", Proc. of the 4th Canadian Marine Geotechnical Conf., St John, pp. 465-476.

Nauroy, J.F., Dubois, J.C., Meunier, J., Marsset, B., Puech, A., Lapierre, F., Kervadec, J.P. and Kuhn, H. (1994a), "Tests in offshore Monaco of new techniques for a better integration of geotechnical and seismic data", Proc. Offshore Technology Conference, Houston, OTC Paper 7375, Vol. 1, pp. 43-52.

Nauroy, J.F., Dubois, J.C., Meunier, J., Puech, A., Colliat, J.L., Poulet, D. and Lapierre, F. (1994b), "The use of VSP techniques in geotechnical boreholes: first tests in offshore Monaco", Proc. BOSS'94 Behaviour of Offshore Structures Conf., Boston, Vol. 1, pp. 111-123.

Robertson, P.K., Campanella, R.G., Gillespsie D. and Rice A., (1985), "Seismic CPT to measure in situ shear wave velocity", Proc. American Society of Civil Engineering, Journal of Geotechnical Engineering, Vol. 112, N° 8, pp. 789-803.

Confidence in seismic characterisation of the ground

G. A. RICKETTS, J. SMITH and B. O. SKIPP
Soil Mechanics Limited, Wokingham, UK

1.0 INTRODUCTION

Geophysical techniques are widely used in site investigation practice and appraisal of their utility is given in a variety of codes. Until recently their principal objective has been to fill in stratigraphic information between conventional boreholes, although sometimes they are used to pinpoint anomalies which can then be bored as part of an interactive problem solving exploration.

Methods applied to stratigraphic or discrete anomaly searches can only function where contrasts between physical properties are present. The geophysicist searches for inhomogeneities and the geologist hopes to translate those inhomogeneities into his own language and hence develop a three dimensional model of the ground. In the absence of such contrasts it is still possible to derive useful physical parameters. There are however fundamental physical limitations to the resolving power of all geophysical techniques and this introduces uncertainty into such model building.

Increasingly geophysical methods are being exploited to yield not only the morphology of the ground but a set of physical properties which can be turned into engineering properties. Examples would include the derivation of the electrical resistivity of resolved formations and hence information on water quality but perhaps the most promising development is the derivation of the small strain shear modulus of the ground which, given appropriate constitutive relationships can be the starting point for the analysis of ground response to static and dynamic action. This development calls for knowledge of the small strain shear wave velocity and bulk density and their variation, ideally in three dimensions, but usually only with depth. For dynamic studies some information is also needed on the material damping capacity of the ground.

This paper will concentrate upon an appreciation of what can be expected of the variety of seismic techniques now available to us. Such techniques are currently most closely associated with investigation where design has to allow for dynamic loading from say a machine or ground shaking from an earthquake.

The paper touches firstly upon some theoretical aspects of the resolution of geophysical inhomogeneities and uncertainties which might be expected in trying to develop a model of the ground. Secondly, the implications of the uncertainty in model morphology and material parameters are explored using simple standard models.

Thirdly the uncertainties which can be expected in deriving shear wave velocity and bulk density and material damping are discussed and the values obtained by different seismic techniques on the same ground profile presented.

Finally a comparison of shear wave velocities obtained by different methods is presented.

2.0 THEORETICAL CONSIDERATIONS
2.1 *Resolution with seismic methods*
With any system that uses wave phenomena for detection or measurement purposes its ability to resolve will depend on the wavelength of the radiation field being used. In the seismic domain the wavelength is a function of the frequency and velocity of the wave type in the materials under study. Wave theory, see for example Backus (1962), predicts that the limit of resolution is determined by the half wavelength of the radiation field. Table 1 gives half wavelength values over a wide band of seismic frequencies and velocities, typical frequencies range from near earthquake values of a few Hz up to ultrasonic frequencies, 1 MHz, used in laboratory testing. Examples of wave and material types are cited for the velocities shown and examples of the source types and techniques are given under the frequency headings. When considering the application of seismic techniques to a problem an understanding and evaluation of the method's ability to resolve the feature or features under investigation is fundamental.

Frequency Hz		10	100	1000	10000	100000
Source examples		Near earthquake frequencies, deep refraction studies usin explosives	Deep reflection surveys using explosives, vibrators and airguns	Shallow penetration marine reflection and crosshole seismics. Sparkers, airguns and boomers	High resolution marine seismic sources such as pingers. Wireline seismic sondes	High resolution echo sounding and laboratory measurements of seismic velocities
Material examples	Velocity m/s	Half wavelength	Half wavelength	Half wavelength	Half wavelength	Half wavelength
S wave - sands P-wave very loose blown sands Less than speed of sound in air	100	5	0 5	0.05	0.005	0 0005
S wave - very stiff clays P wave - dry soils	500	25	2 5	0 25	0 025	0 0025
S wave - weak rocks P wave - dry soils	1000	50	5	0 5	0 05	0 005
S wave - moderate rocks P wave - saturated soils Velocity of sound in water	1500	75	7 5	0 75	0 075	0 0075
S wave - strong rocks P wave - weak rocks - chalk, sandstone	2500	125	12 5	1 25	0 125	0 0125
P wave - strong rocks - metamorphic and igneous	5000	250	25	2 5	0 25	0 025

Table 1. Examples of half wavelengths in different materials

2.2 *Implications of uncertainty*
The ultimate implications of the uncertainties in model geometry and material parameters are in the uncertainty in the static or dynamic deflection of a foundation or the free ground surface predicted and ultimately verified by observation. There has been some discussion on how to assess the closeness of a prediction - in dynamic terms it may be expressed as the confidence which can be expected in prediction of peak spectral amplification from a rock base input to a surface response. It has been claimed that single value estimates of ground layering and material parameters can, with standard linear elastic wave propagation analysis yield maximum ground surface accelerations and rms accelerations within 20-30% of observation see Johnson and Silva (1981).

Different numerical procedures may give rise to different sensitivity to uncertainty and different ground morphologies may also have an impact. The vulnerability to free field resonance of near surface layers has been noted by many authors in both modelling and actuality, see Johnson and Silva (1981) and Henderson et al (1989). However Tinawi et al (1993) note that the spectral amplification (response spectra) for eastern Canadian earthquakes with high frequency content, attributed to soft clays 10-70m, thick was not sensitive to thickness. Field and Jacob (1993) also draw attention to response of the top few metres.

When considering the implications for a soil-structure interaction (SSI) model the tracing through of uncertainties of all kinds to the final outcome is difficult and very dependent on the numerical procedures adopted and the way in which the response motion is characterised, see Skipp (1994). Krauthammmer and Chen (1988) report a case where a 100% change in far field ground stiffness resulted in a 15% change in maximum displacement but only 2% change in peak acceleration. It has been argued by Hadjian (1991) that the outcome of SSI modelling is more dependent on the geotechnical model than the form of numerical analysis adopted, Hadjian (1991).

Robertson (1994) has explored the impact of best estimate and bound assessments of ground stiffness on the response of a safety critical structure and Llambias et al (1993) note how changes in the assessment of soil impedances and strain dependent stiffness degradation have had significant impact on the predicted response of a structure to an earthquake. A increase in rotational, translational and vertical stiffnesses of around 100% leads to an increase of fundamental SSI frequency of 40% for rocking and a similar amount for the first vertical mode.

The adoption of either generic eg Seed and Idriss (1970) shear strain/shear stiffness degradation curves or site soils specific ones has in recent years become an issue (EPRI, 1986). Llambias et al (1993), Eldred and Hight (1995). The effects of cyclic shear strains of .0001 which might be imposed by seismic shaking may reduce the stiffness of a soil by 30% and although there have been attempts to examine such phenomena by seismic methods they are not in widespread use. The phenomenon may however be significant in some cases where surface accelerations are more than 0.3g and its estimation is another uncertainty which has to be viewed in the light of those already stated.

Some appreciation of the relative importance of the elements in a model, and in particular the relative importance of getting the right geometry as against the right material properties and constitutive relationship can be explored, at least initially, by examining the simplest of theoretical models.

The two simplest cases are:

i) A rigid circular disc resting on an isotropic elastic layer of infinite lateral extent itself resting on a rigid base with the disc being exited by continuous sinusoidally varying force. This can be regarded as representing the problem of foundations under dynamic load.

ii) Horizontal strata of isotropic elastic materials, infinite in lateral extent and resting on a semi-infinite isotropic elastic half space and excited by vertically propagating horizontally polarised shear waves. This can be regarded as representing the free field response of a site to an earthquake.

2.2.1 *A simple model of a vibrating foundation.* Considering firstly Case (i), a convenient approach is to make use of the classic analyses of Bycroft (1956) and Warburton (1957). The behaviour of a 5 m diameter rigid disc resting on an elastic layer 5 m thick was examined.

Using Warburton (1957) a magnification factor, to be applied to static deflection, can be derived for a range of ratios of layer thickness and a range of values for mass ratio, b.

As long as the b value is above 5 or thereabouts, the sensitivity to an a change in thickness H can be calculated by applying the magnification factor to a recalculated static deflection. If this is done for an increase of 20% on H it may be shown that the maximum dynamic deflection is increased by a factor of 1.55. (If however, for the same geometry, the mass of the disc is reduced so that b is 5, the comparable increase on maximum dynamic deflection is now, on admittedly a smaller absolute magnitude, is now by a factor x 1.8.)

Since problems are likely to arise at the maximum response (loosely called a "resonance") it is interesting to examine how sensitive that maximum is to changes in the thickness H. Again, looking at vertical motion, the sensitivity can be explored with reference to Warburton (1957). For a mass ratio of 35, for example, dimensionless frequency at resonance changes from 0.55 at H/r of 1 to 0.45 at H/r of 2. For b=5 the corresponding change is from 1.3 to 0.8. The significance of uncertainty in the estimation of H upon the outcome in terms of error in surface vertical dynamic displacement depends upon the shape of the amplitude/frequency response curve. When b is large this is more peaked about the critical frequency, when low the response curve is flat.

Generally the presence of a lower reflecting interface reduces the radiation damping and this is especially so for rocking motion, see Awojobi and Grootenhuis (1965).

Changes in the values of G are reflected linearly in the maximum dynamic amplitude. Effects on the frequency for maximum amplitude are seen in the impact of changes in Vs on the dimensionless frequency term from which that frequency is extracted. This means that the amplitude value scales inversely as Vs squared and the frequency scales to Vs.

Summarising therefore we may say:

a) Error in estimating reflecting boundaries within a depth equal to one or two times the radius of a surface foundation of around 20% can have 60% impact on the maximum magnification under vertical dynamic load excluding any consideration of material damping.

b) Error in estimating the shear wave velocity of 20% can have 40% impact on estimating the maximum magnification under vertical dynamic load excluding any consideration of material damping.

Missing from the scene so far is material damping. This is strongly strain dependent, frequency and stress dependent but is usually much less than the radiation damping term for translational motions of a surface foundation but it does mean that the amplifications derived from elastic assumptions are larger than would in reality be experienced.

2.2.2 *Free surface response.* Sensitivity to uncertainty in assessed shear modulus has been tested by modelling the response of a typical site followed by observing the change in response resulting from modifying the model parameters in such a way as to simulate experimental uncertainties. The question arises as to what effect such an uncertainty has upon the dynamic response to a shaking input and also whether localised uncertainties, say in a stratum where field records are problematic, contribute significantly differently depending on their depth.

The initial model is shown in Fig 1. It represents a sequence of glacial deposits 17 m thick overlying slightly weathered mudstone which in turn overly very strong, fresh limestone. The model is a simplification of a real site, Site D, extensively investigated using crosshole, downhole and wireline logging. Gamma-gamma density and derived S-wave sonic logs are shown alongside crosshole and downhole velocity sections and a simplified stratigraphy and model parameters.

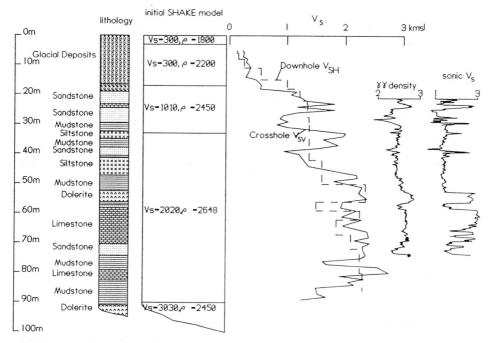

Fig 1. Site D, Summary of Field Data and SHAKE model

The excitation used in the examples that follow was the extensively studied Parkfield event of 27 June 1966, the strong motion record being taken from the Temblor station whose time history is shown in Fig 2. This shaking with a maximum horizontal acceleration of 0.36 g has been applied to the rigid base of the model using the University of California, Berkeley programme SHAKE with the simple assumption of vertically propagating horizontally polarised shear waves. Since the materials assumed were not of a character which would lead to very high shear strain degradation a more advanced model was not considered necessary, Skipp (1993). The computed maximum accelerations through the soil and rock sequence are depicted in Fig 3. The process was then repeated having reduced the shear modulus of a 3 m thick surface sublayer by 30% with the result that the surface acceleration increased by only about 1%. However if a corresponding reduction in shear modulus is applied to a sublayer within the soil profile immediately above rockhead then the reduction in predicted surface acceleration of 29.5%, see Fig 3.

The next issue investigated was the importance of precise knowledge of the shear modulus in the underlying rock. The rockhead sublayer 5 m thick was modified to simulate a shear wave velocity reduction from 1010 m/s to 840 m/s with imperceptible effect at the ground surface again demonstrated in Fig 3.

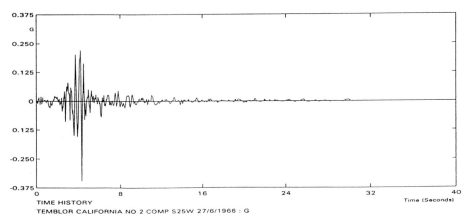

TIME HISTORY

TEMBLOR CALIFORNIA NO 2 COMP S25W 27/6/1966 : G

A second style of site is represented in Fig 4. It is also a simplification of a real site, Site C, where up to 20 m of silt and alluvium with high organic content overlies strong metamorphic rocks. The question addressed here is the implication of error in estimating the depth to an interface. Such an error might arise if inadequate soil/rock profile data are confused by seismic refraction effects as the strong velocity contrast is approached. The initial model comprises 19 m of overburden with a density of 1800 kg/m³ and a shear wave velocity of 135 m/s. A 2 m error in estimating the overburden thickness is then assumed and the model rerun with 17 m of overburden. The 10.5% depth error results in a 11.7% increase in predicted surface acceleration. The depth related responses of the two models are shown in Fig 4.

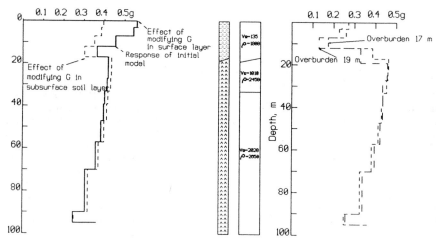

Fig3 Response of Site D   Fig 4. Site C. Effect of uncertainty in rockhead level

The final model study concerns the style of site where a relatively competent surface layer, say made ground, overlies soft material over bedrock. This is a slightly more realistic representation of Site C. Crosshole (Vsv) profiles shot in two opposite directions in the overburden and a downhole (Vsh) profile are shown in Fig 5. Such situations may present problems to imaging from the surface. They are of particular relevance to SSI,see Llambias (1993). The depth related responses resulting from two shear modulus values are shown in

Fig 5. A 36% reduction in shear modulus results in a 25% reduction in predicted surface acceleration.

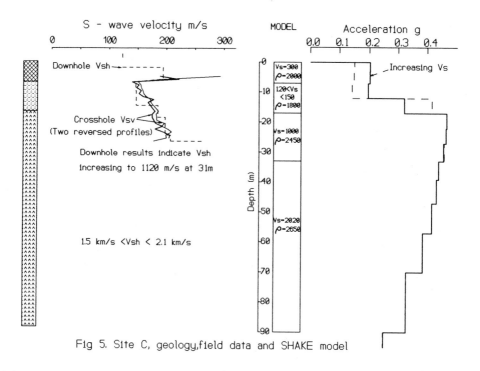

Fig 5. Site C, geology, field data and SHAKE model

3.0 THE EXPERIENCE OF FIELD OBSERVATIONS

3.1 *Estimates of uncertainties in acquisition of Vs*

The SI industry's preferred method of obtaining a shear modulus profile uses crosshole seismic shooting to derive a set of discrete shear wave velocities ascribed to discrete depths beneath the site combined with an estimated density profile derived from either in situ or laboratory measurements or empirically from borehole records. Briefly, at each test depth, the transit time of a shear wave wavelet is measured using a timing device (seismograph) and its velocity is calculated assuming that it has followed a straight raypath between two points in the receiver boreholes whose relative three dimensional coordinates are known from some form of borehole survey. In practical terms the seismic energy source and receivers are designed to enhance the shear wave signal to noise ratio and asymmetry of the source is utilised to improve confidence in identification by observation of phase reversal in the S wave portion of the waveform. The boreholes are sufficiently close together as to allow the justification of the straight line raypath assumption (at least when remote from strongly refracting boundaries) and to minimise the errors in shear wave identification which might arise from dispersion effects. Nevertheless an uncertainty in the calculated modulus profile will derive from experimental error in the basic field measurements and these are briefly addressed below and seen to be of the order of the example perturbations used in the free field behaviour of example sites examined above.

A widely used alternative, usually referred to as downhole shooting utilises a single borehole in which a detector is located and a source activated at one or more source points which are usually at the surface. Variations put the source in shallow pits or in adjacent boreholes. The main difference is that fewer boreholes are needed and the raypaths are near vertical.

Figure 6 gives some examples of some crosshole seismograms showing the received waveforms in two adjacent holes resulting from upward and downward sense motion in the source borehole. It will be noted that identification of particular waves in practice is often not a simple matter and may require the use of expert judgement.

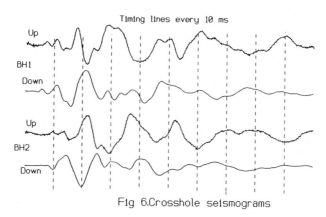

Fig 6.Crosshole seismograms

Assuming a timing uncertainty on each trace of 1 ms and a combined survey error of 0.3 m (both would normally be worst case figures) it follows that the small strain dynamic shear modulus can be estimated to within about 30%.

To illustrate scatter of observation and variation between methods comparisons of shear wave velocities from six sites have been made.

Site A consists of Neogene slightly cemented cohesionless material over Palaeocene clays and Cretaceous Chalk, Site B consists of Jurassic limestones with some mudstones overlying Triassic argillites, Site C has recent alluvium overlying Palaeozoic low grade metamorphics and Site D has glacial deposits on a mixture of Palaeozoic sandstones, limestones and siltstones with igneous intrusions. Site E has typically 20 m of glacial till over sandstone. Site F was originally granular fill in which a number of stone columns have been vibrated in which has resulted in a higher Vsh than in the original state.

3.2 Scatter of data in field observations

The uncertainties in path length, wave identification and timing are inherent to the technique but there are other uncertainties which can be associated with sampling and with the real departures in the ground from lateral homogeneity and isotropy. Some idea of the scatter of determinations of Vs can be seen in Figs 7 and 8, Sites F and E respectively. These results were obtained from shooting downhole in the same volume of ground from multiple source points close to the boreholes.

3.3 Comparison of Vs by various methods

Although crosshole methods are generally favoured on major projects they are often accompanied by up-hole/down hole, VSP, Borehole acoustic logging. Each of these procedures can yield a value of Vs. The values of Vs will often be different.

Powell and Butcher (1991) discuss the differences between the "dynamic" shear modulus derived from shallow surface refraction, seiscone and Rayleigh wave techniques as well as comparing them with small strain direct methods all values being normalised for effective stress. At small strains the differences between values derived from surface refraction and Rayleigh waves were large - around 100% but there was a close match between the results from seiscone and Rayleigh waves.

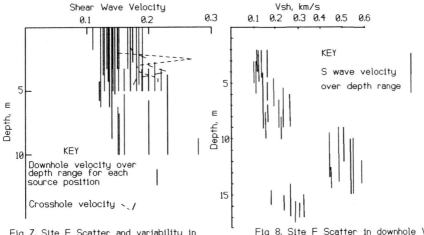

Fig 7. Site E Scatter and variability in crosshole and downhole velocities

Fig 8. Site F Scatter in downhole Vs

At Site A the upper 40m can be regarded as a single unit within which there are no significant compliance contrasts. Uphole horizontally polarised Vsh closely matches vertically polarised crosshole Vsv but is somewhat lower as seen in Fig 9.

The comparisons between Vs as determined from crosshole, uphole and sonic borehole logs at Site B in approximately the same geological units (limestones and argillites) are shown in Fig 10 and 11. The uphole and crosshole velocities in the predominantly limestones are have been examined in three blocks of data. The aggregated histogram shows that the VSh values are significantly smaller than the Vsv values. At this site the shear wave velocity as measured in the well logging was available as a direct measurement and it is clear that these values were much higher, circa 50%, than Vsv derived from crosshole measurements in the limestones although there appeared to be convergence with the crosshole and uphole values in the Triassic argillites as shown in Fig 11.

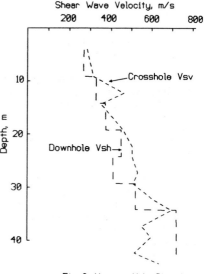

Fig 9. Vsv vs Vsh, Site A

At Site C, referred to above in Fig 5, the comparison between Vsh and Vsv in the Holocene units showed Vsh/Vsv to be about 0.9.

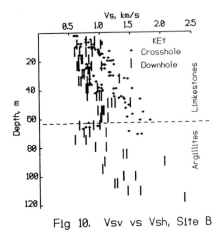

Fig 10. Vsv vs Vsh, Site B

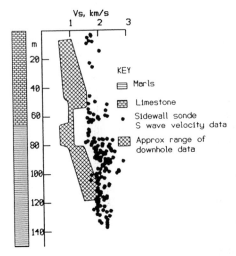

Fig 11. Vs from wireline log vs downhole
data, Site B

At Site D, Fig 1, a comparison can be made between Vsh and Vsv in the relatively uniform (N ranging from about 30 to 80) glacial deposits at one location. Here there was concurrence in the upper 10m of the unit but thereafter until the influence of the bedrock was seen in the Vsv values which remained lower than the Vsh values. Comparisons were also made at this site in a particular massive 12 m thick limestone unit. Here the Vsv values were generally higher than Vsh with an overall ration Vsh/Vsv of 0.91. At this site shear wave velocity values were available from wire line logging but they have ben calculated from P wave velocities using Christiansen's Equation (see Forster and McCann 1979). Values from this expression were generally higher than Vsv by as much as 50%. Taking the rock succession as a whole above the igneous dolerite Vsv tends to be higher than Vsh the ratio Vsv/Vsv being 0.85.

3.4 *Some comment on possible reasons for disparity*
Inspection of these results shows disparities between velocities between crosshole, uphole and sonic log too great to be explicable in terms of a Biot type model. It is tempting to invoke anisotropy, that ubiquitous geophysical and geotechnical property which is so hard to handle properly (King et al, 1994)

Theoretical and empirical ways of reconciling values derived from borehole logs at relatively high frequencies(several kHz) with the frequencies associated with reflection seismic surveying (A few tens of Hz) are being given increasing attention (Mavko and Jizba, 1994; Mavko and Nolen-Hoeksma, 1994). The behaviour in saturated rocks appears to be dominated by large scale processes (Biot, 1956) and local flow(squirt), the former being more important in high porosity rocks and sands. Dispersion (ie frequency dependant velocities) is minimal in dry rocks so it has been suggested 'dry' velocity can be taken as the low frequency velocity (Winkler, 1985). According to Van Dalfsen (1989), using Biot Gassman equations, the effect of increasing the frequency from 10Hz to 200Hz was less than 0.5% in uncemented porous sand and imperceptible for cemented porous sandstone. For P wave velocity the dispersion in uncemented porous sand was calculated to be about 5% between 10Hz and 2000Hz. Most of this work has been carried out in the laboratory but does indicate that well log velocities need to be corrected to compare with seismic velocities at the 5% difference level and that much depends upon permeability.

4.0 DETERMINATION OF DENSITY

In order to calculate shear modulus G the density must be known. Where cores are available this can be accomplished to within a few percent confidence following standard procedures. Spot values derived from core samples are compared with wireline logs in Fig 1. Where soils cannot be recovered as a core, although this is increasingly possible bulk samples may have to be reconstituted and a range of in situ values assessed. In situ testing of granular soils to give an assessment of density state falls within the field of current SI practice. Geophysical techniques which use borehole sondes carrying the means to measure backscatter gamma radiation are available and in principle can achieve a value within a few percent of truth for most rocks and cohesive soils, Fig 12.

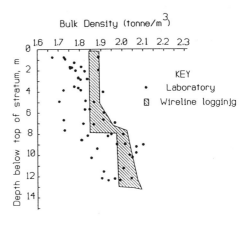

Fig 12, Laboratory vs in situ Density, London Clay

5.0 MATERIAL DAMPING

Material damping is a mainly strain dependant parameter which, over the strain range and strain rates associated with seismic methods can in principle be assessed from any method for which the radiation pattern of a particular wave can be determined or assumed see for example Stokoe ( ). The amplitude decrement with distance from source, or usually from two or more locations on a 'ray' is regarded as composed of a energy spreading term and a material dissipative term. If the geometric spread term is assumed the residual can be expressed as a material damping term. This is an approach familiar to seismologists in their search for Coda Q,(Greensmith and Gutmanis, 1990). In engineering geophysics the ranges at which measurements are often made precludes a clear and unique attribution of motion to a particular wave type. An alternative method using the 'pulse spreading' has attractions where a full wave form such as a high frequency half wave can be traced over a short distance between say adjacent boreholes. Over short distances geometric spread is small but a distinct broadening of the wave can be seen and measured. If the assumption is made that the dissipation can be described as an effective viscosity and damping term can be derived using a formula suggested by Knopoff (1956). In principle an estimate of damping could be obtained from observation of pulse broadening in seismograms such as those shown in Fig 6. Such measurements cannot be given high confidence since much depends on the efficiency of coupling of borehole casing to the soils.

6.0 DISCUSSION AND CONCLUSIONS

Simple numerical exploration suggests that a range of +/- 20% of some 'true' value of soil stiffness would be fit for most geotechnical and foundation dynamic modelling having regard for the limitations of assumptions and numerical procedures. There are however scenarios where uncertainties and their combinations may be significant. These usually involve a thin (ie several m thick) surface layer and free field resonance and have not been addressed in this paper.

There is now a variety of seismic methods which yield shear wave velocities and with density, the small strain shear modulus. In Table 2 these methods are summarised with an attempt to assess their utility and utilisation.

SEISMIC TECHNIQUES FOR GEOTECHNICS

PURPOSE	TECHNIQUE	APPROPRIATENESS	UTILISATION	TECHNIQUE	APPROPRIATENESS	UTILISATION
	NON INTRUSIVE			INTRUSIVE		
MORPHOLOGY/ IMAGING	Surface Refraction	X X X X X	X X X X X	Vertical Seismic Profiling S, P	X X X X X	X X X
	Reflection (Land and Marine)	X X X X	X X X X X	Borehole Tomography S, P	X X X	X X X
	Surface wave dispersion	X X X	X X X X	Borehole Logging	X X X X	X X X X
	Surface Arrays	X X	X X X			
	Inversion of Site Response	X X	X X			
ELASTIC PARAMETERS	Surface Refraction	X X X	X X	Surface - Downhole	X X X X	X X X X X
	Reflection (Land and Marine)	X	X	Cross - Borehole	X X X X X	X X X X X
	Surface wave dispersion	X X X	X X X X	Borehole Tomography	X X X X X	X X X X
	Surface Arrays			Seismic Cone Penetrometer	X X X X X	X X X X
	Inversion of Site Response	X X X	X X X			

The case histories and the very simplified numerical examination presented above suggest that where stratification is present with compliance variations at intervals of the order or smaller than the sampling wave lengths there are significant differences between shear wave velocities emerging from different methods of measurement. Although Vsh and Vsv are close in shallow relatively uniform deposits, the differences - up to 30% - in profiles with marked stratification are greater than can be explained by Biot-Gassman models for saturated poroelastic materials. Long-wave length-equivalent-theory may have to be invoked although account should be taken of general lower reliability of uphole/downhole measurements. This means that the methods used to estimate low frequency parameters from high frequency shear wave measurements in boreholes are not apposite for geotechnical engineering or the modelling of foundation dynamics.

The differences between Vsh and Vsv are also to be expected in some geologically homogeneous units such heavily overconsolidated clays where anisotropy is present.

Within the population of one Vs type scatter may be marked in stratified ground. Notwithstanding the efforts to interpret within a geological unit, the scatter from downhole measurements may amount to +/-50% on the best estimate. Even crosshole results may show sufficient variability over quite short vertical intervals within a single unit to give the appearance of having up to 20% scatter. It is not surprising that USNRC require modelling to incorporate sensitivity studies with a +/- 50% of best estimate ground parameters (USNRC, 1989).

It is difficult to validate best estimates but an appreciation of the error combination - assuming a correct identification of wave type - leads us to expect that a value of Vs can in average circumstances be estimated to within 10-20%.

It is generally accepted that the distribution of punctual determinations of geotechnical parameters is log-normal, see Bruce et al (1993), but where they are derived by wavelength sampling this assumption needs to be confirmed in stratified ground. The data presented here is not sufficient for that purpose.

With all these reservations however, Vs (of appropriate type) can now be measured in commercial practice and is the preferred starting point for dynamic geotechnical deformation analysis, especially if we take heed of Lomnitz's 1994 comment on the remarkable self similarity of soil constitutive relationships.

7.0 REFERENCES

Awojobi, A D and Grootenhuis 1965 Vibration of rigid bodies on elastic media. Proc. Royal Soc. London, A282, 27.

Backus, G.E. 1962 Long wave elastic anisotropy produced by horizontal layering. J.Geophys. Res., 67,4427 - 4440.

Biot 1956 Theory of propagation of elastic waves in a fluid saturated porous solid, 1: Low frequency range: J. Acoustical Society of America, 28,168-178.

Bruce, R L, Prakash, P K and Rogers, C P 1993 Generation of secondary URS and use of structural reliability techniques for removal of conservatism. Nuclear Energy, vol 32 No 4, 249 - 255.

Bycroft, G.M. 1956 Forced vibration of a rigid circular plate on a semi-infinite elastic solid and on an elastic stratum. Phil. Trans. Roy. Soc. London, 248, 327-368.

EPRI(1988) Site response to earthquake ground motion. EPRI Report NP5747, Pao Alto, California.

Field, E H and Jacob, K H 1993 Monte-Carlo simulation of the theoretical site response variability at Turkey Flat California given the uncertainty in the geotechnically derived input parameters. Earthquake Spectra, 9(4), 669-701.

Forster, A and McCann, D M 1979 Uses of geophysical logging techniques in the determination of in situ geotechnical parameters. Trans. 6th European Symp. Soc. Prof. Well Log Analysts, Paper G.

Greensmith, J.I. and Gutmanis, J.C. 1990 Aspects of late Holocene depositional history of the Dungeness area, Kent. Proc. Geol. Ass., 101, 225-237.

Hadjian,A.H., Tseng,W.S., Tang,Y.K., Tang,H.I. 1991 Assessment of soil-structure interaction practice based upon synthesised results from the Lotung experiment - forced vibration tests. In: Proc., 4th U.S. Nat. Conf. Earthq. Eng., Palm Springs, Calif.,3, 845-854.

Henderson, P., Heidebrecht, A.C., Naumosti, N. and Pappin, J.W. 1989 Site response study- methodology - calibration and verification of computer programs, EERG Report 89-01, McMaster Univ.,Toronto, 37pp.

Johnson, L.R., Silva, W. 1981 The effect of unconsolidated sediments upon ground motion during local earthquake. Bull. Seis. Soc. Amer., 71,127-142.

King, M S, Andrea, M and Shans Khansmir 1994 Velocity anisotropy of Carboniferous material. Int J. Rock Mech. and Min. Sci. and Geomech. Abstr. 31, 3 261-263.

Knopoff, L. 1956 The seismic pulse in materials possessing solid friction, I:Plane waves, Bull. Seism. Soc. Amer. 46. 175-183.

Krauthammer, T., Chen, Y, 1988 Free field earthquake ground motion: Effect of various mathematical simulation approaches on soil structure interaction results. Engineering Structures, 10, 85-94.

Llambias, J.M., Shepherd,D.J., Rodweel, M.D. Sensitivity of seismic structure response to interpretation of soils data. Soil Dynamics and Earthquake Engineering 12, 337-342.

Lomnitz, C 1994 Funamentals of Earthquake Prediction, John Wiley,

Mavko, G., Jizba, D. 1994 The relation between P and S wave dispersion in saturated rocks. Geophysics 59, 1, 87-92.

Mavko and Nolen Hoeksma

Powell, T.J.M., Butcher, A.P. 1991 Assessment ground stiffness from field and laboratory tests. Proc. 10th Eur. Reg. Conf. SMFE 1. 153-156.

Robertson, C.I., 1994 Management of dynamic ground uncertainty for nuclear facilities. Risk and Reliability in Ground Engineering Ed:B.O.Skipp, Thomas Telford, London.

Seed, H.B., Idriss, I.M. 1970 Soil Moduli and Damping Factor for Dynamic Response Analysis. EERC 70-10, Univ. of Cal. Berkley.

Skipp, B.O. 1994 Selection and use of ground parameters for soil -structure interaction analysis, 17th European Seminar on Earthquake Engineering, Haifa, Israel September 5-12 1993, Balkema, Rotterdam, 1994.
Stokoe

Tinawi

USNRC (1989) Standard review plan for review of safety analysis reports for nuclear power plant. Section 2.5.2, Rev.No.2, NUREG-0800.

Van-Dalfsen, W. 1989 Acoustics of nearly unconsolidated reservoirs, theory and measurement. TNO Report OC 89-25-A, Delft Netherlands.

Warburton, G. 1957 Forced vibration of a body on an elastic stratum, J. Appl. Mech., Trans ASME,224,55-58.

Winkler, K. 1985 Dispersion analysis of velocity and attenuation in Berea Sandstone. J. Geophys. Res. 90, 6793-6800.

Geophysical surveying methods in a site investigation programme

D. M. McCANN and C. A. GREEN
British Geological Survey, Nicker Hill, Keyworth, Nottingham

The recent revision to BS5930 on Site Investigation Practice has emphasised the increasing role of geophysical methods as part of a site investigation programme. Since the publication of the original version of BS5930 in 1981 significant improvements have been made in many areas of geophysical surveying including:-

1.   the introduction of new methods such as ground probing radar and cross-hole seismic tomography.

2.   the use of the micro-computer as part of the geophysical instrumentation.

3.   the application of the improvement in computer technology in the 80's to enhance the ability to the geophysicist to interpretation of the geophysical data.

In this paper, the changes that have taken place in geophysical surveying methods since 1981 are examined in relation to the points mentioned above. Case histories are used to illustrate the significant improvement in the instrumentation, the collection, processing and display of the geophysical data, the interpretation of the data and finally their integration in the site investigation programme. It is shown that the improvement made over the past decade has resulted in geophysical surveying methods becoming more acceptable to the civil engineers and, hence, more widely used site investigation programme. Finally, future prospects for geophysical surveying methods are assessed and areas where greater application is likely are suggested.

INTRODUCTION

Site investigation is defined as the whole of the process by which information is gathered on the ground as part of the engineering design and construction process. The importance of a comprehensive site investigation for every site is emphasised by Littlejohn et al (1994), who concluded that without a properly procured, supervised and interpreted site investigation the hazards in the ground cannot be known.

The point at which geophysical surveying methods were introduced into the site investigation process is difficult to ascertain but certainly by the early 50's they were in common use at many construction sites. The initial use of these methods was not greeted with much enthusiasm by the civil engineer, who preferred the more trusted approach to site investigation based on boreholes, trial pits, and a number of in situ engineering tests. This scepticism was justified in many ways since the geophysicist's surveying ability was limited by the very rudimentary equipment in operation at the time and the length of time taken to interpret the geophysical data obtained in a survey. The limited efficiency of operation associated with geophysical surveying and a general lack of communication between the geophysicist and the civil engineer gave rise to a general feeling in the industry that the geophysicist represented black box technology, which was not to be trusted.

Early geophysical equipment which was based on thermionic valve technology consequently required considerable electrical power to operate it, and was very heavy and difficult to transport in a field situation. Even the introduction of the transistor did little to reduce the size and weight of the equipment initially but by the late 60's a steady improvement in the standard of the geophysical equipment had taken place. However, it was still largely analogue equipment and the only recording medium for the data was the magnetic tape recorder. In the mid 70's the advent of the micro-processor and digital technology gave rise to a new generation of lightweight highly portable geophysical equipment, which revolutionised the rate at which geophysical surveys could be carried out.

Interpretation of geophysical data in the field was largely based on the use of geophysical methods, which were time consuming and could not be used in an interactive ways to adjust the survey parameters, such as line spacing and station density. Considerable improvement was brought about by the use of the main frame computers introduced in the mid-60s but there was still a time delay between the collection of the geophysical data and its final interpretation in the office. The introduction of the micro-computer in the late 70's coincided with the development of the new generation of geophysical equipment referred to above. It was, therefore, not long before on-line interpretation of geophysical data was introduced in the field with the geophysical information being downloaded to a PC operating in the field laboratory.

The original geophysical contribution to BS 5930 was prepared at a time when the improvements in both equipment and interpretational procedures referred to above were in the transitional stage. The text, thus, summarises how geophysical surveying contributed to site investigation at the end of the 70's and the writer did not have the scope in a document, which merely describes current operational practices, to look at new developments at that time or attempt to visualise the future. It is interesting to note, however, that the authors of the Geological Society Engineering Group Working Party Report on Engineering Geophysics, which was prepared in the mid 80's, were able to describe in some detail the new digital technology, which had improved not only the available geophysical equipment but had also enhanced interpretation procedures as a result of the developments in computing hardware and software. The revision of the standard in 1993 gave an opportunity to trace the development of geophysical surveying during the 80's and speculate a little on likely developments into the next century.

INTEGRATION OF GEOPHYSICAL SURVEYS INTO THE SITE INVESTIGATION PROCESS

While all site investigations are based on traditional methods, such as boreholes and trial pits, in many circumstances a large number may be required to achieve a sample density that is statistically valid to ensure that small but significant targets within the rock mass are located. These targets might be geological ones, such as faults or fractured zones of rock, or man-made problems, such as old mineshafts and adits. Geophysical methods can sample a greater volume of the rock mass and can be used in conjunction with the boreholes and trial pits, which are essential for the provision of groundtruth information to calibrate the interpretation of the geophysical survey data across the site. This enables the geophysical survey to be extended into areas where little or no groundtruth information is available such that more confidence can be placed in the interpretation of the geophysical survey data.

It must be stressed that geophysical survey data on its own merely measures the vertical and lateral variation of a physical properties of the geological strata, such as electrical conductivity and this data can only be interpreted in the light of some knowledge of the likely ground conditions that will give rise to the data set measured. In this respect there are two main approaches to carrying out a geophysical survey:-

1. Measurement of a physical property, such as electrical conductivity on a grid basis over the ground surface. Contouring of this data will locate anomalous zones, which may be associated with the presence of faults, fractured rock, swallow holes and gulls etc. Further investigation of the anomalous areas is required, unless historical information exists which indicates the likely cause of geophysical anomaly.

2. Measurement of a physical property along a horizontal profile, such that details of the vertical variation of that property are determined. In this case the geophysicist attempts to produce a mathematical model of the geological structure which will give rise to the measured geophysical data set. Again, the accuracy of the model is largely dependent on the groundtruth information that is available either from historical sources or from boreholes and trial pits.

It is important to differentiate between these two approaches to geophysical surveying since the production of a simple contoured geophysical map which identifies anomalous ground conditions is far quicker and less expensive to carry out than the more detailed survey which gives rise to a model of both the vertical and horizontal geological structure in three dimensions. It is probable that the lack of appreciation of this one factor has resulted in much of the bad press received in many areas where geophysical methods have been applied, since the cost of obtaining the complete 3D model of the geological structure on a site can be four to five times greater than that for producing a simple contoured geophysical map. As in all other fields the final objective of the geophysical survey must be clearly specified such that it can be designed and costed appropriately to achieve the required objective.

There are five major factors, which need to be considered in the design of a geophysical survey, as follows:-

1. the required depth of penetration into the geological formation.
2. the vertical and lateral resolution required for the anticipated targets.
3. the contrast in physical properties between the target and its surroundings.
4. signal to noise ratio for the physical property measured at the site under investigation.
5. historical information from previous investigations at the site, geological maps, etc.

Careful application of all the above factors to the design of a geophysical survey should result in a specification which either achieves the desired objectives or, more importantly, recommends an alternative approach if no geophysical surveying method is deemed appropriate to the solution of the problem specified.

The revision to BS5930 does draw attention to the desirability of using a geophysical specialist on a site investigation programme where the use of geophysical surveying methods is advocated. This is in line with similar requirements for the use of geotechnical specialists recommended in the report published by the Site Investigation Steering Group (SISG, 1994).

It is also appropriate to consider the use of a geophysical advisor in situation where complex geophysical surveys are being carried out and an independent assessment of the survey data is required. The importance of using personnel with the correct level of expertise is highlighted in SISG (1994) and this factor is of particular importance in specialised areas, such as geophysics.

MODERN DEVELOPMENTS

The original scope of BS5930 listed only five major geophysical methods for land-based surveys. These were:

(a)   Seismic Refraction
(b)   Electrical Resistivity
(c)   Gravity
(d)   Magnetic
(e)   Borehole Geophysical Logging

All five methods are still in common use in the site investigation process but have gained substantially from improvements both in equipment development and interpretational procedures over the last decade.

(a)   Seismic Surveys

The major advance made in seismic surveying has been in the area of signal enhancement where digital methods using micro-processors have replaced the analogue recording techniques used in the previous generation of seismic recorders. Enhancement by computer is based on the averaging of repeated measurements and enables small signals to be faithfully measured. The process effectively increases the signal to noise ratio since the measurements are repeated a number of times, added together or stacked and then divided by the number of measurements, giving an increase in the signal to noise ratio of $N^{\frac{1}{2}}$, where N is the number of repeated measurements. The advantage of the modern seismic recorder is that the seismic data can be accessed directly to a PC either in the field or the laboratory so that rapid processing of the seismic data can take place. Signal enhancement extends the range of a geophone spread to around 100 m with a hammer source.

The use of digital filtering together with high frequency geophones has increased the resolution that can be achieved with shallow seismic reflection surveys. Although considerable success has been reported in the literature (Miller and Steeples, 1994; Goforth and Hayward, 1992) with the development of seismic reflection it is still not in common use in the site investigation process. The main reason for this is that the majority of seismic sources currently used for land-based surveys have pulse widths, which are too long to resolve the fine detail of the near surface geological structure. Attempts to use higher frequency sources to improve the basic resolution have been inhibited by the lack of penetration of the seismic pulse caused by attenuation of the seismic energy in the near-surface layers. However, it is emphasised that progress is being made and the seismic reflection method is in common use on land on more regional engineering studies, where the study of the geological structures down to depths of 300 m is required.

(b)   Electrical Resistivity

The modern equipment for carrying out an electrical resistivity survey is now micro-processor controlled and again significant improvements have been made to the signal to noise ratio by signal averaging. The initial range of digitally controlled equipment interfaced with a data recorder, which was then used to input the data to the PC for interpretation. The later range of equipment developed in the late 80's included computer control of the electrode arrays and automatic processing of the data on-line in the field (Barker 1992; Jackson and McCann 1993).

Undoubtably, the major advance in electrical resistivity surveying was the transition from the manual interpretation of the recorded resistivity curves based on the published curves of Orellana and Moorey (1966) to automatic curve fitting of the data on the computer based on mathematical models of the geological strata. Two approaches can be used namely inversion where geological model is adjusted automatically by the computer until the resistivity curve matches the recorded data and forward modelling where the geological model is inputted normally and adjusted interactively until a suitable fit of the data is achieved. The second approach enables the geophysicist to have more control of the geological model that is developed and far more use can be made of other geological information from the site to optimize the interpretation procedure.

(c)   Gravity

For many years gravity data was used mainly for regional geophysical surveys to examine large scale geological structures. The introduction of the La Coste Romberg Model D in the early 70's which enabled readings to be made down to 10 micro gal resulted in an increasing use of gravity surveys in the detection of cavities and buried mineshafts and adits (Neumann, 1977). This particular instrument still remains the most widely used one for engineering site investigation, although more modern meters with digital outputs have been developed.

Initially gravity data were used to produce contoured maps to locate anomalous zones associated with a density reduction in the near surface material resulting from the presence of a cavity. Significant developments in the mathematical modelling of gravity data enabled the geophysicist to produce 2½ D geological models from data recorded along profiles recorded in the field. A typical example is shown in Figure 1 where the presence of precise information on the depth of the cavity obtained from a borehole is used to constrain the model. The problem of equivalence remains a particular difficulty with the modelling of gravity data since the gravity profile can be modelled by a large number of possible geological solutions and additional information is required to constrain the model in some way.

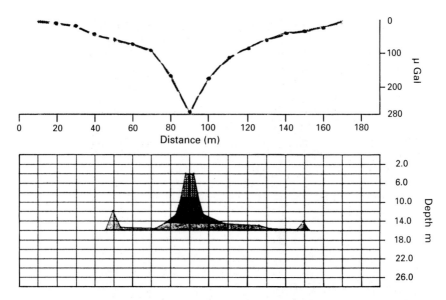

FIGURE 1: Theoretical modelling of an observed gravity traverse across a buried cavern (after McCann et al, 1988)

(d)   Magnetic

The standard instrument in use remains the proton magnetometer and the major improvement made to the instrumentation is the addition of micro-processor control to record the data for downloading to a computer at suitable points in the survey. Modern instruments now include two magnetometers within the system so that measurements of the vertical magnetic gradient can be recorded. This is particularly relevant where rapid changes are observed in the magnetic field resulting from the presence of magnetic bodies, such as oil drums, in the near surface material. Again, significant progress has been made in the mathematical modelling of magnetic data particularly along profiles. Thus, it is common practice to produce 2½ D geological models from the magnetic data and this is often carried out in conjunction with interpretation of gravity data along the same profile. A typical example of a magnetic survey carried out in the reconnaissance mode is shown in Figure 2, which shows a shaded-relief image of total field aeromagnetic data from the Hexham area taken from Evans and Lee (1994). The central part of the image is based on data from a detailed survey (250m line spacing) flown in 1978, while the surrounding area is based on the earlier national aeromagnetic survey (2 km line spacing). Illumination is from the north. The presence of major dykes across the area is more clearly observed in the high resolution survey although their basic trend can be detected on the regional survey. Although this survey is of a regional nature the same image display system can be used on the computer on magnetic and other digital data at the smaller scale of an engineering site.

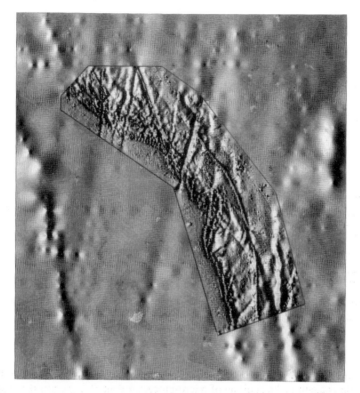

10 km

FIGURE 2: Shaded relief image of total field aeromagnetic data from the Hexham area, (after Evans and Lee 1994)

(e)   **Borehole Geophysical Logging**

Of all the geophysical methods that should be a routine part of the site investigation process, geophysical logging of the boreholes should be considered essential. While the civil engineer is interested in the core taken from the borehole for geological logging and geotechnical testing, the geophysicists sees the site investigation boreholes as access paths to the rock mass for geophysical sensors. The revision to BS5930 has increased the emphasis on the importance of borehole geophysical logging to the site investigation and indicated a number of areas where significant progress has been made.

Increasing use has been made in recent years of gamma logging which measures the variation of natural gamma radiation with depth in a borehole. This is one of the most significant measurements in relation to the determination of lithological changes in the geological structure since clay materials have a high gamma count and sandy materials a low one. The advantage of these measurements is that they can be made in cased boreholes and used for correlation of geological structures across an engineering site. A typical sequence of logs from a site investigation borehole programme is shown in Figure 4. The Fuller's Earth beds are clearly shown dipping to the south on the logs for boreholes A to D; boreholes E and F are to the north of a minor fault with some bifurcation of the Fuller's Earth beds indicated in the appropriate gamma logs.

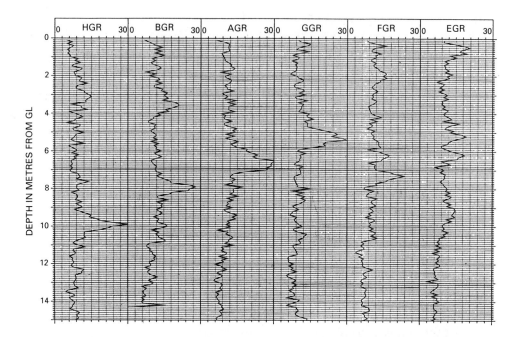

FIGURE 3: Correlation of natural gamma logs in a typical site investigation involving closely spaced boreholes (after Cripps an McCann, under preparation)

Other important geophysical logs for engineering studies are the full-wave train sonic and formation density logs, from which the dynamic elastic properties of the rockmass can be determined. The borehole televiewer and formation microscanner produce respectively ultrasonic and electrical images of the borehole wall and are used extensively in the study of fracture patterns in the rockmass. Increasing use is also being made of high resolution colour television cameras, which under optimum borehole conditions, clear water with few suspended particles, can give a continuous visual recorder of the borehole wall on an on-line basis; close examination of specific zones of interest can be made during the borehole survey.

Probably the most significant development in this area has been the increasing use of cross-seismic measurements for the study of rock mass conditions between the boreholes. An extension of this application has been the increased use of seismic tomography to produce seismic images of the rock mass between boreholes. This approach has produced very interesting information on the position of faults and associated fracture patterns but is still deemed to be a specialist application since survey costs are high.

NEW SURVEYING METHODS

The major advances introduced since BS5930 was published in 1982 are all based on the electromagnetic properties of the ground and these range from the very low frequency (VLF) method through to ground probing radar.

(a)   Very Low Frequency (VLF)

The VLF method is based on the transmission of electromagnetic waves from distant, very low frequency radio station (10 kHz to 30 kHz), such as Rugby in the United Kingdom, which can be used to determine the electrical properties of the earth. The method has been used to date in mineral exploration for the location of mineralised fault zones. However, more recent studies (Fitzgerald et al, 1987) have shown that the method is also applicable to the location of high conductivity zones in a contaminated landfill site associated with the presence of leachate.

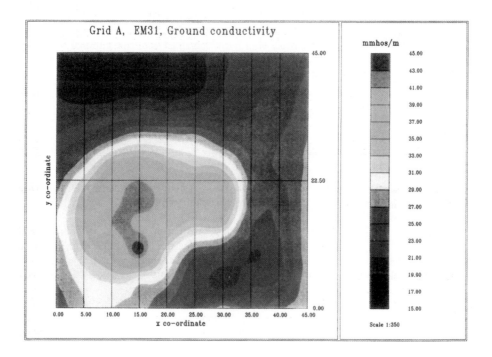

FIGURE 4: Ground conductivity measured with a Geonics EM31 system above a sink-hole resulting from solution in an underlying gypsum bed (Courtesy J P Busby, BGS, personal communication).

(b)   Ground Electrical Conductivity

In this method a transmitter coil is energised with an alternating current and is placed on or above the ground surface. The time-varying electromagnetic field in the transmitter coil induces very small currents in the earth. These currents generate a secondary magnetic field which is sensed together with the primary field by the receiver coil. The intercoil spacing and operating frequency are chosen so that the ratio of the secondary to primary magnetic field is linearly proportional to the apparent ground conductivity. This ratio is measured and a direct reading of apparent ground conductivity is obtained.   Using a fixed separation of 4 m between the transmitter and receiver coils the depth of penetration is limited to less than 6m but the survey can be carried out in a rapid and economic manner by a single operator. A typical example of a ground conductivity over a sink hole resulting from solution in the underlying gypsum beds is shown in Figure 4.   Greater penetration to around 30 m is achieved by moving the two coils apart to a maximum distance of 40 m or by reorientating the coils.

It is important to realise that a ground conductivity survey does not supply the quantitative information on earth layering that can be obtained by resistivity sounding or seismic refraction surveys. However, as the technique is so cost effective it can be considered for the provision of non-quantitative data prior to drilling or for filling gaps between boreholes or resistivity soundings. The constant separation equipment is particularly effective in the location of cavities or buried mineshafts when used in conjunction with a magnetic survey. The measurements compare very closely with results obtained with conventional resistivity profiling and ground conductivity surveys should be carried out in preference to resistivity profiling if the same depth of investigation is required.

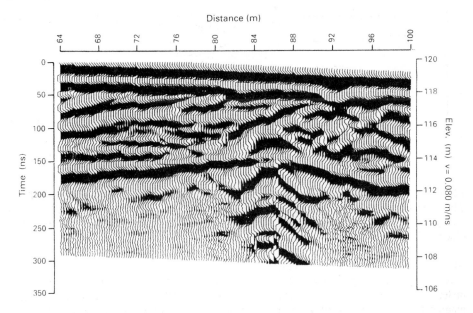

FIGURE 5: Typical ground probing radar over Quaternary deposits with a 50 MHz antenna. (Courtesy P G Greenwood, BGS, personal communication)

(c) Ground probing radar (GPR)

This method has been introduced to site investigation in UK over the past decade. The system consists of a radar antenna transmitting electromagnetic energy in pulse form at frequencies between 25 MHz and 1 GHz. The pulses are partially reflected by the sub-surface geological structures, picked up by the receiving antenna and plotted as a continuous two-way travel time record, a pseudo-geological record section. The vertical depth scale of this section can be calibrated from the measured two-way travel times of the reflected events either by use of the appropriate velocity values of electromagnetic energy through the lithological units identified or by direct correlation with borehole logs.

The depth of penetration achieved by the radar pulse is a function of both its frequency and the resistivity of the ground. For UK soils, the maximum depth of penetration is likely to be between 1m and 4m but useful penetration to greater depths can sometimes be achieved in more resistive geological environments. At the moment ground probing radar is of limited use in normal depth to bedrock determination but future development of the present equipment is likely to include a higher output power from the transmitter and lower frequency antenna. A typical example of a modern ground probing radar survey is shown in Figure 1. The Pulse Echo IV system using a 50 MHz antenna has defined an interbedded sand/gravel sequence overlying bedroom at a depth of about 8m.

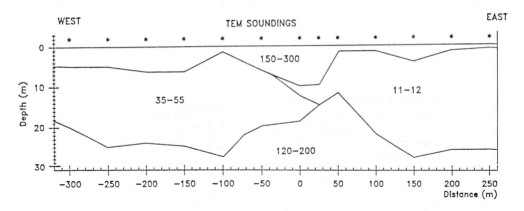

FIGURE 6: Geological model of the margin of a tunnel valley in Suffolk derived from TEM soundings. Resistivities in ohm metres; sandy surface layer 150-300; till in tunnel valley 11-12; clay and weathered chalk 35-55; Chalk bedrock 120-200 (courtesy, J D Cornwell, BGS, personal communication).

(d) Transient Electromagnetic (TEM)

Electromagnetic energy can also be applied to the ground using transient current pulses instead of the continuous waves mentioned above. The collapse of a steady state primary magnetic field will induce eddy currents to flow in a conductive earth, and these in turn will give rise to a transient secondary magnetic field which may be detected in a receiver coil as a time-dependent decaying voltage. The characteristics of this transient decay can be related to the conductivity and geometry of the subsurface geology. Typical TEM systems can provide rapid geoelectric depth scans from a few metres down to several hundred metres and therefore present an attractive alternative to electrical resistivity sounding. TEM is a well established technique for mineral exploration and is increasingly being applied to hydrogeological mapping (especially saline intrusion problems) and to shallow engineering site investigation studies. A typical example of a geological model of the margin of a tunnel valley in Suffolk derived from the interpretation of individual TEM soundings is shown in Figure 6.

CONCLUSIONS AND DISCUSSIONS

The period of time that has elapsed since the publication of the original version of BS5930 is 1981 has coincided with significant advances in digital technology. This has resulted in a whole new generation of geophysical equipment based on microprocessor control that is able to communicate directly with a field based microcomputer for initial interpretation of the geophysical survey data. Indeed, the most modern trend has been the use of the micro-computer itself to control the geophysical field instrumentation and to set the control parameters for the survey. This latter trend has been brought about by the provision of high quality, user friendly software, which can be bought off the shelf for many geophysical interpretation applications.

The development of computing itself has also been subjected to major changes, where the concept of a centralised main-frame computer has largely been replaced by distributed PC's and workstations linked together by either a simple local area network or a more complex regional network. The overall improvement in the interpretation and display of geophysical survey data has brought about a steady improvement in the overall standard of geophysical surveys in the site investigation process. In particular, the provision of high resolution colour plotters has enhanced the display of the contoured geophysical map produced in a reconnaissance survey and given rise to a general improvement in the presentation of geophysical data in its final interpreted form, see for example, figures 2 and 4, produced using the BGS's COLMAP package (Green 1989). The end product is now available in a form that is readily understandable by the civil engineer and has brought about a significant improvement in the relationship between the geophysicist and his client.

What of the future? The revolution in computing technology shows no signs of slowing down. Each generation of computers is significantly more powerful and cheaper than the proceeding one, while the ability of the user to interact with the computer has improved in recent years.

As far as geophysical equipment is concerned general development of new products is largely controlled by economic factors, since demand in the civil engineering industry does not warrant significant investment by geophysical equipment manufacturers. However, the requirements of the hydrocarbons and minerals industries does ensure a steady improvement in the available equipment. In the site investigation area the most likely areas of development appear to be in the electromagnetic methods, where ground probing radar and TEM have demonstrated existing possibilities for the future. High resolution magnetic and ground conductivity surveys coupled with the ability to display the field data in a comprehensible manner are continuing to be developed. Encouraging progress is also being made with the seismic reflection method, but to date it has yet to be used routinely in the UK for shallow site investigation surveys.

To sum up, over the past decade geophysical surveying methods have become an integral part of the site investigation process. On some civil engineering projects they have been used routinely to position the boreholes and trenches used for the geological/geotechnical investigations. In application such as the investigation of contaminated land-fill sites they provide essential information in an indirect form, which can be used to assess the site initially without drill. In the urban environment where drilling is difficult a geophysical survey may well provide the most useful information in the early stages of a site investigation.

The future does look encouraging for geophysical surveying in the site investigation process but it must be carried out as an integral part of the overall investigation and not as an optional extra.

ACKNOWLEDGMENTS

This paper is published with the permission of the Director, British Geological Survey (NERC).

REFERENCES

Baria, R., Jackson, P.D., and McCann, D.M. (1989)
Further developments of a high-frequency seismic source for use in boreholes. Geophysical Prospecting, 37, No. 1, pp 31-52.

Barker, R.D. (1992)
A simple algorithm for electrical imaging of the sub-surface First Break 10 No 2, pp 53-62.

British Standards Institution (1981)
BS5930 - Code of Practice for Site Investigations. British Standards Institution 149p.

Cripps, A.C. and McCann, D.M. (under preparation)
Gamma logging as an aid to borehole studies in the site investigation process.

Evans R. and Lee M.K. (1994)
Imaging the Earth's Crust - geophysics at a regional scale. Geoscientist, 4, No. 4, pp 18-20.

Fitzgerald, L.J., Angers, A.K., and Radvill, M.E., (1987) The Application of VLF Geophysical Equipment to Hazardous Waste site investigations in New England, NUS Corporation.

Geological Society (1988)
Engineering Geophysics. Report by the Geological Society Engineering Group Working Party. Quarterly Journal of Engineering Geology, 21, pp 207-271.

Goforth, T. and Hayward, C. (1992)
Seismic reflection investigation of a bedrock surface buried under alluvium Geophysics, 57, pp 1217-1227.

Green, C. A. (1989) COLMAP: A colour mapping package for 2D geophysical data. British Geology Survey Technical Report WK/89/19.

Jackson, P.D. and McCann, D.M. (1993)
Non-destructive geophysical site investigation on aid to the redevelopment of sites in urban areas with groundwater problems. Proceedings of the International Conference on Groundwater Problems in Urban Areas, Editor W B Wilkinson, Institution of Civil Engineers, 2 - 3 June 1993, Thomas Telford, London, pp 134-148.

Littlejohn, G. S., Cole, K. W. and Mellors, T. W. (1994)
Without site investigation ground is a hazard. Proceedings of the Institution of Civil Engineers, 102, pp 72-78.

McCann, D.M. and Jackson, P.D. (1988)
Seismic Imaging of the Rock Mass, University of Wales Science and Technology Review, No. 4, pp 21-28.

McCann, D. M., Culshaw, M. G., Bell, F. G. and Cripps, J. C. (1988)
Reconnaissance methods for the location of abandoned mineshafts and adits. Proceedings of the Second International Conference on Construction in Areas of Abandoned Mineworkings. Editor Forde, M. C., 28-30 June 1994, Edinburgh University, Engineering Technics Press pp 53-60.2

Miller, R. D. and Steeples, D. W. (1994)
Application of shallow high-resolution seismic reflection to various environmental problems. Journal of Applied Geophysics, 31, pp 65-72.

Neumann, R., (1977) Microgravity methods applied to the detection of cavities. Proceedings of the Symposium on Detection of Subsurface Cavities, Soils and Pavements Laboratory, US Engineering Waterways Experiment Station, Vicksburg, Mississipi.

Orellana, E. and Mooney, H.M. (1966)
Master tables and curves for vertical electrical sounding over layered structure, Madrid, Interciencia, 159p.

SISG (1993)
Site Investigation in Construction, Part 1, Without Site Investigation Ground in a Hazard Report of the Site Investigation Steering Group, Institution of Civil Engineers, Thomas Telford Ltd, 45p.

Practical considerations for field geophysical techniques used to assess ground stiffness

A.P. BUTCHER and J.J.M. POWELL
Building Research Establishment, Watford, Herts, UK.

INTRODUCTION

Recent developments in the understanding of the engineering behaviour of soil has indicated that at very small strains the stiffness of a soil rises to a constant value. These very small strain levels are of the same order of magnitude as the operational strain levels of field geophysical measurements. Based on the fact that the dynamic stiffness of the ground can be calculated directly from the shear wave velocities that can be measured using geophysical techniques this has opened up the possibilities of field geophysical methods to assess the stiffness of the ground.

Field geophysical techniques, which use the propagation of seismic waves, have traditionally been applied to map sub surface features over large areas providing geologists with generalised sections perpendicular to the ground surface. For the engineering application a more detailed profile of information is required along with a greater confidence in the accuracy of the measurements.

This paper concentrates on the practical considerations required to obtain reliable shear wave velocity measurements that can then be used to assess ground stiffness. Three field geophysical techniques are described and typical operational considerations necessary to ensure good data are discussed. Illustrations of the capabilities and shortcomings of each particular method are given by reference to case studies on a range of sub surface conditions including sand, clays and fills of different types. Finally comparisons are made with deformation modulus values derived from other conventional soil tests.

GEOPHYSICAL FIELD TECHNIQUES

Geophysical field techniques to measure seismic waves are either non-intrusive (those which use equipment placed on the surface of the ground) or intrusive (those where either one or both of the source and receiver are placed in the ground). The non-intrusive methods include refraction and Rayleigh wave techniques and the intrusive include downhole and crosshole methods. However, the analysis of refraction measurements requires the assumption of an increasing velocity with depth and a relatively consistent depth profile along the survey line. The actual paths of the refracted waves are not usually known but are assumed to follow strata interfaces or arcs controlled by the depth stiffness profile of the deposit. In contrast, Rayleigh wave, downhole and crosshole techniques use seismic waves travelling in paths which are as controlled as possible and have the potential therefore to yield more reliable data. In this study the Rayleigh wave and the shear wave downhole and crosshole techniques will be outlined and some of the practical considerations necessary to ensure good data will be presented.

Rayleigh waves

Rayleigh waves or surface waves are distortional stress waves that propagate near to the boundary of an elastic half space and were first investigated by Lord Rayleigh (1900). The particle motion in a Rayleigh wave is in an elliptical shape with the long axis of the ellipse perpendicular to the boundary of the half space and the short axis parallel to the half space boundary. The plane of propagation of the wave is in the plane of particle motion.

Continuous Rayleigh waves can be generated by a mass vertically oscillating with simple harmonic motion on the surface of the ground. Pulsed Rayleigh waves can be generated by dropping a heavy weight on to the surface of the ground. Rayleigh waves will then propagate radially outwards from the oscillating mass or point of impact of the weight.

The Rayleigh wave geophysical measurements discussed in this paper were made using equipment and test procedures developed at BRE. The equipment and method of working are fully described by Abbiss (1981) but for completeness are briefly outlined here.
To measure continuous Rayleigh waves a variable frequency Rayleigh wave source is placed on the ground surface and two vertically orientated geophones are placed a distance (d) apart and in a line with the source. The frequency of the source is set and the wave forms received by each geophone are compared and the phase angle change of the waveforms between the geophones measured using a low frequency spectrum analyzer or a frequency response analyzer. By varying the frequency of the source a range of values of frequency and phase angle change can be obtained. Knowing the frequency of the source (n), the distance between the geophones (d) and the measured phase angle change (ø), the wave length (l) and Rayleigh wave velocity (V_r) may be calculated as follows:

$$V_r = n.l \qquad \text{where } \ l = (360/ø).d$$

Since in most soils Rayleigh waves travel at a depth of between one half and one third wavelength below the surface, a reduction of the frequency of the vibrator and consequential increase in wavelength will allow a Rayleigh wave velocity with depth profile to be determined. Rayleigh wave velocity (V_r), in most soils, is about 5% slower than shear wave velocity (V_s) (White 1965) so a correction is made to the Rayleigh wave data to produce a shear wave velocity value.

Practical considerations: The main practical considerations for Rayleigh wave velocity measurements include the selection of the right equipment in order to obtain satisfactory measurable ground response, and the need for on site analysis of the field data.

The signal to noise ratio during Rayleigh wave testing can dictate the depth of penetration. If the ambient noise on a site is at a level close to the source level the spectrum analyzer will not be able to discriminate between the noise and the signal and can result in a scatter of spurious data. To overcome this problem the energy level of the source should be increased until it is significantly higher than the noise. This can be achieved by either increasing the amplifier settings on the signal generator which controls the source or by using a source of higher energy. Signal to noise ratio problems generally become significant as the frequency of the source reduces, since the energy of the source is related to the acceleration imparted to the ground by the source, therefore, the lower the frequency the lower the energy.

Figure 1 shows shear wave velocity data from Rayleigh wave measurements using three different sources. Two of the sources were lightweight (14kg) with different frequency ranges of 0 to 20Hz and 15 to 7000Hz. whilst the third source weighed 1200kg and had a frequency range of 0 to 40Hz. In this case it can be seen that each source, with different energy and frequency characteristics, had a role to play. The higher frequency lightweight source clearly defines the shallow part of the profile (0 to 3m depth) but fails to detect the silty clay beneath the sand which was detected by the low frequency lightweight source. The heavier low frequency range source then gives data for the underlying plastic and lean clay to significant depth. This illustrates the need to match the energy source and frequency range with the likely ground conditions to be investigated.

As the frequency of the source reduces so the wavelength increases and the spacing of the receivers (d) becomes a smaller fraction of the wavelength making measurements of phase changes less accurate. Ideally the receiver spacing should be kept just smaller than the wavelength of the wave to be measured. This is clearly impractical for every reading but for frequencies below about 20Hz with higher energy sources the distance between the receivers can be increased to 2.5m or 5.0m with advantage as shown in Figure 1.

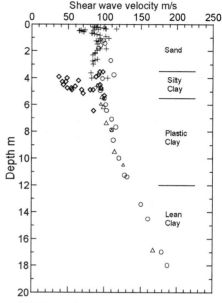

○ 1200kg shaker 5m geophone spacing
△ 1200kg shaker 2.5m geophone spacing
+ 14kg (15 - 7000Hz) 1m geophone spacing
◇ 14kg (0 - 20Hz) 1m geophone spacing

Figure 1: Shear wave velocity vs depth from Rayleigh wave measurements at Museumpark

The distance between the source and the nearest receiver is typically 1m with the BRE Rayleigh wave sources but would be larger for heavier, more powerful sources to avoid the near source effects emanating as seismic noise.

As stated above the effective depth of travel of Rayleigh waves is between one half and one third wavelength. The relationship between depth of travel and wavelength is dependent on the stiffness with depth profile of the soil. In a homogeneous elastic half space the Rayleigh wave velocity would not change with depth whereas in a soil with a uniform increase in shear modulus with depth, the Rayleigh wave velocity would change with depth proportional to the square root of the shear modulus. In reality soils with these characteristics only occur in layers and seldom from the surface but the effects of such stiffness-depth profiles on the Rayleigh wave velocity depth profiles are useful to remember as an aid to interpretation. One of the most difficult stiffness profiles in which to make successful geophysical measurements is where a stiffer layer overlies a less stiff layer. In this case the Rayleigh wave method may, with an initial selection of frequencies, give the same wavelength (related to depth) for certain different frequencies showing as a bunching of data. However, if the increment by which the frequency is changed is reduced the new intermediate frequencies will probably give the missing wavelengths. It can be appreciated therefore, that to ensure the best quality data are obtained, it is essential to calculate the Rayleigh wave velocity and wavelength as the field work proceeds in order to identify the frequencies which need to be selected to 'fill in' the gaps in the wavelength velocity profile.

Research in the USA (Nazarian 1984) has produced computer software which enables forward modelling to match field data with numerical models by iteration. This method produces a stiffness profile which, if present would produce a dispersion curve similar to that recorded in the field. Work at BRE has tended to use the simpler approach of taking the depth of travel as one half a wavelength for clay soils and one third of a wavelength for granular soils and fills. This approach has been justified by comparison of Rayleigh wave profiles with borehole logs (Powell & Butcher 1991, Butcher & Tamm 1994), seismic cone profiles or other *in situ* profiling devices (Butcher & McElmeel 1993).

A Rayleigh wave survey at one location, using a small electromagnetic source, would require a space about 3m by 0.5m in which to set up the equipment, which could all be accommodated in an average size estate car including a generator for electrical power. Larger, more powerful sources obviously need to be transported to site by special vehicle and the geophone spacing would be increased to avoid near source effects. The average survey using a range of frequencies with rapid processing, using a portable computer, to obtain the wavelengths and to check for gaps in the wavelength data would take about one hour after setting up the equipment.

Shear wave techniques

Shear waves are distortional stress waves in that the direction of particle movement caused by the wave is perpendicular to the direction of propagation of the wave. This means that shear waves are reversible in that, if the source action is reversed (in a direction 180° from the first) the resultant wave will be 180 degrees out of phase with the first wave and so will appear as a mirror image of the first wave. This property of shear waves is used in their identification and most shear wave sources are double acting so that reverse shear waves can be produced.

Downhole (seismic cone). The downhole technique comprises a surface source with a receiver or receivers placed at a predetermined depth or depths in the ground. The travel times between the source and receivers are measured and the measurements at successive depths differenced to calculate the wave velocity of the ground between the two depths. The basic assumption is made that the shear waves travel in a straight line from the source to the receiver. The source, to produce shear waves, usually comprises a wooden railway sleeper or an aluminium channel section which is orientated with its long axis at right angles to a radial line on the ground surface from the top of the borehole down which the receiver has been placed. In order to produce shear waves the sleeper or channel is usually trapped against the ground by the weight of a road vehicle and is struck on its end in the direction of its long axis with a heavy hammer incorporating a device to trigger a seismograph or other recording equipment.

The traditional downhole geophysical technique would require the installation of a borehole, with acoustic liner, in which to place the signal receivers. This is avoided with the latest variation of this type of measurement which is to use a receiver or geophone in a static cone penetrometer type body, called a seismic cone (Robertson et al 1986), which can be jacked into the ground to the predetermined depth. When the measurements have been made the seismic cone can be advanced to a new depth and the measurement procedure repeated.

Practical considerations: The practical considerations for downhole seismic wave velocity determinations will be focused on seismic cone testing as a rapid and economical variation of conventional downhole testing. The considerations below are concerned with the test set up and equipment as well as the processing of the data.

The source, as mentioned above, is usually either a wooden railway sleeper or an aluminium channel section beam with steel anvils. The hammers used to strike the beam are usually heavy hand held hammers. The BRE hammers (one for each side of the cone truck), whilst operated by hand, have hammer masses of 10kg and centres of swing fixed to the side of the cone truck). The energy required can be adjusted by the height to which the hammer head is lifted before release. This mechanical hammer system has proved to give consistent energy and a relatively noise free signal with a much higher maximum level of energy than the hand held hammers.

The frequency spectra produced by the BRE hammers striking each of the two suggested sources are given in Figure 2 and show a similar band width but with a higher peak amplitude for the aluminium beam. This characteristic of the aluminium channel reduces dispersion of the signal as it travels through the soil so giving a clearer received signal.

The distance from the point of insertion of the seismic cone into the ground to the shear wave source can have a significant effect on the measured wave travel times and thereby the calculated shear wave velocities. This is because the assumed travel path of the shear waves is in a straight line directly from the source to the receivers in the seismic cone. As the horizontal distance from the seismic cone increases so does the likelihood that the first shear wave arrival will have been refracted from a straight travel path. The outcome would be shorter travel times leading to higher calculated velocities. An illustration of the effect on the shear wave velocity - depth profiles of the horizontal distance between the source and the vertical axis of the seismic cone is given in Figure 3 which shows data from the Madingley site. Above about 7m depth significant differences in the calculated shear wave velocities are evident for horizontal distances between source and receiver of 0.4m and 4.8m. As the source moves closer to the seismic cone axis so the depth affected by the inaccuracies reduces, indicating the desirability of having the source as close to the cone axis as possible.

The received signals can be logged using a multi channel seismograph or a portable computer, with appropriate additional hardware and software. Both systems can give an immediate assessment of the quality of the data and provide storage for later retrieval and further processing but the computer based system allows data processing on site. The computer based system, to record seismic wave data, is typically less than half the cost of the more traditionally used multi channel seismograph based system, which would still require the

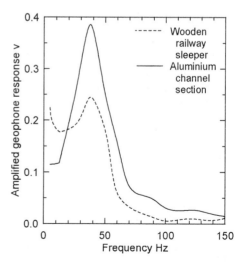

Figure 2: Frequency spectra of the two shear wave sources

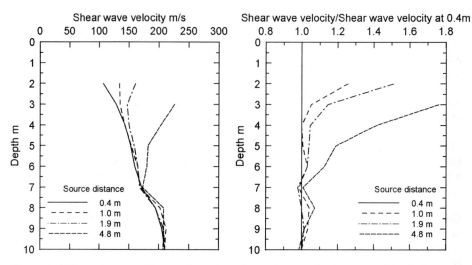

Figure 3: Effect of source distance from insertion point of seismic cone at Madingley

use of a computer to process the data.

The travel times of the shear waves from the source to the receivers need to be obtained from the logged signal records. In many cases ambient noise, existing in the ground, will swamp the shear wave signals to such an extent that the arrival times are difficult to determine. In such cases measurements of the frequency range of the ambient noise can enable a filter, covering the same frequency range as the noise, to be used to clean the signal of the ambient noise. Care must be taken with the use of filters, however, because their action can shift the phase of the signals giving rise to errors. The BRE recording system is based on a portable computer which stores the records in a digital form. The digitised records can then be filtered initially in the positive time direction and then in the negative time direction. Any phase shift that occurs in the positive time filtering is restored during the negative time filtering. If the arrival times are not clear then alternative points to compare are the first positive or negative peaks, or zero amplitude points.

The simplest version of the seismic cone has one set of geophones so that travel times from successive depths are used to calculate the shear wave velocity for that depth increment. This system assumes that the triggering and energy input are the same for each data take at each depth. This system can work well but tends to give scattered data which needs processing further to produce a realistic profile. The further processing involves producing a travel time depth profile and using this to estimate travel time increments, again because of the scatter of data a curve is fitted through the travel time depth profile and the curve used to calculate the velocity increments.

The BRE seismic cone has two groups of geophones instead of the more common single group type. The two groups of geophones can be fixed either 0.5m or 1.0m vertically apart. From a single source activation the dual geophone group seismic cone receives the signals at two depths so allowing the time difference between the arrivals to be measured. By differencing the travel times any errors from the triggering system, to set the recording equipment running are eliminated, as are errors in measuring the successive depths of a single geophone group seismic cone. A comparison of downhole shear wave velocity profiles using both single and dual geophone group seismic cones is shown in Figure 4 taken at the Madingley test bed site (Butcher 1993). It can readily be seen that the dual geophone set gives a much smoother profile with less variability than the single geophone set data. The trigger errors can be related to the position of the trigger, either on the hammer head or on the beam, and the energy generated by the source and its relationship to the required frequency response of the trigger for activation.

The repeatability of the dual geophone group seismic cone is illustrated in Figure 5 which shows a number of profiles taken at Bothkennar. The calculated shear wave velocities form a very narrow band in the soft silty clay deposit with a slight increase in spread as the cone approaches the gravel layer at 18 to 19 m depth.

Crosshole. The crosshole seismic survey technique measures wave travel times between points in the ground at the same depth. In the simplest case this is achieved by placing a seismic source and a geophone receiver at the same depths in adjacent boreholes. To form a good acoustic coupling between the source or geophone with the surrounding ground the boreholes should be lined and grouted to form a continuous bond.

The shear wave sources available for crosshole work are generally of the type using a vertically sliding mass striking a horizontal anvil. These sources produce vertically polarised shear waves which propagate horizontally and radially from the source. Less common are sources which produce shear waves which are horizontally polarised and propagate horizontally. These sources use either a rotary hammer with a vertical axis and anvil or a

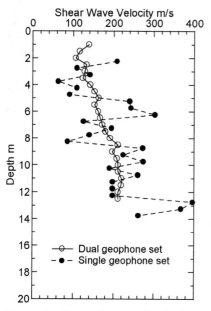

Figure 4: Comparison of single and dual geophone set seismic cone profiles at Madingley

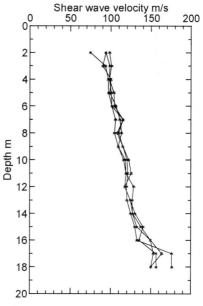

Figure 5: Four seismic cone shear wave velocity depth profiles from Bothkennar

horizontally actuating impact system. In either case the direction of action of the impact must be perpendicular to the required direction of propagation to produce a shear wave.

The best crosshole testing procedure employs three boreholes in a line with the source in an end borehole and receivers in the other two boreholes. The arrival of the wave at each of the receivers will enable the travel times, and hence velocities, between the two receiver boreholes to be calculated. As a check the source should be swapped with the receiver in the borehole at the other end of the line of boreholes and the procedure repeated. The three borehole method avoids problems with triggering delays from the source to the seismograph recording the received signals in a similar way to the dual geophone seismic cone. To obtain a profile of shear wave velocity with depth the source and receivers are placed at the same initial depth, and measurements made; the equipment is then moved to the next depth, measurements taken and the operation being repeated until all the depths of interest are covered.

Practical considerations: The practical considerations for crosshole seismic measurements include the installation of acoustic liners, the verticality of the boreholes, and the effect of increasing stiffness with depth and the use of two seismic cones as receivers.

The installation of acoustic liners is necessary if the boreholes are to remain operational for some time or if instability of some of the layers is suspected. It is essential that a good acoustic coupling is obtained between the surrounding soil and the liner. This can be achieved if the open borehole is partially filled with grout before pushing the acoustic liner, sealed at the base and at joints, into the grout. The grout, when set, should have approximately the same stiffness as the surrounding soil. Then by filling the liner with water it will sink but will require some force to sink to the bottom of the borehole. As the liner sinks the displaced grout will be forced up the borehole so filling in the void between the liner and borehole sides. When the grout has set it is advisable to pump the water out from within the liner before use for geophysical measurements in order to avoid spurious effects of tube waves, which can be set up in the water by the seismic wave sources, and produce seismic waves in the soil which then swamp or can be mistaken for the signals required (Hope 1993).

The verticality of the boreholes can, if not measured, have a significant effect on the measured shear wave velocities. It is not unusual for a borehole to deviate from the vertical by 0.4m in 15m depth which could add, in the worst case, 0.8m to a travel distance. For such a case with a system of boreholes 5m apart at the surface and 5.8m apart at 15m depth in a soil with a shear wave velocity of 100m/s, then the measured travel time at 15m depth would be 0.058s. If the boreholes were assumed vertical, ie a path length of 5m, then the velocity would be calculated as 86.2m/s, which is a significant difference from the real velocity. This 14% error in shear wave velocity would then become a 26% error in stiffness (see below for stiffness calculations). If a seismic cone is used as a receiver the inclinometer should be calibrated to enable the exact position of the receivers to be determined.

In a soil of increasing stiffness with depth the first shear wave arrival is likely to have travelled down to a stiffer, faster shear wave velocity layer, and back up to the receiver so following a curved ray path. This means that the calculated shear wave velocity is higher than reality for the layer at the level of the source and receiver. The only ways to check this are to reduce the borehole spacing or to compare the results with a downhole survey where the distance between the receivers is much smaller so reducing the possibility of curved ray paths. Alternatively the measured travel times and the source and receiver spacings can be used in a correction procedure developed by Hryciw (1989). This procedure can correct for the curved ray path effects due to either a linear or a non-linear increasing seismic velocity with depth.

The crosshole technique can also be achieved by using two seismic cones and a source borehole. With the wave paths extending radially from the source a large area of a site can be assessed by inserting the seismic cones on different projected radial lines. The cost reduction achieved by only installing one cased hole instead of three would be offset by the mobilisation costs of two cone driving trucks. However, a much larger area could be rapidly tested by a much more versatile system.

TYPICAL SHEAR WAVE VELOCITY RESULTS

In the following section typical shear wave velocity results are presented to demonstrate the capabilities of the seismic wave velocity techniques outlined above. The Rayleigh wave velocities have been used to calculate shear wave velocities using the relationship mentioned previously.

The normally consolidated sand site at Holmen, Drammen Norway, has been used by the Norwegian Geotechnical Institute (NGI) as a test bed site and is well documented (Eidsmoen et al 1984). Figure 6 presents shear wave velocity vs depth profiles obtained using Rayleigh wave, downhole (from Eidsmoen et al 1984) and crosshole (from Dyvik 1985) techniques. It is apparent that all three techniques appear to agree on the V_s depth profile even though

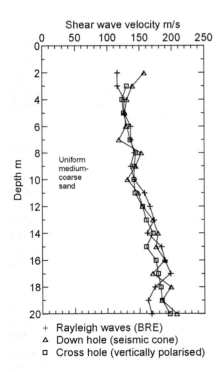

+ Rayleigh waves (BRE)
△ Down hole (seismic cone)
□ Cross hole (vertically polarised)

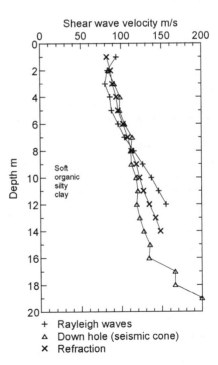

+ Rayleigh waves
△ Down hole (seismic cone)
× Refraction

Figure 6: Shear wave velocity depth profiles from Holmen

Figure 7: Shear wave velocity depth profiles from Bothkennar

both the directions of propagation and polarisation of each of the waves are different.

Figure 7 shows data from the normally consolidated clay on the Bothkennar test bed site. In this case the downhole and Rayleigh wave data closely agree down to 8m depth after which the Rayleigh wave velocities increase more rapidly with depth than the downhole. Also included in Figure 7 is data from a refraction survey which also agrees closely with the downhole data. The shear wave velocity data again appears to be very similar irrespective of the orientation of polarisation and propagation of the shear waves, although following slightly different patterns with depth.

Figure 8 shows data from the BRE test bed site at Chattenden. The soil at Chattenden is a stiff, heavily overconsolidated plastic London clay. The data from Chattenden comprises the Rayleigh wave data forming the lower bound and the horizontally polarised crosshole the upper bound of data. Within these bounds the downhole data is closer to the Rayleigh wave and the vertically polarised crosshole closer to the other crosshole data. In the heavily overconsolidated London clay it is therefore evident that there is a significant difference in the measured shear wave velocities depending on the method of measurement. The reasons for this needs further examination and are probably linked to stress history and the dominant fabric features of the soil. This behaviour has also been observed at two other stiff clay sites (Powell & Butcher 1991, Butcher and Powell 1995).

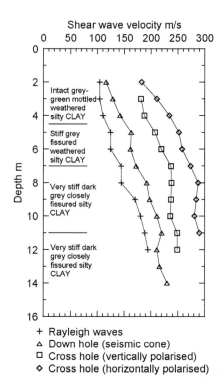

+ Rayleigh waves
△ Down hole (seismic cone)
□ Cross hole (vertically polarised)
◇ Cross hole (horizontally polarised)

Figure 8: Shear   wave   velocity   depth
profiles from Chattenden

Apart from work on natural deposits Rayleigh wave velocity measurements have been carried out on three fill sites, one of refuse fill, one of brick rubble fill and one containing miscellaneous builders fill. In all these cases the only initial means of investigation of the sites was to use a non-intrusive test method. The miscellaneous and brick rubble fill sites were both ground treatment sites where Rayleigh waves were used to detect the effect of ground treatment.

The refuse fill, reported by Butcher & Tamm (1994), highlighted the capability of Rayleigh waves to detect an underlying firm silty clay with its consistent increase in Rayleigh wave velocity with depth as compared with the irregular velocities measured in the refuse. This, used with the practical considerations for Rayleigh wave testing mentioned earlier, enabled the volume of refuse present to be successfully estimated. A typical cross section comparing estimated depths of refuse with those measured by excavation of the refuse is given in Figure 9.

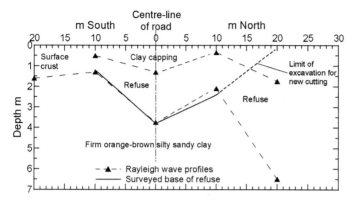

Figure 9: Typical cross section comparing estimated depth of refuse from Rayleigh wave
measurements with surveyed base of refuse upon excavation

The miscellaneous fill comprised of builders debris of concrete waste, bricks and timber with an underlying clay layer. Figure 10(a) gives the pre and post treatment Rayleigh wave velocity depth profiles clearly showing the effect of the rapid impact compaction treatment on the fill. The compaction of the fill, as detected by the Rayleigh wave measurements, was corroborated by measurements made with a magnet settlement gauge given in Figure 10(b) (Butcher & McElmeel 1993).

The brick rubble fill was effectively a 10 to 120mm granular fill, that is without

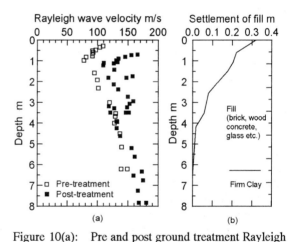

Figure 10(a): Pre and post ground treatment Rayleigh
wave velocity depth profiles
Figure 10(b): Fill settlement caused by the ground
treatment

fines. On this site the Rayleigh wave data enabled the comparison of placed and rolled fill, end tipped and dynamic compaction, end tipped and rapid impact compaction and end tipped prior to treatment. Again the strata underlying the brick rubble fill gave a consistent increase in velocity with depth not evident in the fill (Butcher & McElmeel 1993).

ENGINEERING PROPERTIES FROM SHEAR WAVE VELOCITY RESULTS
In general the data from the above investigations consists of shear wave velocity profiles with depth. These may then be used to calculate dynamic shear modulus (G) depth profiles for the sites as follows:

$$G = \rho(V_s)^2 \text{ kPa}$$

where ρ is the mass density of the soil in Mg/m^3 and V_s is the shear wave velocity in m/s

This relationship shows the influence of mass density on the calculation of dynamic shear modulus although the major influence is the shear wave velocity. Although this paper concentrates on the measurement of, and the likely errors in, shear wave velocity it should be noted that such errors are magnified in the calculation of G because of the $(V_s)^2$ term.

The stress-strain characteristics of the ground are generally non-linear so that the shear modulus varies with strain level. The moduli values calculated from these field geophysical tests are at the very small strain levels of geophysical measurements and are not directly applicable to working strain levels where the moduli may be much less. However, shear modulus - shear strain models (Ramberg and Osgood 1943, Seed and Idriss 1970 and Hardin and Drnevich 1972a&b) have been developed which allow the shear modulus values determined at very small strains to be projected to working strains. Of these models the most popular is Hardin and Drnevich (1972a&b) which will be used here to compare projected shear modulus values from geophysical measurements with shear modulus determinations from other soil tests.

Figure 11 shows shear modulus against shear strain data from the normally consolidated soft clay site at Bothkennar including small strain triaxial and pressuremeter results at appropriate strain levels. The values of shear modulus have been normalised by dividing by the octahedral effective stress (p_o'). The geophysical data covers a fairly narrow range of values which, when projected to higher strains produces an envelope which narrows with increasing strain. The triaxial data follows the lower bound of the projections at higher strains but sits in the middle of the envelope at small strains. The pressuremeter data has a lower bound which agrees with the triaxial data but has a much higher upper bound outside the geophysics projected envelope. In this case of a normally consolidated deposit the projected envelope of shear modulus values from the geophysical measurements agree very well with the results from other tests carried out at larger strains.

Figure 12 shows the normalised shear modulus data from the stiff heavily overconsolidated clay site at Madingley where the geophysical data gives a different projected envelope for each measurement technique. The pressuremeter data lies in the area where the Rayleigh wave and seismic cone envelopes converge with increasing strains. In contrast to the data from Bothkennar (in Figure 14) the spread of G/p_o' values from the different geophysical measurements means that the convergence of the envelopes occurs at a much higher strain level with less agreement at working strain levels. This pattern of behaviour is evident in other stiff overconsolidated clay sites as reported by Powell & Butcher (1991).

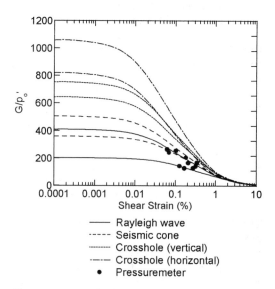

Figure 11: Madingley: Shear Modulus vs Shear Strain

CONCLUSIONS

Three field techniques to measure the shear wave velocity of the ground have been described together with practical considerations which, if implemented, should improve the quality of field data and considerably aid the interpretation of geophysical data for engineering purposes.

From the foregoing the following conclusions can be drawn:

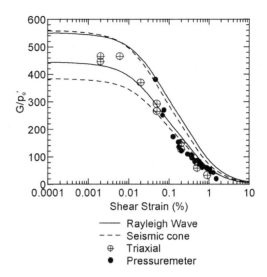

Figure 12: Bothkennar: Shear Modulus vs Shear Strain

The Rayleigh wave technique
- is a rapid method to give good quality seismic wave velocity data on a variety of deposits, especially those which cannot be penetrated by other *in situ* test devices.
- with careful consideration of source energy, frequency selection and geophone spacing can, even with simplified depth approach, provide reliable and repeatable results.
- provided a useful lower bound shear wave velocity profile in the deposits tested.
- was successfully used to detect the presence of fill on the sites tested and the effects of ground treatment in materials in which intrusive *in situ* testing would have been extremely difficult.

Downhole testing using a seismic cone is a rapid test which is more economical than conventional downhole testing and can produce high quality repeatable data by:
- using a device with two geophone groups,
- ensuring the source is within 0.5m of the point of insertion of the seismic cone
- using a portable computer based retrieval and processing system with carefully designed and used noise filters which do not phase shift the signals.

Crosshole seismic wave testing
- must use properly cased, grouted boreholes of known verticality.
- can be carried out more rapidly using a single borehole for the source and two seismic cones (with calibrated inclinometers) for receivers.

The shear wave velocities determined by each method were found to be influenced by the soil properties:
- in the stiff heavily overconsolidated soils the large differences in shear wave velocities, depending on the test method, were found,
- in the lightly overconsolidated clays the differences were less though still in evidence
- in the normally consolidated soils the similar shear wave velocities were found to be relatively independent of the test method.

ACKNOWLEDGEMENTS
The authors would like to thank the Norwegian Geotechnical Institute, in particular T Lunne, for the use of the Rayleigh wave equipment and the Holmen and Museumpark test sites, and also their colleagues at BRE, in particular T.N. Reynolds, who helped with much of the field work.

REFERENCES
Abbiss, C.P. (1981). Shear wave measurements of the elasticity of the ground. *Géotechnique* **31**, No 1, 91-104.
Butcher, A.P. (1993). Discussion - Field measurements of the small strain properties of the Gault. *Geotechnical Engineering in Hard Soils - Soft Rocks*. Vol 3. Ed. A. Anagnostopoulos, F. Schlosser, N. Kalteziotis and R. Frank, Balkema, Rotterdam.
Butcher, A.P. & McElmeel, K. 1993. *Engineered Fills*, pp. 529 - 540. Ed. B.G. Clarke, C.J.F.P. Jones and A.I.B. Moffat. Thomas Telford, London.
Butcher, A.P. & Powell, J.J.M. (1995). The effects of geological history on the dynamic measurement of the stiffness of soils. *Proc. XIth European Conf on Soil Mech and Found Eng*, Copenhagen, May 1995.
Butcher, A.P. & Tamm, W.S.A. (1994). An Example of the use of Rayleigh waves to detect the depth of shallow landfill. *Proceedings of conference on Modern Geophysics in Engineering Geology*, Liege, Geological Society Special Publication - in print.
Dyvik, R. (1985). In situ Gmax versus depth profiles for four Norwegian test sites. Norwegian Geotechnical Institute, Report No 40014-10.
Eidsmoen, T.E., Gillespie, D. and Lunne, T. (1984). Tests with UBC seismic cone at three Norwegian research sites. Norwegian Geotechnical Institute, Report No 59040-1.
Hardin, B.O. and Drnevich, V.P. (1972a). Shear modulus and Damping in soils: Measurement and parameter effects. *J. Soil Mech. and Found. Div. Proc. ASCE*, **98**, SM6, 603-624.
Hardin, B.O. and Drnevich, V.P. (1972b). Shear modulus and damping in soils: design equations and curves. *J. Soil Mech. and Found. Div. Proc. ASCE*, **98**, SM7, 667-691.
Hope, V. (1993). *Applications of Seismic Transmission Tomography in Civil Engineering*. PhD Thesis, Department of Civil Engineering, University of Surrey, Guildford.
Hryciw, R.D. (1989). Ray path curvature in shallow seismic investigations. *J. Soil Mech. and Found. Div. Proc. ASCE*, **115**, No 9, 1268-1284.
Nazarian, S. (1984). *In Situ determination of elastic moduli of soil deposits and pavement systems by spectral analysis of surface waves method*. PhD thesis, The University of Texas at Austin, USA.
Powell, J.J.M. and Butcher, A.P. (1991). Assessment of ground stiffness from field and laboratory tests. *Proc. Xth European Conf on Soil Mech and Found Eng*, Florence. **1**, 153-156.
Ramberg, W. and Osgood, W.T. (1943). Descriptions of stress strain curves by three parameters. Tech Note 902, NACA.
Rayleigh, Lord (1900). On waves propagated along the plane surface of an elastic solid. Scientific papers of Lord Rayleigh, Vol 2, 441-447. London: Cambridge University Press.
Robertson, P.K., Campanella, R.G., Gillespie, D. and Rice, A. (1986). Seismic CPT to measure in situ shear wave velocity. *J. Soil Mech. and Found. Div. Proc. ASCE*, **112** 791-803.
Seed, H.B. and Idriss, I.M. (1970). Soil moduli and damping factors for dynamic response analyses. Earthquake engineering research centre, Univ.of Cal. Berkley, Cal. Rep.No. EERC70-10.
White, J.E. (1965). *Seismic waves; radiation, transmission and attenuation*. McGraw Hill, New York. pp 298.

714

Site investigation for seismically designed structures

P D DAVIS[1], P J L ELDRED[2], J D BENNELL[3], D W HIGHT[4] and M S KING[5]
[1] Nuclear Electric plc, Gloucester, United Kingdom
[2] Soil Mechanics Limited, Wokingham, United Kingdom
[3] University College of North Wales, Menai Bridge, United Kingdom
[4] Geotechnical Consulting Group, London, United Kingdom and Visiting Professor, Imperial College, London, United Kingdom
[5] Oil Industry Professor of Petroleum Engineering, Royal School of Mines, London, United Kingdom

INTRODUCTION

Some structures require a knowledge of the dynamic properties of their foundation materials so that they can be designed to resist seismic events. Nuclear facilities are the most common of such structures, but many others may have to be seismically designed, depending on the assessed seismic hazard at the site and the consequences of failure.

The principal design parameters required for seismic design are the small strain elastic properties and damping ratio of the ground, together with a description of how these parameters vary with shear strain. Traditionally, the small strain elastic properties have been assessed from in situ wave velocity measurements and density tests. The degradation of dynamic stiffness with strain and damping characteristics have been measured by resonant column or cyclic triaxial tests on small samples of the foundation material (reconstituted in the case of sand and undisturbed in the case of cohesive soil). No practical method of in situ measurement of modulus decay and damping characteristics is currently available that can be used for a deep and varied geological sequence.

As dynamic laboratory tests have been difficult to procure in the United Kingdom, much reliance has been placed on published data for modulus decay and damping (eg Seed and Idriss, 1970). Seismic design for civil nuclear power stations in the United Kingdom followed this approach during the 1980s.

Since 1990, Nuclear Electric plc, who own and operate the commercial nuclear power stations in England and Wales, have carried out a thorough review of the dynamic properties of a number of its sites. From a geotechnical viewpoint, there were three main reasons for this:

- In the late 1980s, the results of a number of studies of site response during real earthquakes had been published (eg Electric Power Research Institute, EPRI, 1988). A number of these had suggested that the sites concerned had not exhibited such pronounced non-linearity of response as indicated by the established modulus decay relationships.
- Considerable advances had been made in measuring and understanding the non-linearity of the deformation behaviour of soils under static loading and the reasons for differences between laboratory and field measurements. Many of these advances appear to have applications in deformation response under repetitive dynamic loading.
- Information from modern ground investigations had been acquired during the 1980s for several of Nuclear Electric's sites. These had provided further wave velocity measurements and state-of-the-art measurements of static deformation properties. However, no reliable laboratory measurements of modulus decay and damping on undisturbed specimens had been obtained.

Advances in site investigation practice. Thomas Telford, London, 1996

The basis of the review of the dynamic properties was to be a comprehensive interpretation of all the available site specific information, supplemented wherever possible with published or unpublished information on comparable materials. This paper describes some aspects of the work undertaken for two sites (Sizewell and Hinkley Point). In each case, a set of diverse laboratory and field measurements and literature were synthesised to obtain the best possible estimates of the required dynamic properties.

SEISMIC DESIGN PROCEDURES AND PARAMETERS REQUIRED

The procedures for the seismic design of nuclear structures include the modelling of the transmission of seismic waves up from the bedrock, through the overlying rock and soil strata, into the structure foundations and thence into the structure and equipment. The most important aspect is the vertical propagation of shear waves through the various strata.

As the structure and foundations are set in motion by the seismic waves, energy is radiated back into the ground. This can take the form of surface (Rayleigh) waves and body (shear and compression) waves. In order to model this, the behaviour of the ground in both compression and shear and for both horizontal and vertical propagation of waves is required. In addition, shear waves can have either horizontal or vertical polarisation. A full appreciation of the ground therefore involves obtaining details of its anisotropy.

In practice, values of the small strain shear modulus in a vertical plane and the associated shear damping, as well as the manner in which these properties change with shear strain, are used to model the ground. Properties may be required for the free field (ie those areas of the site not affected by construction loads) and for the stressed material below the structures. In more complex analyses, full or partial inclusion of the anisotropy of the properties may also be required.

THE SITES AND GEOLOGY

Sizewell is situated on the Suffolk coast. There is an operational 'A' Station, the 'B' Station is being commissioned and a 'C' Station is being planned. The study was primarily for the 'B' Station. The geology of the site comprises a sequence of Plio-Pleistocene sand (Crag) and Tertiary clays with subordinate sands (London Clay, Woolwich and Reading Beds and Thanet Beds) overlying the Chalk. The strata are essentially flat over the site, with little dip. A simplified stratigraphy of the site and a summary of relevant test results are included on Fig 1.

Hinkley Point is situated on the Somerset coast. There are operational 'A' and 'B' Stations and the site of a 'C' Station has been investigated. The study was for the 'C' Station. The geology comprises a sequence of Mesozoic mudstones (Lower Lias, Penarth Group and Mercia Mudstone Group), which contain a variable number and thickness of limestone bands. There is a thin cover of superficial deposits. A maximum of about 10 m of the Lower Lias is weathered and the strata dip at approximately 11° to the north. A simplified stratigraphy of the site and a summary of relevant test results are included on Fig 2.

HISTORY OF INVESTIGATION AT THE SITES

Both sites have a long history of investigation because of their phased development. The principal investigations carried out at each site and the techniques utilised are summarised in Tables 1 and 2. These illustrate the increasing sophistication of testing with time.

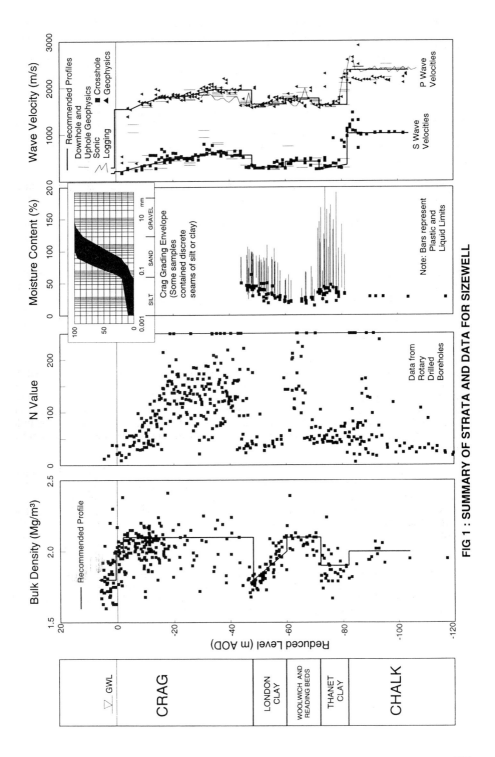

FIG 1 : SUMMARY OF STRATA AND DATA FOR SIZEWELL

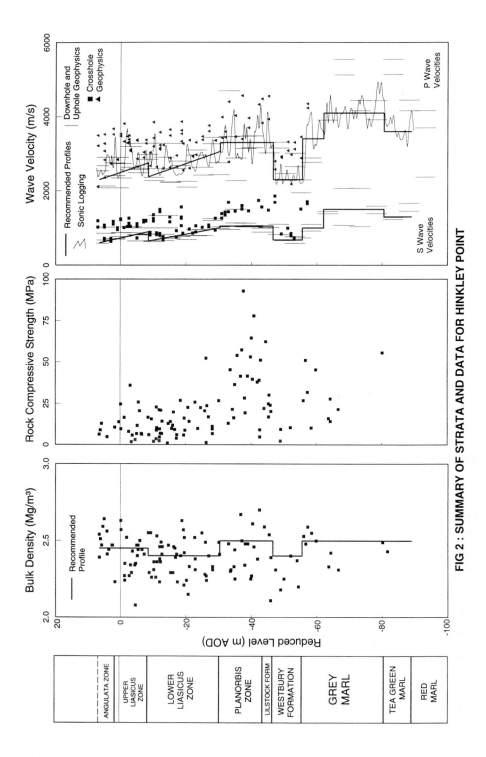

FIG 2 : SUMMARY OF STRATA AND DATA FOR HINKLEY POINT

Date	Principal Techniques Involved
1959	Cable tool boring; inspection shaft with in situ measurements.
1968	Cable tool boring; Bishop sand sampler.
1975	Cable tool boring; experimental wave velocity measurements; pumping tests.
1980	Rotary drilling of soils; cone penetrometer tests, including one test to almost 100 m which penetrated 10 m of Chalk; SPT tests in small diameter rotary drilled holes for liquefaction studies; in situ wave velocity measurements by downhole, uphole and crosshole methods; surface refraction geophysical survey; geophysical logging of boreholes; self boring pressuremeter tests.
1988	Rotary core drilling of granular and cohesive soils; cone penetrometer tests; SPT tests in small diameter rotary drilled holes; in situ wave velocity measurements by crosshole, downhole and uphole methods, including attenuation measurements; geophysical borehole logging*; in situ density tests; nuclear density tests; weak rock self boring pressuremeter tests to 90 m; triaxial tests with local strain measurement.

Date	Principal Techniques Involved
1966	Rotary core drilling; Centex cell tests; Menard pressuremeter tests; trial pits and trenches; vibration trials.
1979	Rotary core drilling; packer permeability tests; pumping tests.
1984	Rotary core drilling; trial trenches; in situ wave velocity measurements by downhole, uphole and crosshole methods; surface refraction geophysical survey; geophysical borehole logging*; borehole plate tests; pressuremeter tests using OYO Elastmeter 100; borehole permeability tests; pumping tests; resistivity survey.
1989	Rotary core drilling; weak rock self boring pressuremeter tests; wave velocity measurements by downhole and crosshole methods; in situ stress measurements using USBM gauge; permeability measurements; unconfined compression tests on rock cores with strain gauge measurement of axial strain; multi-point Westbay piezometers.

* Logs carried out included natural gamma, gamma-gamma density, caliper, neutron porosity, neutron-gamma, focused electric resistivity, spontaneous potential, multi-channel sonic, sidewall sonic, verticality, dipmeter, fluid conductivity, temperature and micro-resistivity.

* Logs carried out included resistivity, natural gamma, gamma-gamma density, caliper, neutron porosity, neutron-gamma, focused electric resistivity, spontaneous potential and multi-channel sonic.

A further site investigation has been carried out in 1994 which post dated the study described in this paper. Testing is still in progress.

TABLE 1 : SITE INVESTIGATION TECHNIQUES EMPLOYED AT SIZEWELL

TABLE 2 : SITE INVESTIGATION TECHNIQUES EMPLOYED AT HINKLEY POINT

PHILOSOPHY AND METHODOLOGY
General aspects

It is recognised that the deformation properties of geological materials relevant to static monotonic loading are dependent on a wide range of factors. Some of the more important factors are geological characteristics, disturbance of the tested material, strain magnitude, drainage conditions, stress history, initial stress state, rate and duration of loading, stress path and anisotropy. Limitations in the testing equipment are also of fundamental importance. In rocks, important geological characteristics affecting stiffness include lithology, weathering and fracture state. It therefore seemed probable that the dynamic deformation characteristics would also be very sensitive to many of the above factors. In order to interpret the available data critically in the light of such influences, a stress path type of approach was adopted, taking account of:

- Stress history and in situ stress state.
- The sequence of construction activities, including dewatering, excavation and structural loading (some of these processes will enhance ground stiffness by increasing effective stresses and others will reduce it by causing yield in soil or opening fractures in rock).
- Ageing of the material below the plant which could lead to an enhancement of stiffness.

Other features specifically considered in the studies were:

- The geological characteristics of the materials and how such characteristics would be altered by sampling and testing and by construction of the plant (for example opening and closing of rock fractures, presence of cementation or structure in sand and clay).
- Differences between the loading conditions experienced in situ by the ground during a real seismic event and the loading applied in a particular type of laboratory or field test.
- Anisotropy of stiffness parameters, both between vertical and horizontal planes and azimuthal anisotropy in the horizontal plane.

As a result of the above influences, the properties beneath the structures were expected to differ from those in the undisturbed free field. The latter correspond more closely to those pertaining at the time of the investigation.

Wave velocities
In order to obtain the required low strain, dynamic properties of the foundation materials, crosshole, downhole and uphole techniques were utilised. The specific methods of testing utilised by University College of North Wales, who carried out most of the testing, are described in Davis (1989). These data were supplemented by surface refraction data (Davis and Bennell, 1988) in order to measure properties of the near surface materials.

Crosshole data, in particular, can be greatly affected by large velocity gradients and by interlayered high and low velocity materials (eg limestone bands within mudstones). Such refraction effects can lead to apparent velocities higher than the true formation velocities. This effect can be overcome to some extent by reducing the borehole spacing and decreasing the depth interval between readings. Downhole and uphole testing avoids some of the problems of layering, although this method only provides average velocities over a depth range as a result of the lower wave frequencies utilised and the longer propagation paths.

At Hinkley Point, the layering was such that the problems with crosshole testing could not be overcome and consequently the downhole and uphole data were concentrated upon. The velocities are therefore less well resolved than for the Sizewell site, which had relatively thick units of reasonably homogeneous material.

Anisotropy
The study of anisotropy requires consideration of the following factors:

- The inherent, or intrinsic, anisotropy of small scale, intact specimens of individual layers of a single material type.
- The structural or mass anisotropy introduced by the layering of material types (eg the mudstones and limestones of Hinkley Point), which may have significantly different stiffnesses at the same strain.
- Anisotropy induced by fractures or discontinuities.
- The anisotropy induced by in situ stress states at all scales, eg where the horizontal stress is very different to the vertical stress and where the horizontal stress varies with azimuth.
- The variation in anisotropy with strain.

In an anisotropic medium, the various in situ methods of measuring the wave velocities can produce different values depending on the orientation of the source and the direction of the transmission path. Thus a horizontally travelling shear wave with vertical polarity (derived from a vertical motion of the source) can result in a different wave velocity to a horizontally travelling shear wave with horizontal polarity (derived from a horizontal motion of the source). The difference between the two gives an indication of the anisotropy of the medium. Incident seismic shear waves are generally of the vertically travelling, horizontally polarised type. Many geological materials are transversely isotropic, with their axis of symmetry perpendicular to the bedding. However, where the horizontal stress varies with azimuth, a more complex state of anisotropy exists.

APPLICATION TO SIZEWELL
Wave velocities and elastic properties at small strains

Profiles of shear and compression wave velocities were derived for the strata based on the crosshole, downhole, uphole and surface refraction measurements carried out as part of the site investigations for both the 'B' and 'C' Stations. Most reliance was put on the crosshole velocities as their quality was generally better. Several aspects also had to be addressed before appropriate values could be recommended. These mainly concerned the changes in stress regime, as follows:

- Some of the wave velocity measurements were carried out in boreholes located in a large pre-existing excavation. Hence the differences in ambient stress had to be allowed for when comparing these results with those from boreholes located at original ground level.
- The excavation and dewatering of the excavation for the foundations will change the ambient stresses during construction.
- The construction of the buildings, backfilling of the excavations and the subsequent rise in groundwater levels will again change the stress state.

The Crag is present immediately below the reactor foundation. It is a very dense, often shelly sand with very high SPT N values (Fig 1). Regional geological evidence suggests that the Crag at Sizewell has not been subject to significant ice loading or removal of overlying sediments. However, it may have been overconsolidated to some extent as a result of natural groundwater lowering during periods of glaciation. It is not noticeably cemented, but the high wave velocity values indicate that it has a much greater stiffness than equivalent reconstituted sands. This factor, along with evidence of a stable but easily disturbed structure and its age of approximately 0.75 to 2 million years, implies a structured or extremely weakly bonded material.

The significance of the structured nature of the Crag is that if strains large enough to destroy the structure occur, there will be a significant reduction in its stiffness and it will tend towards a young, destructured material. Consequently, the possibility of partial or total destructuring had to be allowed for. Simple stress analyses indicated that the strains due to excavation and foundation loading would be sufficient to destructure the Crag to a depth of about 10 m below the reactor foundation, with partial destructuring for a further 25 m. In the absence of test data specific to the Crag, values for the shear modulus and hence shear wave velocity of completely destructured Crag were based on the Hardin and Richart (1963) equation for reconstituted sands.

The effect of the increase in stresses under the plant loading was allowed for by using a power law relationship between shear wave velocity and confining stress of the form of Equation (1):

$$V \alpha \sigma^n \qquad (1)$$

where V is wave velocity, σ is the relevant stress and n is an exponent

A simple stress analysis was used to assess the increase in stresses due to the plant. For reconstituted or destructured sands, recent Japanese research confirms that for shear waves, the value of n is about 0.25, as suggested by Hardin and Richart (1963). For the structured Crag, it was considered that stress increases would have less effect on shear wave velocity and stiffness, so an exponent of 0.125 was adopted. However, fairly wide bounds were applied to the exponent to allow for uncertainty.

In addition to the allowances for stress changes, an allowance was also made for ageing, ie the re-establishment of some of the structure of the Crag after loading. This aspect has been studied in the work of Anderson and Stokoe (1978). An allowance of a 4 to 5% increase in shear modulus per log cycle of time was adopted; the high value being chosen because of the considerable shell content of the Crag. It was applied for an ageing period of 5 to 15 years, giving an increase in the small strain shear modulus of approximately 15%, equivalent to an increase in shear wave velocity of 7½%.

The increase in stresses at the top of the Tertiary strata due to structural loading implied only small changes in shear modulus. Therefore no change in shear wave velocities resulting from the construction of the plant was applied to the Tertiary strata or the underlying Chalk.

The various in situ wave velocity measurements in the Crag showed little evidence of anisotropy of small strain elastic properties. In the Tertiary clays, there was little compression wave anisotropy in either field or laboratory measurements, but laboratory shear wave measurements on unconfined samples yielded horizontal velocities greater than vertical ones, suggesting transverse isotropy of properties. However, the results were widely scattered and may have been influenced by the uncertain but isotropic effective stress state in the unconfined samples.

Modulus decay

No tests had been carried out on undisturbed samples of sand or clay from which the effect of shear strain amplitude on the dynamic shear modulus and damping ratio could be determined. Consequently, it was necessary to investigate indirect evidence and published and unpublished test results on comparable materials.

For Sizewell, the following data were available:

- Unload/reload loops from self boring pressuremeter tests in the Crag and Tertiary strata.
- Triaxial compression tests with limited load cycling and local axial strain measurements on undisturbed rotary core samples of London Clay and Thanet Beds clay.
- Other laboratory stiffness measurements from the London Clay and the Woolwich and Reading Beds clays in the London area (Hight and Jardine, 1993).
- Monotonic and cyclic triaxial, torsional shear, resonant column and other laboratory tests with local strain measurements on reconstituted dense Toyoura Sand (eg Teachavorasinskun et al, 1991).
- Resonant column tests on a range of reconstituted dense sands from other sites carried out at University College of North Wales.
- Resonant column and torsional shear tests on Italian clays carried out in Italy and triaxial tests on the same materials carried out at Imperial College (Georgiannou et al, 1991).
- Summary data linking the shape of the modulus decay curve with the plasticity index of clays (Vucetic and Dobry, 1991).

The modulus decay behaviour of undisturbed, structured Crag was broken down into three elements in order to formulate recommendations:

- The behaviour of the Crag in an unaged, destructured state. This provided a lower limit towards which the structured Crag will tend as the structure is broken down during straining. The destructured behaviour was based on resonant column tests on reconstituted sands, taking account of the change in shape of such curves with increasing mean effective stress. Allowance was also made for pore pressures which might accumulate under cyclic loading.
- The persistence of the structure of the Crag under increasing strains. An estimate was made of the degree to which the structure is broken down at various strains. It was hoped that the self boring pressuremeter tests in the Crag would provide guidance on this aspect. The test results were back analysed using a non linear finite element program to obtain a stress-strain relationship compatible with the shape of the unload/reload curves. However, the resulting behaviour was very similar to modulus degradation relationships for reconstituted sands under monotonic loading. This was considered an underestimate of the response of the Crag structure to straining during a seismic event because:
  - i) Resonant column tests on reconstituted sands exhibit less rapid degradation of shear modulus with shear strain than monotonic loading tests (Teachavorasinskun et al, 1991).

ii) Installation disturbance may have affected the accuracy of the pressuremeter tests in Crag at extremely small strains which correspond to cavity expansion of less than the average grain diameter of the sand.

iii) Considerable significance was attached to the evidence of less marked non-linearity of behaviour of soils in real earthquakes which impose complex, rapid and random fluctuations in shear stress and principal stress directions (EPRI, 1988).

- The modulus decay curve was made compatible with the estimated strength of the Crag, ie the shear stress calculated from shear modulus and shear strain at any point on the curve should not exceed the strength in cyclic simple shear.

For the Tertiary clays, the results from a range of tests on three different Italian clays were utilised. These clays were less plastic than the clays on site, but they were overconsolidated, structured or bonded clays from which block samples had been obtained by careful sampling. The results from these tests produced only a modest scatter. The recommended modulus decay curves were based on these data, with minor differences reflecting the different shear strengths of the Sizewell clays. The Italian clay data were considered the most reliable guidance on dynamic modulus decay for the Sizewell clays because they included resonant column tests and the high quality samples were expected to have preserved much of the material structure and bonding. Such structure was also considered to be present in the Sizewell clays (Hight and Jardine, 1993).

Examples of the recommended modulus decay curves in terms of G/G_{max} are shown on Fig 3.

The self boring pressuremeter tests in the Tertiary clays were analysed by the methods of Jardine (1991) and Muir Wood (1990) to obtain static modulus decay curves. These showed similar behaviour to that exhibited by the triaxial tests with local strain measurement, after reconciliation of the different strain conditions in the two types of test. All of these data were broadly consistent with other published data on the same materials (eg Hight and Jardine, 1993). However, they were not considered appropriate to the complex rapid random shearing of the ground in a seismic event.

Damping ratio
Even less suitable and appropriate data could be obtained from which to estimate material damping characteristics. Only resonant column tests could give a reasonable indication of damping ratio over a range of strains. The adopted approach was therefore to study resonant column data for broadly similar materials and develop relationships between modulus ratio and damping as a function of shear strain for sands and clays. Theoretical models (eg Ramberg-Osgood) provide support for deriving a relationship between the G/G_{max} ratio and damping ratio. From the developed relationships, damping ratio curves were derived from the modulus decay curves.

Examples of the recommended damping ratio curves are shown on Fig 3.

APPLICATION TO HINKLEY POINT
Wave velocities and elastic properties at small strains
Hinkley Point provided a similar range of data and problems, but in a very different geological setting. In particular, there is essentially unweathered rock immediately below foundation level at Hinkley Point. Profiles of shear and compression wave velocities similar to those for the Sizewell site were available.

Some of the aspects which were of particular significance in assessing the dynamic properties of the rock were:

- Stratigraphy. The pronounced strata dip of approximately 11° and the presence of nine different stratigraphical units within 100 m of the surface greatly complicated comparisons of measurements in different boreholes. Measurements had to be related to stratigraphical level and allowances made for depth below ground level or vertical stress. In addition to the contrasts in lithology between many of the stratigraphical units, there was pronounced vertical variation within some units, both within the mudstone and, more importantly, as a result of the presence of limestone bands.

- Anisotropy. All the components of anisotropy listed above required consideration, including variation with azimuth due to variations in the in situ horizontal stress, as determined from measurements with the USBM gauge.
- Discontinuities. These affect the stiffness, strength, anisotropy, Poisson's ratio etc of the rock mass. The permeability results showed a distinct reduction in permeability with depth. However, at depths of greater than approximately 25 m, the reduction in permeability with depth appeared to stop. This was taken to be a dividing point, above which discontinuities were likely to be at least partially open prior to construction.
- Changes in stresses during and after construction. This was similar to the situation at Sizewell, although the complication of an initial excavation was replaced by the problems of dip and stratigraphy.

The recommended free field wave velocity profiles were based on uphole and downhole velocity measurements as these were more relevant to the average properties pertinent to the vertical propagation of seismic waves. Crosshole velocities were generally not used as they were variably affected by the presence of harder limestone bands. Approximate estimates were made of the anisotropy of wave velocities and some of the small strain elastic properties using a synthesis of all wave velocity measurements (including those from geophysical borehole logging), the published data on anisotropy of a number of mudstones and the relative proportions of mudstone and limestone.

The effect of stress changes due to plant loading were then considered. For intact rock specimens, which are generally considered to contain micro-fractures that close under load, the dependence of shear and compression wave velocities on the stress applied to the specimen is governed by the same form of equation used for Sizewell. This type of relationship was also expected to apply to the rock mass. Possible values of the exponent n were investigated by comparing measured velocities at the same stratigraphical level but different depths and hence stresses. A somewhat confusing picture emerged, with calculated values of n in the range 0.04 to 0.22, as compared with 0.01 to 0.15 for intact laboratory specimens. The scatter of results is not surprising given the uncertainties regarding in situ fracture characteristics (spacing, aperture, roughness etc) and indicates the uncertainties in estimating the effect of plant loading on the dynamic properties of the rock mass.

Modulus decay
In the case of Hinkley Point, only the following few suitable data were available for the rock strata:

- Unload/reload loops from weak rock self boring pressuremeter tests.
- Unload/reload loops from unconfined compression tests on strain gauged rock samples.
- Cyclic triaxial and dynamic shear tests on mudstones, carried out in Japan (eg Hara and Kiyota, 1977; Nishi et al, 1989 and Kim, 1992).

As for Sizewell, synthesis of the data was necessary to provide guidance on the appropriate modulus decay. The data from the literature were mainly from broadly similar materials, but it was not possible to match the characteristics as closely as would have been liked. Furthermore, many of the data were for slow rates of loading and were not strictly appropriate to seismic modes of shear. Consequently, judgement had to be used to a great extent to arrive at curves consistent with the available data. The pressuremeter results were found to be too sensitive to membrane and other corrections to provide reliable data at the very small strains of interest.

Examples of the recommended modulus decay curves in terms of G/G_{max} are shown on Fig 4.

Damping ratio

As regards the effect of shear strain on damping, there were even less suitable data than for the modulus decay curves. The damping ratio in saturated porous rocks is controlled by two major factors:

- Frictional dissipation at grain boundaries and across crack surfaces.
- Attenuation due to fluid flow effects.

There is some conflicting evidence on the effect of frequency on damping effects, but the weight of evidence tends to imply little effect for mudstones (Hara and Kiyota, 1977). However, there does appear to be a strong dependence of damping on the stress level. Based on these general guidelines and data from stiff clays (Abbiss, 1986 and Kim et al, 1991), damping ratio curves consistent with the few data and consistent with the modulus decay curves were derived.

Examples of the recommended damping ratio curves are shown on Fig 4.

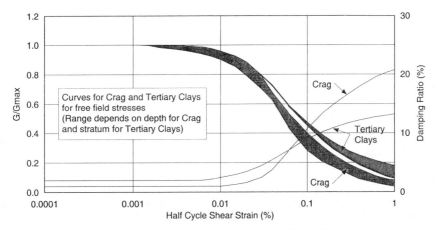

FIG 3 : RECOMMENDED MODULUS DECAY AND DAMPING RATIO CURVES FOR SIZEWELL

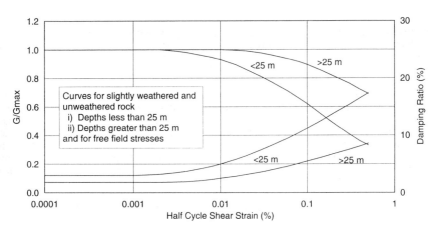

FIG 4 : RECOMMENDED MODULUS DECAY AND DAMPING RATIO CURVES FOR HINKLEY POINT

CONCLUDING REMARKS

We have tried to show in this paper that when tests aimed specifically at obtaining dynamic parameters are not or can not be carried out, some useful estimates can be made from examining the results of the more traditional site investigation tests in conjunction with appropriate published data. We had the advantage in the cases described of having good quality, in situ measurements of the wave velocities. However, even these needed to be assessed and modified, especially when examining the case of loading by the structures.

Space limitations in this paper have precluded the inclusion of as much detail on all aspects of the work as would have been liked, in particular the discussion of uncertainty and bounds. However, detailed consideration was given to bounds on all the data for use in sensitivity analyses.

ACKNOWLEDGEMENT

The authors wish to thank Nuclear Electric plc for allowing them to present the data included in this paper.

REFERENCES

Abbiss C P : 1986 : Shear wave measurements of the elasticity of the ground. Geotechnique, Vol 31, pp 91-204.

Anderson D G and K H Stokoe : 1978 : Shear modulus: A time-dependent soil property. Dynamic Geotechnical Testing, ASTM STP 654, pp 66-90.

Davis A M : 1989 : Determination of dynamic elastic parameters from crosshole testing. Scientific Drilling, Vol 1, pp 54-62.

Davis A M and J D Bennell : 1988 : Resolving overburden characteristics via shear wave propagation. ASEG/SEG Conf, Adelaide, pp 41-44.

Electric Power Research Institute : 1988 : Soil response to earthquake ground motion. EPRI Report NP-5747, Palo Alto, California.

Georgiannou V N, S Rampello and F Silvestri : 1991 : Static and dynamic measurements of undrained stiffness of natural overconsolidated clays. Proc 10th ECSMFE, Florence, Vol 1, pp 91-95.

Hara A and Y Kiyota : 1977 : Dynamic shear tests of soils for seismic analyses. Proc 9th ICSMFE, Tokyo, Vol 2, pp 247-250.

Hardin B O and F E Richart : 1963 : Elastic wave velocities in granular soils. ASCE, Vol 89, SM1, February, pp 33-65.

Hight D W and R J Jardine : 1993 : Small-strain stiffness and strength characteristics of hard London tertiary clays. In Geotechnical Engineering of Hard Soils - Soft Rocks, edited by A Anagnostopoulos, F Schlosser, N Kalteziotis and R Frank, Balkema, Rotterdam, pp 533-552.

Jardine R J : 1991 : Discussion of Strain-dependent moduli and pressuremeter tests (Muir Wood, 1990). Geotechnique, Vol 41, pp 621-626.

Kim D S, K H Stokoe and J M Roesset : 1991 : Characterisation of material damping of soils using resonant column and torsional shear tests. Proc 5th Int Conf Soil Dynamics and Earthquake Engineering, Karlsruhe, pp 189-200.

Kim Y-S : 1992 : Deformation characteristics of sedimentary soft rocks by triaxial compression tests. Dr Engng Thesis, University of Tokyo.

Muir Wood D : 1990 : Strain-dependent moduli and pressuremeter tests. Geotechnique, Vol 40, pp 509-512.

Nishi K, T Ishiguru and K Kudo : 1989 : Dynamic properties of weathered sedimentary soft rocks. Soils and Foundations, Vol 29, No 3, pp 67-82.

Seed H B and I M Idriss : 1970 : Soil moduli and damping factors for dynamic response analyses. Earthquake Engineering Research Center, University of California, Berkeley, Report No EERC 70-10.

Teachavorasinskun S, S Shibuya, F Tatsuoka, H Kato and N Horii : 1991 : Stiffness and damping of sands in torsion shear. Proc 2nd Int Conf Recent Advances Geotech Earthquake Eng and Soil Dynamics, St Louis.

Vucetic M and R Dobry : 1991 : Effect of soil plasticity on cyclic response. ASCE (Geotech), Vol 117, No 1, January, pp 89-107.

The selection and interpretation of seismic geophysical methods for site investigation

M.A. GORDON[1], C.R.I. CLAYTON[1], T.C. THOMAS[2], M.C. MATTHEWS[1]
1 Department of Civil Engineering, University of Surrey, Guildford, U.K.
2 Rust Consulting Limited, Godalming, U.K.

ABSTRACT: During the past decade finite element analyses have shown that the majority of strain levels in the ground at working loads are typically less than 0.1 %. It has therefore become important that stiffness values to be used in geotechnical calculations should be measured at these small strain levels. There is also a growing appreciation of the value of measuring the very small strain stiffness, G_{max}, using seismic methods. There are many types of seismic survey and this paper reviews a selection that have been found, by the writers, to be of value for ground investigations.

INTRODUCTION

Over the past few years there has been a growing use of geophysical methods in the construction industry. In geotechnics seismic geophysical techniques have long been used for logging, to provide correlation between boreholes, but their use to provide data on in-situ ground stiffness has yet to become widespread.

An increasing number of projects in the U.K. require the best estimates to be made of movement around excavations, tunnels or shafts, and as it is now generally accepted that in such situations the strain levels in most of the surrounding ground are less than 0.1 %, small strain (less than 1 % strain) stiffness values are required (Jardine et al., 1986). Until the late 1970's soil stiffness values were obtained in the laboratory from conventional triaxial tests where the axial strains were measured externally. These stiffness values were found to be much lower than those back analysed from field measurements. This discrepancy has been attributed to sampling effects (Burland, 1979), bedding errors and apparatus compliance (Jardine et al., 1984). The advent of local strain measuring devices has meant that small strain stiffness values can be measured in the laboratory and indeed this is now carried out routinely in many soil testing laboratories in the U.K. Although this represents a significant step towards measuring stiffness at working strains, there remains the problem that all laboratory samples are disturbed to a greater or lesser degree.

The value of measuring very-small-strain stiffness, G_{max}, is now more widely accepted, and in the field this can be achieved by seismic geophysical testing (Tatsuoka and Shibuya, 1992). These tests, that are often carried out in addition to more traditional field and laboratory tests (self-boring pressuremeter, local-strain triaxial tests), have the advantage of not being affected by sample disturbance or insertion effects. Obtaining realistic stiffness parameters to use in deformation analyses is one of the most important problems in geotechnical engineering; it is also one of the most difficult. Very small strain stiffness values, G_{max}, from field seismic geophysical tests provide a benchmark against which other stiffness values can be compared. In some cases, for example in highly fractured weak rock, stiffness values from geophysical surveys can be used directly in engineering calculations.

There are many different forms of geophysical tests available. The purpose of this paper is to review the seismic methods that have been found to be of value for ground investigation. It provides an engineering perspective on a range of techniques which have been found to be relatively reliable, and which at the same time are capable of providing valuable geotechnical data. Data from case studies are presented which show the relationship between stiffnesses determined using field geophysics and those determined using other techniques, as well as considering the interrelation of results from a selection of seismic methods.

SEISMIC THEORY

Civil engineers are still reluctant to incorporate seismic surveys as a matter of course into ground investigations, partly due to historically poor success rates. Seismic theory has been developed on the basis of well understood physical principles, for example, elasticity theory (Hooke's law) and wave motion (Huygens' principle), details of which are beyond the scope of this paper, but can be found in many physics text books (Alonso and Finn, 1967, Timoshenko and Goodier, 1951, Whelan and Hodgson, 1978). This suggests that the perceived unreliability of seismic survey techniques when applied in geotechnical engineering stems from inappropriate use and a misunderstanding of what they can provide for the engineer, rather than short comings of the fundamental concepts (Engineering Group Working Party, 1988).

Seismic methods are based on the fact that the velocity of propagation of a wave or impulse in an elastic body is a function of the modulus of elasticity, Poisson's ratio and the density of the material (Hvorslev, 1949). Four types of elastic wave, all of which travel at different velocities, may be propagated when a transfer of seismic energy to the ground occurs. This input of energy can be caused artificially by a falling weight, hammer blow or explosive charge (Clayton et al., 1982), all of which will produce a single elastic impulse. In the case of surface wave geophysics a continuous supply of impulses is provided by a vibrator fixed to the ground. The impulses produce compression (P) waves, shear (S) waves, Rayleigh waves and/or Love waves. P and S waves are known collectively as body waves; Rayleigh and Love waves are termed surface waves. The ground motion resulting from the energy input is detected either by geophones situated at the surface, or, in the case of downhole surveys, by subsurface geophones. Data are collected in the time domain, often using a seismograph, and with the exception of surface wave surveys, the time taken for the seismic wave to travel from source to receiver is measured. A schematic representation of equipment typically used is shown in Figure 1. In the case of surveys using single pulses of energy to generate seismic waves, for example, refraction, crosshole, downhole and seismic cone penetration tests, signals can be stacked in the time domain. This may be necessary if the site is noisey in order to improve the signal to noise ratio.

The petroleum exploration industry has been using seismic surveys since about 1920 and traditionally it has used P waves, although the seismic sources used invariably produce S waves and surface waves as well as P waves. P waves, sometimes referred to as primary waves, are easy to detect as they arrive at the receiver before any other waves. The onset of S waves, or secondary waves, is more difficult to detect. Although S waves have larger

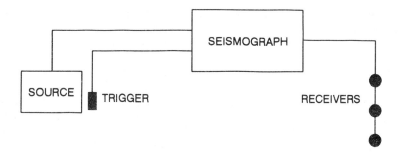

Figure 1 Schematic representation of equipment typically used for seismic surveys.

amplitudes than P waves, their arrival may be masked by the presence of P waves. However, by reversing the polarity of the source input the shear waves themselves reverse, hence if the two traces are superimposed the shear wave arrival can be identified by the first reversal of the wave train. The P wave velocity is controlled by the undrained bulk modulus. This means that in the near surface saturated materials often encountered in this country, the P wave velocity will therefore be a function of the bulk modulus of water. This means that a constant value of about 1500 m/s will be obtained for the P wave velocity, rather than a velocity varying with the stiffness of the soil skeleton. The S wave velocity, on the other hand, is unaffected by drained/undrained conditions and is a function of the shear modulus. The very small strain shear modulus, G_{max}, for an isotropic material, can be found from the relation $G_{max} = \rho \, V_s^2$, where ρ is the bulk density and V_s is the S wave velocity. The bulk density of soils and rocks does not vary a great deal, so variations in S-wave velocity will reflect variations of very small strain stiffness in the ground. This makes the S-wave survey an attractive proposition to the site investigation industry.

SEISMIC SURVEY TECHNIQUES

This section briefly describes five well known seismic survey techniques. Most seismic surveys involve four stages: survey design, data acquisition, data processing and interpretation. At the interpretation stage velocities determined during processing are related to elastic parameters such as shear modulus, G, Young's modulus, E and Poisson's ratio, ν.

Reflection survey

Seismic reflection surveys have traditionally been used to map subsurface structures. They are widely used in the petroleum industry and in geotechnical investigations over water. The method relies on measuring the travel times of P waves that have been reflected back to the surface by geological interfaces between materials of different P-wave velocities. The travel times to the interfaces can be measured from the seismic records but if the depth of the strata is to be found then the P-wave velocity must be known for each layer. At present the seismic reflection survey is unable to detect reliably shallow geological interfaces or weak lithology contrasts and so is rarely used in land based site investigations.

Refraction survey

Traditionally P-wave seismic refraction surveys have been used by geophysicists to provide deep geological sections across sites, but S wave surveys are required at shallow depth and to enable shear moduli to be calculated. S wave energy is usually supplied by placing a heavy object in contact with the ground (for example, railway sleeper) and hitting it horizontally with a sledge hammer. Refraction surveys rely on the critical refraction of waves at a geological interface between two materials having different characteristic wave velocities. Critical refraction can only occur if the velocity in the lower layer, V_l, is greater than the velocity in the upper layer, V_u. This immediately imposes a limitation on the technique in that it can only be used when the wave velocity is greater in each successively deeper layer; it is unable to detect a low stiffness layer underlying a higher stiffness layer. In addition the technique is limited to simple subsurface geometries. Traditionally it has been unable to detect a stratum where the velocity gradually increases with depth because the method of interpretation relies on refraction occurring at layer interfaces. In this country the case of a near surface layer having a stiffness increasing more-or-less linearly with depth is quite common, e.g. London clay. A refraction survey can be used in this situation if the data are fitted to an inverse sinh function which corresponds to a linear increase of velocity with depth (Dobrin, 1960). The refraction paths are now arcs of circles and from the geometry an expression for the travel time of the seismic waves can be deduced. Abbiss has successfully used this alternative interpretation on refraction survey data from chalk at Mundford (Abbiss, 1979).

Parallel crosshole survey
The parallel crosshole survey is a straight forward way of determining horizontal shear wave velocities and of providing information on layering. The method relies on measuring the time taken for S waves to travel between two points at the same depth in the ground, a known distance apart. The shear wave source is usually vertically polarised (e.g. the Bison shear wave hammer) and three component receivers are often used. Prior to the survey three collinear boreholes are drilled to the maximum survey depth, logged and cased using flush coupled plastic casing. The casing is then grouted in place. A verticality survey must be carried out to enable borehole separation at each test depth to be corrected for the drift from true vertical that may occur during drilling.

The parallel crosshole survey is considered to be a reliable field method for determining shear moduli. It has the advantage over the refraction survey in that it can detect low velocity layers providing they are thick compared with the source-receiver spacing. Work has been carried out using a torsional shear wave source which produces horizontally polarised, horizontally propagated shear waves (Hoar and Stokoe, 1978). Torsional shear wave sources are not in general use as yet, but they may offer the opportunity to investigate anisotropy.

Downhole survey and seismic cone penetration test (SCPT)
Boreholes that have been installed for the purpose of carrying out a parallel crosshole survey can also be used for uphole and downhole surveys. Both these types of survey will determine the variation of in-situ S-wave and/or P-wave velocity with depth, and provide additional information about incremental velocities and layering (Ballard and McLean, 1975). They rely on measuring the travel times of the waves between surface and subsurface points. For an S-wave survey, S-wave energy is usually provided in the same way as described in the S-wave refraction survey and ideally a string of three component receivers is used.

In recent years the seismic cone penetration test has increased in popularity. It was originally developed at the University of British Columbia in Canada by Campanella and Robertson (1984). Cone penetration testing has been widely used in geotechnical investigations for some time and although the profile thus obtained may approximate to that of ground stiffness, the introduction of seismic measurements into the test enables the specific determination of G_{max}

(Robertson et al., 1986). The seismic cone test is simply a downhole test without the need for boreholes. It can therefore be carried out much quicker and is subsequently more economical. If two or three cone trucks are available, a crosshole test may also be carried out (Baldi et al., 1988). In general downhole surveys provide more reliable values for average velocities and hence average stiffnesses, whereas crosshole surveys are more able to detect horizontal layering (Pinches and Thompson, 1990). In a strongly-layered strata some of the crosshole waves that are detected may have been refracted waves, leading to an over estimate of velocity and hence stiffness. It is clear that although crosshole and downhole surveys yield valuable information, they should not be used in isolation, but should be interpreted alongside borehole records.

Surface wave survey
The parallel crosshole and downhole surveys previously described are both intrusive, that is, they require boreholes to be drilled or probes to be advanced into the ground (Stokoe et al., 1989). The surface-wave survey is non-intrusive, the testing being performed entirely from the ground surface. The method relies on the dispersive nature of surface-wave propagation in the elastic half space the ground is assumed to be. This means that surface waves of different wavelengths propagate with different velocities, because they penetrate the ground to different depths. Short wavelength, high frequency surface waves penetrate the near surface, while long wavelength, low frequency surface waves penetrate deeper. As stated earlier there are two types of surface waves: Love waves and Rayleigh waves. In practice it is Rayleigh waves that are detected in surface-wave geophysical surveys; Love waves are polarised in a direction perpendicular to that of the receivers and hence will not be detected.

This technique has been used in Europe, America and Asia but has not, as yet, been fully exploited in this country. Although it is limited to relatively shallow depths, it can detect softer materials beneath stiffer materials, and recent research at the University of Surrey has shown that it is capable of detecting a thin, compressible layer sandwiched between two stiffer layers.

CASE STUDIES
London clay
The London clay case study discussed is the site of a major junction improvements scheme. The scheme is approximately 1.8km in length and involves construction of twin bore cut and cover tunnels and extensive retaining walls. Main construction works are due to start in 1995. The scheme is through a built-up urban area of Finchley in North London, with residential properties within 10 metres of 9m high retaining walls.

The site is underlain throughout by over 30m of London Clay with the top 10 to 15m affected by weathering and periglacial action. There are also several old stream channels running across the site and groundwater levels are near surface. Soil stiffness values for Young's Modulus, E, were required for structural design of the tunnels and retaining wall elements, and in addition, for ground movement analyses carried out to assess the effects of the construction on nearby buildings.

Young's Modulus values were obtained in the ground investigation from the following:

- unconsolidated undrained (UU) triaxial tests with local strain measurement (testing was carried out on high quality thin wall samples).

- self-boring-pressuremeter (SBP) tests where stiffness was derived from unload-reload stages.

- a crosshole geophysical survey carried out after the UU and SBP tests.

Pressuremeter testing was carried out in eight boreholes across the site with the UU and geophysical survey carried out in the critical area of site closest to buildings likely to be affected by the construction. Initial attempts to correlate the UU and SBP tests proved difficult with the latter showing a widespread scatter of results and also generally much higher stiffness values. The crosshole geophysical survey was carried out to provide further correlation as it was felt that the very small strain shear modulus, G_{max}, could give a good guide to the upper bound stiffness of London Clay. The designers wanted to assess stiffness values over the strain range of 0.01 % to 0.1 % axial strain, that is the operational strain range for retaining wall behaviour (Mair, 1992).

Stiffness values obtained from the laboratory tests, SBP tests and geophysical survey are compared on Figure 2. Upper and lower bounds are shown for the SBP test results illustrating the scatter mentioned earlier, along with specific values from the SBP test carried out closest to the geophysical survey (SBP3). It was found that stiffnesses at 0.01 % strain, obtained from small strain triaxial testing of thin wall tube samples taken from one of the boreholes, were typically 50 % of G_{max} measured by the crosshole survey(see Figure 2). Laboratory values at 0.1 % strain were typically 20 % of G_{max}. Furthermore, when the laboratory stiffnesses at 0.01 % strain were corrected for the change in mean effective stress that would have taken place as a result of the sampling process, they were equal to or greater than G_{max}. This is clearly impossible. The *in situ* mean effective stresses were calculated using values of *in situ* total horizontal stress derived from self-boring pressuremeter data obtained at various distances (10 - 90 metres) from the crosshole survey site. There are two possible explanations for the corrected laboratory stiffnesses being erroneous. The self-boring pressuremeter may have overestimated the *in situ* horizontal stresses at the locations where it was used, or the *in situ* horizontal stress-depth profile at the crosshole survey site may have been significantly different to that at the pressuremeter locations. Either way values of G_{max} from the crosshole survey provided an invaluable upper bound by which to critically assess stiffness values obtained from other methods.

A surface-wave survey and SCPT were also carried out at the same site. In this case the velocities of the vertically polarised, horizontally propagated S waves in the crosshole survey were almost the same as those of the horizontally polarised, vertically propagated S waves in the SCPT, and both showed good agreement with results from the surface-wave survey (see Figure 3).

Chalk

Over the past five years major field studies have been carried out at three locations on fractured chalk (Matthews,1993). Large diameter (1.8m) plate loading tests were carried out at each location in conjunction with field seismic geophysical testing (surface wave) and small strain laboratory testing on intact chalk. Figure 4 shows the distribution of very small strain shear modulus with depth from the Rayleigh wave tests at each location alongside intact shear moduli values obtained from the small strain laboratory testing. Despite the fact that the stiffnesses will have been measured at much lower strain levels by the geophysical surveys, the intact stiffnesses measured in the laboratory exceeded these by varying degrees. The

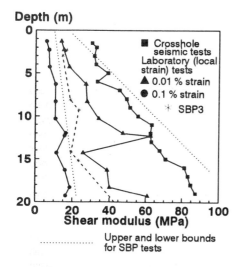

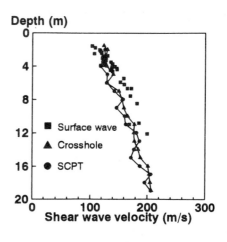

Figure 2 Depth - shear modulus profiles from local strain laboratory tests, crosshole survey and SBP tests.

Figure 3 Depth - shear wave velocity profiles from crosshole survey, surface wave survey and SCPT.

average stiffness of the chalk beneath each plate test was also back calculated from measured plate settlements at the various applied loads. Pre-yield back calculated stiffnesses compared well with very small strain stiffnesses obtained from the surface wave surveys. In most cases the ratio between back calculated stiffness and stiffness from surface wave surveys lay in the range 0.93 - 2.16 for an applied stress level of 200 kPa on the plate (Clayton et al., 1994).

DISCUSSION

Five types of seismic survey have been reviewed and two case studies have been used to illustrate the role that geophysical surveys can play in the site investigation industry by providing data for geotechnical calculations.

In order to be of use in geotechnical calculations, P and S wave velocities measured in seismic surveys must be related to elastic soil parameters, for example, shear modulus, G, Young's modulus, E, and Poisson's ratio, ν. These equations are derived on the basis of the material being isotropic and linearly elastic, a situation that is highly unlikely in reality. At the very least the simplest form of anisotropy, that is , cross anisotropy, is likely to exist, and it is generally now accepted that at small strains soil behaviour is highly non-linear. However, in spite of this, case studies have shown that seismic surveys do provide useful information to the geotechnical engineer.

The dependence of P-wave velocities on the undrained bulk modulus means that from a geotechnical point of view, P-wave surveys will not generally be useful unless the ground is only partially saturated or the skeletal compressibility is less than or equal to the compressibility of the pore water, that is, B is much less than 1. The fact that S-wave velocities are independent of drained/undrained conditions and dependent on the shear modulus, makes S-wave surveys much more useful for the site investigation industry.

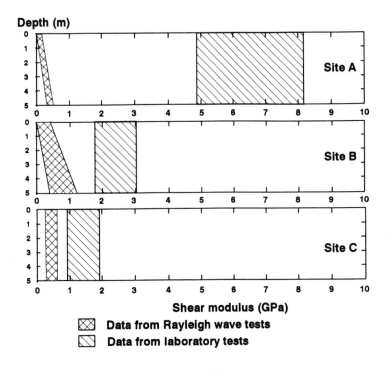

Figure 4 Comparisons of ground stiffness profiles derived from surface wave tests with small strain laboratory stiffness results (after Clayton, Gordon and Matthews, 1994).

In the Finchley case study it was found that the uncorrected laboratory stiffness values at 0.01 % strain were typically 50 % of G_{max} measured by the crosshole survey. At similar strain levels Tatsuoka and Shibuya (1992) compared operational stiffnesses back calculated from ground movements with those obtained from field geophysics and found that operational stiffnesses were about 70 % of G_{max}. A useful feature of stiffnesses obtained from seismic surveys is that they act as a benchmark against which stiffness values from other forms of testing, either *in situ* or laboratory based, can be evaluated and discarded if shown to be unrealistic.

Three seismic surveys were carried out at Finchley and the S wave velocities from all three surveys were about the same. This is not always the case as has been shown by Pinches and Thompson (1990). In a strongly layered site they found crosshole velocities, V_{vh}, to be consistently higher than downhole velocities, V_{hv}. This may be as a result of some of the detected crosshole waves being refracted and travelling along "fast" layers, which would lead to an overestimate of velocity. The velocity of the vertically travelling S waves in the downhole survey will be the average velocity of the formation. In the case of a homogeneous material V_{vh} should equal V_{hv}, hence the ratio of S-wave velocities measured by downhole and crosshole surveys will give an idea of any structural anisotropy present.

In the case of the weak fractured chalk the mass compressibility is lower than that of the intact rock. Here the geophysical survey takes into account the fractured nature of the rock mass, hence the very small strain stiffnesses thus obtained are much closer to operational stiffnesses than those measured on intact specimens in the laboratory. Small strain laboratory tests overestimate operational stiffness because the tests are taking place on intact samples, so the fissured nature of the material is not being taken into account (Clayton et al., 1994). The geophysical surveys therefore provide realistic values of stiffness that may possibly be used directly in geotechnical calculations.

Historically, reflection surveys, although widely used for surveys carried out over water, have not been used much by the civil engineering industry due to their poor performance at shallow depths. This seems likely to change in the future as a result of current research in the field of shallow reflection seismics. Refraction surveys have been widely used in engineering geophysical investigations but the technique suffers from, amongst other things, the inability to detect a soft layer underlying a stiff layer. This problem, as well as that of identifying a soft layer sandwiched between two stiff layers has been overcome by the crosshole and surface wave surveys; the added attraction of the surface wave survey being its non-intrusive nature, enabling surveys to be carried out quickly and economically. The main limitation of the surface wave technique is that its maximum depth of penetration is relatively low. The downhole survey can now be used in conjunction with the *in situ* cone test in the form of the SCPT, an economic way of obtaining S wave velocities without the need for any boreholes.

Laboratory samples, no matter how carefully they are taken will always be affected by the sampling process; seismic geophysical surveys measure stiffness at the right stress level but the wrong strain level. However, the margin by which the operational stiffness is overestimated in seismic surveys is relatively small and values of G_{max} provide an upper bound for stiffnesses values which can be used to check values determined by other methods, field or laboratory. Although not all types of geophysical survey are suitable for all ground investigations, in most cases at least one type of survey will yield valuable information, and therefore S-wave surveys in particular must have a role to play as an integrated part of stiffness investigations.

CONCLUSION

Seismic techniques are based on sound, well established elastic theory and moduli are determined at known, although very small, strains. Stiffness values thus obtained are measured *in situ*, the soil is undisturbed and the stiffnesses are determined for a large volume of soil (Robertson and Ferreira, 1993). Finite element analyses of deformations around civil engineering structures have shown that strain levels at working loads are small (often less than 0.1 %) (Jardine et al., 1986). In the case of soils G_{max} may be 2 or 3 times larger than operational stiffnesses, whilst in the case of weak, highly fractured rock the difference between G_{max} and operational stiffnesses may well be less. Given the relatively large uncertainties associated with other methods of determining soil stiffnesses, whether in the laboratory or using in-situ testing, S-wave seismic surveys must be regarded as a very useful site investigation technique.

ACKNOWLEDGEMENTS
Field work on London clay and chalk formed part of two research projects funded by E.P.S.R.C.
Seismic cone penetration testing at Finchley was carried out by B.R.E.

REFERENCES

ABBISS, C.P. (1981) Shear wave measurement of the elasticity of the ground. Géotechnique, vol.31, no.1, pp.91-104.

ABBISS, C.P. (1979) A comparison of the stiffness of the chalk at Mundford from a seismic survey and a large scale tank test. Géotechnique, vol.29, no.4, pp.461-468.

ALONSO, M., Finn, E.J. (1967) Fundamental University Physics. Volume 2 - Fields and Waves. Addison-Wesley Publishing Company.

BALDI, G., Bruzzi, D., Superbo, S., Battaglio, M., Jamiolkowski, M. (1988) Seismic cone in Po river sand. Proc. First International Symposium on Penetration Testing, ISOPT-1, Orlando. Ed. J. De Ruiter. Vol.2, pp.643-650.

BALLARD, R.F., Jr., McLean, F.G. (1975) Seismic field methods for *in situ* moduli. Proc. Conf. on *In situ* Measurement of Soil Properties. Speciality Conference of the Geotechnical Engineering Division A.S.C.E., Raleigh, North Carolina, vol.1, pp.121-150.

BURLAND, J.B. (1979) Contribution to discussion - Session 4. Proceedings 7th European Conference on Soil Mechanics and Foundation Engineering, Brighton 4:137.

CAMPANELLA, R.G., Baziw, E.J., Sully, J.P. (1989) Interpretation of seismic cone data using digital filtering techniques. Proc. XIIth Int. Conf. on Soil Mech. and Fnd. Engng. Rio de Janeiro, vol.1, pp.195-198.

CLAYTON, C.R.I., Gordon, M.A., Matthews, M.C. (1994) Measurement of stiffness of soils and weak rocks using small strain laboratory testing and field geophysics. International Symposium on Pre-failure Deformation Characteristics of Geomaterials, IS-Hokkaido, Sapporo. In press.

CLAYTON, C.R.I., Simons, N.E., Matthews, M.C. (1982) Site Investigation. Granada Publishing

DOBRIN, M.B. (1960) Introduction to geophysical prospecting. McGraw-Hill, New York.

GEOLOGICAL SOCIETY, Engineering Group Working Party. (1988) Engineering Geophysics. Quarterly Journal of Engineering Geology, vol.21, pp.207-271.

HERTWIG, VON A. (1931) Die Dynamische Bodenuntersuchungsverfahren. Der Bauingenieur, vol.12, pp.457-461, 476-477.

HVORSLEV, M.J. (1949) Subsurface exploration and sampling of soils for civil engineering purposes. The Waterways Experiment Station, Vicksberg, Mississippi.

JARDINE, R.J., Potts, D.M., Fourie, A.B., Burland, J.B. (1986) Studies of the influence of non-linear stress-strain characteristics in soil-structure interaction. Géotechnique, vol.36, no.3, pp.377-396.

JARDINE, R.J., Symes, M.J., Burland, J.B. (1984) The measurement of soil stiffness in the triaxial apparatus. Géotechnique, vol.34, no.3, pp.323-340.

JONES, R. (1958) *In situ* measurement of the dynamic properties of soil by vibration methods. Géotechnique, vol.8, no.1, pp.1-21.

JONES, R. (1962) Surface wave technique for measuring the elastic properties and thickness of roads: theoretical development. Brit. J. Appl. Physics, vol.13, no.1, pp.21-29.

LAVERGNE, M. (1989) Seismic Methods, Graham and Trotman Limited, Éditions Technip.

LEE, S.H.H. (1985) Investigation of low amplitude shear wave velocity in anisotropic material. Ph.D. Thesis, Univ. of Texas at Austin.

MAIR, R.J. (1992) Developments in Geotechnical Engineering Research. Application to Tunnels and Deep Excavations. Unwin Memorial Lecture 1992. Proc. Instn. Civ. Engrs., Civ. Engng., vol.97, issue 1, pp.27-41.

MANCUSO, C., Simonelli, A.L., Vinale, F. (1989) Numerical analysis of *in situ* S wave measurements. Proc. XIIth Int. Conf. on Soil Mech. and Fnd. Engng. Rio de Janeiro, vol.1, pp.277-280.

MATTHEWS, M.C. (1993) The mass compressibility of fractured chalk. PhD Thesis, University of Surrey.

PINCHES, G.M., Thompson, R.P. (1990) Crosshole and downhole seismic surveys in the UK Trias and Lias. Proc. 24th Annual Conference of the Engineering Group of the Geological Society, Field Testing in Engineering Geology, Sunderland. Ed. F.G. Bell, M.G. Culshaw, J.C. Cripps, J.R. Coffey. Geological Society Engineering Geology Special Publication no.6, pp299-307.

ROBERTSON, P.K., Ferreira, R.S. (1992) Seismic and pressuremeter testing to determine soil modulus. Predictive soil mechanics.Wroth Memorial Symposium, ed. G.T. Houlsby, A.N. Schofield, Oxford, pp.562-580.

ROBERTSON, P.K., Campanella, R.G., Gillespie, D., Rice, A. (1986) Seismic CPT to measure *in situ* shear wave velocity. Journal of Geotechnical Engineering, Proc. A.S.C.E., vol.112, no.8, pp.791-803.

ROESLER, S.K. (1979) Anisotropic shear modulus due to stress anisotropy. Journal of the Geotechnical Engineering Division, Proc. A.S.C.E., vol.105, no.GT7, pp.871-880.

SANCHEZ-SALINERO, I., Roesset, J.M., Stokoe, K.H., II. (1986) Analytical studies of body wave propagation and attenuation. Report GR 86-15, University of Texas at Austin.

STOKOE, K.H., II, Rix, G.J., Nazarian, S. (1989) *In situ* seismic testing with surface waves. Proc. XIIth Int. Conf. on Soil Mech. and Fnd. Engng. Rio de Janeiro, vol.1, pp.331-334.

STOKOE, K.H., II, Woods, R.D. (1972) *In situ* shear wave velocity by cross hole method. Journal of the Soil Mechanics and Foundations Division, Proc. A.S.C.E., vol.98, no.SM5, pp.443-460.

TATSUOKA, F., Shibuya, S. (1992) Deformation characteristics of soils and rocks from field and laboratory tests. Report of the Institute of Industrial Science, University of Tokyo, vol.37, no.1.

TELFORD, W.M., Geldart, L.P., Sheriff, R.E., Keys, D.A. (1976) Applied Geophysics. First edition, Cambridge University Press.

TERZAGHI, K. (1943) Theoretical Soil Mechanics. John Wiley and Sons, New York.

TIMOSHENKO, S., Goodier, J.N. (1951) Theory of Elasticity. Second Edition, McGraw-Hill Book Company, Inc.

WHELAN, P.M., Hodgson, M.J. (1978) Essential Principles of Physics. John Murray.

The use of acoustic emission to monitor the stability of soil slopes

DIXON N., HILL R. & KAVANAGH J.
The Nottingham Trent University, Nottingham, England

INTRODUCTION

At present the standard method for assessing the stability of soil slopes is measuring ground deformations. Whether using surface survey markers, or inclinometer tubes installed through the potentially unstable soil mass, the aim is to detect movements as early as possible. Unfortunately the magnitude of the pre-failure deformations which are of interest are often of the same order as the accuracies of the above methods. Therefore, in a large number of cases measurements are taken over a period of time to obtain trends, thus enabling the certainty of ground movement to be established.

Of particular interest are slopes formed in strain softening soils, typically plastic clays and shales, which can experience progressive failure and hence undergo deformation prior to collapse. In these types of material shear deformations in the order of a few millimetres may be sufficient to reduce substantially the shear strength to post peak values, and thus to result in failure. The earlier decreasing stability can be detected the better the chance that remedial measures can be carried out to arrest the ground movements and restore the required factor of safety. An alternative method which could be used to provide an early warning of instability, and which does not necessarily require measurements taken over a period of time would be of significant value.

This paper provides information on an EPSRC funded research project which is presently in progress to develop such a system using measurements of acoustic emission (AE) to detect and assess pre-failure deformations. The basic AE instrumentation and data processing system are outlined, and information is provided on the field monitoring procedures presently being developed. A full scale field trial is described, and the initial results are presented.

BACKGROUND INFORMATION

All materials when stressed emit minute bursts of acoustic emission energy (also called microseismic energy or sub-audible noise). Acoustic emission in soils is believed to be generated from inter-particle friction during the process of shear deformation. Using suitable transducers various properties of the AE can be measured and used to provide information on the stress state of the soil mass.

The main body of research into AE applications for soil assessment was carried out in the United States of America in the 1970's and early 1980's (Koerner et al. 1981), and more recently in Japan (Chichibu *et al.* 1989). A number of these investigations were undertaken to assess fundamental laboratory and field aspects of AE in geotechnical engineering, with a particular emphasis on slope stability monitoring. This work demonstrated clearly that deforming soil produces detectable AE, and that the levels of emission are directly related to the stress state of the soil. In addition, a number of the slope investigations found that the AE monitoring indicated pre-failure deformations before they could be detected using traditional instrumentation.

However, at the present time only the following qualitative system of assessment is available (after Koerner *et al.* 1981):

a)   Soil masses that do not generate AE are probably not deforming and are therefore stable.

b)   Soil masses that generate moderate levels of AE are deforming slightly and are considered to be marginally stable.

c)   Soil masses that generate high levels of AE are deforming substantially and are to be considered unstable.

d)   Soil masses that generate very high levels of AE are undergoing large deformations and can be considered to have failed.

It is the main aim of the present investigation to develop more rigorous methods for site assessment of AE. The intention is that this will lead to a general purpose monitoring technique for early detection of soil slope instability.

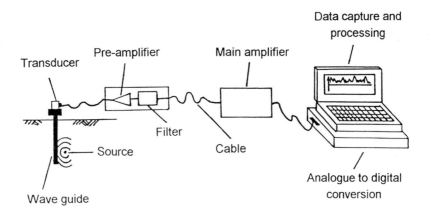

Fig. 1 Schematic of AE monitoring system

AE INSTRUMENTATION

The basic components of an AE monitoring system are shown diagrammatically in Figure 1, and discussed briefly below.

AE senor

A transducer provides the means to convert the mechanical AE wave to a variable electrical voltage, which can be readily processed, analysed and stored. For this investigation a transducer with a resonant frequency of approximately 30 kHz has been selected. This is a compromise response range. If lower frequencies are monitored interference from spurious background noise is encountered. However, if higher frequencies are used problems occur due to the loss of data resulting from attenuation.

Wave guide

The combination of low magnitude and relatively high attenuation of AE signals in soil means that it is not possible to monitor a large volume without providing a low resistance path. It has become standard practice to use metal wave guides to conduct signals from within the soil mass to the AE sensor (Lord *et al.* 1982). Wave guides form a firm mounting location for the transducer at ground level, and if hollow they can provide access into the centre of the soil mass for subsurface monitoring.

Signal Conditioning

A pre-amplifier is used to increase the low level signals from the sensor for transmission through the cable and filters to the main amplifier. A band pass filter is used to restrict the measurement of data to the region where the sensor is most sensitive. The main amplifier which is placed after the filter increases the signal level ready for processing.

Data Capture and Processing

As a direct result of recent increases in computer power and portability, it is now possible to employ digital data capture and real time processing techniques on site. For this study a 486DX 33 MHz laptop with a DAS 1402 analogue to digital converter board, capable of sampling at 100 kHz, have been used. In order to make significant savings on data capture rates, and hence on the volume of data to be stored and processed, the envelope to the upper part of the signal is used rather than the full waveform. The analogue values which define this envelope are converted to digital values and saved to memory. The reduced sampling rates result in a less energy hungry A/D board being suitable, which is of particular importance for a field system. Using this sophisticated system allows direct comparisons between AE phenomena captured on site and under laboratory conditions (Dixon *et al.* 1994).

DATA INTERPRETATION

A range of assessment techniques are available for interpreting AE data. Figure 2 shows a typical AE event and defines a range of parameters which can be used to characterise the signal. The qualitative system of assessment outlined earlier is based on measurement of counts (ring down counts) or events. This is a rather crude method of assessment because not only do different soils generate widely varying levels of detectable AE for the same stress state, considerable problems are encountered in quantifying the influence of the measuring system on the number of counts/events detected. Laboratory studies carried out by Jackson (1986) and Garga & Chichibu (1990) have demonstrated that a more rigorous framework for the assessment of AE can be based on the following criteria.

a) *AE energy per event,* which has been shown to be influenced by strain level. This is proportional to the area under the signal and hence is related to the peak amplitude and duration.

b) *Peak amplitude distributions,* which vary with the mechanism of failure (i.e. mode of particle deformation) and therefore with the magnitude of strain.

c) *Number of counts per event,* which appears to be strain dependant, and is related to frequency and duration of the event.

The above indicate that it may be feasible to use the waveform to distinguish between pre- and post-peak shearing behaviour. It is possible that AE monitoring could be used to provide a direct indication of the stress state of the soil forming a potentially unstable slope.

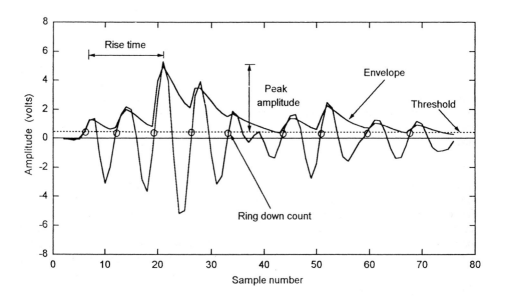

Fig. 2 Typical AE event waveform and envelope with main parameters defined

FIELD MONITORING SYSTEM
Wave Guide Design

Wave guides are normally made from steel as it is readily available, can be easily fabricated to the required cross-section and length, and is an excellent transmitter of AE (i.e. low attenuation of signals). A number of valid wave guide configurations are available. For small slopes it may be possible to drive wave guides to the required depth. However for larger slopes it will be necessary to install the steel rods or tubing in a pre-drilled borehole. This latter system leads to problems selecting the backfill material. Two possible systems are outlined below:

1. *Passive wave guide* - Designed to ensure that the installation does not introduce sources of AE into the soil. Any recorded AE are assumed to emanate from the deforming host soil. Driven systems are likely to be passive as a result of the wave guide being in direct contact with the soil. For systems located in boreholes the annulus around the wave guide has to be backfilled with low AE activity (i.e. quiet) material.

2. *Active wave guide* - As it may prove difficult to obtain quality AE data from quieter cohesive materials, it is proposed that noisy wave guide systems could be used. This is achieved by backfilling the annulus with granular soil. When the host soil deforms, the column of high AE response material will also deform thus producing AE data. Although this data will not relate directly to the stress sate of the host soil, it may be possible to calibrate the system, such that recorded AE behaviour can be related to the magnitude of the general ground deformations.

Wave guide location
It is desirable for the wave guide to be installed to a sufficient depth to ensure that it penetrates any shear surfaces which form during the onset of failure. The main shear zones would be expected to produce high levels of AE. Information related to the position of a slip surface could be obtained by lowering a sensor down a hollow wave guide (i.e. a hydrophone). However, wave guides which do not penetrate the main shear surface can still provide useful information. The large majority of failure surfaces are non-circular and therefore internal shearing occurs within the slide mass. Guides positioned both through and above the main shear surface are capable of transmitting AE from this activity.

Monitoring Philosophy
The mode and frequency of field monitoring require careful consideration. The cost of the prototype AE instrumentation under development is such that it is not practical to provide a dedicated system for each problem slope. Therefore, a monitoring strategy must be based on taking measurements at time intervals. If count/event totals or rates are used, there is the possibility that the most active periods of stress change could be missed. However, there is evidence that the process of stress redistribution during progressive failure, continues to produce AE after such an event occurs. Although measurement of this lower level of activity may result in an overestimation of stability.

A monitoring system based on the assessment of AE event characteristics is better suited for indicating the true condition of the slope. Even if an alteration in stress state is not monitored directly, the characteristics of the AE signal may indicate the new stress conditions within the soil mass, or in particular within the main shear zone.

AE FIELD TRIAL
Introduction
Evaluation of any new instrumentation must include an assessment of it's performance for a range of site conditions. In addition, it has to be compared directly with the established techniques which it is intended to complement. In this instance a comparison is being made between the AE technique and deformation monitoring methods. Instrument installation and monitoring of the first test site are described below. An interpretation of the initial results is provided, and a brief description of the next phase of the field work is also included.

Site Description
The primary test site is situated on the north eastern coast of England at Cowden in North Humberside. The cliffs at this location are 20 metres high and are formed of stiff cohesive till. Failure of these slopes takes the form of rotational sliding which is triggered by marine erosion of the toe. This site was chosen for a number of reasons.

a) Consistently high erosion rates meant that a slope failure was expected to occur in the near future.

b) The Building Research Establishment's coastal cliff test section is situated a hundred metres from the trial site. A slope failure at this location was studied in detail during the early 1980's (Butcher 1991). This has provided information on the failure mechanism of the cliffs, and on the engineering properties of the soil.

c) The site is situated within the boundaries of an operational RAF base, hence vandalism of the instruments was not expected to be a problem.

d) The soil has a high sand content, and hence would be expected to be relatively 'noisy'.

e) Man made background noise would be low as the site is remote.

It was intended to instrument the slope prior to an active winter season (i.e. storm conditions could be expected to produce increased marine erosion). Monitoring of both ground deformations and AE were to be carried out through the winter period, hopefully during the onset of failure.

Instrument Installation
A total of twelve 50 mm diameter steel tubing wave guides (WG) were installed at the locations shown on Figure 3 during September 1993. The lengths of WG required to ensure penetration of all potential shear zones dictated that driven systems could not be used. A number of the WG penetrated to a depth of 20.5 metres below the slope crest in order to intersect any basal shear surfaces (Butcher 1991). The remainder were installed to depths of either 6 or 12 metres. Each WG was installed in a 150 mm diameter borehole, with the annulus between the WG and in situ soil backfilled with either bentonite grout, medium sand or fine gravel. The bentonite was expected to provide low AE when deformed (passive system). Both the sand and gravel were expected to produce relatively high AE (active systems). The boreholes were formed using the continuous flight auguring technique which is ideal for installing instruments in cohesive deposits. In this instance the formation of a constant diameter annulus of backfill material was an important outcome of employing this method.

Groups of WG, with different backfill material, were positioned along the cliff edge and also back from the slope crest. Wave guides 7 and 12 were located back from the slope in order to provide a measure of background noise, and for use in assessing the sensitivity of the AE instrumentation.

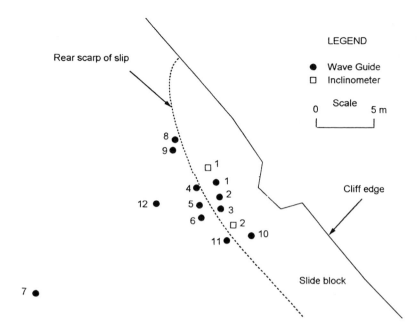

Fig. 3 Plan showing the instrument locations at the Cowden test site

Two aluminium inclinometer casings were installed to a depth of 20.5 metres in 150 mm diameter boreholes. Bentonite/cement grout (4:1) was used as backfill. Aluminium was used in preference to the usual plastic tubing because it was considered that the inclinometer casing might be used as wave guides (i.e. AE attenuation is high in plastic).

Monitoring

The first few sets of inclinometer readings indicated that pre-failure slope movements were occurring. Although AE monitoring was being carried out during this period, the quality of the data and the number of readings taken are inadequate for assessing whether the AE also detected these initial pre-failure deformations. The field AE system was still under development at this early stage. However, monitoring of the ground deformations and AE was continued through the winter of 1993/94 and this has provided useful data. The method used to monitor each WG was to sample the AE signal envelope over a three minute period.

Major slope movements occurred in early April 1994 making the inclinometer casings unusable. Figure 3 shows the position of the landslide rear scarp. It can be seen that it passes through the middle of the instrument array, with some of the WG penetrating the slide block, and a number of others close to, but behind, the main shear surface! The main slide mass will continue to move further down the slope during the next several years. It is intended to monitor the WG during this period. The site still provides an excellent opportunity for assessing and calibrating the field AE system.

Initial Results

Initial interpretation of the AE data has been based on assessment of the area under the signal envelope, which is related to the AE energy. A section of typical signal is shown in Figure 4. It can be seen that events often overlap making it difficult to process data on an individual event basis. Therefore, the results presented here relate to the entire signal recorded for each 3 minute period. Results from WG 7 and 12 have shown that relatively high levels of background AE are detected by the system. In order to remove this background data, each set of readings has been corrected by subtracting the area under the curve obtained from WG 12 for that particular day.

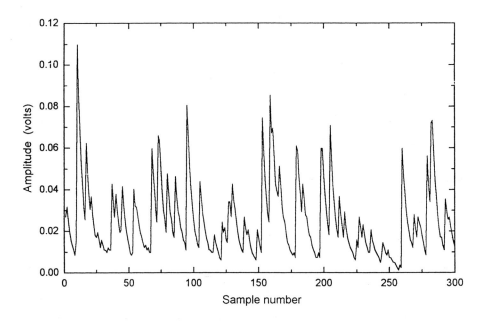

Fig. 4 Typical AE signal envelope obtained from the Cowden site

The main problem encountered when assessing the stability of soil slopes is the sporadic nature of ground deformations. The mechanism of progressive failure is unlikely to result in a smooth increase in the rate of deformation. The 'slip/stick' mode of failure is well known (i.e. active periods of movement can be followed by relatively inactive periods). This type of behaviour would be expected at Cowden because the slope is being destabilised by sporadic marine erosion activity. For this reason the best way of assessing the field data is to use cumulative changes in the main parameters as this will tend to smooth out local variations in the data. Figure 5 shows the cumulative displacement measured by inclinometer casing I2 at a depth of 3.0 metres below ground level, plotted against the number of days since the start of monitoring. It can be seen that the rate of displacement was increasing up to the time when the major failure movements occurred. It should be noted that the rate of displacement between consecutive readings, as indicated by the slope of the line between the points, does not always increase. This data could be misleading if used in isolation.

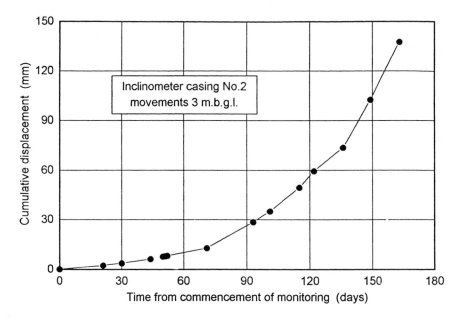

Fig. 5 Cumulative displacement for inclinometer casing I2

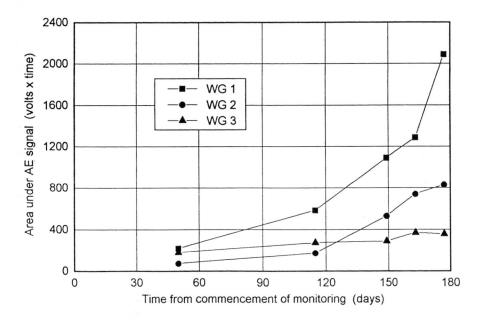

Fig. 6 Corrected cumulative area under AE signals from WGs 1, 2 and 3

Figure 6 shows the cumulative area under AE signals from WGs 1, 2 and 3 after background noise has been removed, plotted against the time since start of monitoring. All three WG are 20.5 metres deep and penetrate through the slide mass. The general trend is an increase in the rate of change of cumulative area (i.e. the graph for WG 1 has a concave shape), which is comparable to the plot of increasing displacement with time (Figure 5). This is not surprising as the area provides an indication of AE energy, which would be expected to increase as the rate of displacement increases. Figure 6 also demonstrates the different behaviour of the three backfill materials. The grout backfill (WG 3) emits the least AE, gravel (WG 2) appears to be the next active and the sand backfill (WG 1) clearly produces the highest levels of AE. These findings are as expected, and confirm that it is deformation of the WG backfill materials which are being detected rather than the host soil.

Assessment of the AE data based on cumulative area under the signal has produced encouraging results. Pre-failure deformations resulted in AE being produced by the active WG systems. In addition, the cumulative area under the signal increased with the rate of deformation as failure was approached. However, due to the fact that the slope started to move before the AE monitoring system was fully operational, it is still not known whether AE can detect the pre-failure deformations before the inclinometer readings.

The next phase of data interpretation is to analyse the peak amplitude distribution of each data set in order to assess whether this gives an indication of the increasing rates of deformation.

Further Field Work
A second field trial is presently in progress in an attempt to monitor AE during the initial onset of failure. A range of WG and two inclinometer casings have been installed behind a stable slope formed in Gault Clay. The slope will be destabilised by excavating the toe, and AE and ground deformation monitoring are to be carried out before, during and after failure. It is hoped that this work will show whether AE can be detected before significant ground deformations are measured. The final phase of the field work is to monitor a wide range of potentially unstable slopes formed in different soils, in order to assess the AE technique for use as a general method for investigating slope stability.

SUMMARY
Acoustic emission techniques for assessing shear deformation of soils have been established by international research carried out over the past 20 years. The most promising area relates to the use of signal characteristics to provide information on the stress state of the soil. The recent rapid increases in computer power and portability, have resulted in the possibility of applying these techniques for field monitoring. Field investigations are in progress to develop an AE measuring system and to compare it's performance with traditional deformation monitoring methods.

Initial results from the first test site have demonstrated that active wave guide systems can be used to detect pre-failure deformations. The energy, indicated by the area under the signal envelope, increases as the rate of deformation increases towards failure. A second field trial is presently in progress to assess whether AE monitoring can detect pre-failure ground deformations before they can be measured using traditional techniques. This is an exciting and relatively new area of research which promises to lead to the development of improved instrumentation techniques for the assessment of soil slope stability.

ACKNOWLEDGEMENTS
A grant from the Engineering and Physical Sciences Research Council has made the field work possible. Mr Kavanagh is supported financially by The Nottingham Trent University and EPSRC. Thanks are due to the Building Research Establishment and the staff at RAF Cowden for allowing access to the test site, Kingston University for the loan of the drill rig used to install the instrumentation, and John Cole for his help during the field work.

REFERENCES
Butcher, A.P. (1991). The observation and analysis of a failure in a cliff of glacial till at Cowden, Holdernes. *Proc. Int. Conf. on Slope Stability*, ICE, Isle of Wight, April 1991, 271-276.

Chichibu, A., Nakamura, K.J.M. & Kamata, M. (1989). Acoustic emission characteristics of unstable slopes. *Journal of Acoustic Emission*, 8, 4, 107-112.

Dixon, N., Hill, R. & Kavanagh, J. (1994). Acoustic emission techniques for assessing deformations in soils. *Int. Symp. on Pre-failure Deformation Characteristics of Geomaterials*, Hokkaido University, Japan, (In Press).

Garga, V.K. & Chichibu, A. (1990). A study of AE parameters and shear strength of sand. *Progress in Acoustic Emission V*, Japanese Soc. for NDI, 129-136.

Jackson, A.L. (1986). *Acoustic emission and yield in sand.* PhD Thesis, University of Aston.

Koerner, R.M., McCabe, W.M. & Lord, A.E. (1981). Acoustic emission behaviour and monitoring of soils. *Acoustic Emission in Geotechnical Practice, ASTM STP 750*, American Soc. for Testing and Materials, 93-141.

Lord, A.E., Fisk, C.L. & Koerner, R.M. (1982). Utilisation of steel rods as AE wave guides. *J. Geotechnical Engng. Div.*, Proc. ASCE, 108.

The use of slimline logging techniques in engineering ground investigation

D. PASCALL, A. D. J. LAW and D. C. CURTIS Arup Geotechnics

Slimline Logging Techniques, which have had widespread usage in the field of exploration geology, have a number of important applications in engineering ground investigations. This paper describes the successful use of such techniques in Triassic sandstones as part of a ground investigation for a possible nuclear power station development at Sellafield, Cumbria.

The aim of the investigation was to provide a clear understanding of the geological structure of the site allowing provision of data for seismic hazard assessment and to provide geotechnical parameters for future structural design. During the planning stages of the investigation the importance of gaining cross-site geological correlation was recognised and Slimline Techniques were chosen as the most appropriate means of providing this given the relatively homogenous and monotonous nature of the sandstones.

The paper describes the investigation and the various Slimline Techniques employed and discusses the relative usefulness of each with respect to the requirements of the investigation. The paper demonstrates a successful correlation between the geophysical signatures obtained from these techniques and the detailed stratigraphy and allowed correlation of sandstone units across the site. The correlation proved to be a key element in the understanding of the geological structure and also allowed confirmation of the recently proposed geological division of the St Bees sandstone into two units, the upper part of which is now known as the Calder Sandstone.

Also described is the use of these techniques in the assessment of basic rock properties such as bulk density and porosity and the estimation of dynamic properties based upon measured seismic velocities. Discussion is made of how these results compared with cross hole survey information.

INTRODUCTION

Between 1991 and 1994 British Nuclear Fuels Plc (BNFL) have arranged for a number of studies to be carried out for siting a nuclear power station to the north of their existing Sellafield complex. Ove Arup and Partners (Arup) were commissioned by BNFL to carry out the Geotechnical Study. The site encompassing about 5 square kilometres comprises three topographical units, the beach and foreshore, the floodplain of the River Ehen and the higher ground to the east of the disused railway line that runs north along the edge of the floodplain. (See Figure 1).

As part of the Geotechnical Study a site investigation was a requirement with the aim of establishing a clear understanding of the geological structure, allowing provision of data for seismic hazard assessment by others, and to obtain geotechnical parameters for future structural design. The investigation which was carried out in two stages between December 1992 and September 1993 included 21 km of seismic surveying, shallow cable percussion boreholes, 1800m of core drilling to depths of 300m, two suites of crosshole and downhole geophysics, cone penetration tests and trial pits. In all the deep cored boreholes geophysical logging was performed using slimline sondes. It is the use of these slimline techniques for stratagraphical correlation that forms the main subject of this paper. In addition the basic rock properties assessed from the slimline logging will be discussed together with the dynamic properties estimated from the measured seismic velocities. A comparison will be made with dynamic properties obtained from the crosshole geophysics.

Advances in site investigation practice. Thomas Telford, London, 1996

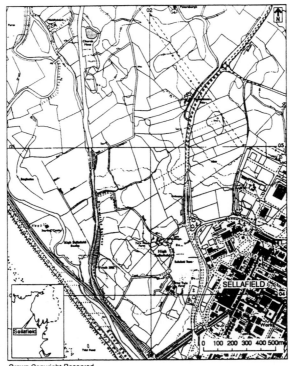

Crown Copyright Reserved

FIGURE 1 THE SITE

GROUND CONDITIONS

The ground conditions comprise a variable thickness of glacial drift over a thick succession of relatively featureless red sandstones. During the planning stages of the investigation the importance of gaining cross-site geological correlation was recognised and slimline techniques were chosen as the most appropriate means of producing this given the relatively homogenous and monotonous nature of the sandstones. Up until recently the sandstones had been termed the St. Bees Sandstone. However, the British Geological Survey (BGS) working on the neighbouring NIREX deep repository, have proposed a geological division of the St. Bees Sandstone into two units. The upper part is now known as the Calder Sandstone. Recognition of the Calder/St. Bees interface proved to be a key element in understanding the geological structure.

SLIMLINE LOGGING TOOLS

The following BPB geophysical sondes were used:-

Sonde code	Measurements
*DD3	Gamma Ray, Linear Density, Bed Resolution Density
*NN1	Gamma Ray, Neutron Porosity
*MS1	Multi-channel Sonic
*RO4	Focused Electric
*RS1	Spontaneous Potential
FT1	Fluid Conductivity Temperature
*DV1	3 Arm Dipmeter Analysis
SS1	Sidewall Wavetrain Sonic
HAV	High Accuracy Verticality
DD2	Cement Bond

Of these sondes those marked with an asterisk were run in all holes drilled for stratigraphical correlation.

The temperature sonde was run in early boreholes but as it only showed a constant temperature since equilibrium conditions had not been achieved when logging took place within 24 hours of the completion of drilling operations. The use of the temperature log was therefore discontinued. The verticality and the cement bond sondes were used in the two suites of boreholes for crosshole seismic measurements, and the sidewall wavetrain sonic for the mechanical properties of the rock.

The DD3 tool contains a natural gamma ray detector at one end. The capability for density measurement is achieved using a gamma ray source located at the lower tip of the tool. Gamma rays omitted from this source are back scattered from the borehole wall and give a measure of bulk density when received at gamma ray detectors placed at differing distances along the tool from the source. The bed resolution detector is the closest at 150mm whereas the long spaced density detector at 480mm from the source gives an averaged density achieved with deeper penetration. The tool is pressed against the borehole wall by a single calliper arm which measures variations in the borehole diameter. The log presents gamma ray intensity in API (American Petroleum Institute) units.

The NN1 tool includes a natural gamma ray detector at the top and a radiation source at the base. This source is Americum/Beryllium which emits fast neutrons. These are absorbed in the surrounding strata by hydrogen atoms which themselves emit gamma rays in proportion to absorption. The gamma ray detectors are spaced at short (250mm) and long (450mm) distances from the source. The ratio of these short and long spaced logs is used to estimate porosity which is plotted in the output on a percentage basis.

The MS1, multi channel sonic sonde measures sonic transit time between a transmitter and four receivers spaced at distances between 600mm and 1200mm from the transmitter. The P or compression wave velocities are calculated from the difference in arrival time at the four receivers and presented for 200mm, 400mm and 600mm transit intervals. The longer spacing gives the best average for the formation velocity while the smaller spacings provide a higher resolution of bed boundaries and structures.

The RO4 focused electric sonde measures the ground resistivity of the formation. The sonde is designed such that a 100mm long sensing electrode is surrounded above and below by two longer guard electrodes. These guard electrodes fulfil the function of directing the current from the sensing electrode perpendicular to the axis of the sonde achieving a high resolution. The log shows resistivity on a logarithmic scale.

The RS1 spontaneous potential tool records the natural electrical potential developed between the borehole fluid and the surrounding formations. Spontaneous potentials are generated by electrochemical electromotive force at the junction of dissimilar materials and by electrokinetic e.m.f's developed when fluid moves through a permeable stratum.

The DV1 dipmeter is a three armed device designed to measure formation dip. It consists of three calliper arms which push micro resistivity pads against the borehole wall. The resistivity curve developed by each pad is compared over a fixed depth interval (approximately 2m) and these comparisons are reported at regular increments of depth, and a dip direction and azimuth is calculated.

The SS1 sidewall sonic sonde measures the primary and shear wave velocities of the formation. Two piezoelectric transducers are clamped to the borehole wall. A sonic pulse is fired from the upper transducer and picked up on the other. The waveform is photographed on an oscilloscope and the P and S or shear wave velocities calculated.

THE TRIASSIC SANDSTONES

The Triassic sandstones in Cumbria have recently been described (Jones and Ambrose in print) in terms of three sandstone formations which together constitute the Sherwood Sandstone Group. The lower St Bees Sandstone comprises primarily of fluviatile channel sandstones. The upper part of the St Bees Sandstone has been renamed The Calder Sandstone which is an alternating sequence of fluviatile and aeolian sandstones. The uppermost sandstone formation, The Ormskirk Sandstone, is regarded by Jones and Ambrose (loc cit) as being of aeolian origin. Offshore in the East Irish Sea Basin (Jackson and Mulholland 1993, Meadows and Beach 1993) the Ormskirk is a mixed fluviatile and aeolian sandstone. Offshore a prominent seismic reflector, the brown reflector, is recognised at the top of the lower part of the St Bees Sandstone with another ephemeral reflector, the yellow, above it. The sandstone above the brown reflector is recognised as a mixed aeolian and fluviatile association.

On-shore the Ormskirk Sandstone outcrops to the southwest of the proposed power station site, and only the lower part of the Calder Sandstone and the upper part of the St Bees Sandstone were cored in the boreholes for the Geotechnical Study.

The fluviatile channel sandstones are characterised by erosively based units up to 5m thick consisting of multistorey and multilateral sand bodies with high width to depth ratios. Some of the channel units display fining upwards sequences with channel tops formed from silty sandstone and sometimes with mudstones. The St Bees sandstone is fine to medium grained micaceous sandstone. Mudstone rip-up clasts are very common and can form a clast conglomerate at channel bases, and are also aligned along the surfaces of foreset beds. The sandstones are cross-bedded in the basal part of each unit which passes up into flat-laminated units which are frequently highly micaceous. The fluviatile units within the Calder Sandstone are medium to coarse grained. In the Calder, fining upwards units are common with the coarse well rounded aeolian sand at the base and occasional mudstone layers at the top. Generally the sandstone is less micaceous than the St Bees but like that formation contains mudstone rip-up clasts.

The aeolian sandstones consist of fine to coarse grained sandstone frequently laminated with alternating fine grained and medium/coarse grained layers. Much of the sandstone is a weak fine grained sandstone in which it was difficult to distinguish the sedimentary structure within the core. Jones and Ambrose (loc cit) have identified dune facies and a variety of interdune facies within this aeolian sequence. The aeolian facies is cross bedded with differing grain size lamination on the foresets. The interdune facies are flat-bedded or low angle cross-bedded sandstones, in part laminated but this is less defined than the laminations in the dune sands.

In carrying out the lithological logging of the Triassic sandstone the core loggers were required to identify the sedimentary features within the sandstones. These sedimentary features were presented separately to the engineering log as a graphical lithological log. Particular attention was paid to the identification of channel bases, micaceous horizons and concentrations of mudstone rip-up clasts. This could be achieved with a high degree of consistency within the fluviatile sandstones, but bed features could not be reliably identified within the aeolian facies, and for these the lithological log only indicates grain size variation.

COMPOSITE LITHOLOGICAL AND GEOPHYSICAL SONDE LOGS

The basic geophysical sonde logs provided by BPB were printed at the same 1:50 scale as the borehole logs. At this scale they are difficult to work with, but it became apparent during the fieldwork that the focused electric log produced a stronger correlation with the logged sedimentary units than either the γ ray or sonic logs. This is illustrated in Figure 2 which shows the γ ray, focused electric and graphical lithological log for typical sections of fluviatile Calder and St Bees Sandstones.

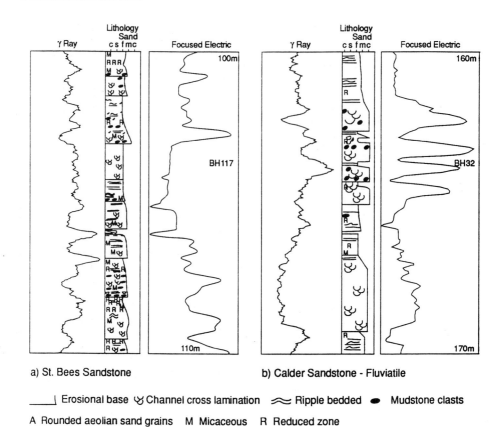

a) St. Bees Sandstone　　　　　　　b) Calder Sandstone - Fluviatile

‿‿| Erosional base　⅋ Channel cross lamination　≈ Ripple bedded　● Mudstone clasts

A Rounded aeolian sand grains　M Micaceous　R Reduced zone

FIGURE 2 COMPARISON OF SONDE AND LITHOLOGICAL LOGS

For report presentation the most useful sonde logs were presented at a compressed scale along with a simplified lithological log that showed the identified sedimentary units, mudstone bands and clasts and aeolian sand sequences (A). Figure 3 shows an example of such a log for the borehole that penetrated the greatest thickness of Calder Sandstone and which was also taken the deepest at 300m. This log shows the clear distinction between the geophysical signatures of the aeolian and fluviatile sandstones. The identified sandstone units in the fluviatile sandstones are clearly picked up by the focused electric though this is best illustrated by the juxtaposition of the lithological and focused electric log for the borehole that penetrated the thickest St Bees sequence (Figure 4).

During the site work logs were available for the NIREX boreholes to the south of Sellafield (NIREX 1992). From these it was apparent that the average gamma ray count was higher in the St Bees Sandstone than in the overlying Calder. In the early boreholes the geophysical sonde logging was carried out incrementally, and it was apparent from the log run to 200m (Figure 3) that the fluviatile sandstones, by then encountered, could not be the St Bees. Although they displayed a more vigorous gamma ray response than the overlying aeolian sandstones the average gamma count was still lower than would be expected for the St Bees. The borehole was then continued to penetrate the lower sandstones as it appeared likely that the fluviatile sandstones first encountered at 150m depth probably corresponded to the yellow seismic reflector even though the NIREX data suggested that this should be lower down.

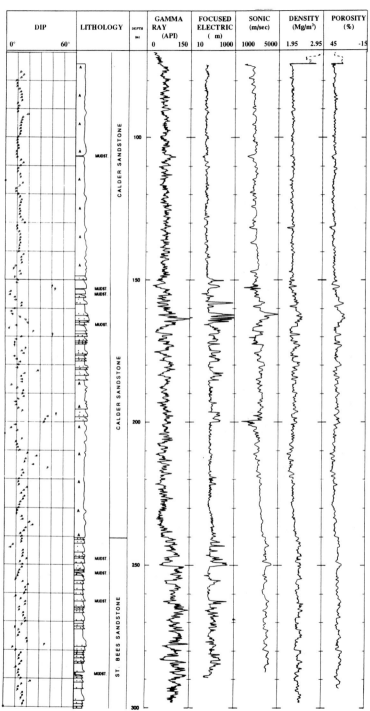

FIGURE 3 SONDE AND LITHOLOGICAL LOG FOR DEEPEST BOREHOLE

It can be seen from the composite log (Figure 3) that along with an increase in gamma ray count there is an increase in density and sonic velocity in the underlying St Bees, although these parameters had been gradually increasing in the basal Calder.

Within the aeolian strata of the Calder, the response on all logs is more muted than in the fluviatile facies. This is particularly the case with the focused electric log. This resistivity log is very peaky in the fluviatile sandstones. For these within the Calder there is a very good correlation between high resistivity peaks and increased sonic velocity and density, and with a decrease in porosity. This correspondence of peaks is not so obvious within the St Bees, but here as shown in Figure 4 there is a good correspondence between resistivity peaks and the identified sedimentary units.

Low resistivity results from the presence of clay material and mica-rich horizons within the sandstones. The clay can either be present as discrete bands or as mudstone clasts within the channel sandstones. Mudstone bands generally correspond with sharp peaks in the gamma-ray log and sharp falls in the resistivity log, as in general the resistivity log takes an inverse form to that of the gamma ray log. The gamma ray log is however much more sensitive to material close to the borehole than is the focused electric log, which sees a wider cylinder of rock. The gamma ray is therefore sensitive to emissions from mudstone clasts lying on foreset beds within the crossbedded channel sandstone. This is one reason why the focused electric log is more responsive to the overall sedimentary structure than the gamma ray log.

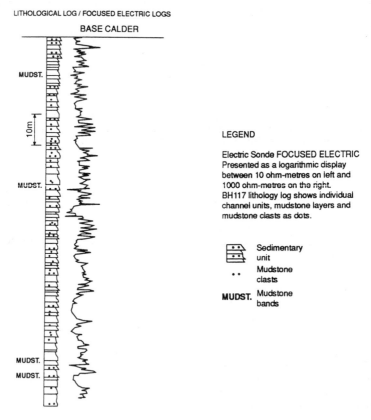

FIGURE 4 LITHOLOGICAL AND FOCUSED ELECTRIC LOG FOR ST BEES

The other reason is that authigenic quartz cement is responsible for a decrease in porosity. This was very obviously the case for the four high resistivity peaks in Figure 2(b), but was observed frequently elesewhere. This phenomena has been observed off-shore within the mixed fluviatile and aeolian Ormskirk Sandstone. Meadows and Beach (1993) observed that within the Ormskirk Sandstone it was only within the fluviatile units that such an authigenic cement occurred, and they suggest that this may be due to the differing provenance of the aeolian and fluviatile material. This suggestion is not fully borne out by the data from this Study where authigenic quartz overgrowths have been observed on rounded aeolian grains, although where this occurs these grains had been reworked in a fluvial environment.

The dipmeter log has been used in conjunction with dip measurements on the core to establish the geological structure of the site. The tadpoles shown on the composite log indicate dip and dip direction and are assessed on the original field data for reliability. The degree of confidence was high apart from the odd spurious high angle dip. The regional dip can be assessed as 10° to the southwest. Higher dips measurements than this have been influenced by false cross-bedding.

BOREHOLE CORRELATION
Using the structural data from the dipmeter and core measurements it was possible to draw up a structural map of the Calder/St Bees interface using the borehole data alone. This could be extended to areas where the Base Calder had not been penetrated, using correlations with the upper and lower fluviatile sandstones as shown in Figure 5. These correlations can be extended beyond the site to the NIREX boreholes in the area. (NIREX 1993).

Over part of the site the St Bees Sandstone was at subcrop and hence the Base Calder could not be used for correlation. Here the drillholes penetrated deeper into the St Bees Sandstone. Using the focused electric log it was possible to discriminate different units with a generally upwards declining resistivity. This permitted the correlation in Figure 6 and allowed the structural Base Calder structural surface to be extended further. The structural map developed using these correlations allowed the fault structure to be positively identified and the correlation of the St Bees strata across one of the faults demonstrated that there had been no vertical movement on this fault since upper St Bees Sandstone time.

ROCK PROPERTIES
Two types of sonde were used to assess sonic velocities, the multichannel sonic and sidewall sonic sondes respectively. The former provides P wave velocities utilising a high voltage pulse transmitter and a series of receivers providing resolutions of between 200 and 600mm giving an essentially continuous record. The latter sonde propagates waveforms at discrete (5m) intervals, the analysis of which can provide both P wave and S wave velocities.

In addition to the use of sondes, crosshole and downhole geophysics was employed as a corroborating method of deriving seismic velocities. These methods allowed P and S wave velocities to be determined at 1m intervals, between boreholes spaced at between 7 and 15m apart. In general good correlation was obtained between P wave velocities measured by both sonde types and between sondes and downhole/crosshole techniques.

The results for Calder Sandstone, gave P wave velocities of about 2500m/sec increasing to 3500m/sec at depth and for St. Bees Sandstone, values increase from about 3000m/sec at rockhead to about 4000m/sec at depth were obtained. Where logging penetrated the Calder/St. Bees interface a sharp increase in velocity was noted, a velocity contrast of 2800 to 3500m/sec being typical.

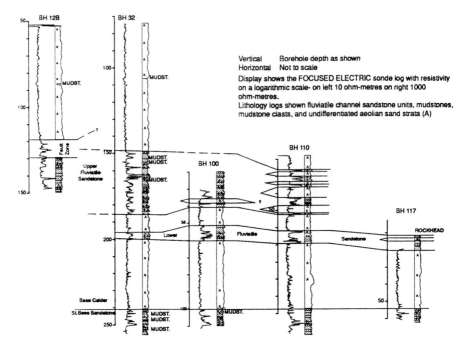

FIGURE 5 CALDER SANDSTONE CORRELATION

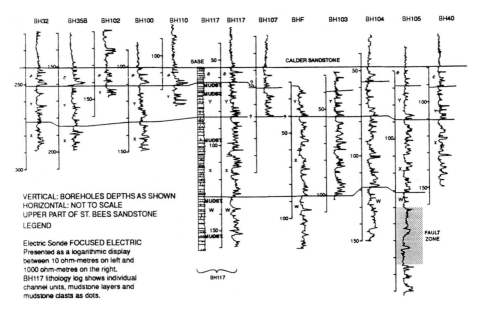

FIGURE 6 ST BEES SANDSTONE CORRELATION

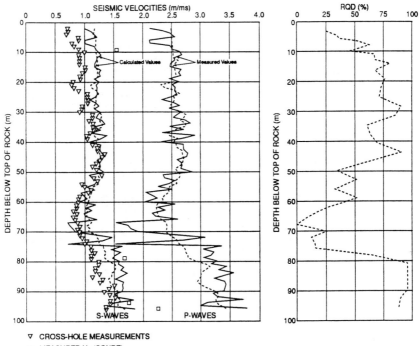

CROSS-HOLE MEASUREMENTS

MEASURED Vs (SONDE)

FIGURE 7 SEISMIC VELOCITIES FOR CALDER SANDSTONE

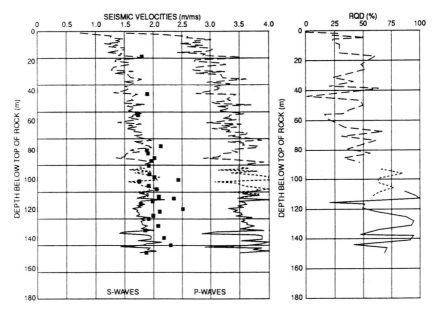

MEASURED Vs (SONDE)

FIGURE 8 SEISMIC VELOCITIES FOR ST. BEES SANDSTONE

759

Correlation of S wave velocities measured by the sidewall sonic sonde and by crosshole methods was reasonable but sonde values were consistently higher than crosshole values. S wave velocities were also calculated using measured P wave velocity and in situ density using the method given by Foster and McCann (1979). Calculated values appeared to form a reasonable fit to the data obtained by the two direct methods, therefore calculated S wave velocities were used in conjunction with measured P wave velocities in the assessment of dynamic moduli using standard relationships.

Figure 7 shows S and P wave velocities obtained from sonde and crosshole techniques together with RQD for Calder Sandstone. A good correlation between seismic velocity and RQD is seen, the zone of reduced RQD corresponding to a fault zone. Figure 8 shows S and P wave velocities for St. Bees Sandstone (sonde value only) and the relevant RQD values, good correlation is again seen.

Measured values of S and P wave velocities for Calder Sandstone correspond reasonably with other published data for the area (Thompson and Leach, 1988). A reduction in the scope of planned work did not allow any rigorous examination of the effect of strain level on shear modulus. Preliminary work would suggest that the ratio of dynamic to static modulus as determined in the laboratory for Calder Sandstone using j (rock mass factor) of 0.5 is approximately 0.05 (for UCS = 1MPa) and 0.12 (for UCS = 10MPa).

For St Bees Sandstone corresponding ratios of 0.13 (UCS = 20 MPa) and 0.5 (UCS = 40 MPa) are suggested. These ratios are generally lower than values quoted by Thompson and Leach (1985), but are based on a limited amount of laboratory strength and modulus data.

CONCLUSIONS

Use of wireline logging proved invaluable in developing the structural geology model for the power station site. Without this logging the initial confidence to considerably extend the deepest borehole may have been lacking. Although it subsequently proved possible to correlate the lithologies of the Calder Sandstone between boreholes, the wireline logging considerably enhanced the correlation, and without the use of the wireline logging of units within the St Bees it could not have been reliably correlated between boreholes. For the stratigraphical correlation the focused electric log was found to be the most valuable. A gamma ray and sonic log are traditional mainstay tools. The superior performance of the focused electric sonde is attributed to high resistivity from secondary cementing of the channel sandstones, and to the fact that it was less influenced by mudstone clasts on crossbeds within channel units.

In addition to the stratigraphical correlation, dynamic moduli values could also be interpreted from the wireline logs.

Considering the large amount of the country underlain by Permo-Triassic strata, for which stratigraphical correlation and quality laboratory testing are problematical, the use of wireline logging should be considered even for relatively small geotechnical investigations. For investigations that are subject to as much external evaluation as those for the nuclear industry they are essential.

REFERENCES

Foster A and McCann D (1979)
Use of Geophysical Logging Techniques in the Determination of In Situ Geotechnical Parameters. Transactions 6th European Symposium SPWLA, Paper G.

Jackson D I and Mulholland P (1993)
Tectonic and Stratigraphic Aspects of the East Irish Sea Basin and Adjacent Areas: Contrasts in their Post Carboniferous Structural Styles. In Parker J R (Ed). Petroleum Geology of Northwest Europe. Proceedings of the 4th Conference. Geological Society, London. Vol. 2 pp 791-808.

Jones N S and Ambrose K (In print)
Traissic Sandy Braidplain and Aeolian Sedimentation in the Sherwood Sandstone Group of the Sellafield Area, West Cumbria. Yorkshire Geological Society Proceedings.

Meadows N S and Beach A (1993)
Controls on Reservoir Quality in the Triassic Sherwood Sandstone of the Irish Sea. In Parker J R (Ed). Petroleum Geology of Northwest Europe. Proceedings of the 4th Conference. Geological Society, London. Vol. 2 pp 823-834.

NIREX (1992)
The Geology and Hydrogeology of Sellafield. UK NIREX Ltd, Report No. 263.

NIREX (1993)
The Geology and Hydrogeology of the Sellafield Area. Interim Assessment. December. UK NIREX Ltd. Report No. 524.

Thompson R P and Leach B A (1985)
Strain-Stiffness Relationship for Weak Sandstone Rock. Proceedings ISSMFE San Francisco, pp 673-676.

Thompson R P and Leach B A (1988)
The Application of the SPT in Weak Sandstone and Mudstone Rocks. Proceedings of Conference on Penetration Testing in the UK. ICE, Birmingham.

ACKNOWLEDGEMENTS
The authors would firstly like to thank BNFL for permission to publish this paper. The main contractor for the site investigation was Norwest Holst Soil Engineering Ltd who sub-contracted the slimline logging and borehole geophysics to BPB Instruments Ltd and Electronic and Geophysical Surveys Ltd respectively. The authors would also like to thank our colleagues in London and Manchester who were involved with the Geotechnical Study.

Use of the resistivity dipmeter for discontinuity assessment and stability design of a major cutting

D G GUY & D A O'CALLAGHAN
Thorburn Colquhoun Limited, Hertford, United Kingdom

1.   INTRODUCTION

Until recently discontinuity surveys for rock mass assessment have relied on good rock exposure for mapping and mechanical downhole borehole techniques. Where the rock mass is closely jointed rock outcrops tend to be weathered to a loose scree surface or obscured by vegetation, and mechanical downhole techniques produce sparse unreliable information. A section of the planned A38 in east Cornwall passing through deep cutting (23m) in metamorphosed Devonian strata presented precisely this problem for slope stability assessment.

The resistivity dipmeter used by the oil industry for many years is now available in a scaled down version suitable for ground investigation boreholes and provided the opportunity to recover continuous discontinuity profiles where otherwise such information would have been sparse or lacking.

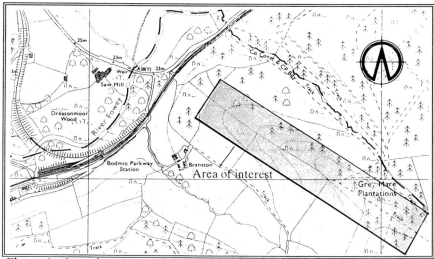

Figure 1 : Location plan, grid ref SX 120640

2. BACKGROUND DATA SOURCES

Information obtained prior to the main ground investigation (Ref 1) indicated that the Staddon Grit Formation and Middle Devonian Slates would be encountered. It was anticipated that rock head would be shallow and that slope stability would be dependent on the properties of the discontinuities of the underlying rock. The surveys indicated that the predominant discontinuity would be a folded penetrative cleavage, which, combined with joints, faulting, shear zones and fracturing had produced closely spaced discontinuities with areas liable to plane sliding, wedge failure and curvi-linear shear failure. The structural mapping of the limited outcrop suggested faulting or major folding in the area of Grey Mare Plantations although neither is indicated on geological sheets for the area.

3. GEOLOGY AND FIELD MAPPING

The geological structure of south-east Cornwall is dominated by the approximate east-west alignment of the numerous fold axes which are a regional feature associated with the Armorican orogeny. The attitude of folds is generally gently inclined to recumbent north-facing. The folds are disrupted by north-south faulting similar to faulting in the Lostwithiel area. This is considered to have occurred as a result of the intrusion of the St. Austell Granite to the west and is associated with the low-grade metamorphism of the area.

The solid geology in the area of Grey Mare Plantations cutting was proved by boreholes to comprise metamorphosed sandstones, siltstones and mudstones of the Lower Devonian Staddon Grit Formation with occasional interbedded tuffs.

Field mapping of outcrops of Staddon Grit within 1km of the proposed centreline was carried out during the main ground investigation to determine the local structure. Two outcrops in the vicinity of Grey Mare cutting were large enough for line surveys to be undertaken. Elsewhere slopes and cuttings were highly weathered and overgrown. Discontinuities were recorded according to type, orientation, persistence, openness and infilling.

4. WEATHERING GRADING

In order to classify the rock weathering a grading system was devised based on the BS 5930 scales I to VI (Ref 2), visual description of the materials and discontinuities, R.Q.D. and block size or core length.

Table 1 : Weathering Grades

Grade	Term	Description
I	Fresh	No visible weathering of the material, slight discolouration of discontinuities. RQD > 80% Block size > 150mm
II	Slightly Weathered	Weathering of the material and discontinuities visible, some fine fragments. RQD < 20% - 80% Block size 100 - 150mm

III	Moderately Weathered	Frequent fines and soil with weathered material. RQD < 20% Block size 50 - 100mm
IV	Highly Weathered	Matrix of fines, soil and weathered material and clay gouge on discontinuities. Block size < 50mm
V	Completely Weathered	Decomposed fines and soil, relict structure and occasional rock fragments.
VI	Residual Soil	Fines and soil without structure, few rock fragments.

The use of RQD or block size enabled unification of gradings from boreholes with those from line surveys at local cuttings. Two locations were surveyed in the vicinity of Grey Mare Plantations, one in a cutting on the Bodmin to Wenford Railway in the River Fowey valley and another further up slope where a farm building was located. Generally elsewhere there was no exposure of rock and natural slopes were vegetated and less than 1 in 1 in gradient.

The weathering of discontinuities was graded in terms of infilling and degree of iron staining and is presented in Table 2.

Table 2 : Weathering of Discontinuities

	Sub-vertical > 80°	Inclined 70° 45° 30°	Sub-horizontal < 10°	Total
Quartz Filled	13%	1% 3% 4% (8%)	6%	27%
Some iron-staining	16%	4% 5% 7% (16%)	5%	37%
Heavy iron-staining	13%	3% 6% 1% (10%)	2%	25%
Clay Filled	3%	1% 2% 1% (4%)	4%	11%

5. THE RESISTIVITY DIPMETER

Resistivity dipmeter analysis was carried out in 17 of the 23 boreholes drilled in the area of Grey Mare cutting to supplement and correlate with the results of the structural mapping survey of surface outcrops and the borehole core logs.

The resistivity dipmeter employed for ground investigation work is 4.5m in length (Fig 2) and suitable for shallow rock drilling. It operates from an all-terrain vehicle in which the logs are produced by a P.C. and printer as the test takes place. The dipmeter comprises a sonde attached to an armoured logging cable which is lowered and drawn up the borehole by a winch mounted on the back of the ATV. At the base of the sonde there are three caliper mounted resistivity pads which produce the resistivity log when in contact with the sides of the borehole. A verticality tool mounted in the sonde corrects for borehole tilt so that true dip is computed. The sonde is drawn up from the base of the hole at a constant rate of 4m per minute simultaneously recording three resistivity traces for correlation, a caliper curve and orientation data. The caliper curve indicates the diameter of the borehole and is a good measure of borehole stability. (Fig 3)

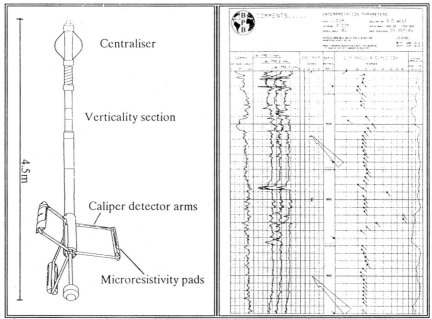

Figure 2 : The resistivity dipmeter. **Figure 3 : Typical dipmeter plot.**

There are three variable input parameters; the correlation interval, step and search angle. These are the length of curve to be analysed in each successive round of correlation, the distance by which the correlation interval is moved from one correlation to the next and the length of curve over which a correlation may be sought for the interval selected respectively. These parameters can be changed at any time after the logging has been completed for additional interpretation.

The paramaters chosen when the logging was carried out were as follows:

Correlation interval	1m and 2m
Search angle	45°-70°
Step	0.5m and 1m

Additional interpretation was carried out after completion of the site work with search angles of 15° and 88° to pick up sub-horizontal and sub-vertical fractures identified in the core.

Data was presented as a tadpole plot with the body representing the true dip magnitude and the tail the dip direction and as stereonet plots with a net diameter of 200mm for direct comparison with surface mapping data.

The resistivity dipmeter will only operate in open, fluid-filled boreholes. Optimum results are achieved in low diameter boreholes ie ≤ 150mm. Where borehole stability is good water may be pumped into the borehole to maintain the water level in the hole for the duration of the test. In highly fractured rock where stability is not guaranteed and permeability is high it is best to attempt to fill the hole with bentonite slurry to stabilise the borehole walls. Where the borehole does not hold water only the section below groundwater level can be logged.

6. ROCK STRENGTH TESTING

All of the rock in the Grey Mare cutting is of the Staddon Grit Formation with a predominance of mudstones and siltstones and occasional sandstone units. The abundant discontinuities throughout the rock mass control the local and overall stability of the eventual cutting slopes although a combination of factors will contribute to the shear strength. (Ref 3)

    i   shear strength of the discontinuities
    ii   shear strength of intact rock

To assess these elements unconfined compressive strength (UCS) and point load index tests were carried out to determine the cohesion. For an assessment of the friction angle and the shear strength envelope shear box tests were also undertaken for gouge filled, open and closed discontinuities for the three rock types.

The cohesion and shear strength are shown in Figs 4 to 6, and illustrate a reasonable distribution in relation to the weathering grades (Ref 4) in the moderately weak to moderately strong scale (Ref 2). The shear strength envelope of results varies from residual clay filled discontinuities to peak strength closed discontinuities with a range from $\phi_r = 20°$ to $\phi_p = 35°$.

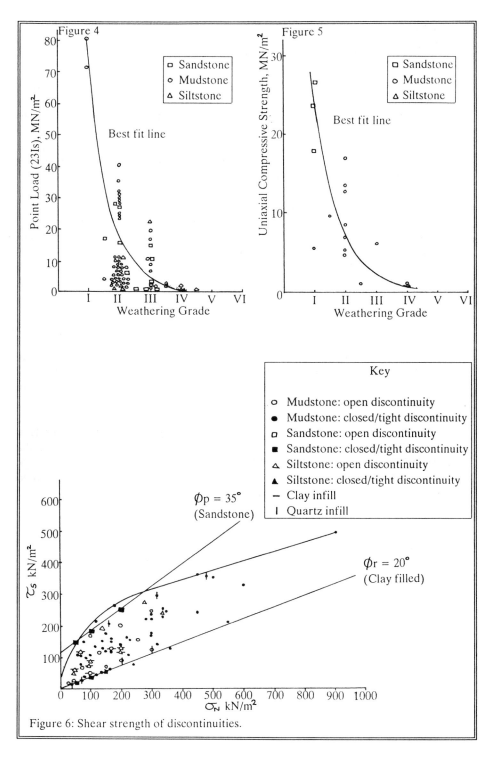

Figure 6: Shear strength of discontinuities.

7. STRUCTURE, DISCONTINUITIES AND PERMEABILITY

The major structure of the area is an Armorican age anticline trending west north west to east south east with the site located on the southern limit. This north-south compression has produced a penetrative cleavage in the Staddon Grit Formation and recumbent north facing folds. Evidence of folding is abundant in the cleavage pattern and a pole plot shows this generally lies on a north-south axis or great circle with the pole probably corresponding with the strike of the anticline (Fig 7). Bedding unit folds are rare and only two examples of north facing recumbent folds were found in the farm cutting and a former railway cutting.

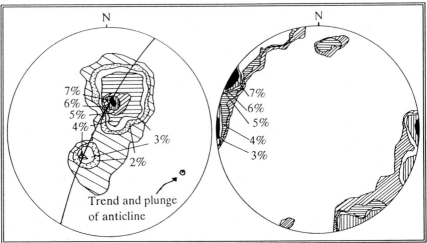

Figure 7 : Cleavage. **Figure 8 : Jointing.**
Contoured plots of discontinuity poles.

While the penetrative cleavage was the dominant structural feature across the whole site, faulting and jointing are also present and affect the stability and design of Grey Mare cutting. The highly fractured and weathered nature of the Staddon Grit and discontinuous sandstone units found in the boreholes are most likely the result of faulting associated with a later north west-south east trending wrench fault. The change of direction from west to south of the river Fowey at the site, and change from a broad flood plain to a narrow gorge probably reflects the intersection of the river with this fault zone.

At least four sets of conjugate and sub vertical joints have been identified (Fig 8). They are almost invariably closely spaced in all but some of the sandstone units and occasionally show movement has taken place in the form of minor kink bands and shear zones displacing the cleavage, as found in the farm cutting.

The condition of the closely spaced cleavage and joint surfaces varied from tight and cemented to open and filled with clay gouge, and generally reflected the weathering grade and rock strength criteria.

For stability purposes the low angle discontinuities measured by the resistivity dipmeter in the boreholes correspond with the penetrative cleavage exposed in the cores and observed and measured in situ in the line surveys of the two local cuttings. Those that are clay filled are of particular interest in that they would give the lowest shear strength which from the shear box tests was $c_r = 0$ kN/m^2 and $\phi_r = 20°$ (Fig 6). The frequency and openness of the sub vertical joint sets means the permeability of the rock mass is often unusually high for low permeability rock types such as mudstone. This was noted on several occasions when water was used to fill the boreholes prior to dipmeter testing and in variable head permeability tests. Values of permeability in the mudstones varied from 10^{-8} m/s to 10^{-5} m/s dependent on the degree of fracture. The sandstone units were highly permeable and some did not hold water. Because of their discontinuous nature, the sandstones are thought to be bounded by normal faults parallel with the north west-south east wrench fault in the Fowey Valley. Where found, groundwater in the form of minor seepages was associated with unfractured mudstone and siltstone layers perched at various levels through the proposed cuttings.

8. **COMPARISON OF THE RESISTIVITY DIPMETER RESULTS WITH BOREHOLE LOGS**
The amount of data produced by the resistivity dipmeter is copious, and the discontinuity dips and orientations are generally in agreement with the sparse information from the geological maps and two surveys of existing cuttings in the Grey Mare Plantations. The highly fractured nature of the Staddon Grit has however resulted in some spurious or questionable results. In order to check the reliability of the resistivity dipmeter information the dip of discontinuities from the core logging has been compared with tadpole plots from the same borehole. Since the core was logged to BS 5930 fractures were not logged individually particularly when extremely to very closely spaced but fracture patterns were recorded. Where core recovery was poor (< 50%) and the core was highly fractured (RQD < 20%) there is little correlation between the resistivity dipmeter results and the core log dips. This is partly due to the broken and often irregular fractures in the core some of which are caused by the drilling, and the small block size (50mm-10mm), from which it would be difficult to judge the true strike and dip of discontinuities. Elsewhere with moderate to good core recovery and RQD, the correlation of observed dip from the core log and the results of the resistivity dipmeter appear a reasonably good fit for inclined surfaces (Fig 9).

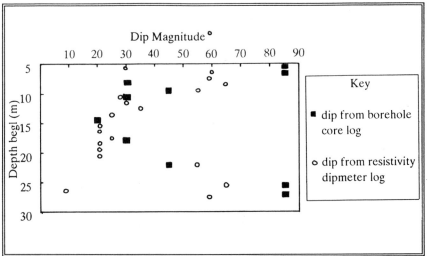

Figure 9 : Correlation of dip from resistivity dipmeter with dip from core log from Borehole 3056.

9.   FAILURE MECHANISMS

The close proximity of a major fault in the River Fowey valley and parallel normal faulting in the Grey Mare valley sides has produced a closely jointed rock mass of weak to moderately strong rocks, with a moderate to high degree of weathering.

The overall stability of slopes in the Grey Mare cutting is considered to be modelled by a curvi-linear shear strength envelope (Ref 5) whereby both rock and discontinuity strength are taken into account and shear takes place along a zone of shear.

Localised minor failure mechanisms within the slope may involve planar blocks or wedges bounded by joints and shearing of the penetrative cleavage.   In particular the competent sandstone and siltstone units have more widely spaced discontinuities and, depending on the final slope design, key stones may require rock bolting where discontinuities daylight in the slope.

The identification of daylighting discontinuities is dependent on the proposed slope angle, which at the current highway boundary limits is a minimum of 22° (1 in 2.5). The great circle to the poles of penetrative cleavage generally falls within this, although dipping in a similar direction to the NE and SW facing slopes. The range of friction angle for the discontinuities is also generally above the minimum slope angle for all but residual clay $\phi_r = 20°$. If the slope angle is steepened to 35° the area of daylighting cleavage discontinuities is significantly increased over the full range of friction angle i.e. $\phi = 20°\text{-}35°$ (Figs 10 and 11).

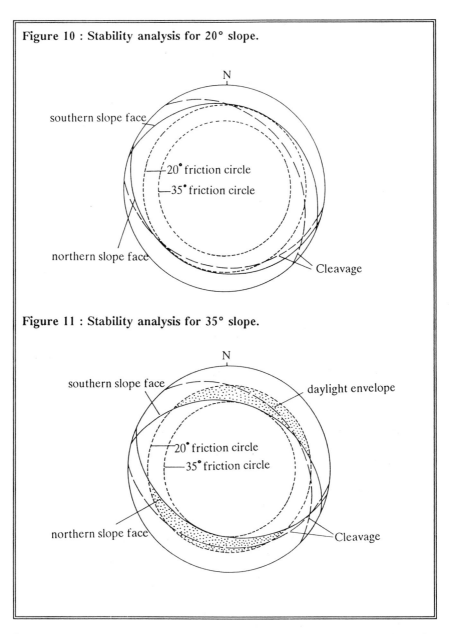

Figure 10 : Stability analysis for 20° slope.

N

southern slope face

20° friction circle

35° friction circle

northern slope face

Cleavage

Figure 11 : Stability analysis for 35° slope.

N

southern slope face

daylight envelope

20° friction circle

35° friction circle

northern slope face

Cleavage

Stereo plots of the joints indicate sub vertical dips with few release surfaces for potential slides or toppling failures and joint surfaces generally parallel to the Fowey Valley (Fig 12).

From Table 1:Weathering of Discontinuities it would appear that potential failure surfaces for slopes between 20° and 35° might vary in daylighting from 4% to 30% in occurrence respectively.

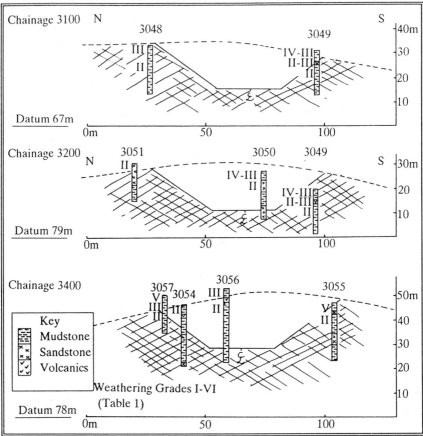

Figure 12 : Sections through Grey Mare Cutting showing daylighting cleavage in 35° slope.

10. ASSESSMENT OF THE RESISTIVITY DIPMETER IN ROCK SLOPE DESIGN

The planned Grey Mare cutting had no natural rock outcrops, limited existing cutting exposure and a paucity of information on the geological maps. Reliance for discontinuity data was placed on the boreholes and obtaining information by remote techniques.

At the time of contract let for the ground investigation the following techniques were reviewed to measure the dip and orientation of the discontinuities:

1. Impression packer - a preliminary survey including 15 tests to 15m depth were carried out. Only 7 tests produced a recognisable impression with only a single identifiable fracture.

2. Core orientation   -   in closely jointed rock overcoring an orientated marker is not a practical proposition due to disturbance and poor rock quality. Results also tend to be scattered and discrete in more favourable formations and limited in number. Carrying out the test is also time consuming.

3. C.C.T.V.   -   the cost of this is relatively inexpensive in the field although interpretation would have been lengthy and costly with a great deal of reliance placed on the individual interpretation, and clarity of camera image.

4. Acoustic televiewer -   This technique was not commercially available but has recently come on to the market. It would have probably been a suitable alternative to the resistivity dipmeter.

For confidence in resistivity dipmeter results cross checking the data with both the known local and regional structure is necessary and where possible surface mapping information and dips from core logs. In highly fractured zones (core recovery $< 50\%$ and RQD $< 20\%$) resistivity dipmeter tests may not be possible if the borehole does not hold water. Where it is possible the results are difficult to confirm and simple curvi-linear analysis may be more appropriate.

Acknowledgements
The Authors are grateful to the Highways Agency for permission to publish this paper and British Plaster Boards Ltd who carried out the resistivity dipmeter logging.

References (1-4)
1. A38 Bodmin to Liskeard Improvement Desk Study (1989). Brian Colquhoun and Partners.
2. BS 5930: Code of Practice for Site Investigations (1981). British Standards Institute.
3. Bray J W, A Study of Jointed and Fractured Rock. Part 1 - Fracture Patterns and their Failure Characteristics. Felsmechanik und Ingenieurgeologie, Vol 5, Nos. 2-3.
4. Hoek E & Bray J W, Rock Slope Engineering,1981 (3rd Edition), IMM.
5. Perry J, A Technique for Defining Non-linear Shear Strength Envelopes, and their Incorporation in a Slope Stability Method of Analysis. Quarterly Journal of Engineering Geology, Vol 27, 231-241, 1994.

Moderator's report on Session 5: geophysical testing

C.R.I. CLAYTON and V.S. HOPE
Department of Civil Engineering, University of Surrey, Guildford, U.K.

INTRODUCTION

The use of geophysics in site investigation practice has a long but somewhat chequered history. Indeed, Terzaghi (1957) wrote:

> *So far as geophysical methods of subsoil exploration are concerned, there can be no doubt about their desirability and merits, because they are extremely cheap; they are even cheaper than geologists.*
>
> *They have only one disadvantage, and that is we never know in advance whether they are going to work or not!*
>
> *During my professional career, I have been intimately connected with seven geophysical surveys. In every case, the physicists in charge of the exploration anticipated and promised satisfactory results. Yet only the first one was a success; the six others were rather dismal failures.*

Geophysicists tend to blame engineers for the lack of success of geophysical methods in site investigation, and *vice versa*. Perhaps both parties have some reason for complaint (Table 1). It is clear that any in-situ testing technique should only be used in those circumstances where it has a very good chance of success. It is, therefore, essential that an informed assessment is made of how appropriate a given geophysical method is to solve a particular site investigation problem. For many civil engineers, geophysical techniques remain shrouded in mystery. Yet geophysical methods depend on the laws of physics, which all civil engineers use daily as the basis of their engineering science. We, as engineers, are certainly as capable of understanding geophysics as are the engineering geologists and geophysics contractors who promote its use.

This review is written, without apology, from the perspective of civil engineers. In general, geophysicists take a quite different view of the use of their techniques in site investigation from that held by engineers. However, the engineer is normally the client and, as has been observed elsewhere, *the customer is always right.*

Advances in site investigation practice. Thomas Telford, London, 1996

Table 1:
Common interdisciplinary problems in geophysical surveys for site investigation.

The problem...	Comment
Expectations of engineers	Engineers expect every tool used during a ground investigation to yield data, preferably in the form of numbers. Many geophysical techniques are interpreted qualitatively.
Poor interdisciplinary understanding	Most engineers do not take the trouble to understand the basic physics of geophysical techniques. Geophysicists don't always understand civil or geotechnical engineering, nor what is of importance to an engineer.
Lack of communication	Engineers are often unclear as to their precise objectives when commissioning a survey, or don't tell their geophysical colleagues what they are. A geophysicist can only give you good advice if you tell him what you are trying to do.
Lack of objective appraisal of chances of success	Geophysicists are often ignorant of the contractual environment in which most engineers work. Our clients do not like to pay for unproductive work. Geophysicists tend to be optimistic about the chances of success. Engineers and geophysicists should establish the previous rate of success of any suggested technique, and also determine whether any site-specific factors may prevent it from working this time.

THE POTENTIAL BENEFITS OF GEOPHYSICS

Many geophysical techniques offer a means to measure and characterise the properties across the full expanse of a site relatively rapidly and inexpensively. Such methods can identify anomalous zones, as well as any regions that are comparatively uniform. Information of this type is of unique value when deciding where to locate most effectively the boreholes and trial pits of the main site investigation.

Most geophysical methods are good at surveying a wide area or volume quickly. In contrast, the conventional site investigation, using boreholes and soil samples, can provide only localised data. It is usual to interpolate, and extrapolate, from these sparse data, to build up a picture of the full volume of ground beneath the site. The results of geophysical surveys can be used to substantiate the conjectures that are inherent in this approach. This is the classic role advocated for geophysics in site investigation: to complement, but never supplant, the main investigation.

The potential applications of geophysics in site investigation are many and diverse. For example, BS5930 (1981) tabulated the following conservative list of possible uses: stratigraphic profiling; identification of buried channels, karstic surfaces, faults and dykes; location of aquifers and saline/potable interfaces; location of deposits of construction materials including gravel banks and clay pockets; determination of (small strain) deformation moduli; monitoring of the effects of ground treatment; assessment of rock rippability; determination of corrosivity potential; identification of buried artefacts, including cables and pipes; identification of sink holes and abandoned mine workings.

AVAILABLE GEOPHYSICAL TECHNIQUES — CATEGORIES

The range of geophysical techniques that are routinely available to a geotechnical engineer is bewilderingly wide and varied. BS5930 (1981) considered four broad classes of land-based geophysical surveys: resistivity, gravimetric, magnetic and seismic. *McCann and Green* suggest that electromagnetic methods should now be added to this list. Each of these broad groupings encompasses numerous distinct techniques. For example, *McCann and Green* cite a number of electromagnetic methods that have developed in the last decade: the very low frequency (VLF) technique; electrical conductivity testing; ground penetrating radar; and the transient electromagnetic method. Recent advances in seismic methods include the seismic cone (*Jacobs and Butcher; Shibuya et al.; Gordon et al.*), surface wave testing (*Butcher and Powell; Gordon et al.*) as well as crosshole seismic transmission tomography.

Geophysical methods can usefully be classified in several ways:
- by ground investigation purpose
- by type of anomaly in the ground
- by prior success rate in ground investigations
- by degree of intrusion
- by cost
- by susceptibility to likely difficulties on site.

Purpose of investigation
Geophysics can be used for all of the general objectives civil engineering site investigation, namely:
- determination of lateral variability
- profiling
- sectioning
- classification
- determination of properties.

Table 1 lists examples of suitable geophysical methods for these jobs.

Lateral subsurface variations cannot be determined easily with conventional investigation techniques such as boring and trial pitting, which are inherently localised. Geophysical methods help to 'fill in' the information that is missing between boreholes or sample points. Surface-based geophysics provides a valuable means of examining near-surface variability. This is extremely valuable when trying to locate point targets or hazards (for example, buried toxic waste drums, dissolution features, old mines, backfilled mineshafts). Benson (1993) has noted that if boreholes are placed randomly (rather than on a grid) 160 boreholes would be required to give a 90% probability of detecting a feature occupying 1% of the site area. For a 100m x 100m site, 1% area would be represented by a relatively large feature 10m x 10m in plan. Given normal borehole spacings there is virtually no chance of locating point features during routine site investigations.

Profiling can be carried out using a number of routine geotechnical methods, for example cone penetration testing. It can also be achieved using geophysics. Geophysical methods are of particular value where direct geotechnical methods are rendered ineffective by, for example, the presence of obstructions in the ground.

Sectioning cannot be carried out using conventional geotechnical procedures, although closely-spaced boreholes or probings may (where finances permit) achieve a similar result. Several geophysical techniques enable cross-sections of the ground to be built-up. A set of such sections can indicate something of the three-dimensional nature of the ground.

Classification is routinely carried out using Atterberg limit and particle size distribution tests. Only a few geophysical tests can differentiate between different types of ground in a way that is of interest to a geotechnical engineer. In site investigations, classification and index testing results may be used to correlate geological horizons between boreholes. This is also readily done using geophysical test results (for example, see *Pascall et al.* and *Nauroy et al.*).

Finally, geophysics can be used to obtain parameters for use in engineering design. As we noted above, electrical geophysical techniques (e.g. resistivity) can be used to determine measures of corrosion potential. More recently, as a result of the growing awareness of the relatively high stiffness of soils and weak rocks at small strains (e.g. Jardine, Symes and Burland, 1984), seismic methods have become recognised as a valuable means by which the stiffness of the ground (albeit at a very-small-strain level) can be investigated, and stiffness anisotropy assessed. Papers by *Butcher and Powell, Davis et al., Gordon et al., Jacobs and Butcher, Ricketts et al.,* and *Shibuya et al.* in this conference testify to the growing interest in the geotechnical community in the use of seismic methods and data.

In this conference we also see geophysics applied in a further area of site investigation, namely field instrumentation and monitoring. *Dixon et al.* report on preliminary trials with active and passive acoustic emission systems, to detect landslip movements in a cliff at Cowden.

Table 2:

Geophysical methods suitable for particular ground investigation objectives

Purpose	Techniques
lateral near-surface variability	ground conductivity (EM and TEM) magnetometry resistivity traversing gravity and micro-gravity
profiling	resistivity depth probing up- or down-hole seismic surface wave
sectioning	ground penetrating radar seismic reflection (off-shore) seismic tomography
classification	ground conductivity (cohesive/granular) resistivity (cohesive/granular, saline intrusion)
correlation	natural gamma logging resistivity logging
determination of properties	seismic refraction up- or down-hole seismic crosshole seismic surface wave
field instrumentation	acoustic emission

Type of anomaly

All geophysical methods rely on there being a suitable, measurable contrast between some physical property of the target of interest and that of the 'background' material. If there is no contrast or only a very slight contrast, or if the contrast cannot be perceived because the target is too small, or too deep, or background noise swamps it, then the feature will not be detected.

Table 2 shows the range of physical properties that are utilised in geophysics. It is important that an engineer appreciates, albeit in a simple way, the principles upon which each technique is founded, because this will suggest how successful a particular method may be in detecting a given target.

778

Table 3:
The range of physical properties exploited in geophysical site investigation

Physical property	Can used to detect or determine...	Works because...	Problems may arise from...
Velocity of seismic propagation	stiffness of ground, rockhead, faults, cavities	velocity properties vary with stiffness and density of soil	traffic noise, construction noise, wind
Electrical resistivity	saline intrusion, pollutant plumes, interfaces between granular and cohesive soil	resistance depends on soil porosity, saturation, and salinity of pore fluid	buried metallic objects (e.g. pipes, cables), overhead power lines, metal fences
Magnetic properties	old mine shafts, dissolution features	magnetic susceptibility of near-surface materials	metallic objects
Gravitational forces	cavities (e.g. old mines), faults	variations of mass density change the local gravitational field	buildings, embankments
Nuclear	marker beds clay content	attenuation of input energy, or natural emission	need to case hole
Electromagnetic (e.g. GPR)	layering, made ground, dissolution features, saline water	dielectric properties of the ground	metallic objects reinforcing mesh

Prior success rate
The success rate of any geophysical technique is difficult to gauge from the literature: just as with engineers, geophysicists prefer to advertise their successes rather than their failures.

The success of a geophysical survey depends on many factors. In general, a successful survey is likely to be one for which:

- the (simple) target was defined explicitly, and there was an appreciation of its properties;
- there was an understanding of the ground conditions, and how the target interacts with them;

- the possible sources of 'background noise' or disturbance to the measurements at the site were recognised;
- the execution and interpretation of the survey was carried out by experienced personnel.

In the authors' experience, some techniques appear more promising than others. Table 3 presents a summary of the techniques which, if the above factors are satisfied, are often worth considering.

Table 4:
Some geophysical techniques which have proved to be useful.

Survey type	Method	Remarks
Lateral, near-surface variability	proton precession magnetometry	detection of dissolution features, buried metallic waste in landfill, or mineshafts
	ground conductivity; resistivity traversing	distinguishing between granular/cohesive soils, detecting saline intrusion and contaminant plumes, detecting infilled dissolution features
	micro-gravity	detecting near-surface cavities
Sectioning	ground penetrating radar	sectioning in granular soils and weathered rocks
	seismic reflection	sectioning over water
	seismic tomography	sectioning for high-budget projects in rock
Determination of geotechnical parameters	surface wave method; crosshole seismics; up- or down-hole seismics	determination of ground stiffness profiles

Degree of intrusion
All geophysical methods can be classified as being one of the following:
- non-contacting
- surface-based
- borehole-based.

Table 4 presents several of the principal geophysical techniques within these categories. Non-contacting methods, in which the instrumentation need not connect with the ground, are quick, non-intrusive and relatively cheap. Surface-based methods, in which the sensors must touch the ground surface, are necessarily more time consuming than non-contacting methods but are not very intrusive. Borehole-based techniques, in which the instruments are deployed within boreholes, are relatively more time-consuming, expensive and are, by their nature, intrusive. In general, it is necessary to case a borehole before lowering any valuable instrumentation within it, in order to preclude collapse of the sides of the bore. A stable borehole is essential when using radioactive geophysical logging sondes, but metallic casing prevents most devices from functioning. Most logging sondes are affected by the borehole diameter, borehole fluid, as well as the borehole casing: it may be found that a borehole intended for other, more routine purposes is unsuitable for geophysical borehole logging.

Table 5:
Categorisation of geophysical methods by degree of intrusion

Degree of intrusion	Examples
Non-contacting	ground conductivity (EM) VLF electromagnetic proton precession magnetometry
Surface-based	resistivity resistivity tomography ground probing radar micro-gravity seismic refraction seismic reflection surface wave nuclear density testing
Borehole-based	seismic up- or down-hole seismic crosshole seismic tomography e.m. tomography geophysical borehole logging

Cost
The cost of a survey is closely allied to the degree of intrusion. Some economies can be achieved by making use of boreholes sunk for other purposes, for example, to obtain samples. Correlation between deep boreholes drilled for other site investigation purposes is generally cost effective (see, for an example, *Pascall et al.*). Other factors which influence the total cost of a survey include the hire cost of the specialist apparatus required, along with its operator, the length of time required on site and the length of time required to process the acquired field data. Table 5 presents a guide to the relative cost of a variety of geophysical surveys. Prices vary from a few hundred pounds per survey, to thousands of pounds per

survey. The relative cost of obtaining the required information by any other route is clearly an important consideration, as is the value of the information to the project itself. *Gordon et al.* report a case where cross-hole seismic results were used to evaluate the quality of self-boring pressuremeter test results. Clearly the geophysical testing, although relatively expensive by the standards of much geophysics, will have been carried out at much less cost than the pressuremeter tests. Both sets of data were vital in obtaining input data for a suite of finite element analyses for a retaining wall complex.

Table 6:
Relative cost of geophysical surveys.

Relative cost	Examples
Low	proton precession magnetometry surface refraction ground conductivity resistivity
Medium	continuous surface-wave testing ground penetrating radar geophysical borehole logging micro-gravity seismic cone testing cross-hole shear-wave seismics
High	crosshole tomography over-water seismic reflection

Susceptibility to "noise" and other difficulties
Since each geophysical technique depends upon a different principal or method of measurement, it will be apparent that each technique will be vulnerable to different technical problems on site. Moreover, each method will be susceptible to potential difficulties to different degrees. For example, noise from passing traffic can cause great problems for a surface-based seismic survey, such as a Rayleigh wave survey. It presents rather less trouble for a deeper, borehole-based seismic survey and gives rise to no problems at all for a natural gamma logging survey. Less expectedly, traffic also causes problems for micro-gravity surveys, as the delicate instrumentation is affected by vibration. Table 3 presented some of the practical problems that can affect geophysical surveys.

It is important, when planning a geophysical survey, to have a detailed knowledge of the site and its surroundings. It is also important to have an adequate understanding of the factors that can affect the viability of the proposed survey. "Active" geophysical techniques, which input energy into the ground (e.g. seismic methods), can use stacking (also termed enhancement) to improve the signal-to-noise ratio (see *McCann and Green*). Mathematical addition of repeated seismic records will enhance systematic signals, but random noise will be self-cancelling. Statistically, the signal-to-noise ratio increases by $N^{1/2}$ where N is the number of stacked records. Another approach is to change the input in a way that helps to distinguish the desired signal from noise. This approach is common for shear wave surveys, where the first arrival is typically within the P-wave train. Here the technique used is to reverse the input direction (for example, see *Butcher and Powell*), and to detect onset of the shear wave on the basis of reversal, or using subtracted (reversed) traces.

Other tactics may be required with "passive" methods, which measure pre-existing energy or field potentials (e.g. micro-gravity). Surveys can be repeated, and may sometimes be configured to minimise noise (for example by aligning equipment). If noise is perceived to be a problematic aspect of a site or its environs, it may be better to use an alternative geophysical method which is relatively unaffected by that type of noise. Where there is no suitable alternative, it may be possible to change the details of the survey. In the case of the seismic techniques, night working is common to avoid high traffic noise levels.

IMPROVING THE CHANCES OF SUCCESS

The one out of seven success rate, quoted from Terzaghi at the beginning of this note, is clearly unacceptably low for site investigation. Although the statistics are hard to obtain, it appears that the rate of success of modern geophysics is better. Nevertheless, geophysical techniques are still perceived by the geotechnical engineering community as being far from reliable. Major technological improvements in field equipment, data acquisition systems and processing facilities have taken place in the last few decades, as noted by the authors contributing papers to this session. These advances mean that, in general, neither the quality of field data nor our ability to manipulate such data present a significant limit on the possible success of a survey. Rather, problems still arise because engineers continue to commission surveys calling for the use of geophysical methods which are unsuited to the investigation problem in hand or the specific conditions obtaining at a site. These conditions are then recognised after the event, by which time the survey has failed.

BS5930 (1981) notes:

> *Geophysics is a specialized subject distinct from geotechnics. Where a geophysical investigation is required, the geotechnical engineer would normally entrust it to an organization specializing in this work. This organisation will usually advise on the details of the method to be used...* (Section 33.1)

The recommendation that a geophysics specialist should be approached to obtain details of the method to be used is important. When drafting a contractual specification for conventional soil investigation procedures, the geotechnical engineer is on familiar ground and can, in any case, usually refer the contractor to the relevant clause of the Standard. In contrast, there are few prescribed guidelines for even the long-established geophysical techniques. It is not unusual for a geophysics specification to be written by an engineer who does not have direct experience of the technique in question, does not take expert advice, and who either omits essential information, or requests what is technically impossible.

It could further be suggested that, as well as advising on the technical details of the method to be used, a specialist advisor should be called in at the earliest stages of a project in order to give expert guidance on the choice of the most appropriate technique. Such an arrangement would have two strongly beneficial effects:

- the geophysicist would gain a clear understanding of what answers the engineer is seeking from geophysics
- it is more likely that a suitable technique will be adopted.

Less tangibly, but no less beneficially, an effective collaborative working relationship would be built between both parties.

CONTRIBUTIONS TO THIS CONFERENCE

The papers presented to this conference give a rather distorted picture of the techniques in use in site investigation, because of some striking omissions. But it should be noted, however, that this conference concerns itself with *Advances* in Site Investigation *Practice*. The majority of the papers deal with seismic geophysics in some way, reinforcing the view that these techniques are rapidly becoming useful and practical (see *Butcher and Powell, Davis et al., Gordon et al., Jacobs and Butcher, McCann and Green, Nauroy et al., Ricketts et al.* and *Shibuya et al.*).

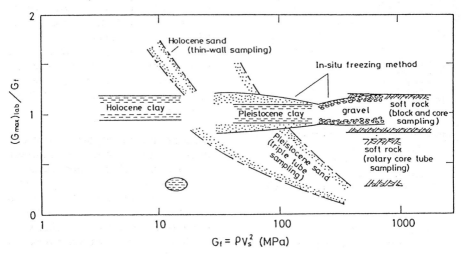

Figure 1   Comparison of shear moduli at very small strains derived from field geophysics and laboratory testing (*Shibuya et al.*)

Shibuya et al. present comparisons between laboratory and field measurements of very small strain shear modulus, G_{max} (Figure 1). In line with previous Japanese experience, they conclude that provided high-quality samples can be obtained, the ratio of G_{max} (lab.) to G_{max} (field) from in-situ shear wave velocity measurement falls within a range of 0.8 to 1.2. By implication they are suggesting that the field geophysics provides the bench mark against which both sampling and laboratory test results are to be judged. This is a startling result for British practitioners, many of whom apparently still believe that field geophysics delivers stiffnesses which are only relevant to dynamic problems, and who consider standard sampling practice as good. However, the results from *Shibuya et al.* should also be judged in the light of results given by *Butcher and Powell* (Figure 2) who find that in stiff clays they obtain differing values of shear wave velocity according to the test technique used. How reliable are our seismic measurements? Are we forgetting the effects of some significant factors (in addition to that of anisotropy, referred to by amongst others *Davis et al.*)?

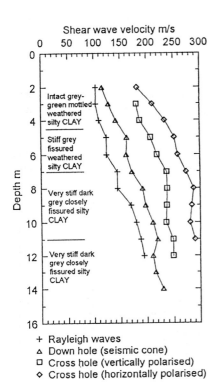

+ Rayleigh waves
△ Down hole (seismic cone)
□ Cross hole (vertically polarised)
◇ Cross hole (horizontally polarised)

Figure 2
Shear wave velocity depth profiles in London Clay, showing a variation depending on the method of determination (*Butcher and Powell*).

Examples of various seismic techniques, and useful advice on how to employ and interpret them, are given by *Butcher and Powell, Davis et al., Gordon et al., Jacobs and Butcher, and McCann and Green* and *Nauroy et al.*. There is clearly now a considerable body of experience in the use of cross-hole, down-hole, seismic-cone and surface-wave seismic methods, and results show the levels of repeatability that can be achieved by those experienced in these techniques. The tests that are described have, quite often, been carried out by geotechnical engineers, indicating that this area is no longer considered to be one where specialist geophysicists are required. We are less certain, however, about how to determine economically the modulus decay curves of the sort shown by *Davis et al.* (Figure 3).

Although the geotechnical community report results which are generally consistent and reliable, heed should be taken of the warnings given by *Ricketts et al.* who examine the confidence which one should have in physical properties measured using geophysics. They show that their measured shear wave velocities may typically contain scatters of up to $\pm 20\%$ for crosshole data, suggesting a scatter in G_{max} of about 50%. Figure 4 gives an example of their data.

Logging techniques are described in two papers. *Pascall et al.* show how a suite of slimline logging tools (including nuclear, sonic, electric, temperature, and verticality) were used at Sellafield in holes up to 300m deep. The results proved particularly useful in correlating rock units between boreholes, which helped with the understanding of the structural geology of the area. The authors suggest that these techniques should be considered even for relatively small geotechnical investigations, although (as noted above) it is our view that these techniques are often impractical because of the need to use casing to maintain borehole stability. *Guy and O'Callaghan* used a resistivity dipmeter (presumably similar to the DV1 three-armed dipmeter used by *Pascall et al.*) to obtain discontinuity data for a proposed deep cutting on the A38 in Cornwall. The results appear promising (Figure 5), but the authors are cautious in their conclusions. They also note the possible use of the acoustic televiewer in this application.

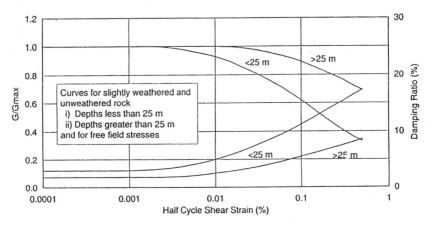

Figure 3
Modulus decay and damping ratios recommended by *Davis et al.* for Hinckley Point.

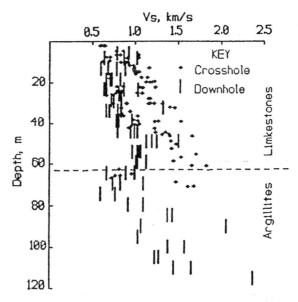

Figure 4
Shear wave velocity determined from downhole and crosshole methods, showing extent of scatter (*Ricketts et al.*)

Dixon et al. report on trials to develop acoustic emission techniques for use in clays (Figure 6). This is done by using sand or gravel backfill around the waveguides used to collect ground noise emitted during landslipping. Although the experiment was in its early stages at the time of writing, the authors find their initial results encouraging, commenting that "this is an exciting and relatively new area of research which promises to lead to the development of improved instrumentation techniques for the assessment of soil slope stability".

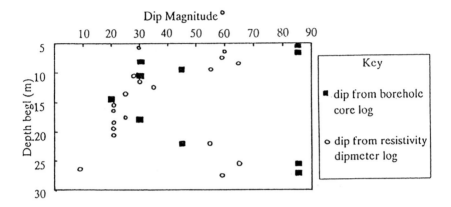

Figure 5
Comparison of dip magnitude derived from the resistivity dipmeter and from core logging (*Guy and O'Callaghan*).

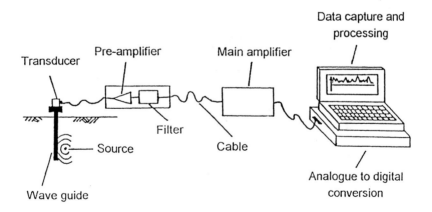

Figure 6
Schematic layout of an acoustic emission monitoring system (*Dixon et al.*)

TOPICS FOR DISCUSSION

- How reliable are G_{max} values, determined either in the field or in the laboratory? Japanese experience strongly suggests that when good sampling and testing practice is followed they should be equal (to within $\pm 20\%$). Why does British experience suggest that there are larger differences? How reliable and repeatable are our shear wave velocities, measured by field seismic test methods? Why do some authors report apparently large differences in V_s when measured using different techniques?

- How important is stiffness anisotropy, and is there experience of the field determination of the five elastic constants required for a simple cross-anisotropic model, using geophysics?

- How should engineers determine by how much G_{max}(field) should be reduced in order to obtain values of the stiffness required at the operational strain levels around civil engineering structures? Is the use of standard modulus decay curves acceptable?

- Are there other geotechnical applications for borehole logging? *McCann and Green* state that "of all the geophysical methods that should be a routine part of the site investigation process, geophysical logging of the boreholes should be considered essential". What are the practical difficulties which prevent this approach?

- Are there further experiences of the use of Acoustic Emission to detect unstable slopes? Are there potential problems due to background noise, since many slopes will be formed for highways or railroads?

REFERENCES

NB: Italicised references in the main text refer to the Conference Proceedings

Benson, R.C. (1993) 'Geophysical techniques for subsurface characterization'
in *Geotechnical practice for waste disposal* (Ed. D.E. Daniel), Chapter 14, pp. 311–357. Chapman and Hall, London.

BS5930 (1981) 'Code of practice for site investigations' British Standards Institute

Jardine, R.J, Symes, M.J., Burland, J.B. (1984) 'The measurement of soil stiffness in the triaxial apparatus' Géotechnique, vol. 34, no. 4, pp.323-340

Terzaghi, K. (1957) Opening address. 4th Int. Conf. on Soil Mech. & Found. Engng, London, vol. III, pp. 55-58

Discussion on Session 5: geophysical testing

Discussion reviewer A. P. BUTCHER
Building Research Establishment

The format of the discussion session was to take presentations prompted by the questions raised in the moderators presentation and to include discussion from the floor.

<u>Dr J-F Nauroy</u> presented work on the possibility of using seismic inversion techniques for extrapolating geotechnical data around a borehole.

Site investigation practice in the offshore oil industry today involves two steps. An initial seismic survey is generally performed, with only single channel acquisition system. This is followed by a geotechnical investigation including one or several boreholes at the location of the future structure. These two surveys are quite distinct and the resulting seismic and geotechnical data are generally difficult to integrate and cross correlate.

In the authors' paper (Nauroy et al 1995) it was shown that for improving offshore site surveys, it is necessary to develop two techniques:

- vertical seismic profiling (VSP) in geotechnical boreholes
- multichannel very high resolution seismic surveying.

These techniques exist already and can be currently used. VSP in particular provides an accurate correlation between seismic reflectors and geotechnical formations. Furthermore, these two techniques allow the processing software, routinely used in oil exploration surveys, to be applied, in particular reservoir assessment tools.

The inversion techniques consist in transforming a seismic section in a map of acoustic impedance by using a wavelet previously calibrated in a borehole located in the same plan as the seismic section.

Initial tests with this technique were made on a seismic section obtained in offshore Monaco. Figure 1 shows the structural model obtained from the interpretation of the seismic section. As a first step, to calibrate the wavelet, a sonic log recorded in a borehole was used, but in the near future, it is planned to use the VSP data obtained by seismic cone testing.

After inversion, a map of acoustic impedance was obtained as shown on Figure 2. If relationships are known between acoustic impedance and standard geotechnical parameters, such as CPT cone resistance for example, a map of such parameters can then be obtained. Work is presently continuing on this subject and initial results are very encouraging.

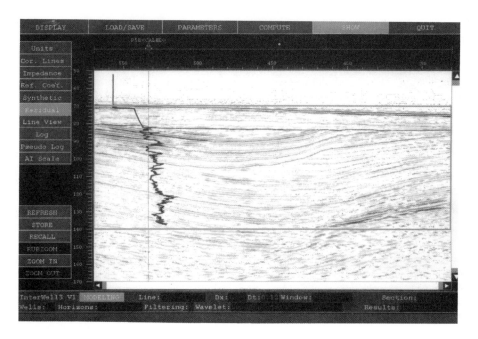

Fig. 1. Structural model obtained from interpretation of seismic section - Monaco site

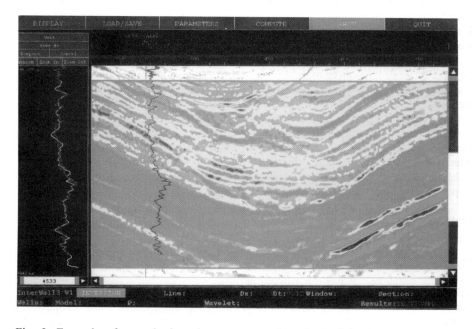

Fig. 2. Example of acoustic impedance cross-section obtained by inverse modelling - Monaco site

The moderator, <u>Professor Chris Clayton,</u> asked 'Will background noise make Acoustic Emissions (AE) impractical?'

In reply <u>Dr N Dixon</u> said he thought not, because although background noise is present, its importance can be reduced and assessed. To ensure that the AE from the deforming soil body can be distinguished from other sources the following conditions are adhered to:

- A transducer with a resonant frequency of $\approx$ 30 kHz is used in conjunction with a low band pass filter (15 kHz). This means that noise from general construction activity can not be detected.

- Covers are used over the waveguides during monitoring to stop wind generated AE (ie. wind blown particles impacting on the waveguide).

- A control waveguide is installed at each site away from the potentially unstable slope. The AE from this waveguide is monitored on each visit to the site.

Any AE detected may be due to the waveguide construction (i.e. movements within the backfill material), and electronic noise generated by the monitoring system. During monitoring of the soil slope, only AE activity greater than that obtained from the control waveguide is considered to be caused by the deforming soil body.

As more experience is gained, the electronic and waveguide construction related AE will be reduced.

<u>Mr P Wilson</u> noted that Dr Dixon had outlined a number of problems (including ambient noise) concerned with acoustic emission measurements but has not seemed to mention any benefits. Perhaps he would elaborate given that all the slips related to road construction would be subject to traffic noise.

<u>Dr Dixon</u> said that a number of field trials, carried out over the past 20 years, have shown that AE can be used to detect pre-failure ground deformations in unstable slopes. In a significant number of cases AE indicated instability before measurements of ground deformations. AE has also been used to indicate the depth of both existing and developing shear surfaces. The aim of the present research is to gain an improved understanding of the role of waveguides, and to improve the design of the field instrumentation and data interpretation techniques. It is believed that this work will lead to the development of a viable field AE system which can be used to give an early indication of instability in soil slopes.

<u>Professor Clayton</u> asked whether the authors were pleased with performance of the resistivity dipmeter:

<u>Ms D. O'Callaghan</u> stated that the authors were pleased with the results obtained but that there were limitations in the use of the dipmeter as it only operates in open, fluid-filled holes and many of the boreholes on this site did not hold water. The logging was carried out in three visits between February and June 1994. On the first visit the boreholes to be logged were filled with water but most had drained by the time they came to be logged and had to be refilled.

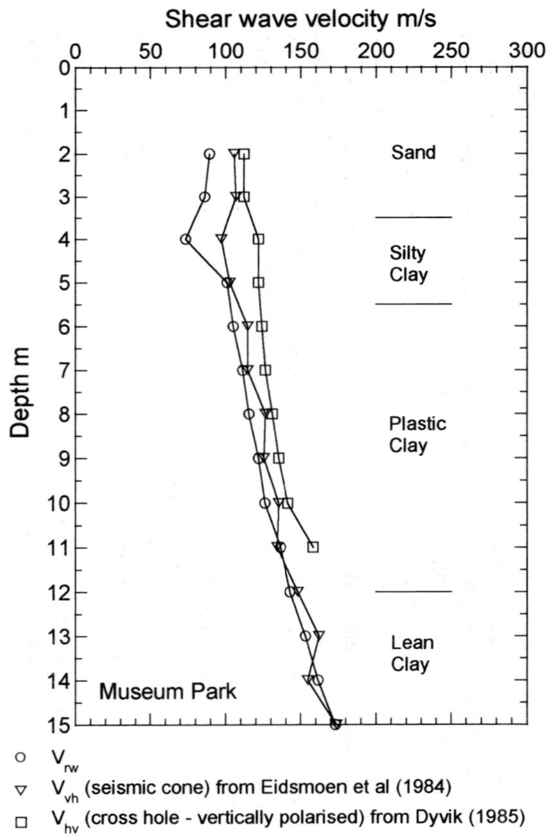

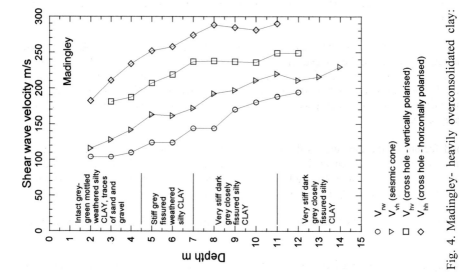

Fig. 4. Madingley- heavily overconsolidated clay: shear wave velocity - depth profile

Fig. 3. Museum Park, Drammen - lightly overconsolidated plastic clay: shear wave velocity - depth profile

On the second visit water was pumped into each borehole and logging took place so that the entire borehole could be logged but this method dislodged rock fragments in one of the boreholes and the sonde became trapped. It was subsequently recovered undamaged by overcoring.

The day before the third and final visit weak bentonite slurry was circulated in the remaining boreholes to be logged and each was filled with water immediately before logging. In boreholes where the water drained away only the section below water was logged. In this particular case it was not critical that whole borehole logs were produced as the purpose was to obtain general structural information to supplement the surface mapping data.

A good feature of the dipmeter is the facility to have additional interpretation performed off-site with different search angles, step or correlation intervals. This was used by the authors to have additional interpretation carried out with different search angles to those used on site without the expense of a return visit to site.

As a contribution to the first discussion topic, <u>A.P. Butcher</u> showed some data obtained by BRE.

Assuming the calculation of G_{max} or G_0 uses the shear wave velocity then the measured shear wave velocity will reflect the shear stiffness of the ground. The first set of data is from a site at Holmen in Drammen, Norway which comprises a medium dense sand. The three seismic wave techniques used, crosshole, downhole and Rayleigh waves, all gave very similar shear wave velocities with depth as shown in Figure 6 of Butcher and Powell (1995). This data agrees well with some of the data reported by Shibuya et al (1995). Looking now at data from Bothkennar (Butcher and Powell (1995) Figure 7), a soft normally consolidated silty clay again very similar shear wave velocity depth profiles were obtained from each measurement technique. The next site to look at is Museum Park, again in Drammen but this is a lightly overconsolidated clay. The plot of shear wave velocity against depth is shown in Figure 3 and here the shear wave velocities from each measurement technique show a distinct pattern with the crosshole consistently higher than the downhole and the Rayleigh waves forming a lower bound. The difference between the measured velocities at any one depth however is not very large. Finally Figure 4 shows data from Madingley on the Gault clay, which is a stiff heavily overconsolidated clay. The pattern formed by the data from Museum Park is well established with again the Rayleigh wave data forming the lower bound and the horizontally polarised crosshole data the upper bound. The vertically polarised crosshole and the downhole fitting between these bounds. These data, with that in Butcher and Powell (1995), show that the measured shear wave velocity, particularly in overconsolidated soils, depends upon the measurement technique used. Now which shear wave velocity should be used to calculate G_{max} or G_0? It is clear that the interpretation of shear wave velocity measurements needs further research but from the data given here appears to be related to stress and fabric anisotropy of the soil.

<u>Professor PK Robertson</u> said that Roesler (1979) found that in calibration chamber work on sands the shear wave velocity was related to the effects of stress anisotropy and those of overconsolidation.

<u>Professor M S King</u> addressed the topic of Stiffness Anisotropy in Earth Materials

1. Importance of anisotropy

Anisotropic stiffness behaviour of near-surface materials is of importance in the following geotechnical applications:
  (a) modelling deformations around tunnels and the foundations of large structures;
  (b) determining the stability against failure of boreholes and tunnels;
  (c) determining principal stress direction and their magnitudes *in situ* in near surface rocks from seismic and differential strain analysis studies;
  (d) interpreting Rayleigh surface-wave, seismic refection, seismic cone and VSP surveys in near-surface materials; and
  (e) the tomographic reconstruction of velocity fields from crosshole seismic tests.

2. Causes of anisotropy

Jardine (1994) suggests several contributory reasons for the stiffness anisotropy exhibited by certain soils and near-surface rocks. These include:
  (a) the micro- and macro-fabrics of clay-rich soils and rocks such as clays and mudrocks;
  (b) the imposition of an anisotropic state of stress on an inherently isotropic granular material; and
  (c) the presence of sets of oriented cracks in earth materials.

The magnitudes of individual stress components will govern the degree of anisotropy in both (b) and (c), with increases in components of normal stress perpendicular to the planes of sets of oriented cracks in (c) leading to a reduction in anisotropy (King et al., 1995).

3. Types of anisotropy

The occurrence in soils or rocks of different magnitudes of stiffness in each of the three principal stress directions leads to the material exhibiting orthorhombic anisotropy. Nine independent elastic constants are then required to describe the relationships between stress and strain (Hearmon, 1961), compared with the two necessary to describe simple isotropy. Computational difficulties and those associated with field and laboratory measurements currently preclude consideration in most cases of orthorhombic anisotropy in geotechnical applications.

Bedded sedimentary materials, particularly clays and mudrocks (King et al., 1994), tend to exhibit hexagonal anisotropy (transverse isotropy), with a vertical axis of symmetry. In this case five independent elastic constants are necessary to describe the material. Transverse isotropy is associated also with sedimentary rocks containing a set of oriented cracks (King et al., 1995). This type of anisotropy is currently capable of analysis and measurement, both in the field and in the laboratory (Thomsen, 1986).

4. Measurement of transverse isotropy

Tatsuoka and Shibuya (1992) demonstrate that the stiffness of soils and soft rocks under static conditions at very low strain levels may be determined from seismic surveys in the field and ultrasonic measurements made under *in situ* stress conditions in the laboratory. Elastic-wave propagation techniques for characterising transverse isotropy with a vertical axis of symmetry require the measurement of five different velocities.

These are normally measured in the field and in the laboratory (King et al., 1994) in the vertical (V_{PV} and V_{SV}) and horizontal (V_{PH}, V_{SH} and/or V_{SV}) directions, and at some intermediate angle (usually on specimens cut at 45° to the vertical in the laboratory, which provide V_{P45} and/or V_{S45}).

In comparison of field and laboratory measurements of P-wave anisotropy (transverse isotropy) at a crosshole seismic test site in near-surface sedimentary rocks, Sams et al. (1993) conclude that intrinsic anisotropy of the rocks is the dominant cause of anisotropy observed in the field. They conclude also that there is little frequency dependence of the intrinsic anisotropy in the ultrasonic to low kHz frequency range. However, Sams et al. (1994) demonstrate that, for the same test site, P- and S-wave velocities exhibit significant frequency dependence (velocity dispersion) in the frequency range ultrasonic (laboratory tests), through sonic (acoustic logging and crosshole seismic surveys) to seismic (VSP surveys), with the velocities increasing with an increase in frequency.

Butcher and Powell (1995) discuss different field techniques for determining S-wave velocities in the field. In particular, they present data from a test site in heavily overconsolidated London clay for Rayleigh surface-wave, downhole seismic cone (V_{SV}), and crosshole techniques with S-waves polarized both vertically (V_{SV}) and horizontally (V_{SH}). It is probable that the approximately 20% difference in velocity observed between the two crosshole methods (V_{SH} and V_{SV}) is due to intrinsic transverse isotropy. The approximately 20% difference between measurements of V_{SV} by the seismic cone and crosshole methods may well be due to velocity dispersion. Rayleigh wave measurements, made at very low frequencies, are observed to yield still lower S-wave velocities.

5. Conclusions

Stiffness anisotropy in near surface soils and rocks is caused by one or more of the following factors: (i) the fabric in clay-rich materials; (ii) the imposition of an anisotropic state of stress on an otherwise isotropic material; (iii) the presence of sets of oriented cracks.

Measurements of stiffness anisotropy by elastic-wave techniques can provide good estimates of that expected under static conditions at very low strain levels. There are, however, indications that elastic-wave velocities measured in tests operating over significantly different ranges of frequency are frequency dependent. This aspect requires further study.

Dr D M McCann responded to the moderators report which used the statement of Terzaghi (1957) that geophysical methods were generally unreliable. This is dealt with in the paper by McCann & Green (1995) which states, in paragraph 2 of the introduction, the following:

> The point at which geophysical surveying methods were introduced into the site investigation process is difficult to ascertain but certainly by the early 50's they were in common use at many construction sites. The initial use of these methods was not

greeted with much enthusiasm by the civil engineer, who preferred the more trusted approach to site investigation based on boreholes, trial pits and a number of in situ engineering tests. This scepticism was justified in many ways since the geophysicist's surveying ability was limited by the very rudimentary equipment in operation at the time and the length of time taken to interpret the geophysical data obtained in a survey. The limited efficiency of operation associated with geophysical surveying and a general lack of communication between the geophysicist and the engineer gave rise to a general feeling in the industry that the geophysicist represented black box technology, which was not to be trusted.

The paper (McCann & Green, 1995) however, emphasises that with modern geophysical equipment and highly sophisticated interpretation software, geophysical methods can be applied rapidly and effectively in the site investigation process and at times can be an essential pre-cursor to the drilling and trenching programme. We agree that these problems still do occur and approach and suggest the following improvements:

The moderator pointed to four areas where problems arose with the use of geophysical methods in a site investigation:

1. Lack of communication
2. Poor inter-disciplinary understanding
3. Optimism by the geophysicist
4. High expectations from the engineer

There is agreement that these problems still do occur and suggest that the following actions are essential:-

1. Improvements in communication by the integration of the geophysical surveys into the site investigation process with close co-operation between the engineers, geologists and geophysicists to address the problem in hand.
2. Use of specialist geophysical advice.
3. Importance of the site visit and a feasibility study of appropriate geophysical surveying method.
4. Specification of the problem.
5. Selection of the most appropriate method(s).
6. Availability of all site geological information for calibration of the geophysical interpretation.
7. Flexibility in the contract to adjust the line and sample station separation in the light of initial survey results.

The moderator also mentioned his reservations on the use of geophysical borehole logging in the shallow site investigation boreholes. It can be argued that a simple natural gamma log in every borehole drilled provides a lithological calibration log which can be used for stratigraphic correlation both at a site and a regional scale. The natural gamma log can be used in both cased and uncased boreholes and also in dry or fluid filled boreholes.

In conclusion, it is considered that the improvement of geophysical methods for use in the site investigation process is such that an appropriate method can usually be chosen for most geological situations. It is, however, essential that a close dialogue is maintained between the geophysical team and the consultant civil engineer/engineering geologist responsible for the site. Provided that the steps mentioned above are taken we feel that Terzaghi would thoroughly approve of the use of geophysical methods in today's site investigation programme.

Why carry out geophysical surveys?

The paper (McCann & Green 1995) describes how geophysical surveys can be integrated into the site investigation process. It emphasises the two basic types of geophysical surveys:

(1) a reconnaissance survey, which is carried out to locate anomalous areas on the site where a borehole or trench survey is required
(2) determination of the geological structure as a 2D or 3D model.

A typical example of (1) is shown in Figure 4 of McCann & Green (1995), which shows the variation of ground conductivity above a sink hole resulting from solution in an underlying gypsum bed.

Theoretical modelling of an observed gravity traverse across a buried cavern in Figure 1 of McCann & Green (1995) is an example of (2) above.

The paper indicated that geophysical methods can be applied rapidly and effectively in many situations in a site investigation and at times can be an essential precursor to the drilling and trenching programme. For instance, a micro-gravity survey could well have been of assistance in the drilling of the borehole, shown in Figure 5, in the Glasgow area. It can be noted that it has not been entirely successful, although one could argue that it has actually located the underground mineshaft. The size of the cavity involved would have been a significant anomaly in a gravity profile (see Figure 1 in McCann & Green 1995) and perhaps a little more consideration of the site engineering capacity would have resulted. Acknowledgement: Reproduced by permission of the Director, British Geological Survey. ©NERC.

A contribution from J C R Arthur said that the loss of credibility of geophysics in engineering programme has been of concern and there is an ongoing offer from our organisation to analyse past failures in an objective, no-blame environment. Engineers have a right, as the employers' agent, to expect value for money and useful results; practising engineering geophysicists need to communicate capabilities of geophysics better and persuade engineers to put planning money "up-front" to:

1. model the expected ground conditions to ensure proposed methodology can deliver required answers to the resolution needed;
2. design an effective programme (e.g. determine adequate borehole depths for tomography/consider geometry of seismic system/establish a survey programme to achieve required results;
3. discuss expectations and plan project effectively within the engineers' overall site investigation programme;
4. construct a realistic contractual relationship for the engineer/consultant/contractor/sub-contractor before documentation is finalised.

Fig. 5. Drilling rig being removed after 'finding' an uncharted mine shaft.
BGS, ©NERC 1994

I Webber made two points regarding the use of acoustic emission monitoring.

Firstly, the technique has been used to predict the onset of a landslip by triggering at certain event levels. This allows an automated approach to landslide prediction.

Secondly, the sources of external "noise" discussed were predominantly surface generated. Other "noise" can be generated by groundwater flow and acoustic emissions have been used for identifying areas of groundwater flow.

Professor J H Atkinson presented some work that he and his co worker V. Jovičić had recently completed at City University in the investigation of anisotropy of the very small strain shear modulus G_0 using bender element techniques in laboratory triaxial tests. In a bender element test shear waves are fired through a sample and G_0 determined from the travel time and distance (Viggiani and Atkinson, 1995). The shear waves are transmitted along the axis of the triaxial sample and the direction of particle motion is normal to the face of the bender element. By changing the orientation of the sample in the apparatus it is possible to investigate different anisotropic shear moduli.

Figure 6. illustrates three samples cut from a homogeneous but anisotropic soil and the shear waves transmitted through these samples by bender elements. One sample is cut vertically; the shear wave velocity is V_{vh} and the corresponding shear modulus is G_{vh}. (The orientation of bender element and the direction of particle motion or polarisation of the shear wave can, of course, be in any direction). Two samples are cut horizontally; in one the bender element is orientated so that the direction of particle motion is in a direction originally vertical and in the other direction of particle motion is in a direction originally horizontal. The shear wave velocities are V_{hv} and V_{hh} and the corresponding shear moduli are G_{hv} and V_{hh} and the corresponding shear moduli are G_{hv} and G_{hh}. (The subscripts are chosen so that the first gives the direction of propagation of the shear wave and the second gives the direction of particle motion both with respect to the in situ directions).

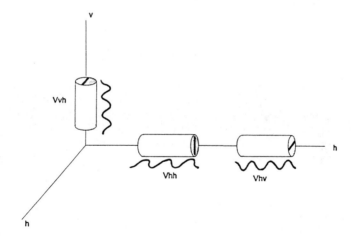

Fig. 6. Orientation of the samples and corresponding notations for shear wave velocities

In samples of an isotropic soil, all at the same stress and overconsolidation ratio, all three shear moduli will be equal. In a homogeneous but anisotropic soil with a vertical axis of symmetry G_{vh} and G_{hv} will be equal but G_{hh} will be different. (In a soil which is layered G_{vh} and G_{hv} may be different).

Bender element tests were carried out on three samples cut from the same core of London Clay. Before the shear waves were fired the samples were isotropically consolidated to a mean stress $p' = 400\text{kPa}$ which is approximately equal to that estimated in situ. The results are given in Table 1. These show that G_{vh} and G_{hv} are approximately equal and G_{hh} is larger by a factor of approximately 1.5. Since the state of stress in the samples was isotropic the differences between the shear moduli must have been due to structural anisotropy in the London Clay.

Table 1

Test	propagation/ polarisation	V_s (m/s)	G_o (MPa)
1	v/h	270	$G_{vh} = 144$
2	h/h	334	$G_{hh} = 220$
3	h/v	263	$G_{hv} = 136$

These results are broadly in agreement with the in situ measurements of G_o reported by Butcher and Powell (1995) in their paper to this Conference. The differences in their values G_{vh} measured in seismic cone tests and G_{hv} measured in cross-hole tests with vertical polarisation may be due to layering evident in situ but not in laboratory samples.

Dr M A Paul presented some work which he and a co-worker, L A Talbot, had completed on the relationship of analogue seismic signatures to geotechnical profiles.

We applaud the work of Nauroy and his co-workers (Nauroy et al., this volume) on the use of high resolution digital data and share his reservations about analogue records. However, there is a large volume of existing analogue data from regional surveys which has potential geotechnical value at the site selection stage. We would like to draw attention to the value of analogue seismic signatures as proxies for geotechnical profiles in well-understood geological settings. The contribution centres on British Geological Survey borehole 88/7,7A which was drilled on the upper continental slope 100km west of Lewis and which has recently been re-examined for geotechnical purposes by the writers (Talbot et al, 1994). The object of this new research has been to relate the profile of seismic facies at the borehole site to corresponding profiles of geotechnical properties and to the regional geological history (Stoker et al 1993).

Figure 7 shows the seismic stratigraphy at the borehole site. Three units were identified by Stoker et al, (1993). The upper, layered unit is a distal glaciomarine mud of Devensian age, (termed the upper acoustic unit of the Upper MacLeod sequence) which is believed to have been deposited by mass flow with some subsequent reworking: sedimentation rates (calculated from radiometric age determinations) were around 150mm/ka. The middle, semi-transparent unit (the lower acoustic unit of the Upper MacLeod sequence) is also a glaciomarine mud, in this case of Anglian age, and probably accumulated by mass flow, although there is evidence of hiatuses during which bioturbation and current reworking occurred. Sedimentation rates in this unit exceeded 500mm/ka on average. The lower, layered unit (the Lower MacLeod sequence) is a mixed sand/mud sequence of late Tertiary to early Pleistocene age which accumulated by along- and down-slope processes and shows evidence of reworking by along-slope currents. The sedimentation rate in this sequence was around 25mm/ka, which is noticeably lower than in the succeeding units.

Figure 8 shows geotechnical and seismic profiles from this borehole. The water content profile records several distinct units which correlate with the seismic stratigraphy. Of note

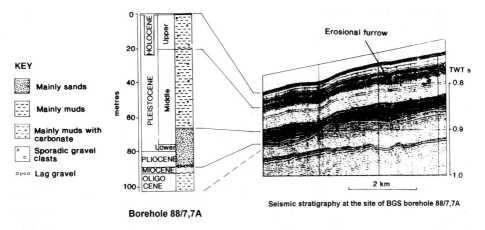

Fig. 7. Seismic stratigraphy and log at borehole 88/7,7A (after Stoker, Hitchen & Graham)

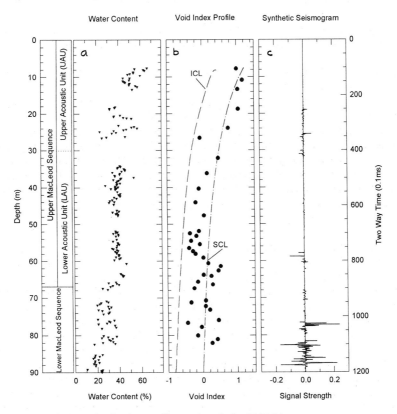

Fig. 8. Geotechnical and seismic profiles at borehole 88/7,7A

is the relatively narrow envelope in the middle unit: this contrasts with the greater scatter in both the layered units above and below. Figure 8 includes a profile of void index: it appears that the values from both the layered units fall close to the sediment compression line (Burland 1990) which characterises marine sediments, whereas in the middle, transparent unit the values fall closer to the intrinsic compression line which characterises remoulded material. Figure 8 also shows a synthetic seismogram constructed from close (200mm) density measurements on this core. The algorithm employed generates both primary and multiple returns. The relative lack of density contrast in the middle unit causes a corresponding lack of strong reflection, so leading to its transparent seismic appearance. In comparison, numerous, small density contrasts in both the uppermost and lowermost units generate multiple reflections which may be expected to give complex multi-layered signatures in the field.

We suggest that there may be a genetic explanation for this pattern which is of general interest. The more rapidly sedimented glaciomarine mud of the lower acoustic unit of the Upper MacLeod (the transparent middle unit) has suffered post-depositional reworking through which it may have lost some of its primary structure. As a result it lies on the ICL and lacks internal density contrasts. By comparison, the units above and below appear to have undergone less post-depositional disturbance and have retained their primary structure (a quantitative study of the electron microfabric of the sediments supports this view (Paul & Talbot, 1995). These sediments lie on the SCL and are characterised by internal density contrasts which generate their signature. A scattering of void index values around the SCL appears to be common in many sediments (e.g. Burland 1990) and may be a normal consequence of the depositional processes by which many muddy sediments are deposited. If so, this explains the multi-layered signatures which are usually associated with these materials.

In conclusion, we believe that water content and void index profiles are sensitive indicators of changing depositional conditions and contain useful information on the sedimentation history of a deposit. This information can aid the production of both local and regional geotechnical models, and can in some cases explain variations in seismic signatures between apparently similar sedimentary units.

Acknowledgement: This work was funded by MTD Ltd and core samples and seismic data were made available by the British Geological Survey.

REFERENCES

Burland J.B. (1990). On the compressibility and shear strength of natural clays. *Geotechnique*, **40,** pp. 329-378.

Butcher, A.P and Powell, J J M. (1995). Practical considerations for field geophysical techniques used to assess ground stiffness. Preprint, *Int. Conf. on Advances in Site Investigation Practice*, London.

Hearmon, R. F. S. 1961. *Applied Anisotropic Elasticity.* Oxford University Press.

Jardine, R. J. 1994. One perspective of the pre-failure deformation characteristics of some geomaterials. Preprint, *Int. Symp. on Pre-failure Deformation Characteristics of Geomaterials*, IS-Hokkaido, Sapporo.

King, M. S., Andrea, M. and Shams-Khansir, M. 1994. Velocity anisotropy of Carboniferous mudstones. *Int. J Rock Mech. Mining Sci. & Geomech.* Abst., **31**, pp 261-263.

King, M. S. Chaudhry, N. A. and Shakeel, A. 1995 Experimental ultrasonic velocities and permeability for sandstones with aligned cracks. *Int. J. Rock Mech. Mining Sci. & Geomech.* Abstr., **32**, pp. 155-163.

McCann, D. M. & Green, C. A. (1995). Geophysical surveying methods in a site investigation programme. Preprint, *Int. Conf. on Advances in Site Investigation Practice*, London.

Nauroy, J-F., Colliat, J.L., Puech, A. Poulet, D., Meunier, J. & Lapierre, F. (1995). Better correlation between geophysical and geotechnical data from improved offshore site investigations. Preprint, *Int. Conf. on Advances in Site Investigation Practice*, London.

Paul, M. A. & Talbot, L. A. (1995). *Geotechnical and seismic analysis of a deep sedimentary sequence on the UK continental margin.* Final report of MTD Research Grant GR/J/07358.

Roesler, S. K., (1979). Anisotropic shear modulus due to stress anisotropy. *Proc. ASCE, J. Geotechnical Engineering Division*, **105**, No GT7, 871-880.

Sams, M. S., Worthington, M. H., King, M. S. and Shams-Khanshir, M. 1993. A comparison of laboratory and field measurements of P-wave anisotropy. *Geophys. Prospecting*, **41**, pp. 189-206.

Sams, M. S., Neep, J. P., Abraham, J., King, M. S. and Worthington, M. H. 1994. *The measurement of frequency dependent attenuation in sedimentary rocks.* Unpublished Final Report: Elastic Wave Propagation and Borehole Seismic Imaging Techniques, Department of Geology, Imperial College, London.

Stoker M. S., Hitchen K. and Graham C. C. (1993). *United Kingdom offshore regional report: the geology of the Hebrides and West Shetland shelves, and adjacent deep water areas.* London: HMSO for the British Geological Survey.

Shibuya, S., Mitachi,T., Yamishita, S. & Tanaka, H. (1995). Recent Japanese practice for investigating elastic stiffness of ground. Preprint, *Int. Conf. on Advances in Site Investigation Practice*, London.

Stoker M. S., Leslie. A. B., W. D. Scott, J. C. Briden, N. M. Hine, R. Harland, I. P. Wilkinson, D. Evans & D. A. Ardus. (1994). A record of late Cenozoic stratigraphy, sedimentation and climate change from the Hebrides slope, NE Atlantic ocean. *J. of the Geological Society, London,* **151,** pp. 235-49.

Talbot, L. A., Paul, M. A. & Stoker, M. S. (1994) Geotechnical studies of a Plio-Pleistocene sedimentary sequence from the Hebrides slope. NW UK continental margin. *Geo-Marine Letters* **14,** pp. 244-51.

Tatsuoka, F. and Shibuya, S. (1992). *Deformation characteristics of soils and rocks from field and laboratory tests.* Report of the Institute of Industrial Science, University of Tokyo, **37,** No. 1., (Serial No. 235).

Thomsen, L. (1986). Weak elastic anisotropy. *Geophysics,* **51,** pp. 1954-1966.

Viggiani, G. & Atkinson, J. H. (1995). Interpretation of bender element tests. *Geotechnique,* **45,** No. 1, 149-154.

Investigation of the fabric of engineering soil using high resolution X-ray densimetry

M.A.PAUL*, G.C.SILLS[†], L.A.TALBOT*, B.F.BARRAS*

*Department of Civil and Offshore Engineering, Heriot-Watt University, Edinburgh, UK

[†]Department of Engineering Science, University of Oxford, Oxford, UK

INTRODUCTION

Natural sediments are complex materials whose fabric usually influences their engineering behaviour. High resolution X-ray densimetry (Been, 1981) allows the non-destructive investigation of this fabric which is of particular value for cores or test specimens that cannot be dissected for fabric studies. Although originally developed for the study of early consolidation in very soft clays, the method has proven valuable in the examination of deep ocean sediments (Edge & Sills, 1989), in the geoacoustic study of continental shelf and slope sediments (Paul & Talbot, 1991) and in fabric studies at the Bothkennar research site (Paul, Peacock & Wood, 1992).

In this paper two case studies are presented which illustrate this work and which together have involved the examination of a total of almost 100m of core samples. The examples are drawn from the coastal and offshore areas of Scotland (Figure 1). The sediments involved are, in the first example, Late Devensian sandy silts and muds which have accumulated by various slope processes on the continental margin west of the Shetland Islands and which have allowed a comparison of the density signatures between gross facies types and, in the second example, estuarine mudflats of Holocene age at the Bothkennar research site near the head of the Firth of Forth in which densimetry has aided the recognition and very detailed description of the engineering facies.

Although drawn from very different environments, these sediments are all fine grained, soft and normally to lightly overconsolidated. In addition, they share a number of characteristics which are suited to the X-ray method. They generally lack non-horizontal fissures and other inclined structures so that the horizontal X-ray beam integrates densities across fairly uniform planes and resolves contrasts well. In all cases but one they have preserved a detailed record of primary bedding and associated boundary structures such as ripples, laminae and inter-penetration of beds which have a clear genetic significance: in some places they are also affected to a greater or lesser extent by bioturbation which has modified or destroyed the visible primary fabric. Thus they each possess features which may be expected to give distinctive density signatures and so are excellent candidates for study by this technique.

X-RAYS IN CORE ANALYSIS

When X-rays are passed through soil, the beam is attenuated by the different soil components and different densities within the soil. This attenuation can be used·to examine an undisturbed core or sample both qualitatively and quantitatively.

The simplest application of X-ray technology is in the production of X-radiographs. In a typical system, the complete or longitudinally sectioned core, or a thin slab, is laid on or close

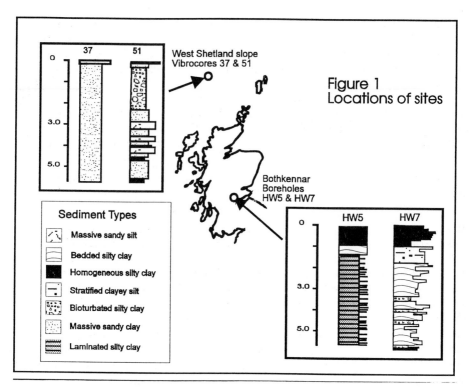

West Shetland slope
Vibrocores 37 & 51

Figure 1
Locations of sites

Sediment Types

Massive sandy silt

Bedded silty clay

Homogeneous silty clay

Stratified clayey silt

Bioturbated silty clay

Massive sandy clay

Laminated silty clay

Bothkennar
Boreholes
HW5 & HW7

HW5   HW7

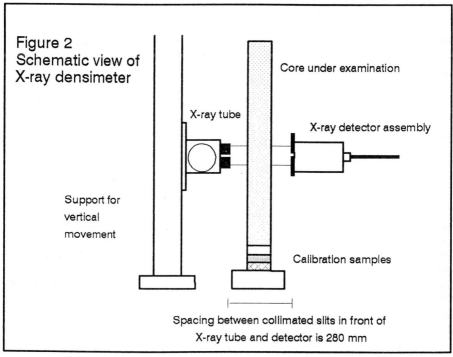

Figure 2
Schematic view of
X-ray densimeter

Core under examination

X-ray tube

X-ray detector assembly

Support for
vertical
movement

Calibration samples

Spacing between collimated slits in front of
X-ray tube and detector is 280 mm

to a length of film and exposed to X-rays. The higher the transmission of X-rays, the darker will be the film or negative. With this technique, it is possible to observe inclusions such as individual clasts or large shells, and features such as cracks. The use of a comparatively thin section of the core allows better resolution of lateral non-uniformities and a more precise identification of the thickness of non-horizontal layers, but the core itself is damaged by being sectioned. This technique provides a qualitative measure of soil variability in the core. Attempts to convert the film contrast to a quantitative measure of density have failed due to the difficulty of controlling the development process. Other problems with precise interpretation of the film arise as a result of parallax, which depends on the focal length of the X-ray tube and the distances between X-ray source, core and film.

For the radiographic results reported in this paper, the cores were sectioned longitudinally to a thickness of about 17mm, the X-ray was run at 55kV and the film negatives have been printed as positives, so that the lighter regions correspond to lower densities.

An alternative method of monitoring the X-ray attenuation uses a detector crystal such as sodium iodide, coupled with a photomultiplier assembly (Been & Sills, 1981). Figure 2 shows a photograph of the system. A finely collimated beam of X-rays at 160 kV is traversed down the core and the output from the photomultiplier is logged. Some variation in count rate occurs due to fluctuations in the intensity of output of the X-rays, so that a suitable integration time for the counting must be set, along with a suitable speed of traverse of the X-ray head. The values chosen for general use leads to a spatial resolution of the order of ± 1mm. The count rate N obtained by such a system can be related directly to density γ through the relationship $N = N_0 e^{-k\gamma}$, where N_0 and k are constants whose values depend on the energy and intensity levels of the X-rays and on the sediment mineralogy. Thus a conversion of count rate to density requires two independent calibration results. These are supplied by calibration samples of known density made up in the same core liner material that is used for the core being analysed.

The accuracy of conversion from count rate to density depends on the uniformity of the core liner as well as on the mineralogical similarity between the calibration samples and the sediment in the core. A realistic accuracy for the density measurement of the cores to be described in this paper is of the order of ± 0.01 Mg m^{-3} although the potential accuracy of the system as applied to long term settling column experiments in the laboratory is of the order of ± 0.002 Mg m^{-3}. The technique is thus well suited to the investigation of a variety of sedimentary structures which involve variations in the packing of the soil framework or gross variations in local density, such as inclusions, primary bedding, grading variations and fabric disturbance (whether natural or as a result of sampling).

CASE STUDIES

Slope deposits on the West Shetland slope

This case study illustrates the identification of gross fabric differences within and between core samples. Plio-Quaternary sediments from the West Shetland slope consist of sediments of both glacial and marine origin; their relative proportions depend on the location and timing of deposition. In this area a stratigraphy has been defined based on seismic sequences, i.e. on the basis of acoustic signature and stratigraphic position (Stoker, Harland, Morton & Graham, 1989; Stoker, Harland & Graham, 1991). X-ray densimetry scans have been carried out on selected shallow vibrocores. Particular attention has been paid to the distinction between acoustically transparent mass flow units and laterally equivalent acoustically layered units since they possess differing density signatures which reflect differences in the depositional processes.

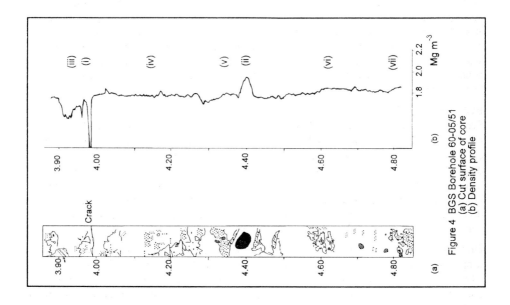

Figure 4 BGS Borehole 60-05/51
(a) Cut surface of core
(b) Density profile

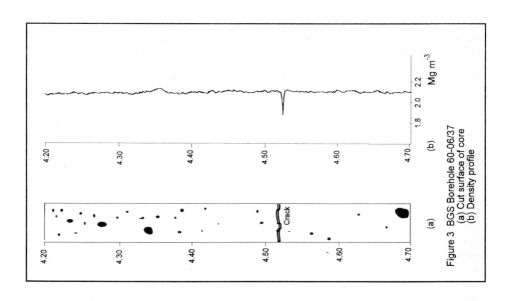

Figure 3 BGS Borehole 60-06/37
(a) Cut surface of core
(b) Density profile

BGS vibrocore 60-06/37, from the acoustically structureless Morrison Sequence, is lithologically uniform in appearance. Figure 3 shows the section from 4.20m to 4.70m below seabed. There is little or no fabric variation with depth (Fig 3a). a uniformity also reflected in geotechnical properties such as water content (Paul, Talbot & Stoker, 1993). On the corresponding density trace (Figure 3(b)) only the larger dropstones and a crack (an artefact) at 4.53m depth are distinguishable from an otherwise uniform background due to the sandy mud. Many dropstones visible in the core section appear too small to show up in the trace, since the sediment itself has a relatively high, almost constant, density of 2.1 Mg m^{-3} and they contribute little additional mass to the overall cross section of the core. This uniform featureless trace is characteristic of nearly all of this core.

BGS vibrocore 60-05/51 is from the acoustically well-layered Faeroe-Shetland Channel Sequence. Figure 4(a) shows the visible fabric features from 3.90m to 4.80m depth. These include: dropstones (e.g. at 4.40m depth); burrows infilled by silt or fine sand and marked by colour changes in the sediment; other general bioturbation is delimited by a fine line on the diagram. Figure 4(b) shows the corresponding density trace from this section of core; the features annotated are described below. The general appearance of the signature reflects the lithofacies changes and the complexity of the fabric; we observe that since many fabric features do not occupy the whole cross section the magnitude of the relative density changes is subdued compared with those described from the other sites.

On the trace, features (i) and (ii) show the position of a crack [artefact] at 3.98m and a large dropstone at 4.40m depth. The area of low density (iii) at the top of the section is produced by a hole (which could also be an artefact). Between 4.00m and 4.30m the density trace is very uniform (iv), a signature typical of a bioturbated area of sediment (see Bothkennar case study), at the base of which there is a rapid fall in density which corresponds to the change in lithofacies from sandy silt to sandy mud. At 4.30m depth the density increases noticeably over a few centimetres (v); although there is no major change of lithofacies, this change correlates with the sandy infilling of a major burrow shown in Figure 4(a). Below the dropstone (ii) the density increases steadily with depth; notably at (vi) corresponding to the bioturbation at 4.60m and at 4.83m (vii) corresponding to the coarse burrow at the base of the section.

In these cores the density signature varies with the seismic sequence from which the sediment was retrieved due to the particular fabric features in these sequences. The uniform, featureless, high density signature which appears characteristic of the Morrison Sequence reflects the accumulation of this sediment by uniform, mass flow processes, whereas the more complex, lower density signature obtained from all vibrocores throughout the Faeroe Shetland Channel Sequence appears to result from the bottom current re-working that has affected the area. Although, in both examples, every detail in the signature can be related to an individual fabric feature, it is considered advisable in the first instance to establish the nature of this relationship by the direct examination of some representative cores or core sections.

The Bothkennar estuarine clay deposits
By contrast with the study above, this case study illustrates the use of the densimeter to investigate very fine details of the soil fabric at scales down to a few millimetres. It is based on the suite of boreholes sunk at the EPSRC Bothkennar research site as part of a major study of the engineering geology and depositional history of the soft clay sequence at the site. The cores from this study have been scanned by the densimeter, thin slabs have been X-ray photographed and split cores have been described in detail after cleaning with an "osmotic" knife (a thin bladed knife connected to a low-voltage DC electric supply which prevents

smearing by re-orientation of the clay particles in the electric field). These procedures have enabled a very detailed comparison to be made of the soil fabric and the corresponding density profile.

The EPSRC site is located on former estuarine tidal flats of Holocene age near the head of the Firth of Forth (Paul, Peacock & Wood, 1992). The uppermost 20m of the geological sequence consists of silty clays which are divisible principally into bedded, laminated and mottled facies: the first is composed of primary sedimentary units, separated by bedding surfaces; in the second the units are eroded and separated by numerous silt laminae; in the last the sediment has undergone post-depositional reworking by organisms which has partially or totally obliterated the primary structure. There is often repetitive interbedding of the bedded and mottled facies at a scale of 10-100mm. The three facies have differing engineering properties (Hight, Bond & Legge, 1992; Little, Muir Wood, Paul & Bouazza, 1992) that are attributed to differences in their fabric: these differences can be clearly seen on high resolution density traces from all the boreholes so far examined.

Figures 5 to 7 show density profiles from the different facies presented together with sediment logs and X-ray photographs. Figure 5 (4.87-5.17m depth in HW7) shows a typical section which includes both bedded and mottled facies. Radiocarbon dating (Paul, Peacock & Wood, 1992) in conjunction with the sea level curve for the Forth valley (Sissons & Brooks, 1992) suggests that the sediments were probably deposited in an inter-tidal to subtidal environment in water of only a few metres depth at most. The X-ray photograph and sediment log show that the lower part of this section (A in Fig. 5) is constructed from individual beds of silty clay each about 5-20mm in thickness, sometimes separated by silt partings and laminae (B). At the top of the section (C) the sediment is bioturbated and the primary sedimentary bedding has been lost. The bioturbation is marked by a visible pattern of mottling, which shows a transition from a large, sparse style to a smaller, denser style over a distance of about 50mm. In the lowermost, bedded section, X-ray attenuation measurements reveal the presence of primary depositional bedding and can resolve density variations within individual beds. It should be noted that the radiographs were exposed and developed in conditions aimed at optimising the show of detail, so that on the photographs the apparent contrast does not necessarily represent equivalent density changes. There is, nevertheless, a broad correlation between the positions of the dark bands on the photograph and the higher density parts of the density profile. A detailed comparison of the density profile with the X-radiograph shows that individual bed contacts are visible (i), as are silt laminae (ii) and that the saw tooth pattern (iii, iv) arises from the upward gradation from higher density clayey-silt to lower density clay within each sedimentary unit. After the transition from bedded to mottled facies (v) the density profile is very uniform, due to mixing by bioturbation as indicated by the dense pattern of small mottling on the cut surface. The onset of mixing occurs over a short distance (vi) which corresponds to the change in the pattern of mottling noted above.

Figure 6 (3.65-3.95m depth in HW5) shows a section in sediments of the laminated facies, which were deposited in a subtidal channel in water of a few metres depth. The core shows a regular, repetitive sequence of fine silt laminae, separated by clay beds often of no more than 5mm thickness. For the most part these beds are horizontal (or nearly so) from one side of the core to the other, with undulating eroded tops and graded bases: there are occasional cross cutting horizons (e.g. at A in Fig. 6) and water escape structures (B). The X-radiograph reveals a wealth of such fabric information down to the limit of its resolution. The density profile is very different from that in Figure 5 and reflects the laminated structure. Again, there is broad correlation between the higher densities and the darker features apparent on the radiograph, although not all the thin, high contrast features correspond to an equally thin layer

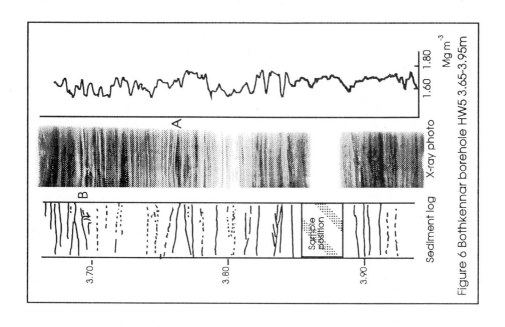

Figure 6 Bothkennar borehole HW5 3.65-3.95m

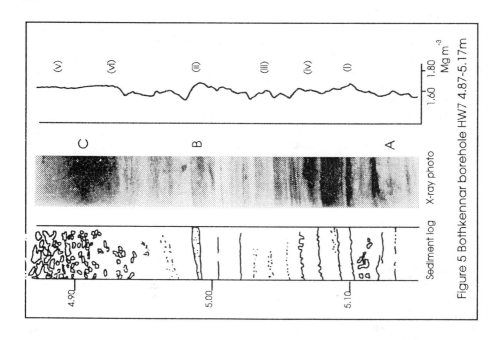

Figure 5 Bothkennar borehole HW7 4.87-5.17m

on the density profile. This may suggest that the layers in the complete core are slightly inclined, while the section used for the radiograph happens to have been taken across the sense of the inclination. Two aspects are noticeable in particular: the magnitudes of the peaks are generally larger than in the bedded facies, and the majority of peaks are symmetrical. The former effect is clearly due to the density of packing in the silt laminae and the latter is due to their internal uniformity; only occasionally do gradations in density produce any discernable saw-tooth pattern.

Figure 7 (2.17-2.47m depth in HW7) shows a section of core from the sub-crust transition (Paul & Barras, 1993). This is composed of sediments belonging to the bedded facies which were deposited under inter-tidal conditions, and which often show evidence of fabric disruption by large burrowing organisms; they also show evidence such as silt clasts, mud balls and flakes which indicate energetic conditions with frequent episodes of minor erosion. Several such features are visible in Figure 7: e.g. a major burrow (A); disturbed and broken laminae (B); water escape (C). Overall, the effect of this disturbance is to introduce small scale lateral variation which reduces the average density contrasts within the section and so smooths the density trace in a manner analogous to, although less complete than, the bioturbation associated with the mottled facies. Comparison of the trace in Figure 7 with those of Figures 5 (lower part) and 6 illustrates this point.

DISCUSSION:

Comparison with conventional methods
Figure 8 shows a quantitative comparison of the value of the density measured by the densimeter with that obtained on this core by a conventional cutting ring at 50mm intervals (dotted line). The data is taken from an earlier study of sediments from the central North Sea (Paul & Jobson, 1991). Three points can be made: (1) the overall shape of the conventional profile matches that of the densimeter trace at the 50-100mm scale, although with considerable loss of detail at finer scales; (2) the trace largely lies within the [two standard deviation] error bound of the conventional method (calculated from previous work on this core (Paul & Jobson, 1987) to be ± 0.08 Mg/m^3) and thus the values are quantitatively comparable, although this is obviously dependent on the accuracy of the calibration; (3) there is a systematic difference between the curves over the lower two thirds of the core. This suggests either a small calibration discrepancy over the whole core, a significant change in sediment mineralogy one third of the way down, or a less than precisely vertical alignment of the core. For the purpose of fabric investigation, these last differences are not considered to be significant: the conventional measurements could in any event be used to recalibrate the trace (either overall or locally) if so required. The overall conclusion is that the densimeter gives results which are directly comparable to conventional methods, is capable of better resolution and, with accurate calibration, of equal or better accuracy.

Resolution of fabric features
These case studies show that the method is able to resolve many small details of the sedimentary fabric. As already noted, this is best shown when the structures involved are sub-horizontal and laterally continuous across the core. Such features are the norm in a bedded, tidal mud such as the Bothkennar clay, which would therefore be expected to produce rich densimeter signatures. Features that are laterally impersistent, such as lenses, clasts, individual shell fragments etc, usually give less well defined features on the density profiles, although they may show up on the radiographs.

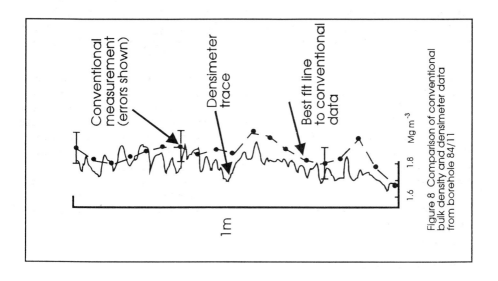

Figure 8 Comparison of conventional bulk density and densimeter data from borehole 84/11

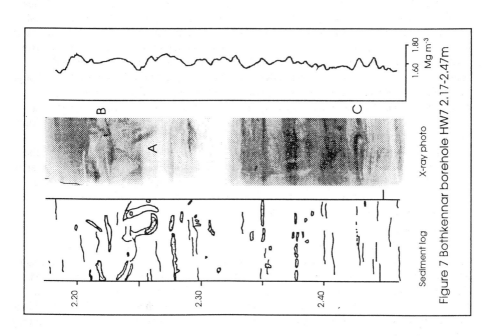

Figure 7 Bothkennar borehole HW7 2.17-2.47m

The profiles show that under good conditions a vertical resolution down to 1-2mm appears possible and also that a density contrast between two adjacent layers of a little as 0.01 Mg m^{-3} detectable within the signatures. Calculation shows that at typical soft sediment water contents this corresponds to a water content difference between the layers of 1% or less. In these sediments, variations of this magnitude appear common over short (millimetre) distances and this may well be the case more generally, although the Authors are unaware of other detailed studies at this fine scale which might confirm this idea.

Application to site investigations and the study of geological processes
The densimeter method has proven very valuable for the examination of whole cores. As described earlier, it has enabled the facies succession to be constructed and the location of particular fabric features to be identified prior to the extrusion and sectioning of the sample. This has allowed the sampling pattern to be decided at an early stage and problematic or difficult fabrics to be identified. When combined with X-radiography a very detailed picture of the soil fabric has been obtained which aids in the explanation of the geotechnical behaviour of individual test specimens.

In addition to its applications to engineering investigations, the densimeter gives novel information about geological processes at the small scale, e.g. density variations within primary depositional bedding; the correlation between the degree of surface mottling and the extent of internal mixing by bioturbation; quantitative data on layer statistics and spatial spectra of laminae and other planar structures. Although it is beyond the scope of this paper to discuss these aspects in detail, the Authors note that these possibilities exist and hope to consider them in later publications.

CONCLUSIONS
These examples have shown that the detailed pattern of density variation within a sediment core can be established quantitatively at the millimetre scale by X-ray densimetry and that the results compare well with those from conventional methods which are commonly used for classification. In the vibrocores from the West Shetland slope, the distinction between gross seismic facies can be explained by the density profiles that arise from visible differences in their respective sedimentology. At Bothkennar, the fine details of the bedded, mottled and laminated facies can be identified in the density profiles and related closely to visible and radiographic features. Thus in each of these examples X-ray densimetry has provided additional information to assist the description and evaluation of geological samples and other site investigation data.

ACKNOWLEDGEMENTS
Sample collection at Bothkennar was undertaken by staff of the Building Research Establishment to whom we express our thanks. The offshore cores were collected by the British Geological Survey as part of the UK Regional Mapping Programme and we are grateful to Mr D.A.Ardus for the opportunity to study them and to BGS staff for helpful discussions. Laboratory assistance was given by Hugh Barras, to whom we express our gratitude and sincere thanks. The financial support of the former SERC (GR/E/74748, GR/H/14151) and MTD Ltd (GR/E/81289) is very gratefully acknowledged.

REFERENCES
Been, K. (1981). Non-destructive soil bulk density measurements by X-ray attenuation. *Geotechnical Testing Journal* **4**: 169-176.

Been, K. & Sills, G.C. (1981). Selfweight consolidation of soft soils: an experimental and theoretical study. *Geotechnique* **31**, 519-535.

Edge, M.J. & Sills, G.C. (1989). The development of layered sediment beds in the laboratory as an illustration of possible field processes. *Q. J. Engng. Geol.* **22**, Part 4, 271-279.

Hight, D.W., Bond, A.J.& Legge, J.D.. (1992). Characterisation of the Bothkennar clay: an overview. *Geotechnique* **42:** 303-347.

Little, J.A., Wood, D.M., Paul, M.A. & Bouazza, A. (1992). Some laboratory measurements of permeability of Bothkennar clay in relation to soil fabric. *Geotechnique* **42**, No. 2, 355-361.

Paul, M.A. & Jobson, L.M. (1987). On the acoustic and geotechnical properties of soft sediments from the Witch Ground basin, central North Sea. *SERC (Marine Technology Directorate) Research Grant Report* Heriot-Watt University: Edinburgh.

Paul, M.A. & Jobson, L.M. (1991). Geotechnical properties of soft clays from the Witch Ground Basin, central North Sea. in Forster, A., Culshaw, M.G., Cripps, J.C., Little, J.A. & Moon, C.F. *Quaternary Engineering Geology.* Geological Society Engineering Geology Special Publication 7, 151-6.

Paul, M.A.& Talbot, L.A. (1991). Thematic geotechnical analysis of sediments from the continental slope northwest of the British Isles. *MTD Ltd Research Grant Report* Heriot-Watt University: Edinburgh.

Paul, M.A., Peacock, J.D. & Wood, B.F. (1992). The engineering geology of the Carse clay at the National Soft Clay Research Site, Bothkennar. *Geotechnique* **42**: 183-198.

Paul, M.A., Talbot, L.A. & Stoker, M.S. (1993). Geotechnical properties of sediments from the continental slope northwest of the British Isles. In: Ardus DA, Clare D, Hill A, Hobbs R, Jardine RJ and Squire JM (Eds) *Offshore Site Investigation and Foundation Behaviour.* Dordrecht: Kluwer Academic Press. 77-106.

Sissons, J.B. & Brookes, C.L. (1971). Dating of the early Postglacial land and sea-level changes in the western Forth valley. *Nature Phys. Sci.* **234**, 124-127.

Stoker, M.S., Harland, R. & Graham, D.K. (1991). Glacially influenced basin plain sedimentation in the southern Faeroe-Shetland Channel, northwest United Kingdom continental margin. *Marine Geology,* **100,** 189-199.

Stoker, M.S., Harland, R., Morton, A.C., & Graham, D.K. (1989). Late Quaternary stratigraphy of the northern Rockall Trough and the Faeroe-Shetland Channel, northeast Atlantic Ocean. *Journal of Quaternary Science.* 4(3) 211-222.

Considerations in the geotechnical testing of contaminated samples

R G CLARK [1] and G P KEETON [2]
[1] C L Associates, Birmingham, United Kingdom
[2] Soil Mechanics Limited, Doncaster, United Kingdom

INTRODUCTION

There continues to be an increasing need for the investigation of contaminated sites either because of a need to remediate the site or because development of the site is proposed or both. For various reasons there may be a need to determine the geotechnical properties of the various strata on the site, such as strength, compressibility and permeability.

In some situations it will be appropriate to remove the contaminated soils from the site in which case geotechnical testing is unlikely to be necessary. However, there is now a move towards either leaving the soils on site without treatment (depending on the type and degree of contamination) or to encapsulate/immobilise the materials. In either case there may be a need to carry out geotechnical laboratory testing of contaminated soil samples.

This poses a number of questions and the authors would suggest that not all these questions are presently being addressed by those who are responsible for either specifying, scheduling or carrying out the tests. These questions include matters of safety, practicality and the validity of the results.

It is recognised that an easy way out of solving these problems might be to not carry out any geotechnical testing but this does not then resolve the question of how geotechnical parameters are to be assessed for design. If assumed parameters are used for design these are likely to be somewhat over conservative due to the uncertainty in the assumptions made or alternatively, of course, an unsafe design could result.

Thus the requirements of testing contaminated samples, including the provision of appropriate funds and time should be recognised in the scoping of an investigation in order to ensure that unnecessary additional costs are not incurred later in a project. The purpose of this paper is primarily to air some of the concerns so that discussion on the topic can thereby be generated.

THE PRESENT SITUATION

At present the proportion of contaminated samples that are being sent to geotechnical laboratories is not high suggesting that either geotechnical parameters are not required for many contaminated sites or, which is more likely, that for those situations where parameters are required engineers are refraining from specifying geotechnical tests.

Nevertheless, many laboratories have experienced the situation where samples have been received at the laboratory without any indication that they are from a contaminated site. This has already lead to the situation where laboratory personnel have inhaled fumes which have caused dizziness, nausea and headaches such that they have had to receive hospital treatment. Cases have also been reported where staff have experienced skin irritation due to handling contaminated samples.

If geotechnical tests are carried out on contaminated samples there is often presently a lack of awareness of the effect that the contamination might be having either on the properties of the material being tested or the test itself. This situation must not continue.

HEALTH AND SAFETY

There is a responsibility on all those who despatch samples to laboratories to provide as much information as possible to the laboratory regarding the condition of the site from which the samples were obtained and/or the nature of possible contamination in the samples. This is particularly relevant in respect of geotechnical laboratories because there is usually a presumption on the part of the laboratory staff that samples sent for geotechnical testing are not contaminated. Under the COSHH regulations there is a legal requirement to provide information relevant to health.

A means of classifying contaminated sites and landfills in respect of the degree of risk to investigation workers is contained within the Site Investigation Steering Group publication entitled "Guidelines for the Safe Investigation by Drilling of Landfills and Contaminated Land" which categorises sites as Green, Yellow or Red depending upon the known or assumed degree of contamination. The categorisation provides a suitable means of also identifying samples. Coloured labels, tapes, tags etc that will not be subject to damage or accidental removal can be used to identify the samples. This is discussed further below.

In view of the need to ensure safe working conditions for laboratory staff and to provide an appropriate testing environment in respect of technical requirements (as discussed later in this paper), consideration may need to be given to setting up a separate laboratory just for the testing of contaminated samples. This, of course, has capital cost implications but could be justified once the industry recognises the need for an overall system for the specifying of contaminated samples.

SAMPLE HANDLING AND STORAGE

Labels should always be on the outside as well as the inside of sample containers in order to reduce the need to handle the samples. Since sample containers may require safety washing prior to transfer to the laboratory it is essential that labels are indelible to the washing medium. Whilst this would usually be water, other substances, eg for removal of oils etc, may be required.

Samples should be protected from accidental damage and exposure. It is therefore appropriate to consider the use of large plastic tubs, or where needed glass containers, as an alternative to bulk bags for known or suspected contaminated samples/sites or to seal the samples inside drums. Sealing inside drums also has the advantage of containing gases or vapours.

All geotechnical laboratories should have appropriate procedures and facilities for the handling and storage of contaminated samples. Preferably the storage should be separate from the storage of uncontaminated samples and such that any fumes given off by the samples do not cause a problem. A means of collecting any fluids which might seep from the samples should also be provided. Restricted access to such an area might also be prudent in order to avoid mistaken selection of samples and hence inappropriate handling and testing procedures being followed and/or personnel receiving unnecessary exposure to any fumes that might be present.

Details of the storage requirements should be given on the sample movement notification sheets to ensure continuity of care. Undisturbed samples should be clearly marked 'top' and 'bottom' and kept the right way up to avoid migration of contamination beyond what may have happened in situ.

SAMPLE PREPARATION

Sample preparation is obviously the stage at which laboratory staff could be most at risk because this is likely to be the first time that the samples will be removed from sample containers, sample bags, etc. Many forms of contamination, but by no means all, can often be detected by olfactory means even if laboratory staff are not familiar with the exact odour given off by each contaminant. But the question needs to be asked as to why laboratory staff should find out the hard way regarding whether samples are contaminated.

On the assumption that the specifier of the geotechnical testing and/or the consignee has indicated that the samples are potentially contaminated then a very sensible procedure, which may appear obvious but is often not done, would be to carry out chemical testing prior to geotechnical testing. Much smaller sub samples are required for chemical testing than for geotechnical testing. Also chemical tests (and biological tests in some cases) lend themselves more readily to hazard control than do geotechnical tests. For example, a GC or GC/MS analysis is easier to control from a chemical hazard point of view than say a compaction test which has the potential to make contaminants air borne.

Fig 1. Simulated photograph of different degrees of protection.

All too often both chemical tests and geotechnical tests are carried out on contaminated samples but because of project programme pressures the results of the chemical tests are not known until after the geotechnical tests have been completed. One of the messages expressed in the SISG guidance document is that all those involved with contaminated sites must appreciate the consequences of the site being contaminated. It is no good ignoring this either for convenience or commercial reasons. The photograph in Fig 1 compares different degrees of protection.

Materials used by the laboratory for sample preparation such as distilled/deionised/tap water, paraffin wax/sealants and chemicals could themselves react with the contamination in the sample. Changes to standard procedures may therefore need to be considered. Any change, however, needs to be balanced against the eventual validity of interpretation of the results.

Exposure to air at various temperatures may affect some less obvious laboratory activities such as sample weighing when for example volatile organic compounds are present and, therefore, the overall laboratory environment should be given full consideration when contemplating the testing of contaminated samples. In this regard a laboratory operating at say 15°C would achieve a different result from a laboratory operating at 25°C even though both are carrying out the test correctly.

LABORATORY EQUIPMENT

Irrespective of whether laboratory equipment is to be used in the main laboratory or in a separate laboratory set aside for the testing of contaminated samples, there are some aspects of equipment which need consideration in terms of the validity of testing. Some of these are dealt with in the sections below.

When the type of contaminants have been defined, consideration should be given to the effects of these on any items of equipment with which they are likely to come into contact and procedures derived to eliminate potentially adverse reactions. Triaxial cell platens for example may be attacked by samples with extreme pH values usually producing gases which inevitably give rise to errors in pressure measurement. Metal containers used for moisture content (a BS requirement for fine grained soils) may react with similar contaminants to give erroneous results. These reactions may only happen under certain conditions of temperature/humidity. Rubber membranes can deteriorate badly in contact with oils and other substances. Test sieves and sample dividers used for screening/wet sieving may react as described above or be clogged by oils/hydrocarbons. It may be preferable to remove such oils prior to sieving/screening but only if the subsequent test validity is acceptable.

It can be noted that different metals are often used to manufacture the same type of apparatus. Therefore, reactions, if they occur, could be different between a number of triaxial cells or a number of oedometers, etc.

It is an obvious but nonetheless essential feature of testing contaminated samples that all equipment, particularly that which has been in direct contact, is thoroughly cleaned to avoid cross contamination of samples. As such there should be precise procedures developed to deal with this, especially where certain items, eg porous stones, are difficult to visually assess whether they are clean. Use, for example, of an ultrasonic bath would be an advisable procedure provided the bath itself is then decontaminated.

PRACTICALITY

Initially it might be thought that certain tests on contaminated samples are not practical. For example the carrying out of a compaction test on a sample that contains asbestos which has the potential to release asbestos fibres into the air. For some laboratories this might indeed not be feasible to carry out in a safe manner, because the test will require the use of breathing apparatus and a separate isolated room which is subsequently decontaminated for asbestos fibres. Some laboratories may feel that they wish to provide this facility.

Other tests may require the use of a fume cupboard such that the operator can handle the sample within the fume cupboard but at the same time be isolated from the atmosphere surrounding the sample. Such techniques are now well established within the nuclear and chemical industries. The cost of such facilities are not necessarily therefore inhibitive but there is a consequential increase in the time required to carry out the test. The extracted fumes from fume cupboards and contaminated water from washing/testing must be appropriately dealt with.

TEST PROCEDURES

It is not proposed in this paper to go through each one of the test methods and discuss how the test should be carried out in the event that the samples are contaminated. Nevertheless, a number of examples would be appropriate.

Classification Tests (BS 1377 Part 2)

For moisture content determination (normally carried out at 105-110°C) where volatile or low flash point contaminants are present it is appropriate to consider longer drying times at lower temperatures to facilitate consistency of data. Consideration should also be given to the effects of contaminants on the use of moisture content as a parameter or manipulation of water content within a test, eg liquid limit. This would be relevant, for example, in the presence of oils, flocculants or substances such as organic compounds that affect the double layer around clay particles.

A particular example here of the potential for procedural changes to allow compliance with likely precautions and production of "valid" results would be to consider the use of a heavy cone for plastic limit tests. This may, or may not, reduce the technical error caused by the effects of contaminants but would certainly allow tests on some materials to proceed whereas the rolling method would prove impossible either due to the material or the use of heavy gloves for protection.

Grading analyses should be pre-appraised for validity in the light of potential;

- Aggregation of particles due to contamination and/or drying either initially or as part of the test; pre-drying of samples should be assessed as to whether it is appropriate and, for cohesive materials especially, total wet sieving with a moisture content control for dry mass should be considered as an alternative.

- Clogging of sieves during wet sieving.

- Chemical reaction with sieve material.

- Chemical or biological activity as a result of copious washing with tap water or addition of dispersant.

- Effects of the use of sodium hexametaphosphate as the dispersing agent. This should perhaps be re-appraised in the light of initial tests.

- Effects of pretreatment with hydrochloric acid for carbonates and/or hydrogen peroxide (H_2O_2) for organics. In effect, all chemical treatments should be re-appraised in the light of the contaminants present. BS 1377 specifically warns of dangers when using H_2O_2 in the presence of manganese oxides or sulphides - especially when heated. Failure to carry out these pretreatments due to precautionary or technical measures may of itself therefore require consideration of the validity of results.

- Effects of the viscosity of the sedimentation liquid. This may have to be reviewed if the contaminants are likely to be soluble in water and corrections applied to calculations. In such cases and where oily contamination exists, extra care may be necessary in the cleaning of the pipette between samplings. If drying and weighing of the pipette specimens constitutes a health and safety problem, then the hydrometer (which can be done to BS on natural material and is "direct" reading) would be the preferable test.

Particle density tests would not appear to require any additional technical procedures, but the results may need an explanation if extreme values result, and often do. Ironically, the test may prove useful in some cases to confirm the presence and, indirectly, the concentration of contaminants as is often done in respect of coal content in colliery spoils. Aggregation would again be a possible factor and an alternative reference liquid to deionised water may have to be considered. It may be appropriate to define from the outset as to whether kerosine is permissible as the liquid.

Chemical Analyses (BS 1377 Part 3)
There must be fundamental questions regarding the validity of geochemical tests carried out on contaminated samples in which, even if the contaminants are known, their potential for interference may not be. All chemical test procedures should be assessed by an analytical chemist/biologist and preferably thereafter any non standard procedures carried out by a chemical laboratory, not a geotechnical laboratory. Any developments/observations arising from these tests would be better identified and reported by a chemist together with his/her professional opinion.

At present, should "normal" testing produce odd results in the geotechnical laboratory the first reaction is to send samples to the analytical chemist for the above checks and opinion. Therefore, once contamination has been confirmed, it would seem preferable for chemical testing to continue in the chemical laboratory rather than to revert back to the geotechnical laboratory.

Compaction Related Tests (BS 1377 Part 4)
The main consideration here is the possible effects on the all metal equipment. Were this to become a problem it could be solved for example with the use of rigid plastic moulds and a plastic coated end to the rammer. Such a mould may seem extreme for quick compaction tests but worth consideration for CBR tests which require the sample to stand for periods prior to the actual test.

Chemical/biological action during the compaction process, particularly if producing gases, may be important from an interpretative point of view since it is conceivable that parameters such as CBR could be affected.

Compressibility and Permeability (BS 1377 Part 5)

Once enclosed within the test cell and immersed in water the test procedures should not require amendment other than if very special precautions are required. For example, samples of extreme pH (<4 or >10) can have a very corrosive effect on the cell metal. Also intermittent anomalous settlement behaviour can occur during a test, due to the sudden escape of gas and subsequent micro collapse of the soil structure or just volume change.

By their nature, compressibility tests may be very sensitive to this type of problem due to the expulsion of water from the specimen under pressure; water which probably holds relatively high concentrations of whatever contaminant is present.

The condition of porous stones after test should be closely examined for any attack from the sample which could have caused deterioration and thus affect the settlement characteristics.

Samples should be described in as much detail as possible, subject to safety precautions, before and after each test such as to identify aggregation or separation of particles before or during the test. Aggregation (chemical bonding) of clay particles can have a very significant affect on the shape and thus laboratory interpretation of time settlement plots, which may not be obvious to those receiving only the evaluated C_v value. Swelling tests should be considered in a similar manner particularly if deionised/distilled water is being introduced to the sample.

For constant head permeability tests the cell and associated materials (rubber tubing etc) should be assessed for potential corrosion by the sample and effluent water recovered and disposed of to a set procedure. Again aggregation and the effects of specimen compaction need to be closely monitored and recorded.

Consolidation and Permeability in Hydraulic or Triaxial Cells. (BS 1377 Part 6)
Shear Strength Tests (Total Stress) (BS 1377 Part 7)
Shear Strength Tests (Effective Stress) (BS 1377 Part 8)

These are all considered together since effectively they all use similar equipment and are affected by contamination in similar ways.

Experience with colliery spoils has shown that high levels of ammoniacal concentrations (from washery) can be transformed into very high gas pressure in the course of a triaxial strength test with the gas, initially held within the sample, rising dramatically to the top of the sample in a pocket by the time the test is concluded. The effect of this on measured values of pore pressure can be considerable.

Reactions between the contaminants and the test equipment have been experienced many times, particularly porous elements and metallic end platens, perspex being recommended as an alternative for the latter. It should be noted that BS 1377 itself recognises the need to use tap water in triaxial cells due to the corrosive effects of deionised/distilled water on some types of cell seals. In recognising that deionised water can cause such problems then it is not difficult to envisage contaminants also doing so.

Probably the least recognised problem relating to this type of laboratory testing, is the effect of pressure upon the physical, chemical and biological structure of samples especially when gas is evolved, not necessarily during application of stress but at some point in the test at which the effective stresses rise or fall to certain unknown limits. It is sometimes misunderstood that the effective stress in a sample is often at its highest at the end of consolidation (not on application of the total stress) and at its lowest at some point during shear, often near failure (normally consolidated samples). The likelihood of physical breakdown increases as consolidation continues and if this occurs the test data will appear erroneous. However, if gases are being formed or concentrated by pressure, the release of this pressure may occur as the effective stress reduces in shear, thus again giving apparently erroneous data.

Similar behaviour to the examples above may be produced by contaminated samples. If the contaminants are known, it may be possible to identify potential laboratory "problems" and facilitate a reasoned approach to interpretation of the results. However, it is essential to balance test behaviour attributable to the sample against laboratory error/equipment malfunction. It is too easy for a rational explanation of one set of circumstances to be used incorrectly to explain others. As such it is helpful for the laboratory, where possible, to be forewarned as to possible anomalous behaviour, such that when events occur they are not hidden by the desire to produce "standard" test results.

It is the authors' belief that many of the concerns regarding contaminated samples in the laboratory could be reduced by having clear and detailed procedures as well as safe precautions. Given both of these it should be possible to derive acceptable geotechnical parameters safely. Acceptability of the data produced in terms of its interpretation will depend greatly upon the liaison between the chemist, the engineer and the laboratory.

VALIDITY OF RESULTS

The next consideration is whether the results of the geotechnical testing of contaminated samples are valid in respect of the subsequent use in any design. There are two considerations that might have a bearing on this. Firstly, if the contamination permanently modifies the physical properties of a soil and the test truly reflects this then it can be argued that the testing will produce results that can be appropriately utilised in design. On the other hand it is possible that the contamination will impair the ability of a particular test to produce a true indication of how a particular property will act in practice.

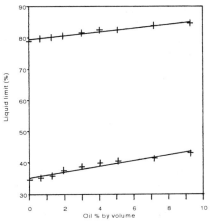

Fig. 2   Effect of oil contamination on the liquid limit of two clays.

Tests carried out by the authors demonstrate the effects of oil contamination on the liquid limit of two clays. The results in Fig 2 show an increase in liquid limit with increase in oil contamination. The question is which index values should be used in correlations to obtain other material properties? Do these other properties respond to the contamination in the same way?

823

In summary, there are three aspects which primarily need to be considered in respect of the effects of contamination on the validity of the results of geotechnical testing. These are :

- Chemical reactions within the samples between the contaminant and the natural solid material or groundwater

- Reactions between the contaminants and either the testing apparatus or the chemical preparations used for some tests

- Reaction of the contaminant or the contaminant/soil combination on exposure to air.

Another point which should be borne in mind is that temperature and humidity amongst other factors could affect the response of a particular contaminant eg. the change in viscosity with change in humidity and/or temperature or the precipitation of salts with change in humidity and/or temperature.

The natural inclination on the part of laboratory personnel might be to clean certain samples prior to testing, eg. the removal of surplus oil, either to avoid contamination of apparatus or just to make the test possible. Note should be taken of the possible consequences of this cleaning in respect of a change to the measured mechanical properties.

Procedures should be established whereby unusual behaviour, pretreatments, precautionary cleaning, suspect results etc., are all appropriately recorded by the laboratory so that they can be considered by the geotechnical engineer/designer and taken into consideration in the assessment of design parameters and the soil behaviour.

There are various examples in the literature which indicate the effects of various contaminants on soil properties. The following discussion does not seek to provide a comprehensive review of the literature but simply to provide examples of some of these effects.

Ho and Pufahl (1987) have published the results of tests to examine the effects of brine contamination on the properties of fine grained soils. They found that the brine had no discernable effect on particle size distribution and specific gravity of the two clays tested but that there was a decrease in the liquid limit and a slight increase in the plastic limit with increase in brine concentration. This was considered to be due to the presence of sodium ions (+ve) in the pore water molecules causing the negative charged clay particles to respond to the brine and thereby result in changes in the physical properties of the soils. Decreases in liquid limit with increase in brine concentration were also found by Chassefiere and Monaco (1983) on smectite-illite marine soils and by Ridley et al (1984) on montmorillonite clays.

Foreman and Daniel (1986) have carried out tests on a number of different clays permeated separately with water, methanol and heptane. The organic compounds methanal and heptane caused significant shifts (both positive and negative) in Atterberg limits and in the case of heptane destroyed the plasticity of the soil.

Ridley et al (1984) report that some soils are susceptible to significant increases in permeability as the electrolyte concentration increases in the pore fluid. Anderson (1982) found large increases in permeability of clays through which organic fluids had permeated. Ho et al (1987) found very little change in permeability due to brine contamination.

Boldt - Leppin et al (1994) carried out tests which showed the effects of gasoline contamination on an organophilic clay. They found a reduction in permeability from 1 x 10^{-7} m/s when water was used as a permeant compared to 1 x 10^{-10} m/s when gasoline was used as the permeant. This was attributed to the swelling of the samples by 9 or 10% due to the gasoline.

Meegoda and Rajapakse (1993) showed that the exposure of saturated clays to organic chemicals causes a change in the soil structure and an increase in permeability once the contaminant has replaced the pore fluid in the soil.

Many authors have demonstrated that the increase in permeability of clays in the presence of organic chemicals can be attributed to a reduction in the thickness of the double layer of water that surrounds the particles of clay. The low dielectric constant of organic chemicals is responsible for reducing the double layer thickness (Fernandez and Quigley, 1988). The reduction in the thickness of the double layer leaves more space for fluid flow and hence an increase in permeability.

Experiments carried out by Hijazi and Pamukcu (1990) indicate, as one might expect, that fuel oil contamination of clay soils reduces significantly the strength and stiffness and increases plasticity. However, they found these effects even at very low concentrations. They also found a slight increase in permeability with the addition of fuel oil.

Al-Tabbaa and Walsh (1994) have examined the physical effects of soil-contaminant interactions in respect of a kaolinite clay contaminated with polyethylene glycol. They used eight different concentrations of polyethylene glycol ranging from 25 to 300 g in 100 ml of water and tested the subjected clay samples for liquid and plastic limits, compaction, unconfined compression strength and permeability. In general they found both increases and decreases in the soil properties as the concentration of contamination increased. For example for concentrations below 100 g/100 ml they found decreases in liquid limit and plastic limit. Above 100 g/100 ml these index properties were above the values with just clean water. The plasticity of the soil increased with increased concentrations. In the compaction test they found a progressive decrease in maximum dry density and increase in optimum moisture content as the concentration of the contaminant increased (see Fig 3). It can be seen that the amount of air in the sample at maximum dry density decreased as the concentration of the contaminant increased.

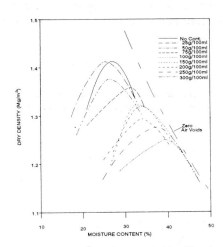

Fig. 3   Dry density vs moisture content for different polyethylene glycol concentrations. (After Al-Tabbaa and Walsh 1994).

Evgin and Das (1992) carried out an assessment of the mechanical behaviour of an oil contaminated sand. They found that oil contamination can cause a significant reduction in the friction angle of both loose and dense sands. The volumetric strains in drained and undrained triaxial tests increased considerably due to oil contamination.

If contaminants can have the effects on physical properties described above then a matter for consideration is whether these changes have come about as a result of a particular sample coming into contact with contamination in the borehole water thus rendering the sample non-representative of the mass of the soil stratum or whether the contamination has permeated through the soil prior to sampling.

Another point for consideration is whether the soil eg. beneath a foundation or as a landfill liner could be subjected to contamination after the laboratory testing has been carried out in which case the properties could change. Examples of this would be the use of a clay liner for a brine pond or the effect of leachate on a clay landfill liner.

TRANSPORTATION AND SAMPLE DISPOSAL

Particular transportation arrangements may need to be made for contaminated samples. As suggested in the SISG publication, reference should be made to the Road Traffic (Carriage of Dangerous Substances in Packages, etc) Regulations 1986 and Amending Regulations 1989, the Classification, Packaging and Labelling of Dangerous Substances Regulations 1984 and the Approved Code of Practice, Packaging and Labelling of Dangerous Substances for Conveyance by Road 1990.

The consequences of geotechnical samples being contaminated does not end with the completion of the testing. Appropriate disposal arrangements will need to be made. Unless the samples are automatically to be taken to a landfill facility licensed to accept special waste, and maybe also even then, it will be necessary to carry out chemical tests on the samples in order to be able to categorise the samples as a particular type of waste. The cost of disposal should be taken into consideration when determining the overall cost of the testing or considered as a separate cost item to testing.

Although the requirements for registration as a waste carrier under EPA 1990 have not yet been fully clarified in respect of the transportation of contaminated samples for laboratory testing, it appears likely that registration will not be required for the transportation from site to laboratory but registration will be required for the disposal of excess arisings from the investigation and also for the final disposal once testing is complete.

CONCLUDING REMARKS

There is presently much uncertainty with regard to the geotechnical testing of contaminated samples. It is hoped that this paper has appropriately aired some of these uncertainties and provided a discussion document to increase awareness of the problems. The intention has been to address the safety (but not necessarily dwelling on this) and the technical implications of carrying out this testing so that methodologies can be developed to deal with both.

ACKNOWLEDGEMENT

The authors which to express their thanks to the Directors of Soil Mechanics and C L Associates for permission to publish this paper. It should be noted, however, that the opinions expressed are those of the authors.

REFERENCES

1. Al-Tabbaa A and Walsh S : 1994 : Geotechnical Properties of a Clay Contaminated with an Organic Chemical, Proceedings of First International Congress on Environmental Geotechnics, Edmonton, Canada.

2. Anderson D C : 1982 : Does Landfill Leachate Make Clay Liners More Permeable. ASCE Civil Engineering, Vol 52, No 9.

3. Boldt - Leppin B, Kozicki P, Hang M D and Kozicki J : 1994 : Use of Organophilic Clay to Control Seepage from Underground Gasoline Storage Tanks, Proceedings of First International Congress on Environmental Geotechnics, Edmonton, Canada.

4. BS 1377 : 1990 : Methods of Test for Soils for Civil Engineering Purposes. British Standards Institution.

5. Chassefiere B and Monaco A : 1983 : On the use of Atterberg Limits on Marine Soils. Marine Geotechnology, Vol 5, No 2.

6. DoE : 1974 : Health and Safety at Work etc Act. HMSO (London).

7. DoE : 1990 : Environmental Protection Act. HMSO (London).

8. Evgin E and Das B M : 1992 : Mechanical Behaviour of an Oil Contaminated Sand. Proceedings of the Mediterranean Conference on Environmental Geotechnology. Usmen and Acar (eds), Balkema, Rotterdam

9. Fernandez F and Quigley R M : 1988 : Viscosity and Dielectric Constant Controls on the Hydraulic Conductivity of Clayey Soils Permeated with Water -soluble Organics. Canadian Geotechnical Journal, Volume 25.

10. Forman D E and Daniel D E : 1986 : Permeation of Compacted Clay with Organic Chemicals. Journal of Geotechnical Engineering, Volume 112, No.7.

11. HSE : 1984 : Classification, Packaging and Labelling of Dangerous Substances Regulations. HMSO (London).

12. HSE : 1986 : Road Traffic (Carriage of Dangerous Substances in Packages, etc) Regulations (and Amending Regulations (1989). HMSO (London).

13. HSE : 1988 : Control of Substances Hazardous to Health Regulations. SI 1988 No. 1657. HMSO (London).

14. HSE : 1990 : Packaging and Labelling of Dangerous Substances for Conveyance by Road. Approved Code of Practice. HMSO (London).

15. Hijazi H and Pamukcu S : 1990 : Possible Re-use of Petroleum Contaminated Soils as Construction materials. Proceedings of the 22nd Mid-Atlantic Industrial Waste Conference, Philadelphia, USA. Published by Technomic Publication Company Inc.

16. Ho Y A and Pufahl D E : 1987 : The Effects of Brine Contamination on the Properties of Fine Grained Soils. Proceedings of Speciality Conference - Geotechnical Practice for Waste Disposal. ASCE. New York.

17. SISG : 1993 : Guidelines for the Safe Investigation by Drilling of Landfills and Contaminated Land. Part 4 of Site Investigation in Construction Series. Thomas Telford.

18. Meegoda N J and Rajapakse R A : 1993 : Short-term and Long-term Permeability of Contaminated Clays. Journal of Environmental Engineering, Volume 119, Number 4, New Jersey Institute of Technology.

19. Ridley K D J, Bewtra J K and McCorquodale J A : 1984 : Behaviour of Compacted Fine - Grained Soil in a Brine Environment. Canadian Journal of Civil Engineering, Vol II.

A hydraulic fracturing test based on radial seepage in the Rowe consolidation cell

C H de A C MEDEIROS, PhD, MSc, Eng.
University of Feira de Santana, Brazil, and University of Newcastle, United Kingdom

A I B MOFFAT, RD, BSc, CEng, MICE, FGS, DL
University of Newcastle, United Kingdom

SYNOPSIS   The hydraulic fracturing mechanism in embankment dam cores has been the subject of considerable research effort and continues to be of great interest to the dam engineer. In general terms, hydraulic fracturing may take place if the total stress on any plane within the core of the dam falls below the hydrostatic pressure acting at that plane.

The paper describes the successful development of a test procedure for the experimental study of hydraulic fracturing in soils based on use of a 250 mm Rowe consolidation cell containing an annular soil sample.

Tests were conducted with specimens partially drained to simulate the field condition associated with initiation of hydraulic fracturing at the upstream core-shoulder interface on initial impounding. Closing pressures for a developed crack were used to define fracturing pressures. Recorded hydraulic fracturing pressures conformed with those predicted from elastic theory.

INTRODUCTION

The mechanism of hydraulic fracturing in soils or rocks due to internal fluid pressure is not yet completely understood, but it is believed that a tensile effective stress is required. The minor principal effective stress, σ'_3, on the plane of the crack must therefore be negative, i.e. the corresponding total stress, σ_3, must be less than the local porewater pressure, u_w, and it is postulated that when the crack is full of water the effective stress on its boundaries must be zero (i.e. $\sigma_3'=0$).

In the case of embankment dams the accepted concept is that hydraulic fracturing may occur when the local hydrostatic pressure exceeds the minor principal total stress σ_3, taking into consideration soil tensile strength, σ_t. Fracturing may initiate on first filling or, less commonly, may occur later as the result of differential deformations or strain incompatibility. This concept of hydraulic fracturing is not yet fully developed and needs to be tested further under controlled laboratory conditions for differing soil characteristics and boundary stress states.

A number of laboratory and field studies of hydraulic fracturing have been conducted in recent years (e.g. Panah and Yanagisawa, 1989; Lo and Kaniaru, 1990; Mhach, 1991; Murdoch,

1993; Atkinson *et al*, 1994, *et al*). The question of how the potential for hydraulic fracturing under prescribed field conditions may be assessed has not yet been conclusively resolved.

THE ROWE CELL HYDRAULIC FRACTURING (HF) TEST:

The Rowe Cell HF Test Concept
A hollow cylindrical specimen confined within a 250 mm dia. Rowe consolidation cell is subjected to internal water pressure P_i, external water pressure, P_o, and vertical axial stress σ_v **(Figure 1)**. Hydraulic fracturing is induced by raising P_i with P_o and σ_v constant, and P_{if} at the initiation of fracturing determined.

The radial and tangential stresses in the specimen, σ_r and σ_θ respectively, attributable to the boundary pressures P_{if} and P_o at fracturing can be calculated from elastic theory (Timoshenko and Goodier, 1965). If an internal pressure increment, ΔP_i is applied at the inner surface of the cavity the stress changes at that surface are given by:

$$\Delta\sigma_r = \Delta P_i \qquad [1]$$
$$\Delta\sigma_\theta = -\Delta\sigma_r = \Delta P_i \qquad [2]$$
$$\text{and } \Delta\sigma_v = \upsilon(\Delta\sigma_r + \Delta\sigma_\theta) = 0 \qquad [3]$$

(The above equations assume a horizontal plane strain condition with constant elastic modulus, E, and Poisson's ratio, υ).

Rowe cell HF Test Stress Conditions:
In executing the Rowe cell HF test the internal water pressure, P_i, is raised incrementally until fracturing occurs, as previously indicated. The corresponding changes in cylinder stresses are calculated bearing in mind that this procedure increases radial stress σ_r and decreases the initial tangential stress σ_θ. Hydraulic fracture occurs when the minor principal effective stress, σ_3', becomes negative (tensile), and exceeds the tensile strength, σ_t, of the soil, thus:

$$\sigma_3' + \sigma_t \leq 0 \qquad [4]$$

Hydraulic fracture may therefore be expected to occur when $P_i = P_{if}$ and exceeds the sum of the initial minor principal effective stress σ_3' (or *closing pressure* (see below)) and the tensile strength σ_t, i.e.:

$$P_{if} \geq \sigma_3' + \sigma_t \qquad [5]$$

The following assumptions have been made for the annular test specimen employed in the Rowe cell test:

$$\sigma_1 = \sigma_r \qquad [6]$$
$$\text{and } \sigma_3 = \sigma_\theta + \sigma_t \qquad [7]$$

Stress-paths may be plotted to assist in studying the mode of failure (tension or shear) and fracture orientation.

The Determination of 'Closing Pressure':
In a number of hydraulic fracturing studies the crack 'closure pressure', i.e. the pressure at which an 'open' fracture will close, is assumed to equal the minor principal total stress σ_3, although this has been questioned (Penman, 1976). Penman found values of σ_3 to be less than the closing pressure obtained using a procedure, where the recorded closing pressure of a crack initiated by hydraulic fracturing was assumed to have the same magnitude as the minor principal stress, σ_3 (Bjerrum and Andersen, 1972). In the Rowe cell HF test procedure it was determined that σ_3 should be derived from the fracture closing pressure rather than the opening pressure. (The latter is normally larger due to the tensile strength of the soil.)

'Closing pressure' may be determined by reducing internal pressure post-fracturing, plotting against apparent specimen permeability and/or flow rate and selecting the point at which the latter parameters decrease dramatically, or by plotting decay in internal pressure against time. The similar approach used by Bozozuk for measuring the *in-situ* minimum principal effective stress, determined when a fracture closes, and referred to as 'closure pressure' was adopted for this programme (Bozozuk, 1974).

In the Rowe cell HF programme initiating and closing pressures were recorded and compared with the tangential stress, σ_θ, calculated from elastic theory as a function of the prevailing internal and external water pressures P_i and P_o.

THE ROWE CELL HF-TEST: PRINCIPLES
The Rowe consolidation cell was originally developed for consolidation or vertical and radial permeability tests on larger soil samples (Rowe and Barden, 1966). In such tests the cylindrical specimen is loaded axially by a vertical stress, σ_v, generated by water pressure within a flexible rubber bag or membrane.

The 250 mm diameter Rowe cell was adapted for the HF tests to accommodate annular samples of 150 mm diameter and 90 mm height with a central axial cavity or drain of 40 mm diameter (**Figure 1**). Control of specimen drainage enabled loading to be applied to the specimen under 'partially drained' conditions. (A 'partially drained' condition was considered to result from the unconstrained development of equilibrium flows, pore pressures and seepage forces within the annular specimen due to the pressure differential between P_i and P_o ($P_i > P_o$).) In this respect the Rowe cell technique therefore models field conditions, since an 'undrained' loading condition and seepage regime are allowed to develop freely.

The pressure regime employed for the HF tests generates a hydraulic gradient and seepage and, with no rubber membrane enclosing the specimen, allows the development of excess internal porewater pressures. The procedure therefore involves pure hydraulic fracturing induced by seepage within a specimen where pressures P_i and P_o are applied in free contact with the surface of the specimen. This approach was based on the observation that internal stress conditions generated by seepage forces are more critical than those registered when applying the same pressure regime via a rubber membrane at the soil-water interface (Bjerrum and Andersen, 1970). The adoption of seepage-induced fracturing in the Rowe cell HF test is therefore a reasonable simulation of conditions at the upstream core-shell interface of an embankment dam on initial filling.

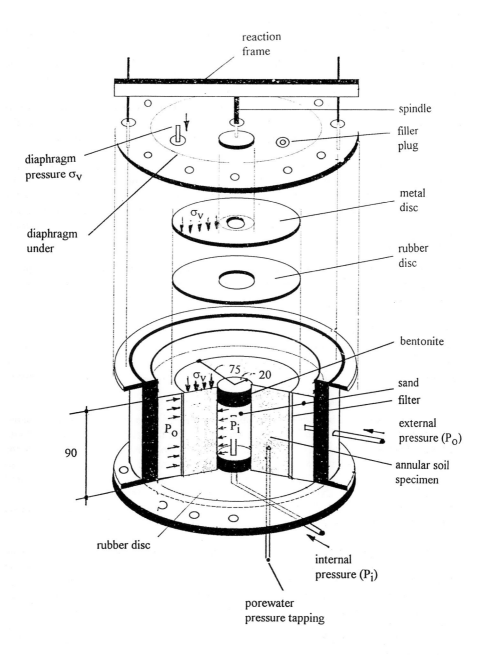

Figure 1: Exploded schematic of Rowe cell HF test:
(boundary stress conditions indicated)

LABORATORY TESTING

A) - FIELD CONDITIONS:

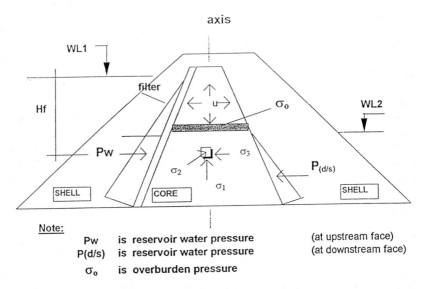

Note:

    Pw   is reservoir water pressure   (at upstream face)

    P(d/s) is reservoir water pressure   (at downstream face)

    σ_0 is overburden pressure

B) - LABORATORY CONDITIONS:

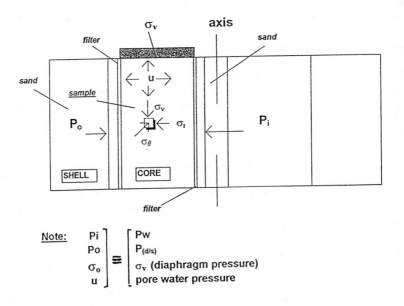

Note: $\begin{bmatrix} Pi \\ Po \\ \sigma_0 \\ u \end{bmatrix} \equiv \begin{bmatrix} Pw \\ P_{(d/s)} \\ \sigma_v \text{ (diaphragm pressure)} \\ \text{pore water pressure} \end{bmatrix}$

Figure 2:   Schematic comparison of field and HF test boundary stress conditions

FIELD VERSUS LABORATORY CONDITIONS

Analysis of Rowe cell stress conditions was performed in terms of effective stress. It was assumed that given the low effective stress levels employed the horizontal effective stress due to the applied vertical effective pressure, σ_v', was sensibly zero. It was also assumed that the hollow cylindrical specimen was subjected to boundary total stresses equivalent to the directly imposed fluid pressures, P_i and P_o. The corresponding changes in effective stresses were determined from monitored internal porewater pressures.

Figures 2A and **2B** allow comparison of the stress conditions existing in the field and within a Rowe cell HF test specimen in the laboratory. The reservoir water pressure, P_w, acting against the upstream face of the prototype core wall (**Fig. 2A**) corresponds to the applied internal water pressure P_i in the cell (**Fig. 2B**). In the laboratory the external water pressure P_o is kept constant, and as a consequence the resultant changes in radial (σ_r) and tangential (σ_θ) stresses within the annular laboratory specimen are those generated by the imposed incremental changes in internal water pressure, ΔP_i. This progressive change is analogous to the filling of the reservoir.

The Rowe cell HF test assumes that $\sigma_1 = \sigma_r$ and $\sigma_3 = \sigma_\theta + \sigma_t$, where σ_t is the tensile strength of the soil (Eqns. 6 and 7). The porewater pressure u_w generated in both field and laboratory is a function of the appropriate internal seepage regime. In the laboratory test u_w is assumed to be a function of the pressure gradient arising from the imposed pressure difference, $P_i - P_o$ inducing radial outward seepage forces in the specimen. The major principal total stress in the prototype dam, σ_1, is represented in the HF-test by the radial stress σ_r as the diaphragm pressure, σ_v, is maintained constant and below the internal water pressure P_i. The external water pressure P_o is similarly kept constant and below the diaphragm pressure σ_v ($P_o \cong 0.90\ \sigma_v$). The boundary stress conditions are such as to justify the assumption of a horizontal plane-strain condition, with no axial strain on the specimen.

THE ROWE CELL HF-TEST: PROCEDURE

Hydraulic fracturing tests have been carried out using a number of different techniques. Alternative theoretical approaches have also been developed, but the principle of the hydraulic fracturing process is considered the same in all cases, i.e. fracturing occurs when minor principal effective stress is progressively diminished (Eqn. 4).

In conducting the Rowe cell HF-tests the constant external water pressure, P_o and diaphragm pressure, σ_v were controlled independently of internal pressure, P_i, until hydraulic fracturing occurred. Three values of diaphragm pressure ranging up to 150 kPa were employed within the development programme. Tests were executed in two phases; pressure equalisation, in Phase 1, followed by fracturing in Phase 2.

Phase 1 - Pressure equalisation phase:

Diaphragm pressure, σ_v, was applied together with external water pressure P_o. Axial stress σ_v was kept 10% greater than P_o to prevent leakage between the diaphragm and the top of the specimen. Internal water pressure P_i was applied to a predetermined value of $P_i = 1.1\ P_o$ and then kept constant. At the final stage of Phase 1, σ_v was thus 10% higher than P_o and equal to P_i (e.g. $\sigma_v = P_i = 110$ kPa and $P_o = 100$ kPa). This boundary stress condition was maintained for 24 hours to allow the specimen to adjust to an equilibrium stress state prior to inducing hydraulic fracturing in Phase 2.

Phase 2 - Hydraulic fracturing:
A volume-change monitoring device (VCD) was used to monitor seepage flow relative to the datum flow value established when σ_v, P_0 and P_i were stabilised in Phase 1.

The porewater pressure, u_w, was monitored at mid-plane within the specimen (**Figure 1**). A linear pressure gradient through the specimen was assumed and when the recorded u_w value approached (Pi + Po), i.e. near the steady-state condition, Pi was raised incrementally and the final test stage to hydraulic fracturing commenced. Internal pressure P_i was raised at one of two incremental rates, 1.0 kPa/minute or 10.0 kPa/minute, until a suddenly enhanced inflow to the specimen, indicated the onset of hydraulic fracturing. Representative loading pattern and measured inflow versus time plots are illustrated on **Figure 3**.

Coincident with the beginning of the hydraulic fracturing phase a solution of sodium fluoresceine dye was injected into the Rowe cell through the volume-change water supply to assist in identifying fracture pattern and orientation within the specimen.

ROWE CELL HF-TEST RESULTS
In developing the Rowe cell HF test an artificial soil composed of 85% PFA + 15% bentonite was employed in the interests of uniformity. This soil mixture had a similar particle size distribution and plasticity to that of the core material of the Teton dam, which failed in the USA in 1976 due to internal erosion believed to have been initiated by hydraulic fracturing in the core trench.

The laboratory hydraulic fracturing pressure, P_{if}, can be determined in the HF test by two alternative techniques:

Test Type I:
In tests of Type I P_{if} is determined by observing a sudden increase in flow into the specimen. This was readily detected by the VCD and recorded automatically. The internal water pressure is then further increased above P_{if} to ensure full crack propagation through the specimen. A representative hydraulic fracturing test result of this type (HF-7) is shown in **Figure 4**.

Test Type II:
For Type II tests the internal water pressure P_i is increased until the dye used to facilitate fracture identification is confirmed as emerging from the outer face of the specimen through a length of transparent tubing connected to the external pressure measuring point. At the instant of soil fracturing, i.e. $P_i = P_{if}$, the valve controlling the internal water pressure is closed, and the subsequent time-dependent decay of P_{if} recorded. A pressure decay test curve as originally suggested for the determination of closing pressure was thus obtained (Bozozuk, 1974). As an example, the internal closing pressure for test HF-37 determined according to this method is shown on **Figure 5**.

A vertical crack was consistently obtained in all test series of Type I and Type II. This can be interpreted as proof of a valid hydraulic fracturing process. The Rowe cell HF technique was therefore considered suitable for further parametric studies, e.g. on the influence of soil plasticity, moisture state, compacted density, etc.

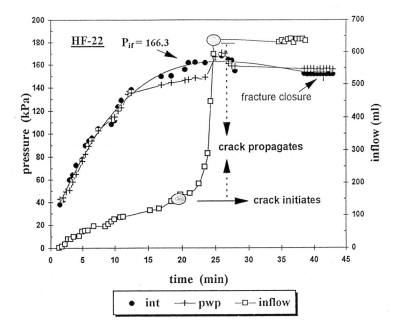

Figure 3: Representative HF test result; plot of inflow v. time (test HF 22)

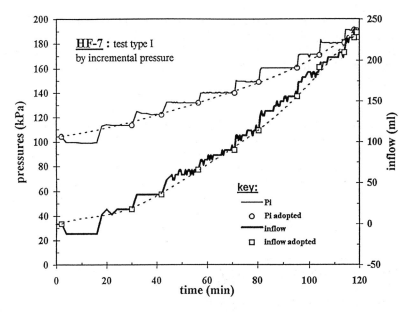

Figure 4: HF Test Type 1 : plots of pressure and inflow v. time (Test HF 7)

CONCLUSIONS

A comprehensive analysis of the hydraulic fracturing process would extend beyond the scope of this paper. The primary objective of the research programme was, however, attained in that a technique for conducting hydraulic fracturing tests which simulated conditions at the upstream core-shell interface of an embankment dam during first filling was developed and proved.

Some analytical simplifications were necessarily employed, and as a consequence it must be recognised that certain important controlling parameters could not be investigated fully within the constraints applicable to the programme. Two such parameters may be highlighted:

i. The influence of consolidation of the annular HF specimen by radial drainage under controlled levels of vertical stress.

ii. Influence of the initial stress conditions within the HF specimen. This factor was not investigated as specimen preparation procedures were not comparable to those used for conventional triaxial tests, where pore pressures and corresponding effective stresses can be measured under a controlled consolidation regime. A similar procedure may, however, be developed for the Rowe cell, since the start and the conclusion of the primary consolidation stage can be established by monitoring porewater pressures. With such a procedure it will be possible to consolidate the HF specimen at representative stress levels prior to the hydraulic fracturing phase.

Conclusions drawn from the initial experimental investigation may be summarised as:

1. The results of 24 HF-tests provide strong evidence in support of the general assumption that hydraulic fracturing occurs if the minor principal effective stress σ'_3 becomes negative and surpasses the tensile strength of the soil σ_t (i.e. $\sigma'_3 + \sigma_t \leq 0$).

2. In all HF tests a tension failure was observed, resulting from a decrease in tangential stress to a negative (tensile) value as a result of seepage forces. (This was subsequently confirmed in terms of stress-path plots.)

3. All specimens fractured with the vertical crack anticipated in terms of the applied boundary stress conditions **(Figure 6)**. The stress-path diagrams of **Figure 7** confirm the validity of the vertical fracture concept.

4. Specimens fractured with an average stress ratio $K = \sigma_3/\sigma_1 = 0.60$. This is compatible with the crack patterns observed in the test programme, as a vertical crack may be formed for $K < 1.0$ (Bjerrum and Andersen, 1972).

5. An additional limited study explored the effect of loading rate on the hydraulic fracturing mechanism but proved inconclusive.

ACKNOWLEDGEMENT

The support of the Brazilian Government through the combined sponsorship of the Universidade Estadual de Feira de Santana (UEFS) and Coordenação de Aperfeicoamento de Pessoal de Nivel Superior (CAPES), is gratefully acknowledged.

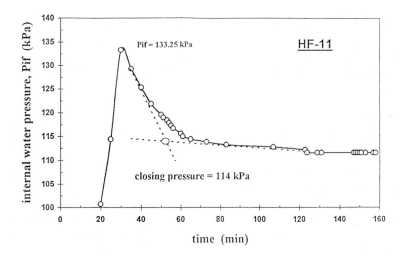

Figure 5: HF Test Type II : determination of closing pressure (Test HF 11)

Figure 6: Specimen with vertical fracture : (Test HF 36)

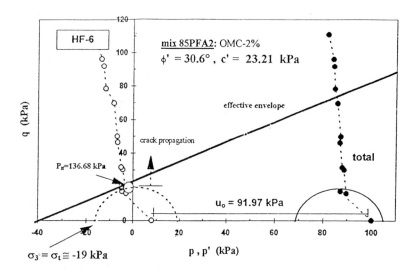

Figure 7: Stress path plots (Test HF 6 : pressure increased post-fracturing to encourage crack propagation)

REFERENCES

Atkinson, J H, Charles, J A and Mhach, H K. (1994) Undrained hydraulic fracture in cavity expansion tests. In Proc. *13th International Conference Soil Mechanics and Foundation Engineering, New Delhi*, **3**, pp 1009-1012.

Bjerrum, L and Andersen, K H. (1970) *Initiation and propagation of cracking in soil caused by permeability tests with excess pore pressure.* Internal Report F382-2, Norwegian Geotechnical Institute, Oslo.

Bjerrum, L and Andersen, K H. (1972) 'In-situ measurements of lateral pressures in clay'. In Proc. *5th European Conf. Soil Mechanics and Foundation Engineering*, Madrid, **1**, pp 11-20.

Bozozuk, M. (1974) 'Minor principal stress measurements in marine clay with hydraulic fracture tests'. In Proc. ASCE Speciality Conf. *Subsurface Exploration for Underground Excavations and Heavy Construction*, pp 333-349.

Jaworski, G W, Duncan, J M and Seed, H B. (1981) Laboratory study of hydraulic fracturing. Journal Geotechnical Division, American Society of Civil Engineers, 107, No. GT6, pp 713-732.

Lo, K Y and Kaniaru K. (1990) 'Hydraulic Fracture in earth and rockfill dams'. Canadian Geotechnique, Jnl., No. 27, pp 496-506.

Mhach, H K. (1991) *'An experimental study of hydraulic fracture and erosion'*. PhD Thesis, City University, London.

Murdoch, L C. (1993) 'Hydraulic fracturing of soil during laboratory experiments' (in 3 parts: Part 1: Methods of observation, Part 2: Propagation, Part 3: Theoretical analysis). Geotechnique, **43**, 2, pp 255-287.

Panah, A K and Yanagisawa, E. (1989) 'Laboratory studies on hydraulic fracture criteria in soil'. Jnl. Japanese Society of Soil Mechanics and Foundations, **29**, No. 4, pp 14-22.

Penman, A D M. (1976) 'Earth pressures measured with hydraulic piezometers'. In Proc. ASCE Speciality Conf. *In-situ measurements of soil properties*, Raleigh, NC., **2**, pp 361-381.

Rowe, P W and Barden, L. (1966) 'A new consolidation cell'. Geotechnique, **16**, 2, pp 162-163.

Timoshenko, S and Goodier, J N. (1968) *'Theory of elasticity'*. McGraw-Hill, New York, 506 pp.

Quick, accurate, consistent measurements of permeability of clays

J. T. ARARUNA Jnr, B. G. CLARKE and A. H. HARWOOD
University of Newcasle upon Tyne, UK

SYNOPSIS

A description of an apparatus to accurately measure the permeability of natural and reconstituted soils is given highlighting the key elements of the equipment. Two features of this flexible wall permeameter are the use of a flow pump to induce an hydraulic gradient and volume change units with coarse and fine settings to ensure the correct resolution is available for the consolidation and permeation stages of a test. A test procedure is described which demonstrates that repeatable results can be obtained and that the time of the test is less than a test in which a differential pressure is set and a flow rate monitored. The apparatus has been used to carry out tests on natural and reconstituted soils to show the effect of consolidation pressure on permeability and support recent work on relationship between permeability and void ratio.

INTRODUCTION

BS 1377: Part 6 1990 describes two techniques to measure coefficients of permeability of cylindrical specimens of cohesive soils. The first, in a hydraulic consolidation cell or rigid wall permeameter, allows vertical or radial flow under a constant hydraulic gradient and vertical effective stress. The second, in a triaxial cell or a flexible wall permeameter, allows vertical flow under a constant gradient and mean effective stress. The flow in both cases is created by the constant difference in pore pressures at the top and bottom of the specimen which are set by adjusting the back pressures. Pressures, inflow and outflow are measured directly.

The volume change apparatus used to measure the flow is often the same apparatus used to measure outflow during consolidation. The resolution needed to measure the change during consolidation is often coarse compared with the small quantities of flow generated during permeation; it is, therefore, necessary to monitor the test for a considerable period to ensure steady state conditions are achieved. Typically, tests on clays last between two weeks and several months (Benson and Daniel, 1990). Alternatively the hydraulic gradient can be increased resulting in a greater flow rate. The disadvantage of this is that the effects of seepage induced consolidation become significant, giving rise to a non-homogeneous specimen and preventing tests at low effective stresses. Outflow could include pore fluid expelled because of seepage induced consolidation. It would be better to use a more sensitive volume change unit for permeation thus allowing low hydraulic gradients.

Daniel (1993) gives a comprehensive review of these and other methods of measuring the permeability of soils. One alternative method, presently used in the UK only in research institutions, is to force permeant through a specimen at a constant rate and measure the

Advances in site investigation practice. Thomas Telford, London, 1996

resulting hydraulic gradient (Olsen et al, 1985). This reduces the time of the test since a constant hydraulic gradient is more easily noticed than a constant flow rate which is a time dependent variable. A more sensitive volume change unit also allows the system to be checked for leaks, measurements of seepage induced consolidation to be made and the outflow to be measured accurately.

A new apparatus has been developed (Araruna Jnr et al, 1994a) which incorporates a flow pump and sensitive volume change units to allow tests to be carried out at low hydraulic gradients, different effective stresses and low flow rates. A description of the apparatus, test procedure and the accuracy of the measuring devices are given, together with examples of results obtained from tests on reconstituted and natural clays with coefficients of permeability of less than 10^{-10} m/s.

TESTING EQUIPMENT
This test is based on the principle that a permeant is forced into the specimen at a controlled rate using a flow pump (1 in Figure 1) while a differential pressure transducer (2) monitors the gradient induced over the length of the specimen. The amount of water expelled from the pore spaces of the specimen is measured by the outflow volume change unit (3).

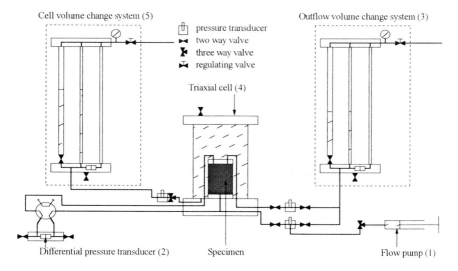

Figure 1   An apparatus to induce flow and accurately measure flow rate and hydraulic gradient

There are three main components to the equipment shown in Figure 1; the flexible wall permeameter, the volume change units and the flow pump. The flexible wall permeameter, a modified triaxial cell (4 in Figure 1) manufactured from clear acrylic, is rated up to 1000 kN/m^2 and can accommodate up to 100 mm diameter specimens.

Independent volume units are used to measure the outflow (3) and changes in the total volume of the soil specimen (5). This allows the degree of saturation of the soil specimen to be checked during the consolidation stage of the test since, for a saturated soil, the outflow should equal the change in volume of the specimen. Any leaks within the system can be identified during this stage.

The volume change unit, described in detail by Araruna Jnr. et al. (1994b), is a dual burette volumetric system. Two flow tubes of different diameters and one dry tube are pressurised internally to the same air pressure. The height of water in the two flow tubes is measured relative to the dry reference tube using a differential pressure transducer which can operate up to 10 kN/m^2. The tubes are 0.8m long and the volume resolution is adjusted by altering the diameter of the tubes. Two flow tubes of different diameter are used so that the larger tube can be isolated during permeation to permit more accurate measurements; in the fine setting the resolution is less than 0.005 ml.

The flow pump consists of a stainless steel syringe mounted in a standard variable-speed driver typical of those used in the medical profession. The pump has a synchronous motor and a linear gear box to give thirty accurate speeds ranging between 1.26×10^{-2} mm/s and 6.30×10^{-7} mm/s. The flow rate is a function of the speed of the pump and the internal diameter of the syringe. An 8 ml capacity syringe, capable of producing flow rates ranging from 4 ml/hr to 0.0002 ml/hr, is suitable for testing soils with a coefficient of permeability of less than 10^{-8} m/s. A larger syringe extends the range to 10^{-6} m/s.

The pressures and flow rate are controlled manually. The variables, including cell pressure, top and bottom back pressure, cell volume, back pressure volume, flow pump rate and head difference, are monitored using transducers connected to an autonomous data acquisition unit linked to an IBM PC compatible computer. Management software enables the operator to exercise control over the system and record all relevant data for subsequent retrieval and analysis.

TESTING PROCEDURE

The test procedure, shown as a flow chart in Figure 2, is based on ASTM D 5084-90, BS 1377: 1990 and good practice. Specimens are saturated using a back pressure prior to consolidation to the required effective stress, which is set by maintaining a constant cell pressure and reducing the back pressure. To reduce the amount of seepage induced consolidation, the induced hydraulic gradient during permeation is kept as small as feasible by reducing the rate of inflow which is then kept constant throughout the permeation stage. Permeation is maintained until the head difference is constant.

Saturation Stage

Tests are carried out on saturated specimens to ensure that the fluid forced into the specimen produces a similar outflow. If a sample were partially saturated some of the inflow fluid would be used to further saturate the specimen and the permeability would vary since it would be influenced by the amount of air present.

It is important to prevent air becoming trapped in the system while setting up the specimen since this will increase the time for saturation. There are two connections to the top and two to the bottom of the specimen to allow deaired water to be flushed through the pipe work and porous disks after the first increments of pressure are applied. A cell pressure of 50 kN/m² and a back pressure of 30 kN/m² are applied to give an effective stress of 20 kN/m² and this is maintained throughout the saturation stage unless swelling occurs.

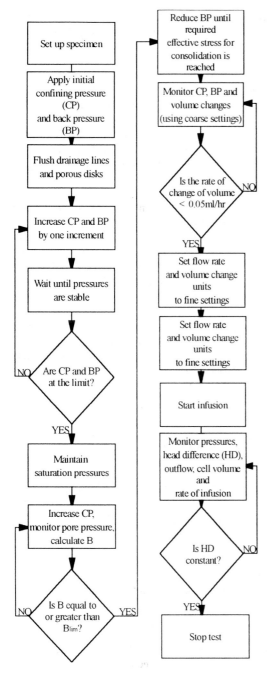

The cell and back pressures are increased in twelve increments of 50 kN/m² over six hours until the cell pressure is 600 kN/m². This pressure was determined by the available pressure supply and maximum cell pressures used in the majority of tests for routine investigations. The full pressures are maintained for twenty four hours to ensure saturation.

It is assumed that a specimen is fully saturated after twenty four hours at a cell pressure of 600 kN/m² and effective stress of 20 kN/m². The change in pore pressure with respect to a subsequent change in cell pressure should be greater than B_{lim}, Skempton's pore pressure parameter for isotropic stress changes which, for soils, is typically 0.95. The actual value of B is determined by increasing the cell pressure to 650 kN/m² with the back pressure isolated. It has been found that the pore pressure is at equilibrium within thirty minutes for all soils.

The change in specimen volume (by measuring the change in cell volume) and the change in pore fluid are measured throughout each stage of a test so that the final volume of the specimen can be calculated from the initial dimensions.

It is important in any test involving flow or pressure to check the systems for leaks which can occur either at pipe fittings or in the membrane. External leaks result in loss of fluid though if the leaks are small, pressures are maintained. Internal leaks can result in either greater flow into the base of the specimen giving increased flow through the specimen and an increase in the hydraulic gradient, or flow into the top of the specimen giving increased outflow. The system is tested at each stage for leaks by comparing the outflow and the specimen volume change.

Figure 1   The test procedure

843

During saturation when a specimen is partially saturated, water should flow into the specimen and the specimen volume should decrease slightly due to consolidation; there should be no changes with a saturated specimen. If the specimen volume increases during saturation it is because the specimen is swelling. This is prevented by increasing the cell pressure. During a B test, when the specimen is isolated, there should be no changes in volume (ignoring elastic compression). If the B value is greater than one, cell fluid at pressure must be leaking into the specimen.

Consolidation Stage

Tests are currently carried out in a modified triaxial cell which allows isotropic consolidation. During consolidation the confining stress is kept constant at 600 kN/m^2 and the back pressure is reduced to reach the required effective stress in order to minimise the effects of compliance. Double end drainage is permitted and consolidation is terminated when there are no further changes in pore and cell volumes. For a fully saturated specimen these changes should be equal if compliance is kept to a minimum; that is the change in volume of the specimen is equal to the amount of pore fluid expelled from the specimen. Figure 3 shows that this is the case for a typical test on reconstituted till. Note that BS1377:1990 only permits single drainage because of the need to monitor changes in pore pressure. The time taken to consolidate and permeate a specimen using double drainage is a quarter of that using single drainage but it is only possible to use double drainage because of the accuracy of the volume change units.

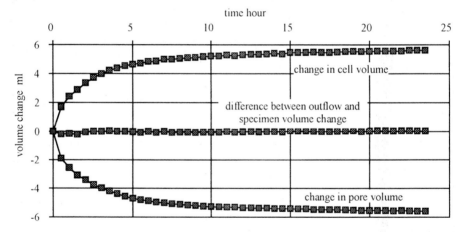

Figure 3   Changes in specimen volume and pore fluid during consolidation of a saturated specimen

Experiments using single drainage showed the times for dissipation of excess pore pressures in specimens of reconstituted clays were similar to the times for changes in volume to cease. Typically for clays with coefficients of permeability ranging between 10^{-8} and 10^{-12} m/sec and 100 mm long specimens, consolidation takes place in 24 to 150 hrs. Both BS1377:1990 and ASTM D 5084-90 state that consolidation is deemed to be complete when the remaining excess pore pressure is 5% of the maximum but for measurements of permeability at low hydraulic gradients it is necessary to ensure full dissipation of excess pore pressure.

Throughout the saturation and consolidation stages of a test the coarse settings of the volume change units are used since volume changes up to 110 ml can be measured. During saturation the change in specimen volume does not necessarily equal the change in pore fluid because of air in the system and compliance of the triaxial cell. An increase in cell pressure will expand the cell and its volume change unit. This system compliance was found to be 3.34×10^{-3} ml/kN/m² which represents about 2% of the full range of the volume change unit. It is possible to correct specimen volume changes for this relatively large compliance during the application of the consolidation pressure but with the cell pressure constant this is unnecessary. The back pressure system compliance is much smaller since the volume change unit on the back pressure line only affects the results. This compliance is reduced to zero by setting the back pressure before consolidation starts.

Permeation
Water is forced into the specimen at a very low and constant rate by the flow pump. Tests on reconstituted clays using different permeants have shown that deaired tap water is acceptable. It is unnecessary to use pore fluid as the quantity of fluid displaced during a test is less than 0.01% of the total volume of pore fluid, that is the measured permeability represents the flow of the pore fluid through the specimen. This is reflected in Table 1 where the permeability is shown to be independent of permeant.

Table 1   Results of tests on reconstituted till showing that the permeant has no effect on the results and the results are consistent

Test	void ratio	bulk density Mg/m³	water content %	consolidation pressure kN/m²	coefficient of permeability x 10^{-11} m/sec	permeant
DOP4	0.55	1.75	19	340	4.08	dist water
DOP5	0.54	1.76	19	310	4.66	dist water
DOP6	0.54	1.76	19	370	3.94	dist water
DOP7	0.49	1.82	18	360	4.57	dist water
DOP8	0.53	1.77	19	360	4.82	dist water
TW1	0.54	1.76	19	360	4.07	tap water
TW2	0.53	1.77	19	360	5.15	tap water
TW4	0.56	1.74	19	360	4.73	tap water
SP 1	0.53	1.77	19	360	3.66	'standard'
SP2	0.55	1.74	19	360	4.62	'standard'
SP3	0 55	1.74	19	360	3.92	'standard'

standard   0.005N CaSO₄ (ASTM 5084-90)

The rate of flow is a function of the speed of the motor and the internal diameter of the syringe. These can be accurately measured and converted to flow rate but, in practice, the actual speed of the piston in the pump is measured with an LVDT and this is converted to volume change by multiplying the rate of displacement by the area of the piston. The piston in the syringe displaces the fluid thus the diameter of the piston rather than the internal bore of the syringe governs the flow. It is easier to measure the diameter of the accurately machined piston than the bore of the syringe. The speed of the piston is precisely controlled after about thirty minutes from starting during which gear backlash is overcome but the speed may vary with back pressure. The speed of the motor at zero back pressure is ±0.01% of the actual speed. The LVDT measures displacement independent of pressure.

LABORATORY TESTING

The quantity of fluid forced into the specimen is fixed by the speed of the motor and the size of the syringe but the induced hydraulic gradient is a function of the permeability of the specimen. Tests have been carried out on soils with coefficients of permeability varying between 10^{-6} and 10^{-12} m/s at hydraulic gradients of less than ten. The motor speed and syringe capacity are varied between soils so that flow rates of between 10^{-3} ml/s and 10^{-7} ml/s can be set thus allowing hydraulic gradients of less than ten to be created.

The head difference across the specimen created by the flow is measured directly together with the quantity of fluid expelled from the specimen. These are used to calculate the coefficient of permeability. Seepage induced consolidation will occur but by creating a head difference of less than 10 kN/m² it is kept to a minimum. Changes in volume of the cell occur but these are due to the unavoidable, small fluctuations in line pressure of the order of 0.05% of the full scale (0.3 kN/m² for 600 kN/m²) rather than changes in volume of the soil specimen.

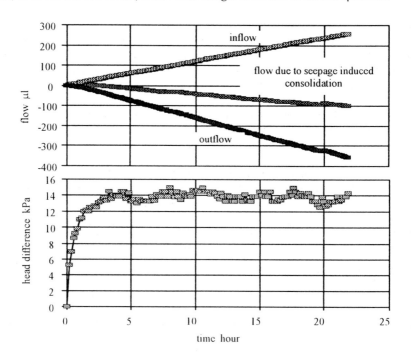

Figure 4   The variation in outflow, inflow and head difference during the permeation
            stage of a test on a specimen of reconstituted till

Data are recorded throughout the permeation stage but for analysis it is assumed that permeation starts when the speed of the motor is constant. The stage is finished when the head difference is reasonably constant. Figure 4 shows a typical set of results from a test on reconstituted till highlighting that the difference between the outflow and the seepage induced consolidation (measured directly) is approximately equal to the inflow and that steady state conditions are reached. Note that the fine settings of the volume change units are used during this stage. The fluctuation in head difference arises because of small variations in line pressure and flow rate, both of which are within acceptable limits (ASTM D 5084-90).

Any leak in the system will affect the results but, if small, it can be accounted for. Once steady state conditions are reached and provided seepage induced consolidation is kept to a minimum, the inflow should equal the outflow. Any leakage within the permeameter will cause an increase in outflow over the inflow. If the cell volume unit shows a reduction in specimen volume, the outflow will increase and exceed the inflow though it will not be clear whether the leak is at the inflow or outflow end of the specimen. Data from the saturation stage are used to decide whether the inflow or outflow should be used to calculate permeability. Note that it is only possible to detect leaks because of the resolution of the volume change units.

DISCUSSION
The developments of the equipment and test procedure were carried out on reconstituted till formed by consolidating a slurry of water and fine particles mixed at 125% to 150% of the liquid limit. The results of tests on eleven samples of this material consolidated to the same void ratio were consistent as shown in Table 1, suggesting that repeatable results can be obtained from tests on similar samples.

Table 2   Results of tests on natural clay and reconstituted till showing time for consolidation and permeation (all tests on 76 mm diameter, 76 mm long specimens)

Soil Type	Consolidation pressure kN/m^2	Hydraulic Gradient	Coefficient of permeability m/sec	Time for Consolidation hr	Time for Permeation
Natural Brown	125	6.18	2.54×10^{-11}	100	18
London Clay	250	34.21	9.95×10^{-12}	132	19
76 mm dia	310	95.89	6.79×10^{-12}	96	37
Reconstituted	50	3.82	1.27×10^{-10}	34	32
Brown London	100	6.93	7.01×10^{-11}	42	19
Clay	200	14.06	3.64×10^{-11}	64	24
76 mm dia	400	34.08	1.57×10^{-11}	65	16
Reconstituted	50	3.49	2.00×10^{-10}	7	16
Brown Till	100	6.23	1.15×10^{-10}	12	16
76 mm dia	200	9.37	7.50×10^{-11}	12	16
	400	15.22	4.80×10^{-11}	14	16

Table 2 is a summary of tests carried out on natural and reconstituted soils showing the soil type, coefficient of permeability, consolidation pressure and times to saturate, consolidate and permeate the specimens.

A low hydraulic gradient together with constant flow rate and accurate measurements permit steady state conditions to be achieved relatively quickly and with little change in pore fluid compared to standard tests using applied pressure to create flow. Thus, it is possible to repeat tests at the same effective stress and carry out tests at different effective stresses on the same specimen in confidence that there is little change to the pore fluid. Problems of long term tests such as creep, bacterial growth and environmental changes are largely avoided. Further, the effects of seepage induced consolidation are minimised as the maximum difference in effective stress across the specimen is less than 10 kN/m².

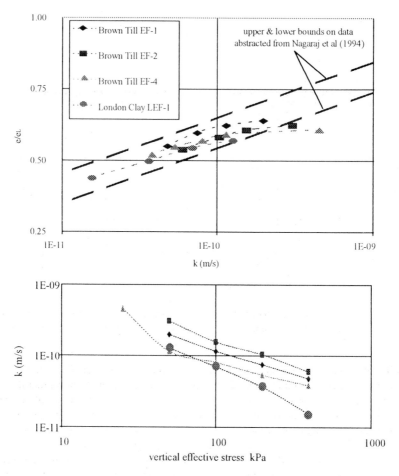

Figure 5   Results of permeation tests on overconsolidated specimens of reconstituted clays consolidated to different stresses plotted as e - $log\ k$ and $log\ k$ - $log\ \sigma'_v$

The results of tests on four specimens of overconsolidated reconstituted clay are shown on Figure 5 where each curve represents different preconsolidation pressures. The void ratios, determined from the initial dimensions of the specimens and measured specimen volume changes, are normalised with respect to the void ratios at the liquid limit as suggested by Nagaraj et al (1994) who showed that results of tests on reconstituted clays lie about a line given by

$$\frac{e}{e_L} = 2.162 + 0.195\ log_{10}k \qquad (1)$$

where e_L is the void ratio at the liquid limit and k is in cm/s. These results also lie about that line but, for a particular specimen, the results at different preconsolidation pressures lie on a curved line of the form

$$\frac{e}{e_L} = \Sigma\, C_n \,(\log_{10} k)^n \tag{2}$$

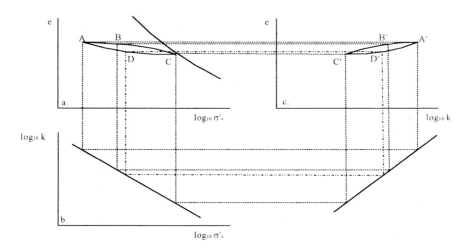

Figure 6   Stress - state - permeability relationship showing (a) compression curve; (b) variation of permeability with stress; and (c) variation of permeability wit void ratio

This is different from most published empirical correlations between e and k which generally indicate a trend toward a limiting value of k at high void ratios. Figure 6 shows that the shape of the e - $log\ k$ plot depends on whether the specimen is on a compression or swelling line. Figure 6a is a typical compression curve of e - $log\ \sigma'_v$ which includes an unload-reload cycle exhibiting hysteresis. Figure 6b, a $log\ k$ - $log\ \sigma'_v$ plot, is shown as linear because the experimental evidence from these (see Figure 5) and other reported tests confirms this assumption. Consider A, B and C on the recompression line in Figure 6a. This is mirrored as A', B' and C' in Figure 6c (e - $log\ k$) which is a similar shaped curve to the results of the tests shown in Figure 5. Note that the tests were carried out on overconsolidated soil in which the consolidation pressure was increased between measurements of permeability. The opposite trend, commonly reported in the literature is likely due to the fact that soils are either on the swelling line or they are undergoing seepage induced swelling during permeation caused by relatively large hydraulic gradients with respect to the consolidation pressures (Harwood et al. 1995). It is only because of the accuracy of the measuring systems used here and the low hydraulic gradients induced can the variation of e with $log_{10}k$ be determined. Note that for practical purposes the variation of e with $log_{10}k$ can be assumed to fall within a band centred around Equation (1).

The void ratio cannot be measured directly and has to be calculated either from the initial dimensions of the specimen and the volume changes that occur during saturation, consolidation and permeation or from the final water content and specific gravity of the soil

849

particles. The effective stress is measured directly, so a relationship between coefficient of permeability and effective stress may be more useful for design purposes. For example, Vaughan (1994) demonstrated that the variation of pore pressure within the core of a dam was attributed, in part, to the variation of the coefficient of permeability with effective stress; Fernanadez and Quigley (1991) suggest that increasing the effective stress reduces the permeability of clay liners. Settlements of structures founded at the surface are often faster than predicted. This could be due, in part, to the fact the in situ permeability is greater than that measured since permeability tests would normally be carried out at effective stresses greater than those in situ near the ground surface.

CONCLUDING REMARKS

Equipment has been developed that permits accurate measurements to be made of volume changes occurring in a clay due to consolidation and permeation. Constant rate of flow tests allow the coefficient of permeability to be determined at hydraulic gradients of less than ten and head differences across the specimen of less than 10 kN/m² which permit the stress-state-permeability relationships for a particular clay to be established with confidence.

Tests on several specimens of reconstituted till showed that consistent results were obtained and these were little affected by the permeant used. Further tests at different consolidation pressures on the same specimens showed, as expected, that the coefficient of permeability varies with void ratio and mean effective stress. The sensitivity of this equipment allowed tests to be carried out at effective stresses as low as 25 kN/m² showing that the coefficient of permeability can vary by an order of magnitude for a stress change of 0 to 100 kN/m².

REFERENCES

ARARUNA Jnr, J T, CLARKE, B G and HARWOOD, A H (1994a) A new apparatus for measuring permeability of soils, 10th Brazilian Conf Soil Mech and Found Engng, Iguassu Falls, Brazil

ARARUNA Jnr, J T, CLARKE, B G and HARWOOD, A H(1994b) A precise, practical and economical volume change measurement device, Geotechnique, to be published

ASTM D5084-90 Measurement of permeability of saturated porous materials using a flexible wall permeameter

BENSON, C H and DANIEL, D E (1990) Influence of clods on permeability of compacted clay, J Geotech Engng, Vol 116, No 8, pp 1231-1248

BS 1377:Part 6:1990 Consolidation and permeability tests in hydraulic cells and with pore pressure measurement

DANIEL D E (1993) State of the Art: Laboratory permeability tests for saturated soils, Proc Conf Permeability and Waste Contaminant Transport in Soils, ASTM STP 1142

FERNANDAZ, F and QUIGLEY, R M (1991) Controlling the destructive effects of clay-organic liquid interactions by application of effective stress

HARWOOD, A H, CLARKE, B G and ARARUNA Jnr, J T (1995) Discussion on 'Nagaraj et al 'Stress-state-permeability relations for overconsolidated clays' Geotechnique, Vol 44, No 2, pp 349-352

NAGARAJ, T S, PANDIAN, N S and NARASIMHA RAJU, P S R (1994) Stress-state-permeability relations for overconsolidated clays Geotechnique, Vol 44, No 2, pp 349-352

OLSEN, H W, NICHOLS, R W and RICE, T L (1985) Low gradient permeability measurements in a triaxial system, Geotechnique, Vol 35, No 2, pp 145-157

VAUGHAN, P R (1994) 34th Rankine Lecture: Assumption, prediction and reality in geotechnical engineering, Geotechnique, to be published

The shear strength and deformation behaviour of a glacial till

A. CHEGINI AND N. A. TRENTER
Sir William Halcrow & Partners Ltd, London

SYNOPSIS

The determination of the strength and deformation characteristics of a Glacial Till offers a considerable challenge to the geotechnical engineer. In many cases, the presence of coarse particle sizes (gravel and cobbles) makes conventional tube sampling difficult and further problems are introduced when the matrix is friable and crumbly, reducing the likelihood of successful sample retrieval.

A major site investigation for a proposed nuclear facility at Chapelcross, Dumfriesshire, provided an opportunity to introduce laboratory testing techniques into site investigation practice which are normally reserved for research and permitted both strength and deformation properties to be determined, despite the sampling difficulties. The paper describes the various techniques adopted and compares the results with established correlations when such are available.

INTRODUCTION

In 1991, a comprehensive geotechnical investigation of the Chapelcross site, Dumfriesshire, was made as part of the nuclear generation studies commissioned by British Nuclear Fuels Ltd. The investigation was required to determine the seismic hazard and all geotechnical aspects affecting design, construction and operation of a nuclear power station and accordingly was designed to the standards stipulated for nuclear power stations and carried out within a quality assured framework. Site work included cable tool boring and rotary core drilling; surface and cross-hole geophysics; down-the-hole nucleonic logging; high pressure dilatometer testing; Westbay piezometer data acquisition; and a pumping test. Laboratory testing included a series of special stress path and resonant column tests.

The site was underlain by a variable thickness of till which proved difficult to sample using conventional U100 equipment and for which only poor correlations between SPT 'N' value and undrained shear strength could be achieved. Consequently a programme of laboratory testing using research techniques was conducted in order to determine strength and deformation properties of the material.

SITE INVESTIGATION PROGRAMME AT CHAPELCROSS

The boreholes sunk were sub-divided into series, each series having a particular primary objective. Thus the series 100 boreholes were sunk to identify the site's geological structure and to place it within a regional context; series 300 boreholes were sunk as control on the seismic refraction survey; series 400 boreholes for geophysical purposes; series 500 and 600 for geotechnical data.

Laboratory routine triaxial and oedometer tests were performed on intact Glacial Till by the Contractor, Norwest Holst Soil Engineering. Special tests using Bishop and Wesley equipment were made on intact and reconstituted Glacial Till by the Geotechnical Engineering Research Centre, City University, and the test types and numbers are given in Table 1 (at end of text). Resonant column tests were performed by Bangor University and the details are also summarised in Table 1.

The reconstituted specimens used in the special tests (Table 1) were prepared from material retained at the end of each intact specimen stress path test, generally in accordance with Gens and Hight (1979); the de-aired slurry was initially one-dimensionally consolidated in a consolidometer under a vertical stress of 88 kPa. Instrumentation comprised piezoceramic bender elements (Shirley 1978, Schultheiss 1981) and Hall effect local strain transducers (Clayton and Khatrush 1986). Axial strain measurements were also made using external displacement transducers, calibrated for apparatus "stiffness" deviations or for initial seating adjustment compliance. The results from the external gauges were considered as control on the local strain devices and for monitoring soil behaviour at large strains.

The low frequency cyclic simple shear and resonant column tests (Table 1) were performed using the Drnevich device in accordance with the guidelines presented by Hardin and Drnevich (1970) and Novak and Kim (1981). The specimens were instrumented using external axial gauges. Bender elements were also incorporated into the resonant column device in order to propagate shear and compressional waves through the specimen length to provide an alternative method for measuring the dynamic shear modulus (G_{max}). The crosshole and refraction geophysical surveys provided information to determine seismic velocities and hence mechanical properties of the till in-situ.

GENERAL CHARACTERISTICS OF TILLS
A description of the properties of some lodgement tills in west central Scotland is given by McGowan et al (1975). They describe how the till's colour reflects the rocks from which it was derived, with grey-black tills from the Carboniferous Limestone in the Glasgow area and red tills from the Upper Coal measures in central Ayrshire. At Chapelcross further south, the tills are also red and red-brown, derived from the Permo Trias. Investigations made by McGowan et al indicate that west central Scotland tills have silt or clay filled discontinuities parallel to the orientation of the drumlin field, with two or more conjugate vertical fissure sets which are often weathered. The authors point out that the tills must be considered as stiff fissured soils, with potentially anisotropic engineering properties.

Atkinson and Little (1988) report that lodgement tills are deposited in a high stress environment within the basal traction zone of an ice sheet. They usually contain a wide range of particle sizes, often with predominant gap grading, and may or may not display evidence of fabric. Unlike sedimented soils, which were subjected to one-dimensional consolidation and swelling, tills may have been reworked by ice movements after deposition and their stress history may depart from this condition.

Table 1 : Summary of Specialist Laboratory Tests Scheduled for Glacial Till

Test Type	Test No.	Comments
Bishop & Wesley undrained stress path tests on undisturbed (intact) 100mm diameter specimens with external and local (Hall effect) strain transducers. Pore pressures measured at base of specimen.	5	Specimens mounted in cells and isotropically consolidated (and overconsolidated) to initial effective stresses in the range 50 to 640 kPa and then sheared along nine undrained and one drained stress paths. Small strain instrumentation mounted locally to allow accurate resolution of the axial strains.
Bishop & Wesley undrained stress path tests on reconstituted 38mm diameter specimens with external, local (Hall effect) strain measurements and very low strain dynamic shear modulus (G_{max}) determinations using bender elements. Pore pressures measured at base of specimen	5	Reliability of stiffness measurements at $\leq 0.01\%$ strain reduced as instruments were affected by ambient conditions. Large differences between the stiffnesses of the two internal and one external transducers reflect inhomogeneity in the till and for the latter type also bedding and trimming disturbance errors.
Resonant column tests with very low strain dynamic shear modulus measurements on 35 and 70mm diameter specimens. Pore pressures measured at base of specimens. External strain and bender element transducers employed.	4	Undrained cyclic and dynamic torsional shear tests were performed using a Drnevich apparatus. The specimens were either isotropically consolidated twice (p_o = 100 and 500 kPa) or isotropically consolidated and then swelled (p'_o = 500 and 100 kPa) and dynamic measurements made during each stage.

Note: Moisture Contents (MC) for all specimens were taken from trimmings at the start of tests. For the stress-path tests MC was also determined for the whole sample but, because of inhomogeneities and initial poor saturation levels, analysis of the test data was based on average of the final MCs.

Stress path and resonant column test specimens proved more difficult to saturate than in the triaxial tests even though much higher back pressures were used. B values ≥ 0.95 were obtained.

In matrix dominant tills, the laboratory approach to determining design parameters is hampered by the difficulty of obtaining reasonably undisturbed samples because of the stiffness of the material and the presence of cobbles and boulders (Gens and Hight 1979). Moreover, during in situ Standard Penetration Tests (SPTs), unrepresentatively high blow counts often result from the presence of coarse clasts in the till which may also prevent the use of pressuremeter and cone penetration devices. Therefore, predictions of till behaviour have often been based on the results of laboratory tests on remoulded material but, as pointed out by Gens and Hight (loc. cit.), care is necessary in sample preparation and in test procedures.

Tills often show strong dilatancy because the matrix is generally densely packed and, in undrained shear, it expands inducing negative pore pressures leading to a high apparent undrained shear strength. But the packing and fabric can be variable at the sample scale and this, together with sample disturbance, may contribute to the large scatter in apparent undrained shear strength generally encountered in tills. This behaviour may be reflected in SPT results, which may show high values due to dilatancy as well as the obstructive effect of larger size clasts.

GLACIAL TILL AT CHAPELCROSS

At Chapelcross, the Devensian age till varied between 0.5 and 13.2m in thickness although, near the centre of the site, a more uniform thickness between 4 and 6m was displayed. Depressions, frequently between 3 and 5m below surrounding rockhead were evident locally and the geophysical surveys produced evidence for a pseudo-channelised feature which could be attributed to differential glacial gouging. The till comprised poorly-sorted, sub-rounded and rounded Permo-Triassic sandstone and siltstone with some igneous clasts in a red brown sandy clay and silt matrix; there was little evidence of fabric and structure. The range of particle size distributions determined for the till at Chapelcross suggests that the Chapelcross till may be classified as matrix dominant. The moisture content of some one hundred and seventy till samples was determined and the modal value was 16%. Most of the till analysed fell into the 'CL' (inorganic clay of low plasticity) sub-group of the Casagrande classification ($\bar{L}L \doteq 30\%$; $\bar{P}L \doteq 16\%$).

RESULTS AND ANALYSIS OF DATA

Undrained Strength

The apparent undrained shear strength (c_u) for the till, measured in routine unconsolidated undrained triaxial compression tests varied appreciably at any one depth, generally lying in the range from 25 to 300 kPa. This variation is typical of overconsolidated tills and may be a result of sample disturbance, dilatancy effects as well as variations in the soil's lithology. The mean liquidity index of the Chapelcross till was -0.04. Since by definition clays are friable and crumbly below their plastic limit, sample disturbance seems inevitable.

The SPT 'N' values and corresponding c_u for adjacent samples are presented in Figure 1, where the plasticity indices or the triaxial test samples are also shown. The results are scattered, and no reliable relationship between N and c_u can be discerned. It is evident that the method proposed by Stroud and Butler (1975), Stroud (1988), for relating cohesion to SPT results is not suitable in this case.

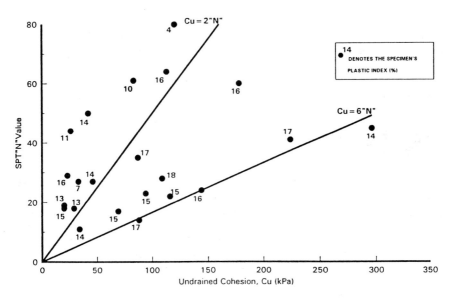

Figure 1 : SPT 'N' vs. Undrained cohesion at corresponding depth for Glacial Till at Chapelcross

Stress Strain Relationships
Figures 2 and 3 show plots of deviatoric stress (q') against corresponding external axial strains for intact and reconstituted stress path tested specimens (Table 1). For the intact specimens, presented on Figure 2, clear peak deviatoric stresses or critical states were not reached, although specimen 5 showed distinct strain softening, probably due to a faint localised shear plane detected on completion of the test; all other specimens, including those reconstituted, exhibited a bulging mode of failure.

Conversely Figure 3 illustrates that for the reconstituted specimens critical states were generally approached for the range of strains investigated, particularly in the case of the two isotropically normally consolidated specimens (6 and 9). Again peak deviatoric stresses are not readily identifiable here even for the overconsolidated specimens. The data were also plotted to show effective stress ratios (q'/p') against axial strains for the intact and the reconstituted specimens in Figures 4 and 5, respectively (Atkinson and Little 1988). These two plots show good similarities for the overconsolidated specimens, where peak stress ratios are defined and signs of approaching critical state at relatively large strains are more clearly exhibited.

Peak and Critical State Parameters
The results of routine consolidated undrained and consolidated drained triaxial tests on 100mm diameter intact specimens of Glacial Till are plotted in s' : t' stress space on Figure 6. It will be observed that most results fall on or about the line corresponding to an effective shear strength: c' = 0 kPa; ø' = 31 degrees. The small scatter is noteworthy (coefficient of correlation, $r^2=0.94$), particularly in view of the fact that the consolidated drained results were also included.

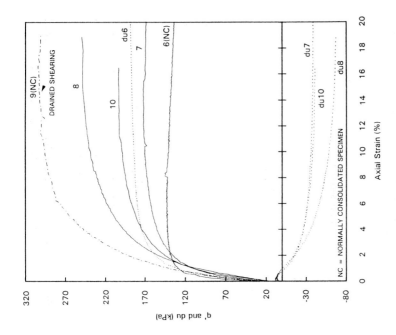

Figure 2 : Deviatoric Stress (q') and Pore Pressure change (du) vs. Axial Strain (100mm Intact Specimens)

Figure 3 : Deviatoric Stress (q') and Pore Pressure change (du)vs. Axial Strain (38mm Reconstituted Specimens)

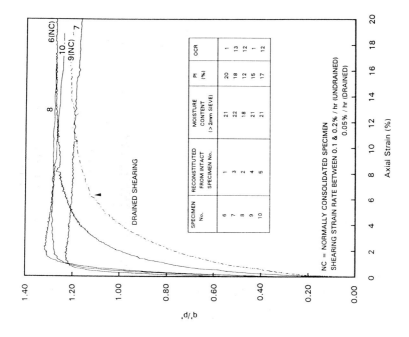

Figure 5 : Effective Stress Ratio (q'/p') vs. Axial Strain (38mm Reconstituted Specimens)

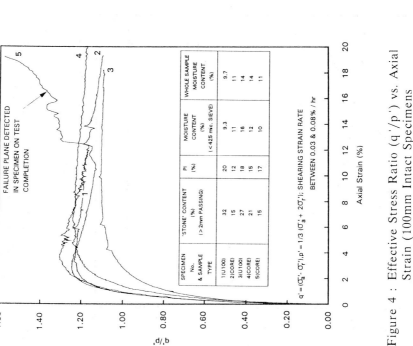

Figure 4 : Effective Stress Ratio (q'/p') vs. Axial Strain (100mm Intact Specimens)

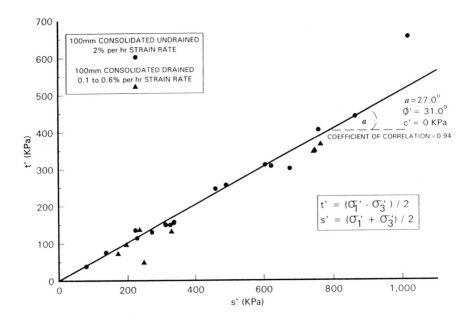

Figure 6 :   Results of Consolidated Drained and Undrained Triaxial Tests on
Intact Glacial Till

As mentioned above, for the reconstituted specimens, tested in the stress path
apparatus, the end of test states are generally close to the critical states. Stress
paths for the reconstituted specimens in q' : p' space are presented on Figure 7
and the critical state points for these specimen all lie close to a critical state line
(CSL) with little scatter. The paths for the intact specimen are also shown on
Figure 7; however, as previously stated, tentative true critical states were only
approached in two cases. Nevertheless, the CSL on Figure 7 is an acceptable
envelope to most paths for both sets of data and the chosen value for the CSL
gradient (M) is equivalent to a critical state friction angle $ø'_{cs}$ = 31 degrees,
equal to the Mohr-Coloumb angle shown on Figure 6. These results indicate that
for the Glacial Till at Chapelcross, critical state parameters may be independent
of the initial state, cementing, moisture content and fabric. It is also evident
that differences between shearing strain rates, quoted on Figures 4 to 6, do not
appear to influence results.

Isotropic Compression and One-Dimensional Consolidation
The one dimensional consolidation of the Chapelcross Glacial Till was measured
in oedometer tests which showed the till to be overconsolidated but with only a
poorly discernable pre-consolidation pressure (determined by the method
described by Casagrande 1936) in the range 400 to 1600 kPa. Isotropic
compression and swelling tests were made on both the intact and reconstituted
specimens and the results are illustrated on Figure 8. In the case of the
reconstituted specimen swelling lines, it is noted that they are in fact curved
rather than linear.

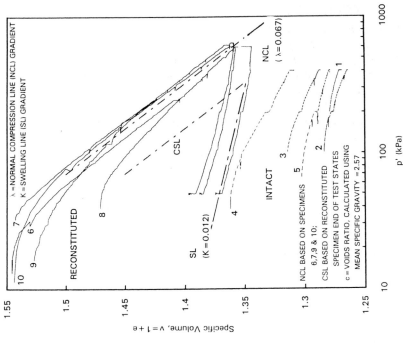

Figure 8 : Isotropic Compression Data for 100mm
Intact and 38mm Reconstituted
Specimens

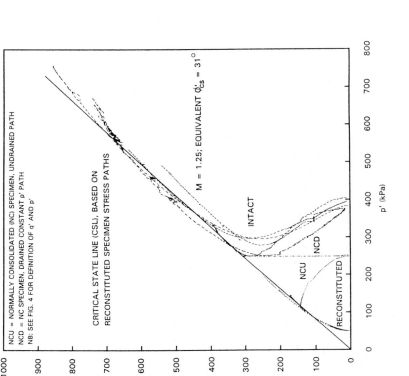

Figure 7 : Stress Paths during Shearing Stage for 100mm
Intact and 38mm Reconstituted Specimens

The results from the above tests were used to define an isotropic, normal compression line (NCL) and an isotropic swelling line (SL), for the reconstituted specimens. The critical state line (CSL) plotted in v : ln p' space is also shown on Figure 8, passing through the end states of the reconstituted specimens with the same gradient as the isotropic NCL. Only specimen 8 appears to indicate a slightly lower gradient, probably because of the specimen's relatively low plasticity index. The chosen compression and swelling parameters given on Figure 8 correspond to an equivalent isotropic compression index, C_c = 0.154 and isotopic swelling index, C_s = 0.028. The corresponding oedometer data are in the range : C_c between 0.05 and 0.15, and C_r (recompression index derived from unload-reload loops) between 0.01 and 0.09. On the basis of these results it is anticipated that the NCL and CSL shown on Figure 8 are probably similar to those likely to be derived for the intact specimens, with due regard to any cementing effects.

Deformation Moduli

It is assumed that the stress-strain behaviour of soil is generally non-linear for states inside a locus, defined by a series of points which may be recognisable as yield points on plots similar to Figures 3 to 6 (Atkinson and Sallfors 1991). In order to examine the non-linear stiffness profile of soils, especially at states inside the yield envelope and over the whole range of strains of practical importance within the state boundary surface (from less than 0.001 per cent to more than 1 per cent), it is necessary to use the equipment discussed below.

The small strain stiffness properties of the Glacial Till at Chapelcross were investigated by in situ crosshole seismic surveys with tests at about one metre depth intervals, by laboratory triaxial stress path tests with bender elements and by laboratory torsional resonant column tests, over a significant strain range

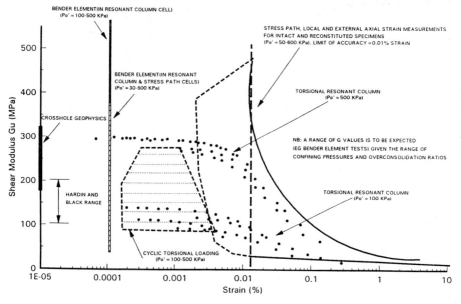

Figure 9 : Summary of all Shear Modulus Determinations for Glacial Till

(Table 1). The results are summarised on Figure 9 which shows values of shear modulus versus shear strain and also an empirical prediction of the G_{max} range of values using the Hardin and Black (1969) relationship for clays. The range of test results for the monotonic stress path triaxial tests, with external and Hall effect transducers on reconstituted and intact specimens is also shown. It can be seen that the resonant column and most bender element tests tend towards the in situ crosshole results at very small strains and at larger strains the resonant column tests approximately correspond to the monotonic tests. Application of the Hardin and Black relationship, using typical parameters for the Glacial Till produces results below the range indicated by in situ crosshole geophysics but accords with some of the laboratory test results.

A summary of the results from the resonant column tests is presented on Figure 10, where the influence of confining pressure, test stress history and specimen diameter is clearly evident. The normalised shear modulus variation (normalised with respect to each specimen's measured initial G_{max}) versus shear strain is demonstrated on Figure 11. The normalised resonant column stiffness data indicates that a well defined value of critical shear strain ϵ_c of less than or equal to 5×10^{-3} per cent exists beyond which non-linear stress strain behaviour is clearly observed. The measured data at Chapelcross presented on Figure 11 show significantly less degradation in modulus with shear strain than would be indicated by Seed and Idriss (1970) for gravelly soils and may in part be attributed to the dependency of ϵ_c on the soil's plasticity index (Georgiannou et al 1991, Dobry and Vucetic 1987). The data on Figure 11 fit well within an empirically determined hyperbolic expression known as the Ramberg-Osgood Model used to describe the variation of shear modulus with strain (Jennings 1964, Macky and Saada 1984).

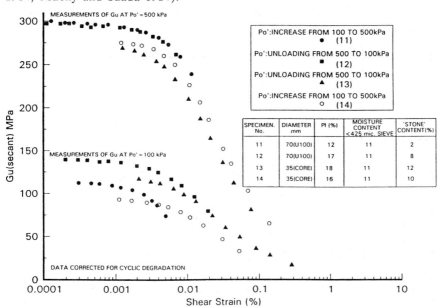

Figure 10 :   Summary of Torsional Resonant Column Undrained Secant Shear Modulus vs. Shear Strain Results for Glacial Till

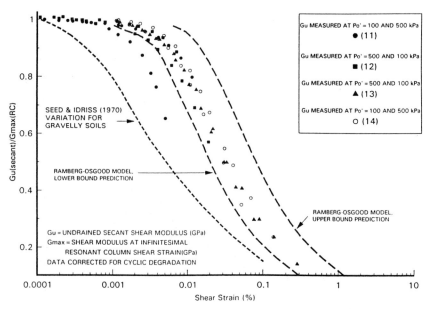

Figure 11 :   Summary of Normalised Torsional Resonant Column Secant Shear Modulus vs. Shear Strain Results for Glacial Till

On Figures 12 and 13, the stiffness profiles normalised with respect to the maximum in situ crosshole G_{max} (320MPa) are illustrated for the stress path tests on the intact and reconstituted specimens; the strain range corresponds to one where reasonably accurate resolution could be made. The figures illustrate the dependency of shear modulus on confining pressure which is quite marked for both intact and reconstituted specimens. They also demonstrate similar results for both intact and reconstituted samples. Gens and Hight (1979) found for a Yorkshire till that ultimate strength properties could be determined from reconstituted samples, irrespective of their consolidation histories, whilst pre-failure behaviour could only be reproduced by samples consolidated aniso-tropically. For the Chapelcross tills, the relatively good agreement between shear modulus for intact and isotropically reconstituted specimens suggests that anisotropic consolidation was not necessary.

The lower bound Ramberg-Osgood prediction is shown on Figures 12 and 13 and it will be seen that most results fall below this line, although the resonant column results fell reasonably well within the upper and lower bound Ramberg-Osgood predictions as demonstrated in Figure 11. However, in making comparisons, it is necessary to take into account the fact that the resonant column tests results are quoted in terms of octahedral shear strains. Had a factor of $\sqrt{3}$ (Georgiannou et al (loc. cit.)) been applied to the axial strain shown in Figures 12 and 13, significantly more comparable results would have been obtained. Furthermore, the normalising parameter applied to the stiffnesses determined in the stress path tests was chosen as the G_{max} value from the in-situ crosshole tests, whereas for the resonant column data G_{max} was derived for each specimen at the start of the test. Precise correlation cannot therefore be expected.

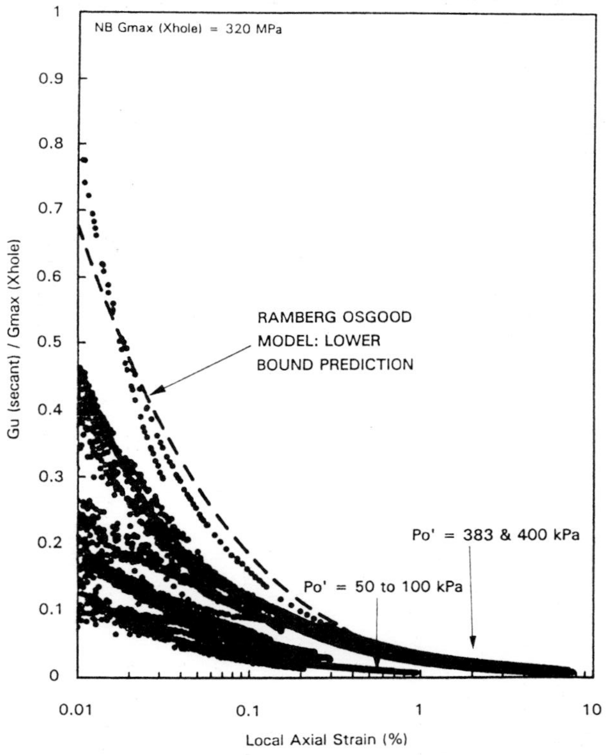

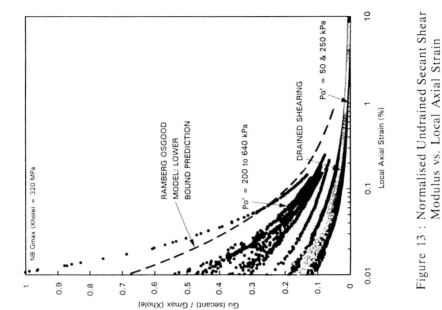

Figure 13 : Normalised Undrained Secant Shear Modulus vs. Local Axial Strain (38mm Reconstituted Specimen)

Figure 12 : Normalised Undrained Secant Shear Modulus vs. Local Axial Strain (100mm Intact Specimen)

A summary graph of the undrained shear moduli corresponding to axial strains of 0.01 and 0.1 per cent, normalised with respect to the initial isotropic effective pressure p'_o is plotted against OCR on Figure 14. The linear regression lines fitted to both sets of data show that normalised stiffnesses increase approximately linearly with OCR plotted to a log scale, with the small strain stiffnesses being most strongly dependent on OCR.

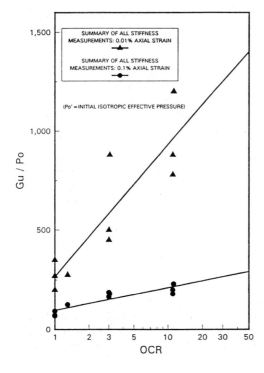

Figure 14:
Variation between Normalised Secant Shear Modulus (Gu) with Over-consolidation Ratio (OCR) for Reconstituted Specimens for Stress Path Tests

CONCLUSION

A site investigation was undertaken at Chapelcross, Dumfriesshire, where the site was underlain by Glacial Till having a widely varying particle size distribution, together with a matrix of low plasticity. The work demonstrated the following:

(i)   The undrained shear strength as measured on samples retrieved by conventional 100mm diameter tube samples varied appreciably, generally in the range 25 to 300 kPa. No useful relationship was found between SPT 'N' value and the undrained shear strength measured on adjoining tube samples.

(ii) Consolidated undrained and consolidated drained triaxial tests made on 100mm diameter tube samples produced an effective shear strength of $c' = 0$ kPa; $\emptyset' = 31$ degrees with little scatter; tests conducted using the Bishop and Wesley stress path apparatus and reconstituted test specimens yielded identical critical state shear strength parameters.

(iii)   Oedometer and isotropic compression and swelling tests were made on intact and reconstituted specimens. Clearly defined normal compression and isotropic swelling lines were obtained for the isotropic compression tests on reconstituted test specimens and the compression and swelling indices, C_c and C_s, respectively, were tolerably close to the range of oedometer values.

(iv)   Deformation moduli were determined in situ using geophysical cross-hole techniques and in the laboratory using stress path and resonant column equipment; intact and reconstituted tests specimens were used and the results were compared with formulations proposed by Hardin & Black, Seed & Idriss and Ramberg-Osgood. The stress path tests yielded very similar results from both intact and reconstituted specimens and the resonant column tests yielded normalised shear modulus values within the upper and lower bound predictions obtained from the Ramberg-Osgood model.

(v)   When plotted against overconsolidated ratio to a log scale, shear moduli normalised with respect to initial isotropic confining pressure demonstrated a linear relationship. This relationship was shown to be particularly sensitive for small strain shear moduli, less so for large strain values.

(vi)   Laboratory testing procedures normally reserved for research can be successfully employed in site investigations in order to determine the strength and deformation properties of a Glacial Till.

ACKNOWLEDGEMENTS

The authors are grateful to British Nuclear Fuels plc and to Sir William Halcrow and Partners Ltd for permission to publish the paper. They are also grateful to their colleagues in the Geotechnics Department at Halcrow and to the staff at the Geotechnical Research Centre, City University, for the many helpful comments made.

REFERENCES

ATKINSON, J. H. & LITTLE, J. A. 1988. Undrained triaxial strength and stress-strain characteristics of glacial till soil. Can. Geotech. J. 25. 428-439.

ATKINSON, J. H. & SALLFORS, G. 1991. Experimental determination of stress-strain-time characteristics in laboratory and in situ tests. Proc. 10th Euro. Conf. Soil Mechanics and Foundation Engineering. Florence. 915-956.

BISHOP, A. W. & WESLEY, L. D. 1975. A hydraulic triaxial apparatus for controlled stress path testing. Géotechnique, Vol, 25, No. 4, 657-670.

CASAGRANDE, A. 1936. The determination of preconsolidation load and its practical significance. Proc. 1st Int. Conf. Soil Mechanics, Cambridge, Mass, Vol. 3. 60-64.

CLAYTON, C. R. I. & KHATRUSH, S. A. 1986. A new device for measuring local axial strains on triaxial specimens. Géotechnique, Vol. 36, No. 4, 593-597.

DOBRY, R. & VUCETIC, M. 1987. Dynamic properties and seismic response of soft clay deposits. Proc. Int. Symp. Geotechnical Engineering of Soft Soils, Mexico City, 2, 51-87.

GENS, A. & HIGHT, D. W. 1979. The laboratory measurement of design parameters for a glacial till. Design parameters in Geotechnical Engineering. Vol 2. Brighton, 57-65.

GEORGIANNOU, V.N., RAMPELLO, S. & SILVESTRI, F. 1991. Static and dynamic measurements of undrained stiffness on natural overconsolidated clays. Proc. 10th Euro. Conf. Soil Mechanics and Foundation Engineering. Florence 91-95.

HARDIN, B. O. & BLACK, W. L. 1969. Vibration modulus of normally consolidated clay: Closure ASCE Vol 95, SM6, 1531-1537.

HARDIN B. O. & DRNEVICH, V. P. 1970. Shear modulus and damping of soils. Technical reports UKY 27-70-CE 2 and 3, College of Engineering, University of Kentucky, Lexington, Kentucky.

JENNINGS, P. C. 1964. Periodic response of a general yielding structure, AXSCE, Vol. 90. EMZ 131-166.

MACKY, T. A. & SAADA, A. S. 1984. Dynamics of anisotropic clays under cyclic and transient loading. Int. Symp. Soils under Cyclic and Transient Loading. Swansea, Vol 1. 315-324.

McGOWAN, A., ANDERSON, W. F. & RADWAN, A. M. 1975. Geotechnical properties of tills in West Central Scotland. Proc. Symp. Engineering Properties of Glacial Materials. Birmingham 81-91.

NOVAK, M. & KIM, T. C. 1981. Resonant column technique for dynamic testing of cohesive soils. Can. Geotech. J. 18, 448-455.

SCHULTHEISS, P. J. 1981. Simultaneous measurements of P and S wave velocities during conventional laboratory soil testing procedures. Marine Geotechnology, Vol 4, No. 4, 343-367.

SEED, H.B. & IDRISS, I. M. 1970. Soil moduli and damping factors for dynamic response analyses, Earthquake Engineering Research Centre, University of California, Berkeley, Report No. EERC.

SHIRLEY, P. J. 1978. An improved shear wave transducer. Journal of the Acoustical Society of America, Vol. 63, No. 5, 1643-1645.

STROUD, M. A. & BUTLER, F. J. 1975. The standard penetration test and the engineering properties of glacial materials, Proc. Symp. behaviour of glacial materials, University of Birmingham.

STROUD, M. A. 1988. The standard penetration test - its application and interpretation. Proc. Symp. Penetration Testing in the UK, University of Birmingham.

The use of local strain measurements in triaxial testing to investigate brittleness of residual soil

L.A. BRESSANI, A.V.D. BICA & F.B. MARTINS
Universidade Federal do Rio Grande do Sul, Porto Alegre, Brazil

INTRODUCTION

One of the objectives of this paper is to highlight the importance of using stress path testing - with local strain measurement - as an adequate tool to identify brittle behaviour of weakly bonded soils. This type of test has been increasingly used during the last ten years to measure soil properties at low strain levels (less than 0,5 %). Conventional triaxial tests - with external strain measurement - are of little use to identify yield at small strains, due to severe bedding errors in the equipment.

Site investigation for slope stability problems usually includes conventional direct shear or triaxial compression testing. The main objective is to evaluate soil strength. Most design methods which are currently applied to slopes in residual soil require the determination of the factor of safety against shear failure. The prediction of slope displacements is not considered critical for design, unless sensitive structures are involved. As a result, stress path testing with local strain measurement is hardly ordered in this context.

The Authors have been investigating residual soils in Southern Brazil over the last four years. The soil originated from the weathering of a local sandstone denoted Botucatu Formation was found to behave according to the framework developed at Imperial College (Vaughan, 1988). Central to this framework is the concept that most residual soils have a weakly bonded structure and a characteristic yield surface (Leroueil and Vaughan, 1990). The Authors' experimental programme has shown that the Botucatu Formation residual soil has an yield surface and either a ductile or a brittle behaviour according to the stress path followed.

The brittle behaviour was highlighted in a slope failure which occurred in this soil. The slope was investigated and it seems that the mechanism has been influenced by the stress path followed to failure.

THE BOTUCATU FORMATION

Most of Southern Brazil is covered by thick basaltic layers. These are underlain by a sandstone of aeolic origin, the Botucatu Formation. In some areas soils derived from this formation daylight. These areas are being increasingly developed, particularly to the north of Porto Alegre, the capital of Rio Grande do Sul state. Geotechnical problems are being noticed in these soils, mainly related to surface erosion and slope stability.

The Botucatu Formation gives origin to residual soil profiles where layer thickness varies according to local weathering conditions. The topsoil layer (A horizon) is generally less than 0.5 m thick. Immediately below is a clay-enriched layer (B horizon) less than 5 m thick. The underlain soil layer (C horizon) is the most important; it may be up to 25 m thick. Its stratification - which is inherited from the parent rock - is visible. Particles in the C horizon are composed mainly of quartz and feldspar. Traces of the parent rock cement - generally due to iron oxides - are also present. Both colour (yellowish to red) and grading (fine to medium sand) are homogeneous with depth in this layer. Figure 1 shows the grading of two sites (Vila Scharlau and Estância Velha) distant about 10 km.

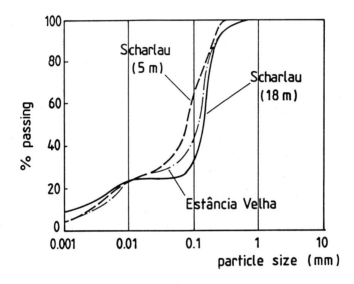

Figure 1 - Particle size distribution of Botucatu sandstone residual soil from two sites.

Block samples were taken from the C horizon at Vila Scharlau site. Martins (1994) carried out triaxial tests using these samples. Figure 2 shows some of the stress paths. It is worthwhile to examine the stress path followed by the test with null radial strain. At low stresses, this test follows a steep stress path up to the point where the yield surface is touched. For increasing stresses, the stress path progresses just bellow the yield surface. For larger stresses, the stress path begins to depart from the yield surface. Its shape then increasingly resembles the rectilinear K_o stress path typical of normally consolidated soils. Similar behaviour has been observed in other structured soils (Maccarini, 1987; Leroueil and Vaughan, 1990; Bressani, 1990). The stress path followed by the test with null radial strain is typical of a weakly bonded soil according to the framework described by Vaughan (1988). At low stresses, the structured soil is not significantly disturbed by the stress path with null radial strain. As a result, its structure is able of sustaining vertical stresses which are higher than for the corresponding de-structured soil at an equivalent lateral stress. However, as the stress path progresses, particle bonds are increasingly broken, and the soil becomes de-structured. The existence of a weakly bonded structure in

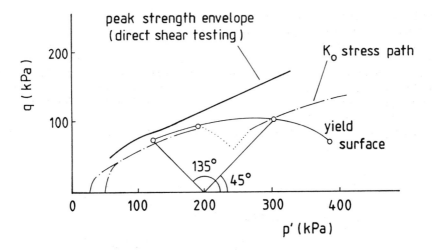

Figure 2 - Triaxial testing of Botucatu Formation residual soil: stress paths used (Martins, 1994).

the Botucatu Formation residual soil is the explanation for the particular shape shown in Figure 2 by the stress path with null radial strain.

The test in which the stress path follows a constant angle of 135° on the *p' x q* diagram deserves closer examination. As shown by the corresponding σ_d x ϵ_a and ϵ_v x ϵ_a curves (Figure 3), the peak strength and the maximum rate of dilation *do not coincide*. The peak

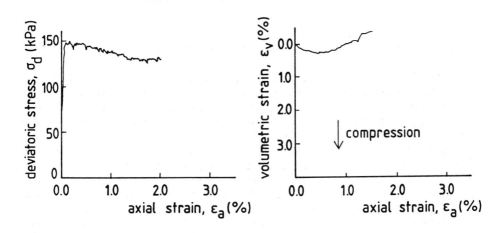

Figure 3 - Triaxial test where the stress path follows a constant angle of 135° on the *p' x q* diagram: σ_d x ϵ_a and ϵ_v x ϵ_a curves (Martins, 1994).

strength is mobilized at an axial strain of approximately 0.1 % whilst the maximum rate of dilation develops at a much higher ϵ_a (about 1.3 %). This behaviour can be interpreted - following Maccarini (1987) - as another evidence that the Botucatu Formation residual soil has a weakly bonded structure. According to Maccarini (1987), structured soils have characteristic yield points which are not related to dilatancy but instead to the breakage of particle bonds.

SMALL-STRAIN YIELD
Stress path tests shown in Figure 2 had yield points developed at $\epsilon_a < 0.5$ %. In order to determine such small strains accurately, all specimens of Botucatu Formation residual soil were instrumented with local displacement transducers. In these tests, local axial displacements were measured with Hall-effect transducers, as recommended by Clayton and Khatrush (1986). Local radial displacements were measured with a Hall-effect calliper (Bressani, 1990). It must be stressed that, if the external displacement measurement is used instead, the identification of the yield point becomes difficult. In addition, the axial strain corresponding to the yield point is wrongly given by this measurement technique. This is clearly shown in Figure 4 for the case of a deviatoric stress path. Bedding and equipment compliance are mainly responsible for such errors (Clayton et al, 1989).

The Botucatu Formation residual soil showed the stiffest response for the stress path which followed a constant angle of 135° on the $p' \times q$ diagram. The yield point developed at a very small axial strain, of about 0.1 % (Figure 3). This result can be compared with previous stress path tests described by Bressani and Vaughan (1989). An artificial soil was used in their tests, which simulated the bonded properties of residual soil. The

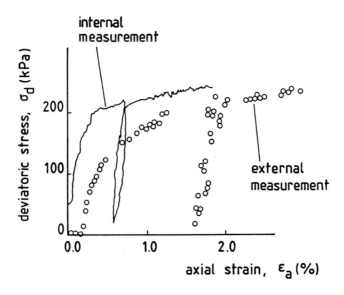

Figure 4 - Triaxial compression test: comparison between local and external displacement measurement (Martins, 1994).

technique of preparation was described in detail by Maccarini (1987) and Bressani (1990). It consisted of mixing sand with kaolin slurry, followed by air drying so that the clay slurry retreats into the sand voids, and firing in an oven. The fired kaolin then formed a stable bond between the sand particles.

Figure 5 shows two stress paths used by Bressani and Vaughan (1989). It is interesting to examine the stress path of test P, in which a confining pressure was first applied, followed by a deviatoric stress which was significantly less than that required to cause yield (point A). The specimen was subsequently failed by increasing the pore pressure while maintaining the deviatoric stress constant. The corresponding σ_d x ϵ_a curve is also shown in Figure 5. Note that for the section C-Y (pore pressure increase under constant

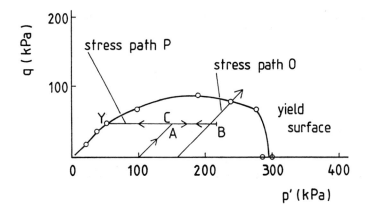

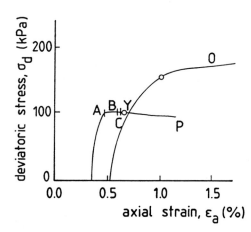

Figure 5 - Triaxial testing of artificial residual soil: stress paths and σ_d x ϵ_a curves (Bressani and Vaughan, 1989).

deviatoric stress) the specimen showed negligible axial strain. After the yield surface was reached (point Y), the specimen failed abruptly.

A comparison can be made between results of test P and the conventional drained compression test O, also shown in Figure 5. After yield (at about $\epsilon_a = 1$ %), test O was able to sustain increasing deviatoric stress with considerable axial strain. It is therefore appropriate to denote soil behaviour in test P as *brittle* in comparison with test O. Similar stress path tests on residual soil from granite [Geotechnical Control Office, *apud* Bressani and Vaughan (1989)] showed the same behaviour with both small axial strain at failure and brittleness. There are therefore considerable experimental evidence that weakly bonded soils (such as the artificial residual soil, the granite residual soil and the Botucatu Formation residual soil) can show brittle behaviour. This behaviour is not only due to cementation but is also stress path dependent. It was observed during tests in which the mean effective stress was reduced while the shear stress either increased or was kept constant.

The stress path from drained compression tests is inappropriate for slope stability problems where failure occurs as a result of increased pore pressure due to heavy rain (Brenner *et al*, 1985). For such reason, the stress path tests described above are of importance for slope stability analysis, specially when brittle weakly bonded soils are involved. The conventional slope stability approach - where only the knowledge of the factor of safety is sought - may not be entirely appropriate in such soils. The main problem is that, in case slope failure is approached in weakly bonded soils, little field warning signs can be expected, such as slope surface displacements or the opening of tension cracks. Therefore, the conventional factor of safety approach must be used with caution when brittle failure has been identified.

THE ESTÂNCIA VELHA SLOPE FAILURE
In July 1993, a slope failure severed the BR-116 single carriageway road near the town of Estância Velha, about 35 km to the north of Porto Alegre. This failure involved both the road fill itself and the underlain Botucatu Formation residual soil. A steep scarp was left behind (Figure 6). Debris spreaded to a large distance below the slope foot (Figure 7). Ten lives were lost and three houses were destroyed.

The slope failed abruptly after two days of heavy rain. About 26 hours before the accident, limited signs of movement were reported; a depression and a small fissure were observed on the pavement surface near the slope crest. The slope failed without further warning. Water was observed to seep in large quantity from the scarp, specially from two holes which seemed to be the result of previous erosion piping. In the debris flow no intact block of fill or residual soil could be seen. The failed soil mass seemed to have liquefied instead. The soil resembled to some extent specimens of Botucatu Formation residual soil at the end of those stress path tests in which brittle failure took place.

A limited post-failure site investigation was carried out due to the requirement of fast re-construction. This investigation revealed that the failure surface crossed the fill material, the B horizon soil and to some extent the C horizon soil. No water level could be measured down to a depth of 20 m below the road centerline one week after the accident. During debris removal other holes were found in the C horizon, which also seemed to be the result of previous piping. It seems that significant pore pressures developed at the site

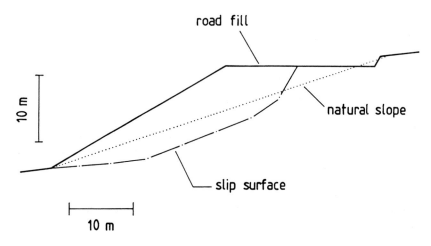

Figure 6 - Estância Velha slope failure: cross-section.

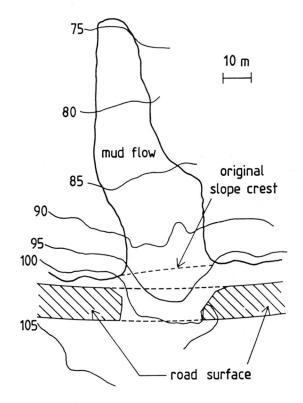

Figure 7 - Estância Velha slope failure: site plan.

only immediately before failure. As a result, it is likely that soil stresses varied according to the described stress path tests with structured soil.

CONCLUSION
Weakly bonded soils show brittle behaviour during stress path tests in which the mean effective stress is reduced while the shear stress either increase or is kept constant. Such behaviour will be of importance for slope stability problems where failure is induced by pore pressure increase. A kinematic mechanism consistent with this brittleness happened in a slope failure involving Botucatu Formation residual soil. Site investigation for such problems should include stress path triaxial testing with local strain measurement to identify brittleness.

ACKNOWLEDGEMENTS
The work of F.B. Martins was supported by Conselho Nacional de Desenvolvimento Científico e Tecnológico (CNPq) of Brazil.

REFERENCES
Brenner, R.P., Tam, H.K. and Brand, E.W. (1985). Field stress path simulation of rain-induced slope failure. *Proc. XI Int. Conf. Soil Mech. Fndn Engng, San Francisco* **2**, pp. 991-996.

Bressani, L.A. (1990). Experimental properties of bonded soils. *Ph.D. Thesis*, University of London, London, U.K.

Bressani, L.A. and Vaughan, P.R. (1989). Damage to soil structure during triaxial testing. *Proc. XII Int. Conf. Soil Mech. Fndn Engng, Rio de Janeiro* **1**, pp. 17-20.

Clayton, C.R.I. and Khatrush, S.A. (1986). A new device for measuring local strains on triaxial specimens. *Geotechnique* **36**, No. 4, pp. 593-597.

Clayton, C.R.I., Khatrush, S.A., Bica, A.V.D. and Siddique, A. (1989). The use of Hall effect semiconductors in geotechnical instrumentation. *Geotechnical Testing Journal* **12**, No. 1, pp. 69-76.

Leroueil, S. and Vaughan, P.R. (1990). The general and congruent effects of structure in natural soils and weak rocks. *Geotechnique* **40**, No. 3, pp. 467-488.

Maccarini, M.M. (1987). Laboratory studies of a weakly bonded artificial soil. *Ph.D. Thesis*, University of London, London, U.K.

Martins, F.B. (1994). Triaxial equipment automation for the study of bonded tropical soils. *M.Sc. Dissertation*, Universidade Federal do Rio Grande do Sul, Porto Alegre, Brazil (in Portuguese).

Vaughan, P.R. (1988). Characterizing the mechanical properties of in-situ residual soil. *Proc. 2nd Int. Conf. on Geomechanics in Tropical Soils, Singapore* **2**, pp. 469-487.

Recent Japanese practice for investigating elastic stiffness of ground

S. SHIBUYA [1], T. MITACHI [1], S. YAMASHITA [2] and H. TANAKA [3]
1) Hokkaido University, Sapporo, Japan.
2) Kitami Institute of Technology, Kitami, Japan.
3) Port and Harbour Research Institute, Yokosuka, Japan.

1.INTRODUCTION

Recent case histories in geotechnical engineering indicate that the ground strains induced under working loads are mostly 0.1 % or even less. (Burland, 1989, Tatsuoka and Kohata, 1994). Therefore, it is important to understand the stiffness of geomaterials under states of stress/strain which are far from failure. This statement is equally applicable to monotonic loading problems as well as dynamic or cyclic loading problems.

In–situ seismic survey is a common practice for exploring the G_f profile of ground, where G_f represents the shear modulus defined as $\rho_t V_s^2$ (ρ_t : total soil mass, V_s : in–situ seismic shear wave velocity). The shear strain level involved in the survey is very small of the order of 0.0001% (10^{-6}). The investigation making use of borehole(s) is common. However, seismic cone is a competent piece of tool owing to not only the versatility in obtainable information but also the better performance when used in soft soils. The huge experimental data indicates that the stiffness at very small strains (less than about 0.002%) as measured in the laboratory can be regarded as elastic in an engineering sense. The pseudo–elastic shear modulus, G_{max} , is a fundamental soil property which is little influenced by shearing rate, loading frequency, type of loading (monotonic and cyclic), number of cycles, etc (Tatsuoka and Shibuya, 1992, Shibuya et al, 1992, 1995). Provided that a) quality samples are supplied, and b) in–situ stress/strain conditions are successfully duplicated in the laboratory, G_{max} should be close to G_f at the relevant depth.

The paper aims to provide an overview of recent Japanese case records of site investigation, each in which the agreement between G_f in–situ and G_{max} in the laboratory is examined. Moreover, a first–hand case record of this kind obtained by the authors is described.

2.MEASUREMENT OF PSEUDO-ELASTIC STIFFNESS – A LITERATURE REVIEW

Table 1 summarizes recent case records in which the pseudo–elastic stiffness of ground is measured comparatively by in–situ seismic surveys and laboratory tests. A total of thirteen case records are conveniently classified in terms of the geomaterial type, together with the

Table 1   Comparisons of pseudo-elastic shear moduli between laboratory and in-situ seismic type measurements – review of recent Japanese case records

	Reference	Type of Geomaterial	Sampling method	Labo. test	Consolidation in the labo.	G_f (MPa)	$\dfrac{(G_{max})_{labo}}{G_f}$
①	Shibuya & Mitachi(1994)	Holocene clay	thin-wall	CTS	$\sigma'_c = \sigma'_{v\ in-situ}$ $\sigma'_c = 0.67 \times \sigma'_{v\ in-situ}$	3~15	1.0~1.2
②	Koga et al (1994)	Holocene clay	thin-wall	RC	$\sigma'_c = \sigma'_{v\ in-situ}$	8~12	0.9~1.3
③	Ohneda et al (1984)	Holocene clay	thin-wall	RC	$\sigma'_c = 0.67 \times \sigma'_{v\ in-situ}$ ($K_0 = 0.5$ assumed)	10~20	≒ 0.3
		Pleistocene clay				40~60	0.9~1.2
④	Mukabi et al (1994)	Holocene clay	thin-wall	MTXC	Anisotropic $\sigma'_v = \sigma'_{v\ in-situ}$ ($K_0 = 0.5$ assumed)	≒ 40	0.9~1.1
		Pleistocene clay				50~200	0.9~1.1
⑤	Yasuda, S. et al(1994)	young sand fill	thin-wall	CTS	$\sigma'_c = \sigma'_{v\ in-situ}$	10~20	1.2~1.6
			triple			70~300	0.2~0.5
⑥	Tokimatsu & Ohara(1990)	Holocene sand	thin-wall	CTX	$\sigma'_c = \sigma'_{v\ in-situ}$	20~100	0.3~1.5
		Pleistocene sand	triple-tube			100~400	0.1~1.0
		Both above	in-situ freezing			20~400	0.9~1.1
⑦	Koga et al (1994)	Holocene sand	In-situ freezing	CTS	$\sigma'_c = \sigma'_{v\ in-situ}$	30~100	0.8~1.25
		Pleistocene sand	triple tube			200~500	0.3~0.7
⑧	Tanaka, Y. et al(1994)	gravel	in-situ freezing	CTX	$\sigma'_c = \sigma'_{v\ in-situ}$	250~350	0.8~1.1
⑨	Yasuda, N. et al(1994)	gravel	in-situ freezing	CTX	$\sigma'_c = \sigma'_{v\ in-situ}$	250~800	0.9~1.2
⑩	Tatsuoka et al (1993) Tatsuoka & Kohata(1994)	sedimentary soft rocks	core boring	MTXC	$\sigma'_c = \sigma'_{v\ in-situ}$	300~ 2,000	0.8~1.2
			rotary core tube				0.3~0.7
⑪	Koga et al (1994)	sedimentary rock	core boring	CTX	$\sigma'_c = \sigma'_{v\ in-situ}$	≒ 700	≒ 0.6
⑫	Tatsuoka & Shibuya(1992) Shibuya et al (1992)	cement-treated sandy soil	block	CTX	$\sigma'_c : 0.05 \sim 0.25$ MPa	≒ 1,000	0.8~0.95 0.9~1.1
⑬	Toki et al (1994)	sedimentary soft rock	core boring	CTX	$\sigma'_c = \sigma'_{v\ in-situ}$ Anisotropic $K = 0.33$	≒ 1,000	0.85~1.1 0.9~1.05

σ'_c : isotropic consolidation pressure, unless otherwise stated.

$\sigma'_{v\ in-situ}$: in-situ effective overburden pressure.

$(G_{max})_{labo}$:shear modulus at shear strain levels less than 0.002% measured in the laboratory.

G_f : shear modulus from in-situ shear wave velocity.

CTS: torsional shear test, CTX: cyclic triaxial test, RC: resonant column test, and MTXC: monotonic triaxial compression test.

geological background that was sited; i.e., Holocene clay (soft clay), Pleistocene clay (stiff clay), Holocene sand, Pleistocene sand, gravel and soft rock. It should be noted that except for two cases of ③ and ④, the intact samples were isotropically consolidated to the stress level equal to the in-situ effective overburden pressure, $\sigma'_{v(in-situ)}$. In addition to it, G_{max} of triaxial specimens of Pleistocene clay (④), gravel (⑧) and soft rocks (⑩, ⑫ and ⑬) was obtained from 'local' axial strain measurement, for which bedding error at the specimen ends was successfully avoided (e.g., Burland and Symes, 1982, Jardine et al, 1984, Goto et al, 1991).

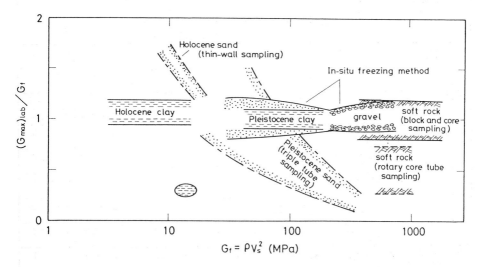

Fig.1 Comparison of shear moduli at very small strains between in-situ and laboratory measurements (refer Table 1).

Fig.1 shows the ratio, $(G_{max})_{labo}/G_f$, with respect to G_f. The following may be noted:

i) The $(G_{max})_{labo}/G_f$ values fall within a narrow band between 0.8 and 1.2 for most of clays retrieved by thin-wall samplers, sands and gravels by in-situ freezing method (Yoshimi et al, 1989, 1994) and soft rocks by block and core sampling, and

ii) the $(G_{max})_{labo}/G_f$ value is far greater than unity for loose sands (G_f <50 MPa) retrieved by thin-wall samplers, however, it was noticeably smaller than unity for dense sandy soils (50 MPa< G_f <400 MPa) recovered by a triple-tube sampler.

As already pointed out by Tokimatsu and Oh-hara (1990) and Tatsuoka and Shibuya (1992), data points outside the band may have involved significant effects of sample disturbance in the laboratory measurements of the stiffness. As far as results of sand and gravels are concerned, densification must have occurred duing the sampling of loose sandy soil. On the other hand, a triple-tube sampler which is commonly used for sampling dense sandy soil would have caused a significant modification of the in-situ fabric, resulting in breakage of interparticle bonding, reduction in density, etc. It may be surmised that these effects brought about an apparent increase and decrease of the stiffness, respectively, as measured in the laboratory. Based solely on the data presented in Table 1, the in–situ freezing method of sampling appears to be competent in avoiding disturbance in the course of sampling and transportation of sandy soil.

The above insights become more definite, if;–

i) the applicability of in–situ freezing technique is examined for very loose sandy deposit with some fine particles, and
ii) the performance of laboratory tests is more rigorously testified to the exact link to G_f by recompressing the intact samples to in–situ anisotropic states of stress.

The investigation which will be described in the following is currently on–going in an attempt to challenge these unknown issues.

3. A CASE RECORD OF OFFSHORE RECLAIMED ISLAND IN TOKYO BAY
3.1 Test Site

Higashi–Ohgijima island is a man–made island located offshore Kawasaki City in Tokyo Bay (Fig.2). The test borehole records show that the site consists of the top sand layer down to 14 m below the ground surface, which overlies Holocene deposits of silt and clay. The reclamation was completed only a few decades ago. The water level is about 2 meters below the ground surface. In addition, the overconsolidation ratio (OCR) of the original ground ranges between 1.0 and 1.2, and the primary consolidation due to the reclamation has been fully completed for the entire depth examined down to 30 meters below the ground surface (Tanaka H. et al. 1994).

Table 2 Gradings of top sand layer

Depth (m)	D_{max} (mm)	D_{50} (mm)	U_c	F.C. (%)
3.0	9.50	0.341	2.58	4.5
5.0	9.50	0.249	2.31	6.3
7.0	4.75	0.223	2.49	7.9
10.4	4.75	0.234	2.93	9.1

The material used for the reclamation is Sengenyama sand. Table 2 shows the representative gradings of some samples retrieved by in–situ freezing method. The mean diameter is 0.25 mm on average with the coefficient of uniformity ranging from 2.3 to 2.9. It contains considerable amount of fines (<0.074 mm) the content of which ranges from 2 to 10 percent along the depth.

It should be mentioned that the stratigrafy is typical of many reclaimed islands in Tokyo Bay. The SPT N–value ranges between 2 and 4 for both the sand and clay layers (Iai and Kurata, 1991).

3.2 Overview of In–Situ and Laboratory Tests
3.2.1 In–Situ seismic cone test and soil sampling

The G_f profile was defined by every 1 meter in depth by means of a downhole seismic cone test (Tanaka et al, 1994). The sand samples were retrieved using different sampling techniques; the in–situ freezing method and the conventional technique using two kinds of tube samplers (ie,a fixed–piston thin–wall sampler and a triple–tube sampler), both having

the inner diameter of 76 mm. The retrieved
samples were immediately put into a refrigerator,
and these were delivered to the laboratories. It
should be mentioned that the in–situ freezing
sampling took four full days to fulfil the entire
process. The clay samples were retrieved by the
fixed–piston thin–wall sampler.

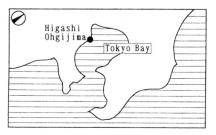

Fig.2 Test site

Fig.3 shows the variation of total density, ρ_t,
with depth, which was measured from the freezed
sand column with 15 cm in diameter extracted
from the ground. In this figure, the solid and
open circles represent the measurements from
the sizes of block with 25 cm and 10cm long,
respectively. Note that the scatter is considerable
due possibly to the variations of the fines
content (see Table 2) and random inclusion of
coarse grains. It should be mentioned that ρ_t
of 1.85 g/cm^3 was used for the calculations of
G_f and in–situ effective overburden pressure,
σ'_v.

3.2.2 Laboratory Cyclic Loading Tests
The G_{max} value of the soil samples was
obtained comparatively from two kinds of
laboratory tests; a cyclic torsion shear (CTS) test
and a cyclic triaxial (CTX) test. The testing
procedure has been described in detail by
Shibuya et al (1995). The descriptions of the
CTX apparatus developed at Hokkaido University
are given in Toki et al (1994).

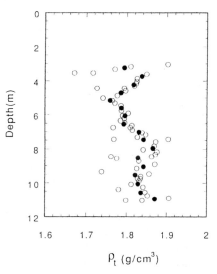

Fig.3 Total density in sand fill

The CTX specimen had the dimension of 5 cm in diameter and 10–12 cm in height,
whereas a hollow cylindrical specimen, having the inner/outer radii of 1.5 cm/3.5 cm and
the height of 7.0 cm, was prepared for the CTS test. The sand specimens were each
isotropically recompressed to the corresponding σ'_v (in–situ). Most of the clay
specimens was recompressed to the anisotropic state of stress with σ_h'/σ'_v (in–situ) =
0.5 (σ'_h : effective stress in the horizontal direction). In the CTX test, the specimen was
anisotropically consolidated by maintaining σ_h'/σ'_v constant at 0.5. On the ther hand,
the CTS specimen was initially subjected to an isotropic stress state with a half of σ'_v
(in–situ), and σ'_v was subsequently increased to σ'_v (in–situ). The recompression
paths were imposed aiming at closely reproducing in–situ stress states, which is based on a
postulation of K_o value of 0.5 for the clay deposit in a state of normal consolidation.

The specimen deformation was mesured at right above the top cap using proximity
transducers the use of which made the stiffness to be examined over a range of shear

Table 3 Results of cyclic loading tests in the laboratory

Depth (m)	Sampler	ρ_s (g/cm³)	LL (%)	I_P	Test	σ'_{vc} (kPa)	e_0	e_c	K	G_{max} (MPa)	f (Hz)
3.05	IFM	2.737	——	—	CTS	45.1	0.727	0.711	1.0	49.8	0.5
3.2	IFM	2.737	——	—	CTX	46.1	0.742	0.734	1.0	34.9	0.1
4.05	IFM	2.708	——	—	CTS	53.9	0.906	0.897	1.0	23.0	0.5
5.05	IFM	2.702	——	—	CTS	61.8	1.003	0.987	1.0	32.5	0.5
5.2	IFM	2.702	——	—	CTX	62.8	1.078	1.068	1.0	33.0	0.1
5.6	IFM	2.702	——	—	CTX	66.7	1.039	1.028	1.0	36.9	0.1
	TTS						0.926	0.919		37.6	
6.6	IFM	2.729	——	—	CTX	74.5	1.004	0.998	1.0	38.6	0.1
	TTS						0.791	0.777		52.9	
7.05	IFM	2.729	——	—	CTS	78.5	0.879	0.858	1.0	42.0	0.5
7.2	IFM	2.729	——	—	CTX	79.4	0.952	0.927	1.0	44.0	0.1
8.1	IFM	2.729	——	—	CTX	87.3	0.962	0.946	1.0	45.4	0.1
	TWS						0.927	0.919		56.4	
9.2	IFM	2.681	——	—	CTX	97.1	0.967	0.941	1.0	50.8	0.1
10.0	IFM	2.681	——	—	CTX	103.0	0.954	0.936	1.0	52.3	0.1
	TWS						0.840	0.819		57.8	
10.4	IFM	2.681	——	—	CTS	108.9	1.022	0.994	1.0	59.5	0.5
18.4	TWS	2.688	52.1	24.6	CTX	169.7	1.157	1.128	0.5	34.2	0.1
					CTS	113.1	1.266	——	0.75	32.5	0.5
20.4	TWS	2.684	55.6	30.6	CTX	184.4	1.208	1.194	0.5	34.8	0.1
					CTS		1.255	——	1.0	47.2	0.5
22.4	TWS	2.674	66.2	35.7	CTX	198.1	1.557	1.565	0.5	27.8	0.1
					CTS		1.564	——	0.5	36.6	0.5
25.4	TWS	2.685	83.1	46.4	CTX	215.7	1.453	1.515	0.5	30.5	0.1
					CTS		1.528	——	0.5	29.8	0.5
28.4	TWS	2.678	97.6	57.7	CTX	232.4	2.283	2.290	0.5	23.7	0.1
					CTS		2.318	——	0.5	21.6	0.5
31.4	TWS	2.655	100.2	60.6	CTX	248.1	2.226	2.126	0.5	28.6	0.1
					CTS		2.254	——	0.5	24.4	0.5
34.4	TWS	2.684	67.4	37.8	CTX	263.8	1.438	1.480	0.5	42.5	0.1
					CTS		1.286	——	0.5	61.4	0.5

σ'_{vc} : effective vertical stress at the end of consolidation.
e_0 : initial void ratio
e_c : void ratio at the end of consolidation.
f   : loading frequency , $K = \sigma'_h / \sigma'_v$
TWS : thin-wall sampler ($\phi = 76$ mm)
TTS : triple-tube sampler ($\phi = 76$ mm)
IFM : in-situ freezing method ($\phi = 150$ mm)

strain, between 1×10^{-5} (or less) and 1×10^{-2}. In the CTS test, the peak-to-peak shear modulus (equivalent shear modulus), G_{eq}, is determined from the hysteresis loop of the shear-stress and shear-strain relationship. In the undrained CTX test, G_{eq} was estimated as $E_{eq}/3$ (E_{eq} : equivalent Young's modulus) (Shibuya et al, 1995, Toki et al, 1994). The hysteretic damping ratio, h, is also obtained directly from the stress-strain relationship.

4. PRESENTATION OF LABORATORY TEST RESULTS

The tests performed are summarized in Table 3, in which void ratio at two instants;ie, when trimmed, e_0, and at the start of cyclic shear, e_c, is shown, together with the G_{max}

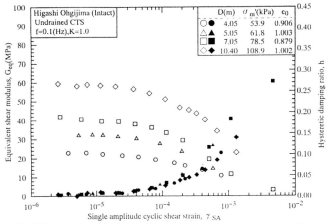

Fig.4 Results of CTS tests on intact sand specimens retrieved by in-situ freezing method of sampling.

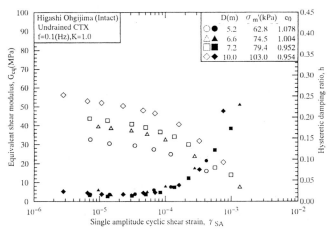

Fig.5 Results of CTX tests on intact sand specimens retrieved by in-situ freezing method of sampling.

value. Figs.4, 5 and 6 show the variations of G_{eq} and h with respect to single amplitude cyclic shear strain, γ_{SA}. The following may be noted;–

i) the $G_{eq} - \gamma_{SA}$ relationship exhibits the initial plateau for γ_{SA} less than 0.001% and 0.002% for the sand and clay samples, respectively,

ii) the damping ratio in these small strain regions is very small, and therefore

iii) G_{max} is determined as the first data point in each cyclic loading test (see Table 3).

It should be mentioned that the results of CTX tests on the clay samples are presented in the separate paper of Shibuya (1995).

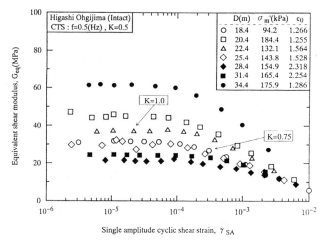

Fig.6 Results of CTS tests on intact clay specimens retrieved by thin-wall sampling

5. SOME REMARKS ON TEST RESULTS

A comparison is made for shear moduli of the sand at very small strains (Fig.7). The different data sets shown in these figures are; a) $G_f - \sigma'_v$ relationship from in-situ seismic cone test performed by Tanaka H. et al (1994) (open triangles), b) $G_{max} - \sigma'_c$ relationship from laboratory CTX and CTS tests on the intact specimens retrieved by in-situ freezing method of sampling (solid squares and solid circles, respectively) and c) $G_{max} - \sigma'_c$ relationship from CTX tests performed on the tube samples (open circles). In Fig.7, these relationships are directly compared against the effective stresses with which the shear moduli are measured.

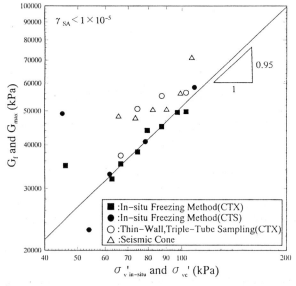

Fig.7 Shear moduli at very small strains with effective stresses

Some remarks to be made are;–

i) the $G_f - \sigma'_v$ (in–situ) relationship forms the upper bound in this diagram,
ii) the good agreement is seen of the results between the CTX and CTS tests, both using in–situ freezed samples,
iii) the stress dependency of G_{max} is expressed by the power of 0.95, and
iv) the G_{max} values of tube samples are noticeably higher compared with those of in–situ freezed samples.

In an attempt to see the effects of difference in density amongst the laboratory samples, the shear moduli are devided by the void ratio function, $f(e) = (2.17 - e)^2 / (1+e)$ (Hardin and Richart, 1963). In Fig.8, each e_c value was used for G_{max} of the laboratory specimen (see Table 3), whereas the correction was made for G_f by using the mean value of 1.013 (see Fig.3). The following may be noted;–

i) the power for G_{max} reduced to the value of 0.73, and
ii) the data points of the tube samples stayed within the scatter of those of the in–situ freezed samples.

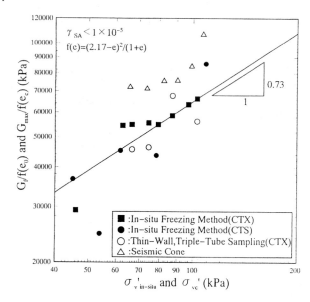

Fig.8 Corrected shear moduli at very small strains with effective stresses

It may be said that the higher values of G_{max} of the tube samples are probably due to the densification of the samples. Since more comprehensive laboratory investigation is still underway, the authors would not intend to say more about the results shown in Fig.8, until;–

a) effects of fines content on G_{max}, for example as previously investigated by Iwasaki and Tatsuoka (1977), are quantitatively evaluated for the material, and

b) the density profile is more precisely determined by taking account of the effects of coarse grains.

Fig.9 shows comparisons between G_f and G_{max} for the entire depth examined. In this figure, the change of density observed during the consolidation of the laboratory specimens is also examined by showing the e_c/e_0 value. The ratio, G_{max}/G_f is less than unity for the intact samples; it ranges between 0.75 and 0.98 for the in-situ freezed sand samples, and between 0.6 and 0.75 for the clay samples recompressed using K equal to 0.5. Provided that G_f is the most reliable measure, the underestimation of the small strain stiffness in the laboratory may be due to; a) underestimation of in-situ K_0 value for the clay (for example, G_{max}/G_f is approximately unity for the specimen recompressed using K=1.0), b) insufficiency of time allowed for secondary compression of the laboratory specimens, and c) effects of sample disturbance (this is unlikely at least for the clay samples tested, since e_c/e_0 is close to unity). Obviously, further study is needed to clarify every of these effects till the final conclusion is drawn.

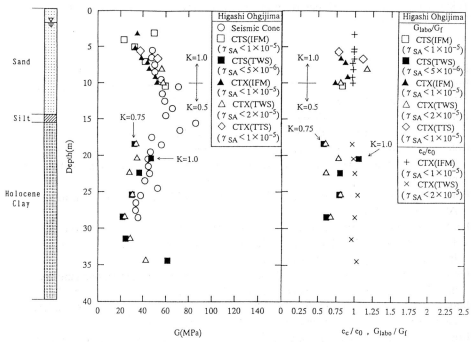

Fig.9 Profile of shear moduli at very small strains with depth and change in density of laboratory specimens during recompression

6. CONCLUSIONS

The current Japanese practice; thin-wall tube sampling is used for cohesive soils, and in-situ freezing method of sampling for granular materials, coupled with isotropic recompression to in-situ effective overburden pressure, results in a reasonable laboratory estimate of pseudo-elastic shear modulus of natural ground. The value of G_{labo}/G_f both determined at very small strains (less than 0.002%) fall within a narrow range between 0.8 and 1.2.

The performance of in-situ freezing method of sampling was well demonstrated in the first-hand case record presented in this paper, in which the small strain stiffness of very loose sand deposit containing some fines was measured by a in-situ seismic survey and laboratory cyclic loading tests. It is also demonstrated that the undrained cyclic triaxial test can be a reliable testing method to measure the small strain shear modulus of soft soils.

ACKNOWLEDGEMENTS
The authors are grateful to Professor S. Toki, Hokkaido University, for the continuous encouragement and support in carring out this project. Mr M. Nakajima, Dia Consultant Co Ltd., Messrs F. Fukuda, T. Furukawa, H. Inahara, A. Hosomi and D. Sugimoto, Hokkaido University, are greatly appreciated for help in performing the laboratory cyclic tests. Valuable discussion on the test results is given by Prof. F. Tatsuoka, University of Tokyo. The laboratory work described in this paper has been partly supported by TEPCO Research Foundation.

REFERENCES
Burland, J.B. and Symes, M.J. (1982). A simple axial displacement gauge for use in the triaxial apparatus. Geotechnique, Vol.32, No.1, pp.62–65.

Burland, J.B.(1989). The ninth Bjerrum Memorial Lecture: small is beautiful– stiffness of soils at small strains. Canadian Geotechnical Journal, Vol.26, pp.499–516.

Goto, S., Tatsuoka, F., Shibuya, S., Kim, Y–S. and Sato, T. (1991). A simple gauge for local small strain measurements in the laboratory. Soils and Foundations, Vol.31, No.1, pp.169–180.

Hardin, B.O. and Richart, F.E. (1963). Elastic wave velocities in granular soils. Jour. ASCE, Vol.89, No.33–65.

Iai, S. and Kurata, E. (1991). Pore pressures and ground motions measured during the 1987 Chiba–Toho–Oki Earthquake. Technical Note of the Port and Harbour Research Institute, No.718, pp.3–18 (in Japanese).

Iwasaki, T. and Tatsuoka, F. (1977): Effects of grain size and grading on dynamic shear moduli of sands. Soils and Foundations, Vol.17, No.3, pp.19–35.

Jardine, R.J., Symes, M.J. and Burland, J.B. (1984). The measurement of soil stiffness in the triaxial apparatus. Geotechnique, Vol.34, No.3, pp.323–340.

Koga, Y., Matsuo, O. and Sugawara, N. (1994). In situ measurement of shear moduli of soils and its evaluation. Pre–failure Deformation of Geomaterials (S. Shibuya, T. Mitachi and S. Miura edns), Balkema, Vol.1, pp.213–216.

Mukabi, J.N., Tatsuoka,F., Kohata, Y., Tsuchida, T. and Akino, M. (1994). Small strain stiffness of Pleistocene clays in triaxial compression. Pre–failure Deformation of Geomaterials (S. Shibuya, T. Mitachi and S. Miura edns), Balkema, Vol.1, pp.189–195.

Ohneda, H., Umehara, Y., Higuchi, Y. and Irisawa, K. (1984). Enginering properties of marine clay in Osaka Bay (Part 4); Dynamic stress–strain and strength properties. Technical Note of the Port and Harbour Research Institute, Ministry of Transportation, No.498, Sept., pp.115–136 (in Japanese).

Shibuya, S., Tatsuoka F., Teachavorasinskun S., Kong X.J., Abe F., Kim Y–S. and Park, C–S. (1992). Elastic deformation properties of geomaterials. Soils and Foundations, Vol.32, No.3, pp.26–46.

Shibuya, S. and Mitachi, T. (1994). Small strain shear modulus of clay sedimentation in a state of normal consolidation. Soils and Foundations, Vol.34, No.4 (in press).

Shibuya, S., Mitachi T., Fukuda F. and Degoshi T. (1995). Strain rate effects on shear modulus and damping of normally consolidated clay. Geotechnical Testing Jour., Vol.18, No.1 (in press).

Shibuya, S. (1995). Characterization of small strain shear modulus of Tokyo Bay clay. Proc. XI European Regional Conference on SMFE, Copenhagen (in press).

Tanaka H., Tanaka, M., Iguchi, H. and Nishida, K. (1994). Shear modulus of soft clay measured by various kinds of tests. Pre–failure Deformation of Geomaterials (S. Shibuya, T. Mitachi and S. Miura edns), Balkema, Vol.1, pp.235–240.

Tanaka, Y., Kudo, K., Nishi, K. and Okamoto, T. (1994). Shear modulus and damping ratio of gravelly soils measured by several methods. Pre–failure Deformation of Geomaterials (S. Shibuya, T. Mitachi and S. Miura edns), Balkema, Vol.1, pp.47–54.

Tatsuoka, F. and Shibuya, S. (1992). Deformation characteristics of soils and rocks from field and laboratory tests. Keynote paper, Proc. 9th Asian Regional Conference on SMFE, Vol.2, pp.101–170.

Tatsuoka, F., Kim, Y–S., Kohata, Y. and Ochi, K. (1993). Measuring small strain stiffness of soft rocks. Proc. Inter. Symposium on Hard Soils–Soft Rocks, Athens, pp.809–816.

Tatsuoka, F. and Kohata, Y. (1994). Stiffness of hard soils and soft rocks in engineering applications. Keynote Paper, Pre–failure Deformation of Geomaterials (S. Shibuya, T. Mitachi and S. Miura edns), Balkema, Vol.2 (in press).

Toki, S., Shibuya, S. and Yamashita, S. (1994). Standardization of laboratory test methods to determine the cyclic deformation properties of geomaterials in Japan. Keynote Paper, Pre–failure Deformation of Geomaterials (S. Shibuya, T. Mitachi and S. Miura edns), Balkema, Vol.2 (in press).

Tokimatsu, K. and Oh–hara, J. (1990). 8.2 In–situ freezing sampling. Tsuchi–to–Kiso, Proc. of JSSMFE, Vol.38, No.11, pp.61–68 (in Japanese).

Yasuda, N., Ohta, N. and Nakajima, A. (1994). Deformation characteristics of frozen specimens sampled from riverbed gravel. Pre–failure Deformation of Geomaterials (S. Shibuya, T. Mitachi and S. Miura edns), Balkema, Vol.1, pp.41–46.

Yasuda, S., Masuda, T., Nagase, H., Oda, S. and Morimoto, I. (1994). A study on appropriate number of cyclic shear tests for seismic response analysis. Pre–failure Deformation of Geomaterials (S. Shibuya, T. Mitachi and S. Miura edns), Balkema, Vol.1, pp.197–202.

Yoshimi, Y., Tokimatsu, K. and Hosaka, Y. (1989). Evaluation of liquefaction resistance of clean sands based on high–quality undisturbed samples. Soils and Foundations, Vol.29, No.1, pp.93–104.

Yoshimi, Y., Tokimatsu, K. and Hosaka, Y. (1994). In situ liquefaction resistance of clean sands over a wide density range. Geotechnique, Vol.44, No.3, pp.479–494.

The measurement of strength, stiffness and in situ stress in the Thanet Beds using advanced techniques

JGA JOHNSON[1], RL NEWMAN[2], TS PAUL[3], DS PENNINGTON[4]
[1]Ove Arup and Partners, Sheffield, UK
[2]Ove Arup and Partners, London, UK
[3]Ove Arup and Partners International, Singapore
[4]Department of Civil Engineering, University of Bristol, UK (formerly Ove Arup and Partners, UK)

ABSTRACT: The paper describes a site investigation of the Thanet Beds, a dense fine Tertiary sand, underlying Central London. The paper presents an interpretation of the data on the strength, stiffness and *in situ* stress in the Thanet Beds. In addition to conventional site investigation and laboratory tests, good quality samples were obtained using rotary coring techniques. Measurements of the stiffness at low strains were made in triaxial apparatus using Hall Effect Gauges. *In situ* tests using the Weak Rock Self Boring Pressuremeter were carried out and the results interpreted to corroborate the stiffness and strength data, and enable an approximate evaluation of the *in situ* stress in the deposit.

INTRODUCTION

The CrossRail Project is a proposed new railway line being developed jointly by London Underground Ltd (LUL) and the British Railways Board (Network South East). It will link counties on the east and west sides of London via 11 km of new 6 m diameter twin tunnels beneath central London. The high profile nature of the scheme and the sensitivity of the overlying urban environment to ground movement induced by tunnelling meant that the site investigation assumed an almost unique significance. Consequently a comprehensive programme of fieldwork and laboratory testing (including a number of advanced field and laboratory tests) amounting to some £1.2M was implemented over a period of nine months during 1992.

The proposed alignment of the tunnels included partial or near intersection of the Thanet Beds in the area of Farringdon and Liverpool Street stations. The aim of investigation in the Thanet Beds was to obtain reliable design parameters, particularly for use in ground stiffness models for assessment of ground movement. The basic philosophy of the CrossRail Investigation was to combine high quality sampling and description of the strata with both sophisticated and traditional field and laboratory testing.

FIELD WORK

Traditionally investigation of this stratum has relied upon light cable percussive methods recovering disturbed samples, combined with the Standard Penetration Test (SPT) to assess density and other parameters. For the CrossRail Project, SPT's were carried out in approximately half of the boreholes penetrating the Thanet Beds, and piezometers were installed to monitor ground water levels. The more advanced aspects of the investigation are discussed below.

Rotary Coring

Rotary cored holes were drilled to depths of up to 75 m using wireline techniques and triple tube core barrels incorporating semi-rigid plastic liners, with a variety of combinations of drill bit and flushing media. The purpose of specifying this modern drilling technique was to obtain high quality continuous core for detailed assessment of the lithology, close examination of the fabric of the soil, and recovery of good quality samples for laboratory testing. In the Farringdon Station area a top drive drilling rig was employed using Geobore wireline 'S' size triple tube core barrels with a variety of impregnated diamond set, tungsten carbide and combination bits, and IDS Hypergum mud flush. In the Liverpool Street area coring was carried out by SK6L wireline techniques using GS 550 polymer mud flush. At both locations, the core barrels consisted of the SWF size (146 mm) double tube systems fitted with semi-rigid removable plastic core liners to permit the continuous recovery of core of approximately 100 mm diameter.

Sampling

Core samples were selected from particular locations and these were sealed with clingfilm, waxed and protected by plastic casing within an hour of being removed from the ground to aid retention of the integrity of the sample and minimise disturbance. In some areas the combination of high quality rotary coring together with the high relative density and weakly interlocked nature of these beds enabled relatively undisturbed samples to be retrieved. This allowed carefully trimmed specimens of undisturbed material to be prepared for triaxial and shear box testing. In other areas it was not possible to obtain undisturbed samples and the material was therefore reconstituted in the laboratory to produce specimens which replicated field densities as far as practically possible.

Pressuremeter Testing

The Weak Rock Self-Boring Pressuremeter (WRSBP) instrument was used to test the Thanet Beds. This instrument has a membrane of neoprene rubber approximately 4.9 mm thick (reinforced with nylon fibres) and can operate up to a maximum cavity pressure of 20 MN/m^2. The WRSBP used a twin rod drilling system to install the instrument at the required test level. The thrust from an attendant rotary rig was transferred to the instrument via the non-rotating outer rods while the cutter was turned by the rotating inner rods. A full face cutter was used rather than 'fish tail' or rock roller bits used on the ordinary self-boring pressuremeter in order to penetrate the very dense material. The tests consisted of a loading stage, which included a minimum of three unload/reload loops, and a final unloading stage.

CLASSIFICATION PROPERTIES OF THE THANET BEDS

Classification tests on rotary cored samples were carried out to assist in the evaluation of geotechnical parameters through an understanding of their fundamental properties.

Petrographic Analyses

Petrographic analyses showed that the mineralogy of the Thanet Beds consists of mainly monocrystalline quartz (75%), with minor constituents being chert (5%), glauconite (7%), polycrystalline quartz, pyrite, feldspar, albite and muscovite and traces of organic matter. All specimens were reported as uncemented although there were traces of grains with rounded overgrowths suggesting reworked quartz sandstone. The shape of the dominant quartz was described as subangular to subrounded with fine to very fine grains. Measurements of porosity in specimens from samples deemed to be undisturbed, indicated pore sizes typically of 50 to 150 μm in diameter with corresponding estimates of total porosity ranging from 25 to 29%.

Particle size distributions

Grading analyses carried out on specimens from rotary cored samples indicated silt contents varying from 7 % to 20 %, while the percentage of material less than 0.2 mm diameter (fine sand) was in excess of 78 %. The Thanet Beds show a gradual 'coarsening upwards' grading as indicated in Figure 1. Silt and clay contents are generally 10 ± 5 % except over the top metre where they are up to about 22 %.

Angle of Repose

The angle of repose was determined by pouring samples of the oven dried sand onto a dry horizontal glass plate. The mean angle of repose was determined to be 33°. The value of 33° is comparable with values determined using the relationship between particle roundness and critical state angle proposed by Youd (1973).

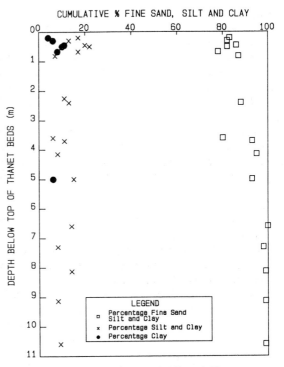

Figure 1: Percentage Fine Sand, Silt and Clay vs. Depth

Maximum and Minimum Void Ratio

The minimum void ratio was determined by the vibrating hammer method in a CBR mould, while the maximum void ratio was determined accordingly to Kolbuszewski (1948) using a 2 litre water bottle. Typical values of maximum and minimum void ratios measured were 1.32 and 0.66 respectively.

In situ Relative Density.

The *in situ* relative density (D_r) of rotary core samples was determined using Equation 1.

$$D_r = \left(\frac{\rho_{d\,in\,situ} - \rho_{d\,min}}{\rho_{d\,max} - \rho_{d\,min}} \right) \frac{\rho_{d\,max}}{\rho_{d\,in\,situ}} \times 100 \ (\%) \ \dots\dots\dots\dots\dots\dots\dots\dots\dots\dots\dots\dots\dots\dots\dots\dots\ 1$$

Typical minimum and maximum dry densities ($\rho_{d\,min}$, $\rho_{d\,max}$) measured in the laboratory were 1.12 and 1.60 Mg/m^3 respectively The *in situ* dry density, ($\rho_{d\,insitu}$), was measured by trimming a block of material from the centre of good quality core. A mean value of 1.5 Mg/m^3 was determined using Equation 1. Values of D_r ranging from 75 to 93%, with a mean of 84%, were estimated, indicating *in situ* relative densities in the dense to very dense range. This trend is consistent with the SPT data which show a range in $(N_1)_{60}$ values, (Skempton 1986), of 50 to 150 . For $(N_1)_{60}$ values in the range 50 to 90, and assuming the Thanet Beds are lightly overconsolidated, Stroud (1988) suggests relative densities between 75 and 100 %.

IN SITU HORIZONTAL STRESS

The results of WRSBP tests were used to evaluate the *in situ* horizontal stress in the Thanet Beds. In addition to being required directly in design of foundations and underground works, σ_{h0} is used in the assessment of angles of shearing resistance and dilation, and in relating shear moduli to appropriate stress levels. As stated by Mair and Wood (1987), the *in situ* total horizontal stress (σ_{h0}) in sands can, in principle, be determined from the lift-off pressure in much the same way as for clays. In practice, it is extremely difficult to avoid disturbance in the sand immediately adjacent to the pressuremeter during installation. The values of σ_{h0} assessed from lift-off pressures are considered to be lower-bound values. Fahey and Randolph (1984) suggest that even if perfect drilling could be carried out, the state of stress would be altered because of the shear stress between the sand and the instrument as it is installed.

An assessment of σ_{h0} was therefore carried out using two alternative procedures based on i) inspection of initial stage of loading curve (Fahey and Randolph, 1984), and ii) Yield Point Analysis (Whittle, 1991).

Inspection of initial loading curve (Fahey and Randolph, 1984)
The stiffness of soil on unloading is generally greater than on initial reloading. Drilling disturbance causes the radial stress to reduce and hence the deviatoric stresses to increase. When pressuremeter expansion starts, the radial stress increases and the deviatoric stresses reduce to zero before increasing. This change in stress path from unloading to reloading can be marked by a noticeable change in slope of the pressure-strain curve, which may represent the position of σ_{h0} if drilling disturbance has only created elastic stress changes.

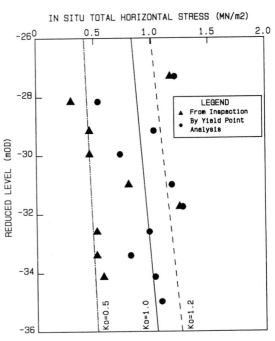

Figure 2: *In Situ* Horizontal Stress in Thanet Beds from Pressuremeter

The results of this analysis are shown in Figure 2, and indicate, that for the majority of results (6 in total), σ_{h0} lies between 400 and 600 kN/m², while 3 results lie in the range 800 to 1200 kN/m².

Yield Point Analysis (Whittle, 1991)
The sand reaches failure in a pressuremeter test when the stress ratio is:

$$\frac{\sigma_\theta'}{\sigma_r'} = \frac{\sigma'_{\theta 0} + \Delta\sigma'_\theta}{\sigma'_{r0} + \Delta\sigma'_r} = \frac{1 - \sin\phi'_{pp}}{1 + \sin\phi'_{pp}} \quad \dots\dots\dots\dots\dots\dots\dots\dots\dots\dots\dots\dots\dots\dots 2$$

where, ϕ_{pp} is the peak plane strain angle of friction, σ_θ' is the circumferential effective stress, and σ_r' is the radial effective stress. Since initially $\sigma'_{\theta 0} = \sigma'_{r0}$ and $\Delta\sigma_r' = -\Delta\sigma_\theta'$, it follows that:

$$\sigma_r' = \sigma'_{r0}(1+\sin\phi'_{pp}) \dotfill 3$$

The yield point, σ_r', can be detected from a log plot of effective cavity pressure against cavity strain as the point at which the plot becomes linear and at which the sand is assumed to be failing at constant stress ratio. The gradient of the plot is used to assess ϕ'_{pp} and hence σ'_{h0} $(= \sigma'_{r0})$ can be determined from Equation 3. The results of this analysis are shown in Figure 2 and indicate σ'_{h0} in the range of 800 to 1200 kN/m².

The majority of the results from the Fahey and Randolph approach lie close to a K_0 profile of 0.5, while the results from the Yield Point Analysis generally lie closer to a K_0 profile of unity. The latter is considered to represent more closely the condition which should be anticipated for design. This view is endorsed by the back analysis data from pile tests in the deposit reported by Troughton and Platis (1989).

STRENGTH

SPT

The $(N_1)_{60}$ values range from 50 to 150. There was no discernible trend with depth, and no apparent difference between the more silty material within the upper metre of the stratum and the consistently fine sand in the lower 9 to 10 m. A mean $(N_1)_{60}$ value of approximately 80 would indicate, according to correlation's given by Stroud (1988), peak plane strain angles of friction (ϕ'_{pp}) of about 44°.

Shear Box Tests

The effective stress strength parameters of the Thanet Beds in direct shear were assessed using a standard 60 mm square shear box. Specimens were sheared at a drained rate of 0.0081 mm/min to total relative displacements of up to 8.8 mm. This enabled both the peak and constant volume (critical state) angle of friction to be determined.

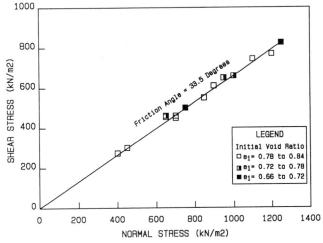

Figure 3: Constant Volume Angle of Friction - Shear Box

Figure 3 shows the variation of shear stress at constant volume conditions with the normal effective stress (σ'_n) applied to the specimen. The best fit line shows a constant volume angle of friction (ϕ'_{cv}) of 33.5 ± 1°, which compares well with the angle of repose measurement of 33°. There is no apparent dependence of ϕ'_{cv} on σ'_n for the range of normal effective stress range investigated. The peak angle of friction $[\phi'_p = \tan^{-1}(\tau_p/\sigma'_n)]$ of the material is 35.5 ± 2°, and there appears to be no curvature in the failure envelope for σ'_n ranging from 400 to 1200 kN/m².

The dependence of ϕ'_p on the initial voids ratio (e_i) is illustrated in Figure 4. Close inspection of the data plotted shows that, while ϕ'_{cv} is relatively independent of the initial void ratio, ϕ'_p for the dense samples ($e_i \approx 0.68$) is about 4 degrees higher than ϕ'_{cv} whilst the loosest samples ($e_i \approx 0.80$) give ϕ'_p values that are comparable with ϕ'_{cv}.

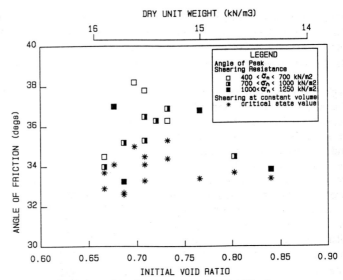

Figure 4: Variation of Angle of Friction with Void Ratio

Triaxial Tests

Triaxial tests were carried out on good quality 100mm diameter specimens obtained from rotary core samples from the Thanet Beds. Five tests were carried on specimens considered to be reasonably representative of the *in situ* structure and density of the Thanet Beds. All were initially isotropically consolidated to mean effective stresses between 350 kN/m² and 1000 kN/m². Three specimens then followed conventional drained stress paths, and two followed constant p', (p'=(σ₁+2σ₃)/3), stress paths. The specimens were sheared at strain rates of between 0.039 and 0.042 % per hour. The stress paths followed in each test are summarised on Figure 5. The results show peak angles of friction (ϕ'_{pt}) of the order of 42° and constant volume angles of friction (ϕ'_{cv}) of the order of

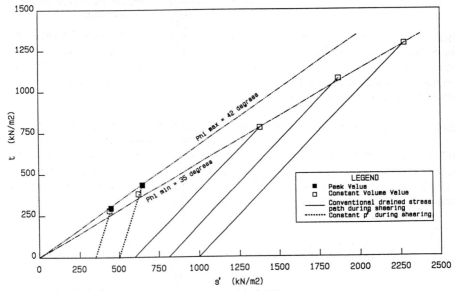

Figure 5: Stress Paths for Consolidated Drained Triaxial Tests

37° for specimens sheared at p′ <500 kNm². For tests carried out at mean effective stresses in the range 1350 kN/m² < p′ < 2250 kN/m², the constant volume angle of friction mobilised is 35°: These determinations of ϕ'_{cv} are higher than the plane strain results determined in the shear box as expected. No peak angles of friction were observed in this range.

Self-Boring Pressuremeter Tests
The angles of friction and dilation determined from the pressuremeter tests are for plane strain conditions and consequently the values are generally larger than those obtained from conventional triaxial tests. The two methods of analysis used to evaluate the peak plane strain angle of friction (ϕ'_{pp}) and peak angle of dilation (ν) were based on Hughes *et al.* (1977) and Manassero (1989). The derivation of ϕ'_{pp} and ν assumes no cavity expansion occurs until the cavity pressure exceeds σ_{h0}. As installation of the WRSBP causes expansion of the cavity to take place prior to reaching the actual *in situ* horizontal stress σ_{h0}, the datum for strain used in these analyses is the point at which the cavity pressure reaches σ_{h0}, i.e. (ϵ_c) current = (ϵ_c)measured -(ϵ_c) at σ_{h0}.

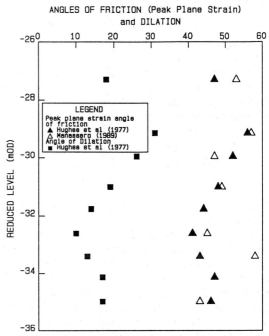

Figure 6: Peak Plane Strain Angles of Friction and Dilation from Pressuremeter

Hughes et al. (1977) The peak plane strain angles of friction (ϕ'_{pp}) and dilation (ν) derived by this method are shown in Figure 6. The trends of ϕ'_{pp} and ν are similar. Above -32 mOD, there are two particularly high values giving ϕ'_{pp} of 52° and 56° and ν of 26° and 31°. If these two high values are neglected the average $\phi'_{pp} \approx 47.5°$ and ν $\approx 18.5°$. Between -31 mOD and -35 mOD, average ϕ'_{pp} and ν values are 44° and 14° respectively.

Manassero (1989) The peak angles of friction derived using the Manassero method are also plotted in Figure 6. The trend and range in results is more variable than the Hughes *et al.* (1977) analysis, with ϕ'_{pp} varying from 43° to 58°. Manassero (1989) states that the method of Hughes *et al.* (1977) is likely to under-estimate ϕ'_{pp} due to the simplified constitutive rule adopted. In this context, it is notable that the majority of results obtained using the Manassero method exceed those obtained using the Hughes method.

Summary
Lade and Lee (1976) related ϕ'_{pp} from the pressuremeter to ϕ'_{pt} from triaxial tests by the following expression:

$$\phi'_{pt} = \frac{\phi'_{pp} + 17°}{1.5} \quad\quad\quad\quad\quad 4$$

If ϕ'_{pp} from the WRSBP in taken to vary generally from 44° to 47.5°, Equation 4 predicts ϕ'_{pt} of 40.5° to 43°, which is in good agreement with actual ϕ'_{pt} values of 42° for $p' < 500$ kN/m². In comparison with shear box tests. The angle of friction measured in the shear box is lower than the peak plane strain angle of friction (ϕ'_{pp}), as expected Jewell (1989).

STIFFNESS

SPT Data

Based on linear elastic back analyses of the settlements of footings and rafts founded on over-consolidated sands, Stroud (1988) proposed that for a given average stress ratio beneath a foundation represented by q_{net}/q_{ult}, the mean drained vertical Young's Modulus (E'_v) of the stratum could be correlated directly with the mean SPT N value. Stroud's correlation's for overconsolidated sands are listed in Table 1. The mean uncorrected SPT N value was approximately 150. For a q_{net}/q_{ult} of 0.1, which is typical of working foundations, $E'_v = 375$ MN/m².

q_{net}/q_{ult}	E_v'/N (MN/m²)
0.05	4.0
0.10	2.5
0.20	1.4

Table 1: Stroud's correlation for overconsolidated sands

Triaxial Tests

Hall effect gauges, mounted at the mid height of the triaxial stress path tests described previously, were used to measure the local axial strains (ε_a), allowing accurate resolution (to 0.003 %) of the secant shear stiffness at low strains.

The values of E'_v normalised by the mean effective stress in samples prior to shearing, p'_0, are plotted against the axial strain (ε_a) on Figure 7. From Figure 7 it is evident that for the conventional drained tests, (E'_v/p'_0) reduces from between 700 and 1200 at $\varepsilon_a = 0.01$ % to 350 ±50 at $\varepsilon_a = 0.1$ %, while the constant p' tests (E_v'/p_0') reduces from 700 ±100 at $\varepsilon_a = 0.01$ % to 425 ±75 at $\varepsilon_a = 0.1$ %.

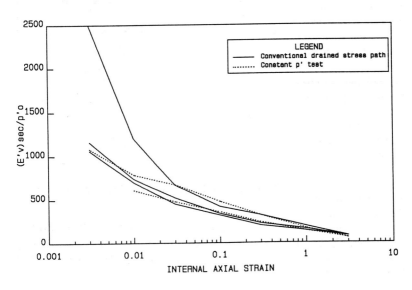

Figure 7: E'_v (secant) / p'_0 for Triaxial Compression

Pressuremeter

The elastic shear modulus was measured by performing small unload-reload loops during a pressuremeter test in much the same ways as for clays. If the soil is perfectly elastic in unloading, then the unload-reload cycle will have a gradient 2 x G_{ph}, (where G_{ph} is the unload-reload pressuremeter modulus), so that:

$$G_{ph} = \frac{1}{2} \frac{d\psi}{d\varepsilon_c} \qquad\qquad\qquad\qquad 5$$

where ε_c = cavity strain and Ψ = cavity pressure.

The procedure for performing the unload-reload loops to take account of creep deformation comprised holding the cavity stress constant and recording cavity strain until creep strain had apparently ceased. At this point the unload-reload loop was performed. The values of G_{ph} resulting from this analysis for a range of cavity strains from 0.04 to 0.1 % are shown in Figure 8. The range in G_{ph} values from the unload-reload loops is large and for the upper 3 tests between -27mOD and -30 mOD, G_{ph} values lie between 400 and 600 MN/m². From -30.5 to -34 mOD the values increase and range from 450 to 615 MN/m². The deepest test at -35 mOD gave G_{ph} ranging from 329 to 375 MN/m².

It is assumed that the Thanet Beds are relatively free draining and that no excess pore pressure develops during the test. It would therefore follow that the mean effective stress would increase as the test progresses. As a consequence G_{ph} values measured in this way would increase as the position of the unloading-reloading loop moves further from the start of the test. This explains the large range in values seen in Figure 8. For instance, tests at -30 and -31 mOD, in which the unload-reload loops have been evaluated over approximately the same range of strain, show a consistent increase in G_{ph} with successive loops.

The moduli obtained from unload-reload loops should be corrected for the average effective stress and strain

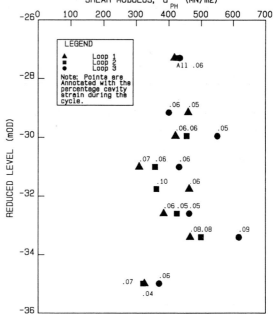

Figure 8: Reload Shear Modulus from Pressuremeter Tests

level existing around the pressuremeter cavity. Bellotti *et al.* (1989) provide a consistent framework for correcting G_{ph} by calculating average values of the mean plane strain effective stress S_{av}' and shear strain amplitude of the cycle (γ_{av}) within the plastic zone that exists at the start of the unload-reload loop.

Bellotti *et al.* (1989) proposed to correct G_{ph} by the following:

$$G^c_{ph} = G_{ph}\left(\frac{S_0'}{S_{av}'}\right)^n \qquad \dots\dots\dots\dots\dots 6$$

where $S_0' = 0.5\,(\sigma_r' + \sigma_\theta') = \sigma_{h0}'$, and σ_r' and σ_θ' are the radial and circumferential effective stresses respectively. $S_{av}' = \sigma_{h0}' + \alpha\,(P_c' - \sigma_{h0}')$ where P_c' is the effective cavity pressure at the start of the unloading loop and α is a reduction factor which is function of p_0', P_c' and ϕ'_{pp}. Bellotti *et al.* found that $\alpha = 0.2$ for Ticino sand, whereas for Thanet Beds α has been found to be 0.18, (Newman *et al.*,1991).

The exponent n value was assessed from a plot of log G_{ph} against log S_{av}'. The gradient of the best fit line through the data confirmed that n is close to 0.5. Equation 6 therefore becomes:

$$G^c_{ph} = G_{ph}\left[\frac{\sigma_{h0}'}{\sigma_{h0}'+0.18(P'_c - \sigma_{h0}')}\right]^{0.5} \qquad \dots\dots\dots\dots\dots 7$$

The values of G_{ph}^c assessed from Equation 7 are plotted in Figure 9. In comparison with Figure 8, the effect of stress level correction proposed by Bellotti *et al.* is a reduction in the range of values between individual loops as they are normalised to the *in situ* mean plane strain stress. Values of G_{ph}^c for changes in cavity strain which, on average, are 0.06 % range from 300 to 400 MN/m². Except for 3 tests at -30, -31 and -35 mOD which gave a mean G_{ph}^c value of 300 MN/m², a mean value through the remaining data is approximately 350 MN/m². The results do not indicate a systematic increase with depth as a function of the current mean effective stress. The SPT results similarly do not show an apparent increase.

Bellotti *et al.* also suggest that the average shear strain (γ_{av}) existing in the plastic zone around the probe at the start of the unload-reload loop is given by:

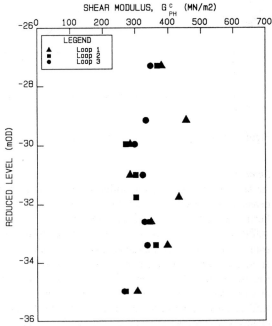

Figure 9: Reload Shear Modulus from Pressuremeter, corrected to *In Situ* Mean Effective Stress Level

$$\gamma_{av} = \beta\gamma_L \qquad \dots\dots\dots\dots\dots 8$$

where, β is a reduction factor which is a function of p_0', P_c' and ϕ'_{pp} which was found to be typically 0.5, (Bellotti *et al.* 1989), and γ_L is the shear strain amplitude of the loop around the cavity, i.e. $\gamma_L = 2\,\varepsilon_c$. The shear strain is given by:

$$\gamma_L = \sqrt{(2/9\,(\varepsilon_r - \varepsilon_z)^2 + (\varepsilon_r - \varepsilon_\theta)^2 + (\varepsilon_z - \varepsilon_\theta)^2)} \qquad \dots\dots\dots\dots\dots 9$$

where, ε_r, ε_θ and ε_z are the radial, hoop and vertical strains. The vertical strain, ε_z, is zero while in the elastic region $\varepsilon_r = -\varepsilon_\theta$ and at the cavity wall the cavity strain ε_c is equal to the hoop strain. The shear strain therefore reduces to:

$$\gamma_L = \frac{2\varepsilon_c}{\sqrt{3}} \quad or \quad \frac{\varepsilon_c}{\gamma_{av}} = \sqrt{3} \quad\text{...} \quad 10$$

As stated by Jardine (1992), this amounts to a transformed strain approach. Assuming elastic conditions and no change in mean effective stress, a more useful term G_{sh}, the secant shear modulus in the horizontal direction, can be introduced. Hence G_{ph} vs. ε_c plots can be transformed to G_{sh} vs. γ_{av} plots. The results of this analysis are shown in Figure 10.

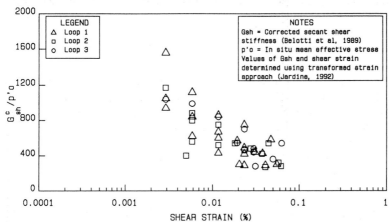

Figure 10: G_{sh}^c/p_o' from Pressuremeter Tests

Summary
As seen from Figure 7, the stiffness ratios E'_v/p_0 from the triaxial tests appear low for the dense Thanet Beds and are comparable to values for London Clay, (Jardine et al 1984). Furthermore, the degree of non-linearity of 0.56, indicates a relatively moderate reduction in stiffness with increasing strain level. In comparison, the pressuremeter results indicate an equivalent E'_v/p_0' ratio (assuming isotropic, linear elastic behaviour, i.e. $E_v' \approx 2.5\ G_{sh}^c$) of 3000 at an approximate axial strain of 0.05 %, which is substantially higher than the corresponding triaxial values. The apparently low values of stiffness from triaxial tests may be due to damage of particle contacts during sampling and extrusion and the formation of a different and 'less stiff' contact structure during subsequent consolidation (prior to shearing). Ageing and other effects are probable reasons for the stiffer structure shown by the material *in situ* and as measured in the pressuremeter tests.

CONCLUSIONS

The site investigation for the CrossRail Project included clusters of concentrated high quality site investigation at key locations to obtain detailed information on the ground profile and representative geotechnical design parameters. At these locations, both routine and sophisticated laboratory and in-situ tests enabled development of non-linear stiffness models for assessment of ground movements.

Use of simple, cost effective classification tests, e.g. angle of repose, petrographic analyses, maximum and minimum voids ratio, in conjunction with more advanced testing, provided confirmation of the reliability of the data.

High quality rotary coring of the Thanet Beds using modern drilling techniques provided continuous and detailed logs of the ground conditions, which are essential to the assessment of tunnelling conditions, as well as representative samples for testing. Traditional cable percussive drilling in this deposit fails to provide the same level of information. The measurements made *in situ* and on samples revealed that:

- the WRSBP can provide credible estimates of σ_h and K_0, although great care is needed in the choice of the method of analysis adopted to evaluate test results;
- peak angles of friction obtained from the WRSBP are comparable with those from triaxial tests on intact specimens. (Credible values of angles of dilation could not be measured in the shear box);
- Critical State angles assessed from mineralogy, grain shape and shear box agree well. It would appear that the strength parameters are less sensitive to the alterations of the contact-structure by sampling;
- stiffness as measured in the triaxial tests appear low for the dense Thanet Beds. This may be due to damage of particle contacts during sampling and specimen preparation.. The WRSBP gave substantially higher stiffness values. The SPT correlations give values which are more comparable with those of the WRSBP, but investigations for major projects may benefit from a comprehensive non-linear elastic model made possible by the use of this advanced *in situ* technique.

The WRSBP is an advanced *in situ* test and is the only means of providing a direct measure of the *in situ* strength and stiffness of dense cohesionless deposits at their correct density, stress level and stress history. The accuracy of these measurements is, however, highly dependent on the amount of drilling disturbance during installation of the WRSBP. Moreover, as an advanced *in situ* technique, sophisticated methods of assessment and interpretation are required to produce reliable design values.

ACKNOWLEDGEMENT

The authors would like to thank the CrossRail Project Team for the opportunity to present the data given in the paper and the assistance of colleagues in Arup Geotechnics. Acknowledgement is also extended to Foundation and Exploration Services, Soil Mechanics Ltd, Cambridge Insitu, City University, Reading University and Surrey Geotechnical Consultants.

NOTATION

γ_{av}........ shear strain amplitude of loading cycle

ε_a......... axial strain

ε_c......... cavity strain

ν.......... angle of dilation

σ_θ'....... circumferential effective stress

σ'_{h0}...... *in situ* horizontal effective stress

σ'_{r0}...... *in situ* radial effective stress

σ'_n....... normal effective stress

σ_{r0}'...... *in situ* radial effective stress

ϕ'_{cv}...... constant volume angle of friction

ϕ'_p....... peak angle of friction

ϕ'_{pt}....... peak drained angle of friction from triaxial tests

ϕ_{pp}........ peak plane strain angle of friction

$(N_1)_{60}$... SPT blowcount N corrected for rod length, borehole diameter and effective stress

D_r........ relative density

E'_v....... drained Young's modulus in vertical direction

e_i......... initial void ratio

G_{ph}....... pressuremeter shear modulus

K_0 .... coefficient of earth pressure at rest

mOD ... metres above Ordnance Datum

p'......... $=(\sigma'_1+2\sigma'_3)/3$),

p'_0....... *in situ* mean effective stress

q_{net}....... net loading on shallow foundation

q_{ult}....... ultimate capacity on shallow foundation

S_{av}' ...... mean plane strain effective stress

s' .... $= (\sigma'_1 + \sigma'_3)/2$

t' .... $= (\sigma'_1 - \sigma'_3)/2$

REFERENCES

Bellotti, R., Ghionna, V., Jamiolkowski, M., Robertson, R. & Peterson, R. (1989). Interpretation of moduli from self-boring pressuremeter tests in sand. Géotechnique, Vol. 39, No. 2, pp. 269-292.

Bolton, M.D. (1986). The strength and dilatancy of sands. Géotechnique, Vol. 36, No. 1, p.65.

Fahey, M. and Randolph, M.F. (1984). Effect of disturbance on parameters derived from self-boring pressuremeter tests in sand. Géotechnique, Vol. 34, No. 1, pp. 81-97.

Hughes, J.M.O, Wroth, C.P. and Windle, D. (1977). Pressuremeter tests in sands. Géotechnique Vol. 27, No. 4, pp. 453-477.

Jardine, R.J. (1992). Non-linear stiffness parameters from undrained pressuremeter tests. Canadian Geotechnical Journal, June.

Jardine, R.J., Symes, M.J. and Burland, J.B. (1984). The measurement of soil stiffness in the triaxial apparatus. Géotechnique. Vol. 34, No. 3, pp. 323-340.

Jewell, R.A. (1989). Direct shear tests on sand. Géotechnique. Vol. 39, No. 2, pp. 309-322.

Kolbuszewski, J. (1948). An experienced study of the maximum and minimum porosity's of sands. Proc. 2nd Int. Conf. Soil Mech. Foundn. Eng. Rotterdam. Vol 1, pp 158-165

Lade, P.V. and Lee, K.L. (1976). Engineering properties of soils. Report UCLA ENG-7652. University of California at Los Angeles.

Mair, R.J. and Wood, D.M. (1987). Pressuremeter Testing Methods and Interpretation. CIRIA. Project 335. Pub. Butterworth, London. ISBN 0-408-02434-8.

Manassero, M. (1989). Stress-strain relationships from drained self-boring pressuremeter tests in sand. Géotechnique. Vol. 39, No. 2, p.293.

Newman, R.L., Chapman, T.J.P., and Simpson, B. (1991). Evaluation of pile behaviour from pressuremeter tests. Xth. Eur. Conf. SMFE, Florence.

Rowe, P.W. (1962). The stress dilatancy relation for static equilibrium for an assembly of particles in contact. Proc. Royal Society. Vol. 269, Series A, pp. 500-527.

Skempton, A.W. (1986). Standard Penetration Test procedures and the effects in sands of overburden pressure, relatively density, particle size, ageing and overconsolidation. Géotechnique, Vol. 36, No. 3, pp. 425-447.

Stroud, M.A. (1988). The Standard Penetration Test - its application and interpretation. Proc. Conf. on Penetration Testing in the UK. Birmingham. Thomas Telford. London.

Troughton, V.M., Platis, A. (1989) The effects of changes in effective stress on a base grouted pile in sand. Proc. Int. Conf. on Piling and Deep Foundations, London, Balkema, Vol. 1, pp.445-453

Whittle, R. (1991). Yield point analysis for self-boring pressuremeter tests in sands. Reported by R.L. Newman in Soil Mechanics Ltd. Pressuremeter for Design in Geotechnics. November.

Wroth, C.P., Randolph, M.F. and Houlsby, G.T. (1979). A review of the engineering properties of soil with particular reference to the shear modulus. Cambridge University Engineering Department. Internal Report CUED/D-Soils TR 75.

Youd, T.D. (1973). Factors controlling maximum and minimum densities of sands. Proc. Sym. Eval. Density. ASTM STP 523, pp. 98-112.

Moderator's report on Session 6: laboratory testing

R. J. JARDINE
Department of Civil Engineering, Imperial College of Science, Technology and Medicine, London, UK

SYNOPSIS
The theme of Session VI is recent advances in laboratory methods and their practical application. This Moderator's report consists of an introductory discussion on the rôle of laboratory testing, a brief review of recent developments (drawn from conferences, journal papers and personal experience) and some specific comments on the issues raised by the eight Session VI papers. Site investigation is a multi-faceted activity and reference is also made to several papers from earlier Sessions which report interesting laboratory work. Topics for Discussion Session VI are suggested at the end of the report.

ROLE OF LABORATORY AND THE THEMES OF PAPERS PRESENTED
Laboratory testing is a key feature of almost all geotechnical site investigations. As detailed below and summarised in Table 1, it is performed in practice for at least six main purposes:

Purpose 1 - Strata Identification
Strata identification and profiling are the first vital steps in site investigation. **Clark and Keeton,** and **Paul et al** address this question in their Session VI papers. The issue is also raised in Session III by **Norbury and Gosling,** and by **Hight** in his Moderator's report. X-ray radiographs have long been used (at the NGI and elsewhere) to help select test specimens from uncut cores, so avoiding unrepresentative discontinuities or inclusions. **Paul et al's** paper goes on to show how X-ray photographs and densimeter measurements can identify detailed fabric changes, with density variations of 0.01 Mg m^{-3} being resolved at a millimetre scale. These images and traces provide specific signatures that can be indicative of sediment type, geological origin and post depositional history.

Purposes 2 to 4 - Conventional Testing
Sessions I and II discuss how improvements have been made recently in the management, interpretation and manipulation of data from conventional laboratory tests. Although few papers concentrate on the testing techniques involved, **Clark and Keeton** describe in Session VI how routine procedures may have to be modified when dealing with contaminated ground. The important issues raised include: health and safety in the laboratory during sample preparation and testing (even water content determinations can be dangerous if volatile fluids are present!); the potential for chemically active soils to be altered during routine work; and the effects that various industrial waste products can have on basic soil properties. Noting that some tests cannot be carried out on soils in their contaminated states, they question the validity of data obtained from modified, or cleaned-up, samples.

Purpose 5 - Physical modelling
Recent specialist meetings such as Centrifuge '94 (Leung et al 1994) show that physical

Advances in site investigation practice. Thomas Telford, London, 1996

modelling remains an actively developing area. However, the Session VI paper by **Medeiros and Moffat** provides the sole example presented at this Conference. Their paper reports a study on the hydraulic fracture of fill material used in earth dams which was carried out in a modified Rowe consolidation cell.

	Purpose	Notes and examples
1	Strata identification, profiling, assessment of environmental hazard	Sample description and fabric studies; composition and index properties; chemical composition
2	Parameters for empirical design procedures	CBR for road pavements; MCV tests to assess compactability; Su's for pile capacity; Eu_{50} for P-Y curves, etc
3	Assessment of potential construction materials	Compaction; grading; water content; durability; permeability; Su value etc
4	Parameters for idealised models	Tests to evaluate 'elastic' moduli, or estimate λ, κ, Γ (or N), G, M for Modified Cam Clay analyses
5	Physical models to assist design and in-situ test evaluation	Centrifuge; soil tank tests; shaking table; calibration chamber etc
6	Element tests to characterize true mechanical behaviour of geomaterials	Stiffness; yield; flow; strength Brittleness; post-rupture behaviour; loss of 'structure' and sensitivity Cyclic and dynamic responses Diffusion; permeabilities to of liquids and gases Consolidation; creep; effects of strain rate

Table 1. The rôle of laboratory testing in site investigations

Purpose 6 - Geotechnical element testing
Five of the eight papers in Session VI are devoted to laboratory element tests on (principally) soils. The earlier contributions by **Pickles and Everton, Tatsuoka et al, Maddison et al and Davis et al** (presented in Sessions I, III, IV and V respectively) also refer to the practical application of 'advanced' laboratory testing. These proceedings, and other recent conferences (such as the 1991 Florence European Conference of the ISSMFE, the 1992 Wroth Memorial Symposium, the 1994 New Delhi International Conference of the ISSMFE or the 1994 IS-Hokkaido Symposium held at Sapporo, Japan) testify to the renewal of interest in laboratory soil element testing.

ELEMENT TESTING
The purpose of element testing is to measure, as accurately as possible, the response of a soil or rock element to changes in stresses, strains, pore-pressures or hydraulic gradients. The essential aims are to study a sufficient number of representative samples intensively in tests that simulate the expected field conditions as closely as possible. To do this requires control over:

- initial soil state
- initial stresses
- imposed stress/strain/pore pressure changes
- sequence and rate of changes
- field drainage conditions

The data obtained are then generalised to cover the whole area of interest, using sample descriptions, index and in-situ tests to identify and account for variations in ground conditions. Laboratory and in-situ testing techniques have complementary strengths and weaknesses, as summarised in Table 2. Good site investigations try to integrate the two approaches!

Control parameter	Laboratory element test	In-situ test method
Initial soil fabric and state; sample disturbance	Good only if advanced sampler is used. Particularly difficult with sands and weakly bonded soils	Good with low disturbance tools. May be assessed indirectly from high disturbance penetration tests
Initial stress state	Good with stress path equipment	Generally poor
Applied stress path	Good, depending on apparatus capability	Generally poor
Applied strain rates and creep	Generally good	Generally poor
Drainage conditions	Generally good	Generally poor
Uniformity of stresses and strains	Generally good, can be measured reliably	Generally poor, hard to measure
Frequency of testing	Low, time consuming and costly	Can be very high, as in CPT testing
Representativeness	Often poor, key layers can easily be missed	Can be very good, as in CPT testing

Table 2. Strengths and weaknesses of field and laboratory methods for measuring soil properties

BACKGROUND RESEARCH AND TECHNICAL DEVELOPMENTS
The practical application of geotechnical element testing has developed principally through a gradual and selective assimilation by industry of the ideas and techniques generated in research. Reviewing the papers at this Conference, research appears to have been influential in three main areas: laboratory equipment and testing techniques; data interpretation; improving sampling and understanding sample disturbance. The following paragraphs considers each of these topics in turn, focusing mainly on developments in soil mechanics.

Developments with equipment and techniques
Static stress-strain and strength measurements
Laboratory equipment for measuring static stress-strain and strength behaviour developed rapidly in the 1970's. Stress path triaxial and true triaxial testing became feasible (see Hambly 1969, Green 1971, Bishop and Wesley 1974, Tatsuoka 1988) and electrical transducers for measuring force, displacement and pressure became common. Further progress has been made since with more advanced shear testing apparatus, including the Directional Shear cell, Hollow Cylinder and True Triaxial apparatus (see Hight et al 1983, and Arthur 1988).

Figure 1 illustrates the stress conditions that can be developed in a well designed Hollow Cylinder Apparatus, showing how the four control parameters (inner and outer hydraulic pressures, vertical load and torque) allow the values of σ_1, σ_2, σ_3 and α (the angle of σ_1 to the vertical) to be varied as required, permitting a far wider range of stress paths to be followed than is possible in either conventional, or stress path, triaxial cells.

The development of local on-sample instrumentation has greatly increased the value of soil element tests. Hight (1982) described how sub-miniature pressure transducers can be mounted at mid-height on a sample's surface to measure local pore pressure. The consolidation of clay samples may then be monitored more precisely and rates of stress change applied during drained stages can be controlled so that specified hydraulic gradients are maintained (such as excess pore pressure $< 0.05p'$ along a given stress path). In undrained tests, the central pore water pressures can be measured without having to wait for the entire sample and external measurement system to reach hydraulic equilibrium. Reliable data are thus available from an earlier stage and tests can be accelerated if required. The errors due to stress non-uniformity can also be minimised.

Ridley and Burland (1993) describe probes which can measure strongly negative pore water pressures (up to 1,500 kPa below atmospheric), making tests on unsaturated samples possible without the need for an elevated air pressure. **Ridley and Burland** present a paper in Session IV on the in-situ use of their 'suction probe'. Such devices can also be used to check suctions immediately after sampling, giving an indication of sample quality. Filter paper techniques have also been used more widely to measure soil suctions; Chandler et al (1992).

Devices incorporating LVDT sensors, electrolevel inclinometers, electromagnetic proximity transducers, Hall effect gap sensors and strain gauged flexible elements have been designed to measure the strains developed locally on the surface of soil samples; see Burland and Symes (1982), Jardine et al (1984), Clayton and Khatrush (1986), Ackerley et al (1987), Goto et al (1991) and Lo-Presti (1994). Research with such equipment has shown that conventional 'external' measurements of soil strains are often subject to very large errors as a result of apparatus compliance, bedding at the sample ends, filter paper compression, tilting of specimens (due to eccentric load contacts) and non-uniform conditions at the sample ends; Jardine et al (1984) and Tatsuoka and Shibuya (1992). Figure 2 shows the local strain and pore pressure instrumentation scheme adopted by Smith et al (1992) for tests on soft Bothkennar clay.

Steps can be taken to minimise external strain measurement errors, but local transducers are recommended for most cases where accurate stiffness data are required, particularly when working with the most common UK designs of triaxial apparatus (Lo Presti et al 1994).

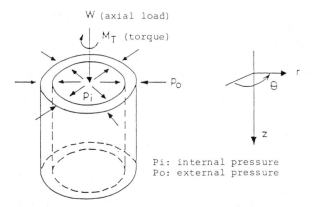

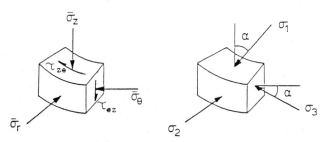

(a) Hollow cylinder sample

(b) Stresses on an element   (c) Principal stresses on an element

Figure 1.   Stress system within a Hollow Cylinder Apparatus designed for controlled principal stress rotation tests.

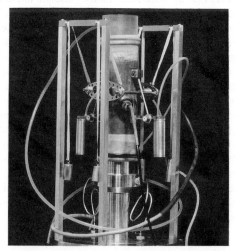

Figure 2.   Example of local instrumentation of 38mm diameter triaxial sample: pendulum type inclinometers for axial strain; LVDT arrangement for radial strain; and sub-miniature transducer for mid-height pore pressure. After Smith et al (1992).

Local strain sensors resolve strains to between 0.0001% and 0.005%, depending on the transducer design, the laboratory environment, the signal conditioning system and interpretation method employed. Lo-Presti et al (1994) describe comparative trials on four different types of sensor. A set of LVDT sensors gave the best resolution, accuracy and drift characteristics (0.3μm, 1μm and less than 0.5μm/day), but these operated over a working range of only 2mm ($\approx 2\%$ strain for their test arrangements). Electrolevel Devices designed at Imperial College to work over a 30% strain range (with 130mm gauge length) on Hollow Cylinder samples show good long term stability (better than 0.002%/day), but with scatters of 0.002 to 0.004%, depending on the amount of electrical noise. Signal smoothing is required to quantify stiffness at strains smaller than 0.01%.

The Session VI papers by **Bressani et al, Chegini and Trenter, Johnson et al and Shibuya et al** all describe practical applications of stress path testing and 'small-strain' stiffness measurements. **Tastsuoka et al, Maddison et al, Davis et al** and others also refer to the same techniques in earlier Sessions. A later section of this report discusses their findings in more detail.

Cyclic and dynamic tests

Cyclic experiments can be considered as a of sub-set stress path testing. Low frequency tests can be performed relatively easily in automatically controlled stress path triaxial or torsional shear equipment, taking full advantage of the static testing developments described above. However, if the cycling period is to less than perhaps 100 to 10s, then more powerful actuators, faster control systems and quicker data acquisition arrangements may be required. No attempt is made here to summarise recent developments in large strain cyclic and dynamic testing; readers are referred instead to Tatsuoka (1988), Brown and O'Reilly (1991) or Toki et al (1994).

Cyclic and dynamic tests have long been used to investigate small-strain stiffness characteristics; Richart et al (1970). Such tests are not common in the UK, although research with Resonant Column (RC) and Bender Element techniques has been carried out at University College, North Wales since the 1970's and commercial testing has been carried out by Fugro for more than 10 years. The recent interest in static behaviour at small strains has led more research groups to try out the same methods. For example, Dyvic and Madshus (1985) report on direct comparisons between the elastic shear stiffnesses (G_{max}) measured by RC and BE methods, while Vigianni (1992) introduced bender elements into stress path triaxial equipment and re-opened the debate on the interpretation of wave velocities from the BE signals. Porovic and Jardine (1994) describe RC tests performed within a computer controlled Hollow Cylinder Apparatus (HCA) which improves the stress and strain uniformity within test specimens. They also refer to parallel BE oedometer tests and data from static HCA and Triaxial (TC) experiments. Generally good agreement was seen between RC and static HCA tests and Figure 3 illustrates the trends seen in BE, RC and TC tests on reconstituted Ham River sand. Note that the strain range over which G_{max} applies rarely extends far beyond $\gamma \approx 0.001\%$, and that the various test methods tend to give more divergent results at larger strains.

Care must be taken when interpreting such data. Experiments where P waves and differently polarised S waves are passed through prismatic samples, held under various stress conditions, show that soils and soft rocks manifest both induced and inherent elastic stiffness anisotropy; King et al (1994), Jamiolkowski et al (1994). Field test data also indicate anisotropy; see for

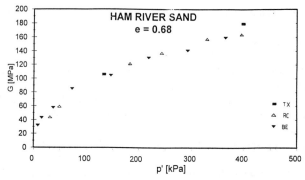

Figure 3. G_{max} determinations for normally consolidated Ham River sand by Triaxial (TX), Resonant Column (RC) and Bender Element (BE) tests; after Porovic and Jardine (1994)

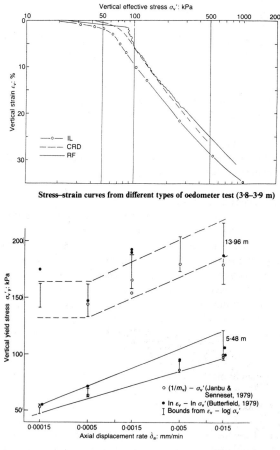

Figure 4. Effects of strain rate and test type on compressibility and vertical effective yield stress of Bothkennar clay; after Hight et al (1992).

example **Davis et al's** paper in Session V.

Permeability measurements
The flow pump technique advocated by Olsen et al (1985) is probably the most significant recent development in laboratory permeability testing. This technique involves introducing fluid into, and withdrawing it from, samples whilst monitoring their hydraulic gradients with sensitive differential pressure transducers. Hydraulic gradients can be kept low with this system and test times can be reduced greatly. The arrangement can be applied in axial or radial flow experiments and the flow pumps may be either automatic medical syringe systems, or ram-driven geotechnical pressure sources. The Session VI paper by **Araruna et al** describes how such a system provided quick and reliable measurements for two UK clays. Flow pumps were also used in the hydraulic fracture study described by **Medeiros and Moffat.**

Systems for test control and data manipulation
Following the introduction of data loggers, automatic systems for controlling stress path experiments became feasible in the mid 1970's. The introduction of stepping motors and powerful personal computers has reduced the cost of such systems and increased their flexibility greatly, as reviewed by Toll and Ackerley (1988), Tatsuoka (1988), Baldi et al (1988). Although little is reported in the literature, the constant development and up-grading of the associated software now constitutes a major part of research into soil element behaviour.

Increasing sophistication in test interpretation
The Critical State framework set out by Schofield and Wroth (1968) has become increasingly widely understood and used, as may be seen in the papers by **Bressani et al, Chegini and Trenter and Johnson et al**. Many practitioners now distinguish carefully between peak and ultimate states and try to use 'Critical State' ideas to link together the shear and volumetric behaviours of the geomaterials they encounter.

Recognition of soil brittleness and non-linear behaviour
Attention has also focused on features of behaviour that were not captured in the original Critical State models. For example, the significance of residual strength has long been recognised (Skempton 1964) and accounted for in slope stability studies and other applications. Although not mentioned in any of the Session VI papers, research into rate effects and interface shear behaviour (Lemos 1985, Tika-Vassilikos 1992) and the effects of pore water chemistry has produced some new results which practitioners should find useful. For example, Bond and Jardine (1991) showed that the rate dependent interface shear characteristics play a key rôle in determining the shaft capacity of piles in clays, whilst tests on sands show trends contrary to those incorporated into standard pile design procedures; Jardine et al (1992).

Following Vaughan (1988) the more general characterisation of brittle, sensitive and bonded soils has also improved. **Bressani et al** report further stress path testing studies into the yiedling and failure of model and natural brittle residual soils in their Session VI paper.

More evident at this Conference is the impact of the research into pre-failure deformation characteristics reported by Jardine et al (1984), (1991), Burland (1989), Atkinson and Sallfors (1991), Jamiolkoswki et al (1991), Tatsuoka and Shibuya (1992) and the multitude of papers given at the IS-Hokkaido conference (edited by Shibuya et al 1994) .

Moving away from stress-strain behaviour to consider the flow of fluids through soils, there is now a wide spread recognition that permeability can be both anisotropic and vary strongly with void ratio or stress state (see Rowe 1972, Olsen et al 1985). Vaughan (1994) has demonstrated that variable permeability can have a strong influence in many geotechnical problems and the Session VI paper by **Araruna et al** adds to the discussion on laboratory permeability measurements in clays and their interpretation.

Rate effects, creep, stress-strain and strength anisotropy
As mentioned earlier, the transfer of ideas from research to practice is selective. Some important research findings appear to have had less influence on practice than might have been expected. One example is an appreciation of the effects of strain rate and creep. It is now clear that compressibility, strength and stiffness are all strongly rate dependent (see for example Leroueil et al 1985, Hight et al 1987). Figure 4 illustrates the rate sensitivity of soft Bothkennar clay, showing (i) the effect of oedometer axial displacement rate on the vertical yield stress, σ'_{vy} and (ii) the tendency of CRS, Restricted Flow (RF) and other 'advanced' procedures to give e - log σ'_v curves which plot well above those given by the standard Incremental Loading (IL) procedure. CRS and RF tests may give non-conservative foundation designs, unless a specific allowance is made for the difference between the field and laboratory strain rates.

The anisotropic nature of shear strength and deformation behaviour is another facet of soil behaviour that has been recognised long ago (see for example Ladd et al 1977), but has yet to influence practice significantly. Figures 5 and 6 illustrate the effects further, showing (i) the inherent anisotropy of ϕ' for a typical quartz sand, and (ii) the total anisotropy in undrained shear strength of three normally consolidated soils when tested from K_0 stress conditions in a large Hollow Cylinder Apparatus.

Improving samplers and understanding sampling effects
It is now clear that many soil properties are altered by conventional tube sampling processes. For example, **Shibuya et al** show that freeze sampling methods are essential with natural sands if their fabric and properties are to be preserved. With clays significantly higher strengths have been reported when block quality (carved) samples and high quality rotary cores are tested, as emphasised in the Session III Moderator's report by **Hight**. The potential effects of sampling are discussed further **Tatsuoka et al** (Session III) and in the Session VI papers by **Shibuya et al, Johnson et al, Chegini and Trenter and Bressani et al**.

TRENDS IN PRACTICAL APPLICATIONS OF ADVANCED TESTING
Reviewing the recent literature and the papers presented at this Conference shows that three main developments have taken place in the application of advanced soil element testing:

1.   Practitioners are commissioning more sophisticated soil element tests. In particular, this Conference has shown how advanced triaxial tests, often involving anisotropic consolidation and local measurements of pore water pressure and strains, have been carried out in connection with a wide range of civil engineering projects.

2.   There is a perceived need in design parameter selection to move away from relying on data from 'model' soils, index tests and published correlations. The ability and

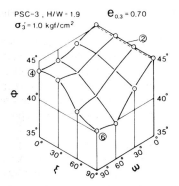

Figure 5.   Drained anisotropy in ϕ' of dense air-pluviated Toyoura sand in plane strain compression. Note $\zeta =$ angle of normal to bedding plane to $(\sigma_1 - \sigma_2)$ plane; $\varpi =$ angle of bedding plane normal to σ_1 direction.

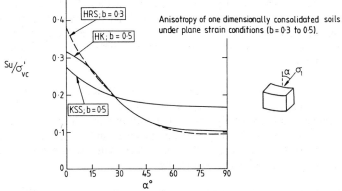

Figure 6.   Anisotropy in undrained strength of Ham River sand (HRS), an artificial clay KSS (Kaolin-Silt-Sand) and an artificial clayey sand (HK). After Menkiti 1994.

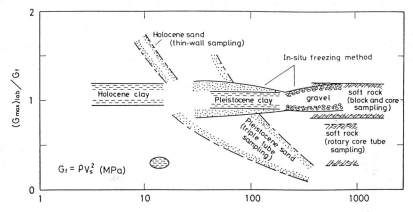

Figure 7.   Comparisons between laboratory and seismic field (G_f) measurements of shear modulus at very small strains, after Shibuya et al (Session VI, this Conference).

motivation to sample and test natural soils has improved; detailed laboratory investigations of the properties of 'real' soils are becoming more common on a regional and a site specific basis.

Examples include the recent interest in testing carbonate sands, soft sedimentary rocks, natural 'structured' soft clays, weakly bonded sands and tropical soils. Contributions made at this Conference include the overviews given by **Shibuya et al** and **Tatsuoka et al** of how a broad spectrum of Japanese geomaterials behave at very small strains; the studies of various Tertiary, Plio-Pleistocene and Holocene deposits from the London basin described by **Pickles and Everton, Davis et al and Johnson et al; Davis et al's** work on Mesozoic UK mudstones; the paper by **Chegini and Trenter** on Devensian glacial till from Dumfriesshire, Scotland; **Bressani et al's** investigation of Botucatu residual soil from Southern Brazil..

3.   It is also clear that practitioners are aware of potential sampling problems and are keen to maintain an integrated approach to site investigations. Comparisons between G_{max} data from field seismic, static laboratory and dynamic laboratory tests are presented at this Conference by **Tatsuoka et al** (Session III), **Davis et al** (Session V) and, in Session VI, by **Shibuya et al and Trenter et al**. Weak Rock Self-Boring Pressuremeter tests were used by **Johnson et al** (Session VI) to check their laboratory studies.

SPECIFIC ISSUES FOR DISCUSSION

Potential differences between laboratory and field behaviour
Having shown that sophisticated laboratory testing is now feasible for many projects, the essential question is whether the data obtained represent in-situ behaviour reliably. Here comparisons with full scale field behaviour provide the acid test; Jardine et al (1991), Hight and Higgins (1994). In Session VI, **Bressani et al** link their laboratory observations of residual soil behaviour to a recent flowslide in southern Brasil. Sessions IV and V of this Conference are devoted to in-situ testing. Some overlap exists in the cases cited below where comparisons were made between laboratory and in-situ test data. As mentioned earlier, over-simple comparisons between seismic, dynamic and static tests, and between laboratory and in-situ behaviour, can lead to incorrect conclusions. Most geomaterials appear to be anisotropic over the full strain range. Laboratory and field tests often employ different stress paths and shearing directions. The relative magnitude of σ_2 can also be influential.

Comparisons with seismic tests
Shibuya et al present close comparisons between laboratory and field seismic G_{max} values. As shown in Figure 7, generally good agreement was seen in clay soils, confirming the broad trends observed for UK sites (see Clayton et al 1994 and Jardine 1994). **Chegini and Trenter** also report that laboratory and in-situ seismic G_{max} data sets tended to overlap for their insensitive low plasticity till. However, the rotary techniques adopted in Japan appear to cause some damage in sedimentary soft rocks. In addition, in-situ freezing during sampling seems to be essential for sands. An interesting question is whether corrections made on the basis of in-situ void ratio and aging time compensate adequately for sample disturbance in sands. Conversely, is it valid to try to reconstruct the full, in-situ, stiffness-strain characteristics of a deposit on the basis of field seismic measurements of G_{max}?

Comparisons with other tests
The difficulties of interpreting stiffness, strength and yielding characteristics from in-situ tests flow from the basic limitations outlined in Table 2. The glacial till study reported by Chegini and Trenter proved again that 'quick' UU tests can show greater scatter than the SPT blow count, reflecting in part the effects of the percussive U100 sampling procedure. Appropriately consolidated samples provided far more consistent and meaningful information, but it was not possible to be sure that the interpreted strength and yielding parameters reflected conditions in-situ.

Johnson et al presented a non-linear elastic plastic interpretation of pressuremeter tests in Thanet sand (a London Tertiary deposit), concluding that their laboratory stress path tests on rotary core samples under-estimated the material's in-situ shear moduli at comparable strains and effective stresses. Similar conclusions emerged from an earlier investigation at another Thanet sand site, although the discrepancies between laboratory and pressuremeter data appeared to be smaller (Love 1991). Hight and Jardine (1993) report that rotary samples of a very stiff plastic clay from the Thanet bed formation at Sizewell gave more credible laboratory stiffness data.

Johnson et al also interpreted ϕ' values and dilation angles, ν, from their pressuremeter data. As with shear stiffness, many assumptions were required to make direct comparisons. The Authors concluded that compatible peak ϕ' values were obtained in triaxial and pressuremeter tests. Shear box experiments seemed to under-estimate peak ϕ' and give unrealistic dilation rates. No correlation was made between triaxial and pressuremeter ν values; broad agreement was found between $\phi_{critical\ state}$ values estimated from three types of laboratory test.

Laboratory model tests which involve markedly non-uniform stress conditions may suffer from the same difficulties in interpretation as in-situ tests. In this respect, there are several parallels between the assumptions needed to analyse the hydraulic fracture experiments described by **Medieros and Moffat** and those required in pressuremeter analysis.

Following from the above, themes for the Discussion Session might well include:

- How can the practical limitations of laboratory procedures be established?
- How can testing techniques and interpretive methods be improved?
- When is the use of advanced testing appropriate?
- How can we combine the respective strengths of different testing methods in practice?

Other issues for discussion
While laboratory element tests on saturated soils has been discussed in several papers, other potentially interesting issues have received less attention. There is scope for redressing any imbalance in the Discussion Session. Issues that might be raised include:

- The recording and quantification of soil fabric
- New techniques required for environmental engineering problems
- Apparatus and procedures for testing unsaturated soil
- The effects of rate and creep
- Cyclic loading
- The characterisation of soil anisotropy

- Soil-interface shear tests for soil-structure interaction problems
- Developments in data acquisition and test control software
- Advances in testing rocks
- Laboratory physical model tests

REFERENCES

Ackerley, S K, Hellings, J E and Jardine R J (1987)
Discussion on Clayton, C.R.I. and Khatrush, S.A. (1987) . A new device for measuring local axial ,strains on triaxial specimens. Geotechnique, 37, No.3, pp 413-417.

Arthur, J R F (1988)
Cubical devices: Versatility and Constraints. Advanced triaxial testing of soil and rock. ASTM STP 977. Eds Donaghe, Chaney and Silver, pp 743-765.

Atkinson, J H and Sallfors, G (1991)
Experimental determination of soil properties (stress-strain-time). General Report. Xth ECSMFE, Florence, Balkema, Vol 3, pp 915-958.

Baldi, G, Hight, D W and Thomas, G E (1988)
A re-evaluation of conventional Triaxial Test Methods. Advanced triaxial testing of soil and rock. ASTI STP 977. Eds Donaghe, Chaney and Silver, pp 219-263.

Bishop, A W and Wesley, L D (1974)
A hydraulic triaxial apparatus for controlled stress path testing. Geotechnique, Vol.25, No.4, pp 657-670.

Bond, A. J. and Jardine, R. J. (1991)
The effects of installing displacement piles in a high OCR clay. Geotechnique, Vol 41, No. 3, pp 341-363.

Brown, S F and O'Reilly, M (1991)
Editors, The Cyclic Loading of Soils, Blackie & Son, Glasgow.

Burland, J B and Symes, M (1982)
A simple axial displacement gauge for use in the triaxial apparatus. Geotechnique 32, 1, pp 62-65.

Chandler, R J, Crilly, M S and Montgomery-Smith, G (1992)
A low cost method of assessing clay desiccation for low-rise buildings. Proc. ICE, Vol 92, May, pp 82-89.

Clayton, C.R.I. and Khatrush, S.A. (1986)
A new device for measuring local axial strain on triaxial specimens Geotechnique 36, No 4, pp 593-597

Clayton, C R I, Gordon, M A and Matthews, M C (1994)
Measurement of stiffness of soils and weak rocks using small strain laboratory testing and field geophysics. Proc. IS-Hokkaido, Vol 1, Balkema, Rotterdam, pp235-240.

Dyvik, R and Madshus, C (1985)
Laboratory measurements of G_{max} using bender elements. ASCE National Convention, Detroit, October 1985.

Goto, S, Tatsouka, F, Shibuya, S, Kim, Y S and Sato, T. (1991)
A simple gauge for local measurements in the laboratory. Soils and Foundations, Vol. 31, No 1, pp169-180.

Green, G E (1971)
Strength and Deformation of sand measured in an independent stress control cell. Proc. Roscoe Memorial Symposium, Cambridge, 285-323.

Hight, D W (1982)
A simple piezometer probe for the routine measurement of pore pressures in triaxial tests on saturated soils. Geotechnique, Vol.32, No.4, pp 396-402.

Hight, D W, Gens A and Symes, M J (1983)
The development of a new hollow cylinder apparatus for investigating the effects of principal stress rotation in soils. Geotechnique, 33, No.4, pp 355-384.

Hight, D. W., Jardine, R. J. and Gens, A. (1987)
The behaviour of soft clays. Embankment on soft clays, Public Works research Center, Athens, Ch.2, pp 33-158.

Hight, D W, Bond, A J and Legge, J D (1992)
Characterisation of Bothkennar clay: an overview. Geotechnique, 42, No 2, pp 303-348.

Hight, D W and Jardine, R J (1993)
Small strain stiffness and strength characteristics of hard London Tertiary clays. Proc. Int. Symp. on Hard Soils - Soft Rocks, Athens, Greece, Vol 1, Balkema, Rotterdam, pp 533-522.

Hight, D W and Higgins, K G (1994)
An approach to the prediction of ground movements in engineering practice: background and application. Keynote Lecture, Proc. IS-Hokkiado, Vol 2, In Press.

Jamiolkowski, M, Leroeuil, S and Lo Presti, D C F (1991)
Design parameters from theory to practice. Geo-coast '91 Int. Conf., Yokohama, Port and Harbour Research Institute, Yokosuka, Vol. 2, pp 877-917.

Jamiolkowski, M, Lancellotta, R and Lo-Presti, D C F (1994) Remarks on the stiffness at small strains of six Italian clays. Keynote Lecture, Proc. IS-Hokkaido, Vol 2, In Press.

Jardine, R J, Symes, M J and Burland, J B (1984)
The measurement of soil stiffness in the triaxial apparatus. Geotechnique 34, No.3, pp 323-340.

Jardine R. J., St John, H. D., Hight, D. W. and Potts, D. M. (1991) Some practical applications of a non-linear ground model. Proc. Xth ECSMFE, Florence, Vol. 1, pp 223-228.

Jardine, R J, Everton, S J and Lehane, B M. (1992)
Friction coefficients for piles in cohesionless materials. Offshore site investigations and foundation behaviour. Proc. SUT Int Conf, Kluwer, Dordrecht, pp 661-680.

Jardine, R J (1994)
One perspective on the pre-failure deformation characteristics of some geomaterials. Keynote lecture. Proc. International Symposium on pre-failure deformation characteristics of geomaterials. Hokkaido, Japan, Volume II, In Press.

King, M S, Andrew, M and Shams Khanshir, M (1994) Velocity anisotropy in Carboniferous mudstones. Int. Journ. Rock Mech. Min. Sci. & Geomech. Abstr., Vol 31, No 3, p 261-263.

Ladd, C C, Foot, R, Ishihara, K, Schlosser, F and Poulos, H G (1977) Stress-deformation and strength characteristics. State of the Art report, Proc. 9th ICSMFE, Tokyo, 2, pp 421-494.

Lemos, L. J. (1985)
The effects of rate on the residual strength of soil. PhD Thesis, University of London, Imperial College.

Leroueil, S, Kabbaj, M, Tavenas, F and Bouchard (1985)
Stress-strain-strain rate relation for the compressibility of sensitive natural clays. Geotechnique, 35, No.2, pp 159-180.

Leung, C F, Lee, F H and Tan, T S (1994)
Editors, Proc. Int Conf Centrifuge 94, Singapore, Balkema, Rotterdam.

Lo Presti, D C F, Pallara, O, Costanzo, D and Impavido, M (1994) Small strain measurements during triaxial tests: many problems, some solutions. Proc. IS-Hokkaido, Balkema, Rotterdam, Vol 1, pp 11-16.

Love, J P (1991)
Personal Communication

Menkiti, C O (1994)
The general behaviour of a clay and a clayey-sand with particular reference to the effects of principal stress rotation. PhD Thesis, University of London (Imperial College).

Olsen, H W, Nichols, R W and Rice, T L (1985)
Low gradient permeability measurements in a triaxial system. Geotechnique, Vol 35, 2, pp 145-157.

Porovic, E and Jardine R J (1994)
Some observations on the static and dynamic shear stiffness of Ham River sand. Proc. IS-Hokkaido, Vol I, pp. 25-30, Balkemna, Rotterdam.

Richart, F E, Woods, J D and Hall, J R (1970)
Vibrations of soils and foundations. New Jersey; Prentice-Hall.

Ridley, A M and Burland, J B (1993)
A new instrument for the measurement of soil moisture suction. Geotechnique, Vol 43, 2, 321-324.

Rowe, P W (1972)
The relevance of soil fabric to site investigation practice. 12th Rankine Lecture, Geotechnique Vol 22, 2 pp 195-300.

Schofield, A N and Wroth, C P (1968)
Critical state soil mechanics, McGraw-Hill, London.

Skempton A.W. (1964)
Long-term stability of clay slopes, 4th Rankine Lecture, Geotechnique, 14, No. 2, pp77-102.

Smith P.R., Jardine R.J. and Hight D.W. (1992)
On the yielding of Bothkennar clay. Geotechnique, 42, No 2, pp 257-274.

Tatsuoka, F (1988)
Some recent advances in triaxial apparatus for cohesionless soils. Advanced triaxial testing of soil and rock. ASTI STP 977. Eds Donaghe, Chaney and Silver, pp 7-81.

Tatsuoka, F and Shibuya, S (1992)
Deformation characteristics of soils and rocks from field and laboratory tests. Keynote lecture, 9th ACSMFE, Bangkok. Vol 2, pp 101-170

Tika-Vassilikos, T. (1991)
Clay-on-steel ring shear tests and their implications for displacement piles. Geotech. Testing J. Am. Soc. Testing and Materials, Vol. 14, No 4, pp 457-463.

Toki, S, Shibuya, S and Yamashita, S (1994)
Standardisation of laboratory test methods to determine the cyclic deformation properties of Geomaterials in Japan. Proc. IS-Hokkaido, Vol. 2, In Press.

Toll, D G and Ackerley, S K (1988)
The development of two data acquisition systems for a geotechnical laboratory. Proc. Int. Symp. on New Concepts in Laboratory and Field Tests in Geotechnical Engineering, Rio de Janeiro.

Vaughan, P R (1988)
Characterising the mechanical properties of in-situ residual soil. Proc. 2nd Int. Conf on Geomechanics in Tropical Soils, Singapore, Vol 2, pp 469-487.

Vaughan, P R (1994)
Assumption, prediction and reality in geotechnical engineering, 34th Rankine Lecture, Geotechnique, 44, No 4, pp 573-609.

Viggiani, G (1992)
Small strain stiffness of fine grained soils. PhD Thesis, City University, London.

Discussion on Session 6: laboratory testing

Discussion reviewer J. P. LOVE
GCG

Professor King of Imperial College spoke about the development of a device for the true triaxial testing of rock specimens, which could also be used for cemented sands and clays. Each of the three principal stresses may be varied independently under servo-control in the range 0 - 115MPa, with pore pressures to 5MPa possible (Figure 1). Stress is tranmsmitted to each of the six faces of the cubic specimen via 5mm thick faceplates matching approximately the elastic properties of the materials being tested. Magnesium has been found a suitable material for use with soft sandstones. Deformation of the specimen is measured by three pairs of LVDTs. Feedback from the LVDTs may be used for servo-control of the specimen deformation. Each of the six transducer holders contains stacks of piezoelectric transducers capable of producing or detecting pulses of compressional (P) or two shear (S) waves polarised at right angles propagating in the three principal directions. The bandwidths of the P and S wave pulses are in the range 350-850kHz. The system permits the measurement of the deformation and elastic-wave velocities and attenuation in each direction. Nine components of velocity are measured: three compressional and six shear. The redundancy in S wave velocity measurements provides the opportunity to confirm that the state of stress within the specimen is indeed homogenous. Permeability and electrical conductivity may be measured in the vertical direction. Tests have been performed on Penrith Sandstone specimens, during which aligned cracks were introduced by increasing σ'_1 and σ'_2 in unison while maintaining σ'_3 low. The three deformations, all nine components of velocity and permeability in the vertical direction were measured during this cracking cycle. Finally, the three principal stresses were increased together in order to close the cracks formed. The same set of measurements were repeated during the crack closing cycle. There was an excellent correlation between permeability in the direction of the set of oriented cracks and the S wave velocity in a direction perpendicular to the set. The system will next be used to model the behaviour of soft sandstones adjacent to a borehole during drilling.

Mr Jovicic of City University spoke about the use of Bender Elements to measure the stiffness of soils. Bender elements consist of two bonded piezoelectric plates (10mm x 13mm x 0.5mm) which polarise in opposite directions when a voltage is applied. If one end of the element is fixed, polarisation will make the opposite end move and the element will bend. Piezoelectricity works in both ways so a bender element will generate a voltage if forced to deform. Bender elements may be built into a triaxial cell. The free ends of the transmitter and receiver elements protrude into the triaxial sample. Input to the transmitter element is obtained from a function generator. The waves passing through the specimen are detected by the receiver element and are observed on a digital oscilloscope together with

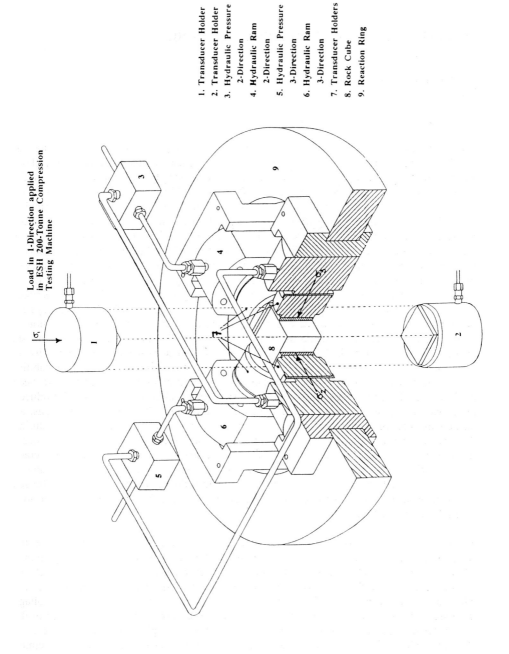

1. Transducer Holder
2. Transducer Holder
3. Hydraulic Pressure 2-Direction
4. Hydraulic Ram 2-Direction
5. Hydraulic Pressure 3-Direction
6. Hydraulic Ram 3-Direction
7. Transducer Holders
8. Rock Cube
9. Reaction Ring

Load in 1-Direction applied in ESH 200-Tonne Compression Testing Machine

Figure 1   True triaxial testing apparatus

the input signal (Figure 2). The measurement of time delay between characteristic points can be taken directly off the screen (Figure 3). Data from the oscilloscope are digitally stored and used for further processing. The waves originating from a bender element are a combination of shear waves and near field waves. These near field waves decay rapidly with distance from the source and in the far field are negligible. Recent work at City University has indicated that near field effects are significant when the travel distance is less than 2.5 times the wave length. In bender element tests it is preferable to use a sine wave rather than a square wave so that the wave length is greater than the critical value. Providing that the measurement of arrival time is made in the far field the receiver will

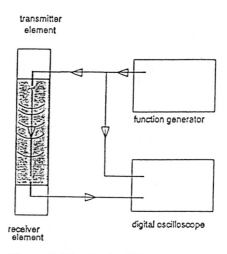

Figure 2 The Bender Element set up

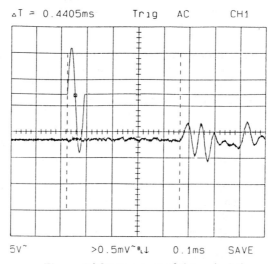

Figure 3 Measurement of the arrival time

capture the transverse motion which corresponds to the shear wave. Knowing the transmitter-receiver tip to tip distance (L) and measured propagation time (T_a) the velocity of the shear wave is easily determined from $V_s = L/T_a$. Assuming that soil is a porous elastic material the shear modulus may be calculated from $G = \rho V_s^2$, where ρ is the total mass density of the saturated soil. Current research at City University is investigating variations of G_0 with stress, overconsolidation ratio and water content for soils and soft rocks. The bender element method has been used routinely in testing fine grained materials, sands and soft rocks. Tests have been carried out with cell pressures in the range 50 - 5000kPa resolving stiffnesses in the range 50 - 3000MPa. The bender element method has the advantage of being less expensive and relatively easier to use and interpret than resonant column tests.

Dr Shibuya of Hokkaido University discussed the effects of sample disturbance on laboratory derived values of G_{max} (contained in **Shibuya et al**). Using data obtained from the Higashi-Ohgishima man-made island in Tokyo Bay, he compared values of G_{max} derived in the laboratory from reconsolidated samples with the values of G_f obtained from a seismic cone survey in the field. As reported in the paper, the ratio of G_{max}/G_f for the clay samples was found to lie in the approximate range 0.6 to 0.8. Assuming the field data (ie G_f) to be the more reliable, the poor match between the two sets of data was attributed to (i) possible differences in strain level, γ (ii) possibly the wrong value of K_0 used for lab reconsolidation (a value of 0.5 was assumed) (iii) the effects of sample disturbance. These three areas were therefore systematically examined using Bender Element tests performed on two intact clay samples. The following was found. (i) The laboratory test data for Cyclic triaxial tests ($\gamma \approx 10^{-5}$), Cyclic torsional shear tests ($\gamma \approx 10^{-5}$) and the Bender Element tests ($\gamma \approx 10^{-5}$) all gave indistinguishable results, despite the order of magnitude range of shear strain, hence the magnitude of γ was concluded not to be an issue. (ii) The measured value of K_0 was confirmed to be 0.5. (iii) When $G/G_{t=tp}$ was plotted against $\log(t/t_p)$ it was found that for $\sigma'_v = \sigma'_{v\ insitu}$, the gradient of the plot was not zero as it should have been for no sample disturbance. In fact there was no discernible difference in the gradient when the tests were repeated at $2\sigma'_{v\ insitu}$. From this it was deduced that the clay samples were grossly disturbed. In Figure 9 of the paper, the values of e_c/e_0 were all close to unity, however. This suggests that the ratio of e_c/e_0 is not a sensitive parameter for diagnosing the potential degree of sample disturbance, but that plotting $G/G_{t=tp}$ vs $\log(t/t_p)$ is.

Mr Johnson of Ove Arup & Partners highlighted some results from the paper by **Johnson et al**. He compared measurements of stiffness for the Thanet Sands obtained from triaxial tests in the laboratory with those from SBP (Weak Rock Self-Boring Pressuremeter) tests in the field. Details may be found in the paper. In summary, the laboratory data showed that typically $E'_v/p'_0 = 500$ at a strain of 0.05%. Although three of the five laboratory samples were undisturbed and the other two reconsolidated, there was no appreciable difference between the two. One of the "undisturbed" samples does indicate higher stiffness, but in general the stiffness observed was lower than anticipated. The SBP data was then discussed in comparison. Interpretation of the SBP required many assumptions. Manassero's method was used to establish the stress paths followed by elements of soil at the cavity wall, as shown on Figure 4. The analysis relied on correcting the cavity strain using an estimate of the insitu horizontal stress, which is not straightforward. The stress paths indicated peak plane strain angles of friction of 40°-60°, which were comparable with the results obtained in triaxial conditions (35°-42°) allowing for the different stress fields, which gave confidence in the analysis. However, the SBP gave comparative values of E'_v/p'_0 some three to four times higher than for the triaxial tests (see Figure 5, noting the

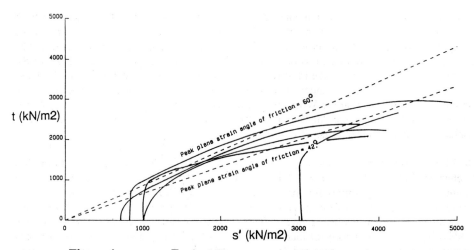

Figure 4 **Typical Pressuremeter Stress Paths**

Results based on analysis by Manassero (1989)

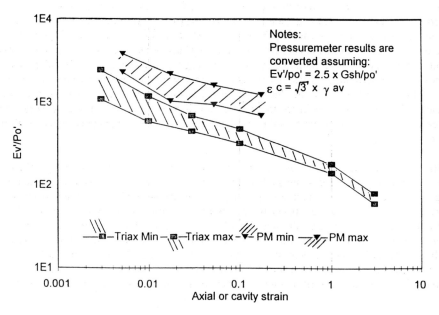

Figure 5 **Triaxial and Pressuremeter Stiffness**

A comparison assuming a strain transformation

strain transformation), which is in contrast to the data presented earlier by the speaker from Exploration Associates for tests on clays. The following explanations were offered for the mis-match: (i) sample disturbance (ii) the in situ horizontal stiffness (as measured by the SBP) greater than vertical stiffness (as measured by the laboratory tests). The results of the pressuremeter are considered the most representative in this case, on the basis that the dense Thanet Sand is known to support relatively large loads with small settlements. Data presented by Crova et al at the Wroth Memorial Symposium supports this view. This does not invalidate the importance of laboratory tests, however, since it is important to validate results by means of other tests.

Mr Eldred presented the following comments prepared by Mr Eccles, both from Soil Mechanics Ltd, on the paper by **Johnson et al**. This paper reported maximum and minimum dry density values for the Thanet Beds of 1.60 and 1.12 Mg/m^3 respectively. Figure 6 presents results of dry density measurements on wireline core samples which we have collected from a number of sites. These samples were taken from the liner very soon after coring. The dry density of many of these samples is in excess of the maximum dry density of Thanet Beds as measured by Johnson et al in the laboratory. Since the stratum is known to have a relatively uniform composition, this large variation in measured densities can be attributed to sample disturbance. This has occurred even when sand samples taken from wireline core have been carefully preserved using techniques similar to those presented by **Hepton**. Disturbance of sand samples could be due to a number of factors including drilling disturbance and those suggested by Johnson et al. However, vibration of the core in its liner during transportation between the drilling rig and the on-site core handling and sampling facilities (up to 4 miles on CrossRail) is probably one of the most significant factors causing the reduction of density. As noted by **Hepton**, there is a clear annulus between the core liner and the core, hence the sand is not restrained after its initial recovery until the sampling is complete. Johnson et al present results which show stiffnesses measured in the triaxial apparatus to be lower than those measured using the SBP. This difference could be due to the triaxial tests being carried out on disturbed samples. Figure 7 shows stiffness to be proportional to dry density. There is more than a three fold increase in stiffness as the dry density of the tested specimens increases from 1.47 to 1.64 Mg/m^3.

Mr Blight of Sir William Halcrow & Partners also commented on the paper by **Johnson et al**. The paper presented some sophisticated analyses of pressuremeter tests carried out using the WRSBP. The maximum operating cavity pressure of this pressuremeter was reported to be 20MPa. But the paper by **Beckwith et al** describes the use of a SBP in both the London Clay, the sands, silts and clays of the Woolwich & Reading Beds and the Thanet Sands. The maximum operating cavity pressure for this instrument was 4MPa which gave the advantage of increased sensitivity combined with lower drilling disturbance. However, the disadvantage of using such an instrument in the Thanet Sands was the relatively high membrane attrition rate. The membrane had often to be replaced after every test.

Dr Allen of Exploration Associates, in response to the last speaker, stated that the SBP could readily be fitted with a more arduous membrane.

Dr Toll of Durham University discussed computer control of geotechnical testing. The School of Engineering at Durham can now do a full range of stress path testing. This

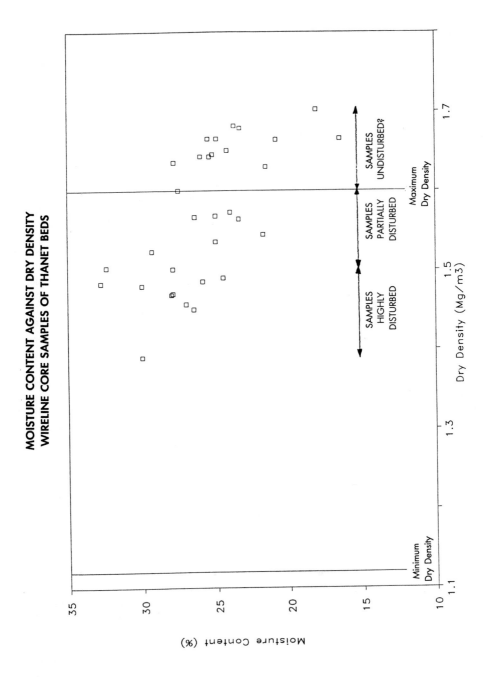

**MOISTURE CONTENT AGAINST DRY DENSITY
WIRELINE CORE SAMPLES OF THANET BEDS**

Figure 6

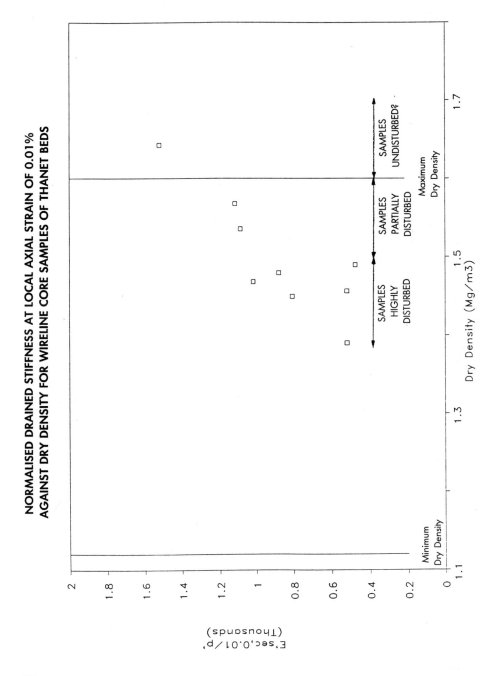

Figure 7

includes anisotropic consolidation, K_o consolidation, cyclic loading and probing tests. Advances in the computer control of testing are now such that the rate of testing may be either (i) stress rate controlled (ii) displacement controlled (useful for obtaining post rupture behaviour) (iii) excess pore water pressure controlled or (iv) "true" strain rate controlled (based on feedback from local strain devices). Dr Toll then provided a brief description of his program.

Mr Harwood of Newcastle University presented some additional information on the flow pump and the volume change meter used by **Araruna et al**. The flow pump (Figure 8) employed was a medical-type syringe driver with a stainless-steel syringe designed and built at Newcastle University, based on one described by Aiban and Zuidarcic (1989). The syringe driver was chosen since it delivers very low, constant flowrates with little dependence on fluid pressure. The syringe comprises a one-piece stainless-steel body incorporating static O-ring seals and an integral flushing port. The piston, also stainless-steel, has a very accurately ground diameter, which improves overall accuracy when compared with relying on the less accurately machinable bore. An LVDT permits independent measurement of flow rate. The volume change meter (Figure 9) is based on the burette principle described by Tatsuoka (1981) and employs a DPT to measure the height of fluid in two flow tubes, the larger of which can be isolated using the selection valve as discussed in the paper. The volume change meter was also designed and built in the University and is robust and self-contained, allowing its possible use on site.

Dr Ridley of Imperial College discussed the laboratory testing of unsaturated soils. For saturated soils the two stress variables (σ and u_w) uniquely characterise the soil behaviour. In unsaturated soils there are also two stress variables; the net total stress (σ-u_a) and the soil suction (u_a-u_w), where u_a is the pore air pressure. However, they cannot be combined to form a single "effective stress", the magnitude of which is independent of their absolute values, that will uniquely define every aspect of the soil behaviour. Therefore when investigating the behaviour of an unsaturated soil it is necessary to independently measure and control the net total stress and the soil suction, whilst measuring the strength and/or the volume change. The main difference between a saturated soil and an unsaturated soil is found in the pore water pressure. In an unsaturated soil the absolute pore water pressure can be several atmospheres negative. Standard laboratory pore water pressure transducers have proved inadequate to measure negative water pressures in excess of one atmosphere. Consequently the measurement and control of pore water pressures in unsaturated soils has been very difficult. Currently popular ways of *measuring* soil suction are the axis translation technique (Figure 10) and the filter paper method (Figure 11). In addition a new device, the "suction probe" tensiometer, capable of measuring suctions up to 1500kPa, has recently been developed at Imperial College and is the subject of **Ridley and Burland**. In recent years, however, a new technique has also emerged for *controlling* soil suction. Water is extracted from a soil sample osmotically, and does not require elevated air pressures. Water can be caused to move through a filter membrane if the solutions on either side of the membrane have different osmotic potentials and the membrane is impermeable to the salt in one of the solutions. Polyethylene Glycol (PEG) has a large molecular size and will not pass through Visking Dialysis membrane. Therefore circulating PEG through the drainage system of the triaxial cell and placing such a membrane between the soil sample and the drainage system will cause water to be extracted from the soil at the ambient atmospheric air pressure. The pore water pressure that is generated in the soil using this approach will be negative and the tension can be of the order of 1500kPa which

Flow pump

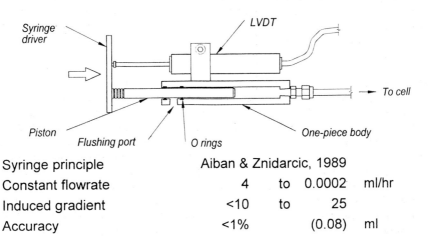

Syringe principle	Aiban & Znidarcic, 1989		
Constant flowrate	4 to 0.0002		ml/hr
Induced gradient	<10 to 25		
Accuracy	<1%	(0.08)	ml

Figure 8

Volume change meter

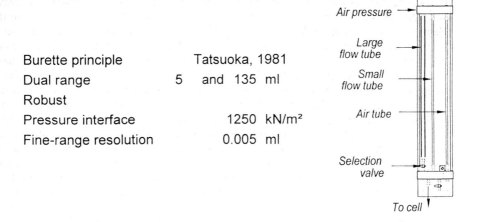

Burette principle	Tatsuoka, 1981
Dual range	5 and 135 ml
Robust	
Pressure interface	1250 kN/m²
Fine-range resolution	0.005 ml

Figure 9

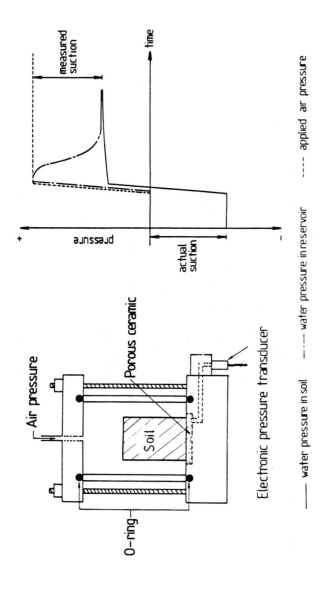

— water pressure in soil   —·— water pressure in reservoir   ---- applied air pressure

Figure 10   Pressure plate apparatus and the axis translation technique of suction measurement

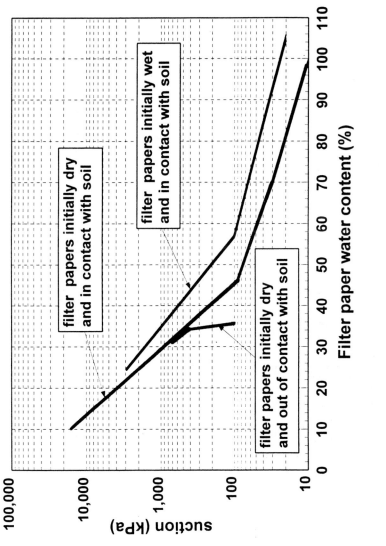

Figure 11   Seven day calibrations for Whatman No.42 filter papers

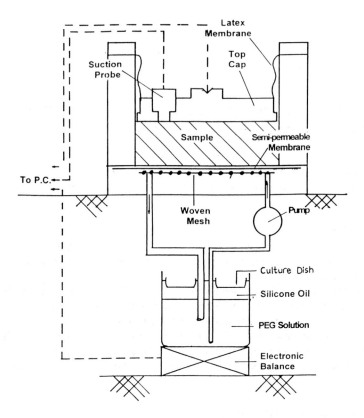

Figure 12   A new osmotically controlled oedometer system (after Dineen & Burland, 1995)

is significantly higher than that achievable using the alternative axis translation technique (400kPa). Recent research at Imperial College (described in a paper by Dineen and Burland recently submitted to the 1st Conf. on unsaturated soils, Paris, Sep 1995) has combined the method of PEG suction control with the new tensiometer in an oedometer system (Figure 12) for investigating the swelling behaviour of soils. Measurements of the sample's volume, moisture content and suction can be continuously logged on a PC. This system is now ready for inclusion in a triaxial stress path cell and should pave the way for the elemental testing of unsaturated soils under conditions which simulate their natural environment.

Dr Paul of Heriot-Watt stated that improved photography of soft clays could be obtained by using an "osmotic knife". This is a thin bladed knife which runs off a DC power supply and cleans up the sample by causing the soil particles to separate on the surface.

Dr Smart of Glasgow University stated that there were now available automatic methods of image analysis for fabric (structure) seen in electron and optical micrographs, ordinary photographs, aerial photographs and remote sensing. These are expected to develop in scope before the next conference on Site Investigation.

Closing address

The role of in situ testing in geotechnical engineering — thoughts about the future

M. JAMIOLKOWSKI and N. MANASSERO
Technical University of Torino and Studio Geotecnico Italiano of Milan

INTRODUCTION

Seven years ago the first author, jointly with Prof. P.K. Robertson [Jamiolkowski and Robertson (1988)] have attempted to foresee the future of the penetration tests in the site investigation practice.

Looking back at this attempt it appears that the crystal ball used by the authors has worked quite effectively. Infact, the scenario which was pictured in 1988 looks consistent with what has emerged during this conference relatively to penetration testing.

This positive experience has encouraged the writers to undertake a similar challenge in the attempt to envisage the future role of the insitu methods in the geotechnical investigation practice.

Within this scope it is worth recalling the main aims of insitu tests as formulated by Wroth (1984).

· Soil profiling and characterization

· Measurement of a specific ground property relevant to geotechnical design

· Direct dimensioning of foundations

· Quality control of constructions

· Monitoring of performance

The use of insitu tests should be observed in combination with laboratory testing methods hence only a harmonious and properly planned use of both can lead to a comprehensive and cost-effective geotechnical site characterization.

When comparing insitu versus laboratory tests the following points should be kept in mind:

· The continuous evolution in the two areas of soil testing renders mutable with time their relative merits.

· The remarkable progress made in the eighties in laboratory testing e.g. *undisturbed sampling techniques, increasing reliability in stress-strain measurements* contributes to a more rational use of insitu tests.

The development of new insitu techniques, the improvement of the existing ones and the rationalization of interpretation methods confer them a vital role in any modern and cost-effective soil investigation programme, e.g. *multipurpose penetration probes, new geophysical methods.*

As to the capacity of the insitu methods to satisfy multipurpose needs of a comprehensive soil characterization programme, they can be grouped as follows:

A. Tests aimed at stratigraphic profiling, classification and assessment of the spatial variability of soil deposits, e.g. Static Cone Penetration Test (CPT) with pore pressure measurement (CPTU)

B. Tests aimed at obtaining a specific ground property, e.g. evaluation of the undrained shear strength s_u from Field Vane Test (FVT).

C. Multipurpose tests aimed at measuring a number of ground properties yielding positively redundant information, e.g. Marchetti Flat Dilatometer (DMT) or Pressiocone (CPMT).

The discussion that follows will take into account the above classes.

PRESENT DEVELOPMENT TRENDS

Present development trends of insitu testing are listed in the following:
1. Evolution of non destructive geophysical testing techniques as applied to the broad spectrum of problems in geotechnical engineering.

In this area, seismic tomography, georadar, acoustic emission and the use of surface waves, which has received a new impulse by the development of advanced numerical interpretation techniques, are of major importance as far as geotechnical characterization of natural soil deposits and earth structures are concerned [Dixon et al. (1995), Gordon et al. (1995), Mc Cann and Green (1995) and Nauroy et al. (1995), Rechtien et al. (1994), Mc Gee and Ballard (1995)].

Particularly, the recent progresses in the spectral analysis of surface waves (SASW) rendered the application of this method especially useful and cost-effective for the characterization of soil deposits, see Nazarian and Stokoe (1983, 1984), Stokoe et al. (1994), Santamarina (1994).

The key aspects of such method are briefly summarised in the following [Jamiolkowski et al. (1995)]:
- In terrestrial applications Rayleigh waves are propagated by means of impulsive mechanical sources and are detected by a pair of transducers located at different distances (r1 and r2) from the source, see Fig. 1. Waveforms with a high frequency content (short wavelength) propagate in the shallower strata; the deeper strata are investigated by generating waveforms with longer wavelengths - the signals at the receivers are ditized and recorded by a dynamic signal analyser. The Fast Fourier Transform (FFT) is computed for each signal. Thus the cross power spectrum and coherence function can be computed.
The coherence function indicates for which frequency interval meaningful measurements have been obtained. On the other hand, the phase angle of the cross power spectrum enables ones to determine the travel time between the two receivers and consequently the Rayleigh wave velocity. The wavelengths (L_R) corresponding to the frequency (f) dependent Rayleigh wave velocities (V_R) are computed as $L_R = V_R (f) / f$. It is thus possible to plot the surface wave phase velocity versus wavelength (field dispersion curve). The above mentioned data analysis is performed in real time.

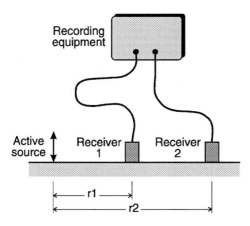

Fig. 1 - Basic configuration for SASW test.

- It is assumed that $V_R / V_S \cong 0.92$; in the conventional steady state Rayleigh wave method the velocity profile is obtained by assuming that the depth corresponding to each determined velocity is equal to 1/2 or 1/3 of the corresponding wavelength [Abiss and Viggiani (1994)].

This is an approximate hypothesis especially in layered soils. In the SASW method a theoretical dispersion curve is computed for a given velocity profile and compared to that experimentally determined. The soil profile and the velocities assigned to each stratum are adjusted until the theoretical curve matches the experimental one.

Experimental (Gauer 1990) and theoretical (Manesh 1991) studies have shown that it is possible to use the SASW technique offshore propagating the Scholte Waves. Difficulties in test interpretation can arise for very stiff seafloor see Stokoe et al. (1994).

The SASW potentiality can be highlighted by the following examples:
- Figure 2 reports the comparison between the shear wave velocity (V_s) as measured in cross-hole and SASW tests respectively.
  Such results refer to the sand and gravel deposit at the Sicilian shore along the axis of the proposed suspended bridge over the Messina Strait, see Crova et al. (1992).
  The agreement yielded by the two testing methods appears excellent.
- Figure 3 shows an application of SASW by Stokoe (1995) aimed at obtaining the V_s profile of an existing old toxic waste landfill. The potential of the method in detecting the different layers is clearly evidentiated.
- The use of SASW offshore appears particularly interesting. Figures 4 and 5 present examples of the related test arrangements while Figure 6 reports the results of an offshore SASW test performed at the bottom of the Venice harbour as compared against the Vs measured using seismic cone [Luke et al. (1994)].

In the meantime, current practice and recent innovations have consolidated the important role gained by the borehole seismic techniques, such as cross-hole, down-hole, up-hole and seismic cone tests, in the everyday geotechnical site investigation practice both on and off shore [Campanella et al. (1986), Stokoe (1980), Jamiolkowski et al. (1994, 1995), Woods and Stokoe (1985)].

The use of these techniques is mostly aimed at soil profiling, assessment of small strain stiffness i.e. shear (G_o) and constrained (M_o) moduli and via a comparison between field and laboratory measured V_s, at the evaluation of the quality of the undisturbed sampling.

Besides the continuous relevant progresses made in the hardware and software for the acquisition and processing of the seismic insitu tests results, bore-hole techniques benefit from

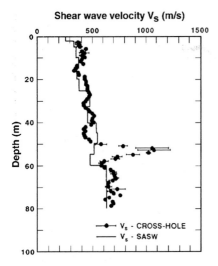

Fig. 2 - Shear Wave Velocity from SASW vs. That from Cross-Hole Test in Sand and Gravel Messina Straits Crossing.

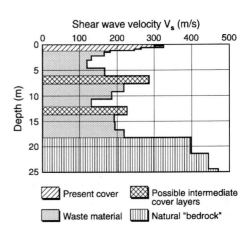

Fig. 3 - Shear Wave Velocity Profile Determined by SASW Testing at a Landfill

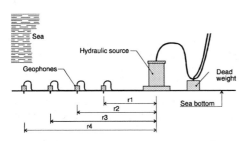

Fig. 4 - Test Arrangement for Offshore SASW.

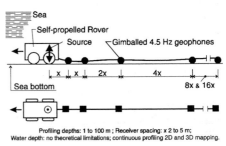

Profiling depths: 1 to 100 m ; Receiver spacing: x 2 to 5 m;
Water depth: no theoretical limitations; continuous profiling 2D and 3D mapping.

Fig. 5 - Offshore SASW

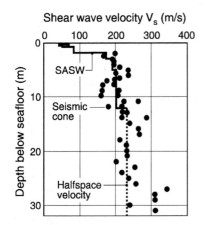

Fig. 6 - Comparison between underwater SASW and seismic cone performed in Venice harbour, water depth 7m.

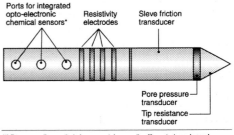

Integrated opto-electronic (IO) chemical sensors utilizing wave interference spectra to detect chemical pollutants in quantities PPM are mounted on CPT tip.

(*) Benzene, xilene, ethyle benzene, toluene, oil spills, petroleum loosed contaminants.

Fig. 7 - Integrated Optics Geoenvironmental Cone Penetrometer.

the following factors:
- The recognition that the reliability and the quality of the measured body waves velocities can be highly improved when their assessment is made on the basis of the true interval method [Patel (1981)]. This imposes the use of at least three holes in case of cross-hole tests and of two receivers at spacing of ~ 1 m when down-hole and up-hole techniques are used [Patel (1981), Rice (1984), Laing (1985)].
- The development of mechanical sources able to generate the shear waves (V_s^{hh}) having both propagation and particles vibration direction located on horizontal plane.
  This allows to estimate the initial elastic anisotropy of the deposit and opens some perspectives for the evaluation of the insitu horizontal stress [Woods and Stokoe (1985), Stokoe (1985), Jamiolkowski and Lo Presti (1994), Sully and Campanella (1995)].

What stated above may well exemplify the present, impetuous trend towards the application of the geophysical methods in the geotechnical site investigation practice. This tendency encompasses all the three groups of insitu techniques classified according to their targets, as reported in the introduction.

In addition, geophysical methods are gaining more and more importance in the nondestructive quality control of engineered geotechnical constructions such as, buried structures, pavements, piles, earth structure, etc., e.g. Stokoe and Jackson (1995), Stokoe et al. (1995), Jalinoos et al. (1994), Olson et al. (1995), Rits (1992), Rechtien et al. (1993), Lee et al. (1995), Bay et al. (1995).

2.   Another field worthy of attention within this contex is the environmental geotechnics. In this area, numerous static cones prototypes have been developed in the nineties in Canada, Italy, Japan, UK and USA aimed at assessing the presence and the extend of pollutants in the subsoil.

A typical example of the technology presently available in this field is the Geo-Environmental Cone described by O'Neill et al. (1995), allowing to evaluate, in real time, the physio-chemical parameters ([1]) of pore fluid and soil skeleton in addition to the information gained from the CPTU.

New advanced Site Characterization and Analysis Penetrometer System (SCAP) is being presently developed by the US Army Corps of Engineers with the participation of many US universities and private organizations, see Mayne and Burns (1994) and Robataille (1994). This Geo-Environmental CPTU tip, see Fig. 7, incorporates a variety of highly sophisticated sensors able to detect physio-chemical parameters of the soil skeleton and of the pore fluid. The use of the integrated opto-electronic sensors in combination with the laser induced fluorescence technique allows to discover and quantify a large number of pollutants present in the subsoil. The system under development appears extremely flexible allowing to conform to many site specific needs.

The SCAP once developed and validated will largely contribute to a rapid and reliable screening of the contaminated sites, supporting the subsequent remediation activities.

All Geo-Environmental cones are the typical example of insitu methods which fall within those belonging to group C.

3.   Remaining in the area of more conventional insitu methods it should be mentioned the development of the tests belonging also to group C aimed at measuring simultaneously more than one parameter, as much as possible independent among them. Because of the redundancy of the information which such tests can supply, their use increase the potential of the insitu techniques in the attempt of characterizing the ground property.

---

(1) e.g. pH, redox potential, dissolved oxigen, electrical resistivity, temperature, etc.

933

Typical examples of this trend are the Cone Pressuremeter Tests (CPMT). This recently developed tool [Campanella and Robertson (1986), Whiters et al. (1986)] combines CPTU and pressuremeter (PM) features. It ensures simple installation, high reproducibility of test results, continuous logging and well established links with foundation design practice, proper of CPTU. In the meantime stopping the penetration at the selected depths the pressuremeter tests can be performed obtaining pressure vs. displacement curve matching the expansion of a cylindrical cavity into the surrounding soil. The information yielded by the expansion stage of the test allows the assessment of the ultimate limit cavity pressure (p_u), unload-reload shear modulus (G_{ur}) together with other three recorded during the penetration stage of the test: the cone resistance q_c, the local mantle friction f_s and the penetration pore pressure u. Therefore, the Cone-Pressuremeter tests offer great potential for the ground property characterization of the investigated deposit and in the direct use of test results in foundation design.

Electrical resistivity cone combining the features of the CPTU and the allowance to measure the vertical or horizontal electrical resistivity ρ_v of the penetrated soil deposit is an additional example of the recently developed insitu techniques see Campanella et al. (1990). The assessment of ρ_v results in the mapping of the contaminated zone present in the subsoil and allows, in case of sand deposits, to distinguish, in first approximation, the strata exhibiting the contractive behaviour from those which appears to have dilatant response [Kokan (1992)].

Other multipurposes probes worth a remark are the Lateral Stress Cone (LSC) and the Vibrocone. However, their development have not reached yet such a stage at which the use of the obtained results in practice can be outlined.

The LSC [Campanella et al. (1990), Huntsman (1985), Jefferies et al. (1987), Sisson (1990)] in addition to q_c, f_s and u allows to measure the total horizontal stress σ_h and the corresponding pore pressure acting on the penetrometer shaft well above the tip. The obtained results are looked with interest in relation to problems such as the semi-empirical approach of the initial insitu horizontal stress estimate and the assessment of the horizontal stress involved in the evaluation of shaft resistance of displacement piles.

The Vibrocone throughout the comparison of the conventional static q_c profile and that obtained at the adjacent location with the cone tip subject to a vibration of known characteristics, offers the possibility of attempting to distinguish the deposits that are likely to liquefy from those that are not.

In addition, the already mentioned SCPT, thanks to the measurements of V_s besides to the usual information yielded by the CPTU adds a great potential to the ground properties characterization by means of insitu tests. Some applications of this test will be mentioned in the following section of this paper.

As to the tests belonging to group B, aimed at determining a given ground property, among many new developments and improvements of the existing methods, the following deserve special attention when thinking ahead:

- The new tensiometer developed by Ridley and Burland (1995) and presented in a paper submitted to this conference. It allows to measure reliably and in a short time the soil suction insitu up to 1.5 MPa.

  The suction represents a basic parameter framing the mechanical behaviour of partially saturated soil which plays a particularly important role in predicting their swelling potential. It can be of great help when excavating deep tunnels and shafts in highly OC clays and allows to assess the quality of undisturbed samples.

- The Self-Boring Electrical Resistivity Probe test (SBEPT) developed by Bellotti et al. (1994), combines the self-boring insertion which minimize the disturbance with the simultaneous measurements of the horizontal (ρ_h) and vertical (ρ_v) electrical resistivity of the surrounding soil.

The device is mainly aimed at assessing the insitu porosity (n) of sands via its link with the average formation factor F_{av} defined by Arulmoli (1985) as follows:

$$F_{av} = \frac{F_v + 2\,F_h}{3}$$

where: F_v, F_h = vertical and horizontal factors respectively, defined as the measured electrical resistivities normalized with respect to that of the pore fluid (ρ_w).

The formation factors may be correlated to the porosity via the Archie (1942) law, i.e.

$$F_v = n^{-mv}, \quad F_h = n^{-mh}, \quad F_{av} = n^{-ma}$$

being: m = exponents depending on shape of the soil grains.

Typical laboratory established correlations between F_v or F_h and n are reported in Fig. 8. Once such correlations are available for a given sand the porosity can be assessed via insitu tests using SBEPT. An example [Morabito (1994)] of these tests is reported in Fig. 9 referring to the quaternary Po river sand at San Prospero site in North Italy. Although the lack of laboratory tests on undistorted samples obtained using the freezing or impregnation techniques do not allow to assess the absolute reliability of the measured porosities, however they represent the upper bound of those inferred from the correlations between relative density and penetration tests results. Taking into account that these correlations are based on CPT's and SPT's performed using freshly deposited sand, the obtained results look qualitatively good if framed within the scenario of aged sand as that existing at the site in question.

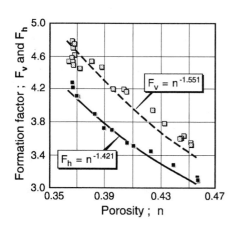

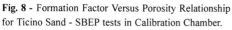

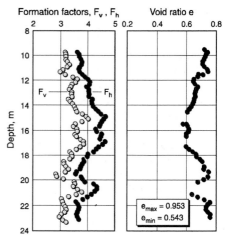

Fig. 8 - Formation Factor Versus Porosity Relationship for Ticino Sand - SBEP tests in Calibration Chamber.

Fig. 9 - SBEP tests results in Po River Sand.

In addition, the ratio F_h/F_v or the electrical index A defined by Arulanandan and Kutler (1978) as follows:

$$A = \left(\frac{F_h}{F_v}\right)^{0.5}$$

are a measure of the fabric anisotropy of the tested soil.

An interesting example of a device falling within the category B and aimed at measuring single soil parameter, in this case the hydraulic conductivily of the fine grained soils, is the newly developed self-boring permeameter [Znidaric and Piccoli (1995)].

This device by using the flow pump technique overcomes the problems which arise when using the self-boring permeameter developed in Canada [Tavenas et al. (1990)] and allows the full authomation of the test and a very precise flow control.

The additional pore pressure transducer installed on the body of the proble offers, at least in principle, the possibility to assess the anisotropy of the soil with respect to the hydraulic conductivity.

USE AND INTERPRETATION

If on one side modern insitu techniques offer a broad spectrum of advanced tests aimed at the characterization and property assessment of different soil deposits, on the other hand their interpretation particularly for the property assessment is not always straightforward.

Basically, the approaches used in the interpretation of insitu tests can be grouped as follows:

A. Direct

Insitu test results are correlated directly to the desired response of the foundation system, i.e. bearing capacity, settlements, etc., see for example: Bustamante and Gianeselli (1982), Reese and O'Neill (1988), Schmertmann (1972), Burland and Burbidge (1985), Berardi and Lancellotta (1991), Ghionna et al. (1993, 1994).

B. Indirect

Insitu tests are interpreted to obtain the physical and mechanical properties of the ground such as: stiffness, shear strength, hydraulic conductivity etc.

Within this class the following three conditions can be pictured depending on the amount of empirism involved:

B.1 Soil elements surrounding the insitu device are subject to a similar effective stress path (ESP). In this circumstance adopting an appropriate constitutive relationship, a specific ground property reflecting the behaviour of a soil volume element can be inferred.

Unfortunately very few insitu tests comply with the above condition. The most convincing examples are the seismic borehole methods allowing to evaluate the small strain stiffnesses G_o and M_o.

With some reservations, within this sub-category self-boring pressuremeter and permeability tests can also be classified.

B.2 The soil elements implicated follow different ESP's. Making some assumption about the soil stress-strain behaviour, a ground property reflecting an overall response of the soil volume involved is obtained. A typical example is represented by the interpretation of plate loading and Menard Pressuremeter or Cone-Pressuremeter Tests for stiffness and strength.

B.3 The soil surrounding the insitu device is subject before the test to a large straining. All penetration tests (CPT, DPTU, DMT, SPT) fall within this sub-category when interpreted for stiffness and strength. In this case, only empirical correlations between ground properties and tests results are possible.

In undertaking a brief review of the most promising trends in use and of the interpretation of insitu tests results priority is given to seismic tests, especially to those performed in the boreholes.

For such tests, the improvement of the testing equipment and data acquisition system combined with a better understanding of their basic significance resulting from large scale laboratory tests [Lee and Stokoe (1986), Mok (1987) , Lee (1993), Stokoe et al. (1994), Jamiolkowski et al. (1995a)] has opened new perspectives for the interpretation and use of the results in geotechnical design.

At present, it is widely accepted that the velocity of the seismic waves in soils depends on factors that are illustrated in the following relationships [Roesler (1979), Stokoe et al. (1985)]

Shear wave velocity

$$: V_s = C_{ij} \ (k\text{-}e) \ p_a^{\ (1\text{-}ni\text{-}nj)} \ (\sigma'_i)^{ni} \ (\sigma'_j)^{nj} \qquad [LT^{-1}]$$

Compression wave velocity

$$: V_p = C_{ii} \ (k\text{-}e) \ p_a^{\ (1\text{-}2ni)} \ (\sigma'_i)^{2ni} \qquad [T^{-1}]$$

where:

C	=	dimensional anisotropic material constant, reflecting soil type and fabric	$[LT^{-1}]$
k	=	dimensionless constant	$[-]$
e	=	current void ratio	$[-]$
σ'_i, σ'_j	=	current effective stresses in direction of wave propagation and particles motion respectively	$[FL^{-2}]$
p_a	=	atmospheric pressure	$[FL^{-2}]$
ni, nj	=	experimental stress exponents	$[-]$

Considering that the small strain elastic soil stiffnesses are linked to the wave velocities as follows:

$$G_o = \rho \, V_s^2 \ [FL^{-2}]; \qquad\qquad M_o = \rho \, V_p^2 \ [FL^{-2}]$$

being:

G_o	=	initial shear modulus	$[FL^{-2}]$
M_o	=	initial constrained modulus	$[FL^{-2}]$
ρ	=	soil density	$[FL^{-3}]$

the influence of the factors appearing in the relationship for the body waves velocity on the elastic stiffness of soils can be delineated as follows [Roesler (1979), Yu and Richart (1984), Stokoe et al. (1985), Lee and Stokoe (1986)]:

$$G_o = S_{ij} \, F(e) \, p_a^{\ (1\text{-}2ni\text{-}2nj)} \ (\sigma'_i)^{2ni} \ (\sigma'_j)^{2nj}$$

$$M_o = S_{ii} \, F(e) \, p_a^{\ (1\text{-}4ni)} \ (\sigma'_i)^{4ni}$$

being:

F(e)	=	empirical void ratio function
S	=	dimensionless anisotropic material constants depending on the soil type and fabric.

The above formulae describe the influence of the current soil state [1] on its initial stiffness. Such formulae which express the link between M_o and the velocity of the constrained compression

(1) Combination of soil fabric, σ' and e.

937

wave V_p are also valid when an unconstrained compression wave is propagated whose velocity is related to the initial Young modulus E_o of the tested soil.

With the above background in mind and assuming that the soil deposits when subject to the propagation of the seismic body waves behave as a transversely isotropic elastic body, the following remarks can be made:

- During the conventional cross-hole tests the propagated shear wave velocity $V_s = V_s^{hv(1)}$ while during the down-hole or up-hole tests $V_s = V_s^{vh}$. On the basis of the above exposed relationships it results that if the postulated cross-anisotropic elastic model holds, $V_s^{vh} = V_s^{hv}$ and consequently the two tests in question yield the initial shear modulus acting on vertical plane G_o^{hv} and G_o^{vh} which, in principle, should be equal. This holds under the condition that $ni = nj$, what generally is, in first approximation, confirmed by laboratory experiments.

- Only in a cross-hole test during which V_s^{hh} is propagated using the special source able to generate horizontally polarized shear waves [ISMES (1993), Fuhriman (1993)] the shear modulus acting on horizontal plane G_o^{hh} can be assessed and consequently the small strain anisotropy evaluated.

TABLE 1a - Small strain anisotropy of two pluvially deposited sands.

Sands (*)	e	σ'_h/σ'_v	G_{hh}/G_{vh}	M_h/M_v	E_h/E_v	
Predominantly silica Ticino river Sand	0.778	0.5	0.96	0.83	0.81	Medium dense
		1.0	1.20	1.20	1.22	
		1.5	1.25	1.55	1.52	
		2.0	1.44	1.88	1.86	
$D_{50}=0.55$ mm Bellotti et al.(1995)	0.617	0.5	1.13	1.05	NA	Very dense
		1.0	1.15	1.31	NA	
		1.5	1.25	1.40	NA	
Carbonatic Kenya sand	1.573	0.5	1.09	0.76	0.76	Medium dense
		1.0	1.13	1.28	1.23	
		2.0	1.29	1.98	1.85	
$D_{50}=0.13$ mm	1.315	0.5	1.05	0.94	0.97	Very dense
		1.0	1.24	1.25	1.29	
		2.0	1.40	2.00	1.92	

(*) Seismic tests performed in Calibration Chamber; NA = not available.

---

(1) The first apex refers to the propagation direction while the second to the particles motion direction, being h for horizontal and v for vertical.

TABLE 1b - Small strain anisotropy of two natural clays.

	σ'_h/σ'_v	OCR	G_{hh}/G_{vh}		σ'_h/σ'_v	OCR	G_{hh}/G_{vh}
	0.55	1	1.43		0.55	1	1.27
Panigaglia	0.55	1	1.36	*Upper Pisa*	0.55	1	1.34
	0.71	1.7	1.43	*Clay(**)*	1.49	1.5	1.29
Clay(**)	0.87	2.5	1.59		2.62	2.6	1.40
	1.24	5.1	1.55		4.01	4.0	1.48
PI=44%	1.97	12.8	1.79	PI=41%	7.96	8.0	1.56
e_0=1.6	2.85	26.9	2.08	e_0=1.843	16.39	16.4	1.77

(**) Seismic tests performed in laboratory on high quality undisturbed samples, Jamiolkowski et al. (1994)

Tables 1a and 1b report the data regarding the initial elastic anisotropy of two reconstituted sands and of two natural clays obtained from seismic tests performed by Jamiolkowski et al. (1995).

The results obtained allow to figure out the relevance of the fabric anisotropy alone when referring to the case of $\sigma'_h/\sigma'_v = 1$ as compared to the initial one for all values of stress ratios $\sigma'_h/\sigma'_v \neq 1$. In this case, both the fabric and the stress induced anisotropy are reflected in the measured stiffness.

- When interpreting the results of the insitu seismic tests it should be kept in mind that the velocity of the body waves is influenced by aging, even slight cementation and other structure forming processes being the part of the early diagenesis phenomena. This on one side explains why the V_s or V_p of natural deposits of some age differ from that of the same soil reconstituted in laboratory with the same e and σ' as it has insitu. On the other hand, what stated above corroborate the idea that a comparison between insitu and in laboratory measured values offers insight into the quality of the undisturbed samples.

Table 2 reports the data collected by Jamiolkowski et al. (1995) and quantifying the influence of aging on the G_0 by means of the following empirical formula [Anderson and Stokoe (1978), Mesri (1987)]:

$$G_0(t) = G_0(t_p)\left[1 + N_G \log\left(\frac{t}{t_p}\right)\right] \quad [FL^{-2}]$$

$G_0(t)$	=	initial shear modulus at $t > t_p$	$[FL^{-2}]$
$G_0(t_p)$	=	as above at $t = t_p$	$[FL^{-2}]$
t	=	any generic time larger that t_p	$[T]$

t_p = time to the end of primary compression   [T]

N_G = dimensionless parameter indicating the rate of increment of G_o per log cycle of time.

TABLE 2 - Influence of aging on initial shear modulus.

Soil	d_{50} [mm]	PI [%]	N_G [%]	Notes
Ticino sand	0.54	-	1.2	Predominantly silica
Hokksund sand	0.45	-	1.1	Predominantly silica
Messina sand and gravel	2.10	.	2.2 to 3.5	Predominantly silica
Messina sandy gravel	4.00	-	2.2 to 3.5	Predominantly silica
Glauconite sand	0.22	.	3.9	50% Quartz 50% Glauconite
Quiou sand	0.71	-	5.3	Carbonatic
Kenya sand	0.13	.	12	Carbonatic
Pisa clay		23-46	13 to 19	
Avezzano silty clay		10-30	7 to 11	
Taranto clay		35-40	16	

- The formulae relating the $G_o(V_s)$ and $M_o(V_p)$ to e, σ' and S, indicate that in the geostatic stress field G_o is function of the two principal effective stress components, $G_o^{vh}=f(\sigma'_v, \sigma'_h)$, $G_o^{hh}=f(\sigma'_h \cdot \sigma'_h)$, while M_o and E_o depend only on one principal effective stress component, $M_o^v = f(\sigma'_v)$, $E_o^v = f(\sigma'_v)$, $M_o^h = f(\sigma'_h)$, $E_o^h = f(\sigma'_h)$.
  These experimental evidences should be kept in mind when normalizing the initial stiffnesses with respect to the effective stress state.

- Another important issue in the normalization of the initial stiffnesses is related to the definition of the void ratio function F(e).
  Recent experimental results published by Jamiolkowski et al. (1994, 1995) and Shibuya et al. (1995) indicate that for many sands and clays this function can be satisfactory approximated by a simple relationship: $F(e) = e^{-x}$ with x ranging in a relatively narrow boundary, between 1.1 and 1.5.

- Based on the experimental evidence that $V_s^{vh} - V_s^{hv}$ is function of σ'_v and σ'_h while V_s^{hh} depends on σ'_h only, allows at least in principle, to postulate that when both horizontally and vertically polarized shear waves are measured during cross-hole tests one can attempt to estimate the coefficient of the earth pressure at rest $K_o = \sigma'_{ho}/\sigma'_{vo}$ [Stokoe (1985), Jamiolkowski and Lo Presti (1994), Sully (1991), Sully and Campanella (1995)].

Referring to the previously exposed formulae relating the shear wave velocity to the current void ratio and effective stress state one obtains:

$$K_o = \left(\frac{C_{vh}}{C_{hh}} \cdot \frac{V_s^{hh}}{V_s^{vh}} \right)^{\frac{1}{nh}}$$

However, the use of this approach at present is hampered by the need to know, a priori, the ratio of C_{vh}/C_{hh} which reflects the fabric anisotropy of the tested deposit. This information should be obtained by means of another insitu test.

The already mentioned SBEPT [Bellotti et al. (1994)] offers some possibility with this respect if one assumes that the ratio of the formation factors $F_h/F_v - C_{vh}/C_{hh}$.

Moreover, an important obstacle to the use of seismic tests for the estimate of K_o lies in the fact that typically $7 \leq 1/nh \leq 9$ which requires extremely high reliability and precision in the measurements of V_s and C_{ij} values, which at present might result not affordable.

- The use of the seismic tests offers new possibilities in the evaluation of the undrained behaviour of sands insitu.

Cunning et al. (1994) and Robertson et al. (1995) combining the insitu measured V_s and the reference critical state parameters λ_{ss}, Γ of a given sand have proposed the method for the evolution of the insitu state parameter ψ [Schofield and Wroth (1968), Been and Jefferies (1985)] being:

$\lambda_{ss} =$ slope of the Steady State Line (SSL) in $\ln \sigma'_m$ vs. e plane
$\Gamma =$ intercept of SSL on the void ratio axis at $\sigma'_m = 1$ kPa
$\sigma'_m =$ current mean effective stress.

This in turn allows to plot for a given sand in the V_s vs. σ'_v plane descriminating the contractive versus dilative boundary as shown in Fig. 10 which can be of great help in the evaluation of the susceptibility of the material insitu to static or cyclic liquefaction.

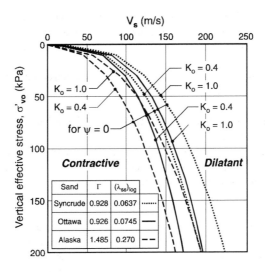

Fig. 10 - Contractive versus dilative boundary from seismic tests.

In the above figure the dilative behaviour is linked to $\psi < 0$ while the contractive one requires that $\psi \geq 0$.

Fear and Robertson (1994) have also worked out a procedure allowing the estimation of the undrained steady shear strength τ_{us} of sands using the insitu measured V_s, again in combination with known values of λ_{ss} and G.

- Finally, the ongoing research activities attempt to evaluate the small strain material damping directly from in situ seismic tests, e.g. Campanella and Stuart (1990), Mok (1989), Fuhriman (1993).

Other insitu techniques that are progressing with interesting future prospectives are the pressuremeter devices. Many kinds of pressuremeter probes are currently used in different countries [Clarke (1995)].

The common features of all these devices is that at least in principle they allow, during the test, to model the expansion of a cylindrical cavity into the surrounding soil. This requires that the length to ratio of the probe should be sufficiently large, at least between 8 and 10 in order to meet this postulate. Once it is fulfilled the test results can be interpreted within the frame of the above scheme referring to an a priori adopted constitutive equation.

The differences among various probes are mostly related to the way they are inserted into the ground i.e.: predrilled hole (PMT), pushed-in (CPMT) or self-bored (SBPT).

Especially, the latter one, allowing an insertion with a minum disturbance, makes possible quite rigorous theoretical interpretation of the test results and corresponds to one of the very few insitu methods yielding under ideal conditions the information about the magnitude of the geostatic total horizontal stress σ_{ho}.

The insertion into the granular soil of the other two types of probes (PMT, CPMT) on the contrary causes a significant change in the state of the surrounding soil which renders the theoretical interpretation of the test more difficult.

The use of the three mentioned types of pressuremeter offers many valuable options in the ground characterization some of which are listed in Table 3.

An application of the pressuremeter test, susceptible of further developments is the following: evaluation of shear stiffness, of special interest in coarse grained soil in which the expansion test occurs in drained conditions.

With this respect the usual approach is to refer to the small unload-reload loops as shown in Fig. 11, determining the secant modulus G_{ur} as evidentiated in Fig. 12. This stiffness is judged more reliable than the one G_i representing the initial part of the expansion curve whose magnitude results strongly influenced even by small disturbance.

The reliability of the measured unload-reload shear modulus G_{ur} is dependent on the equipment features and on the test procedure [Fahey and Jewell (1990), Fahey (1991)] among which the most important are:
· compliance of the strain measuring system
· hysteresis effects due to creep
· loop depth criterion.

The value of G_{ur} measured in coarse grained soils represents the drained stiffness at intermediate strain level $1 \cdot 10^{-4} \leq \gamma \leq 1 \cdot 10^{-3}$ relatively insensitive to soil disturbance caused by probe insertion.

Its use in the geotechnical deformation analyses has not been yet properly clarified within the frame of the soil stress-strain behaviour [Bellotti et al. (1989), Byrne et al. (1990), Fahey and Carter (1993)].

TABLE 3 - Applicability of pressuremeter tests for ground properties characterization.

Drained Expansion-Coarse Grained Soils

	Parameter	SBP	PM	CPM
p_u	Ultimate cavity stress	A	A	A
G_{ur}	Unload-reload shear modulus	A	U	U
ϕ'_{PS}	Plane strain angle at shearing resistance	A	U	U
ψ	State parameter	N	N	$U^{(1)}$
D_R	Relative density	N	N	$A^{(1)}$
K_o	Coefficient of earth pressure at rest	A	N	$U^{(1)}$
-	Direct use in foundation design	N	A	$U^{(2)}$
-	Quality control of soil improvement	N	A	A

Undrained Expansion-Fine Grained Soils

	Parameter	SBP	PM	CPM
p_u	Ultimate cavity stress	A	A	A
G_{ur}	Unload-reload shear modulus	A	U	U
s_u	Undrained shear strength	A	A	A
C_h	Coefficient of consolidation for horizontal flow	A	N	N
K_o	Coefficient of earth pressure at rest	A	N	U
-	Direct use in deep and shallow foundation design	N	A	$U^{(2)}$
-	Quality control of soil improvement	N	A	A

(1) *Application which requires further validation* A = *Appropriate*; N = *Not appropriate*
(2) *Appropriate rules not yet developed* U = *Semi-emperical, reliability uncertain*

The progresses with this respect require that the following aspects of the problem should properly be considered:
- Representative mean effective stress around the expanding cavity at which the G_{ur} is measured
- Location of the G_{ur} on the G vs γ cyclic degradation curve
- Tie with the stiffness of stress-strain curve from monotonic tests

·   Link of G_{ur} to the initial shear stiffness G_o
·   Anisotropic nature of the soil deformability, i.e. $G_{ur} = G_{hh}$.

What evidentiated above applies also to fine grained soils where the uncertainty related to the drainage conditions existing during the test [Fioravante et al. (1994)] adds extra complexity to the problem.

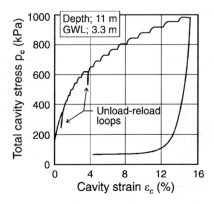

Fig. 11 - Self-Boring Pressuremeter Test in Po River Sand San Prospero Site

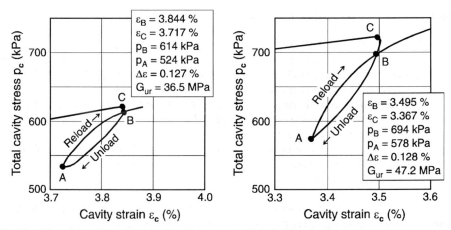

Fig. 12 - Unload-Reload Shear Modulus from Pressuremeter Tests Po River Sand

The attempts to assess G_o from the values of G_{ur} obtained from SBPT's carried out in sands have been presented by Bellotti et al. (1989), Byrne et al. (1990), Fahey (1991) and Ghionna et al. (1994a, 1995).

The approach followed by these latter authors considers a non-linear unload-reload loop departing from the vergin τ versus γ curve. In order to link the measured G_{ur} to G_o, Ghionna et al. (1994a) make the following two assumptions:

-     The non-linearity of soil within the unload reload loop can be matched by a preselected stress-strain relationship, i.e. Ramberg and Osgood (1943), Duncan and Chung (1970), Fahey and Carter (1993).

- Masing's (1936) rule applies when referring the reloading branch of a loop to the backbone curve.

- Using Duncan and Chang (1970) hyberbolic stress-strain relationship, Ghionna et al. (1994a) have estimated G_o from the G_{ur} measured during the SBPT's performed in sands both in the CC (Ticino sand) and in the field (Po river sand). The values of G_o obtained have been compared against those resulting from seismic tests using cross-hole techniques, see Fig. 13.

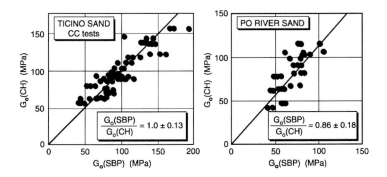

Fig. 13 - G_o from Cross-Hole Seismic Tests Versus that Inferred from SBPT's

- Another application of the SBPT's subject to further development is related to the use of the SBPT's for the determination of the consolidation characteristics of the fine grained soils insitu. Clarke et al. (1979) have introduced the strain holding test (SHT), while Fahey and Carter (1986) have illustrated the stress holding tests(PHT).
  Both tests, together with the already mentioned self-boring permeameter probe offer a great potential for a reliable estimate insitu of the coefficient of consolidation referred to the horizontal pore water flow.
  However, in order to maximise this possibility, many practical and theoretical problems for performing and interpreting such tests have still to be overcome, Fioravante et al. (1994), Fahey and Goh (1995).

In addition, to the two above mentioned applications, considered of priority interest, the SBPT, at least in principle, offers the possibility to assess the entire shear stress τ versus shear strain γ relationship in sands (drained) and in fine grained soils (undrained), see [Palmer (1972), Ladanyi (1972), Baguelin (1972), Manassero (1989)].

Despite the availabiliy of these solutions the research has not yet advanced enough to indicate if such approaches are adequately warranted.

Within the area of the pressuremeters it is imperative to mention the use of the PMT's results in the foundation design [Gambin and Frank (1995)]. With this respect the rules which have been developed in the last thirty years in France, covering the broad spectrum of foundation problems, represent an excellent example of direct use of insitu tests results in the geotechnical design and deserve further development and validation in the future.

Also, the further development of the Cone-Pressuremeter represents an attractive and powerful alternative for ground properties characterization by means of insitu tests. This example of the multipurpose device opens interesting perspectives in the interpretation of its results especially

in coarse grained soils where the expansion occurs in drained conditions. Recently in addition to the parameters like plane-strain peak of shearing resistance ϕ'_{ps}, and G_{ur}, semi-empirical interpretation procedures have been proposed aimed at assessing K_o, D_R, and the state parameter ψ. These approaches combining the penetration and expansion phases of the test results, lead to interesting applications like those proposed by:
-    Schnaid and Houlsby (1992) aimed at the assessment of D_R
-    Manassero (1991) attempting to evaluate K_o
-    Yu et al. (1994) indicating the possibility for estimating ψ.
   The other important area of application of insitu methods in geotechnical engineering is linked to the use and interpretation of the conventional penetration testing like SPT, CPT, and DMT.

   Because of lenght constraint this topic is not covered here. Updated information with this respect can be found in many of the papers presented at this conference and in those submitted to the CPT '95, to be held on October 1995 in Linköping.

REFERENCES

Abiss, C.P. and Viggiani, G. (1994), "*Surface Wave and Damping Measurements of the Groud with a Correlator*", Proc. XIII ICSMFE, New Delhi, Vol. 3.

Anderson, D.G. and Stokoe, K.H. II (1978), "*Shear Modulus: A Time Dependent Soil Property, Dynamic Geotechnical Testing*", ASTM STP 654.

Archie, G.E. (1942), "*The Electrical Resistivity Log as an Aid in Determining some Reservoir Characteristics*", Transactions of the American Institute of Mineral Metallurgy Engineering, 146.

Arulanandan, K. and Kutter, B. (1978), "*A Directional Structure Index Related to Sand Liquefaction*", ASCE Speciality Conference on Earthquake Engineering a Soil Dynamics, Pasadena.

Arulmoli, K., Arulanandan, K. and Seed, H.B. (1985), "*New Method for Evaluating Liquefaction Potential*", JGED ASCE No. 1.

Baguelin, F., Jézéquel, J.F., Lemée, E., Le Méhauté, A. (1972), "*Expansion of Cylindrical Probes in Cohesive Soils*", JSMFE DIV. ASCE SM 11.

Bay, J.H., Stokoe, K.H. and Jackson, J.D. (1995), "*Development of Preliminary Investigation of a Rolling Dynamic Deflectometer*", Accepted for publication in the Transport Research Board.

Been, K. and Jefferies, M.G. (1985), "*A State Parameter for Sands*", Geotechnique, No. 2.

Bellotti, R., Ghionna, V.N., Jamiolkowski, M., Robertson, P.K. and Peterson, R.W. (1989), "*Interpretation of Moduli from Self-Boring Tests in Sand*", Geotechnique, No. 2.

Bellotti, R., Benoit, J. and Morabito, P. (1994), "*A Self-Boring Electrical Resistivity Probe for Sands*", Proc. of XIII ICSMFE, New Delhi.

Berardi, R. and Lancellotta, R. (1991), "*Stiffness of Granular Soil from Field Performance*", Geotechnique No. 1.

Byrne, P.M., Salgado, F.M. and Howie, J.A. (1990), "*Relationship between the Unload Shear Moduli from Pressuremeter Tests and the Maximum Shear Modulus for Sand*", Proc. III Int. Symposium on Pressuremeters - British Geotechnical Society - Oxford, U.K.

Burland, J.B. and Burbidge, M.C. (1985), "*Settlement of Foundations on Sand and Gravel*", Proc. I.C.E., Part 1, London, U.K..

Bustamante, M. and Gianeselli, L. (1982), "*Pile Bearing Capacity Prediction by Means of Static*

Penetrometer", Proc. ESOPT II, Amsterdam.

Campanella, R.G. and Robertson, P.K. (1986), "*Research and Development of the UBC Cone Pressuremeter*", Proc. III Canadian Conference on Marine Geotechnical Engineering - University of St. Johns - New Foundland.

Campanella, R.G., Robertson, P.K. and Gillespie, D.G. (1986), "*Seismic Cone Penetration Test*", Proc. ASCE In-Situ '86, Blacksburg, Va.

Campanella, R.G. and Stewart, W.P. (1990), "*Seismic Cone Analysis Using Digital Signal Processing for Dynamic Site Characterization*", 43rd Canadian Geotechnical Engineering Conference, Quebec, Canada.

Campanella, R.G. and Weemees, I. (1990), "*Development and Use of an Electrical Resistivity Cone for Groundwater Contamination Studies*", Soil Mechanics Series No. 140, Dept. of Civil Engineering, The University of British Columbia.

Campanella, R.G., Sully, J.P., Greig, J.W. and Jolly, G. (1990), "*Research and Development of Lateral Stress Piezocone*", Draft submitted to TRB Symposium on Measurement of Lateral Stress, FHWA, Washington DC.

Clarke, B.G., Carter, J.P. and Wroth, C.P. (1979), "*In Situ Determination of the Consolidation Characteristics of Saturated Clay*", Proc. VII ECSMFE, Brighton, U.K.

Clarke, B.G. (1995), "*Pressuremeters in Geotechnical Design*", Chapman and Hall.

Crova, R., Jamiolkowski, M., Lancellotta, R. and Lo Presti, D.C.F. (1992), "*Geotechnical Characterization of Gravelly Soils at Messina Site*", Selected Topics, Proc. Wroth Memorial Symposium, Thomas Telford, London, U.K.

Cunning, J.C. "*Shear Wave Velocity of Cohesionless Soils for Evaluation of In-Situ State*", MSc Thesis, Dept. of Civ. Engng., Univ. of Alberta, Edmonton, Alberta, Canada.

Dixon, N., Hill, R. and Karanagh, J. (1995), "*The Use of Acoustic Emission to Monitor the Stability of Soil Slopes*", Proc. I.C.E. Int. Conf. on Advances in Site Investigation Practice. London.

Duncan. J.M. and Chung, C.Y. (1970), "*Non-Linear Analysis of Stress-Strain in Soils*", JSMFE Div., ASCE 96, SM5.

Fahey, M. and Lee Goh, A. (1995), "*A Comparison of Pressuremeter and Piezocone Methods of Determining the Coefficient of Consolidation*", Proc. 4th Intern. Symposium - The Pressuremeter and Its New Avenues, Sherbrooke, Canada.

Fahey, M. and Jewell, R.J. (1990), "*Effect of Pressuremeter Compliance on Measurement of Shear Modulus*", Proc. 3rd Int. Symposium on Pressuremeters, Oxford , U.K.

Fahey, M. (1991), "*Measuring Shear Modulus in Sand with the Self-Boring Pressumeter*", Proc. of X ECSMFE, Florence, Italy, Balkema, Rotterdam.

Fahey, M. and Carter, J.P. (1993), "*A Finite Element Study of the Pressuremeter Test in Sand Using a Non-Linear Elastic Plastic Model*", Canadian Geotechnical Journal, No. 2.

Fear, C.E. and Robertson, P.K. (1994), "*Estimation of the Ultimate Undrained Steady Shear Strength of Sand Using Shear Wave Velocity*", Draft of a paper submitted for publication in Canadian Geotechnical Journal.

Fioravante, V. et al. (1994), "*An Analysis of Pressuremeter Holding*", Geotechnique No. 2.

Fuhriman, M.D. (1993), "*Cross-Hole Seismic Tests at two Northern California Sites Affected by the 1989 Loma Prieta Earthquake*", M. Sc. Thesis, The University of Texas at Austin.

Gambin, M.P. and Frank, R.A. (1995), "*The Present Design Rules for Foundations Based on Ménard PMT Results*", Proc. 4th Intern. Symposium - The Pressuremeter and Its New Avenues, Sherbrooke, Canada.

Gauer, R.C. (1990), "*Experimental Study of Applying the Spectral-Analysis-of-Surface-Waves Method Offshore*", M. Sc. Thesis, The University of Texas at Austin.

Ghionna, V.N., Jamiolkowski, M. and Lancellotta, R. (1993), *"Base Capacity of Bored Piles in Sands from In Situ Tests"*, 2nd Int. Geotechnical Seminar "Deep Foundation and Auger Piles", Ghent.

Ghionna, V.N., Jamiolkowski, M., Pedroni, S. and Salgado, R. (1994) *"The Tip Displacement of Drilled Shafts in Sands"*, Proc. ASCE Symposium "Settlement-94", College Station, Texas.

Gionna, V.N., Karim, M., Pedroni, M. (1994) *"Interpretation of Unload-Reload Modulus from Pressuremeters Test in Sand"*, XIII ICSMFE, New Delhi.

Ghionna, V.N., Jamiolkowski, M., Pedroni, S. and Piccoli, S. (1995), *"Cone Pressuremeter Tests in Po River Sand"* Proc. 4th Int. Symposium - The Pressuremeters and Its New Avenues, Sherbrooke, Canada.

Gordon, M.A., Clayton, C.R.I., Thomas, T.C. and Matthews, M.C. (1995) *"The Selection of and Interpretation of Seismic Geophysical Methods for Site Investigation"*, Londra, Proc. I.C.E. Int. Conf. on Advances in Site Investigation Practice. London, U.K.

Houlsby, G.T. and Nutt, N.R.F. (1992), *"Development of Cone Pressuremeter"*, Proc. Wroth Memorial Symposium, Oxford. U.K.

Huntsman, S.R. (1985), *"Determination of In Situ Lateral Pressure of Cohesionless Soils by Static Cone Penetration"*, Ph.D. Thesis, Univ. of California at Berkeley.

ISMES (1993), *"Sviluppo e Sperimentazione di Attrezzature per Indagini Geofisiche e di Tecniche per Indagini Non Distruttive e Applicazioni in Situ - Controlli Sonici non Distruttivi (Radar, Termografia, Onde di Taglio)"*, Report, Prog. DGF-6761, Doc. N. RAT-DGF-004.

Jefferies, M.G., Jonsson, L., Been, K. (1987), *"Experience with Measurement of Horizontal Geostatic Stress in Sand during Cone Penetration Test Profiling"*, Geotechnique, No.4.

Jalinoos, F., Olson, L.D., Aonad, M.F. and Balch, A.H. (1994), *"Acoustic Tomography for Qualitative Nondestructive Evolution (QNDE) of Structural Concrete Using a New Ultrasonic Scanner Source"*, Proc. of QNDE Conf. Snowmass, Co.

Jamiolkowski, M. and Robertson, P.K. (1988), *"Future Trends for Penetration Testing"*, Proc. Conf. on Penetration Testing in the UK, Birminghan.

Jamiolkowski, M., Lancellotta, R. and Lo Presti, D.C.F. (1994), *"Remarks on the Stiffness at Small Strains of Six Italian Clays"*, Theme Lecture Proc. IS on Prefailure Stress-Strain Behaviour of Geomaterials, Sapporo, IS Hokkaido.

Jamiolkowski, M. and Lo Presti, D.C.F. (1994), *"Validity of In Situ Tests Related to Real Behaviour"*, Proc. XIII ICSMFE, 1994, New Delhi.

Jamiolkowski, M., Lo Presti, D.C.F. and Pallara, O. (1995), *"Role of In Situ Testing in Geotechnical Earthquake Engineering"*, Proc. III Intern. Conf. on Recent Advances in Geotechnical Earthquake Engineering and Soil Dynamics, St. Louis.

Kokan, M.J. (1992), *"Dilatancy Characterization of Sands Using the Resistivity Cone Penetration Test"* M.Sc. Thesis, University of British Columbia, Vancouver.

Ladanyi, B. (1972), *"In Situ Determination of Undrained Stress-Strain Behaviour of Sensitive Clays with the Pressuremeter"*, Canadian Geotechnical Journal, No 3, Technical Note.

Laing, N.L. (1985), *"Sources and Receivers with the Seismic Cone Test"*, Civil Engineering Department, M.S.Sc. Thesis, University of British Columbia, Vancouver.

Lee, N.K.J., Stokoe, K.H., Rits, M.P. and Young, Y.K. (1995), *"Long-Term Evaluation of Subgrade Moduli Using Permanently Embedded Geophones"*, Accepted for publication in the TRB, Washington D.C.

Lee, S.H. and Stokoe, K.H. II (1986), *"Investigation of Low Amplitude Shear Wave Velocity in Anisotropic Material"*, Geotechnical Engineering Report GR 86-6, University of Texas at Austin.

Luke, B.A., Stokoe, K.H. II and Piccoli, S. (1994), *"Surface Wave Measurements to Determine*

In Situ Shear Wave Velocity Profiles: On Land at Pontida and Offshore at Treporti, Italy", Geotechnical Engineering Report GR94-1, Geotechnical Engineering Center, Civil Engineering Department, University of Texas at Austin.

Manassero, M. (1989), *"Stress-Strain Relationship for Drained Self-Boring Pressuremeter Tests in Sands"*, Géetechnique, No. 3.

Manassero, M.(1991), *"Assessment of the Earth Pressure at Rest Coefficiente (K_o) in Sand Deposits"*, Proc. X ECSMFE - Florence.

Manesh, M.S. (1991), *"Theoretical Investigation of the Spectral-Analysis-of-Surface-Waves (SASW) Thecnique for Application Offshore"*, Ph.D. Thesis, University of Texas at Austin.

Masing, G. (1926), *"Eigespannungen und Vereestigung bei Messing"*, Proc. II Intern. Congress of Applied Mechanics, Zurich.

Mayne, P.W. and Burns, S.E. (1994), *"Development of an Integrated Optics Geoenvironmental Cone Penetrometer for Detecting and Mapping Soil and Groundwater Contaminants"*, Proc. Workshop on Advancing Technology for CPT for Geotechnical and Geoenvironmental Site Characterization, Dept. of Civil Engineering, The University of Texas at Austin.

Mc Cann, D.M. and Green, C.A. (1995), *"Geophysical Surveying Methods in a Site Investigation Programme"*, I.C.E. Int. Conf. on Advances in Site Investigation Practice. London.

Mc Gee, R.G. and Ballard, R.F. (1995), *"A Technique to Assess the Characteristics of Bottom and Subbottom Marine Sediments"*, Technical Report DRP-95-3 Prepared for U.S. Army Corps of Engineers, Washinghton, DC.

Mesri, G. (1987), *"Fourth Law of Soil Mechanics"*, Proc. Int. Symposium on Geotechnical Engineering of Soft Soils, Mexico.

Mok, Y.J. (1987), *"Analytical and Experimental Studies of Borehole Seismic Methods"*, Ph.D. Thesis, University of Texas at Austin, Austin, TX.

Morabito, P. (1994), Personal Communication.

Nauroy, J.F., Colliat, J.L., Puech, A., Poulet, D.. Meuniev, J. and Lapierre, F. (1995) *"Better Correlation between Geophysical and Geotechnical Data from Improved Offshore Site Investigations"* Proc. I.C.E. Int. Conf. on Advances in Site Investigation Practice. London.

Nazarian, S. and Stokoe, K.H. II and Hudson, W.R. (1983), *"Use of the Spectral Analysis of Surface Waves Method for Determination of Moduli and Thickness of Pavement Systems"*, Transportation Research Board, No. 930, TRB, Washington D.C..

Nazarian S. and Stokoe, K.H. II (1983), *"Use of the Spectral Analysis of Surfac Waves for Determination of Moduli and Thickness of Pavement Systems"*, Transportation Research Recod, No. 954, TRB, Washington, D.C.

Nazarian, S. and Stokoe, K.H. II (1984), *"In Situ Wave Velocities from Spectral Analysis of Surface Waves"*, Proc. of the 8th WCEE, San Francisco.

Olson, L.D., Lew, M., Phelps, G.C., Murthy, K.N. and Ghadioli, B.M. (1994), *"Quality Assurance of Drilled Shaft Foundations with Nondestructive Testing"*, Proc. Int. Conf. on Design and Construction of Deep Foundations, Orlando.

O'Neill, D.A., Piccoli, S. and Della Torre, A. (1995), *"Capabilities of the Envirocone Test System for Rapid Deep Contaminated Site Characterization"*, Proc. I.C.E. Int. Conf. on Advances in Site Investigation Practice. London.

Palmer, A.C. (1972), *"Undrained Plane-Strain Expansion of a Cylindrical Cavity in Clay: a Simple Interpretation of the Pressuremeter Test"*, Geotechnique, No. 3.

Patel, N.S. (1981), *"Generation and Attenuation of Seismic Waves In Downhole Testing"*, Geotechnical Engineering Thesis GT81-1, Department of Civil Engineering, University of Texas at Austin.

Powell, J.J.M. and Shields, C.H. (1995), *"Field Studies of the Full Displacement Pressuremeter in Clays*", Proc. 4th Int. Symposium, The Pressuremeter and its New Avenues, Sherbrooke, Canada.

Ramberg, W. and Osgood, W.T. (1943), "*Description of Stress-Strain Curves by Three Parameters*", Technical Note 902, National Advisory Committee for Aeronautics.

Rechtien, R.D., Greenfield, R.J., Ballard, R.F. (1993), "*Cross-Borehole Seismic Signature of Tunnels*", Fourth Tunnel Detection Symposium on Subsurface Exploration Technology, Golden, Co.

Reese, L.C. and O'Neill, M. (1988), "*Drilled Shafts: Construction Procedures and Design Methods*", ADSC - TC4 Publication.

Rice, A.H. (1984), "*The Seismic Cone Penetrometer*", Civil Engineering Departmente, M.A.Sc. Thesis, University of British Columbia, Vancouver.

Ricketts, G.A., Smith, J. and Skipp. B.O. (1995), "*Confidence in Seismic Characterization of the Ground*", Proc. I.C.E. Int. Conf. on Advances in Site Investigation Practice. London.

Ridley, A.M. and Burland, J.B. (1995), "*A Pore Water Pressure Probe for the In Situ Measurement Suctions*" Proc. I.C.E. Int. Conf. on Advances in Site Investigation Practice. London.

Rits, M.P. (1992), "*Development of an In-Situ Method for Continuous Evaluation of the Risilient Modulus of Pavement Subgrade*", M.Sc. Thesis, The University of Texas at Austin.

Robataille, G. (1994), "*Site Characterization and Analysis Penetrometer System - SCAPS*", Proc. Workshop on Advancing Technology for CPT for Geotechnical and Geoenvironmental Site Characterization, Dept. of Civil Engineering, The University of Texas at Austin.

Robertson, P.K., Sasitharan, S., Cunning, J.C. and Sego, D.C. (1995), "*Shear Wave Velocity to Evaluate In Situ State of Ottawa Sand*", JGED, ASCE No. 3.

Roesler, S.K. (1979), "*Anisotropic Shear Modulus Due to Stress-Anisotropy*", JGED, ASCE, GT 7.

Sack, D.A. and Olson, L.D. (1994?), "*In-Situ Nondestructive Testing of Buried Precost Concrete Pipe*", Olson Engineering Inc., Golden, Co.

Santamarina, J.C. (1994), "*Tomographic Inversion in Geotechnical Engineering*", International Report, Univ. of Waterloo, Ontario.

Schmertmann, J.H. (1972), "*Effects of In Situ Lateral Stress on Friction Cone Penetrometer Data in Sands*", Fugro Soundeering Symp.

Schnaid, F. and Houlsby, G.T. (1992), "*Measurements of the Properties of Sand in a Calibration Chamber by the Cone Pressuremeter Test*", Geotechnique No. 4.

Schofield, A. and Wroth, P. (1968), "*Critical State Soil Mechanics*" McGraw-Hill, New York.

Shibuya, S. and Tanaka, H. (1995), "*Estimate of G_{max} in Quaternary Clay Deposits*", Submitted for Publication to Soils and Foundation.

Sisson, R.C. (1990), "*Lateral Stress on Displacement Penetrometers*", Ph.D. Thesis, Univ. of California, Berkeley.

Soliman, A.A. and Fahey, M. (1995), "*Measurement and Interpretation of Shear Modulus in SBP Tests in Sand*", Proc. 4th Int. Symposium, The Pressuremeter and its New Avenues, Sherbrooke, Canada.

Stokoe, K.H. II (1980), "*Field Measurement of Dynamic Soil Properties*", Proc. of 2nd ASCE Conference on Civil Engineering and Nuclear Power, Knoxville, Tennessee, Vol. II.

Stokoe, K.H. II, Lee, S.H.H. and Knox, D.P. (1985), "*Shear Moduli Measurement Under True Triaxial Stresses*", Proc. of Conf. ASCE Convention Advances in the Art of Testing Soils Under Cyclic Loading Conditions, Detroit.

Stokoe, K.H. II, Wright, S.G., Bay, J.A. and Roësset, J.M. (1994), "*Characterization of Geotechnical Sites by SASW Method*", Proc. XIII ICSMFE New Delhi.

Stokoe, K.H. II (1995), Personal Communication.

Sully, J.P. (1991), "*Measuremets of In Situ Lateral Stress During Full-Displacement Penetration Tests*", Ph.D.Thesis, Univ. of British Columbia, Vancouver.

Sully, J.P. and Campanella, R.G. (1995), "*Evaluation of In Situ Anisotropy from Cross-Hole and Down-Hole Shear Wave Velocity Measurements*", Geotechnique, No. 2.

Tavenas, F., Diene, M. and Leroueil, S. (1990), "*Analysis of the In Situ Constant-Head Permeability Tests in Clays*", Canadian Geotechnical Journal, No. 3.

Whiters, N.J., Schaap, L.H.J., Kolk, K.L. and Dalton, J.C.P. (1986), "*The Development of the Full Displacement Pressuremeter*", Proc. of 2nd Int. Symposium on The Pressuremeter and Its Marine Applications, University of Texas, ASTM STP 950.

Woods, R.D. and Stokoe, K.H. II (1985), "*Shallow Seismic Exploration in Soil Dynamics*", Richart Commemorative Lectures, ASCE.

Wroth, C.P. (1984), "*The Interpretation of In-Situ Tests*", XXIV Rankine Lecture, Geotechnique No. 4.

Yu, P. and Richart, F.E. Jr. (1984), "*Stress Ratio Effects on Shear Modulus of Dry Sand*", JGED, ASCE, GT 3.

Yu, H.S., Shnaid, F. and Collins, I.F. (1994), "*Analysis of Cone Pressuremeter Tests in Sands*", Research Report - Dept. Civil Engineering and Surveying - University of Newcastle - New South Wales, Australia.

Znidovic, D. and Piccoli, S. (1995), "*Field Measurements of Hydraulic Conductivity*", Proc. IV Int. Symposium on Field Measurements in Geomechanics. ISMES, Bergamo

Zuidberg, H.M. and Post, M.L. (1995), "*The Cone Pressuremeter: An Efficient Way of Pressuremeter Testing*", Proc. 4th Int. Symposium, The Pressuremeter and its New Avenues, Sherbrooke, Canada.